Methods in Enzymology

Volume XXV

ENZYME STRUCTURE

Part B

METHODS IN ENZYMOLOGY

EDITORS-IN-CHIEF

Sidney P. Colowick Nathan O. Kaplan

Methods in Enzymology

Volume XXV

Enzyme Structure

Part B

EDITED BY

C. H. W. Hirs

DIVISION OF BIOLOGICAL SCIENCES
INDIANA UNIVERSITY
BLOOMINGTON, INDIANA

Serge N. Timasheff

GRADUATE DEPARTMENT OF BIOCHEMISTRY
BRANDEIS UNIVERSITY
WALTHAM, MASSACHUSETTS

1972

ACADEMIC PRESS New York and London

ACADEMIC PRESS, INC.
111 Fifth Avenue, New York, New York 10003

United Kingdom Edition published by
ACADEMIC PRESS, INC. (LONDON) LTD.
24/28 Oval Road, London NW1 7DD

LIBRARY OF CONGRESS CATALOG CARD NUMBER: 54-9110

PRINTED IN THE UNITED STATES OF AMERICA

Table of Contents

Section I. Amino Acid Analysis and Related Procedures

Section II. End-Group Analysis

Section III. Chain or Subunit Separation

Section IV. Cleavage of Disulfide Bonds

Section V. Cleavage of Peptide Chains

Section VI. Separation of Peptides

Section VII. Sequence Determination

Section VIII. Modification Reactions

Section IX. Specific Modification Reactions

Contributors to Volume XXV, Part B

Article numbers are in parentheses following the names of contributors. Affiliations listed are current.

RICHARD P. AMBLER (10, 21), *Department of Molecular Biology, University of Edinburgh, Edinburgh, Scotland*

RUTH ARNON (50), *Section of Chemical Immunolgy, The Weizmann Institute of Science, Rehovot, Israel*

M. Z. ATASSI (49), *Wayne State University, Detroit, Michigan*

P. J. G. BUTLER (14), *Medical Research Council, Laboratory of Molecular Biology, Cambridge, England*

PAUL X. CALLAHAN (22, 23), *Biochemical Endocrinology Branch, Environmental Biology Division, Ames Research Center, National Aeronautics and Space Administration, Moffett Field, California*

JOHN CANN (11), *Department of Biophysics and Genetics, University of Colorado Medical Center, Denver, Colorado*

K. L. CARRAWAY (56), *Department of Biochemistry, University of California, Berkeley, California*

HENRY B. F. DIXON (33), *Department of Biochemistry, University of Cambridge, Cambridge, England*

R. F. DOOLITTLE (18), *Department of Chemistry, University of California, San Diego, California*

STANLEY ELLIS (22, 23), *Biochemical Endocrinology Branch, Environmental Biology Division, Ames Research Center, National Aeronautics and Space Administration, Moffett Field, California*

ROBERT FIELDS (33, 38), *Department of Biochemistry, University of Cambridge, Cambridge, England*

ANGELO FONTANA (34, 40), *Institute of Organic Chemistry, University of Padua, Padua, Italy*

MARINA J. GORBUNOFF (43), *Graduate Department of Biochemistry, Brandeis University, Waltham, Massachusetts*

WILLIAM R. GRAY (8, 26a), *Biology Department, University of Utah, Salt Lake City, Utah*

FRANK R. N. GURD (34a), *Department of Chemistry, Indiana University, Bloomington, Indiana*

A. F. S. A. HABEEB (37, 49, 51), *The University of Alabama in Birmingham, School of Medicine, Birmingham, Alabama*

B. S. HARTLEY (14), *Medical Research Council, Laboratory of Molecular Biology, Cambridge, England*

FRED C. HARTMAN (59), *Biology Division, Oak Ridge National Laboratory, Oak Ridge, Tennessee*

C. H. W. HIRS (1), *Division of Biological Sciences, Indiana University, Bloomington, Indiana*

H. R. HORTON (39), *Department of Biochemistry, North Carolina State University at Raleigh, Raleigh, North Carolina*

M. J. HUNTER (54), *Biophysics Research Division and Department of Biological Chemistry, University of Michigan, Ann Arbor, Michigan*

ADAM S. INGLIS (5), *Division of Protein Chemistry, Commonwealth Scientific and Industrial Research Organization, Parkville, Victoria, Australia*

MICHAEL H. KLAPPER (46, 47), *Department of Chemistry, Ohio State University, Columbus, Ohio*

IRVING M. KLOTZ (46, 47), *Department of Chemistry, Ohio State University, Columbus, Ohio*

WILLIAM KONIGSBERG (13, 17, 26), *Department of Molecular Biophysics and Biochemistry, Yale University, New Haven, Connecticut*

D. E. KOSHLAND, JR. (39, 56), *Department of Biochemistry, University of California, Berkeley, California*

RICHARD A. LAURSEN (27), *Department of Chemistry, Boston University, Boston, Massachusetts*

ALBERT LIGHT (20), *Department of Chemistry, Purdue University, Lafayette, Indiana*

TEH-YUNG LIU (4, 5) *Biology Department, Brookhaven National Laboratory, Upton, Long Island, New York*

ROBERT E. LOVINS (25), *Department of Biochemistry, University of Georgia, Athens, Georgia*

M. L. LUDWIG (54), *Biophysics Research Division and Department of Biological Chemistry, University of Michigan, Ann Arbor, Michigan*

J. KEN MCDONALD (22, 23), *Biochemical Endocrinology Branch, Environmental Biology Division, Ames Research Center, National Aeronautics and Space Administration, Moffett Field, California*

JAMES M. MANNING (2), *The Rockefeller University, New York, New York*

NORBERT P. NEUMANN (31), *Division of Biochemistry, Ortho Diagnostics Raritan, New Jersey*

A. NIEDERWIESER (6), *Kinderspital Zurich, Steinwiesstrasse 75, 8032, Zurich, Switzerland*

ROBERT F. PETERSON (12), *Eastern Utilization Research and Development Division, Agricultural Research Service, U.S. Department of Agriculture, Philadelphia, Pennsylvania*

J. J. PISANO (3), *National Heart and Lung Institute, National Institutes of Health, Bethesda, Maryland*

DAVID PRESSMAN (35), *Department of Biochemistry Research, Roswell Park Memorial Institute, Buffalo, New York*

VERNON N. REINHOLD (19), *Department of Chemistry, Massachusetts Institute of Technology, Cambridge, Massachusetts*

FRANK F. RICHARDS (25), *Department of Internal Medicine, Yale University Medical School, New Haven, Connecticut*

J. F. RIORDAN (36, 41, 42, 44, 45), *Department of Biological Chemistry, Harvard Medical School, Boston, Massachusetts*

OLIVER A. ROHOLT (35), *Department of Biochemistry Research, Roswell Park Memorial Institute, Buffalo, New York*

DEBDUTTA ROY (17), *Department of Molecular Biophysics and Biochemistry, Yale University, New Haven, Connecticut*

WALTER A. SCHROEDER (9, 15, 16, 24), *California Institute of Technology, Pasadena, California*

ERNESTO SCOFFONE (40), *Institute of Organic Chemistry, University of Padua, Padua, Italy*

MICHAEL SELA (50), *Section of Chemical Immunology, The Weizmann Institute of Science, Rehovot, Israel*

ELLIOTT SHAW (58), *Biology Department, Brookhaven National Laboratory, Upton, New York*

GEORGE R. STARK (7, 28, 29, 53), *Department of Biochemistry, School of Medicine, Stanford University, California*

B. L. VALLEE (36, 41, 42, 44, 45), *Department of Chemistry, Harvard Medical School, Boston, Massachusetts*

E. W. WESTHEAD (32), *Biochemistry Department, University of Massachusetts, Amherst, Massachusetts*

FREDERICK H. WHITE, JR. (30, 48), *National Heart and Lung Institute, National Institutes of Health, Bethesda, Maryland*

PHILIP E. WILCOX (55), *Department of Chemistry, Harvard University, Cambridge, Massachusetts (Deceased)*

FINN WOLD (57), *Department of Biochemistry, University of Minnesota, Minneapolis, Minnesota*

J. A. YANKEELOV, JR. (52), *Department of Biochemistry, University of Louisville, School of Medicine, Louisville, Kentucky*

Preface

"Enzyme Structure," the eleventh volume in this series, first appeared in 1967. Although the objectives sought in its publication have largely been realized, it had been apparent almost from the beginning that to maintain its usefulness a periodic revision and updating of the text would be necessary. The original volume, moreover, contained some notable omissions. In particular, the scope of the section on conformational changes was restricted deliberately in the expectation that ultimately another volume would be published in which physical methods generally would be covered in a more comprehensive fashion.

This volume should be regarded as the first of three supplements to Volume XI. It is designed to update the first nine sections of Volume XI and for that reason retains the organization of that volume. In addition to a number of new procedures not previously covered, it includes revised and updated versions of most of the articles from Volume XI which experience has shown to be those most frequently referred to in the literature. The last section on methods suitable for the investigation of conformational changes will be substantially enlarged upon in two additional volumes presently in preparation.

We acknowledge with pleasure the generous cooperation of the contributors to this volume. Their suggestions for its improvement have been of especial value. We also wish to thank the staff of Academic Press for their many courtesies.

C. H. W. Hirs
Serge N. Timasheff

METHODS IN ENZYMOLOGY

EDITED BY

Sidney P. Colowick and Nathan O. Kaplan

VANDERBILT UNIVERSITY
SCHOOL OF MEDICINE
NASHVILLE, TENNESSEE

DEPARTMENT OF CHEMISTRY
UNIVERSITY OF CALIFORNIA
AT SAN DIEGO
LA JOLLA, CALIFORNIA

I. Preparation and Assay of Enzymes
II. Preparation and Assay of Enzymes
III. Preparation and Assay of Substrates
IV. Special Techniques for the Enzymologist
V. Preparation and Assay of Enzymes
VI. Preparation and Assay of Enzymes (*Continued*)
Preparation and Assay of Substrates
Special Techniques
VII. Cumulative Subject Index

METHODS IN ENZYMOLOGY

EDITORS-IN-CHIEF

Sidney P. Colowick   Nathan O. Kaplan

VOLUME VIII. Complex Carbohydrates
Edited by ELIZABETH F. NEUFELD AND VICTOR GINSBURG

VOLUME IX. Carbohydrate Metabolism
Edited by WILLIS A. WOOD

VOLUME X. Oxidation and Phosphorylation
Edited by RONALD W. ESTABROOK AND MAYNARD E. PULLMAN

VOLUME XI. Enzyme Structure
Edited by C. H. W. HIRS

VOLUME XII. Nucleic Acids (Parts A and B)
Edited by LAWRENCE GROSSMAN AND KIVIE MOLDAVE

VOLUME XIII. Citric Acid Cycle
Edited by J. M. LOWENSTEIN

VOLUME XIV. Lipids
Edited by J. M. LOWENSTEIN

VOLUME XV. Steroids and Terpenoids
Edited by RAYMOND B. CLAYTON

VOLUME XVI. Fast Reactions
Edited by KENNETH KUSTIN

VOLUME XVII. Metabolism of Amino Acids and Amines (Parts A and B)
Edited by HERBERT TABOR AND CELIA WHITE TABOR

VOLUME XVIII. Vitamins and Coenzymes (Parts A, B, and C)
Edited by DONALD B. MCCORMICK AND LEMUEL D. WRIGHT

VOLUME XIX. Proteolytic Enzymes
Edited by GERTRUDE E. PERLMANN AND LASZLO LORAND

VOLUME XX. Nucleic Acids and Protein Synthesis (Part C)
Edited by KIVIE MOLDAVE AND LAWRENCE GROSSMAN

VOLUME XXI. Nucleic Acids (Part D)
Edited by LAWRENCE GROSSMAN AND KIVIE MOLDAVE

VOLUME XXII. Enzyme Purification and Related Techniques
Edited by WILLIAM B. JAKOBY

VOLUME XXIII. Photosynthesis (Part A)
Edited by ANTHONY SAN PIETRO

VOLUME XXIV. Photosynthesis and Nitrogen Fixation (Part B)
Edited by ANTHONY SAN PIETRO

VOLUME XXV. Enzyme Structure (Part B)
Edited by C. H. W. HIRS AND SERGE N. TIMASHEFF

VOLUME XXVI. Enzyme Structure (Part C)
Edited by C. H. W. HIRS AND SERGE N. TIMASHEFF

Section I

Amino Acid Analysis and Related Procedures

[1] Automated Sample Injection in Amino Acid Analyzers

By C. H. W. Hirs

Automated sample injection affords a significant technical advantage in situations where a large number of relatively similar samples are to be subjected to routine analysis in a liquid chromatographic system. This is particularly true in applications involving automatic amino acid analyzers.

All the injection procedures developed thus far are based on one of two fundamental elements: a sample loop, in which the sample is stored prior to chromatography as a solution; or a sample cartridge, in which the substances to be chromatographed are stored in the adsorbed state on a suitable support from which they are eventually eluted when the analysis is commenced. Automated injection of samples becomes feasible with satisfaction of the additional technical requirement for some device with which a succession of sample loops or cartridges can be brought, in an appropriately programmed fashion, into line with the eluent stream that is delivered to the separation system.

Sample Loops

Sample loops were used in gas–liquid chromatography long before their application in liquid chromatography was explored. Crestfield,[1] in studies on the chromatography over IRC-50 columns of bovine pancreatic ribonuclease and its carboxymethylated derivatives, found that reproducibility of the chromatograms was markedly improved if the samples were injected from a sample loop into a column the elution of which was not interrupted. The sample loop he used was a simple arrangement of four T-tubes, a coil of 1-mm i.d. Teflon tubing in which about 1–2 ml of liquid could be stored, and some pinch clamps (Fig. 1). Normally, clamps 3 and 4 were kept closed, 1 and 2 open. The loop could be filled by opening clamps 5 and 6. The sample was drawn into the coil by applying suction from a syringe at one end, thus displacing the liquid previously in the coil. When all the sample was aspirated, transfer to the coil was made quantitative with several rinses of solvent. Clamps 5 and 6 were now closed, 3 and 4 opened, and finally 1 and 2 closed. After a time sufficient to assue complete transfer of the contents of the sample loop into the column, clamps 1 and 2 were opened

[1] A. M. Crestfield, *Anal. Chem.* **35**, 1762 (1963).

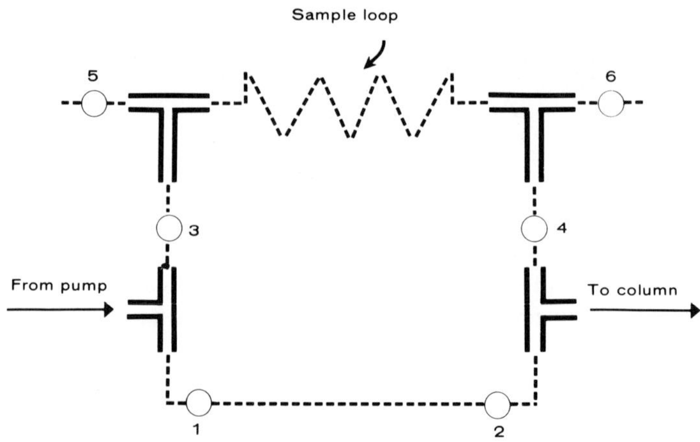

FIG. 1. Sample injection loop described by A. M. Crestfield [*Anal. Chem.* **35**, 1762 (1963)]. The dashed lines indicate connections made with 1-mm i.d. Teflon tubing. Locations of pinch clamps are indicated by circles. For a description of the manner in which the T-fittings were made, the original article should be consulted.

again, and 3 and 4 were closed to prepare for the loading of the next sample.

With the advent of reliable micro multiport valves, the operations just described could be much simplified. A versatile valve for purposes of this type is manufactured by Chromatronix, Inc.[2] It can be used to advantage for the manual addition of samples to an amino acid analyzer from a sample loop.

The first fully automatic sample injection device for liquid chromatography was developed in the author's laboratory by Dr. A. M. Murdock and accommodated 8 sample loops.[3] A drawing of the machine is shown in Fig. 2. It was employed initially to automate analytical chromatography of reaction mixtures produced in the chemical modification of ribonuclease, but was soon incorporated into an amino acid analyzer.[4] A technical description of the device is beyond the scope of this article; an indication of how it operates must suffice.

Samples of up to 1 ml in volume can be accommodated in coils of ~1-mm Teflon tubing (AGW 22) which may be filled through valves that give access through a T to both the upper and lower ends of each coil. The coils are mounted on a rotating center section. Seals between

[2] Chromatronix, Inc., 2743 Ninth Street, Berkeley, California 94710. R60SV Sample injection valve.

[3] A. L. Murdock, K. L. Grist, and C. H. W. Hirs, *Arch. Biochem. Biophys.* **114**, 375 (1966).

[4] N. Alonzo and C. H. W. Hirs, *Anal. Biochem.* **23**, 272 (1968).

the center section and the fixed ends are provided by silicone rubber O rings. In the original machine, the rotating center section and the fixed plates were machined from 316 stainless steel. Experience has shown that for amino acid analysis PVC plastic is equally satisfactory mechanically and superior in resistance to corrosion.

Precise alignment of the center section with the ports in the fixed ends is assured with a Geneva movement through which the center section is rotated by a suitable motor controlled with limit switches. When a sample loop is brought into line with the buffer stream from the pump, the solution it contains is displaced into the column and a coil from which the sample has been displaced is left filled with buffer. During the movement associated with sample change, buffer from the pump escapes through a bypass, thus preventing development of excess pressure at the pump. Samples of predetermined volume may be aspirated into the coils with a calibrated syringe and a quantitative transfer achieved by a solvent rinse.

In amino acid analyzer applications it is advantageous to replace the sample loops with a sample tube of appropriate capacity, usually either 1.0 or 1.5 ml, of the type illustrated in Fig. 3. Such tubes can be fabricated conveniently from Nylon or Teflon stock. The closure is identical in dimensions to an SB 12/5 standard ball joint. The recess in the inner member accommodates a suitable O ring of Buna N with which a reliable seal is assured. Successful use of such sample tubes requires that the sample solution be made up in buffer that contains 20% by volume of polyethylene glycol (Polyethylene glycol 400; Fisher). The sample solution is pipetted directly into the tube and is then overlayered with sample buffer devoid of polyethylene glycol.

In practice, the tube is opened and the contents are removed by aspiration, most conveniently with the aid of a 25-ml syringe equipped with a length of 1-mm polyethylene tubing. It is advisable to rinse the closure several times with distilled water and to aspirate the rinses away each time. Care should be taken to avoid overfilling the tubes with sample buffer when preparing for sample addition, because the excess forced out when the tube is closed acts as a trap for ammonia in the laboratory air. This is a significant consideration, particularly in urban areas. The ammonia trapped around the closure tends to contaminate the next sample to be applied unless great care is exercised in cleaning the closure at that point. It is a good precaution to forbid smoking around the analyzer, especially when samples are being pipetted into the sample tubes. With good technique it is possible to limit the access of adventitious ammonia to about 0.01 μmole.

When sample tubes are used in place of the loops, the microvalves

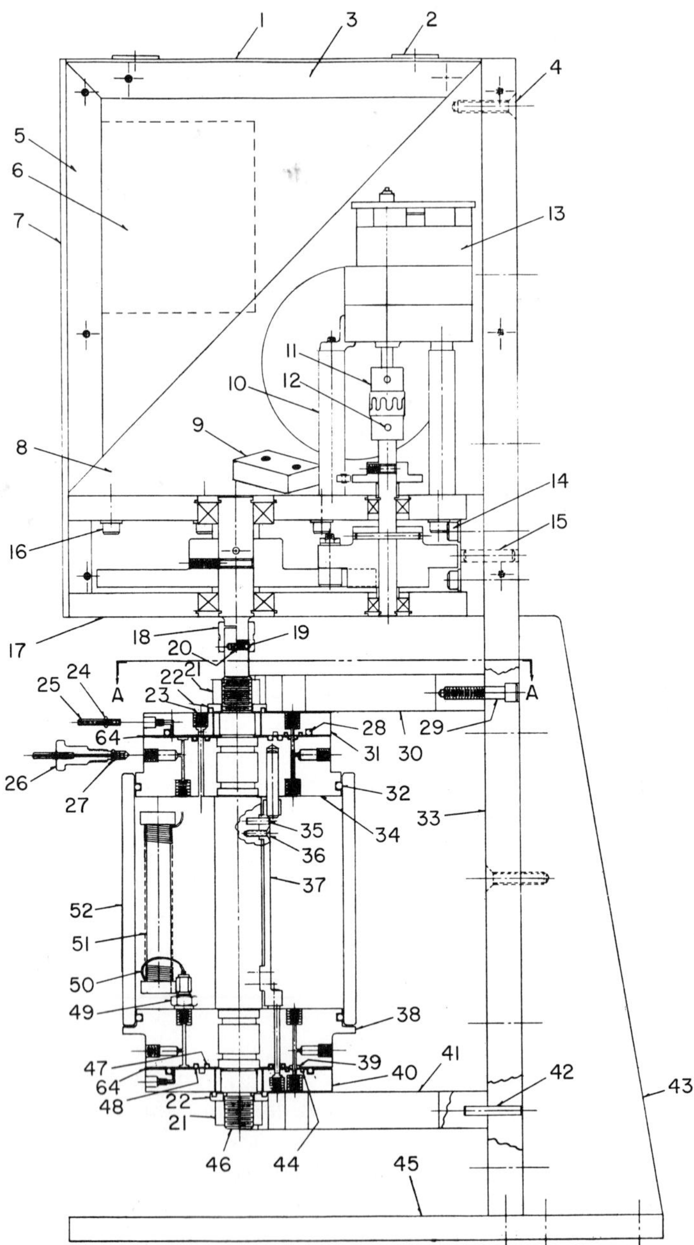

FIG. 2. Sample injector with storage capability for 8 samples described by A. L. Murdock, K. L. Grist and C. H. W. Hirs [*Arch. Biochem. Biophys.* **114,** 375 (1966)]. Vertical section through the revolving shaft. One of the Teflon coils that

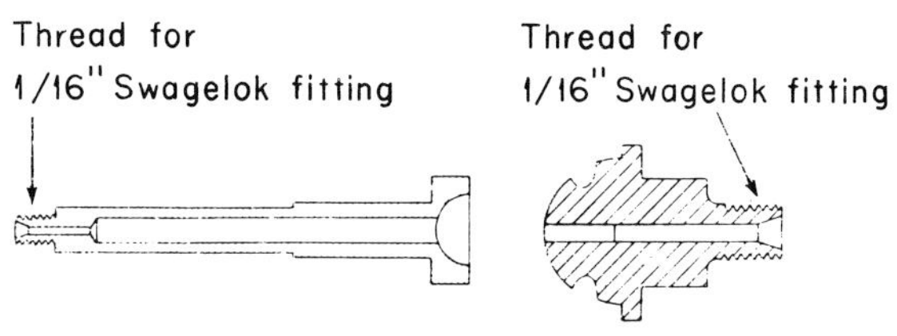

FIG. 3. Storage tube. The tube is fabricated of Nylon. The closure has the same radius as a standard SB 12/5 joint. The connection at the end of the tube is machined to be identical to a Swagelok 1/16-inch tube fitting.

are not needed and can, in fact, be dispensed with entirely. In a more recent modification of the instrument, Mr. Nicholas Alonzo, in the author's laboratory, has modified the configuration of the end plates in such a fashion as to leave the sample tubes stationary. A sample changer of this general type is now distributed by Chromatronix, Inc. and accommodates 20 sample loops. It is used as an accessory to the amino acid analyzer distributed by the BioCal organization. A rotary sample changer with a capacity for 80 sample loops is featured in the Beckman Model 120C amino acid analyzer. The respective technical bulletins from these companies should be studied for detailed descriptions of how to load the sample loops.

Sample Cartridge

The cartridge principle was introduced by Eveleigh and Thomson[5] and has been incorporated into amino acid analyzers distributed by the Technicon Chromatography Corporation and by the Perkin-Elmer Corporation. In the Technicon instrument the cartridges (filled with a suitable cation-exchange resin) are stored on a turntable and are suc-

[5] J. W. Eveleigh and A. R. Thomson, *Biochem. J.* **99**, 49P (1966).

accommodate samples (50) is shown wound on a spool (51). One of the microvalves that provide access to these coils is shown in cross section (26, 27). The tip of the valve is made of Viton and is sealed with an O ring. Separate valves are provided at the top and bottom of each coil. Diametrically opposite the valve and coil shown in the drawing may be seen the channels that would be associated with a second coil and a second set of valves. These channels are shown in apposition with ports into which are screwed fittings that connect to the pump and to the chromatograph column. A plastic cylinder (52) encloses the space in which the coils are accommodated. Port (23) and a similar port at the bottom give access to this space and permit circulation of refrigerated water around the coils. Seals between the stationary ends and the moving center section are provided by O rings (28, 39, 44, 47). Fitting (25), sealed by O ring (24), is used to drain the space between the upper end pieces. A similar fitting is provided for the lower end pieces.

cessively rotated into position above the column in which the amino acid separation is achieved. Here they are coupled to the column and the pumping system which delivers the eluting buffer. The arrangement has the advantage of simplicity, but limits flexibility of operation in equipment in which selection of more than one type of separating column may be desirable. The Perkin-Elmer machine operates in such a fashion that the individual sample cartridges are selected by a rotary valve arrangement and is thus similar to the equipment described in the preceding section. There is no doubt that sample addition with a sample tube, such as that shown in Fig. 3, or with a resin-filled cartridge is technically less complex, and on this account probably more reliable in a routine analytical context, than addition with a sample loop.

Sample Injection

The process of injection of the sample into the analyzer column may be accomplished in one of two ways. The simpler of the two is to permit injection by the same solvent pumping system as will be used in the subsequent analysis. This approach is subject to certain limitations, the most important of which is the pressure developed during the analysis in the column inlet system. The type of rotary microvalves used in sample changers of the type illustrated in Fig. 2 cannot withstand pressures in excess of about 250 psi for protracted periods without excessive wear on the O rings and consequent leakage. Thus, if the sample changer is to be used directly in line with the pumping system, elution pressures must be kept below 200 psi. With certain resins and at higher flow rates of the kind required in accelerated procedures, this can be a serious limitation. A second limitation, already mentioned in connection with the Technicon system, is that in-line configurations of the sample changer restrict flexibility of operation. For example, in automatic regeneration of the column after an analysis, the sample changer must remain immobilized while this step is completed.

A second, more flexible injection procedure was introduced by Mr.

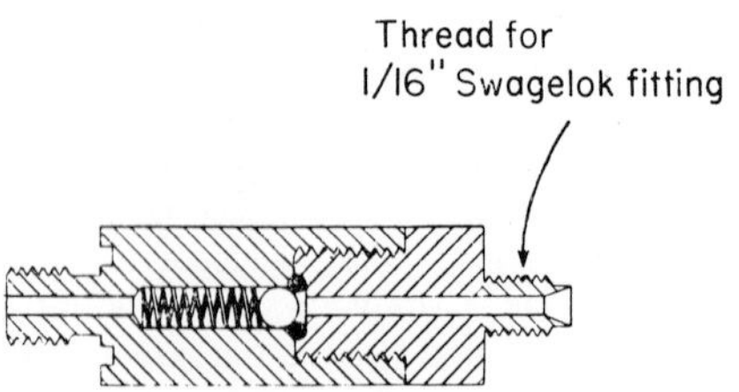

FIG. 4. Microbore check valve. The seat is a 1/16-inch silicone rubber O ring. The ends are machined to be compatible with a Swagelok 1/16-inch tube fitting.

Nicholas Alonzo in the author's laboratory several years ago.[4] In this, a separate pumping system is required for the sample injection step. The sample is applied from the sample tube through a line which connects with the buffer line to the column just where this enters the column closure. Isolation of the sample injection and buffer inlet lines from cross effects is achieved by placing a microbore check valve in each. Such check valves can be fabricated of stainless steel to the dimensions shown in Fig. 4. The springs have a tension sufficient to require a pressure of about 10 psi before flow is possible.

The use of a separate injection line overcomes the difficulties that result from use of the simpler configuration. A disadvantage is the requirement for an additional pump.

A question frequently asked about equipment for automated sample injection is: How much holdup is there between samples? In our experience, very little, if any. Clearly, the most important consideration in this regard is that the equipment be tested periodically to ensure that, in fact, holdup or holdover is negligible. The simplest test is to run a series of blanks after an analysis of a sample that is recognized to have been overloaded.

[2] Determination of D- and L-Amino Acids by Ion-Exchange Chromatography

By JAMES M. MANNING

Principle

Amino acid enantiomers are converted into diastereomeric dipeptides, and these derivatives are separated on an amino acid analyzer.[1] The dipeptides are prepared by a modification of the *N*-carboxyanhydride (NCA) procedure of Hirschmann *et al.*[2]

L-Leucine NCA is used for coupling with most of the acidic and neutral amino acids; the basic amino acids are coupled with L-glutamic acid NCA to give dipeptides which can be eluted from the column without inconvenient retardation. The method of Hirschmann *et al.* for the preparation of dipeptides offers several advantages: it is rapid, and it does not require blocking groups for the amino acids; the yield is 90–

[1] J. M. Manning and S. Moore, *J. Biol. Chem.* **243,** 5591 (1968).

[2] R. Hirschmann, R. G. Strachan, H. Schwam, E. F. Schoenewaldt, H. Joshua, H. Barkemeyer, D. F. Veber, W. J. Paleveda, Jr., T. A. Jacob, T. E. Beesley, and R. G. Denkewalter, *J. Org. Chem.* **32,** 3415 (1967).

L-Amino acid NCA (HN—C(=O)—O—C(=O)—C(H)(R)—, cyclic) + DL-Amino acid ($H_2NCHR'COOH$) $\xrightarrow[\text{2 min}]{\text{pH 10.4, 0°}}$ $\xrightarrow{H^+,\ \text{pH 2–3}}$ L-D-Dipeptide ($R—C(NH_2)(H)—CONH—C(R')(H)—COOH$) + L-L-Dipeptide ($R—C(NH_2)(H)—CONH—C(H)(R')—COOH$)

L-Amino acid NCA

DL-Amino acid

L-D-Dipeptide

L-L-Dipeptide

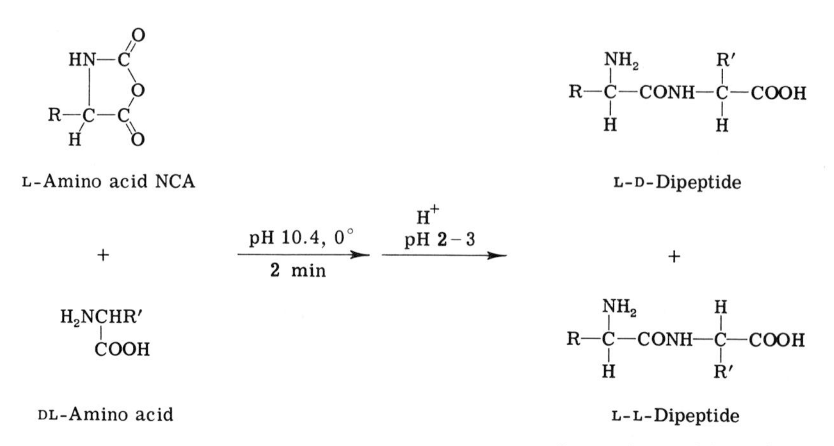

95%. The derivatives can be applied directly to the amino acid analyzer, and the dipeptide pairs can be separated, in many instances, with the eluents that are used for routine amino acid analysis. An important feature of the method is that the synthesis of the dipeptides proceeds without detectable racemization.[3] The ease of preparation of the derivatives together with the sensitivity of current amino acid analyzers are the basis of this procedure for the precise determination of the D- and L-isomers of amino acids.

With hydrolyzates of proteins, the amount of racemization that occurs during acid hydrolysis must be determined before the amounts of D- and L-amino acids in the protein can be established. Use of tritiated HCl provides a convenient means for measurement of the racemization that takes place during the hydrolysis. This technique is valid only for those amino acids whose side-chain hydrogen atoms are not exchanged under these conditions.[4]

Procedure

Reagents. Most of the L-amino acids were samples from CalBiochem and were found to be stereochemically pure by the method described here. The D-amino acids were obtained from Cyclo Chemical Corporation. The DL-amino acids were purchased from various commercial sources.

The initial samples of L-leucine NCA and of L-glutamic acid NCA were generously donated by Dr. R. Hirschmann and Dr. R. G. Denke-

[3] R. G. Denkewalter, H. Schwam, R. G. Strachan, T. E. Beesley, D. F. Veber, E. F. Schoenewaldt, H. Barkemeyer, W. J. Paleveda, Jr., T. A. Jacob, and R. Hirschmann, *J. Amer. Chem. Soc.* **88**, 3163 (1966).

[4] J. M. Manning, *J. Amer. Chem. Soc.* **92**, 7449 (1970).

walter of Merck and Company, Rahway, New Jersey. These compounds were chemically and stereochemically pure within the limits of detection of the present method. L-Leucine NCA is now available from Pilot Chemical Company, Watertown, Massachusetts, and L-glutamic acid NCA can be obtained from Cyclo Chemical Corp., Los Angeles, California. However, the quality of these preparations of L-glutamic acid NCA is variable because the synthesis of this amino acid NCA requires absolutely dry conditions. If necessary, the L-glutamic acid NCA can be prepared by the method of Hirschmann *et al.*,[5] who have devised the most satisfactory method for the synthesis.

The NCA's are stored in 4-ml screw-cap glass vials (Kimble 609106), each containing 100–200 mg. The vials are kept at —20° over Drierite in a larger bottle, which is allowed to come to room temperature before it is opened; NCA's that have been stored unopened for 9 months at —20° show no sign of decomposition as judged by the yield of dipeptide. However, repeated opening of the same vial leads to gradual decomposition of the anhydride, especially during humid weather. Batches which have been adequately stored give good yields of dipeptides (about 90%) and a clear reaction mixture; when deterioration of the NCA has occurred, the yields of dipeptides are lower than 90% and the reaction mixture is cloudy.

The borate buffer (0.45 *M*, 500 ml) used for preparation of the dipeptides is made up by dissolving boric acid (13.9 g) in 497 ml of boiled, distilled water, and adding 50% NaOH (approximately 3.0 ml) to bring the solution to pH 10.2 at 25° (corresponding to pH 10.4 at 0°).

Preparation of Derivatives. The method of dipeptide synthesis is a scale-down (about 1:1000) of the procedure of Hirschmann *et al.*[2] These investigators showed that, with the exception of the ϵ-NH_2 group of lysine, the —SH group of cysteine, and the —OH group of tyrosine, the only functional group of amino acids which reacts with an NCA under their conditions is the α-NH_2 group. The problems presented by the side reactions with these other groups are dealt with below.

The amount of an individual amino acid taken for the derivatization step is usually 20 μmoles; 1 μmole can be used if the sample is in limited supply. The sample is weighed into a Pyrex test tube, 25 × 150 mm, and dissolved in 2.0 ml of the borate buffer. L-Leucine NCA (MW 157) or L-glutamic acid NCA (MW 173) in 10–20% molar excess is weighed into a stoppered weighing tube (Kimble 46465). With lysine, a 5-fold molar excess of the NCA is taken in order to attain the maxi-

[5] R. Hirschmann, H. Schwam, R. G. Strachan, E. F. Schoenewaldt, H. Barkemeyer, S. M. Miller, J. B. Conn, V. Garsky, D. F. Veber, and R. G. Denkewalter, *J. Amer. Chem. Soc.* **93,** 2746 (1971).

mum yield of the disubstituted derivative, which is the compound subjected to chromatographic analysis.

The amino acid solution is cooled to 0° in an ice bath and brought into the cold room (4°) along with the weighing tube. The vigorous agitation essential for efficient coupling is accomplished on a microscale by placing the test tube on a Vortex mixer (Scientific Industries, New York) at maximum speed. The NCA is dumped from the weighing tube into the swirling solution; to ensure complete solution of the NCA, the reaction tube is removed from the mixer at 30-second intervals and replaced within 2 seconds. After 2 minutes, 1.0 *M* HCl (0.80 ml) is added rapidly to stop the reaction. The resulting solution should have a pH of 1–3 (pH paper). If the mixture is slightly cloudy, it can be filtered through a Millipore filter (0.22 μ, No. GSWP01300) in a Swinny adapter (No. XX3001200). At this point the peptide solution may be stored in the freezer.

Chromatographic Analyses. The separations are carried out with an automatic amino acid analyzer of the design of Spackman, Stein, and Moore[6] with the use of a column, 0.9 × 62 cm, packed with Beckman-Spinco AA-15 resin. The temperature of the column is maintained at 52° except when otherwise noted, and the flow rate through the column is 50 ml per hour; the effluent–ninhydrin mixture remains in the heating coil for 9 minutes. The relative elution positions of the dipeptides on columns which are packed with other custom resins and which operate at another temperature may be different than those found under our conditions. A recorder with a 0 to ∞ or a 0 to 0.10 absorbance scale may be used, and the load of dipeptide that is applied is calculated accordingly. All buffers are 0.20 *N* sodium citrate[7]; the pH 3.25 and pH 4.25 buffers are those ordinarily used for amino acid analysis. Dilution of 0.38 *N*, pH 5.28, sodium citrate, used for the analyses of the basic amino acids, with distilled water to 0.20 *N* results in a buffer of pH 5.42. The pH 4.68 buffer is prepared by mixing equal volumes of the pH 4.25 and the pH 5.42 buffers. The pH 2.80 and pH 3.10 buffers are prepared by addition of concentrated HCl to the pH 3.25 buffer, and the pH 5.10 buffer is similarly prepared from the pH 5.42 buffer.

For chromatographic analysis, an aliquot of the peptide solution is diluted with one-half or more of its volume of the pH 2.2 citrate buffer normally used for samples to be added to the ion-exchange column.

Separation of L-D *and* L-L *Dipeptides.* The pairs of dipeptides prepared from each of 21 amino acids have been separated by ion-exchange

[6] D. H. Spackman, W. H. Stein, and S. Moore, *Anal. Chem.* **30,** 1190 (1958).

[7] S. Moore, D. H. Spackman, and W. H. Stein, *Anal. Chem.* **30,** 1185 (1958).

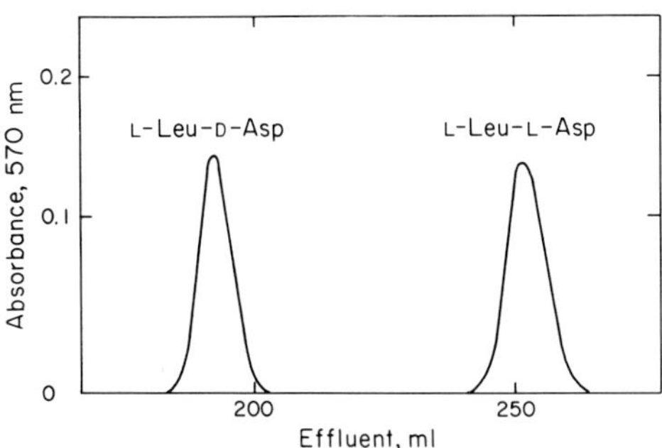

FIG. 1. Separation of L-Leu-D-Asp and L-Leu-L-Asp on the amino acid analyzer. Column: 0.9 × 62 cm. Eluent: 0.20 N sodium citrate, pH 3.25, at 52°.

chromatography on the amino acid analyzer. In every case complete separation of the diastereomers is achieved; typical chromatograms are presented in the first two figures. The elution pattern obtained at pH 3.25 with L-Leu-D-Asp and L-Leu-L-Asp is shown in Fig. 1. Unreacted aspartic acid is eluted at 63 ml, well ahead of the dipeptides; excess leucine from the NCA is eluted at 345 ml. The chromatographic pattern obtained at pH 5.42 with the L-glutamyl derivatives of a basic amino acid, DL-histidine, is shown in Fig. 2. Unreacted histidine is eluted from the column much later.

The elution conditions for the dipeptides are described in Table I. The dipeptide from each D- or L-amino acid was prepared separately, and its position with a given eluent was determined. Many of the L-D and L-L dipeptides are successfully separated by eluents routinely used on the amino acid analyzer. Occasionally, it is necessary to use a different buffer; for example, pH 3.10 was needed to separate L-Leu-D-Glu from free leucine. For the separation of all but two of the dipeptides

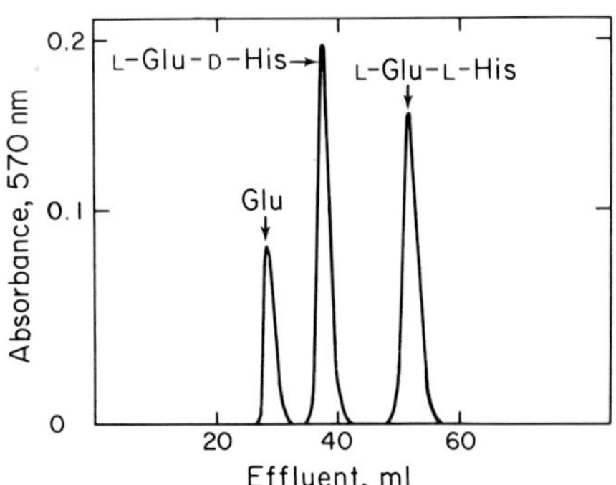

FIG. 2. Separation of L-Glu-D-His and L-Glu-L-His on the amino acid analyzer. Column: 0.9 × 62 cm. Eluent: 0.20 M sodium citrate, pH 5.42 at 52°.

TABLE I

Elution Conditions and Operational Color Values of Dipeptides

Dipeptide	Eluent pH	Peak position (ml)	Operational color value (relative to leucine as 1.00)
L-Leu-D-Asp	3.25	190	0.69
L-Leu-L-Asp	3.25	251	0.83
L-Leu-D-Asn	3.25	193	0.67
L-Leu-L-Asn	3.25	291	0.73
L-Leu-D-Ser	3.25	243	0.90
L-Leu-L-Ser	3.25	320	0.89
L-Leu-D-Thr	3.25	267	0.74
L-Leu-L-Thr	3.25	291	0.88
L-Leu-D-alloThr	3.25	242	0.86
L-Leu-L-alloThr	3.25	293	1.00
L-Leu-D-Glu	3.10	382	0.59
L-Leu-L-Glu	3.10	453	0.75
L-Leu-D-Gln	3.25	223	0.78
L-Leu-L-Gln	3.25	323	0.84
L-Leu-D-Pro	4.25	120	0.49
L-Leu-L-Pro	4.25	162	0.83
L-Leu-D-Ala	4.25	122	0.87
L-Leu-L-Ala	4.25	148	0.97
L-Leu-D-Val	4.25	164	0.77
L-Leu-L-Val	4.25	185	0.95
L-Leu-D-Met	4.25	185	0.73
L-Leu-L-Met	4.25	213	0.89
L-Leu-D-Ile	5.42	68	1.05
L-Leu-L-Ile	5.42	56	1.06
L-Leu-D-alloIle	5.42	65	0.86
L-Leu-L-alloIle	5.42	53	1.02
L-Leu-D-Leu	5.42	70	ca. 0.83[a]
L-Leu-L-Leu	5.42	60	ca. 1.06[a]
L-Leu-D-Cys(CM)	2.80	452	0.58
L-Leu-L-Cys(CM)	2.80	422	0.66
L-Leu-D-Tyr	5.10	93	0.37[b]
L-Leu-L-Tyr	5.10	119	0.49[b]
L-Glu-D-Phe[c]	4.25	108	0.91
L-Glu-L-Phe[c]	4.25	155	0.90
$N^{\alpha,\epsilon}$-(di-L-Glu)-D-Lys	4.25	187	1.31
$N^{\alpha,\epsilon}$-(di-L-Glu)-L-Lys	4.25	336	1.34
L-Glu-D-His	5.42	38	0.95
L-Glu-L-His	5.42	52	0.98
L-Glu-D-Arg	5.42	56	0.95
L-Glu-L-Arg	5.42	93	0.93
L-Glu-D-Trp	4.68	84	0.89
L-Glu-L-Trp	4.68	128	0.94

[a] Since L-leucine NCA was used, the amount of unreacted leucine could not be calculated; a yield of 95% was assumed by analogy with the L-isoleucine derivatization.

[b] The operational color value can be increased by hydrolyzing *N*,*O*-dileucyltyrosine at pH 12.5 as described in the text.

[c] Chromatography was carried out at 40°.

reported in Table I, the chromatographic column was maintained at 52°. However, the differential effect of temperature on the mobility of some dipeptides with respect to amino acids was sometimes used in order to effect a separation; for example, L-Glu-L-Phe and phenylalanine overlap at 52° but are separated when the chromatography is carried out at 40°.

The L-D dipeptides prepared from 17 of the 21 amino acids are eluted before the corresponding L-L dipeptides under the conditions given in Table I. It seems likely that the relative chromatographic positions of the diastereomers are related to the pK values of the α-carboxyl groups of the dipeptides. Thus, L-Leu-D-Met is eluted before L-Leu-L-Met when the eluent is pH 4.25 sodium citrate as described in Table I. However, when the eluent is pH 5.42 sodium citrate, L-Leu-L-Met is eluted from the resin before L-Leu-D-Met.

Ninhydrin Color Values of Dipeptides. The color values have been obtained from the chromatograms with the practical objective of determining the proportion of D isomer in a sample containing the D and L antipodes of an amino acid. The operational color values listed in Table I serve this purpose; they are calculated as follows. The difference between the amount of amino acid taken for the coupling and the amount of unreacted amino acid remaining, as determined by chromatography, is taken as the "difference yield" of the dipeptide. In most cases these yields are about 90%. On the basis of this yield of peptide, in micromoles, and the area of the dipeptide peak, the color value per micromole is calculated and expressed relative to a color yield of 1.00 for L-leucine.

The applicability of this calculation for practical analyses was checked by first determining the operational color value for the L-D and the L-L dipeptides individually, and then applying these constants to the analysis of a sample of the DL-amino acid subjected to the derivatization procedure. With 17 racemic amino acids, the D-amino acid content was found to be 50.0 ± 1.5% in each instance.

The true color value can be measured by isolating the analytically pure L-D or L-L dipeptide and determining its color yield. For example, both L-Leu-L-Val and L-Leu-L-Met have a true color value of 1.04 relative to that of leucine. The corresponding operational color values in Table I are lower because of the formation of ninhydrin-negative products during the derivatization. Hirschmann *et al.* observed that side reactions in the synthesis gave about 6% of ninhydrin-negative hydantoic acid under their conditions; in our scaled-down procedure, there is an average difference of about 10% between the operational color values and the true color values. The use of operational color values involves the assumption that the proportion of the L-D and L-L dipeptides does not change appreciably with small variations in the side products of the coupling reaction. The validity of this assumption is supported by the satisfactory results in the analyses of DL mixtures cited above and the analysis of known mixtures of D- and L-amino acids described below.

Tyrosine presents a special case. Hirschmann *et al.* have pointed out that the hydroxyl group of tyrosine may react with an NCA at pH 10.5. The operational color values for the tyrosine peptides (Table I) are about half of the expected values. If the reaction mixture is incubated at pH 12.5 for 4 hours at 37°, the amounts of the L-D and L-L dipeptides increase (the operational color value of L-Leu-L-Tyr increases from 0.49 to 0.77), a result that is consistent with the conclusion that *N,O*-dileucyl-tyrosine is formed; this disubstituted derivative is retained on the column under the conditions of the chromatography. Hydrolysis of the disubstituted derivative at pH 12.5 before the acidification step may be a generally desirable procedure in an analysis for D-tyrosine, along with the use of a higher molar excess of the NCA.

The high color value of the lysine derivatives results from the formation of the disubstituted product. This structure was verified by isolating the derivative and showing that it contains glutamic acid and lysine in a 2:1 ratio.

The color values of the L-D dipeptides are lower than those of the corresponding L-L dipeptides in most cases. Yanari[8] has also observed this type of difference in color yield for pairs of isolated L-D and L-L dipeptides. These results reflect differences in the rates of reaction or side product formation when ninhydrin reacts with structurally different peptides. Similar variations in color value are also observed with different amino acids. The occurrence of different, but reproducible color yields does not affect the application of the ninhydrin reaction for analytical purposes. These differences may result from incomplete reaction of the dipeptides with ninhydrin during the 9-minute heating period in the 100° reaction coil under the conditions used here; the color values in Table I may not be strictly applicable to other analyzers if the buffer flow rate differs from the 50 ml per hour rate used in these experiments. If it is not convenient to adopt this flow rate, it will be necessary to establish operational color values for a given analyzer when maximum precision is needed.

Sensitivity for Analysis of Free D- *and* L-*Amino Acids.* The method can be used to determine the stereochemical purity of free amino acids. Since the dipeptide pairs are so well resolved, it is possible to load the column heavily with respect to one diastereomer in order to determine

[8] S. Yanari, *J. Biol. Chem.* **220**, 683 (1956).

very small amounts of a possible stereochemical impurity. The sensitivity as well as the accuracy of the method are illustrated by the analysis of a solution of alanine in which D-alanine was added at a final concentration of 0.003%. After preparation of the diastereomeric dipeptides, the solution was applied to the amino acid analyzer; the added D-alanine was quantitatively recovered. This result also shows that the L-leucine NCA used for the coupling was stereochemically pure to this degree. In general, for most of the amino acids, the sensitivity is about 0.01%. If, for example, one analyzes 20 μmoles of an L-L dipeptide, 2 nmoles of an L-D dipeptide can be easily detected on analyzers having the expanded scale. However, at this high load, it is advisable to run the effluent to drain during elution of the L-L dipeptide so as not to clog the reaction coil with the product of the ninhydrin reaction. With the dimethyl sulfoxide-ninhydrin reagent,[9] the product of the ninhydrin reaction is more soluble than that formed with the Methyl Cellosolve reagent.

Determination of D- and L-Amino Acid Residues in Peptides and Proteins

Racemization during Acid Hydrolysis. The interpretation of the results from an analysis of a free amino acid for D- and L-amino acid content is unequivocal so long as the sample has not been exposed to any conditions that could have caused racemization. With a peptide or a protein, the sample must first be hydrolyzed to liberate the constituent amino acids. Hydrolysis in 6 N HCl at 110° is usually the method of choice. Since some racemization of amino acids occurs during acid hydrolysis of a peptide or a protein, the amount of this racemization must be determined before the configurations of the amino acid residues in the peptide can be rigorously established. Treatment of the corresponding free amino acids with hot acid has been employed as a control for measurement of the racemization of amino acid residues during acid hydrolysis of a peptide.[1] However, this control may not always be adequate because the amino acid residues in some sequences of a peptide may be subject to increased racemization during hydrolysis; for example, in bacitracin A Ile-1 is completely epimerized during the hydrolysis,[10] and the phenylalanyl residue in L-Phe-L-Ser is racemized more than free L-phenylalanine during hydrolysis.[1]

When the stereochemical purity of a synthetic L-peptide is being evaluated by a study of an acid hydrolyzate of the material, the natural

[9] S. Moore, *J. Biol. Chem.* **243**, 6281 (1968).

[10] L. C. Craig, W. Hausmann, and J. R. Weisiger, *J. Amer. Chem. Soc.* **76**, 2839 (1954).

L-peptide is the best control for correction for the racemization that occurs during the hydrolysis. When the natural material is not available for comparison, it has been difficult to establish unequivocally the configurations of the amino acids in the synthetic peptide because there has not been a method for measuring increased racemization of some amino acid residues during acid hydrolysis. The same consideration applies to the examination of a natural peptide for D- and L-amino acid content.

Since racemization occurs concomitantly with removal of the α-hydrogen atom of an amino acid, hydrolysis with tritiated HCl provides a convenient means for measuring directly the amount of racemization that occurs during the acid hydrolysis of a peptide. Correction for the racemization that occurs during the hydrolysis is made for those amino acids that have side-chain hydrogen atoms which are not exchangeable under these conditions. The specific radioactivity of the amino acid is determined with a flow-cell scintillation counter attached to the amino acid analyzer. A standard graph is constructed from the results of an experiment in which pure L-alanine is heated for varying periods in [^{3}H]HCl. The specific radioactivity of the amino acid is plotted against the percentage of D-alanine formed. This graph is then used as a standard in other experiments for determination of the amount of isomer formed during the hydrolysis as calculated from the specific radioactivity of the amino acid. The *total* amounts of D- and L-amino acids in an acid hydrolyzate of a peptide or a protein are measured chromatographically after preparation of the diastereomeric dipeptides. The difference between these values is a measure of the amount of each enantiomer in the peptide.

Hydrolysis in Tritiated HCl.[4] The sample is dissolved in 100 μl of concentrated HCl in a Pyrex test tube (10 × 75 mm) which had been washed with hot $H_2SO_4:HNO_3$ (3:1). Tritiated water (100 μl) (1 Ci/g; New England Nuclear Corp. No. NET-001E) is added, and the contents are mixed; this operation and the sealing of the tubes should be carried out in a hood. The tubes are not frozen or evacuated before they are sealed; hydrolysis is carried out in an oven at 110°. Evacuation of the tubes is omitted so that the concentration of tritiated HCl will be the same for each sample. Each new batch of tritiated water is standardized in terms of disintegrations per minute per milliliter with a flow-cell scintillation counter attached to the amino acid analyzer as described below.

The presence of oxygen in the hydrolysis mixture leads to oxidation of methionine during hydrolysis; 10% of the methionine is converted to the sulfoxides, which are eluted just before aspartic acid, and 5% of the methionine is oxidized to the sulfone. Methionine sulfone, which is eluted just before threonine, incorporates tritium during heating in [^{3}H]HCl. This interferes with the determination of the small amount of radioactivity in threonine. If the peptide for analysis contains both methionine and threonine, this problem can be solved by addition to the hydrolysis mixture of 2-mercaptoethanol (1 μl), which prevents oxidation of methionine.[11]

After the heating in [^{3}H]HCl the amino acid solutions are transferred to 18 $\times$ 150-mm tubes and are concentrated to dryness in a rotary evaporator at 45–50° in a hood. The [^{3}H]HCl is collected in a Dry Ice: acetone-cooled trap and transferred to a bottle for disposal. The residues are washed twice with 1–2 ml of H_2O. More than 99% of the [^{3}H]HCl is removed during the evaporation; the exchangeable tritium on the carboxyl and the amino groups of the amino acids is quantitatively removed during the ion-exchange chromatography.

Determination of Incorporation of Tritium into Amino Acids. A portion of the hydrolyzate (corresponding to 0.1–1.0 μmole of each amino acid) in 0.20 *N* sodium citrate, pH 2.2, is chromatographed on the amino acid analyzer. The 0–∞ absorbance scale is used for this amount of amino acid. The acidic and neutral amino acids are eluted from a 0.9 $\times$ 62-cm column with 0.20 *N* sodium citrate, pH 3.25 and pH 4.00. With the pH 4.25 buffer, which is used for routine analyses, some radioactive material is eluted at the position of the buffer breakthrough, and this interferes with the determination of labeled methionine. The basic amino acids are eluted from a 0.9 $\times$ 13-cm column with 0.38 *N* sodium citrate, pH 5.26. A scintillation flow-cell (Nuclear-Chicago Model 6350,877260,8704,8437 Chroma/Cell Detector assembly with a 2-ml cell) is connected between the bottom of the ion-exchange column and the mixing manifold to record the amount of tritium in the effluent (0.8% counting efficiency as determined with tritiated H_2O). Specific radioactivity is defined as disintegrations per minute (dpm) per micromole of amino acid. The large amount of tritium used in the initial mixture for hydrolysis is necessary in order to attain a minimum counting rate of 500 cpm in the flow-cell apparatus for those amino acids which are only slightly racemized during the acid hydrolysis. If a flow-cell scintillation counter is unavailable, the divider pump on most amino acid analyzers can be used to divert an aliquot of the amino acids into a fraction collector. The amount of radioactivity can then be determined separately in a liquid scintillation counter.

Preparation of a Standard Graph with L-*Alanine.* Samples of

[11] H. T. Keutmann and J. T. Potts, Jr., *Anal. Biochem.* **29**, 175 (1969).

L-alanine (6.2 μmoles) are heated for various times in 6 N [^{3}H]HCl. The extent of ^{3}H incorporation (dpm/μmole) into alanine increases linearly with time and is proportional to the amount of D-alanine found (Fig. 3) as determined by the relative amounts of L-Leu-D-Ala and L-Leu-L-Ala after derivatization. The same total amount of radioactivity was found in each diastereomer for each of the heating times. The specific radioactivity of alanine, extrapolated to 100% racemization, is lower by a factor of 3.5 than the specific radioactivity of the initial [^{3}H]HCl which was used for the hydrolysis; this difference is attributed to an isotope effect. Correction for this isotope effect need not be applied, however, if an empirical standard graph (Fig. 3) is derived which expresses the relationship between the observed incorporation of tritium and the amount of D-isomer formed. Variations in the equipment available to a user or differences in experimental conditions, particularly temperature of hydrolysis, could lead to results different from those obtained in Fig. 3 in terms of the amount of radioactivity incorporated. The user should determine his own standard graph. It is particularly important to measure the specific radioactivity of each new batch of tritiated water so that the amount of tritium used for each hydrolysis may be related to that which was used in the determination of the standard graph.

The other amino acids which incorporated tritium exclusively at the α-carbon atom concomitant with racemization were determined as follows: the free L-amino acid was heated in 6 N [^{3}H]HCl as described for L-alanine in Fig. 3. After chromatography on the amino acid analyzer

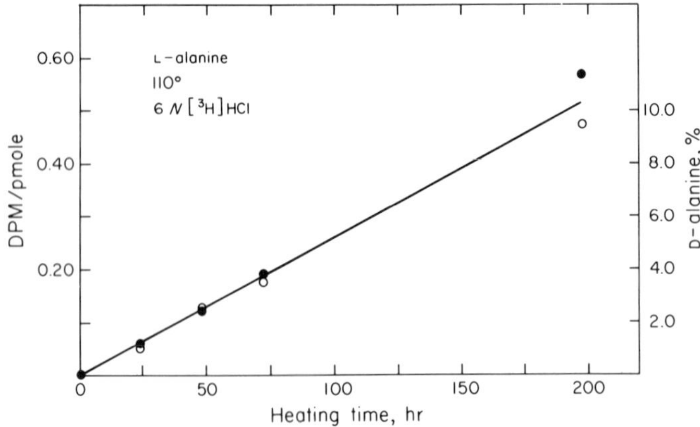

FIG. 3. Relationship between the amount of tritium incorporated (●——●) and the amount of D-alanine formed (○——○) during heating of L-alanine in 6 N [^{3}H]HCl at 110°.

TABLE II
RACEMIZATION OF FREE L-AMINO ACIDS UNDER THE CONDITIONS OF ACID HYDROLYSIS.[a] COMPARISON OF THE TWO METHODS OF ANALYSIS

L-Amino acid	Percent D-isomer as determined by	
	^{3}H-incorporation	L-D dipeptide
Alanine	1.4	1.0
Valine	0.2	0.7
Isoleucine	0.4	1.0[b]
Leucine	0.9	1.3
Serine	0.6	0.4
Threonine	0.9	0.5
Lysine	1.8	3.0
Arginine	1.4	1.6
Methionine	2.7	2.2
Proline	2.3	2.2

[a] Hydrolysis was carried out for 22 hours at 110°.
[b] Epimerization of L-isoleucine at the α-carbon atom affords D-alloisoleucine, which can be separated on the amino acid analyzer.

the amount of D-isomer formed was calculated from the specific radioactivity of the amino acid with Fig. 3 as a standard graph. The true amount of D-isomer present was determined by measurement of the relative amounts of diastereomeric dipeptides after derivatization with the appropriate NCA; the results of the two methods were then compared. In Table II are listed the amino acids that gave values for percentage of D-isomer within a range of about 1% by the two methods of analysis. Thus, hydrolysis with [^{3}H]HCl can be used for measurement of the racemization of these amino acids during acid hydrolysis of a peptide or a protein. Aspartic acid, glutamic acid, phenylalanine, tyrosine, and histidine incorporate more tritium than can be accounted for as the actual amount of D-isomer found. This is most likely due to incorporation of tritium at positions in the molecule other than the α-carbon atom.[4,12-15] Tritium is not completely removed from the labile positions of these amino acids when the mixture is treated with 6 N HCl for 24 hours at 110°. Thus, the amount of racemization during acid hydrolysis in tritiated HCl may be determined for 10 of the 16 amino acids that are usually present in a hydrolyzate of a peptide or a protein.

[12] S. Ratner, D. Rittenberg, and R. Schoenheimer, *J. Biol. Chem.* **135**, 357 (1940).
[13] D. Rittenberg, A. S. Keston, R. Schoenheimer, and G. L. Foster, *J. Biol. Chem.* **125**, 1 (1938).
[14] D. H. Meadows, O. Jardetzky, R. M. Epand, H. H. Ruterjans, and H. A. Scheraga, *Proc. Nat. Acad. Sci. U.S.* **60**, 766 (1968).
[15] C. H. W. Hirs, W. H. Stein, and S. Moore, *J. Biol. Chem.* **211**, 941 (1954).

Those amino acid residues which cannot be analyzed by this method can be studied in another way. When one of these residues is suspected of undergoing increased racemization during hydrolysis, the segment of the peptide chain in which it occurs can be synthesized and the amount of racemization that takes place during the hydrolysis can be studied in a separate experiment. This approach is tedious, but it was useful in the study of the amount of racemization of Phe-5 during hydrolysis of bradykinin.[1]

Application to the Analysis of Synthetic and Natural Peptides. As a practical test for correction of racemization during acid hydrolysis, synthetic L-bradykinin, a nonapeptide having the sequence L-Arg-L-Pro-L-Pro-Gly-L-Phe-L-Ser-L-Pro-L-Phe-L-Arg, which was synthesized by the solid phase method,[16] was chosen for study. The *total* amounts of D-amino acids in the acid hydrolyzate of this peptide were determined by analysis of the diastereomeric dipeptides prepared from the acid hydrolyzate. The fact that bradykinin contains only five different amino acids, four of which have optical isomers, makes the separation of the L-leucyl and L-glutamyl dipeptides a relatively simple chromatographic problem. The chromatograms (Figs. 4 and 5) are included as an example of how one can use the two parameters, pH and temperature, to find chromatographic conditions for the resolution of all the L-D dipeptides prepared from the acid hydrolyzate of a simple peptide. For a hydrolyzate of insulin A chain, which contains 10 different amino acids, D-leucine NCA (available from Pilot Chemical Corp.) was used for resolution of the D-D dipeptides prepared from a hydrolyzate of this peptide. With hydrolyzates of proteins that contain a full complement of amino acids, the problem of separating the diastereomers is likely to be more formidable. Thus far, complete separations of the diastereomeric dipeptides have been achieved for the D-isomers of serine, allothreonine, aspartic acid, and glutamic acid.[1] Another approach, which is generally applicable, is simply to isolate the amino acid by a preliminary chromatogram[17] and then to apply the derivatization to that sample.

The conditions for acid hydrolysis and the preparation of the hydrolyzate for the derivatization are illustrated for bradykinin. The peptide (5–10 μmoles) is heated in 1–2 ml 6 N HCl in an evacuated, sealed tube for 22 hours at 110°. The acid hydrolyzate is concentrated to dryness in a rotary evaporator at 40°. A few milliliters of water are added, and the hydrolyzate is again concentrated to dryness. The residue is then dissolved in 2.0 ml of the sodium borate buffer, and the solution

[16] R. B. Merrifield and J. M. Stewart, *Nature (London)* **207**, 522 (1965).
[17] C. H. W. Hirs, S. Moore, and W. H. Stein, *J. Amer. Chem. Soc.* **76**, 6063 (1954).

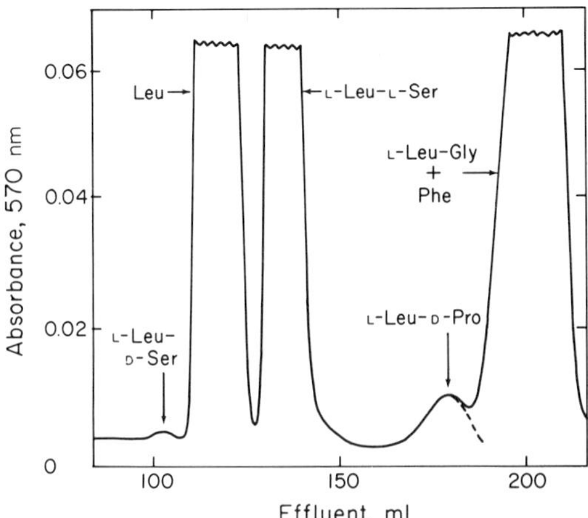

FIG. 4. Separation of L-leucyl dipeptides prepared from a hydrolyzate of synthetic L-bradykinin. Eluent: 0.20 N sodium citrate, pH 4.00, at 35°.

is adjusted to pH 10.2–10.3 with 1 N NaOH; the alkali is added dropwise with rapid mixing. The solution is cooled to 0° in an ice bath and the NCA is added as described above for a single amino acid. With bradykinin, one-half of the hydrolyzate is mixed with L-leucine NCA and the other half is treated with L-glutamic acid NCA. Two chromatographic systems are used for the separation of the dipeptides. A buffer of pH 4.00 is used to separate the L-leucyl dipeptides of serine, proline, and glycine, as shown in Fig. 4; the chromatographic column is maintained at 35° for this analysis. This combination of pH and temperature gives the best separations of the L-leucyl dipeptides in this mixture. The more retarded L-leucyl dipeptides of phenylalanine and arginine are not eluted under these conditions. The dipeptides of phenylalanine and arginine are separated as the L-glutamyl derivatives, as shown in Fig. 5. With these dipeptides, it is convenient to introduce a temperature and a buffer shift after the emergence of L-Glu-L-Phe. The material which is eluted at 185 ml, at the buffer breakthrough, consists mainly of isoglutamine formed by reaction of L-glutamic acid NCA with ammonia. The small peaks eluted just after L-Glu-D-Arg are reproducible and are probably side products of the coupling reaction with the NCA. However, they are formed in insignificant amounts compared with the dipeptides and are therefore neglected in the calculations. The load of the L-L dipeptides (1.5–2.0 μmoles) is sufficiently high to provide accurate data for the small amounts of L-D dipeptides present. Excellent separation of the dipeptides is obtained even at this high level of ninhydrin-positive material. The data in Table III under the heading

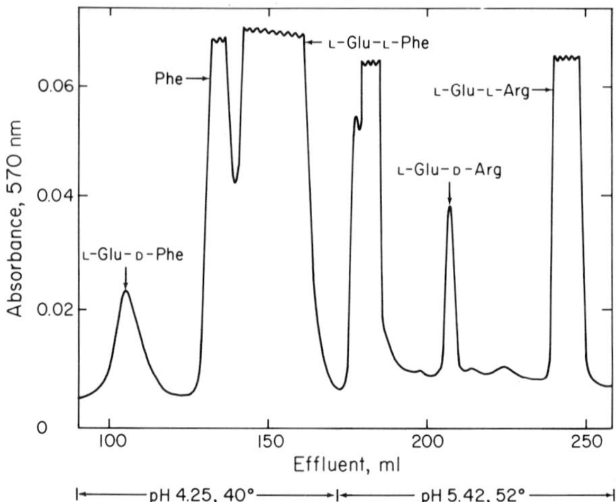

FIG. 5. Separation of L-glutamyl dipeptides prepared from a hydrolyzate of synthetic L-bradykinin.

L-D dipeptide show the total amounts of D-isomers of serine, proline, and arginine. This value includes the amount of D-isomer formed by racemization during acid hydrolysis as well as the amount which may have been present in the synthetic peptide.

The amount of D-isomer formed by racemization during acid hydrolysis is determined as follows: The synthetic peptide (1–2 μmoles) is dried in a desiccator over NaOH pellets and then heated for 22 hours at 110°

TABLE III
CORRECTION FOR RACEMIZATION OCCURRING DURING ACID HYDROLYSIS OF PEPTIDES[a]

Peptide	L-Amino acid	Percent of D-isomer as determined by	
		^{3}H incorporation	L-D dipeptide
Synthetic L-bradykinin	Serine	0.1	0.3
	Proline	2.8	2.3
	Arginine	1.7	1.6
Natural bacitracin A	Isoleucine	17.4	15.7[b]
	Leucine	5.8	6.8
	Lysine	2.2	3.0

[a] Hydrolysis was carried out for 22 hours at 110°.
[b] Epimerization of L-isoleucine affords D-alloisoleucine which can be separated on the amino acid analyzer.

in 200 μl of 6 N [^{3}H]HCl. A portion of the hydrolyzate is chromatographed on the amino acid analyzer equipped with the flow-cell scintillation counter as described above. Results for D-serine, D-proline, and D-arginine are calculated from the specific radioactivity of each amino acid (Table III) with Fig. 3 as the standard graph; the amounts of the D-isomers thus found are formed during the acid hydrolysis. Since the values obtained are the same as those for the total amount of each D-amino acid in the hydrolyzate, as determined by the amount of L-D dipeptide, the conclusion is that these residues in the original synthetic peptide were stereochemically pure. The amount of D-phenylalanine formed during acid hydrolysis could not be determined from the tritium incorporation because there was labeling in the side chain. However, evidence for the stereochemical purity of the phenylalanyl residues of synthetic bradykinin has been obtained[1] by preparation of phenylalanyl dipeptides corresponding to sequences in bradykinin and examination of the amount of racemization that takes place during hydrolysis of the peptides.

Another use of this method is for detecting labile hydrogen atoms which result from interactions between amino acid residues in a peptide. In natural bacitracin A, one isoleucine residue, Ile-1, forms a thiazoline ring with the adjacent cysteine residue, Cys-2. This ring formation results in the removal of the α-hydrogen atom of this particular isoleucine residue during acid hydrolysis to give 0.5 equivalent of alloisoleucine.[10] The total amounts of D-isomers of leucine and lysine are determined by analysis of the L-leucyl and L-glutamyl dipeptides, respectively, prepared from a hydrolyzate of natural bacitracin A; D-alloisoleucine is determined directly on the amino acid analyzer. Typical results are shown in Table III. The amounts of D-isomers of these amino acids that were formed by racemization during the hydrolysis was determined after hydrolysis in tritiated HCl. These results are also given in Table III under the heading "^{3}H incorporation." Since the peptide contains three L-isoleucine residues, the complete epimerization of one of them should give rise to 50% of 33.3% = 16.7% of the D-allo-isomer in the hydrolyzate; this is close to the value found (Table III). The result from tritium incorporation into isoleucine is also consistent with this course for the hydrolysis.

The single leucine residue of bacitracin A (Leu-3) is of the L-configuration. The amount of D-leucine in the hydrolyzate, determined as the diastereomeric dipeptide, is 5–6 times higher than that expected from the racemization of free L-leucine during acid hydrolysis (Table II). The amount of D-leucine agrees with that calculated from the amount of tritium incorporated and indicates that the increased racemization of leucine occurred during the hydrolysis. Craig *et al.*[18] noted the low optical rotation of the leucine isolated from a hydrolyzate of bacitracin A. The increased racemization of Leu-3 during acid hydrolysis may be due to some interaction with Cys-2. The data indicate that the lysine residue in bacitracin A is not racemized any more than the corresponding free amino acid during acid hydrolysis.

Sensitivity for Analysis of D- *and* L-*Amino Acid Residues in Proteins.* Since the amounts of D- or L-amino acid residues in a protein are determined by the difference between the total amount of D- or L-isomer in the hydrolyzate and that amount which was formed during the hydrolysis, the sensitivity of the method is not as great as that for the analysis of a given amino acid. The precision of the analysis is the sum of the precision of the chromatographic determination of diastereomers ($\pm 0.5\%$) plus the precision of the tritiated HCl method ($\pm 0.3\%$); the limits of precision are therefore about 1%.

Comments

The method finds its simplest application in the determination of the configuration of a single free amino acid. It possesses one main advantage over the enzymatic methods presently available: the configurations of amino acids that are poor substrates for a given enzyme can readily be determined by this technique. The method should be useful to the peptide chemist for determination of the stereochemical purity of the amino acids used as starting materials in a synthesis.

For determination of the stereochemical purity of a synthetic peptide, this technique is more sensitive than one based upon optical rotation, since the intrinsic rotation of the peptide may sometimes be low. In addition, the use of optical rotation is limited to those cases where the natural peptide is available for comparison. The present approach is more definitive than one in which the synthetic product is assumed to be stereochemically pure if the coupling of two amino acid derivatives as a model proceeds without racemization. Stereochemical purity can be demonstrated in some instances by hydrolysis of the synthetic peptide with aminopeptidases which are specific for L-amino acid residues. However, the occurrence of certain sequences, such as Arg-Pro-Pro- in bradykinin, precludes the use of enzymatic hydrolysis to establish the stereochemical purity of some peptides. For natural peptides and proteins, knowledge of the amount of racemization which has occurred during acid hydrolysis of the sample is prerequisite before the configurations of the amino acid residues can be rigorously established. The use of tritiated HCl for the hydrolysis has provided a solution to this

[18] L. C. Craig, W. Hausmann, and J. R. Weisiger, *J. Biol. Chem.* **199**, 865 (1952).

problem for 10 of the 16 optically active amino acids that are usually found in the hydrolyzate of a peptide or a protein.

Acknowledgments

The author is indebted to Dr. Stanford Moore and Dr. William H. Stein, in whose laboratory this work was carried out, and to Miss Wanda Jones for her expert assistance.

[3] Gas-Liquid Chromatography (GLC) of Amino Acid Derivatives

By J. J. PISANO

Of the many amino acid derivatives that may be analyzed by GLC, three are of particular interest in the present context: the amino acid phenylthiohydantoins (PTH's), the *N*-trifluoroacetyl (TFA) *n*-butyl esters, and the trimethylsilyl (TMS) derivatives. The PTH analysis will be presented in detail because the method is fully developed and in use in several laboratories determining amino acid sequences of polypeptides by Edman's procedure. The other two derivatives have been used for the assay of free amino acids. However, the methodology has only recently been delineated to the point where it may be generally useful. At this time an easy choice between the more established *N*-TFA *n*-butyl ester and the simpler TMS derivative cannot be made. Hence, only an outline of each method will be presented with some critical comments.

GLC methods for the separation of D and L amino acids deserve mention. An interesting application is the recent analysis of amino acids in the Murchison meteorite. The D and L enantiomers of certain amino acids were determined as the *N*-TFA-D-2-butyl esters.[1] Optically active stationary phases have also been used to separate stereoisomers of amino acids.[2]

Amino Acid PTH's and Amino Acid Sequence Analysis

Compared to the established thin-layer chromatographic procedure for identifying the PTH's,[3] the GLC method[4] is superior in resolving

[1] K. A. Kvenvolden, J. G. Lawless, and C. Ponnamperuma, *Proc. Nat. Acad. Sci. U.S.* **68**, 486 (1971).

[2] A convenient source of references is the bibliography section of *Journal of Chromatography*.

[3] P. Edman, *in* "Protein Sequence Determination" (S. B. Needleman, ed.), p. **211**. Springer-Verlag, Berlin and New York, 1970.

power, sensitivity, speed of analysis, and ease of quantitation.[5] In order to realize the full potential of the method, some attention to detail is required, particularly in the preparation of the support and the selection of the proper stationary phases.

Materials

Several manufacturers offer suitable instruments. However, metal injection ports and columns cause destruction of PTH's; hence glass columns and direct, i.e., on-column, injection are recommended. In order to accommodate the large number of samples that are formed in the automated Edman degradation,[3,6] it is desirable to have the capacity for the analysis of two samples simultaneously. This may be accomplished most economically with a double column oven, two hydrogen flame detectors, electrometers and recorders, and one temperature programmer. With this instrument package and simultaneous injection, one has the capacity of two chromatographs, but saves the cost of a separate column oven and temperature programmer.

The support, Chromosorb W, 100–120 mesh, the stationary phases DC-560, SP-400, OV-210, and OV-225, and the silylating reagents, *N*,*O*-bis(trimethylsilyl)acetamide or bis(trimethylsilyl)trifluoroacetamide and dichlorodimethylsilane may be purchased from various supply houses (e.g., Supelco, Inc., Bellefonte, Pennsylvania).

Preparation of Support and Column

The preparation of the support[7] has been modified to include a Na_2CO_3 prewash. In a 1-liter beaker is mixed 50 g of Chromosorb W and approximately 500 ml of 0.5 *M* Na_2CO_3. After soaking overnight generated fine particles are removed by decanting several times, distilled water being used. Concentrated HCl (500 ml) is added to the almost neutral support, which is allowed to stand over a period of 16–24 hours with occasional swirling. Fine particles are again removed by decantation, using distilled water. The support is dried at 140° and, while still warm, a 25-g portion is added to a 1-liter flask. A 5% solution (v/v) of dichlorodimethylsilane in toluene (200 ml) is added, and the mixture is degassed by the use of a suitable, trapped aspirator. The support is

---

[4] J. J. Pisano and T. J. Bronzert, *J. Biol. Chem.* **244**, 5597 (1969).

[5] A useful introduction to gas chromatography is: H. M. McNair and E. J. Bonelli, "Basic Gas Chromatography." Varian Aerograph, Walnut Creek, California, 1969.

[6] P. Edman and G. Begg, *Eur. J. Biochem.* **1**, 80 (1967).

[7] E. C. Horning, W. J. A. VandenHeuvel, and B. G. Creech, *Methods Biochem. Anal.* **11**, 69 (1963).

gently swirled 2 or 3 times during the 10–15-minute reaction period. The reagent is decanted, and the support is rinsed 3 times with toluene, care being taken that the support is always wet with toluene. This prevents reaction of the silylated support (which contains unreacted Si-Cl groups) with atmospheric moisture. The deactivation is effected by the addition of 300 ml of anhydrous methanol. The preparation is allowed to stand for 10–15 minutes, then the methanol is decanted and the support is rinsed with methanol until the supernatant is clear. Finally, the support is filtered on a sintered-glass funnel, rinsed with acetone, air-dried, and finally dried thoroughly in an oven at 140°.

The filtration technique[7] is used to coat the support with stationary phase. The amount of stationary phase used is generally expressed as its percentage (w/v) in the solvent. Thus a 10% DC-560 packing is prepared with 10 g of DC-560 made up to 100 ml with acetone. Typically, 75 ml of solvent containing stationary phase is added to 5 g of support in a 125-ml filter flask. The mixture is degassed by gentle swirling under reduced pressure (aspirator), filtered on a 150-ml sintered-glass funnel until apparently dry, and transferred to a dish and thoroughly dried at 140°. The volume of coating solution is not critical, but should be sufficient to allow transfer of the support (with swirling) to the sintered-glass funnel.

No single stationary phase tested has been found to be totally acceptable for the separation of all the PTH's in a mixture. However, two, SP-400 (like DC-560) and OV-225 (like XE-60) have certain unique features.[8] A blend of these phases plus OV-210 retains these features and provides the best single column prepared to date. This blend (CFC) consists of equal volumes of acetone solutions of SP-400 (11%), OV-210 (8%), and OV-225 (1%). Dilution of this blend with acetone, 1 part acetone, 2 parts blend, is recommended because it shortens the analysis time about 10 minutes and there is less baseline rise (due to column bleed) during temperature programming. Columns[9] 4 feet × 2 mm i.d. are usually conditioned with an initial helium[10] flow of 150 ml/minute and temperature of 50°. After 30 minutes the temperature

[8] J. J. Pisano and T. J. Bronzert, *Anal. Biochem.*, in press (1971).

[9] Columns are filled by gravity flow or with the aid of house vacuum. While filling they are gently tapped to ensure even packing. A vibrator should not be used as it may damage the support or give columns with prohibitively slow flow rates; e.g., <100 ml of helium per minute at 50° with an inlet pressure of 40 psi. Glass wool plugs at the column inlet and outlet are silylated with dichlorodimethylsilane, like the support.

[10] Helium is superior to the more common argon and nitrogen carrier gases. Resolution is best and flow rates may be varied over a wider range without affecting efficiency.

is raised at the rate of 0.5° per minute until it reaches 300°. It is held at this temperature for 16 hours or until a suitably low baseline rise (e.g., 2%) is obtained in the temperature-programmed analysis. The isothermal period and rate of temperature rise may be varied as long as volatile impurities are gradually exhausted.

Analyses are usually performed with the injector port temperatures at 250–270°. At lower temperatures (e.g., 200°) volatilization of asparaginyl, glutaminyl, tyrosyl, lysyl, histidyl, and tryptophanyl PTH's is incomplete; use of temperatures higher than 270° results in the decomposition most notably of glutaminyl, lysyl, and histidyl PTH's. Most derivatives decompose significantly near 300°. The hydrogen flame detector must be operated above the column temperature; 280–300° is suitable.

Standard Solutions

The derivatives obtained in an Edman degradation are readily soluble in solvents such as ethyl acetate and ethylene dichloride. However, some crystalline standards are not soluble in these solvents to the extent of at least 1 mg/ml. These are: the asparaginyl derivative, which is dissolved in 2:1 pyridine:ethyl acetate (4 mg/ml); glutaminyl PTH, in 1:4 pyridine:ethyl acetate (4 mg/ml); aspartyl and glutamyl PTH's in 1:4 pyridine:ethyl acetate (1 mg/ml); and histidyl PTH in 1:1 water:methanol (4 mg/ml). Methanol alone can be used for these derivatives, but less concentrated solutions (0.5 mg/ml) are obtained. All solutions are stored in screw-cap vials at 0–4°. Pyridine-containing samples are stable for only 1 or 2 days. Ethyl acetate or methanol solutions are stable for several weeks.[11]

Silylation of Phenylthiohydantoins

Amino acid PTH's differ markedly in their chemical and chromatographic properties (see table). Some, i.e., the aspartyl, glutamyl, and cysteic acid PTH's, must be converted to more volatile derivatives; others, i.e., lysyl, seryl, threonyl and *S*-carboxymethylcysteinyl (SCMC) PTH's, although measurable by direct injection, often have much better chromatographic properties (i.e., higher yields and better peak symmetry) when they are also converted to a suitable derivative. The derivative of choice is that formed by reaction with *N,O*-bis(trimethyl-

[11] Dioxane containing 0.01% w/v diethyldithiocarbamate may prove to be the best single solvent (T. J. Bronzert and J. J. Pisano, unpublished). It dissolves all phenylthiohydantoins except the histidine derivative. Solutions are stable at room temperature for several weeks. A possible limitation is the interference of the antioxidant with the silylation reaction. This is under study.

Grouping of Amino Acid PTH Derivatives According to Gas Chromatographic Behavior[a]

Group I	Group II	Group III
Alanine	Asparagine	Aspartic acid
Glycine	Glutamine	*S*-Carboxymethylcysteine
Valine	Tyrosine	Cysteic acid
Leucine	Histidine	Glutamic acid
Isoleucine	Tryptophan	Lysine
Methionine		Serine
Proline		Threonine
Phenylalanine		

[a] Group I amino acids are most volatile and generally give symmetrical peaks. Members of group II are least volatile. Histidine, asparagine, and glutamine show the greatest tendency to adsorb to the column packing, give tailing peaks, and low responses. Group III derivatives include those which must be silylated before analysis (aspartic, glutamic, and cysteic acids) and others which, when silylated, have significantly better chromatographic properties.

silyl)acetamide[12] which readily silylates PTH's at sites containing exchangeable protons.[4,13,14] The variable reactivities and stabilities of the derivatives, however, makes it necessary to compromise on the conditions of silylation.[4] Most convenient is "on column" silylation, in which the PTH is withdrawn into a 10-μl syringe containing an equal volume of reagent. In this situation, the PTH's should not be dissolved in solvents that will react with the reagent (e.g., methanol). Ethyl acetate and ethylene dichloride, acetonitrile, and dimethyl formamide are suitable solvents. Of the group II derivatives (see the table) the aspartyl, glutamyl, and lysyl PTH's appear to be completely silylated. The seryl and threonyl derivatives, however, barely react. They are treated with an equal volume of reagent in a sealed tube at 50° for 15 minutes.[4] Analysis of the hydroxy amino acids obtained from an automated degradation may not be improved by silylation when dithiothreitol[15] is used as a protective reagent because it yields interfering substances. This is of no consequence because the reducing agent so improves the yield and stability of the hydroxy amino acids that silylation is unnecessary. The same laboratory is investigating other reducing agents

[12] J. F. Klebe, H. Finkbeiner, and D. M. White, *J. Amer. Chem. Soc.* **88**, 3390 (1966).

[13] J. J. Pisano, *in* "Theory and Applications of Gas Chromatography in Industry and Medicine" (H. S. Kroman and S. R. Bender, eds.), p. 147. Grune and Stratton, New York, 1968.

[14] R. E. Harman, J. L. Patterson, and W. J. A. VandenHeuvel, *Anal. Biochem.* **25**, 452 (1968).

[15] M. A. Hermodson, L. H. Ericsson, and K. A. Walsh, *Fed. Proc. Fed. Amer. Soc. Exp. Biol.* **29**, 728 (1970) (Abstract).

(e.g., butanedithiol and ethanethiol) which might afford better protection and no interference.

Quantitation

Although identification of the cleaved amino acid is the primary aim of a sequential degradation, the yield is also sometimes important. The GLC method is capable of quantitatively determining the amount of sample injected, but the significance of the value obtained must be ascertained in terms of prior handling of the sample and the stability of the particular derivative. Thus, it is well known that in the Edman degradation several derivatives, especially seryl, threonyl, asparaginyl, glutaminyl, tryptophanyl, histidyl, and SCMC PTH's are often obtained in low yields that are difficult to control. Quantitation of the stable derivatives is obviously more useful in following the efficiency of the sequence analysis.

The high resolving power as well as the ease of quantitation inherent in the GLC method make it possible to analyze small peptides by an additive procedure. In this approach an aliquot of the reaction mixture is removed after each cleavage step and analyzed after conversion.

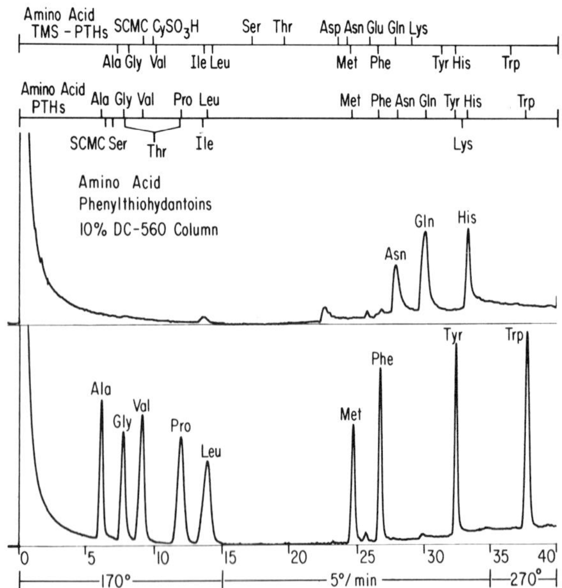

FIG. 1. Separation of PTH's on a 4-foot × 2 mm i.d. 10% DC-560 column. Sample size: 2 μg each of asparaginyl, glutaminyl, histidyl, and tryptophanyl PTH's; all others, 1 μg each. Full scale deflection is 3×10^{-10} A with a 5 mV recorder. Helium flow: 65 ml per minute.

Since some of the released derivative remains in the reaction vessel, the chromatogram reveals all prior steps of the degradation together with the newly released derivative.[16] The principal advantage of this approach is that extraction steps that often result in serious loss of peptides are omitted.

Chromatographic Results

A column in general use (Fig. 1) and particularly suited for group I derivatives[8] is one prepared with 10% DC-560 (or SP-400, which is purified DC-560). However, columns made with the CFC blend are suited for all the PTH's, and they provide superior resolution. Only the tyrosyl and lysyl PTH's are poorly separated, but their silyl derivatives are completely resolved (Fig. 2). The behavior of the prolyl derivative on all columns is unique in that its position, relative to the other derivatives, is temperature dependent. If a newly prepared CFC column does not adequately resolve prolyl PTH from valyl or glycyl PTH, the initial isothermal temperature can be either raised to move prolyl PTH closer to glycyl PTH or lowered to move it closer to valyl PTH. Sub-

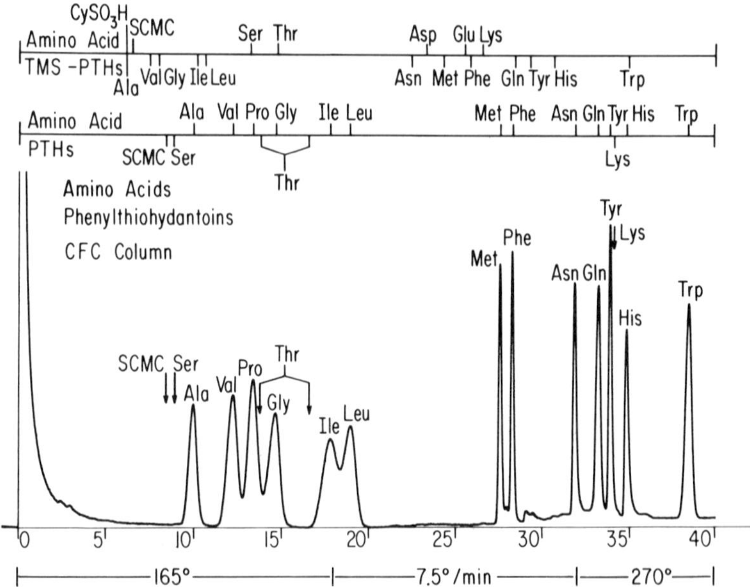

FIG. 2. Separation of PTH's with the CFC blend. Sample size 0.5 μg each of methionyl and phenylalanyl PTH's; all others same as Fig. 1 (i.e., 1–2 μg). Helium flow 105 ml per minute. Electrometer setting and column size same as Fig. 1.

[16] D. E. Vance and D. S. Feingold, *Nature* (*London*) **229**, 121 (1971).

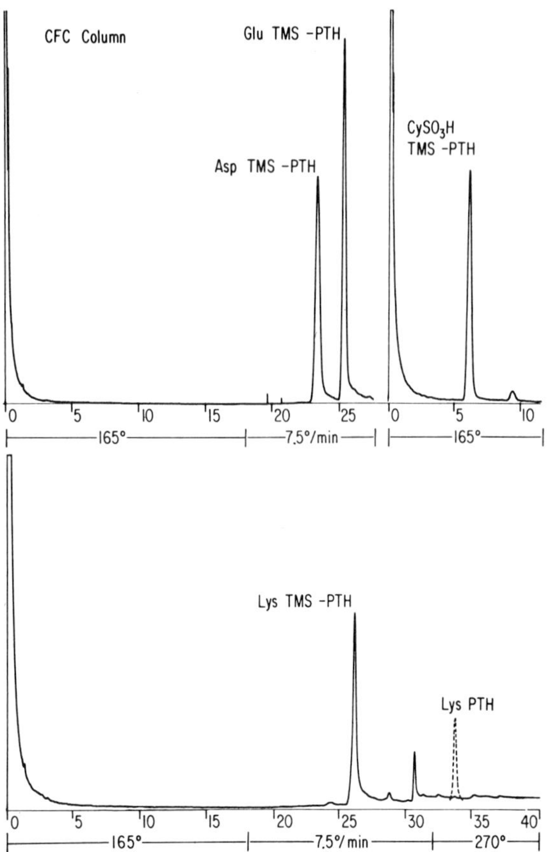

FIG. 3. Separation on the CFC column of the trimethylsilyl products of 2 μg of the aspartyl, glutamyl, and lysyl PTH's and 4 μg of cysteic acid PTH. The dashed peak is 4 μg of lysyl PTH before silylation. Helium flow: 105 ml per minute.

sequent adjustment of the helium flow rate will then position the group I peaks to any desired time without affecting column efficiency. Figures 3 and 4 show the silyl derivatives of the group III compounds.

Methylthiohydantoins

Several laboratories have recently considered substituting the more volatile methylisothiocyanate for phenylisothiocyanate.[17-21] The ad-

[17] A. Dijkstra, H. A. Billiet, A. H. Van Doninck, H. Van Velthuyzen, L. Matt, and H. C. Beyerman, *Rec. Trav. Chim. Pays-Bas* **86**, 65 (1967).

[18] F. F. Richards, W. T. Barnes, R. E. Lovins, R. Salomone, and M. D. Waterfield, *Nature (London)* **221**, 1241 (1969).

[19] M. D. Waterfield and E. Haber, *Biochemistry* **9**, 832 (1970).

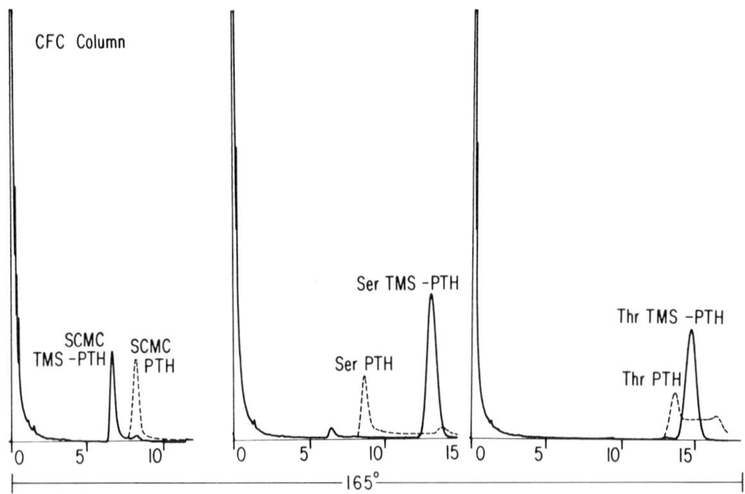

FIG. 4. Same as Fig. 3: 1–2 μg of SCMC, seryl, and threonyl PTH's silylated. Dashed peaks were obtained before silylation.

vantages of such a substitution are not yet established, but it may be noted that the chromatographic properties of the methylthiohydantoins (MTH's) are comparable to those of the phenyl derivatives.[8] A column containing the stationary phase OV-225 is better than the CFC blend and previously reported stationary phases[16–21] for resolving the MTH's. Seryl MTH is troublesome. The standard is unstable and tends to polymerize, but serine obtained from a degradation has been identified.[16,19]

Application of the Method

In the manual Edman degradation,[3] all the PTH's except histidyl, arginyl, and cysteic acid PTH, are recovered in ethyl acetate. The latter remain in the aqueous phase. The ethyl acetate solution is concentrated in a nitrogen stream to less than 0.5 ml. This solution is quantitatively transferred to a vial[22] and evaporated to dryness. The residue is redissolved in a small measured volume of ethyl acetate, usually at a concentration of 0.1–1.0 μg/μl. Although, as noted above, some crystalline amino acid thiohydantoin standards are poorly soluble in this solvent, the samples obtained from an Edman degradation will dissolve with brief warming of the vial at 40–50°. A 1–8-μl aliquot is

[20] J. E. Attrill, W. C. Butts, W. T. Rainey, Jr., and J. W. Holleman, *Anal. Lett.* **3**, 59 (1970).
[21] D. E. Vance and D. S. Feingold, *Anal. Biochem.* **36**, 30 (1970).
[22] Kontes Glass Co., Vineland, New Jersey, Catalogue No. K749000.

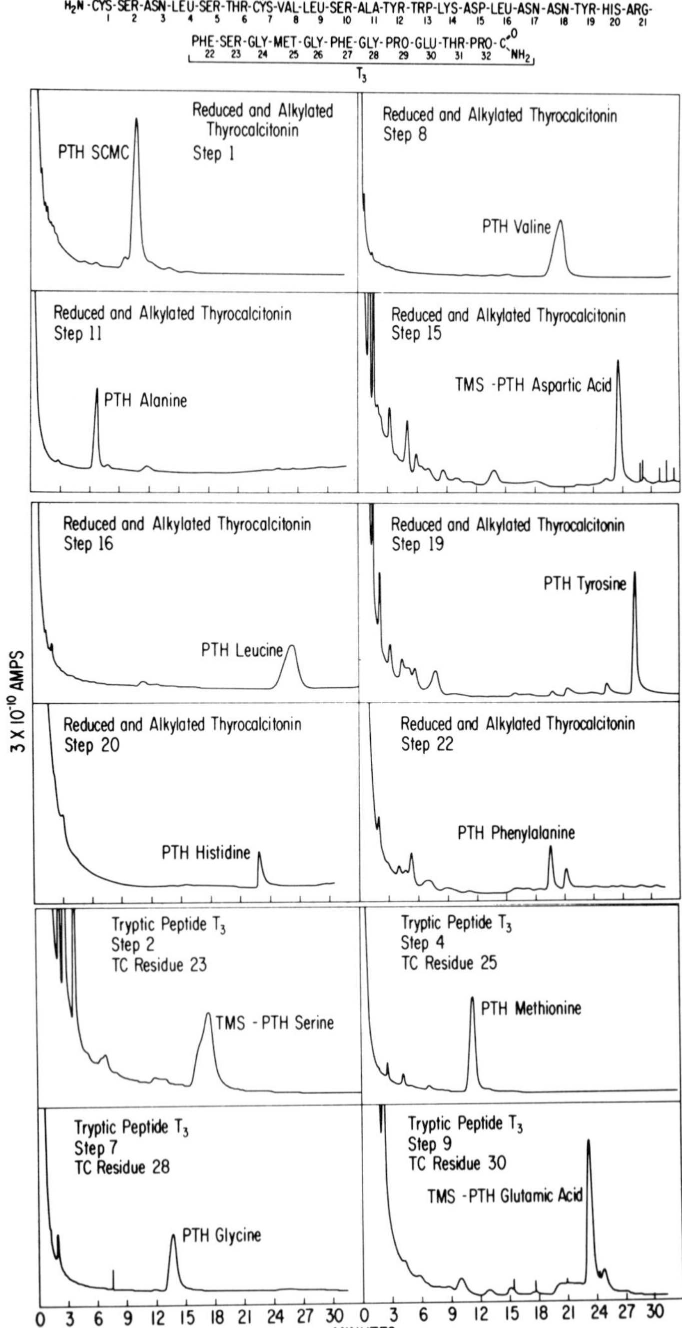

H2N-CYS-SER-ASN-LEU-SER-THR-CYS-VAL-LEU-SER-ALA-TYR-TRP-LYS-ASP-LEU-ASN-ASN-TYR-HIS-ARG-
1 2 3 4 5 6 7 8 9 10 11 12 13 14 15 16 17 18 19 20 21
PHE-SER-GLY-MET-GLY-PHE-GLY-PRO-GLU-THR-PRO-C=O NH2
22 23 24 25 26 27 28 29 30 31 32
T3
Reduced and Alkylated Thyrocalcitonin Step 1
PTH SCMC
Reduced and Alkylated Thyrocalcitonin Step 8
PTH Valine
Reduced and Alkylated Thyrocalcitonin Step 11
PTH Alanine
Reduced and Alkylated Thyrocalcitonin Step 15
TMS -PTH Aspartic Acid
Reduced and Alkylated Thyrocalcitonin Step 16
PTH Leucine
Reduced and Alkylated Thyrocalcitonin Step 19
PTH Tyrosine
Reduced and Alkylated Thyrocalcitonin Step 20
PTH Histidine
Reduced and Alkylated Thyrocalcitonin Step 22
PTH Phenylalanine
Tryptic Peptide T3 Step 2 TC Residue 23
TMS - PTH Serine
Tryptic Peptide T3 Step 4 TC Residue 25
PTH Methionine
Tryptic Peptide T3 Step 7 TC Residue 28
PTH Glycine
Tryptic Peptide T3 Step 9 TC Residue 30
TMS -PTH Glutamic Acid
3 X 10-10 AMPS
0 3 6 9 12 15 18 21 24 27 30 0 3 6 9 12 15 18 21 24 27 30
MINUTES

injected. With a dual instrument; a silylated sample is analyzed at the same time by withdrawing 1–4 μl of *N,O*-bis(trimethylsilyl)acetamide into the syringe followed by an equal volume of the sample. The second injection follows the first by only a few seconds. If no identification has been made, the aqueous phase is examined. Arginine PTH may be identified by spot tests for the guanidine group.[22a] Histidine[23] may also be identified by gas chromatography after extraction from the aqueous phase[4] adjusted to pH 8.5.

Two examples of the application of the technique may be cited. The complete covalent structures of bovine thyrocalcitonin[24] (manual degradation) and parathyroid hormone[25] (manual and automated) were determined utilizing the GLC method for the identification of the Edman derivatives. The chromatograms (Fig. 5) are nearly free of overlap and extraneous peaks, thus allowing unequivocal identification of the cleaved amino acid at each step in the sequence. Most contaminants appear close to the solvent, and none coelute with any of the silylated PTH's. The shoulder on the serine peak (residue 23) is due to incomplete silylation of the sample.

The gas chromatographic method appears to be gaining in popularity, but most applications employ the previously reported DC-560[4] or the similar SP-400. Surprisingly little has been reported with the XE-60 (OV-225) column, which complements the DC-560 column in that it clearly distinguishes isoleucyl and leucyl PTH's and gives more symmetrical peaks for group II derivatives. These two columns were used extensively in the author's laboratory in early applications of the technique[4] but now the CFC blend is used exclusively for the determination of all PTH's except arginine.[8]

An automatic sample applicator (several are commercially available) offers the potential of unattended gas chromatographic analysis, but it would appear that the immediate answers often desired would limit this approach to some extent. Another factor to consider is the possible deterioration of sample in the sample holder due to impurities introduced by reagents used in the sequential degradation. However, crystalline

[22a] S. Yamada and H. A. Itano, *Biochim. Biophys. Acta* **130,** 538 (1966).

[23] F. Sanger and H. Tuppy, *Biochem. J.* **49,** 463 (1951).

[24] H. B. Brewer, Jr. and R. Ronan, *Proc. Nat. Acad. Sci. U.S.* **63,** 940 (1969).

[25] H. B. Brewer, Jr. and R. Ronan, *Proc. Nat. Acad. Sci. U.S.* **67,** 1862 (1970).

FIG. 5. Selected steps from the Edman degradation of bovine thyrocalcitonin. Either the DC-560 or CFC column was employed. The aspartyl, glutamyl, and seryl residues were identified after silylation. Histidyl PTH in step 20 was recovered from the water layer in the conversion step of the Edman procedure. From H. B. Brewer, Jr. and R. Ronan, *Proc. Nat. Acad. Sci. U.S.* **63,** 940 (1969) and J. J. Pisano, T. J. Bronzert, and H. B. Brewer, Jr., *Anal. Biochem.*, in press.

standard PTH's are stable overnight and give results that are indistinguishable from manual injections with a Barber-Colman instrument.

Comments on the Procedure

In the preparation of the support, fine particles present initially or generated by the carbonate and acid washes must be removed in the preparation of efficient columns. As many as 25 decantations after the base and the acid soakings and 3 decantations with methanol after silylation may be necessary. The dried support should be handled carefully, as it is easily crushed, exposing unsilylated sites that cause peak tailing and adsorption of sample.

It is not uncommon for the first few samples to tail on a new column. However, after a few analyses, the column stabilizes and may be used for months or even a year if not abused by the use of excessive temperatures or dirty samples. The seriousness of the problem is determined by the increase in baseline rise during temperature programming and the decrease in resolution, yield, or loss of the derivatives. Such contaminated columns are usually discarded, although removal of the top inch or so of packing will occasionally improve the column.

Some phenylthiohydantoins decompose during GLC analysis, but this decomposition is remarkably reproducible and is not a limitation in quantitation. Yields are determined by comparing peak areas to standards run preferably on the same day. As little as 0.02 μg of many derivatives is adequate, but 0.2 μg may be required for detection of asparaginyl, glutaminyl, histidyl, and lysyl PTH's. Lysyl and SCMC PTH's decompose significantly, and the observed peaks probably represent degradation products. Asparaginyl, glutaminyl, and histidyl PTH's also decompose to some extent. They also tend to adsorb to the packing more than the other derivatives, as evidenced by tailing peaks, and they are the first lost when columns become contaminated or stripped of the coating of stationary liquid phase. Seryl PTH and to a lesser extent threonyl PTH may be dehydrated during the conversion step of the Edman degradation or in the injection port of the gas chromatograph. When crystalline standards are injected, the peak obtained for serine is the anhydro compound,[26] whereas threonine only partially dehydrates. A threonine sample obtained from a degradation, on the other hand, gives predominantly the anhydro compound. Comparing crystalline standards, the approximate peak area ratios observed for the unstable derivatives with reference to alanine (1.0) are: serine and threonine,

[26] J. J. Pisano, W. J. A. VandenHeuvel, and E. Horning, *Biochem. Biophys. Res. Commun.* 7, 82 (1962).

0.5; histidine, glutamine, and SCMC, 0.4; asparagine, 0.3; and lysine, 0.1. All others are near 1.0.

GLC Analysis of Amino Acids

When GLC methods for amino acid analysis were first contemplated some 15 years ago, the now classical ion-exchange procedure took 24 hours and required about 100 nmoles of each amino acid. Current methods routinely require less than 2 hours and a tenth of the sample.[27] With little instrumental modification, analyses using 1 nmole have been reported.[28] While these advances were being made with the ion-exchange procedure, a superior GLC method has not been developed. However, very recent developments now make it likely that the GLC approach will be a worthy alternative to the classical ion-exchange procedure. Analysis of 0.1 nmole of sample in about 1 hour is now feasible. The lower cost of GLC analysis is also an important factor.

Since it is not the purpose to review the field (for reviews see references cited in footnotes 13, 29–33), the most recent and relevant work is given and early important contributions are regrettably omitted.

The Problem

One may reasonably ask why, after 15 years and numerous attempts, is there no established GLC procedure for amino acids. The answer is that most investigations were superficial and never dealt systematically with the chemical and analytical problems inherent in the method.

Since free amino acids are not sufficiently volatile for direct GLC analysis, they must be converted to suitable derivatives. The derivative(s) must be applicable to all the protein amino acids, easily prepared, both quantitatively and reproducibly, and be manageably stable at the submicrogram level. The difficult amino acids have proved to be arginine, histidine, cysteine, cystine, and tryptophan, and to a lesser extent serine, threonine, hydroxyproline, tyrosine, lysine, and methionine. The amides, asparagine and glutamine, are normally

[27] D. H. Spackman, Vol. XI [1] p. 3.

[28] P. B. Hamilton, Vol. XI [2] p. 15.

[29] H. M. Fales and J. J. Pisano, *in* "Biochemical Applications of Gas Chromatography" (H. A. Szymanski, ed.), p. 39. Plenum Press, New York, 1964.

[30] B. Potteau, *Bull. Soc. Chim. Fr.* 3747 (1965).

[31] B. Weinstein, *Methods Biochem. Anal.* **14**, 203 (1966).

[32] W. J. McBride and J. D. Klingman, *in* "Lectures on Gas Chromatography" (L. R. Mattick and H. A. Szymanski, eds.), p. 25. Plenum, New York, 1967.

[33] K. Blau, *in* "Biomedical Applications of Gas Chromatography" (H. A. Szymanski, ed.), p. 1. Plenum, New York, 1968.

determined as their amino acid counterparts as they are hydrolyzed in the commonly used esterification procedures.

Of the dozens of derivatives that have been studied, two appear superior. They are the *N*-trifluoroacetyl *n*-butyl esters,[34,35] and the trimethylsilyl derivatives.[36]

N-Trifluoroacetyl n-Butyl Esters

The formation of the derivative is effected in two steps using *n*-butanol and then trifluoroacetic anhydride (TFAA) as follows:

$$RCH(NH_2)COOH \xrightarrow[100°/15\ min]{C_4H_9OH/HCl} RCH(NH_2)COOC_4H_9 \cdot HCl \xrightarrow[150°/5\ min]{TFAA/CH_2Cl_2} RCH(NHCOCF_3)COOC_4H_9 \quad (1)$$

The butyl esters are superior to the corresponding methyl esters because the latter are so volatile that some are subject to serious losses during the workup.[37,38] The problems of poor solubility of amino acids in butanol and slow rate of ester formation have been overcome by ultrasonic mixing of the amino acid suspension in *n*-butanol which is made 3 *N* in HCl, prepared by bubbling dry acid into the alcohol.[39] Esterification is complete in 15 minutes at 100° except for isoleucine, which is about 85% converted in this time. Tryptophan is partially decomposed, but again about 85% is recovered. These two amino acids are best measured by correcting for losses determined by simultaneously analyzing a calibration mixture. With protein hydrolyzates which contain no tryptophan a 35-minute esterification time may be employed for complete conversion of isoleucine without detrimental effects on the other amino acids. Asparagine and glutamine are not mentioned. Their conversion to the corresponding acids during the esterification and acylation procedure is likely.

Acylation of the amino acid ester hydrochlorides has been troublesome due to the difficulty of acylating the guanidinium moiety of arginine and to the instability of both acylated hydroxyamino acids and the *N*-acylimidazole moiety of histidine. The problems have been overcome by carrying out the acylation in a sealed tube (Teflon-lined screw cap tube) at 150° for 5 minutes and then directly injecting an aliquot of the reaction mix without further handling. Cysteine values have been reported

[34] E. Bayer, *in* "Gas Chromatography" (D. H. Desty, ed.), p. 333. Academic Press, New York, 1958.

[35] C. Zomzely, G. Marco, and E. Emery, *Anal. Chem.* **34,** 1414 (1962).

[36] K. Rühlmann and W. Giesecke, *Angew. Chem.* **73,** 113 (1961).

[37] R. W. Zumwalt, D. Roach, and C. E. Gehrke, *J. Chromatogr.* **53,** 171 (1970).

[38] C. W. Gehrke, K. Kuo, and R. W. Zumwalt, *J. Chromatogr.* **57,** 209 (1971).

[39] D. Roach and C. W. Gehrke, *J. Chromatogr.* **44,** 269 (1969).

but without comment on the extent of its oxidation during derivatization. Usually no more than a 50 *M* excess of TFAA is used. Greater quantities are likely to interfere in the chromatography because TFAA elutes before all the amino acids, and it tails badly.[37]

Precautions

Long experience in the preparation of *N*-TFA *n*-butyl esters has led to specific recommendations that must be followed to make the method applicable to all the amino acids.[37,38] Thus, it has been known for some time that the derivatives of the hydroxy amino acids and cysteine are particularly unstable.[40–42] Water must be excluded during derivatization and chromatography. Furthermore, all reagents and gases must be pure and metal contaminants avoided (particularly for methionine). Samples are best analyzed with all-glass columns using on-column injection.

Considerable effort has gone into the selection of columns. The main search has been directed at obtaining suitable resolution without destruction of the sensitive amino acids. Polyester phases[43] and hydrated diatomaceous earth supports[44] have been implicated in the destruction of the hydroxy and mercapto amino acid derivatives. No single column has been found, but two, a polyester and a mixed polysiloxane, used in parallel, permit the assay of all derivatives (Fig. 6).

Results

Pitfalls in the GLC method have largely been overcome recently.[37,38] Analyses of free amino acids in bovine plasma as well as human urine and of hydrolyzates of corn and soybeans have been reported using an older transesterification method in which the butyl esters were prepared from methyl esters.[37] These specialists have obtained results that compare well with those they obtained by classical ion-exchange procedures. There is every reason to believe that equally good results will be obtained with the new direct esterification with butanol. A total analysis could then be accomplished in about an hour,[38] which is half the time required by ion-exchange methods.

Trimethylsilyl Derivatives of Amino Acids

Silylation of amino acids[12,36] is now readily accomplished in a single step employing the powerful silylating reagents *N,O*-bis(trimethylsilyl)

[40] F. Weygand and H. Rinno, *Chem. Ber.* **92**, 517 (1959).
[41] A. Darbe and K. Blau, *Biochim. Biophys. Acta* **100**, 298 (1965).
[42] S. Makisumi and H. A. Saroff, *J. Gas Chromatogr.* **3**, 21 (1965).
[43] A. Darbe and K. Blau, *Biochim. Biophys. Acta* **126**, 591 (1966).
[44] D. Roach and C. W. Gehrke, *J. Chromatogr.* **43**, 303 (1969).

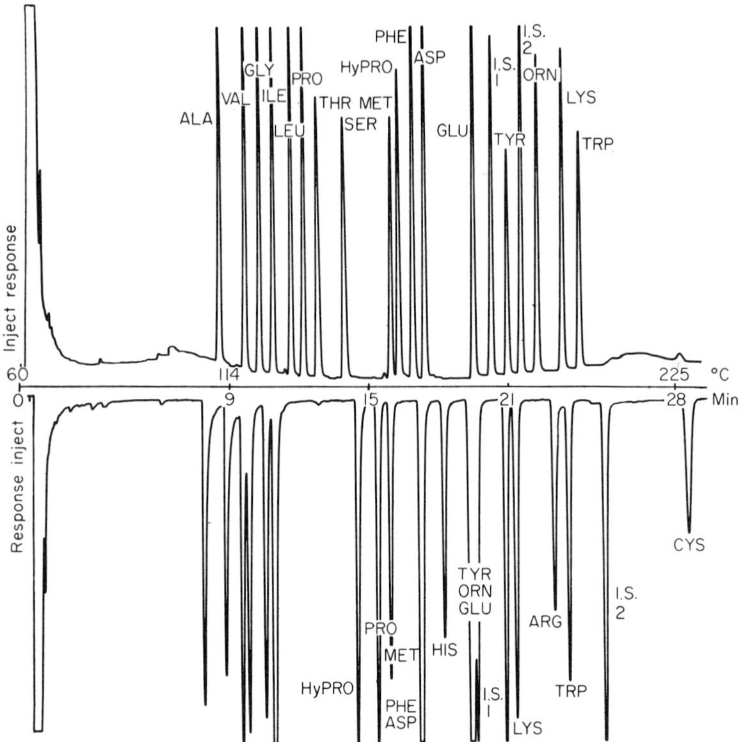

FIG. 6. Separation of *N*-TFA *n*-butyl esters of amino acids on a 1.5 m × 4 mm i.d., 0.65 w/w% EGA column. Support: acid-washed Chromosorb W 80–100 mesh. Sample: approximately 0.6 μg each. Internal standard, I.S. 1: tranexamic acid. Lower chromatogram, column: 1.5 m × 4 mm i.d. containing 2.0 w/w% OV-17, 1.0% w/w% OV-210 on Supelcoport, 100–120 mesh. I.S. 2: *n*-butyl stearate. From C. W. Gehrke, K. Kuo, and R. W. Zumwalt, *J. Chromatogr.* **57**, 209 (1971).

acetamide,[12] or *N,N*-bis(trimethylsilyl)trifluoroacetamide (BSTFA).[45] The typical reaction is:

$$\mathrm{RCH(NH_2)COOH + 2CF_3CON[Si(CH_3)_3]_2 \xrightarrow[150°/2.5\ hr]{CH_3CN} RCH[NHSi(CH_3)_3]COOCSi(CH_3)_3} \quad (2)$$

A systematic study of the silyl products of all the amino acids has not been undertaken, but the results from several laboratories indicate that trimethylsilyl-proton exchange also occurs with the hydroxy, mercapto, imidazole, indole, and guanidinium moieties of the amino acids and presumably the amide nitrogens of asparagine and glutamine under the

[45] C. W. Gehrke, H. Nakamoto, and R. W. Zumwalt, *J. Chromatogr.* **45**, 24 (1969).

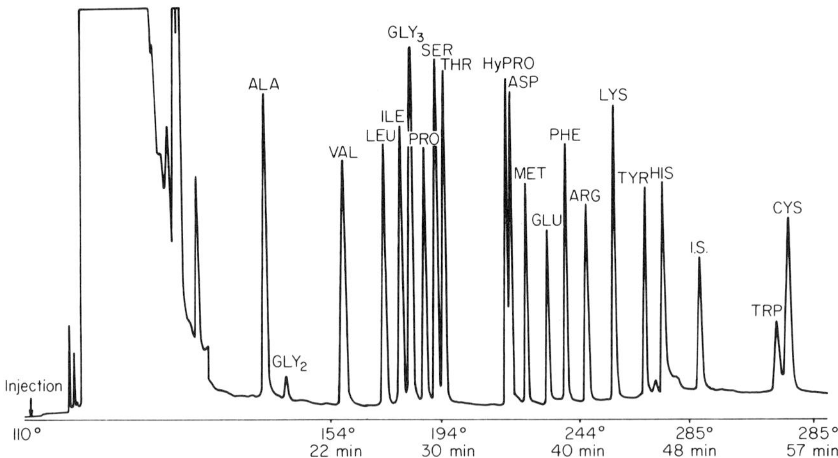

FIG. 7. Separation of TMS amino acids on a 6 m × 2 mm i.d. 10 w/w% OV-11 column. Support: Supelcoport 100/120 mesh. Sample: 1 μg each amino acid. I.S.: phenanthrene. Carrier gas: N_2. Silylation conditions: BSTFA —CH_3CN (1:1) 150° for 2 hr. Slightly inferior resolution was obtained on a 2 m column. From C. W. Gehrke and K. Leimer, *J. Chromatogr.* **57**, 219 (1971).

appropriate conditions.[12,14,45-51] The α- and ε-amino groups of glycine and lysine, respectively, exchange both protons when reacted in polar solvents at high temperatures.[45,49] BSTFA has been investigated in a variety of solvents which markedly influence the rate and products of the reaction.[49,51] Silylated amino acids are quite reactive and labile. They are themselves powerful silylating reagents, a fact that may not be fully appreciated, but must be considered when crude samples are tested and relatively large amounts of acetamide are generated. The silylated amino acids could silylate the acetamide and contaminants.

Working with amino acid standards and hydrolyzates of ribonuclease and casein, impressive results have been obtained[46] (Fig. 7). It is possible that the single-step silylation procedure will supplant the *N*-TFA *n*-butyl ester method, which is a more cumbersome 2-step operation.

Comment

In considering the relative merits of the GLC method to established ion-exchange procedures, an important factor is the sample preparation

[46] C. W. Gehrke and K. Leimer, *J. Chromatogr.* **57**, 219 (1971).
[47] E. D. Smith and H. Sheppard, *Nature (London)* **208**, 878 (1965).
[48] E. D. Smith and K. L. Shewbart, *J. Chromatogr.* **7**, 704 (1969).
[49] K. Bergstrom, J. Gurtler, and R. Blomstrand, *Anal. Biochem.* **34**, 74 (1970).
[50] E. D. Smith, J. M. Oathout, and G. T. Cook, *J. Chromatogr. Sci.* **8**, 291 (1970).
[51] C. W. Gehrke and K. Leimer, *J. Chromatogr.* **53**, 201 (1970).

prior to derivatization. Most biological samples require a separation of amino acids from interfering substances on a cation-exchange resin. Some samples such as urine require a clean-up on both cation and anion-exchange resins. In less time than it takes to accomplish this, the samples could be analyzed directly by the established ion-exchange methods. Clean-up of hydrolyzates of most pure proteins and peptides is not necessary, and it would seem that the GLC method has its greatest applicability in this area of analysis. However, it remains to be seen if the nonspecialist can realize the full potential of the GLC method, particularly at the nanogram level,[52] where the problems of derivatization, handling, stability, and chromatography (e.g., adsorption to the column packing) are more likely to be manifest.

[52] R. W. Zumwalt, K. Kuo, and C. W. Gehrke, *J. Chromatogr.* **57,** 193 (1971).

[4] Determination of Tryptophan[1]

By Teh-Yung Liu

As the precision of analytical methods for the determination of amino acids has increased, the limiting factor in determining the amino acid composition of a protein has become the extent to which the composition of the hydrolyzate is a true reflection of the composition of the parent protein. The two major problems associated with acid hydrolysis utilizing 6 *N* HCl are the destruction of labile amino acids, serine, threonine, and tryptophan, and the slow hydrolysis of some peptide linkages between bulky, sterically hindered amino acids. With the exception of tryptophan, these problems have been circumvented by extrapolating values obtained from 22- and 72-hour hydrolyzates. The quantitative estimation of tryptophan has usually been carried out by methods that avoid acid hydrolysis. Examples include the spectrophotometric procedure of Goodwin and Morton,[2] the colorimetric procedure of Spies and Chambers[3] and that of Barman and Koshland,[4] and the alkaline hydrolysis procedure of Drèze.[5] More recently, Matsubara and Sasaki[6] reported

[1] Research carried out at Brookhaven National Laboratory under the auspices of the U.S. Atomic Energy Commission.

[2] T. W. Goodwin and R. A. Morton, *Biochem. J.* **40,** 628 (1946).

[3] J. R. Spies and D. C. Chambers, *Anal. Chem.* **21,** 1249 (1959).

[4] T. E. Barman and D. E. Koshland, Jr., *J. Biol. Chem.* **242,** 5771 (1967).

[5] A. Drèze, *Bull. Soc. Chim. Biol.* **42,** 407 (1960).

[6] H. Matsubara and R. M. Sasaki, *Biochem. Biophys. Res. Commun.* **35,** 175 (1969).

a procedure for the analyses of tryptophan in proteins by acid hydrolysis in 6 *N* HCl containing 4% thioglycolic acid. This procedure, however, cannot be applied to glycoproteins.

The method described here was developed by Liu and Chang[7] and utilizes *p*-toluenesulfonic acid as the catalyst for hydrolysis rather than HCl. We investigated the suitability of *p*-toluenesulfonic acid for acid hydrolysis of proteins, glycoproteins, and peptides after we noted that it was more effective and less destructive than HCl as a catalyst for methanolysis in studies of polysaccharides from meningococcus.[8] The application of this catalyst for methanolysis and the determination of sialic acid in glycoproteins are described elsewhere.

Analytical Methods

Reagents

p-Toluenesulfonic acid, monohydrate, reagent grade, purchased from the Baker Chemical Co., Phillipsburg, New Jersey. For recrystallization, the acid, reagent grade, is first dissolved in the smallest quantity of water and, after filtering, 3 volumes of concentrated hydrochloric acid is added. The acid crystallizes well. It is collected on a sintered-glass funnel and washed with concentrated HCl. The crystallization is repeated once. The crystals melt at 104–105°. To remove HCl, the crystals are heated at 45° in a desiccator *in vacuo* over NaOH pellets. The final product must be free from HCl contamination as determined by the chloride test using a 1% solution of $AgNO_3$ in 50% HNO_3.

3-(2-Aminoethyl)indole: Reagent grade product from Eastman Organic Chemicals, Rochester, New York, can be used without further purification. 3-(2-Aminoethyl)indole can be prepared from its HCl salt supplied by the manufacturer in the following manner: 3(2-aminoethyl)indole hydrochloride, 1.967 g (10 mmoles), is dissolved in water (24 ml). The aqueous solution is chilled in an ice bath, chloroform (50 ml) is added, and the mixture is shaken with a 2 *N* aqueous NaOH solution (5 ml) in a separating funnel. The chloroform layer is washed twice with 10-ml portions of water and dried over $MgSO_4$. Evaporation of the solvent under reduced pressure gives crystalline 3-(2-aminoethyl)indole (1.5 g).

Test amino acid mixtures: Spinco amino acid standards (1.0 ml

[7] T. Y. Liu and Y. H. Chang, *J. Biol. Chem.* **246**, 2842 (1971).

[8] T. Y. Liu, E. C. Gotschlich, F. Dunne, and E. K. Jonsson, *J. Biol. Chem.* **246**, 4703 (1971).

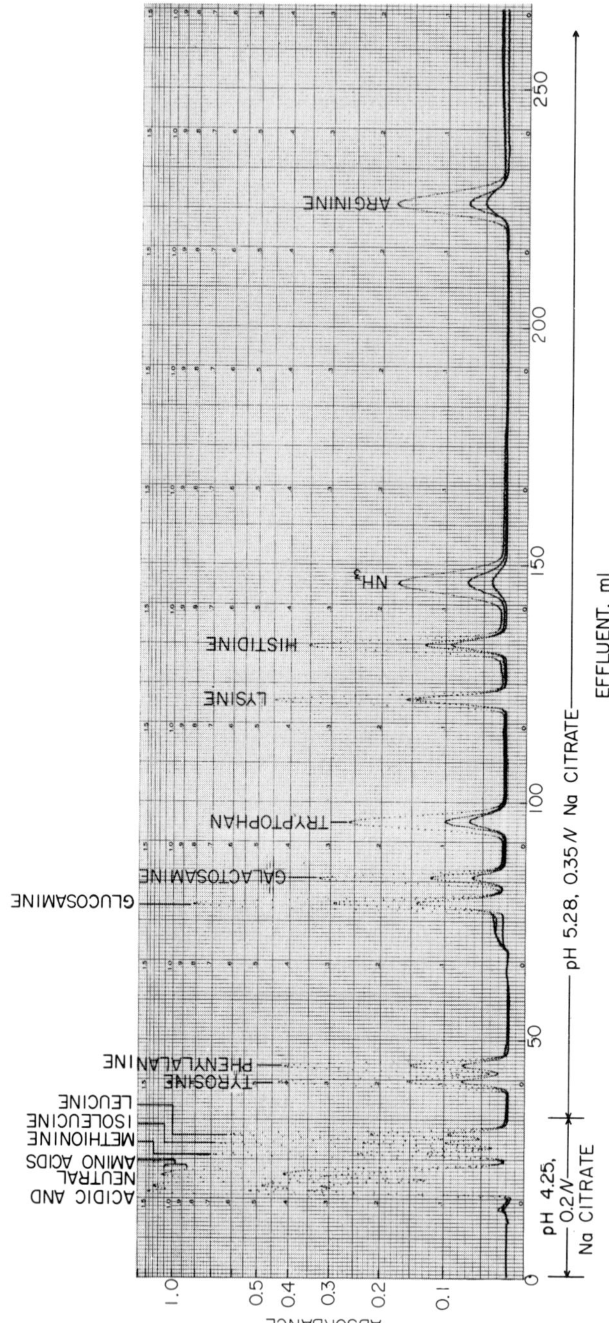

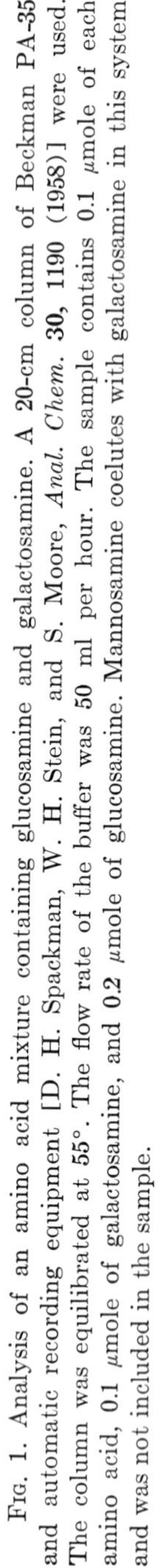

FIG. 1. Analysis of an amino acid mixture containing glucosamine and galactosamine. A 20-cm column of Beckman PA-35 and automatic recording equipment [D. H. Spackman, W. H. Stein, and S. Moore, *Anal. Chem.* **30,** 1190 (1958)] were used. The column was equilibrated at 55°. The flow rate of the buffer was 50 ml per hour. The sample contains 0.1 μmole of each amino acid, 0.1 μmole of galactosamine, and 0.2 μmole of glucosamine. Mannosamine coelutes with galactosamine in this system and was not included in the sample.

containing 2.5 μmoles of each amino acid) are mixed with 2.5 μmoles of tryptophan (Mann Research Laboratory, New York, New York) and lyophilized. The residues are made up to a volume of 10 ml with 3 *N* *p*-toluenesulfonic acid in 0.2% 3-(2-aminoethyl)indole (Eastman Organic Chemicals), and 1.0-ml aliquots of this solution are used in the hydrolyses. The solution should be used within an hour or two.

Hydrolysis. Hydrolysis is carried out in heavy-walled ignition tubes (Corning 9860, 18 × 150 mm) which have been washed with H_2SO_4:HNO_3 (3:1), rinsed in deionized water, and oven-dried. The protein or peptide, 2–3 mg, is hydrolyzed under vacuum (20–30 μ) at 110° for 22, 48, and 72 hours, with 1 ml of 3 *N* *p*-toluenesulfonic acid containing 0.2% 3-(2-aminoethyl)indole.

At the end of hydrolysis, 2.0 ml of 1 *N* NaOH is added, the solution is quantitatively transferred to a 5.0-ml volumetric flask, made to a volume of 5.0 ml with water, and filtered (Millipore Co., Swinny adapter XX30 01200, GSWP 01300, GS 0.22 μ, 13 mm). One-milliliter aliquots are used for analysis on the amino acid analyzer.[9] The dimethyl sulfoxide ninhydrin reagent of Moore[10] is used.

Chromatography. Tryptophan may be determined quantitatively by the standard short-column procedure of Spackman *et al.*[9] using a 10-cm column of Beckman PA-35 resin if galactosamine, mannosamine, and glucosamine are not present in the sample. Galactosamine and mannosamine coelute with tryptophan in this system. In some instances the acid decomposition product of tryptophan is eluted just ahead of tryptophan, where glucosamine is normally eluted. The presence of galactosamine, mannosamine, and glucosamine can readily be detected by extending the elution of the long column (60 cm) with the pH 4.25 buffer for an additional 60 ml beyond the position of phenylalanine.

The system illustrated in Fig. 1 shows how tryptophan and hexosamines can be determined on a 20-cm PA-35 column using elution with pH 4.25 and pH 5.28 buffers. In this system, the elution volumes are: phenylalanine, 42 ml; tyrosine, 46 ml; glucosamine, 82 ml; mannosamine or galactosamine, 90 ml; tryptophan, 98 ml; lysine, 125 ml; histidine, 136 ml; and arginine, 250 ml. The acid decomposition product of tryptophan, if present, will be eluted at 94 ml. This product is known to be present in hydrolyzates of tryptophan-containing proteins when 6 *N* HCl is used for hydrolysis, but it was found to be absent from all the protein hydrolyzates when the current procedure was employed. The amino acid

[9] D. H. Spackman, W. H. Stein, and S. Moore, *Anal. Chem.* **30,** 1190 (1958).
[10] S. Moore, *J. Biol. Chem.* **243,** 6281 (1968).

analyzer constant for tryptophan is 78 ± 2 for an instrument for which the lysine constant is 100 ± 2.

Comments

Calculations. In Fig. 2 the rate of decomposition of tryptophan during hydrolysis of a standard amino acid mixture and some proteins is plotted against the time of hydrolysis. The values obtained for tryptophan at the three times of hydrolysis (Table I) are extrapolated to zero time assuming first-order kinetics. By this procedure, the values obtained for a standard amino acid mixture and for ribonuclease A plus tryptophan were 99 ± 2% of the expected values. Moreover, the slope of the line suggests that, under these hydrolytic conditions, there is very little destruction of tryptophan even after prolonged hydrolysis time. The yields of tryptophan in the amino acid mixture after 22, 48, and 72 hours of hydrolysis are, respectively, 99.8, 96.3, and 98.2%. The recovery of tryptophan from a mixture of tryptophan and bovine ribonuclease A

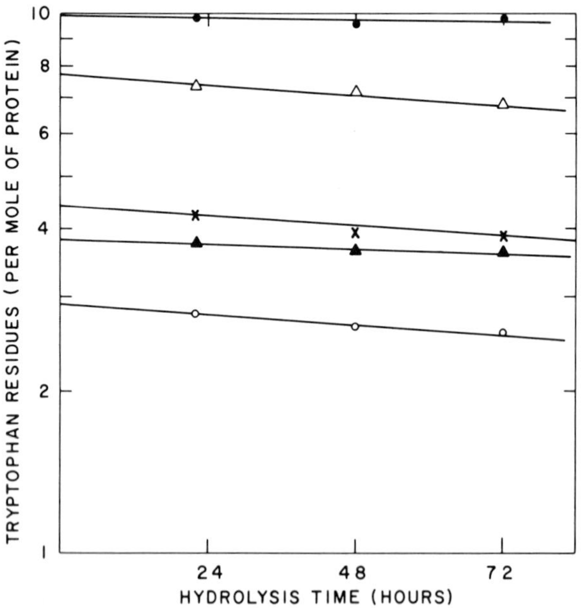

FIG. 2. Rate of decomposition of tryptophan in an amino acid mixture and in proteins during acid hydrolyses with 3 *N* *p*-toluenesulfonic acid containing 0.2% 3-(2-aminoethyl)indole; ●, amino acid mixture (theoretical value is 10); ○, bovine pancreatic deoxyribonuclease A (theoretical value is 3.0, see Table II); ▲, trypsinogen (theoretical value is 4.0, see Table II); △, chymotrypsinogen A (theoretical value is 8.0, see Table II); ×, streptococcal proteinase (theoretical value is 5.0, see Table II).

TABLE I
YIELDS OF TRYPTOPHAN AFTER HYDROLYSIS WITH 3 N p-TOLUENESULFONIC ACID[a]

Sample	Hours of hydrolysis			Extrapolated value[b]
	22	48	72	
Spinco amino acid mixture plus tryptophan[c]	99.8	96.3	98.2	100
Ribonuclease A plus tryptophan[d]	98.4	97.4	96.3	100

[a] Values are expressed as percentage of theoretical values. Duplicate analyses by ion-exchange chromatography were performed on 0-, 22-, 48-, and 72-hour hydrolyzates. Averages of the two analyses are presented. For computation of the percentage of recovery, the zero time value (sample without heating at 110°) was used as 100%.

[b] Values obtained by extrapolating 22-, 48-, and 72-hour values to zero time using semilogarithmic plot (see Fig. 2).

[c] Prepared by adding 2.5 μmoles of tryptophan to 2.5 μmoles of Spinco amino acid standard mixture followed by lyophilization. The residue was dissolved in 10 ml of 3 N p-toluenesulfonic acid in 0.2% 3-(2-aminoethyl)indole. One-milliliter aliquots of this solution were used for each hydrolysis.

[d] Ribonuclease A (20 mg) was mixed with 2 μmoles of tryptophan, and the mixture was dissolved in 10 ml of 3 N p-toluenesulfonic acid in 0.2% 3-(2-aminoethyl)indole. One-milliliter aliquots of this solution was used for hydrolysis.

after 22, 48, and 72 hours of hydrolysis are, respectively, 98.4, 97.4, and 96.3% (Table I).

The method has been applied to a variety of purified proteins, glycoproteins and peptides. The results are documented in Tables II and III. High recoveries of tryptophan were obtained for lysozyme, chymotrypsinogen, trypsinogen, myoglobin, streptococcal proteinase and its zymogen, tyrocidine C, and a bradykinin-potentiating pentapeptide. In each instance the recovery of tryptophan in a 22-hour hydrolyzate was $90 \pm 3\%$. To obtain more accurate figures, the values obtained for 22, 48, and 72 hours were extrapolated to zero time, assuming first-order kinetics. This provided some compensation for the small variations that were observed in the extent of the decomposition of tryptophan from one protein sample to another and may give the most accurate results. Since the numbers of tryptophan residues in these proteins have been well established, it is possible to compare the values calculated from the analyses with the number of residues to be expected.

The use of a longer time of hydrolysis is also required to ensure complete hydrolysis. When 6 N HCl was used, it was noted that some amino acids are liberated unusually slowly. For instance, the isoleucine in ribonuclease A is released very slowly because of the presence of an

TABLE II
ANALYSES OF TRYPTOPHAN IN PROTEINS BY ACID HYDROLYSIS WITH 3 *N* TOLUENESULFONIC ACID CONTAINING 0.2% 3-(2-AMINOETHYL)INDOLE

Sample	Hours of hydrolysis 22	48	72	Extrapolated value[a] Residues	Yield (%)
	Residues				
Bovine deoxyribonuclease A[b]	2.80	2.65	2.57	2.92 (3)	97
Bovine trypsinogen[c]	3.74	3.69	3.65	3.81 (4)	95
Sperm whale myoglobin[d]	1.87	1.87	1.89	1.87 (2)	94
Bovine chymotrypsinogen A[e]	7.45	7.19	6.79	7.85 (8)	98
Thrombin[f]	8.02	7.68	7.69	8.20 (9)	91
Streptococcal proteinase[g]	4.29	3.99	3.87	4.40 (5)	88
Zymogen of streptococcal proteinase[h]	4.22	4.08	3.68	4.70 (5)	94
Egg white lysozyme[i]	5.52	5.45	5.16	5.90 (6)	98
Tyrocidine C[j]	1.85	1.70	1.62	1.90 (2)	95
Bradykinin-potentiating pentapeptide, V-3-A[k]	0.92	0.86	0.85	0.95 (1)	95
Bovine insulin[l]	0	0	0	0	—

[a] The theoretical integral residue numbers are given in parentheses.

[b] Calculations are based on the content of 9 lysine residues per mole of protein. This calculation gave, for a 22-hour hydrolyzate, histidine 5.74, arginine 11.1, and glucosamine 2.80. The extrapolated value for glucosamine from the 22-, 48-, and 72-hour values is 3.0. For this deoxyribonuclease preparation (Lot 9LB, WDPC), Drs. W. Patterson and A. L. Baker, Research Laboratory, Worthington Biochemical Corporation, reported a value of 2.94 residues of tryptophan as determined by the method of H. Edelhoch [*Biochemistry* **6,** 1948 (1967)].

[c] Based on 15 lysine residues [K. A. Walsh and H. Neurath, *Proc. Nat. Acad. Sci. U.S.* **52,** 884 (1964)].

[d] Based on 19 lysine residues [A. B. Edmundson, *Nature (London)* **205,** 883 (1965)]. This calculation gave, for a 22-hour hydrolyzate, histidine 11.65, and arginine 3.86.

[e] Based on 14 lysine residues [J. R. Brown and B. S. Hartley, *Biochem. J.* **101,** 214 (1966)]. This calculation gave, for a 22-hour hydrolyzate, histidine 1.94, and arginine 4.04.

[f] Based on 24 lysine residues (see J. Magnuson, *in* "Structure-Function Relationships of Proteolytic Enzymes," P. Desnuelle, H. Neurath, and M. Ottesen, eds., p. 138. Munksgaard, Copenhagen, 1970; also, private communication from Drs. G. Glover and E. Shaw, Brookhaven National Laboratory). This calculation gave, for a 22-hour hydrolyzate, histidine 7.78, arginine 21.8, and glucosamine 3.39. The extrapolated value for glucosamine from the 22-, 48-, and 72-hour values is 4.2.

[g] Based on 17 lysine residues [T. Y. Liu and S. D. Elliott, *in* "The Enzymes" (P. Boyer, ed.). Academic Press, New York, vol. 3, p. 609 (1971)]. This calculation gave, for a 22-hour hydrolyzate, histidine 7.82, and arginine 8.76.

[h] Based on 29 lysine residues [T. Y. Liu, N. P. Neumann, S. D. Elliott, S. Moore, and W. H. Stein, *J. Biol. Chem.* **238,** 251 (1963)]. This calculation gave, for a 22-hour hydrolyzate, histidine 7.51, and arginine 10.85.

[i] Based on 6 lysine residues per mole of protein [R. E. Canfield, *J. Biol. Chem.* **238,**

isoleucyl-isoleucyl sequence. Similar sequence situations exist in other proteins, and this is a general problem.

The data given in Table III indicate that the release of valine and isoleucine from the three proteins after 22 hours when 3 *N* *p*-toluenesulfonic acid was used in hydrolysis was slower than when 6 *N* HCl was used. However, after 48 and 72 hours of hydrolysis, the values obtained for these two amino acids were quite similar by the two methods. The yields of serine and threonine were found to be somewhat higher with *p*-toluenesulfonic acid than with 6 *N* HCl. The yield of cystine and cysteine for the three proteins listed in Table III are quite similar by both the *p*-toluenesulfonic acid procedure and the 6 *N* HCl procedure. For more accurate determination of the sum of cystine and cysteine, an independent determination should be made by determining the cysteic acid content of the performic acid-oxidized protein[11,12] or *S*-sulfosulfenyl cysteine content of the hydrolyzate after treatment with sodium tetrathionate.[13] The recovery of proline by the present procedure is slightly higher compared with the theoretical values, possibly owing to interference by cysteine which, if not oxidized before analysis, will emerge at the position of proline.

Analysis in Glycoproteins. Successful analyses were also obtained with two glycoproteins: bovine pancreatic deoxyribonuclease A and thrombin, the former containing 4–5% carbohydrate,[14] and the latter containing 5–6% carbohydrate.[15] Well characterized glycoproteins with higher percentage of carbohydrate were not available for our studies. Analyses, therefore, were performed on a mixture of chymotrypsinogen A and different amounts of α-methylmannoside and *N*-acetylglucosamine. These two carbohydrates are common constituents of glycoproteins. The results indicate that, with increased amounts of added carbohydrate, the

[11] S. Moore, *J. Biol. Chem.* **238,** 235 (1963).

[12] C. H. W. Hirs, *J. Biol. Chem.* **219,** 611 (1956).

[13] A. S. Inglis and T. Y. Liu, *J. Biol. Chem.* **245,** 112 (1970).

[14] P. A. Price, T. Y. Liu, W. H. Stein, and S. Moore, *J. Biol. Chem.* **244,** 917 (1969).

[15] S. Magnuson, *in* "Structure-Function Relationships of Proteolytic Enzymes" (P. Desnuelle, H. Neurath, and M. Ottesen, eds.), p. 138. Munksgaard, Copenhagen, 1970.

2698 (1963)]. This calculation gave, for a 22-hour hydrolyzate, histidine 0.99, and arginine 10.96.

[j] Based on 1 lysine residue [M. A. Ruttenburg, T. P. King, and L. C. Craig, *Biochemistry* **4,** 11 (1965)].

[k] Based on 1 lysine residue [S. H. Ferreira, D. C. Bartelt, and L. J. Greene, *Biochemistry* **9,** 2583 (1970)].

[l] Bovine insulin is devoid of tryptophan [F. Sanger and E. O. P. Thompson, *Biochem. J.* **53,** 353 (1953)].

TABLE III

Amino Acid Compositions of Proteins after Acid Hydrolysis with 3 *N* Toluenesulfonic Acid

	Lysozyme[a]				Ribonuclease A + tryptophan[a]				Streptococcal proteinase[a]			
Amino acid	22-Hour hydrolyzate[b]	48-Hour hydrolyzate[b]	72-Hour hydrolyzate[b]	Average or extrapolated value	22-Hour hydrolyzate[b]	48-Hour hydrolyzate[b]	48-Hour hydrolyzate[b]	Average or extrapolated value	22-Hour hydrolyzate[b]	48-Hour hydrolyzate[b]	72-Hour hydrolyzate[b]	Average or extrapolated value
Tryptophan	5.5	5.4	5.2	5.9[c]	0.98	0.97	0.96	0.97[c]	4.3	4.0	3.87	4.5[c]
Lysine	6.0 (6.0)	6.0 (6.0)	6.0 (6.0)	6.0 (6.0)	10.0 (10.0)	10.0 (10.0)	10.0 (10.0)	10.0 (10.0)	17.0 (17.0)	17.0 (17.0)	17.0 (17.0)	17.0 (17.0)
Histidine	1.0 (0.60)	0.97 (0.89)	0.98 (0.88)	1.00 (0.91)	3.68 (3.84)	3.89 (3.89)	3.93 (4.11)	3.80 (3.90)	7.60 (8.21)	7.71 (8.30)	8.00 (8.40)	7.80 (8.40)
Arginine	11.3 (9.01)	11.0 (10.1)	10.8 (10.6)	11.0 (10.7)	3.92 (4.00)	3.95 (4.00)	3.87 (4.20)	3.90 (4.00)	8.80 (8.90)	8.60 (8.30)	8.50 (8.10)	8.60 (8.10)
Aspartic acid	21.4 (21.4)	21.3 (21.0)	21.4 (20.9)	21.4 (21.0)	15.1 (15.4)	15.1 (15.5)	15.1 (14.8)	15.1 (15.2)	39.1 (40.1)	37.7 (39.5)	39.3 (39.0)	36.3 (39.5)
Threonine	6.95 (6.76)	6.56 (6.17)	6.22 (5.88)	7.25[c] (6.9)[c]	9.60 (9.48)	10.1 (9.20)	9.25 (8.84)	9.8[c] (9.80)[c]	11.1 (10.9)	10.6 (0.70)	9.60 (9.50)	11.2[c] (11.5)[c]
Serine	10.3 (8.87)	9.79 (7.28)	9.03 (5.92)	10.4[c] (9.3)[c]	14.1 (13.6)	13.4 (12.4)	12.6 (10.9)	14.9[c] (15.2)[c]	23.5 (22.0)	22.0 (18.0)	21.1 (16.5)	24.8[c] (25.0)[c]
Glutamic acid	4.90 (4.84)	4.96 (4.79)	5.00 (4.84)	5.00 (4.8)	11.6 (12.1)	11.6 (11.8)	12.0 (12.2)	11.7 (12.1)	27.5 (29.0)	27.5 (28.2)	27.8 (28.2)	27.8 (28.5)
Proline	1.84 (2.06)	1.71 (1.98)	2.27 (2.09)	1.9 (2.0)	4.07 (3.98)	4.12 (3.80)	4.01 (3.80)	4.51 (3.90)	15.8 (15.5)	15.5 (15.0)	16.1 (15.0)	15.7 (15.2)
Glycine	11.9 (12.1)	11.9 (12.0)	12.0 (12.1)	11.9 (12.0)	2.92 (3.05)	3.07 (3.22)	3.21 (3.09)	3.00 (3.10)	36.5 (38.2)	36.5 (37.0)	36.1 (37.8)	36.4 (37.7)
Alanine	12.0 (12.0)	12.0 (12.0)	12.0 (12.0)	12.0 (12.0)	12.0 (12.0)	12.0 (12.0)	11.90 (12.0)	12.0 (12.0)	21.7 (22.3)	21.7 (21.5)	21.7 (21.9)	21.7 (21.9)
Half-cystine	7.64 (7.36)	6.45 (6.55)	6.20 (6.22)	7.9[c] (7.7)[c]	6.37 (7.77)[d]	5.83 (8.25)[d]	4.67 (8.21)[d]	7.71[c] (8.10)[d]	— (1.0)[e]	—	—	—
Valine	4.73 (5.59)	5.54 (5.70)	5.90 (6.01)	5.9[f] (5.9)[f]	8.31 (8.90)	9.0 (8.92)	8.88 (8.64)	8.91[f] (8.80)[f]	18.4 (20.1)	20.9 (21.6)	20.8 (21.3)	20.8[f] (21.3)[f]
Methionine	1.59 (1.89)	1.78 (1.92)	2.02 (1.91)	2.00 (1.9)	3.63 (3.71)	3.88 (3.66)	4.06 (3.84)	3.81 (3.70)	4.70 (4.81)	4.66 (4.60)	4.60 (5.00)	4.61 (4.80)
Isoleucine	4.71 (5.78)	5.21 (5.84)	6.02 (5.88)	6.0[f] (5.9)[f]	1.57 (2.18)	2.09 (2.54)	2.65 (2.81)	2.71[f] (2.81)[f]	11.1 (13.0)	12.3 (13.5)	12.9 (13.3)	12.9[f] (13.3)
Leucine	7.75 (7.84)	7.91 (7.85)	7.74 (7.88)	7.8 (7.9)	2.0 (2.01)	2.07 (1.93)	2.25 (1.94)	2.00 (2.00)	17.0 (17.0)	17.0 (17.0)	17.0 (17.0)	17.0 (17.0)
Tyrosine	2.96 (2.94)	2.96 (2.90)	3.22 (3.01)	3.04 (3.0)	5.95 (5.70)	5.81 (5.54)	6.20 (5.36)	5.91 (5.90)	18.0 (17.7)	17.9 (17.3)	18.0 (17.3)	18.0 (17.4)
Phenylalanine	2.81 (2.91)	2.94 (2.77)	3.12 (2.88)	2.96 (2.8)	3.04 (2.95)	2.86 (2.91)	2.97 (2.97)	3.00 (2.90)	11.7 (12.4)	11.7 (11.7)	11.9 (12.1)	11.8 (12.1)

[a] The values shown in parentheses were obtained after acid hydrolysis with 6 *N* HCl and are recalculated from values given in the literature. Lysozyme [see R. Canfield, *J. Biol. Chem.* **238,** 2698 (1963)]. Lysine was taken as 6.0 residues in calculations for the short column, and alanine was taken as 12.0 residues for the long column in order to compensate for a slight difference in flow rate through the two columns. Ribonuclease A [see A. M. Crestfield, S. Moore, and W. H. Stein, *J. Biol. Chem.* **238,** 622 (1963)]. Lysine was taken as 10.0 residues in calculations for the short column and alanine was taken as 12.0 residues for the long column. Streptococcal proteinase [see T. Y. Liu and S. D. Elliott, *in* "The Enzymes" (P. Boyer, ed.). Academic Press, New York, vol. 3, p. 609 (1971)]. Lysine was taken as 17.0 residues in calculations for the short column, and leucine was taken as 17.0 residues for the long column.

[b] Average of at least two analyses.

[c] The 22-, 48-, and 72-hour values for tryptophan, serine, threonine, and half-cystine were extrapolated to zero time to correct for decomposition during hydrolysis.

[d] Determined as *S*-carboxymethylcysteine.

[e] Determined as cysteic acid after performic acid oxidation.

[f] The 72-hour values are taken for valine and isoleucine, which are slowly liberated.

TABLE IV
EFFECTS OF CARBOHYDRATE CONCENTRATION ON THE YIELDS OF TRYPTOPHAN FROM BOVINE CHYMOTRYPSINOGEN A ON HYDROLYSIS WITH 3 N TOLUENESULFONIC ACID IN 0.2% 3-(2-AMINOETHYL)INDOLE

Carbohydrate content[a] (%)	Hours of hydrolysis		
	22	48	72
	Residues[b]		
0	7.45 (93.1)	7.19 (89.9)	6.79 (84.9)
12	6.86 (85.8)	6.75 (84.4)	6.65 (83.1)
30	5.70 (71.3)	5.64 (70.5)	5.48 (68.5)

[a] The carbohydrates consist of N-acetylglucosamine and α-methylmannoside in a ratio of 1:4.5.

[b] The numbers given in parentheses are percentage recovery based on 8 tryptophan residues per mole of protein.

destruction of tryptophan increases (Table IV). Thus, in the 22-hour hydrolyzate, when the amount of carbohydrates changed from 0 through 12, to 30% of the weight of the protein, the recovery of tryptophan decreased correspondingly from 94, through 86, to 72%. The use of longer hydrolysis times, 48 or 72 hours, did not result in substantial decrease in the recovery of tryptophan within a given sample, and extrapolation of 22-, 48-, and 72-hour values did not yield values comparable with the expected integral values. However, when the logarithm of the yields of tryptophan were plotted against the percentage of carbohydrate content, as shown in Fig. 3, a straight line was obtained. Thus, if the carbohydrate content of a glycoprotein is known, the amount of tryptophan obtained chromatographically can be divided by a factor derived from the curve shown in Fig. 3 to give a corrected value.

Matsubara and Sasaki recently reported a procedure for the analysis of tryptophan in proteins by acid hydrolysis in 6 N HCl containing 2–4% thioglycolic acid.[6] The recovery of tryptophan for some proteins reported by these authors are comparable with the present results. However, when glucose was present in the protein sample, the protective effect of thioglycolic acid was lost and no tryptophan could be recovered from the hydrolyzate. The method described here has been applied to two glycoproteins and to a mixture of carbohydrate and protein and was found to be successful in the determination of tryptophan in these samples.

Conclusion. The present procedure provides a new method for the determination of tryptophan in proteins and glycoproteins. It gives tryp-

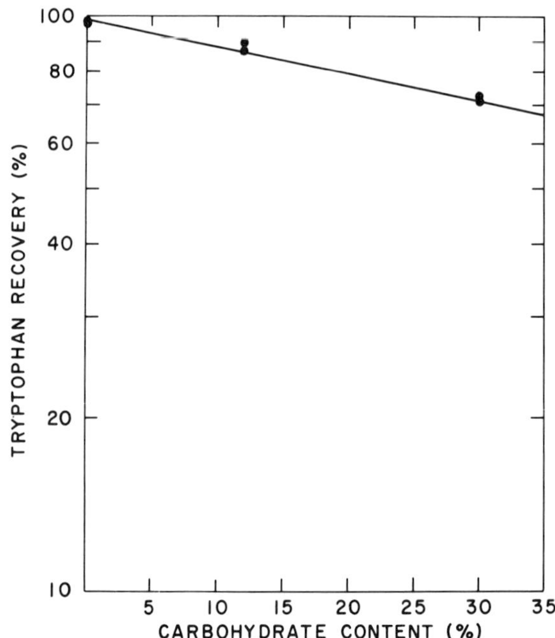

FIG. 3. The effect of carbohydrate concentration on the recovery of tryptophan from chymotrypsinogen A on acid hydrolysis in 3 *N* *p*-toluenesulfonic acid containing 0.2% 3-(2-aminoethyl)indole.

tophan values for proteins in agreement with those obtained by other methods reported in the literature not using acid hydrolysis. In addition, the recovery of all other amino acids was comparable to that obtained from a 6 *N* HCl-hydrolyzate. Since in this procedure the hydrolyzate can be placed on the amino acid analyzer column for analysis without prior removal of the solvent, as is required when 6 *N* HCl is used, the procedure using *p*-toluenesulfonic acid should be more convenient than the 6 *N* HCl method for routine amino acid determinations and has the exclusive advantage of permitting a reliable analysis for tryptophan.

We have studied the behavior of other acids as substitutes for *p*-toluenesulfonic acid. These include benzenesulfonic acid, methanesulfonic acid, and trifluoromethanesulfonic acid. These acids were used in the hydrolysis of lysozyme, chymotrypsinogen, trypsinogen, and myoglobin. The yields of tryptophan and other amino acids, with the exception of valine and isoleucine, were identical (±3%). Recoveries of valine and isoleucine with trifluoromethanesulfonic acid were 6–8% higher than with the remaining sulfonic acids for 22-hour hydrolyzates. After hydrolysis for 72 hours, however, all the above acids give identical recoveries of these more slowly released amino acids. Reagent grade *p*-toluenesulfonic acid as well as the other acids can be used directly in the procedure without additional purification. It should be noted that the toxicity of trifluoromethanesulfonic acid has not yet been reported and therefore constitutes a potential hazard.

[5] Determination of Cystine and Cysteine as *S*-Sulfocysteine[1]

By TEH-YUNG LIU and ADAM S. INGLIS

Cystine and cysteine have been considered to be too unstable during acid hydrolysis to permit the amounts of these amino acids to be estimated quantitatively with an automatic amino acid analyzer. Hence, the determination of these amino acids is usually made either by analyzing for cysteic acid after performic acid oxidation[2,3] or for *S*-carboxymethylcysteine after reduction and carboxymethylation in 8 *M* urea.[4] However, there have been indications that extensive degradation of cystine does not in fact occur during acid hydrolysis and that cystine is probably converted to closely related derivatives. Such derivatives, with suitable modification, might be amenable to determination on the analyzer. One advantage of such an approach for the analysis of cysteine and cystine residues in proteins is that a special hydrolyzate would not be required for the analysis.

The reversible masking of reduced cystine residues in proteins with sodium tetrathionate[5] prompted tests on the applicability of the reagent to a study of the sulfur-containing amino acids in protein hydrolyzates. The procedure to be described was developed by Inglis and Liu[6] for the determination, of both cysteine and cystine in acid hydrolyzates as *S*-sulfocysteine by reduction of cystine with dithiothreitol followed by treatment of the reaction solution with excess sodium tetrathionate.

[1] Research carried out at Brookhaven National Laboratory under the auspices of the United States Atomic Energy Commission.

[2] C. H. W. Hirs, *J. Biol. Chem.* **219,** 611 (1956).

[3] S. Moore, *J. Biol. Chem.* **238,** 235 (1963).

[4] A. M. Crestfield, S. Moore, and W. H. Stein, *J. Biol. Chem.* **238,** 622 (1963).

[5] T. Y. Liu, *J. Biol. Chem.* **242,** 4029 (1967).

[6] A. S. Inglis and T. Y. Liu, *J. Biol. Chem.* **245,** 112 (1970).

Analytical Methods

Reagents

Amino acid mixture, a standard solution containing 1.25 μM cystine and 2.5 μM aspartic acid.

The amino acid mixture is used to check the procedure and to establish the color factor for *S*-sulfocysteine on a Beckman-Spinco Model 120C amino acid analyzer. The color values obtained for both amino acids are identical to those obtained when the derivitization procedure is applied to the calibration mixture for the analyzer. The presence of a dip in the baseline just preceding the peak from *S*-sulfocysteine may make it desirable to integrate the peak by hand rather than with an electronic integrator. The aspartic acid is used as an internal standard. The dimethyl sulfoxide ninhydrin reagent of Moore[7] is used.

Hydrolysis. Hydrolysis is carried out in heavy-walled test tubes, cleaned first in $H_2SO_4:NHO_3$ (3:1), rinsed in deionized water, and oven-dried. The protein, 2–5 mg, is hydrolyzed under vacuum at 110° for 22 hours with 1 ml of constant-boiling HCl (twice redistilled from all-glass apparatus). When tryptophan is present, sodium tetrathionate (0.5–1.0 mg, about twice the molar quantity of tryptophan) is added to the hydrolysis tube before the addition of HCl. Tetrathionate in the hydrolyzate will react with dithiothreitol in the reduction step, and the quantities of reagents to be used subsequently should be selected accordingly.

The hydrolyzate is concentrated by rotary evaporation at 37°, and the residue is dissolved in 4 ml of deionized water. Since 0.1-μmole amounts of individual amino acids are optimal with a Spinco Model 120C analyzer, the quantities taken generally allow duplicate runs and the analysis of the untreated hydrolyzate. The latter is desirable for establishing whether other ninhydrin-positive substances are present in the position of *S*-sulfocysteine.

Reduction. One milliliter of the hydrolyzate is placed in a test tube (the heavy-walled hydrolysis tube may be used), followed by 3 ml of deionized water, 0.1 ml of pyridine, and 1 ml of dithiothreitol solution (approximately 4 mM). The tube is flushed with N_2 for 1 minute, sealed with a silicone rubber stopper, shaken on a Vortex shaker, and kept at 37° for 1 hour.

Oxidation. At the end of the reduction period, about 200 μmoles (60 mg) of sodium tetrathionate ($Na_2S_4O_6 \cdot 2H_2O$) is added to the test tube, and the mixture is allowed to stand for 16 hours at 25°. Oxygen is then bubbled through the solution for 7 minutes. After at least 1 hour, the solution (or a suitable aliquot thereof) is dried by rotary evaporation, diluted with pH 2.2 buffer (5–10 ml), and applied to the 60-cm column of the analyzer.

Preparation of Sodium Tetrathionate.[8] Highly pure samples of sodium tetrathionate may be conveniently prepared by the dropwise addition of a concentrated solution of $Na_2S_2O_3 \cdot 5H_2O$ (250 g in 150 ml of H_2O) to an alcoholic solution of iodine (127 g of I_2 and 50 g of NaI in 500 ml of 90% ethanol). The reaction mixture is vigorously stirred and maintained below 20° during the addition of thiosulfate. Precipitation of tetrathionate begins when approximately one-half of the thiosulfate has been introduced. The reaction is considered complete when the color of the iodine solution attains a yellowish blue. One liter of absolute ethanol and 500 ml of anhydrous ethyl ether are then added. After precipitation appears to be complete, the deposit is collected on a Büchner funnel and washed with small portions of absolute ethanol to remove excess iodine. The precipitate is pressed down well and dissolved in an equal weight of H_2O. The solution is filtered by gravity through filter paper, and the filtrate is allowed to fall directly into 1 liter of absolute ethanol. The precipitate so formed is collected on a Büchner funnel, washed well with absolute ethanol, and dried at room temperature to constant weight *in vacuo* over $CaCl_2$. The yield is about 100 g of a white, crystalline material. Sodium tetrathionate crystallizes from ethanol–water as the hydrated salt, $Na_2S_4O_6 \cdot 2H_2O$. The product is unstable at room temperature in the presence of moisture and therefore should be kept in a desiccator over $CaCl_2$ at 4°.

Preparation of S-Sulfocysteine Dihydrate. To 3.06 g (10 mmoles) of $Na_2S_4O_6 \cdot 2H_2O$ in 30 ml of water is added 1.21 g of cysteine (10 mmoles). The reaction mixture is stirred at room temperature for 60 minutes and then kept at 4° for 18 hours. The precipitate is removed by centrifugation at 4° at 12,000 *g*. To the clear supernatant solution, absolute ethanol is added to a concentration of 70% (v/v); the solution is allowed to stand at 4° for 18 hours. Crystals of *S*-sulfocysteine start to appear when the vessel is scratched with a glass rod and are collected by filtration. The product is recrystallized from 70% ethanol and dried over phosphorus pentoxide in a desiccator.

Comments

A typical chromatogram obtained when a lysozyme hydrolyzate was treated with dithiothreitol followed by tetrathionate is shown in Fig. 1.

[7] S. Moore, *J. Biol. Chem.* **243,** 6281 (1968).

[8] A. Gilman, S. F. Philips, E. S. Koelle, R. P. Allen, and E. St. John, *Amer. J. Physiol.* **147,** 115 (1946).

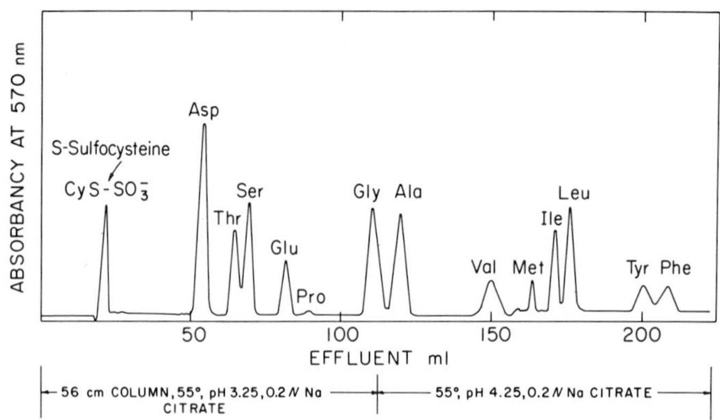

FIG. 1. Effluent curve obtained by ion-exchange chromatography of a sample of egg white lysozyme hydrolyzate treated with dithiothreitol and then with sodium tetrathionate.

After treatment with dithiothreitol followed by sodium tetrathionate, a mixture of cystine and aspartic acid gives only two peaks on the long column of an amino acid analyzer—the aspartic acid peak and one in the elution position of cysteic acid. The color yield of *S*-sulfocysteine is much lower than that of cysteic acid. It is convenient to relate the color factor to that of aspartic acid; *S*-sulfocysteine gives a color yield which is 0.745 ± 0.02 times that for aspartic acid under our conditions, but the color factor should be checked for each analyzer. The product with ninhydrin also shows a relatively high absorption at 440 nm, as indicated by the relative areas of the peaks at 570 nm and 440 nm. The ratio of these areas is only 2.3:1 for *S*-sulfocysteine, whereas it is 6:1 for cysteic acid.

With respect to the specificity of the procedure, it should be stressed that any ninhydrin-positive substance eluted without retardation on the amino acid analyzer will interfere and produce high results. The possibility of interference can be checked by running the untreated sample as a blank determination. If this is not convenient, the area of the 440-nm peak should be calculated and the ratio of this area to the area calculated for the 570-nm absorption, when compared with the ratio in a standard analysis, should give an indication of the possible presence of interfering substances. In this connection, when serine and threonine (0.5 μmole) were hydrolyzed in the presence of tetrathionate (3 mg) and the resulting hydrolyzate was treated with dithiothreitol followed by tetrathionate, sulfoserine and sulfothreonine were not found to be present on the chromatogram. The conclusion is that under the experimental condition described, serine and threonine do not form substances that interfere in the analysis of *S*-sulfocysteine by this procedure.

ANALYSIS OF HYDROLYZATES FOR HALF-CYSTINE

Protein	Percentage recovered as *S*-sulfocysteine[a]	
	HCl hydrolysis	HCl hydrolysis with $Na_2S_4O_6$
Bovine pancreatic ribonuclease A[b]	98	95
Streptococcal proteinase[c]	78	100
Egg white lysozyme[d]	97	97
Egg white lysozyme[e]	N.D.	100
Wool I[f]	90	101
Wool II[g]	83	102

[a] Average of duplicate analyses of 22-hour hydrolyzates.

[b] Based on 8 half-cystine residues [C. H. W. Hirs, *J. Biol. Chem.* **219,** 611 (1956)].

[c] Based on 1 half-cystine residue [T. Y. Liu and S. D. Elliott, *J. Biol. Chem.* **240,** 1138 (1965)].

[d] Based on 8 half-cystine residues [R. E. Canfield, *J. Biol. Chem.* **238,** 2691 (1963)].

[e] Hydrolyzed for 72 hours.

[f] Merino wool which contains 23 μmoles of cysteine and 450 μmoles of cystine per gram of protein. Calculations are based on amperometric determinations with methylmercuric iodide as 100% [S. J. Leach, *Aust. J. Chem.* **13,** 547 (1960)].

[g] Reduced merino wool which contains 845 μmoles of cysteine and 58 μmoles of cystine per gram of protein. Calculations are based on amperometric determination with methylmercuric iodide as 100% [S. J. Leach, *Aust. J. Chem.* **13,** 547 (1960)].

The procedure provides a new method for the determination of half-cystine residues in proteins. It gives results comparable to those of the performic acid method (see the table). Since the same hydrolyzate can be used as a control on the procedure and the chemical reactions are carried out on the free amino acids in the same solution, conditions are optimal for reproducible results. If for some reason the reactions are not quantitative, this will be evident from the appearance of cystine on the chromatogram. In routine practice, small differences are found for the factor obtained for standard cystine solutions; for the most accurate results, it is recommended that a control be run in parallel with the unknown solution.

For the analysis of proteins that contain both 3-carboxymethylhistidine and cystine, the present procedure is particularly advantageous. Formerly, the protein was reduced and carboxymethylated before amino acid analysis in order to convert cystine to *S*-carboxymethylcysteine, which emerges before aspartic acid and, therefore, does not interfere with the determination of 3-carboxymethylhistidine. With the present procedure, this extra step—reduction and carboxymethylation of protein—will not be necessary.

The addition of tetrathionate before hydrolysis may lead to low tyrosine values, as exemplified by the result with ribonuclease. However, the results of experiments on proteins containing both thiol groups and tryptophan indicate that the addition of tetrathionate is essential for correct results. Streptococcal proteinase, with one half-cystine residue and five tryptophan residues per molecule, yielded only 78% of the expected amount of *S*-sulfocysteine if tetrathionate was not added to the hydrolysis tube, while similarly two Merino wool samples, which contained 23 and 845 μmoles of thiol per gram of wool protein, respectively, yielded only 90 and 83% of the expected values. Tetrathionate and *S*-sulfocysteine are unstable in hot acid, but there is evidence from the present work that during acid hydrolysis in 6 *N* HCl tetrathionate reacts with tryptophanyl residues, thereby preventing reaction of this amino acid with cysteine. That the cysteine may then be determined quantitatively after hydrolysis as *S*-sulfocysteine using the procedure described herein reopens the possibility of determining the half-cysteine residues in modified proteins, for example, *S*-sulfoproteins, after acid hydrolysis.

[6] Thin-Layer Chromatography of Amino Acids and Derivatives

By A. Niederwieser

The popularity of thin-layer chromatography (TLC) for the analysis of amino acids and derivatives is astonishing. There are other techniques which give, in general, better qualitative *and* quantitative results. Automatic ion-exchange chromatography has progressed considerably (for a review see Benson and Patterson[1]); a modern analyzer quantitates 2 nmoles of a protein hydrolyzate within one single hour (Technicon TSM 1). Meanwhile also gas chromatography has been developed to a powerful tool for the analysis of amino acid derivatives (for a critical review see Coulter and Hann[2]) and can be combined with mass spectrometry if quite unequivocal results are necessary. The reasons for the unbroken popularity of TLC are the modest experimental requirements, its versatility, and the capability for ultramicroanalysis, which com-

[1] J. V. Benson and J. A. Patterson, *in* "New Techniques in Amino Acid, Peptide, and Protein Analysis" (A. Niederwieser and G. Pataki, eds.), p. 1. Ann Arbor-Humphrey Science Publ., Ann Arbor, Michigan, 1971.

[2] J. R. Coulter and C. S. Hann, *in* "New Techniques in Amino Acid, Peptide, and Protein Analysis" (A. Niederwieser and G. Pataki, eds.), p. 11. Ann Arbor-Humphrey Science Publ., Ann Arbor, Michigan, 1971.

pensate for an occasional ambiguity in the interpretation of a result. There is a great demand for qualitative routine analyses which can be done by TLC: comparison of peptide and protein hydrolyzates, control of the course of hydrolysis, end-group determination, identification in sequence analysis, and also screening for inborn errors of amino acid metabolism.[3]

Since the appearance of the first edition of this volume, TLC of amino acids and derivatives has expanded further. Only selected papers can be mentioned here, omitting many technical details. It is supposed that the reader is well acquainted with the special requirements for TLC and knows the importance of careful standardization of the technique, including the following points: quality of layer material, layer thickness and homogeneity, layer activation and relative humidity, amount of sample applied, narrow starting point or line, effective drying of the applied sample, quality of solvent (aging), chamber saturation, chamber volume and geometry, temperature, distances of starting point and solvent front from immersion line, and drying of the layer prior to TLC in the second dimension. Such details and problems have been discussed in our article in Volume 11[4] and, more thoroughly, in monographs.[5-8]

Layers

General Remarks on Layer Materials

A great variety of different layer materials is offered. Precoated thin-layers on glass plates, glass fiber, plastic, or alumina are available, based on silica gel, cellulose, polyamide, aluminum oxide, or other materials, with or without fluorescent agents. It is convenient to use these precoated layers, particularly for serial analyses. In general, similar results are obtained using precoated or self-made thin-layers. Unfortunately, some precoated layers show less resolution or a much slower rate of solvent migration than the self-made thin-layers. Naturally, the

[3] A. Niederwieser and H.-C. Curtius, *Z. Klin. Chem. Klin. Biochem.* **7**, 404 (1969).

[4] M. Brenner and A. Niederwieser, Vol. 11, 39.

[5] G. Pataki, "Techniques of Thin-Layer Chromatography in Amino Acid and Peptide Chemistry." Ann Arbor-Humphrey Science Publ., Ann Arbor, Michigan, 1969; G. Pataki, "Dünnschicht-Chromatographie in der Aminosäure- und Peptid-Chemie." Walter De Gruyter, Berlin, 1966.

[6] "Thin-Layer Chromatography" (E. Stahl, ed.). Academic Press, New York, 1965.

[7] J. G. Kirchner, "Thin-Layer Chromatography." Wiley (Interscience), New York, 1967.

[8] "Progress in Thin-Layer Chromatography and Related Methods" (A. Niederwieser and G. Pataki, eds.), Vols. I-III. Ann Arbor-Humphrey Science Publ., Ann Arbor, Michigan, 1970, 1971.

same precautions regarding layer activation and humidity in the laboratory are necessary. Because of the long storage time usual for precoated layers, it is often necessary to clean them prior to use. This is best done by a "chromatographic" run in chloroform–methanol (1:1, v/v) or in the solvent used afterward and then drying first for 5 minutes at room temperature, then for 10 minutes at 110°. Fluorescent additives in the layer are necessary for detecting ultraviolet (UV)-absorbing substances. In general, they are superfluous if a photographic record of the chromatogram is made with UV light.

Preparation of the Layer

Silica Gel G. A slurry is prepared from 25 g Silica Gel G and 50 ml of water in a closed 250-ml Erlenmeyer flask with vigorous shaking for 30 seconds. The suspension is poured at once into the clean spreader (the under edges of the spreader in particular must be free of grease), which is carefully moved over the clean plates to be coated (slit width 0.25 mm). Usually, an aligning tray is used in which can be arranged five plates (20×20 cm) in a row and one auxiliary plate (5×20 cm) at each end. In this case the coating should take about 5 seconds. The plates are air-dried overnight. The thickness of the dry layer is about 0.15 mm. Some authors recommend drying of the layer in a current of air; this must be done very carefully, because the layer is easily disturbed when freshly spread. The plates should not be touched until their surface has become dull (10–15 minutes); then they can be handled more safely.

If the relative humidity of the laboratory atmosphere exceeds 65%, a short activation is recommended. The plates are heated for 10 minutes at 110° and immediately covered with a suitable glass plate or aluminum foil that leaves an open margin for sample application, which should be started while the plate is warm. Chromatography is begun as soon as the plate has cooled down to room temperature. Cooling takes at least 20 minutes; it may be accelerated by a thick metal block support or best by a cooling device of the type provided in the BN chamber (see p. 48 in Vol. 11[4]).

Cellulose MN 300. A slurry is prepared from 15 g Cellulose MN 300 (Machery & Nagel, Düren, Germany), 70 ml of water, and 10 ml of ethanol. Vigorous mixing in an electrical blender for 3–5 minutes is necessary to ensure an even mixture. The suspension is spread immediately on clean glass plates, as described above (silica gel G). The plates are air-dried overnight.

Polyamide. Five grams of polyamide for TLC (Woelm, Eschwege, Germany) is mixed with 45 ml of chloroform–methanol (2:3, v/v) in an Erlenmeyer flask and spread as described above. Ready-to-use foils, coated on *both* sides (15 × 15 cm), are available from Cheng-Chin Trading Co., Ltd., No. 75 Sec. I. Hankow St., Taipei, Taiwan.

Ion-Exchange Resin. The ion-exchanger must be brought into the appropriate form prior to use, by treatment with acid or base and washing with distilled water on a funnel provided with a porous disk. If a polystyrene resin is employed, cellulose is added as binder. For instance, 20 g of Dowex 50-X8, 400 mesh (H^+ form), 10 g of cellulose MN 300, and 65 ml of water are mixed with an electrical blender. The suspension is spread immediately as described above.

Twin-Layer Ion-Exchange Cellulose. A thin-layer consisting of two different materials, a narrow zone of a material A along a border of the plate and a material B on the rest of the plate area is useful for trapping unwelcome or disturbing substances at the starting point[9] or to enable two-dimensional TLC on different stationary phases. These techniques of discontinuous (and continuous) gradient TLC have been reviewed recently.[10]

Twin layers are prepared by filling the two different suspensions into a spreader which is divided into two compartments by a plastic barrier. For instance, the cylinder of a Desaga spreader is divided by a Teflon disk, 30 mm in diameter and about 1 mm thick. Both compartments are filled to an equal height. During spreading the position of the cylinder must be such that air can enter both compartments.

Free Amino Acids

Layer Material

Free amino acids are mostly separated on silica gel or on (microcrystalline) cellulose. Besides these materials, Supergel,[11] Celite,[12] Silica Gel–kieselguhr mixture,[13] aluminum oxide,[14,15] Silica Gel H–cellulose MN 300 (2.5:15, w/w),[16] Silica Gel G–cellulose MN 300 G (4:10, w/w),[17,18]

[9] J. A. Berger, G. Meyniel, J. Petit, and P. Blanquet, *Bull. Soc. Chim. Fr.* **11**, 2662 (1963); *C. R. Acad. Sci.* **257**, 1534 (1963).

[10] A. Niederwieser, *Chromatographia* **2**, 23, 362 (1969).

[11] J. Huber, W. Schaknies, and J. Rückbeil, *Pharmazie* **18**, 37 (1963).

[12] B. Shasha and R. L. Whistler, *J. Chromatogr.* **14**, 532 (1964).

[13] E. Bancher, H. Scherz, and V. Prey, *Mikrochim. Acta* **1963**, 712.

[14] M. Mottier, *Mitt. Geb. Lebensmittelunters. Hyg.* **49**, 454 (1958).

[15] L. Rastekiene and T. Prauskiene, *Liet. TSR Mokslu Akad. Darb. Ser. B* 5 (1963); see *Chem. Abstr.* **60**, 6217*h* (1964).

[16] B. Granroth, *Acta Chem. Scand.* **22**, 3333 (1968).

[17] N. A. Turner and R. I. Redgwell, *J. Chromatogr.* **21**, 129 (1966).

[18] G. Molnár, *Clin. Chim. Acta* **27**, 535 (1970).

TABLE I
SOLVENTS FOR TLC OF FREE AMINO ACIDS

Combination for two-dimensional TLC	No.	Solvent
1.	1	*n*-Propanol–water (70:30, v/v)[a]
2.	2	*n*-Butanol–glacial acetic acid–water (4:1:1, v/v)[a–c]
2.	3	Phenol–water (75:25, w/w)[a,c]
	4	*n*-Propanol–34% w/w ammonia (70:30, v/v)[a]
	5	Methyl ethyl ketone–pyridine–water–glacial acetic acid (70:15:15:2, v/v)[d]
1.	6	Chloroform–methanol–17% w/w ammonia (2:2:1, v/v)[e–g]
	7	96% Ethanol–water (70:30, v/v)[a,h]
	8	Chloroform–*n*-butanol–methanol–glacial acetic acid–water (100:60:20:20:15, v/v)[i]
1.	9	Methanol–pyridine–water (20:1:5, v/v)[g]
	10	Acetone–pyridine–conc. ammonia–water (30:45:5:20, v/v)[j]
2.	11	2-Propanol–formic acid–water (75:12.5:12.5)[j]
1.	12	2-Propanol–formic acid–water (40:2:10, v/v)[k,l]
2.	13	*t*-Butanol–methyl ethyl ketone–ammonia (d = 0.88)–water (50:30:10:10, v/v)[k]
1. 1.	14	2-Propanol–butanone–1 *N* hydrochloric acid (60:15:25, v/v)[m]
2.	15	*t*-Butanol–butanone–propanone–methanol–water–ammonia (d = 0.88) (40:20:20:1:14:5, v/v)[m]
2.	16	2-Methyl-2-butanol–butanone–propanone–methanol–water–ammonia (d = 0.88) (50:20:10:5:15:5, v/v)[n]
	17	*n*-Butanol–glacial acetic acid–water (60:15:25, v/v)[o]
	18	*n*-Butanol–pyridine–water (60:60:60, v/v)[o]
	19	2-Propanol–water–ammonia (d = 0.88) (200:20:10, v/v)[o]
	20	*n*-Propanol–1 *N* acetic acid (3:1, v/v)[o]
	21	*n*-Propanol–0.2 *N* ammonia (3:1, v/v)[o]
	22	*n*-Butanol–acetone–diethylamine–water (70:70:14:35, v/v)[o]
1.	23	Cyclohexanol–acetone–water–1-dimethylaminopropanol-2-diethylamine (10:5:5:1:1, v/v)[p]
2.	24	*t*-Butanol–glacial acetic acid–water (5:1:1, v/v)[p]
	25	*n*-Butanol–acetone–diethylamine–water (10:10:2:5, v/v)[l,q]
	26	2-Butanol–methyl ethyl ketone–dicyclohexylamine–water (10:10:2:5, v/v)[l,q]
	27	Phenol–water (75:25, w/w); gas phase equilibrated with 3% ammonia[l,q]

[a] M. Brenner and A. Niederwieser, *Experientia* **16,** 378 (1960).
[b] *n*-Butanol–glacial acetic acid–water (4:1:1, v/v) and (3:1:1, w/w) are practically identical.

DEAE-cellulose,[19,20] starch,[21] and polyamide (for aromatic amino acids and derivatives)[22] have been proposed.

Solvents, R_f Values, and Maps for Identification

Many solvents known from paper chromatography can be employed directly or with minor modifications. For silica gel and cellulose, the solvents of Table I are recommended. The corresponding R_f values are compiled in Tables II and III and should only be regarded as a guide. Even greater variations are observed on different cellulose materials, as shown in Table III (cellulose MN 300 compared with Avicel). These variations are probably caused by secondary effects, such as variable vapor impregnation of the layer and evaporation of the solvent from the layer due to different migration rates of the solvent. R_f values related to a certain amino acid remain more constant. Nevertheless, R_f values are very helpful for special separation problems. For instance, it can be seen from Tables II and III that leucine and isoleucine are best separated on silica gel in solvent 5 and on cellulose in solvents 13, 15, 19. One-dimensional TLC is very economical and is hence especially suited for serial analyses. The chromatograms can be evaluated rapidly by a scanner (or by eye). The resolution may be sufficient for screening tests, purity tests, reaction control, etc., but it is not suitable for separating complex amino acid mixtures. Resolution increases somewhat if multiple TLC is employed. For multiple TLC on cellulose, *n*-butanol–

[19] P. de la Llosa, C. Tertrin, and M. Jutisz, *J. Chromatogr.* **14**, 136 (1964).
[20] R. Vercaemst, V. Blaton, and H. Peeters, *J. Chromatogr.* **43**, 132 (1969).
[21] S. M. Petrović and S. E. Petrović, *J. Chromatogr.* **21**, 313 (1966).
[22] J. D. Sapira, *J. Chromatogr.* **42**, 136 (1969).

[c] D. Horton, A. Tanimura, and M. L. Wolfrom, *J. Chromatogr.* **23**, 309 (1966).
[d] M. Brenner, A. Niederwieser, and G. Pataki, *in* "Dünnschicht-Chromatographie" (E. Stahl, ed.), pp. 403–452. Springer-Verlag, Heidelberg, 1962.
[e] M. Brenner and A. Niederwieser, Vol. 11, p. 39.
[f] A. R. Fahmy, A. Niederwieser, G. Pataki, and M. Brenner, *Helv. Chim. Acta* **44**, 2022 (1961).
[g] E. Bujard and J. Mauron, *J. Chromatogr.* **21**, 19 (1966).
[h] E. J. Shellard and G. H. Jolliffe, *J. Chromatogr.* **31**, 82 (1967).
[i] F. Fussi and G. F. Fedeli, *Boll. Chim. Farm.* **107**, 711 (1968).
[j] H. H. White, *Clin. Chim. Acta* **21**, 297 (1968).
[k] K. Jones and J. G. Heathcote, *J. Chromatogr.* **24**, 106 (1966).
[l] E. von Arx and R. Neher, *J. Chromatogr.* **12**, 329 (1963).
[m] C. Haworth and J. G. Heathcote, *J. Chromatogr.* **41**, 380 (1969).
[n] J. G. Heathcote and C. Haworth, *J. Chromatogr.* **43**, 84 (1969).
[o] I. Smith, L. J. Rider, and R. P. Lerner, *J. Chromatogr.* **26**, 449 (1967).
[p] M. N. Copley and E. V. Truter, *J. Chromatogr.* **45**, 480 (1969).
[q] R_f values, see reference in footnotes *e* and *l*.

TABLE II
hR_f VALUES OF FREE AMINO ACIDS ON SILICA GEL

		Silica gel G (Merck)									Silica gel F_{254} precoated plates (Merck)
		Unbuffered						Buffered[a] (pH)			Unbuffered
								3	5	8	
		hR_f				hR_{Leu}[b]		hR_f			hR_{Phe}[c]
Abbreviation	Amino acid / Solvent:	1	2	3	4	5[d]	6	7	7	7	8
Ala	Alanine	37	22	29	39	51	72	47	32	24	29
β-Ala	β-Alanine	26	22	30	30	35	51	—	—	—	—
Abu	α-Amino-*n*-butyric acid	—	27	—	—	59	75	48	53	45	—
Aib	α-Aminoisobutyric acid	—	27	—	—	61	73	—	—	—	—
β-Aib	β-Aminoisobutyric acid	—	25	—	—	48	72	—	—	—	—
γ-Abu	γ-Amino-*n*-butyric acid	—	27	—	—	45	51	47	41	24	—
ε-Aco	ε-Amino-*n*-caproic acid	—	34	—	—	68	70	—	—	—	—
Acy	α-Amino-*n*-caprylic acid	65	59	69	58	143	107	—	—	—	—
Arg·HCl	Arginine·HCl	2	6	19	10	19	17	19	22	3	4
Asn	Asparagine	—	14	—	—	43	66	47	29	13	—
Asp	Aspartic acid	33	17	6	9	18	32	47	33	21	7
$CySO_3H$	Cysteic acid	50	10	4	17	53	61	49	41	27	—
Cys·HCl	Cysteine·HCl	—	—	—	—	—	—	53	49	62	0
$(Cys)_2HCl$	Cystine·HCl	32	9	12	27	18	70	24	20	10	—
Gln	Glutamine	—	15	—	—	47	74	53	39	24	—
Glu	Glutamic acid	35	24	10	14	32	50	51	49	31	15
Gly	Glycine	32	18	24	29	45	61	42	35	12	17
His	Histidine	20	5	32	38	18	74	21	18	15	4
$(Hcys)_2$	Homocystine	—	17	—	—	28	—	—	—	—	—
Hylys	Hydroxylysine	—	—	4	—	—	19	—	—	—	—
Hyp	Hydroxyproline	34	16	38	28	50	63	49	42	25	24

Ile	Isoleucine	53	43	49	52	92	97	—	—	—	94
Leu	Leucine	55	44	48	53	100	100	41	51	53	95
Lys·HCl	Lysine·HCl	2	3	9	18	11	26	13	13	1	0
Met	Methionine	51	35	49	51	92	—	64	43	52	79
$MetO_2$	Methionine sulfone	—	—	—	—	66	76	—	—	—	—
1-Mehis	1-Methylhistidine	—	—	—	—	9	—	—	—	—	—
Nle	Norleucine	57	45	52	53	102	102	63	53	57	—
Nval	Norvaline	50	36	42	49	77	88	—	—	—	—
Orn·HCl	Ornithine·HCl	—	4	—	—	14	24	—	—	—	—
Phe	Phenylalanine	58	43	55	54	109	106	69	47	57	100
Sar	Sarcosine	22	12	37	34	32	62	—	—	—	—
Ser	Serine	35	18	20	27	47	60	49	31	13	16
Tau	Taurine	—	18	—	—	79	76	—	—	—	—
Thr	Threonine	37	20	26	37	51	75	51	40	25	28
Trp	Tryptophan	62	47	63	55	122	102	68	54	57	—
Tyr	Tyrosine	57	41	47	42	107	77	49	50	55	60
Val	Valine	45	32	40	48	72	85	56	49	45	68
Literature reference		*e*	*e*	*e*	*e*	*f*	*g*	*h*	*h*	*h*	*i*

[a] Silica gel layer prepared[h] by suspension in McIlvaine's universal buffer.[j]
[b] hR_f value related to leucine.[f,g]
[c] hR_f value related to phenylalanine.[i]
[d] One-dimensional TLC in the BN-chamber[k] with overflow technique.[l] Duration of run 4.5 hours.[f]
[e] M. Brenner and A. Niederwieser, *Experientia* **16,** 378 (1960).
[f] M. Brenner, A. Niederwieser, and G. Pataki, *in* "Dünnschicht-Chromatographie" (E. Stahl, ed.), pp. 403–452. Springer-Verlag, Heidelberg, 1962.
[g] M. Brenner and A. Niederwieser, Vol. 11, p. 39.
[h] E. J. Shellard and G. H. Jolliffe, *J. Chromatogr.* **31,** 82 (1967).
[i] F. Fussi and G. F. Fedeli, *Boll. Chim. Farm.* **107,** 711 (1968).
[j] T. C. McIlvaine, *J. Biol. Chem.* **49,** 184 (1921).
[k] Available from D. Desaga GmbH, Heidelberg, Germany; C. A. Brinkman, Westbury, Long Island, New York.
[l] M. Brenner and A. Niederwieser, *Experientia* **17,** 237 (1961).

TABLE III
hR_f Values of Free Amino Acids on Cellulose

Amino acid	Cellulose MN 300									Cellulose MN 300 (C) Avicel microcryst. (A)												Avicel	
	hR_f in Solvent No.:																						
										17		18		19		20		21		22			
	6	9	10	11	12	13	14	15	16	C	A	C	A	C	A	C	A	C	A	C	A	2	3
α-Amino adipic acid	29	61	—	—	—	—	62	3	—	—	—	—	—	—	—	—	—	—	—	—	—	—	—
Alanine	64	63	34	56	55	10	57	23	18	43	36	41	34	16	13	40	33	35	25	37	27	37	38
β-Alanine	46	46	29	56	—	—	46	19	—	45	39	33	27	9	6	30	24	31	21	37	29	—	—
α-Amino-*n*-butyric acid	—	—	42	62	64	16	65	29	—	53	45	48	41	24	20	47	41	45	34	47	37	—	—
α-Aminoisobutyric acid	—	—	—	—	—	—	74	30	—	—	47	49	42	25	21	49	43	48	37	46	34	—	—
β-Aminoisobutyric acid	—	—	42	70	—	—	60	26	—	56	48	42	33	18	13	42	33	47	33	47	38	—	—
γ-Amino-*n*-butyric acid	53	48	35	64	—	—	51	23	—	53	45	34	27	11	8	32	25	45	30	44	34	44	64
ε-Amino-*n*-caproic acid	—	—	—	—	—	—	76	23	—	—	—	—	—	—	—	—	—	—	—	—	—	—	—
α-Amino-*n*-caprylic acid	—	—	—	—	—	—	93	81	—	89	89	80	78	73	66	90	87	88	91	76	73	—	—
Arginine	34	4	12	24	24	3	19	6	8	22	18	7	4	4	3	21	14	19	9	5	3	17	49
Asparagine	32	24	22	20	—	—	21	14	—	22	16	22	16	5	5	25	18	14	11	22	19	—	—
Aspartic acid	21	46	10	34	41	0	48	1	2	27	21	31	21	1	0	17	15	18	14	14	9	19	18
Citrulline	36	35	25	32	—	—	34	12	—	28	23	32	23	4	2	27	20	20	16	20	12	—	—
Cysteic acid	31	47	24	11	11	5	53	8	6	14	12	36	31	2	1	21	15	13	9	26	19	—	—
Cysteine	—	—	14	9	8	3	12	5	5	—	—	—	—	—	—	—	—	—	—	—	—	8	56
Cystine	24	22	14	9	0	0	6	3	2	13	9	18	13	1	1	17	13	6	4	17	10	—	—
Glutamic acid	26	62	13	46	52	1	56	1	2	41	28	40	33	2	2	22	16	37	22	15	9	28	23
Glutamine	39	34	25	27	29	4	31	13	—	33	22	33	22	6	4	31	20	21	14	23	15	—	—
Glycine	42	46	25	37	36	7	37	16	14	33	24	30	22	8	7	29	24	22	15	27	23	25	32
Histidine	48	27	42	16	15	9	11	26	21	23	18	29	23	7	6	32	28	13	6	38	33	17	60
Hydroxyproline	46	46	32	37	42	7	48	17	12	—	—	—	—	—	—	—	—	—	—	—	—	—	—

Isoleucine	90	81	60	85	76	45	90	63	51	71	68	62	58	44	36	65	69	65	60	62	56	70	71
Leucine	85	82	60	85	78	52	90	69	55	73	71	64	61	50	40	68	74	69	64	64	59	71	71
Lysine	56	5	28	20	20	4	16	17	13	18	17	5	1	4	2	18	14	12	7	32	28	17	24
Methionine	81	66	52	67	66	36	78	51	41	59	50	57	52	24	18	55	55	53	46	56	48	55	76
Methionine sulfoxide	54	45	32	32	—	—	39	20	—	30	23	31	26	7	6	31	31	23	16	22	20	—	—
Methionine sulfone	67	49	46	33	—	—	50	27	—	32	25	41	35	9	8	35	36	26	18	37	33	—	—
1-Methylhistidine	63	34	—	—	—	—	14	18	—	21	17	29	25	9	8	36	33	12	7	27	22	—	—
3-Methylhistidine	63	44	31	16	—	—	12	14	—	29	21	36	33	8	6	37	32	20	10	31	21	—	—
Norleucine	—	—	—	—	—	—	92	73	—	—	—	—	—	—	—	—	—	—	—	—	—	—	—
Norvaline	—	—	—	—	—	—	—	—	—	63	57	56	49	35	31	58	60	60	53	57	49	—	—
Ornithine	44	3	24	15	15	4	11	15	—	15	15	6	2	3	5	16	15	10	5	26	24	—	—
Proline	75	64	42	54	54	17	58	30	24	45	38	42	33	20	17	41	39	39	29	39	33	41	89
Phenylalanine	89	71	63	72	69	51	82	67	55	67	62	64	59	36	32	64	67	59	60	65	59	64	77
Sarcosine	—	—	35	48	—	—	48	24	—	38	31	36	29	15	15	35	32	30	23	35	31	—	—
Serine	49	46	46	33	36	13	39	27	23	32	25	32	25	7	7	30	28	23	15	43	40	22	24
Taurine	71	52	55	23	22	17	41	33	—	28	22	42	36	17	15	42	40	29	22	50	42	18	22
Threonine	51	53	68	44	45	45	51	61	48	38	33	38	36	14	8	41	41	32	24	64	54	32	34
Tryptophan	68	50	61	56	60	42	70	54	49	58	50	64	61	22	19	55	51	49	42	63	54	47	52
Tyrosine	62	60	61	57	60	21	72	35	27	53	47	62	59	15	12	47	45	49	44	48	38	45	52
Valine	85	72	49	74	69	30	79	44	35	61	55	53	50	32	23	57	49	58	49	56	44	58	73
Chamber saturation			+	+	−	−	+	+	+	TLC in a 6-plate frame												+	+
Length of run (cm)			13	13	13	13	13	13		15	12	15	12	15	12	15	12	15	12	15	12	12	12
Duration of run (h)			1.5	2	3	2.5	2.5	2.5		5	4.5	5	4.5	5	4.3	5	4.3	5	4.3	2.5	3	1.5	2
Literature reference	*a*	*a*	*b*	*b*	*c*	*c*	*d*	*d*	*e*	*f*	*f*	*f*	*f*	*f*	*f*	*f*	*f*	*f*	*f*	*f*	*f*	*g*	*g*

[a] E. Bujard and J. Mauron, *J. Chromatogr.* **21,** 19 (1966).

[b] H. H. White, *Clin. Chim. Acta* **21,** 297 (1968).

[c] K. Jones and J. G. Heathcote, *J. Chromatogr.* **24,** 106 (1966).

[d] C. Haworth and J. G. Heathcote, *J. Chromatogr.* **41,** 380 (1969).

[e] J. G. Heathcote and C. Haworth, *J. Chromatogr.* **43,** 84 (1969).

[f] I. Smith, L. J. Rider, and R. P. Lerner, *J. Chromatogr.* **26,** 449 (1967).

[g] D. Horton, A. Tanimura, and M. L. Wolfrom, *J. Chromatogr.* **23,** 309 (1966).

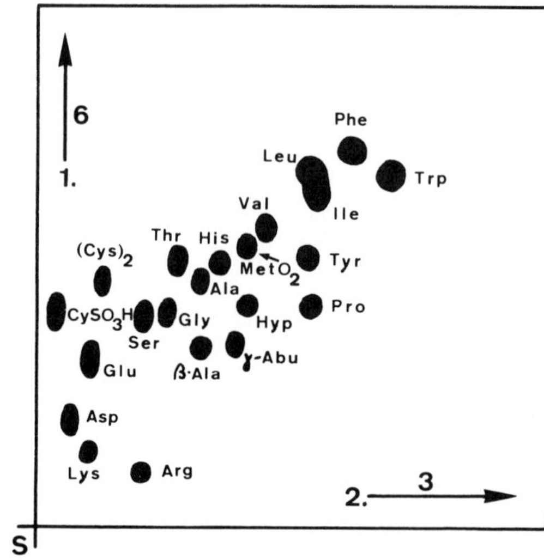

FIG. 1. Two-dimensional thin-layer chromatography (TLC) of free amino acids on Silica Gel G in solvents 6 and 3 (see Table I) with chamber saturation. Detection with ninhydrin. Methionine migrates more or less with leucine, but is visible as methionine sulfone after performic acid oxidation; then cystine will be oxidized to cysteic acid. Load: 1 μg per amino acid, applied in 0.5 μl 0.1 N hydrochloric acid. S: start. Abbreviations as in Table II. Compare A. R. Fahmy, A. Niederwieser, G. Pataki, and M. Brenner, *Helv. Chim. Acta* **44**, 2022 (1961).

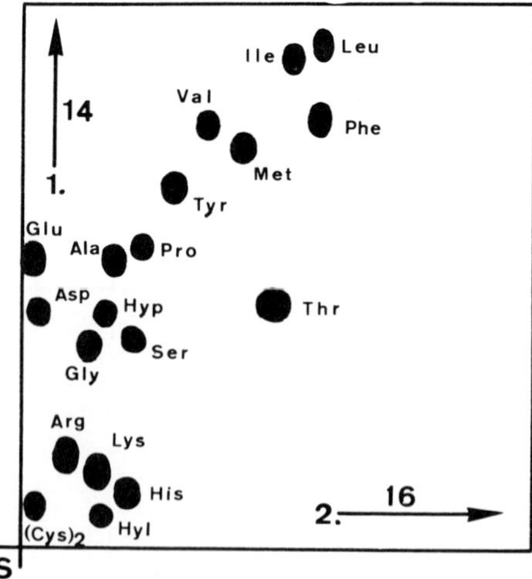

FIG. 2. Two-dimensional TLC of free amino acids on cellulose MN 300 in solvents 14 and 16 (see Table I) with chamber saturation. Load: about 1 μg per amino acid. Detection with ninhydrin–cadmium acetate reagent. S: start. Map according to J. G. Heathcote and C. Haworth, *J. Chromatogr.* **43**, 84 (1969).

acetic acid–water (60:20:20) twice,[23] ethanol–water (87:13) three times,[24] and *n*-butanol–88% formic acid–water (75:15:10, v/v) twice,[18] have been proposed.

Resolution is much better on two-dimensional chromatograms provided that carefully selected solvents are employed. Solvents suitable for a two-dimensional combination are indicated in Table I. A separation of amino acids on Silica Gel G is shown in Fig. 1 and on cellulose MN 300 in Figs. 2 and 3. The separation technique shown in Fig. 2 was worked out by Heathcote and Haworth for the quantitative determination of amino acids in protein hydrolyzates.[25] The map of Fig. 3 was obtained by Copley and Truter[26] on an Amberlite CG 120 (H^+ form)–cellulose MN 300 twin layer to which a mixture of potassium chloride and 22

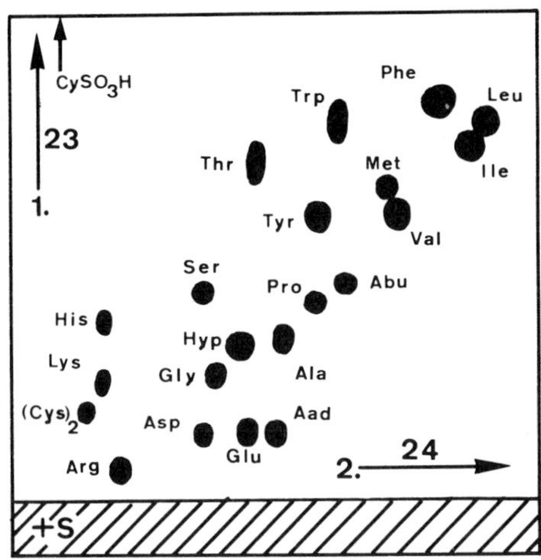

FIG. 3. Two-dimensional TLC of free amino acids and an 87-molar excess of potassium chloride on a cellulose MN 300–Amberlite CG 120 (H^+ form) twin layer in solvents 23 and 24 (see Table I) without chamber saturation. The hatched area where the start S is located is coated with Amberlite CG 120–cellulose MN 300 (3:0.45, w/w), and the unmarked area with cellulose MN 300 alone. After application of the salt-containing sample, the chloride is washed away by permitting the ascent of water to the top of the plate (65 minutes, 16 cm). Subsequently, the chromatogram is dried at 40° for 2 hours, and then chromatography is started. First dimension, 16 cm in 200 minutes; second dimension, 16 cm in 210 minutes. Detection with ninhydrin. Map according to M. N. Copley and E. V. Truter, *J. Chromatogr.* **45,** 480 (1969).

[23] R. B. Mefferd, R. M. Summers, and J. G. Fernandez, *Anal. Lett.* **1,** 279 (1968).

[24] H. J. Bremer, W. Nützenadel, and H. Bickel, *Monatsschr. Kinderheilk.* **117,** 32 (1969).

[25] J. G. Heathcote and C. Haworth, *J. Chromatogr.* **43,** 84 (1969).

[26] M. N. Copley and E. V. Truter, *J. Chromatogr.* **45,** 480 (1969).

amino acids (87:1, molar) was applied. Despite an enormous excess in salt, a perfect separation is obtained. Twin-layer techniques for routine desalting and subsequent one-dimensional TLC of urinary amino acids have been described by Kraffczyk and Helger.[27] A two-dimensional technique for separation of urinary amino acids was described recently, using, first, pyridine–dioxane–ammonia–water (35:35:15:15) twice, and, second, *n*-butanol–acetone–acetic acid–water (35:35:10:20) twice.[28]

Detection

Numerous reagents and the relevant literature, especially on specific reagents, have been tabulated[29] (see also footnote 30). "Non-destructive" detection of amino acids using 1-fluor-2,4-dinitrobenzene has been described.[31] This procedure, although not as sensitive as detection with ninhydrin, allows further investigation of the resulting DNP-amino acids, if necessary. Ninhydrin and isatin reagents are described below.

Reagents

Ninhydrin–acetic acid: 500 mg of ninhydrin, 5 ml of glacial acetic acid, and 100 ml of *n*-butanol. The chromatogram is sprayed and kept for 30 minutes at 80° or 10 minutes at 110°.

Ninhydrin–collidine–copper(II) nitrate (Moffat-Lytle reagent,[32] modified): I: 1 g of ninhydrin, 20 ml of 2,4,6-collidine, 100 ml of glacial acetic acid, 500 ml of absolute ethanol; II: 1% $Cu(NO_3)_2 \cdot 3H_2O$ in absolute ethanol. Spray: mix 15 ml of I and 1 ml of II immediately before use. The mixture is unstable. The sprayed chromatogram is kept for 15–30 minutes at 80° or 5–10 minutes at 110°.

Ninhydrin–cadmium acetate:[25] 0.5 g of cadmium acetate, 50 ml of water, 20 ml of glacial acetic acid, 430 ml of propanone. Immediately before use dissolve 100 mg of ninhydrin in 50 ml of this solution. The sprayed chromatograms are kept at 60° for 15 minutes.

Isatin–cadmium acetate for detection of proline and hydroxyproline.[25] Dissolve 100 mg of isatin in 50 ml of the above cadmium

[27] F. Kraffczyk and R. Helger, *Z. Anal. Chem.* **243**, 536 (1968).

[28] F. Kraffczyk, R. Helger, and H. Lang, *Clin. Chem.* **16**, 662 (1970).

[29] M. Brenner, A. Niederwieser, and G. Pataki, *in* "Dünnschicht-Chromatographie" (E. Stahl, ed.), pp. 403–452. Springer-Verlag, Heidelberg, 1962.

[30] "Chromatographic and Electrophoretic Techniques" (I. Smith, ed.), 3rd ed., Vol. 1. Heinemann, London, 1969.

[31] G. Pataki, J. Borko, and A. Kunz, *Z. Klin. Chem. Klin. Biochem.* **6**, 458 (1968).

[32] E. D. Moffat and R. I. Lytle, *Anal. Chem.* **31**, 926 (1959).

acetate solution immediately before use and keep the sprayed plate at 90° for 10 minutes.

Note on Quantitative Analysis

Automatic ion-exchange chromatography[1] according to Spackman *et al.*[33] should be used for quantitative analysis of amino acids. If this is not possible, gas chromatography[2,34,35,35a] or TLC may be employed. The compounds produced by the reaction of amino acids with ninhydrin can be eluted from a thin-layer chromatogram and measured photometrically.[36-38] Visual comparison of spot size[39] and intensity between an unknown and a standard, both applied in systematically varied amounts and chromatographed under identical conditions, is still considered to be useful. The best thin-layer chromatographic method is, however, *in situ* measurement by transmission or by reflection densitometry[40,41] using a scanner. Excellent results have been obtained recently by Heathcote and Haworth.[25]

Phenylthiohydantoins (PTH-Amino Acids)

Phenylthiohydantoins are obtained when reaction products of peptides or proteins with phenyl isothiocyanate are properly degraded. Their separation and identification are essential steps in the N-terminal and sequential analysis of peptide primary structures by stepwise degradation (Edman degradation).[42]

Preparation of PTH-Amino Acids

Collections of PTH derivatives of all protein amino acids are commercially available. PTH-amino acids may be prepared from free amino

[33] D. H. Spackman, W. H. Stein, and S. Moore, *Anal. Chem.* **30,** 1190 (1958).

[34] K. Blau, *in* "Biomedical Applications of Gas Chromatography" (H. A. Szymanski, ed.). Plenum, New York, 1968.

[35] C. W. Gehrke, D. Roach, R. W. Zumwalt, D. L. Stalling, and L. L. Wall, "Quantitative Gas-Liquid Chromatography of Amino Acids in Proteins and Biological Substances." Analytical Biochemistry Laboratories, Inc., Columbia, Missouri, 1968.

[35a] C. W. Gehrke and K. Leimer, *J. Chromatogr.* **53,** 201 (1970).

[36] K. Esser, *J. Chromatogr.* **18,** 414 (1965).

[37] R. Bondivenne and N. Busch, *J. Chromatogr.* **29,** 349 (1967).

[38] M. E. Clark, *Analyst* **93,** 810 (1968).

[39] S. J. Purdy and E. V. Truter, *Chem. Ind.* (*London*) **1962,** 506 (1962).

[40] M. M. Frodyma and R. W. Frei, *J. Chromatogr.* **15,** 501 (1964); **17,** 131 (1965).

[41] R. W. Frei, *in* "Progress in Thin-Layer Chromatography and Related Methods" (A. Niederwieser and G. Pataki, eds.), Vol. II, pp. 1–61. Ann Arbor-Humphrey Science Publ., Ann Arbor, Michigan, 1970.

[42] P. Edman, *Acta Chem. Scand.* **4,** 283 (1950).

acids according to procedures described elsewhere.[43-51] Melting points,[45] optical rotatory dispersion (ORD) measurements of optically active PTH-amino acids,[52] and infrared (IR) spectra[49] also of microgram amounts,[53] have been reported. A great number of procedures for the Edman degradation[42] of peptides and proteins have been described (see, e.g., references in footnotes 54–58): paper strip method,[59,60] silica gel thin-layer method,[61] radioisotope method using [^{35}S]phenylisothiocyanate[62-64] or [^{14}C]phenylisothiocyanate.[65] The processes involved in the Edman degradation have been automated in the protein sequenator.[66]

Solvents, R_f Values and Identification Maps

Layers of silica gel, aluminum oxide, cellulose, and polyamide can be employed for TLC of PTH-amino acids. Useful solvents are compiled in Table IV and the corresponding R_f values in Table V. It is not as yet possible to separate all PTH-derivatives of protein amino acids on a single chromatogram. However, the Edman degradation theoretically delivers only one single phenylthiohydantoin at a time, and, practically

[43] P. Edman, *Acta Chem. Scand.* **4**, 277 (1950).
[44] J. Sjöquist, *Biochim. Biophys. Acta* **41**, 20 (1960).
[45] J. Sjöquist, *Ark. Kemi* **11**, 129 (1957).
[46] D. Ilse and P. Edman, *Aust. J. Chem.* **16**, 411 (1963).
[47] E. Cherbuliez, J. Marszalek, and J. Rabinowitz, *Helv. Chim. Acta* **43**, 87 (1960); **46**, 1445 (1963).
[48] A. L. Levy and D. Chung, *Biochim. Biophys. Acta* **17**, 454 (1955).
[49] L. K. Ramachandran, A. Epp, and G. McConnell, *Anal. Chem.* **27**, 1734 (1955).
[50] P. Edman and K. Lauber, *Acta Chem. Scand.* **10**, 466 (1956).
[51] V. Ingram, *J. Chem. Soc.* **1953**, 3717 (1953).
[52] D. Djerassi, K. Undheim, R. C. Sheppard, W. G. Terry, and B. Sjöberg, *Acta Chem. Scand.* **15**, 903 (1961).
[53] M. Murray and G. F. Smith, *Anal. Chem.* **40**, 440 (1968).
[54] J. Sjöquist, B. Blombäck, and P. Wallén, *Ark. Kemi* **16**, 425 (1961).
[55] J. Sjöquist, *Ark. Kemi* **14**, 291 (1959).
[56] J. Sjöholm, *Acta Chem. Scand.* **18**, 889 (1964).
[57] H. Fraenkel-Conrat, J. I. Harris, and A. L. Levi, *Methods Biochem. Anal.* **2**, 393 (1955).
[58] S. Eriksson and J. Sjöquist, *Biochim. Biophys. Acta* **45**, 290 (1960).
[59] P. Edman, *Proc. Roy. Aust. Chem. Inst.* **24**, 434 (1957).
[60] J. Jentsch, *Z. Naturforsch. B* **23**, 1613 (1968).
[61] T. Wieland and U. Gebert, *Anal. Biochem.* **6**, 201 (1963).
[62] E. Cherbuliez, B. Baehler, J. Marszalek, A. R. Sussmann, and J. Rabinowitz, *Helv. Chim. Acta* **46**, 2446 (1963); **47**, 1350 (1964).
[63] E. Cherbuliez, A. R. Sussmann, and J. Rabinowitz, *Pharm. Acta Helv.* **36**, 131 (1961).
[64] W. G. Laver, *Biochim. Biophys. Acta* **53**, 469 (1961).
[65] D. Brandenberg, *Hoppe-Seyler's Z. Physiol. Chem.* **350**, 741 (1969).
[66] P. Edman and G. Begg, *Eur. J. Biochem.* **1**, 80 (1967).

TABLE IV

SOLVENTS FOR THIN-LAYER CHROMATOGRAPHY (TLC) OF PTH-AMINO ACIDS

Combination for two-dimensional TLC	No.	Solvent
1.	28	Chloroform[a,b]
1.	29	Chloroform–methanol (90:10, v/v)[a,b]
2.	30	Chloroform–formic acid (100:5, v/v)[a,b]
2.	31	*n*-Heptane–ethylene chloride–formic acid–propionic acid (90:30:21:18, v/v) upper phase[a,b]
	32	Chloroform–methanol–formic acid (70:30:2, v/v)[b]
	33	Chloroform–isopropanol–water (28:8:1, v/v)[c]
	34	Ethyl acetate–pyridine–water (7:2:1, v/v)[c]
	35	Chloroform–ethyl acetate–water (6:3:1, v/v)[c]
	36	Heptane–*n*-butanol–90% formic acid (40:40:20, v/v)[d,e]
	37	*o*-Xylene. Sheet impregnated with acetone–formamide (4:1) by dipping and air-drying before application of the sample [e,f]
	38	*n*-Butyl acetate–propionic acid–formamide. The butyl acetate is saturated with water, 3% v/v propionic acid is added, and this mixture is subsequently saturated with formamide. The thin-layer is impregnated with acetone–formamide (4:1) by dipping and air-drying before application of the sample[e,f]
	39	Heptane–ethylene chloride–propionic acid (58:25:17, v/v)[g]
	40	Multiple TLC: 1. Solvent 39, 2. (in the same direction) *n*-Heptane–*n*-butanol–75% formic acid (50:30:9, v/v)[g]
	41	Chloroform–ethanol (98:2, v/v)[c]
	42	Chloroform–isopropanol–formic acid (35:4:1, v/v)[c]
2. 2.	43	90% Formic acid–water (1:1)[h]
1.	44	*n*-Heptane–*n*-butanol–glacial acetic acid (40:30:9)[h]
1.	45	Carbon tetrachloride–glacial acetic acid (9:1)[h]
	46	Benzene–glacial acetic acid (9:1)[h]
	47	*n*-Hexane–2-butanone (3:1)[h]

[a] M. Brenner, A. Niederwieser, and G. Pataki, *in* "Dünnschicht-Chromatographie" (E. Stahl, ed.), pp. 403–452. Springer-Verlag, Heidelberg, 1962.

[b] M. Brenner, A. Niederwieser, and G. Pataki, *Experientia* **17,** 145 (1961).

[c] E. Cherbuliez, B. Baehler, J. Marszalek, A. R. Sussmann, and J. Rabinowitz, *Helv. Chim. Acta* **46,** 2446 (1963); **47,** 1350 (1964).

[d] J. Sjöquist, *Acta Chem. Scand.* **7,** 447 (1953).

[e] K. Dus, H. DeKlerk, K. Sletten, and R. G. Bartsch, *Biochim. Biophys. Acta* **140,** 291 (1967).

[f] P. Edman and J. Sjöquist, *Acta Chem. Scand.* **10,** 1507 (1956).

[g] J.-O. Jeppsson and J. Sjöquist, *Anal. Biochem.* **18,** 264 (1967).

[h] K.-T. Wang, I. S. Wang, A. L. Lin, and C.-S. Wang, *J. Chromatogr.* **26,** 323 (1967).

TABLE V
hR_f VALUES OF PTH-AMINO ACIDS ON THIN-LAYER CHROMATOGRAMS

		Silica gel G								Eastman chromagram type K 301 R silica gel					Aluminum oxide (Fluka)			Polyamide[a]				
		hR_f in Solvent No.																				
Abbreviation	PTH-Amino acid	28	29	30	31	32	33	34	35	36	37[b]	38[b]	39[c]	40[c]	28	41	42	43	44	45	46	47
Ala	Alanine	16	68	39	11	100	41	—	45	—	20	79	33	66	51	66	—	68	60	43	67	47
Arg	Arginine	0	1	0	0	24	—	—	—	—	—	—	0	17	0	0	0	90	40	0	6	0
Asn	Asparagine	0	23	7	0	100	3	67	2	26	—	14	3	34	0	0	32	—	—	—	—	—
Asp	Aspartic acid	0	1	13	0	70	0	17	0	49	—	11	7	51	0	0	20	65	25	3	15	0
$CySO_3H$	Cysteic acid	—	—	—	—	—	—	—	—	—	—	—	0	8	—	—	—	—	—	—	—	—
$(Cys)_2$	Cystine	—	—	—	—	—	—	—	—	—	—	—	—	—	—	—	69	—	—	—	—	—
Gln	Glutamine	1	28	8	0	100	2	65	2	28	—	22	4	38	0	0	36	—	—	—	—	—
Glu	Glutamic acid	1	4	17	0	75	1	25	0	49	—	26	12	52	0	0	32	51	12	0	31	0
Gly	Glycine	10	56	33	5	90	30	—	30	—	8	64	25	61	28	45	78	71	65	30	63	29
His	Histidine	1	29	0	2	100	1	58	0	—	—	—	0	15	0	0	4	—	—	—	—	—
Hyp	Hydroxyproline	—	—	—	—	—	—	—	—	—	—	—	—	—	—	—	—	73	14	0	0	0
Ile	Isoleucine	40	77	57	37	100	64	—	66	—	67	—	57	76	78	80	—	44	76	60	76	75
Leu	Leucine	40	77	60	37	100	69	—	70	—	70	—	60	77	78	80	—	44	74	63	80	72
Lys	Lysine	12	71	34	3	100	28	—	25	—	—	87	15	58	28	62	—	29	39	19	55	3

Met	Methionine	33	75	51	13	100	52	—	51	—	45	—	40	70	58	76	—	51	63	48	72	50
MetO	Methionine sulfoxide	1	40	12	1	100	—	—	—	—	—	—	—	—	—	—	—	—	—	—	—	—
Phe	Phenylalanine	28	74	50	18	100	56	—	53	—	64	—	44	72	60	74	—	38	65	55	66	56
Pro	Proline	60	82	65	21	100	75	—	61	—	82	—	47	72	91	91	—	61	75	85	90	77
Ser	Serine	—	—	—	—	—	6	79	7	—	—	33	5	47	0	0	48	—	—	—	—	—
Thr	Threonine	4	45	15	0	100	11	83	14	—	38	45	10	50	0	0	51	46	63	51	71	56
Trp	Tryptophan	13	62	39	10	100	—	—	—	—	15	—	23	65	28	52	—	24	40	14	48	22
Tyr	Tyrosine	3	47	21	1	100	19	—	35	—	—	75	11	58	0	8	64	43	34	4	16	4
Val	Valine	32	74	55	23	100	56	—	61	—	51	—	50	74	72	76	—	57	70	53	74	53
MPTU	Monophenylthiourea	12	54	31	3	—	—	—	—	—	—	—	30	64	—	—	—	70	65	67	52	4
DPTU	Diphenylthiourea	43	76	67	22	—	—	—	—	—	—	—	—	—	—	—	—	37	65	32	70	23
Chamber saturation		+	+	+	+	+	+	+	+	−	−	−	+	+	+	+	+					
Literature reference		*d*	*d*	*d*	*d*	*e*	*f*	*f*	*f*	*g*	*g*	*g*	*h*	*h*	*f*	*f*	*f*	*i*	*i*	*i*	*i*	*i*

[a] ε-Aminocaprolactam resin CM 1011, Toyo Rayon Co., Tokyo, Japan.[i]

[b] Layer impregnated with acetone-formamide (4:1) and air-dried.[g]

[c] hR_f values calculated from figures given by Jeppsson and Sjöquist.[h]

[d] M. Brenner, A. Niederwieser, and G. Pataki, *in* "Dünnschicht-Chromatographie" (E. Stahl, ed.), pp. 403–452. Springer-Verlag, Heidelberg, 1962.

[e] M. Brenner, A. Niederwieser, and G. Pataki, *Experientia* **17,** 145 (1961).

[f] E. Cherbuliez, B. Baehler, J. Marszalek, A. R. Sussmann, and J. Rabinowitz, *Helv. Clim. Acta* **46,** 2446 (1963); **47,** 1350 (1964).

[g] K. Dus, H. DeKlerk, K. Sletten, and R. G. Bartsch, *Biochim. Biophys. Acta* **140,** 291 (1967).

[h] J.-O. Jeppsson and J. Sjöquist, *Anal. Biochem.* **18,** 264 (1967).

[i] K.-T. Wang, I. S. Wang, A. L. Lin, and C.-S. Wang, *J. Chromatogr.* **26,** 323 (1967).

speaking, only a few substances have to be separated. Often (in smaller peptides) only a few PTH-amino acids are expected to come out. The rather poor resolution of one-dimensional TLC may suffice under such conditions.

Jeppsson and Sjöquist[67] proposed a combination of two solvents on Eastman chromatogram sheets type K 301 R; it entails a single application of sample, which is chromatographed first in solvent 39. All except the water-soluble derivatives (i.e., PTH-arginine, PTH-histidine, and PTH-cysteic acid) move. After drying of the chromatogram at 110° for 15 minutes, the PTH-derivatives of Trp, Tyr, Asn, and His appear as yellow spots, and of Gly, Thr, and Ser as pink spots. Only PTH-Val moves as a single spot. The nonresolved phenylthiohydantoin group of Tyr, Thr, and Glu is distinguishable after rechromatography in the same solvent 39 and in the same direction. The nonresolved groups of Trp and Gly; Asn and Gln; Arg, His, and cysteic acid are separated by rechromatography in solvent 40 in the same direction. The group of Pro, Phe, and Met is resolved by paper chromatography in solvent II of Sjöquist.[44]

Running three one-dimensional chromatograms simultaneously on Silica Gel G in solvents 33, 34, 35 allows analysis for 15 PTH-amino acids. PTH-asparagine and PTH-glutamine are not resolved.

Two-dimensional TLC leads to a better resolution and is of advantage especially when the sample contains several PTH derivatives. According to Brenner *et al.*[29] two two-dimensional chromatograms and one one-dimensional chromatogram are run simultaneously as shown in Fig. 4. Separation is complete except for PTH-glycine and monophenyl-

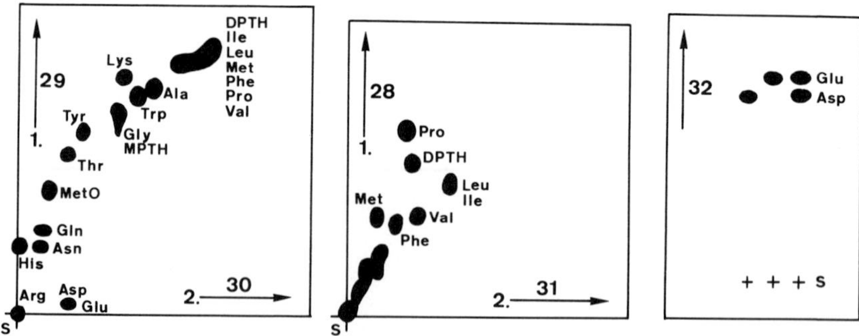

FIG. 4. Simultaneous separation of PTH-amino acids on two two-dimensional chromatograms and one one-dimensional thin-layer chromatogram on Silica Gel G in solvents 29/30, 28/31, and 32, according to M. Brenner, A. Niederwieser, and G. Pataki [*Experientia* **17**, 145 (1961)]. PTH-Leucine and PTH-isoleucine are not resolved. S: start. For solvents and abbreviations, see Tables IV and V. Load: about 0.5 μg of each substance.

[67] J.-O. Jeppsson and J. Sjöquist, *Anal. Biochem.* **18**, 264 (1967).

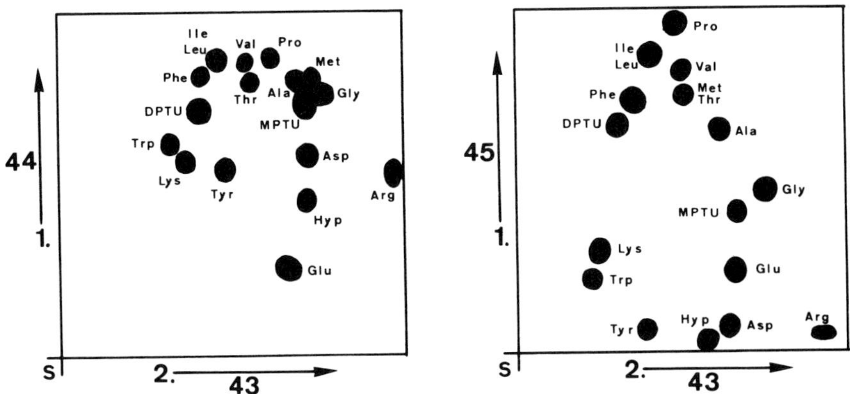

FIG. 5. Two-dimensional separations of PTH-amino acids on polyamide thin-layers (ε-polycaprolactam resin CM 1007 s, Toyo Rayon Co., Tokyo, Japan) according to K.-T. Wang, I. S. Wang, A. L. Lin, and C.-S. Wang [*J. Chromatogr.* **26**, 323 (1967)]. PTH-Leucine and PTH-isoleucine are not resolved. PTH-Methionine may remain obscured by PTH-threonine and monophenylthiourea, respectively. For solvents and abbreviations, see Tables IV and V. S: start. Load: 8 μg of each substance.

thiourea on the one hand, and PTH-leucine and PTH-isoleucine on the other hand. PTH-Glycine is detectable by its very characteristic color reaction with ammonia.[68] Identification of PTH-leucine and of PTH-isoleucine is possible after elution of the spot, hydrolysis in 6 *N* HCl for 12 hours, and TLC of the free amino acid in solvent 5 (Table I) with overflow technique.

All PTH-amino acids except PTH-leucine and PTH-isoleucine are resolved by two two-dimensional chromatograms on polyamide, as shown in Fig. 5. Simultaneous presence of PTH-methionine and PTH-threonine will obscure PTH-methionine due to the presence of monophenylthiourea. A disadvantage of this technique on polyamide is that 8 μg of PTH-amino acid are necessary for detection.

To get correct spot patterns in two-dimensional TLC, a standard mixture of all PTH-amino acids must always be run simultaneously under the same chromatographic conditions. If a PTH-derivative cannot be identified unambiguously at once, it should be eluted from the chromatogram, transferred to another thin-layer, and run again between the corresponding reference substances in an appropriate solvent. The transfer, also of minute quantities, is easily accomplished by sucking the adsorbent around the spot in a glass tube (inner diameter about 2 mm) whose other end has been reduced to a capillary and which contains a small cotton plug. The substance is eluted with a few drops of ap-

[68] G. Schramm, J. W. Schneider, and F. A. Anderer, *Z. Naturforsch. B* **11**, 120 (1956).

propriate solvent (e.g., chloroform–methanol–acetic acid, 50:50:1, v/v) directly onto the new thin-layer. The eluting solvent is added with a capillary pipette.

Further identification of a PTH-derivative is possible by IR spectroscopy. This needs about 2 μg of PTH-amino acid.[53]

Detection of Phenylthiohydantoins

PTH-amino acids are visible in UV light at 270 nm on layers containing a fluorescent agent. On silica gel layers the detection limit is about 0.1 μg. The chlorine/toluidine test[29] is still more sensitive, but the substance will be destroyed. Detection is also possible with iodine.[69] Polychromatic detection has been described recently[70] on Eastman chromagrams type K 301 R, using a spray of 100 mg of ninhydrin, 5 ml of collidine, and 95 ml of absolute ethanol. After 15 minutes at 110°, characteristic colored spots appear. PTH-Valine remains invisible. This test seems to be limited to the layer material cited.

PTH-Glycine specifically develops a dark red permanent color if the layer is moistened by a water spray and subsequently exposed to ammonia vapors.[68] The detection limit on Silica Gel G is about 0.08 μg.[29]

N-(2,4-Dinitrophenyl) Amino Acids (DNP-Amino Acids)

Abderhalden *et al.*[71] introduced dinitrophenylation for N-terminal labeling of proteins as early as 1923. Later on, Sanger applied this technique for the sequence analysis of insulin,[72] replacing 2,4-dinitrochlorobenzene by 2,4-dinitrofluorobenzene, a more reactive agent. Other 2,4-dinitrobenzene derivatives have been proposed (see below), but none of them has gained common use. Because of their very strong adsorption near 370 nm, the DNP derivatives are easily detectable in visible and in UV light (*O*-DNP-tyrosine can be recognized only in the UV). They can be used for further purification and for peptide mapping.

Substitution of a proton in the amino group of amino acids with a dinitrophenyl residue destroys the zwitterionic character of all protein amino acids except arginine (a guanidino group does not react), and the resultant derivatives are acids which are less polar than the amino acids. The DNP-amino acids can be separated easily from interfering salts and other polar substances. The conversion of free amino acids into DNP derivatives is, therefore, often applied for determination of amino

[69] K. Toczko and Z. Szweda, *Bull. Acad. Polon. Sci. Ser. Sci. Biol.* **14,** 757 (1966).
[70] G. Roseau and P. Pantel, *J. Chromatogr.* **44,** 392 (1969).
[71] E. Abderhalden and W. Stix, *Hoppe-Seyler's Z. Physiol. Chem.* **129,** 143 (1923).
[72] F. Sanger, *Biochem. J.* **39,** 507 (1945).

acids in salt solution (seawater[73]) and in biological specimens like blood,[74] urine,[75-77] sperm,[78] and excrements.[79]

A disadvantage of the method for serial analyses or ultramicro analysis, the need for liquid-liquid extraction of the derivatives, has been removed recently because DNP-amino acids can be adsorbed from aqueous solution on Porapak Q, a *neutral* polystyrene resin, from which they are eluted with acetone–water (80:20, v/v).[80] By liquid–liquid extraction (a partition between ether and about 0.1 *N* hydrochloric acid) the DNP-amino acids are fractionated into "ether-soluble" and acid-soluble derivatives. Not extractable with ether are DNP-arginine, DNP-cysteic acid, mono-DNP-cystine, α-DNP-histidine, im-DNP-histidine, ε-DNP-lysine, and *O*-DNP-tyrosine. Di-DNP-histidine as well as many DNP-peptides are found in both phases.

Like all other derivatives used for determination of N-terminal residues, DNP-amino acids are sensitive to light.[81-84] They must be stored in the dark and handled and chromatographed in indirect light.

Preparation

Collections of DNP-amino acids are commercially available. A review on the many different preparative techniques has been written by Biserte *et al.*[85] As already mentioned, the liquid–liquid extraction now can usually be replaced by adsorption on Porapak Q.[80]

Dinitrophenylation of Amino Acids and Peptides

The following procedure may be scaled down if necessary. An aqueous solution, 2 ml, containing 50 μmoles of amino acids or peptide is adjusted to pH 8.8. Then, 0.5 ml of 1 *M* carbonate buffer, pH 8.8, and 200 μl of 10% w/v 2,4-dinitrofluorobenzene (DNFB) in absolute ethanol are

[73] K. H. Palmork, *Acta Chem. Scand.* **17,** 1456 (1963).
[74] G. Pataki and M. Keller, *Z. Klin. Chem. Klin. Biochem.* **1,** 157 (1963).
[75] D. Walz, A. R. Fahmy, G. Pataki, A. Niederwieser, and M. Brenner, *Experientia* **19,** 213 (1963).
[76] W. Bürgi, J. P. Colombo, and R. Richterich, *Klin. Wochenschr.* **43,** 1202 (1965).
[77] K. Figge, *Clin. Chim. Acta* **12,** 605 (1965).
[78] M. Keller and G. Pataki, *Helv. Chim. Acta* **46,** 1687 (1963).
[79] F. Tancredi and H.-C. Curtius, *Z. Klin. Chem. Klin. Biochem.* **5,** 106 (1967).
[80] A. Niederwieser, *J. Chromatogr.* **54,** 215 (1971).
[81] B. Keil, V. Tomasek, and J. Sedlachova, *Chem. Listy* **46,** 457 (1952).
[82] B. Pollara and R. von Korff, *Biochim. Biophys. Acta* **39,** 364 (1960).
[83] D. W. Russel, *Biochem. J.* **87,** 1 (1963).
[84] D. W. Russel, *J. Chem. Soc.* 894 (1963).
[85] G. Biserte, J. W. Holleman, J. Holleman-Dehove, and P. Sautière, *J. Chromatogr.* **2,** 225 (1959); **3,** 85 (1960).

added. The glass vessel is closed after warming it with hot water (to avoid subsequent development of pressure within the vessel), and the suspension is well shaken for 90 minutes at 40°C in the dark. The excess of DNFB may then be extracted from the alkaline medium with ether, and the dissolved ether removed from the aqueous phase by leaving it 5 minutes *in vacuo*. Alternatively, the reaction mixture may be filtered directly through a column of 1 g of Porapak Q, 150–200 mesh[86] prewashed successively with acetone, water, and 1 *M* NaCl. The polystyrene resin is washed successively with 5 ml of 1 *M* NaCl, 2 × 5 ml of 1 *N* HCl, and 2 × 5 ml of 5% acetic acid. The first 5-ml portion of the 1 *N* HCl is used to rinse the reaction vessel before it is transferred to the column. The DNP derivatives, which are strongly adsorbed, are then eluted with up to 20 ml of acetone–water (80:20, v/v). Again, small portions of this solvent are employed for rinsing of the reaction vessel, and these washings are transferred quantitatively to the column. At a slow flow rate, the yellow zone can be collected in a few milliliters of solution.

Dinitrophenylated proteins usually are insoluble in water. After reaction they can be washed and centrifuged several times with distilled water and ether. Hydrolysis occurs in 6 *N* HCl for 4–16 hours. If DNP-proline or DNP-glycine are expected, only 4–6 hours are chosen because these derivatives are partially destroyed.

Adsorption of DNP Derivatives from Protein Hydrolyzates

DNP derivatives are quantitatively separated from free amino acids by adsorption on Porapak Q.[80] Only tryptophan and about 20% of phenylalanine are lost into the DNP fraction. The following procedure can be scaled down, if necessary.

The hydrolyzate, 2 ml, containing up to 70 μmoles of DNP-amino acids in 6 *N* HCl, is diluted 10-fold with distilled water and slowly filtered through a column (10 mm inner diameter) containing 300 mg of Porapak Q, prewashed successively with acetone, water, and 0.1 *N* HCl. The resin is then washed with 15 ml of 0.1 *N* HCl (or 5% acetic acid). The combined eluates contain the free amino acids less tryptophan and about 20% of phenylalanine and may be evaporated *in vacuo*. The DNP derivatives adsorbed on Porapak Q are then eluted with up to 10 ml of acetone–water (80:20, v/v) and the solution is evaporated to dryness in a stream of nitrogen at 50°. (Small portions of the 5% acetic acid and acetone–water, respectively, may previously be used to rinse the hydrolyzate vessel and are quantitatively transferred to the column.)

Removal of Dinitrophenol. Dinitrophenol is formed during dinitrophenylation by hydrolysis of DNFB. Its amount relative to the sub-

[86] Product of the Waters Associates, Framingham, Massachusetts.

TABLE VI
SOLVENTS FOR THIN-LAYER CHROMATOGRAPHY (TLC) OF DNP-AMINO ACIDS

Combination for two-dimensional TLC										No.	Solvent
1.	1.	1.	1.							48	Toluene–pyridine–2-chloroethanol–0.8 *N* ammonia (100:30:60:60, v/v)[a] upper phase for TLC.[b,c] The Silica Gel G thin-layer is impregnated for at least 16 hours by the vapors of the lower phase in a tank lined with filter paper out of contact with the thin layer. The vapor-impregnated thin-layer is covered immediately with a glass plate after removal from the tank, leaving a margin of about 1.7 cm for spotting. After spotting (not more than 1 μl) the cover plate is carefully removed and TLC is started at once.[b,c]
				1.	1.	1.	1.			49	Toluene–pyridine–2-chloroethanol–25% ammonia (50:15:35:7, v/v)[d]
			2.				2.			50	Chloroform–benzyl alcohol–glacial acetic acid (70:30:3, v/v)[b,c]
		2.				2.				51	Chloroform–*t*-amyl alcohol–glacial acetic acid (70:30:3, v/v)[b,c]
										52	*n*-Propanol–34% ammonia (70:30, v/v)[b,c]
	2.				2.					53	Benzene–pyridine–glacial acetic acid (80:20:2, v/v).[b,c] Overflow technique
2.				2.						54	Chloroform–methanol–glacial acetic acid (95:5:1, v/v).[b,c] Overflow technique
								1.		55	Benzene–glacial acetic acid (4:1, v/v).[e,f] Overflow technique
										56	Carbon tetrachloride–glacial acetic acid (4:1, v/v).[e,f] Overflow technique
								2.	2.	57	90% formic acid–water (1:1, v/v)[e,f]
									1.	58	*n*-Butanol–glacial acetic acid (9:1, v/v)[e,f]

[a] G. Biserte and R. Osteux, *Bull. Soc. Chim. Biol.* **33,** 50 (1951).
[b] M. Brenner, A. Niederwieser, and G. Pataki, *Experientia* **17,** 145 (1961).
[c] A. Niederwieser, Doctoral Dissertation, University of Basle, 1962.
[d] D. Walz, A. R. Fahmy, G. Pataki, A. Niederwieser, and M. Brenner, *Experientia* **19,** 213 (1963).
[e] K.-T. Wang, I. S.-Y. Wang, S.-S. Yuan, and P.-H. Wu, *J. Chin. Chem. Soc. (Taipei)* **15,** 59 (1968).
[f] K.-T. Wang and B. Weinstein, *in* "Progress in Thin-Layer Chromatography and Related Methods" (A. Niederwieser and G. Pataki, eds.), Vol. III, p. 177. Ann Arbor-Humphrey Science Publ., Ann Arbor, Michigan, 1971.

stance to be dinitrophenylated increases with increasing dilution of the reaction mixture. A large excess of dinitrophenol may disturb the chromatography. Dinitrophenol can be removed by repeated sublimation[87] or by adsorption on aluminum oxide[88] or silica gel.[89]

Solvents, R_f Values and Identification Maps

Silica gel, Supergel,[90] cellulose,[91,92] and polyamide[93,94] can be used for TLC of DNP-amino acids. The behavior of DNP-amino acids on Silica Gel G has been systematically investigated.[29,95] Solvents for TLC on silica gel and polyamide are given in Table VI, and the corresponding

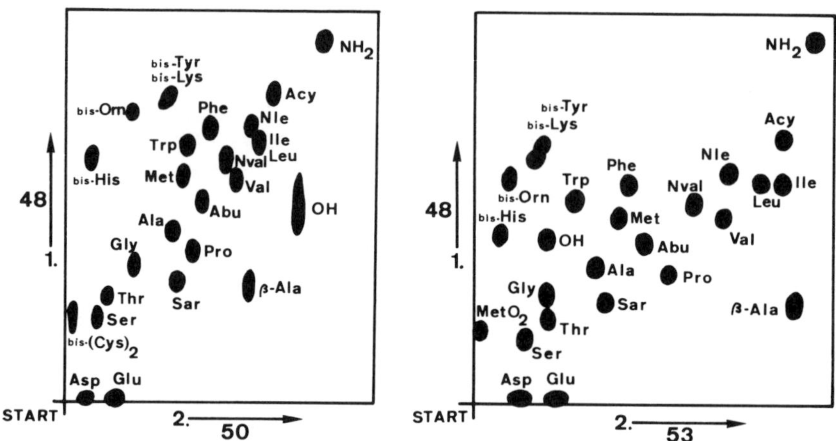

FIG. 6. Two-dimensional separations of DNP-amino acids on Silica Gel G thin-layers in solvents 48/50 and 48/53, respectively (see Table VI). TLC in solvent 53 is done with overflow technique [M. Brenner and A. Niederwieser, *Experientia* **17**, 237 (1961)] during 3 hours in the BN chamber (available from D. Desaga, Heidelberg, Germany; C. A. Brinkman, Westbury, Long Island, New York). Similar results are obtained replacing solvent 48 by solvent 49. Load: 0.1–10 μg of each DNP-amino acid. Detection is best in UV-light (~370 nm). Abbreviations as in Table VII. According to M. Brenner, A. Niederwieser, and G. Pataki, *Experientia* **17**, 145 (1961); and A. Niederwieser, Doctoral dissertation, University of Basle, 1962.

[87] G. L. Mills, *Biochem. J.* **50**, 707 (1952).
[88] F. Turba and G. Gundlach, *Biochem. Z.* **326**, 322 (1955).
[89] F. S. Steven, *J. Chromatogr.* **8**, 417 (1962).
[90] S. Fittkau, H. Hansen, I. Marquardt, H. Diessner, and U. Kettmann, *Z. Physiol. Chem.* **338**, 180 (1964).
[91] G. Brandner and A. J. Virtanen, *Acta Chem. Scand.* **17**, 2563 (1963).
[92] R. L. Munier and G. Sarrazin, *Bull. Soc. Chim. Fr.* **1965**, 2959 (1965).
[93] K.-T. Wang and J.-M.-K. Huang, *Nature (London)* **208**, 281 (1965).
[94] K.-T. Wang and I. S.-Y. Wang, *J. Chromatogr.* **27**, 318 (1967).
[95] A. Niederwieser, Doctoral Dissertation, University of Basle, 1962.

R_f values in one-dimensional chromatography are compiled in Table VII. The R_f values of DNP-amino acids usually change if larger amounts (>5 μg) are applied, because of a tendency of these acids for self-association. A few percent glacial acetic acid in the solvent reduces association and tailing and increases the R_f value. To get sufficient reproducibility, the added acetic acid must be measured exactly. DNP-Amino acids are best identified by two-dimensional TLC on silica gel (Fig. 6) or on polyamide (Fig. 7). In the solvents used in Fig. 6, the bis-DNP derivatives of lysine and tyrosine move together. They are resolved by rechromatography in solvent 54 or in chloroform–methanol–acetic acid (98:2:1, v/v); this separation is dependent on the relative humidity. For transfer suck the adsorbent around the spot into a capillary column of about 2–3 mm inner diameter containing a minute cotton plug, and elute with a few drops of acetone–acetic acid (100:1, v/v) directly onto the start of the new thin-layer. The acid-soluble DNP derivatives remain in the system of Fig. 6 near the start. They can be identified by rechromatography in solvent 52 or in chloroform–benzyl alcohol–methanol–water–15 N ammonia (30:30:30:6:2, v/v).[96] DNP-Arginine and ϵ-DNP-lysine can also be resolved easily by electrophoresis.[85] ϵ-DPN-lysine, δ-

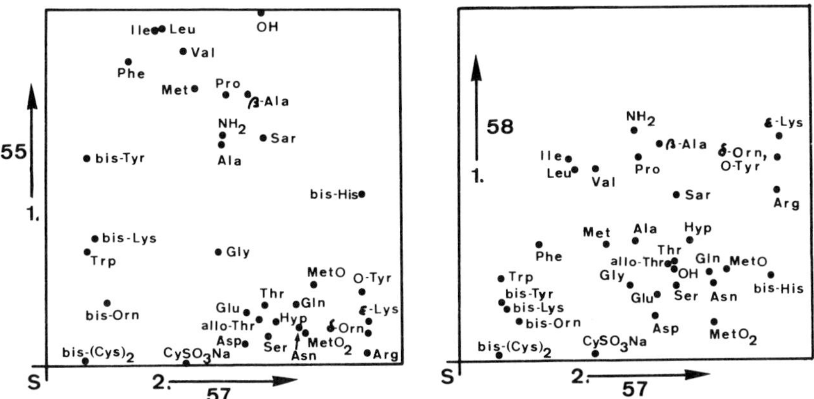

FIG. 7. Two-dimensional map of DNP-amino acids on polyester-supported polyamide in solvents 55/57 and 58/57, respectively (see Table VI). TLC in solvent 55 is done with overflow technique. Abbreviations as in Table VII. S: start. According to Y.-T. Lin, K.-T. Wang, and I. S-Y. Wang, *J. Chin. Chem. Soc.* (*Taipei*) **13**, 19 (1966); and K.-T. Wang and B. Weinstein, *in* "Progress in Thin-Layer Chromatography and Related Methods" (A. Niederwieser and G. Pataki, eds.), Vol. III, p. 177. Ann Arbor-Humphrey Science Publ., Ann Arbor, Michigan, 1971.

[96] J.-M. Ghuysen, E. Bricas, M. Leyh-Bouille, M. Lache, and G. D. Shockman, *Biochemistry* **6**, 2607 (1967).

TABLE VII
hR_f VALUES OF DNP-AMINO ACIDS IN THIN-LAYER CHROMATOGRAPHY (TLC)

Abbreviation	Substance	Silica gel G						Polyamide[a]			
		hR_f				hR_{Leu}[b]		hR_{OH}[c]		hR_f	
		in Solvent No.									
		48[d]	50	51	52	53	54	55	56	57	58
Abu	DNP-α-amino-n-butyric acid	46	72	73	—	52	85	—	—	—	—
Ala	DNP-alanine	34	54	60	—	33	66	62	42	49	33
β-Ala	DNP-β-alanine	27	71	73	—	98	95	75	44	56	60
Arg	DNP-arginine	—	—	—	43	—	—	3	1	90	47
Asn	DNP-asparagine	—	—	—	—	—	—	10	5	70	21
Asp	DNP-aspartic acid	2	13	9	—	5	6	6	4	55	12
bis-$(Cys)_2$	bis-DNP-cystine	12	3	1	—	0	2	1	1	11	1
mono-$(Cys)_2$	mono-DNP-cystine[e]	—	—	—	29	—	—	—	—	—	—
$CySO_3H$	DNP-cysteic acid	—	—	—	29	—	—	0	0	38	1
Gln	DNP-glutamine	—	—	—	—	—	—	17	10	70	24
Glu	DNP-glutamic acid	1	26	31	—	12	12	15	10	56	18
Gly	DNP-glycine	27	32	40	—	18	38	32	16	48	21
bis-His	bis-DNP-histidine	53	11	8	65	4	16	48	15	88	23
α-His	α-DNP-histidine	—	—	—	57	—	—	—	—	—	—
Hyp	DNP-hydroxyproline	—	—	—	—	—	—	12	8	64	33
Ile	DNP-isoleucine	64	83	81	—	107	101	94	90	31	56
Leu	DNP-leucine	66	82	80	—	100	100	94	90	32	53
bis-Lys	bis-DNP-lysine	74	56	60	—	13	73	35	14	14	15
ϵ-Lys	ϵ-DNP-lysine[e]	—	—	—	44	—	—	12	5	90	62
Met	DNP-methionine	55	70	69	—	43	81	77	62	41	32
MetO	DNP-methionine sulfoxide	—	—	—	—	—	—	22	10	75	26
$MetO_2$	DNP-methionine sulfone	17	—	—	—	3	10	9	5	72	10

NH_2	2,4-dinitroaniline	90	90	72	—	128	101	64	42	49	64
Nle	DNP-norleucine	69	82	80	—	90	100	—	—	—	—
Nval	DNP-norvaline	56	77	76	—	70	95	—	—	—	—
OH	2,4-dinitrophenol	41	100	83	—	21	102	100	100	60	26
bis-Orn	bis-DNP-ornithine	70	34	40	—	6	46	17	9	17	11
δ-Orn	δ-DNP-ornithine	—	—	—	—	—	—	9	5	90	56
Phe	DNP-phenylalanine	67	75	74	—	46	86	85	62	23	33
Pro	DNP-proline	29	65	67	—	59	84	75	51	50	46
Sar	DNP-sarcosine	23	56	57	—	35	65	63	41	61	46
Ser	DNP-serine	15	11	11	—	10	8	8	5	61	20
Thr	DNP-threonine	20	17	15	—	14	11	16	10	60	27
allo-Thr	DNP-allo-threonine	—	—	—	—	—	—	13	9	59	27
Trp	DNP-tryptophan	65	69	69	—	25	61	32	17	12	23
bis-Tyr	bis-DNP-tyrosine	76	58	60	—	16	65	58	24	12	16
O-Tyr	*O*-DNP-tyrosine[f]	—	—	—	49	—	—	20	10	88	Tailing
Val	DNP-valine	53	79	77	—	81	98	88	76	38	53
Chamber saturation		+	+	+	+					+	+
Literature reference		*g*	*g*	*g*	*g*	*g*	*g*	*h*	*h*	*h*	*h*

[a] Polyamide on sheets of polyester.

[b] hR_f value related to DNP-leucine. Values obtained in the BN-chamber[i] with overflow technique.[j]

[c] hR_f value related to 2,4-dinitrophenol. Values obtained with overflow technique.

[d] Solvent 49 gives similar results.[k]

[e] Brown color with ninhydrin.

[f] Violet color with ninhydrin.

[g] M. Brenner, A. Niederwieser, and G. Pataki, *Experientia* **17,** 145 (1961).

[h] K.-T. Wang and B. Weinstein, *in* "Progress in Thin-Layer Chromatography and Related Methods" (A. Niederwieser and G. Pataki, eds.), Vol. III, p. 177. Ann Arbor-Humphrey Science Publ., Ann Arbor, Michigan, 1971.

[i] Available from D. Desaga GmbH, Heidelberg, Germany; C. A. Brinkman, Westbury, Long Island, New York.

[j] M. Brenner and A. Niederwieser, *Experientia* **17,** 237 (1961).

[k] D. Walz, A. R. Fahmy, G. Pataki, A. Niederwieser, and M. Brenner, *Experientia* **19,** 213 (1963).

DNP-ornithine, α-DNP-lysine, and α-DNP-ornithine can be separated on silica gel G in chloroform–methanol–acetic acid–water (65:25:13:8, v/v).[97]

For end-group determination, it is best to apply the sample (1–2 μg of each DNP-amino acid) together with a standard mixture (0.2 μg of each DNP-amino acid) to get a two-dimensional pattern where the N-terminal DNP-amino acid is usually unequivocally indicated by the increased intensity of its spot. *O*-DNP-Tyrosine and ε-DNP-lysine can be present, if the peptide sequence contained lysine or tyrosine. *im*-DNP-Histidine is unstable.

Detection and Quantitation

Owing to very strong absorption near 370 nm, DNP-amino acids can be detected on silica gel with a sensitivity of 0.1 μg on two-dimensional and about 0.02 μg on one-dimensional thin-layer chromatograms (silica gel). For small amounts, photographic recording is recommended (e.g., on Agfa Copex Ortho film; illumination with UV light at about 350 nm). Quantitation is possible by elution of the spots with 4% sodium bicarbonate, followed by centrifugation and colorimetry at 365 nm, and at 385 nm in the case of DNP-proline (see, e.g., Dellacha and Fontanive[98]). Much easier is *in situ* densitometry using a scanner.[99] By radioactive labeling with [^{3}H]- or [^{14}C]dinitrofluorobenzene[100] or [^{14}C]diazomethane (giving DNP-amino acid methyl esters),[101] analyses in the range of picomoles are possible.

1-Dimethylaminonaphthalene-5-sulfonyl Amino Acids (DANS-Amino Acids)

1-Dimethylaminonaphthalene-5-sulfonyl chloride (DANS-Cl, dansyl chloride) was introduced by Weber[102] in 1952 for fluorescence labeling in size and shape investigations of macromolecules.[103] Dansyl chloride reacts with nucleophiles, such as thiol groups (to give the corresponding *S*-sulfonyl derivatives), amino groups (to give the corresponding sulfonamides), phenols (to give phenol esters), and hydroxyl ion (from water, to give DAN-sulfonic acid). Aliphatic hydroxyl compounds react

[97] M. Guinand, J.-M. Ghuysen, K. H. Schleifer, and O. Kandler, *Biochemistry* **8**, 200 (1969).
[98] J. M. Dellacha and A. V. Fontanive, *Experientia* **21**, 351 (1965).
[99] G. Pataki, *Chromatographia* **1**, 492 (1968).
[100] G. B. Gerber and J. Remy-Defraigne, *Anal. Biochem.* **11**, 386 (1965).
[101] K. Heyns and R. Hauber, *Hoppe-Seyler's Z. Physiol. Chem.* **348**, 357 (1967).
[102] G. Weber, *Biochem. J.* **51**, 155 (1952).
[103] H. R. Horton and D. E. Koshland, Jr., Vol. 11, p. 856.

only to a small extent, with the exception of choline.[104,105] Gray and Hartley[106] introduced dansyl chloride for end-group labeling in peptide and protein chemistry. Used in a fashion analogous to 2,4-dinitrofluorobenzene (see above), it possesses several distinct advantages over dinitrophenylation: dansylation proceeds faster than dinitrophenylation and owing to the strong fluorescence of dansyl derivatives (excitation at 340 nm, fluorescence at 510 nm), is more sensitive, up to a factor of 10–20 when judged by TLC. Owing to the smaller amounts necessary, often a better chromatographic resolution is achieved. A disadvantage is that dansyl amino acids are partially fragmented to DANS-NH_2, CO, and the aldehyde or ketone with one less carbon atom than the parent amino acid.[107] For this reason, hydrolysis of a dansylated protein should be interrupted after 4 hours, giving in general an optimal recovery of the N-terminal dansyl amino acid.[108] Nevertheless, quantitative determinations of N-terminal amino acids are possible in the range of 10^{-9} mole[108] and 10^{-10} mole[109] using thin-layer chromatographic separation. Dansylation has been applied also for peptide mapping,[110,111] for characterization of transfer RNA in the picomole range (radioactive labeling)[112] and for identification of C-terminal amino acids after hydrazinolysis of peptides.[113] It is very interesting that biologically active peptides can be recovered from their dansyl derivatives by Birch reduction,[114] although prolyl peptide bonds are partially split.

Preparation of Dansyl Derivatives

Collections of DANS-amino acids are commercially available. The original method of Hartley and Massey[115] is suitable for the preparation of larger quantities. The dansylation reaction was studied thoroughly by Gros and Labouesse.[108] The following procedures are based on these results.

[104] N. Seiler, *Hoppe-Seyler's Z. Physiol. Chem.* **348**, 601 (1967).
[105] N. Seiler, L. Kamemikova, and G. Werner, *Collect. Czech. Chem. Commun.* **34**, 719 (1969).
[106] W. R. Gray and B. S. Hartley, *Biochem. J.* **89**, 59P (1963); **89**, 379 (1963).
[107] D. J. Neadle and R. J. Pollitt, *Biochem. J.* **97**, 607 (1965).
[108] C. Gros and B. Labouesse, *Eur. J. Biochem.* **7**, 463 (1969).
[109] G. Schmer, *Hoppe-Seyler's Z. Physiol. Chem.* **348**, 199 (1967).
[110] C. Gerday, E. Robyns, and C. Gosselin-Rey, *J. Chromatogr.* **38**, 408–411 (1968).
[111] J. Langner, *Hoppe-Seyler's Z. Physiol. Chem.* **347**, 275 (1966).
[112] V. Neuhoff, F. von der Haar, E. Schlimme, and M. Weise, *Hoppe-Seyler's Z. Physiol. Chem.* **350**, 121 (1969).
[113] P. Nedkov and N. Genov, *Biochim. Biophys. Acta* **127**, 544 (1966).
[114] Z. Tamura, T. Tanimura, and H. Yoshida, *Chem. Pharm. Bull.* **15**, 252 (1967).
[115] B. S. Hartley and V. Massey, *Biochim. Biophys. Acta* **21**, 58 (1956).

Dansylation of Peptides or Amino Acids.[108] To 1–10 nmoles of peptide or amino acids in 10 μl of water are added 20 μl of 50 mM bicarbonate buffer, pH 8.3, and 30 μl of 10 mM DANS-Cl in acetone. The actual starting pH of the mixture should be 9.5. The final concentrations should be 5 mM DANS-Cl and 10–100 μM peptide or amino acids. The reaction is stopped after 30 minutes at room temperature by addition of 10 μl of 0.1 M NaOH. The reaction mixture is lyophilized, and 0.1 ml of 5.7 N HCl is added for hydrolysis.

Dansylation of Proteins.[108] To 1–10 nmoles of protein in 300 μl of water, 250 mg of urea (ammonia and cyanate free) is added to give a final volume of 500 μl. Then 150 μl of 0.4 M phosphate buffer, pH 8.2, 250 μl of dimethyl formamide, and 100 μl of 0.2 M DANS-Cl in acetonitrile are added. The starting pH should be 9.4 (half an hour later it will be 8.7). After 30 minutes at room temperature, the dansylated protein is precipitated by addition of 10 ml of 10% trichloroacetic acid. After centrifugation for 10 minutes at 3000 rpm the precipitate is washed twice with 1 N HCl and then hydrolyzed in 0.5 ml of 5.7 N HCl.

A limited labeling of reactive groups, for instance in an enzyme, can be attained by replacement of 8 M urea by water.[108,115–117]

Hydrolysis of DANS-Peptides or Proteins.[108] Because of destruction of DANS-amino acids, hydrolysis in 5.7 N HCl for only 4 hours at 110° gives an optimal recovery, except for DANS-derivatives of valine, leucine, and isoleucine, which require 18 hours of hydrolysis.

Removal of Excess DANS-OH. Generally it is unnecessary to remove DANS-OH, the hydrolysis product of DANS-Cl; it will be necessary if a large excess of DANS-Cl is used. Removal is effected, in the case of proteins, by gel filtration, dialysis against distilled water, or protein precipitation. In the case of peptides or amino acids, the reaction mixture is brought to dryness, dissolved in 0.01 N acetic acid, and filtered through a small column of Dowex 50 $\times$ 8 (H^+ form and equilibrated with 0.01 N acetic acid).[109] DANS-OH elutes quickly. Then the DANS-derivatives are eluted with a freshly prepared mixture of acetone–water–conc. ammonia (25:68:7, v/v).

Solvents, R_f Values, and Identification Maps

Thin-layers of silica gel have been used most often for the separation of DANS-amino acids, but good separations are obtained also on polyamide.[110,112,118] Moreover, cellulose[119] and aluminum oxide[120] have been

[116] M. Kasuya and H. Takashima, *Biochim. Biophys. Acta* **99,** 452 (1965).
[117] R. D. Hill and R. R. Laing, *Biochim. Biophys. Acta* **132,** 190 (1967).
[118] K. R. Woods and K.-T. Wang, *Biochim. Biophys. Acta* **133,** 369 (1967).
[119] M. S. Arnott and D. N. Ward, *Anal. Biochem.* **21,** 50 (1967).
[120] Z. Deyl and J. Rosmus, *J. Chromatogr.* **20,** 514 (1965).

proposed. DANS-amino acids are similar in polarity to the DNP-amino acids. It is therefore not surprising that solvents which had been described earlier for TLC of DNP-amino acids also can be used for the DANS-derivatives. Of the many solvents described for TLC of DANS-amino acids, only a few are given in Table VIII. R_f values are compiled in Table IX. One-dimensional TLC will often be enough for identification of a DANS derivative. When a particular amino acid cannot be identified, the adsorbent around the spot is sucked into a micro column (2–3 mm inner diameter) containing a minute cotton plug, and the derivative is eluted with a few droplets of an appropriate solvent directly onto the start of a new thin layer. For this elution a solvent is used in which the substance moves in TLC with an hR_f value greater than about 80. In the search for a solvent for rechromatography, Table IX will be useful. Further solvents for one-dimensional TLC have been recommended, e.g., by Nedkov and Genov,[121] namely modifications of solvents 68 and 66: chloroform–ethyl acetate–methanol–acetic acid (9:15:4.5:0.2, v/v) and ethyl acetate–isopropanol–conc. ammonia (8:20:6, v/v), and, most recently, by Stehelin and Duranton.[122] For the separation of DANS derivatives that cannot be extracted from aqueous solution by ether (those of arginine, cysteic acid, and the mono-derivatives of tyrosine, histidine, and lysine) the latter authors recommend the solvent toluene–2-chloroethanol–conc. ammonia–water (30:50:2:2, v/v); compare solvents 48, 49, 64, and 81.

A more straightforward identification is possible by two-dimensional TLC, especially if the sample is chromatographed together with a standard mixture containing all DANS derivatives. The concentration of the standards should be such that the spots just can be recognized; the concentration of the sample should be about 10 times higher. The sample spot will stand out on a background of the characteristic pattern. Useful solvent combinations are indicated in Table VIII. Examples of two-dimensional patterns are shown in Figs. 8–10. The separation in Fig. 8 was obtained on Silica Gel G. The pattern of Woods and Wang[118] (Fig. 9) was obtained on polyamide,[123] and it can be recognized that it affords a particularly good separation. Solvent 76 is more effective than solvent 77 for separation of DANS-glutamic acid and DANS-aspartic acid. The pattern of Fig. 10 resulted from a combination of TLC on cellulose and electrophoresis.

After two-dimensional TLC, a derivative also can be eluted and

[121] P. Nedkov and N. Genov, *Biochim. Biophys. Acta* **127**, 541 (1966).

[122] D. Stehelin and H. Duranton, *J. Chromatogr.* **43**, 93 (1969).

[123] K.-T. Wang and B. Weinstein, *in* "Progress in Thin-Layer Chromatography and Related Methods" (A. Niederwieser and G. Pataki, eds.), Vol. III. p. 177. Ann Arbor-Humphrey Science Publ., Ann Arbor, Michigan.

TABLE VIII

SOLVENTS FOR THIN-LAYER CHROMATOGRAPHY (TLC) OF DANS-AMINO ACIDS

Combination for two-dimensional TLC			No.	Solvent
			48	See Table VI. No vapor impregnation necessary[a]
		2.	50	See Table VI[a]
			51	See Table VI[b]
	1.	1.	53	See Table VI. No overflow[b,c]
			59	Chloroform–*t*-amyl alcohol–formic acid (70:30:1, v/v)[b]
			60	Chloroform–*t*-amyl alcohol–acetic acid (70:30:0.5, v/v)[b]
	2.		61	*n*-Butanol saturated with 0.2 *N* sodium hydroxide[c]
			62	Chloroform–*n*-butanol (97:3, v/v)[c]
		1.	63	Benzene–pyridine–glacial acetic acid (80:20:5, v/v)[d]
2.		2.	64	Toluene-2-chloroethanol–25% ammonia (6:10:4, v/v)[d]
			65	Chloroform–glacial acetic acid (10:4, v/v)[d]
	1.	1.	66	Methyl acetate–isopropanol–conc. ammonia (45:35:20, v/v)[e]
		2.	67	Chloroform–methanol–glacial acetic acid (75:25:5, v/v)[e]
	2.		68	Chloroform–ethyl acetate–methanol–glacial acetic acid (30:50:20:1, v/v)[e]
			69	Chloroform–ethanol–glacial acetic acid (38:4:3, v/v)[a]
			70	Chloroform–*t*-butanol–glacial acetic acid (6:3:1, v/v)[a]
			71	Petroleum ether–*t*-butanol–glacial acetic acid (75:15:15, v/v)[f]
			72	Petroleum ether–2,4,6-collidine–methyl ethyl ketone–glacial acetic acid (75:5:24:3, v/v)[f]
	1.		73	Diethyl ether–methanol–acetic acid (100:5:1, v/v)[g]
		1.	74	Ethyl acetate–methanol–acetic acid (20:1:1, v/v)[g]
	2.	2.	75	Methyl acetate–ammonia (d = 0.88) (19:1, v/v)[g]
	2.		76	Hexane–*t*-butanol–glacial acetic acid (3:3:1, v/v)[h]
		2.	77	Benzene–glacial acetic acid (9:1, v/v)[h]
	1.	1.	78	Water–90% formic acid (100:1.5, v/v)[h]
			79	Chlorobenzene–glacial acetic acid (9:1, v/v)[h]
			80	Pyridine–glacial acetic acid–water (0.4:0.8:98.8, v/v)[i]
1.			81	Toluene-2-chloroethanol–25% ammonia (6:10:0.5, v/v)[i]

[a] Z. Deyl and J. Rosmus, *J. Chromatogr.* **20,** 514 (1965).
[b] D. Morse and B. L. Horecker, *Anal. Biochem.* **14,** 429 (1966).
[c] M. Cole, J. C. Fletcher, and A. Robson, *J. Chromatogr.* **20,** 616 (1965).
[d] C. Gros, *Bull. Soc. Chim. Fr.* **1967,** 3952 (1967).
[e] N. Seiler and J. Wiechmann, *Experientia* **20,** 559 (1964).
[f] B. Mesrob and V. Holeyšovský, *J. Chromatogr.* **21,** 135 (1966).
[g] K. Crowshaw, S. J. Jessup, and P. W. Ramwell, *Biochem. J.* **103,** 79 (1967).

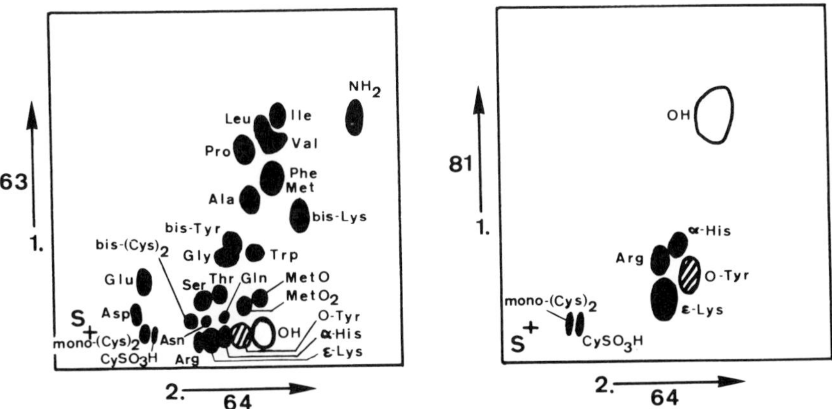

FIG. 8. Two-dimensional TLC of DANS-amino acids on Silica Gel G according to C. Gros and B. Labouesse [*Eur. J. Biochem.* **7**, 463 (1969)]. For solvents and abbreviations see Tables VIII and IX, respectively. S: start. Detection in UV-light. *O*-DANS-tyrosine fluoresces pink, and DANS-OH blue. The derivatives shown at the right are water soluble and can be separated by extraction of all others with ether.

rechromatographed in a third solvent, if necessary. Artifacts may arise (also in one-dimensional TLC), e.g., by oxidation of methionine to methionine sulfoxide or sulfone, or by loss of one DANS residue from a bis-DANS derivative. It should be emphasized further, that DANS derivatives are light sensitive.

Detection and Quantitation

The chromatograms can be evaluated while still damp, immediately after TLC, under UV light (360 nm). The fluorescence intensity decreases rapidly after drying. DANS-amino acids show a strong yellow fluorescence, and DANS-OH a blue-green fluorescence. With increasing dielectric constant of the medium, the emission maximum is shifted to a longer wavelength and the quantum yield usually decreases.[124] Fluorescence intensity of DANS derivatives is low in the protonated form and strong in alkaline media.[125] A considerable increase in sensitivity is, therefore, obtained by spraying the chromatogram with a solution of

[124] R. F. Chen, *Arch. Biochem. Biophys.* **120**, 609 (1967).
[125] N. Seiler and M. Wiechmann, *Z. Anal. Chem.* **220**, 109 (1966).

[h] K.-T. Wang and B. Weinstein, *in* "Progress in Thin-Layer Chromatography and Related Methods" (A. Niederwieser and G. Pataki, eds.), Vol. III, p. 177. Ann Arbor-Humphrey Science Publ., Ann Arbor, Michigan, 1971.
[i] M. S. Arnott and D. N. Ward, *Anal. Biochem.* **21**, 50 (1967).
[j] C. Gros and B. Labouesse, *Eur. J. Biochem.* **7**, 463 (1969).

TABLE IX
hR_f VALUES OF DANS-AMINO ACIDS IN ONE-DIMENSIONAL THIN-LAYER CHROMATOGRAPHY (TLC)

		Silica gel G																			Polyamide				Cellulose MN 300
		hR_f		$hR_{DANS\text{-}NH_2}$[a]					$hR_{DANS\text{-}Leu}$[b]			hR_f													
		In solvent No.																							
Abbreviation	Substance	48[c]	50[c]	51	53	59	60	61	62	63	64	65	66[c]	67[c]	69[c]	70[c]	71	72	73	74	75	76	77	78	79	80
Ala	DANS-alanine	0	45	89	47	96	29	37	—	69	95	—	57	86	83	82	less than 10	less than 13	37	53	49	46	33	54	40	79
β-Ala	DANS-β-alanine	—	—	—	—	—	—	—	—	—	—	—	52	—	—	—			53	63	39	—	—	—	—	—
Arg	α-DANS-arginine	9	0	0	0	0	0	16	—	0	65	—	32	12	0	3			0	0	9	6	2	88	1	63
Asn	DANS-asparagine	—	5	2	3	10	3	15	—	—	—	—	37	—	22	—			1	7	20	—	—	—	—	—
Asp	DANS-aspartic acid	0	0	2	6	60	0	4	—	10	45	—	4	53	27	3			1	9	0	16	3	51	5	87
bis-Cys	Bis-DANS-cystine	—	—	—	4	—	—	40	—	—	—	—	—	—	—	—			0	4	10	tail	2	2	2	—
$CySO_3H$	DANS-cysteic acid	—	0	—	0	—	—	4	—	0	45	—	10	12	30	—			—	—	—	0	0	19	0	96
bis-$(Cys)_2$	Bis-DANS-cystine	—	17	—	2	—	—	35	—	10	65	—	—	—	0	—			0	4	10	—	—	—	—	13
mono-$(Cys)_2$	Mono-DANS-cystine	—	—	—	—	—	—	—	—	0	50	—	—	—	—	—			—	—	—	—	—	—	—	—
Gln	DANS-glutamine	—	5	7	3	14	0	15	—	—	—	—	45	—	27	—			1	9	28	—	—	—	—	—
Glu	DANS-glutamic acid	0	15	42	16	85	8	5	—	24	46	—	10	62	52	46			7	29	2	21	5	49	9	86
Gly	DANS-glycine	21	30	65	32	93	12	32	—	44	88	—	52	77	67	60			21	41	41	34	17	53	23	74
α-His	α-DANS-histidine	—	—	—	0	—	—	8	—	—	—	—	—	—	—	—			—	—	—	—	—	—	—	—
bis-His	bis-DANS-histidine	9	0	59	24	91	9	32	—	0	83	—	46	17	—	—			—	—	—	16	3	88	4	76
im-His	*imidazole*-DANS-histidine	—	—	—	0	—	—	20	—	—	—	—	—	—	—	—			—	—	—	—	—	—	—	—
Hyp	DANS-hydroxyproline	25	9	—	—	—	—	—	—	—	—	—	40	70	52	43			10	31	21	40	14	62	64	—
Ile	DANS-isoleucine	65	62	100	87	100	89	65	19	100	110	116	67	94	93	97	26	27	59	61	60	64	50	27	58	50
Leu	DANS-leucine	57	55	100	83	99	63	63	10	100	100	100	63	94	91	88	26	24	63	63	55	61	45	24	53	50
bis-Lys	bis-DANS-lysine	25	26	93	40	96	33	54	—	63	—	—	69	93	88	3	36	58	27	55	66	20	21	11	30	44
ε-Lys	ε-DANS-lysine	—	—	8	0	5	4	26	—	0	72	—	34	—	—	—	—	—	—	—	—	—	—	—	—	58

Met	DANS-methionine	4	9	92	52	99	33	43	—	77	—	—	—	—	36	15	<10	<13	37	53	52	43	33	30	41	63
$MetO_2$	DANS-methionine sulfone	—	17	—	9	—	—	20	—	5	—	—	57	62	46	—	<10	<13	—	—	—	—	—	—	—	85
1-MeHis	bis-DANS-1-methylhistidine	—	—	—	—	—	—	—	—	—	—	—	—	—	—	—	—	—	0	0	37	—	—	—	—	—
3-MeHis	bis-DANS-3-methylhistidine	—	—	—	—	—	—	—	—	—	—	—	—	—	—	—	—	—	0	0	17	—	—	—	—	—
NH_2	Dansylamide	—	—	—	—	—	—	—	—	—	—	—	94	94	—	—	9	39	—	—	—	40	38	54	38	—
OH	DAN-sulfonic acid	—	—	—	—	—	—	—	—	0	100	—	60	36	—	—	0	—	0	3	57	0	0	33	0	90
bis-Orn	bis-DANS-ornithine	—	—	—	—	—	—	—	—	—	—	—	70	93	—	—	—	—	19	47	67	—	—	—	—	—
δ-Orn	δ-DANS-ornithine	—	—	—	—	—	—	—	—	—	—	—	34	—	—	—	—	—	—	—	—	—	—	—	—	—
Phe	DANS-phenylalanine	4	62	94	59	99	42	51	7	82	—	—	64	94	88	97	16	13	43	58	57	45	36	22	45	41
Pro	DANS-proline	4	58	93	70	96	50	24	7	91	—	—	46	94	88	97	less than 10	less than 13	34	51	33	57	58	41	64	70
Ser	DANS-serine	4	82	20	15	52	2	25	—	20	—	—	46	54	83	24			5	18	31	24	5	64	8	83
Thr	DANS-threonine	7	15	25	18	57	2	28	—	28	85	—	52	62	52	33			7	20	40	31	9	64	12	83
Trp	DANS-tryptophan	—	48	87	34	95	25	45	—	44	100	—	52	86	59	—			35	59	49	17	12	10	17	29
bis-Tyr	bis-DANS-tyrosine	13	48	82	46	92	23	48	—	46	88	—	69	93	91	92			30	53	67	32	45	3	55	64
O-Tyr	*O*-DANS-tyrosine	—	—	—	0	—	—	33	—	5	90	—	51	—	—	—			—	—	—	—	—	—	—	11
Val	DANS-valine	0	62	100	77	97	75	54	13	97	—	89	64	94	91	88	19	24	60	60	57	59	48	39	54	68
Chamber saturation		?	?	?	+	?	?	+	+	?	?	?	+	+	?	?	+	+	+	+	+	—	—	—	—	S-chamber
Literature reference		*d*	*d*	*e*	*f*	*e*	*e*	*f*	*f*	*g*	*g*	*g*	*h*	*h*	*d*	*d*	*i*	*i*	*j*	*j*	*j*	*k*	*k*	*k*	*k*	*l*

[a] hR_f value related to DANS-amide.

[b] hR_f value related to DANS-leucine.

[c] Values estimated from figures in the original paper.

[d] Z. Deyl and J. Rosmus, *J. Chromatogr.* **20,** 514 (1965).

[e] D. Morse and B. L. Horecker, *Anal. Biochem.* **14,** 429 (1966).

[f] M. Cole, J. C. Fletcher, and A. Robson, *J. Chromatogr.* **20,** 616 (1965).

[g] C. Gros, *Bull. Soc. Chim. Fr.* **1967,** 3952 (1967).

[h] N. Seiler and J. Wiechmann, *Experientia* **20,** 559 (1964).

[i] B. Mesrob and V. Holeyšovský, *J. Chromatogr.* **21,** 135 (1966).

[j] K. Crowshaw, S. J. Jessup, and P. W. Ramwell, *Biochem. J.* **103,** 79 (1967).

[k] K.-T. Wang and B. Weinstein, *in* "Progress in Thin-Layer Chromatography and Related Methods" (A. Niederwieser and G. Pataki, eds.), Vol. III, p. 177. Ann Arbor-Humphrey Science Publ., Ann Arbor, Michigan, 1971.

[l] M. S. Arnott and D. N. Ward, *Anal. Biochem.* **21,** 50 (1967).

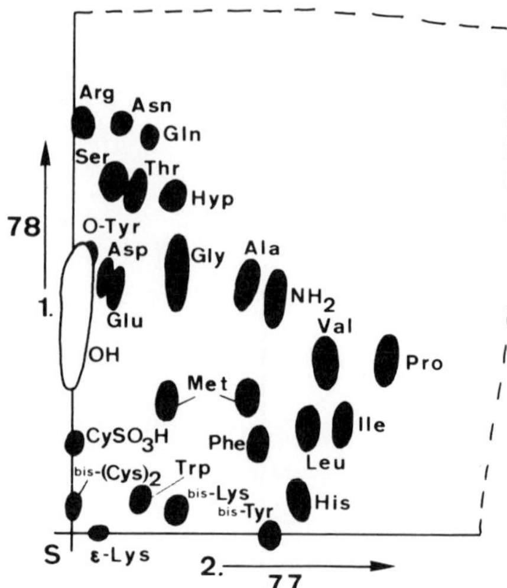

FIG. 9. Two-dimensional TLC of DANS-amino acids on polyamide layers (Cheng-Chin Trading Co., Taipei, Taiwan) according to K. R. Woods and K.-T. Wang [*Biochim. Biophys. Acta* **133**, 369 (1967)]. For solvents and abbreviations, see Tables VIII and IX, respectively. S: start.

triethanolamine in isopropanol.[125] Drying of the sprayed chromatogram causes a further increase in quantum yield. However, such an alkaline spray is advisable only if no rechromatography is to be performed. Quantitative determinations are best done by *in situ* measurements di-

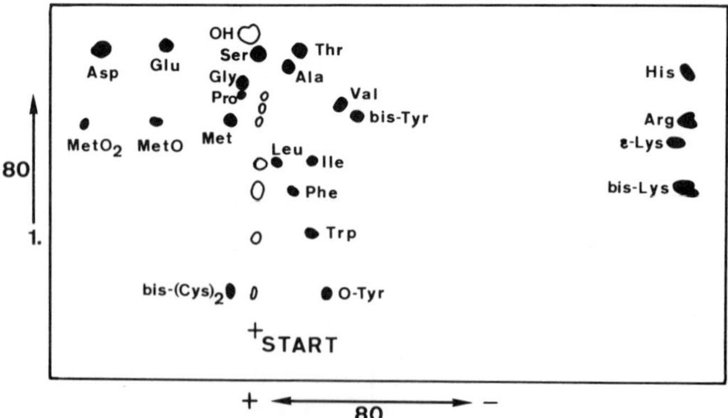

FIG. 10. Two-dimensional separation of DANS-amino acids by TLC combined with high-voltage electrophoresis on cellulose MN 300, according to M. S. Arnott and D. N. Ward [*Anal. Biochem.* **21**, 50 (1967)]. Chromatography and electrophoresis are done using the same solvent (No. 80, see Table VIII). Spots not identified are artifacts. For abbreviations, see Table IX.

rectly on the thin-layer chromatogram using a scanner,[99,125–127] whereby standard deviations of about ±3–5% are achievable. Naturally, it is also possible to elute the spots, e.g., with chloroform–methanol–acetic acid (7:2:2, v/v) and to measure the fluorescence of the solution[108,109,128] (for fluorescence measurement an alkaline medium would be better). As indicated already, a quantitative determination of 10^{-4} μmole N-terminal amino acids in immunoglobulins has been achieved[109] with a precision of about ±10%. Using microchromatograms (3 × 3 cm polyamide sheet), application of the solution under a stereomicroscope, two-dimensional TLC, and scintillation counting of the minute spots or autoradiography followed by microdensitometry, Neuhoff *et al.*[112] succeeded in the quantitative characterization of tRNA's in the picomole range, when the aminoacylation of the tRNA was performed with ^{14}C-labeled amino acids.

Further Amino Acid Derivatives

Instead of phenyl isothiocyanate or dinitrofluorobenzene, many other reagents have been used for stepwise peptide degradation or end-group determination. Some of them are listed below.

Further Derivatives for Stepwise Peptide Degradation

Methylthiohydantoins. These derivatives are formed when methyl isothiocyanate is employed[129] in the Edman degradation. Methylthiohydantoins can be separated by TLC on silica gel,[130] or by gas chromatography.[131] The latter technique can be combined with mass spectrometry to identify the compounds unambiguously. Mass spectrometry of methylthiohydantoins was described as early as 1964.[132]

4-Sulfophenylthiohydantoins (SPTH-Amino Acids). 4-Isothiocyanatobenzenesulfonic acid when used in the Edman degradation gives a sulfonated phenylthiocarbamyl peptide which can be separated easily from unreacted peptide by simple ion-exchange chromatography on DEAE-Sephadex (Cl^- form). Evaporation of the acid eluate splits off the N-terminal residue, which appears as the SPTH-amino acid. Chromatography on DEAE-Sephadex separates the SPTH derivative from the

[126] H. Zürcher, G. Pataki, J. Borko, and R. W. Frei, *J. Chromatogr.* **43,** 457 (1969).
[127] N. Seiler and M. Wiechmann, *Hoppe-Seyler's Z. Physiol. Chem.* **349,** 588 (1968).
[128] K. Crowshaw, S. J. Jessup, and P. W. Ramwell, *Biochem. J.* **103,** 79 (1967).
[129] V. M. Stepanov and V. F. Krivtzov, *Zh. Obshch. Khim.* **35,** 53 (1965).
[130] V. M. Stepanov and J. I. Lapuk, *Zh. Obshch. Khim.* **36,** 40 (1966).
[131] D. E. Vance and D. S. Feingold, *Anal. Biochem.* **36,** 30 (1970).
[132] V. A. Puchkov, V. M. Stepanov, N. S. Wulfson, A. M. Zyakun, and V. F. Krivtzov, *Dokl. Akad. Nauk SSSR* **157,** 1160 (1964).

shortened peptide. In this way, Birr *et al.*[133] achieved an unambiguous identification of the degradation products using TLC.

Fluorescein Thiohydantoins. Fluorescein isothiocyanate, a well-known reagent for fluorescent labeling of proteins, has also been employed for Edman degradation. The amino acid derivatives may be separated by TLC on Silica Gel G.[134] However, many side reactions were found to occur.[135]

N-Thiobenzoylamino Acid Anilides (TBA-Amino Acids). TBA-amino acids, first prepared in 1955,[136] have become of renewed interest in connection with a new stepwise degradation of polypeptides.[137] The terminal *N*-thiobenzoyl derivative of a peptide gives on treatment with trifluoroacetic acid a 2-phenylthiazol-5(4*H*)-one trifluoroacetate, which reacts with aniline in boiling toluene to give the TBA-amino acid. R_f values of these derivatives in TLC on precoated Silica Gel F_{254} plates in 8 solvents have been reported recently.[138]

Further Derivatives for End-Group Determination

2,4,6-Trinitrophenylamino Acids (TNP-Amino Acids). 2,4,6-Trinitrobenzene-1-sulfonic acid (TNBS) was introduced by Satake *et al.*[139,140] for the qualitative and quantitative assay of amino acids and peptides. It has the advantage over dinitrofluorobenzene of being water soluble, which results in homogeneous reaction mixtures. Furthermore, less hydrolysis (to give picric acid) is observed. The resulting TNP-amino acids can be separated by two-dimensional TLC on Silica Gel G according to Nitecki *et al.*[141]

2,4-Dinitro-5-aminophenyl Amino Acids (DNAP-Amino Acids). 2,4-Dinitro-5-fluoroaniline reacts more readily than dinitrofluorobenzene with amino acids and peptides.[142] The resulting DNAP-amino acids can be diazotized[143] to form dyes. DNAP-Amino acids can be separated by

[133] C. Birr, C. Reitel, and T. Wieland, *Angew. Chem.* **82,** 771 (1970).

[134] D. E. Morse and B. L. Horecker, *Arch. Biochem. Biophys.* **125,** 942 (1968).

[135] D. E. Morse, Doctoral Dissertation, Albert Einstein College of Medicine, New York, 1967.

[136] J. B. Jepson, A. Lawson, and V. D. Lawton, *J. Chem. Soc.* **1955,** 1791 (1955).

[137] G. C. Barrett, *Chem. Commun.* **1967,** 487 (1967).

[138] G. C. Barrett and A. R. Khokhar, *J. Chromatogr.* **39,** 47 (1969).

[139] T. Okuyama and K. Satake, *J. Biochem.* (*Tokyo*) **47,** 454 (1960).

[140] K. Satake, T. Okuyama, M. Ohashi, and T. Shinoda, *J. Biochem.* (*Tokyo*) **47,** 654 (1960).

[141] D. E. Nitecki, I. M. Stoltenberg, and J. W. Goodman, *Anal. Biochem.* **19,** 344–350 (1967).

[142] E. D. Bergmann and M. Bantov, *J. Org. Chem.* **26,** 1480 (1961).

[143] R. S. Ratney, *J. Chromatogr.* **11,** 111 (1963).

TLC on Eastman Kodak Silica Gel sheets according to Deyl *et al.*[144] or on Silica Gel G according to Ratney *et al.*[145]

3,5-Dinitropyridyl, Nitropyridyl, and Nitropyrimidyl Amino Acids. By substitution of the ortho group of 1-fluoro-2,4-dinitrobenzene by an aza function to give a 2-fluoronitropyridine, the fluorine atom becomes more reactive toward amino groups in peptides.[146] In fact, even 2-chloro-3,5-dinitropyridine reacts readily with amino acids and peptides.[147] The aza group accelerates hydrolysis of the adjacent peptide bond, permitting gentler conditions of hydrolysis. Furthermore, most of the resulting amino acid derivatives can be converted to the original amino acids. TLC of 3,5-dinitro-2-pyridyl amino acids,[148-150] 3-nitropyridyl amino acids,[151] 5-nitropyridyl amino acids,[151] and 5-nitropyrimidyl amino acids[148] has been described.

[144] Z. Deyl, L. Schinkmannová, and J. Rosmus, *J. Chromatogr.* **30**, 614–617 (1967).

[145] R. S. Ratney, M. F. Godshalk, J. Joice, and K. W. James, *Anal. Biochem.* **19**, 357 (1967).

[146] A. Signor, L. Biondi, A. M. Tamburro, and E. Bordignon, *Eur. J. Biochem.* **7**, 328 (1968).

[147] A. Signor, L. Biondi, M. Terbejovich, and P. Pajetta, *Gazz. Chim. Ital.* **94**, 619 (1964).

[148] C. Di Bello and A. Signor, *J. Chromatogr.* **17**, 506 (1965).

[149] A. Serafini-Fracassini, *Biochim. Biophys. Acta* **170**, 289 (1968).

[150] S. Takahashi and L. A. Cohen, *Biochemistry* **8**, 864 (1969).

[151] E. Celon, L. Biondi, and E. Bordignon, *J. Chromatogr.* **35**, 47 (1968).

Section II

End-Group Analysis

[7] Use of Cyanate for Determining NH_2-Terminal Residues in Protein

By GEORGE R. STARK

Principle[1]

The amino groups of proteins can be carbamylated completely in aqueous solution with KNCO in the presence of a denaturant. If the carbamylated protein is heated in acid, formation of hydantoins that correspond to the NH_2-terminal residues is catalyzed. These hydantoins can be isolated and then quantitated by hydrolysis to amino acids:

$$^{+}NH_3CHRCONHCHR'\cdots COO^- + {}^-NCO \xrightarrow{pH\ 8} NH_2CONHCHRCONHCHR'\cdots COO$$

$$\xrightarrow{H^+} \text{Hydantoin (HC(R)—CO, HN—C(=O)—NH ring)} + {}^{+}NH_3CHR'\cdots COO^-$$

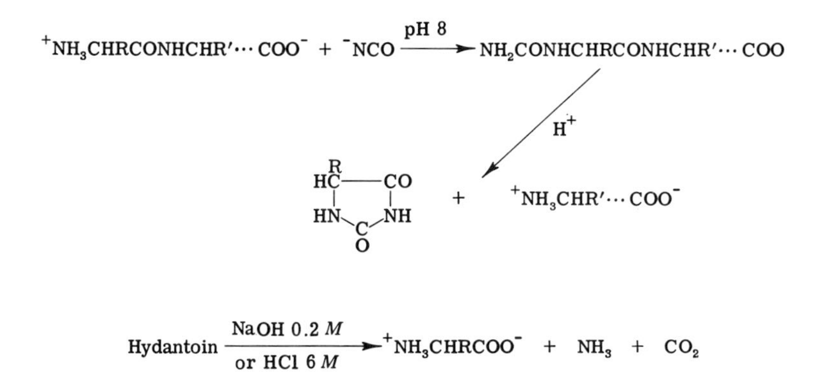

$$\text{Hydantoin} \xrightarrow[\text{or HCl } 6\,M]{\text{NaOH } 0.2\,M} {}^{+}NH_3CHRCOO^- + NH_3 + CO_2$$

Alternatively, the hydantoins can be determined directly.[2]

The cyanate method offers several important advantages. Since KNCO is very soluble in water, addition of cyanate to the amino groups of a protein can be performed as a homogeneous reaction in denaturing solvents, such as 8 *M* urea or 6 *M* guanidinium chloride, in which the protein remains soluble throughout the course of the reaction. The reaction goes well at neutral or mildly alkaline pH. All the hydantoins are formed in high yield during the acid-catalyzed cyclization. The results can be analyzed quantitatively in two ways: (1) Most results up to now have been obtained by hydrolyzing the hydantoins to amino acids, which can then be analyzed with excellent precision, sensitivity, and resolution. The major problems associated with this mode of analysis are the destruction in part of serine and threonine during the hydrolysis and the appearance of small quantities of spurious end groups. How-

[1] G. R. Stark and D. G. Smyth, *J. Biol. Chem.* **238,** 214 (1963).

[2] P. Hagel and J. J. T. Gerding, *Anal. Biochem.* **28,** 47 (1969).

ever, all the other hydantoins can be hydrolyzed in excellent yield, and corrections for spurious end groups are obtained readily from blank determinations. (2) Recently, Hagel and Gerding[2] have developed a sensitive method for direct analysis of hydantoins in which the problems associated with hydrolysis are circumvented.

Procedure

Preparation of Reagents and Reference Hydantoins. Reagent grade urea, KNCO, and serotonin creatinine sulfate may be used without purification. *N*-Ethylmorpholine should be redistilled so that the product is colorless and then stored in the cold. Pure reference hydantoins can be prepared readily by treating the isoionic amino acid with a slight excess of KNCO overnight at room temperature, and then heating the mixture to 100° with an equal volume of concentrated HCl.[1] However, in most instances it is unnecessary to prepare pure samples; amino acids, singly or in any combination, can be converted to the corresponding hydantoins by the procedure given below.

Preparation of Carbamyl Proteins or Large Peptides. Mix 2 ml each of *N*-ethylmorpholine and water, and bring the solution to pH 8 with glacial acetic acid. Dissolve 2.4 g of solid urea in enough of the above buffer to give a final volume of 5 ml (8 *M* in urea). Add 2.5 ml of this solution and then 250 mg of KNCO to a solution of 50 mg of protein or peptide in 2.5 ml of 8 *M* urea. Alternatively, the carbamylation can be carried out in guanidinium chloride or mixtures of this reagent with urea. The choice should depend upon which reagent or combination of reagents denatures a particular protein most effectively. Allow the carbamylation to proceed overnight at 50°.[3]

Carbamyl proteins can be separated from the reagents by any of several simple means. Gel filtration on Sephadex is feasible sometimes, but most often the carbamyl protein is too insoluble. Dialysis of the reaction mixture against distilled water will usually cause the modified protein to precipitate. The contents of the dialysis sac can be lyophilized, or precipitation can be completed by adding a little acetic acid and excess acetone. If gel filtration is to be carried out in an acidic solvent,

[3] The time and temperature of the reaction and the concentration of cyanate are chosen to ensure complete carbamylation of ϵ-NH_2 groups, which have a half-life of 2–3 hours under the conditions given. However, α-NH_2 groups react much more rapidly, with a half-life of 3–4 minutes. If the amino-terminal residue is not lysine, 1 hour should suffice for its complete carbamylation. G. Markus, E. A. Barnard, B. A. Castellani, and D. Saunders [*J. Biol. Chem.* **243,** 4070 (1968)], have found that carbamylation of ribonuclease is complete in 16 hours at 40°.

cyanate should be removed first in a preliminary dialysis, to avoid evolution of CO_2.

Cyclization to Hydantoins. The cyclization is performed on 0.5 μmole of carbamyl protein or peptide, contained in 1–3 ml of 50% acetic acid, in a heavy-walled Pyrex tube, 18 × 150 mm (Corning Catalog No. 9860). Add a volume of concentrated HCl equal to that of the acetic acid, evacuate the tube, and carefully deaerate the solution.[4] Seal the tube and immerse it in a bath of boiling water for 1 hour. Cool and open the tube, evaporate the contents to dryness, and add 1 ml of water to the residue.

If the protein contains tryptophan, 50 mg of serotonin creatinine sulfate should be included during the cyclization to protect NH_2-terminal tryptophan from destruction.

Isolation of Hydantoins. An outline of the fractionation scheme that follows is given in Fig. 1; the chromatographic separations are illustrated in Fig. 2. Transfer the solution of hydantoins quantitatively, with the aid of three 1-ml rinses of water, onto a column of Dowex 50-X2 (10 × 0.9 cm, 200–400 dry mesh). Pour the column in a tube (20 × 0.9 cm) that has a coarse sintered-glass disk. Before each use, the column should be washed with, in turn, 1 *M* NaOH, water, 6 *M* HCl, and finally water. The height of the column should be 10 cm after the final water wash, measured under 10 psi of air pressure. Develop the column

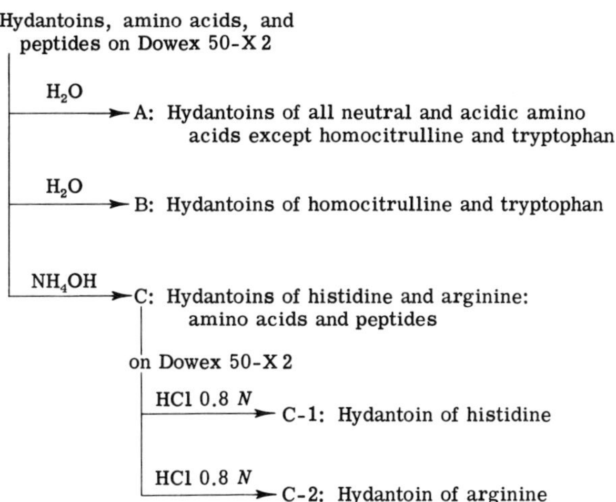

FIG. 1. Fractionation scheme for separation of hydantoins.

[4] See the procedure for sealing hydrolysis tubes, Vol. VI [117].

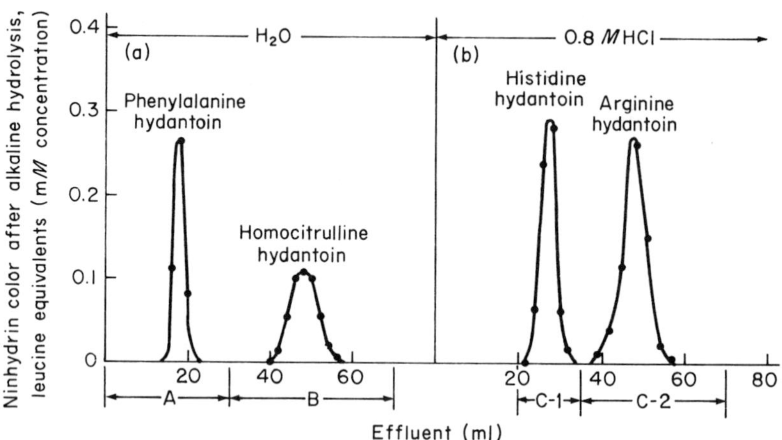

FIG. 2. Illustration of chromatographic separation of hydantoins.

with water at 20–25° at up to 100 ml per hour. Collect as fraction A the first 30 ml of effluent, including the water displaced by the sample; collect as fraction B 30 through 70 ml.[5] Evaporate these solutions to dryness. Wash the column with 40 ml of 1 *M* NH_4OH (fraction C) and evaporate this solution to dryness.

In order to hydrolyze pyrrolidonecarboxylic acid (see Discussion), add 5 ml of 3 *M* HCl to the residue obtained from fraction A and heat the solution at 100° for 30 minutes. Remove the acid by evaporation, dissolve the residue in approximately 2 ml of water, and transfer it quantitatively to a Dowex 50-X2 column (5 × 0.4 cm) with an additional 1 ml of water. The resin should be in the H^+ form and equilibrated with water. Develop the column with 6 ml of water; concentrate the effluent to dryness, and reserve the residue for hydrolysis.

Dissolve the residue from evaporation of fraction C in 1.0 ml of water[6] and place it all, with three 1-ml rinses of water, on a separate column of Dowex 50-X2 in 0.8 *M* HCl (10 × 0.9 cm; the column should measure 10 cm under 10 psi of air pressure). Wash the column before each use with, in turn, 1 *M* NaOH, water, 6 *M* HCl, and finally 0.8 *M*

[5] If serotonin has been included during the cyclization, its decomposition products leave a dark residue that is removed from the Dowex 50 column only very slowly by subsequent washing with NaOH and HCl. This residue does not interfere, however, with the subsequent performance of the column.

[6] E. Gaetjens, H. S. Cheung, and M. Bárány [*Biochim. Biophys. Acta* **93,** 188 (1964)], in an investigation of the NH_2-terminal residues of myosin, found it useful to heat fraction C in 3 *M* HCl for 30 minutes at 100° before chromatography in order to hydrolyze pyrrolidonecarboxylyl peptides, although this is not usually necessary with proteins of smaller size.

HCl. Develop the column at room temperature (20–35°) with 0.8 *M* HCl at approximately 60 ml per hour. Collect the effluent in two fractions, 20 through 35 ml as fraction C-1, and 35 through 70 ml as fraction C-2. Evaporate these solutions to dryness and reserve the residues for hydrolysis.

Hydrolysis of Hydantoins and Determination of Resulting Amino Acids. The procedure of Spackman, Stein, and Moore[7] was used for all the amino acid analyses described here, and the preparation of hydrolyzates described below is appropriate for this method of analysis. If another method is used, the procedure may require modification.

FRACTION A. Dissolve the residue in 2.0 ml of freshly prepared 0.2 *M* NaOH. Transfer 1.8 ml to a hydrolysis tube, 18 × 150 mm. Evacuate and seal the tube[3] and heat for 24 hours at 110°. After hydrolysis, neutralize the NaOH with 1 ml of 1 *M* HCl and then evaporate the solution to dryness in the tube. Add 2.0 ml of citrate buffer, pH 2.2,[7] pour the well-stirred solution into a centrifuge tube, and centrifuge the precipitated silicates. Withdraw 1.8 ml of the supernatant solution for determination of the neutral and acidic amino acids. If the analysis reveals NH_2-terminal glutamic acid, it may be desirable to perform a 96-hour acid hydrolysis of fraction A, since an excellent yield of this amino acid is so obtained.

FRACTION B. Dissolve the residue in 2.0 ml of 0.2 *M* NaOH. Transfer 1.8 ml to a hydrolysis tube, and evacuate and seal the tube. Hydrolyze the sample and prepare it for analysis exactly as fraction A was prepared. Chromatograph it on the short column only, through the position of lysine.

FRACTION C. Dissolve the residues from fractions C-1 and C-2 in 2.0 ml of 6 *M* HCl. Transfer 1.8 ml of each to a hydrolysis tube, and evacuate and seal the tubes. Hydrolyze the samples for 96 hours at 110°. Remove the HCl by evaporation. Dissolve the residues in 1 ml of 0.2 *M* NaOH and evaporate the solutions nearly to dryness to remove excess NH_3, which would otherwise interfere with the subsequent amino acid analysis. Add 1 ml of 1 *M* HCl and evaporate the solution to dryness. Add 2.0 ml of pH 2.2 buffer. Chromatograph 1.8 ml on the short column through the position of histidine for fraction C-1 and of arginine for fraction C-2. As when protein hydrolyzates are analyzed,[7] the first 10 ml of effluent must not be mixed with ninhydrin; large quantities of amino acids, derived from neutral peptides, are present and, if allowed to react with ninhydrin, cause precipitation in the reaction coil of the amino acid analyzer.

[7] See amino acid analysis, Vol. VI [117].

BLANKS. Small amounts of amino acids not derived from NH_2-terminal residues usually appear in the amino acid analysis. These may be neglected, but more accurate information can be secured with the aid of a blank determination, in which a sample of the protein that has not been exposed to either cyanate or urea is put through the procedure. The quantities of amino acids so obtained are then subtracted from those obtained from the carbamylated protein.

A blank end-group determination is always needed when a protein containing serine *O*-phosphate or threonine *O*-phosphate is analyzed, e.g., pepsin or pepsinogen. Peptides containing phosphoamino acids, formed during the cyclization, pass through Dowex 50 unretarded, as does cysteic acid. Such peptides, upon subsequent hydrolysis, contribute amino acids distinguishable from those derived from hydantoins only by the use of a blank.

If the end groups of a protein such as phosvitin, which contains large

TABLE I

RECOVERIES OF AMINO ACIDS AFTER END-GROUP DETERMINATION[a]

Amino acid	Recovery (%)
Lysine	84
Histidine	89[b]
Arginine	76[b]
Tryptophan	77[c]
Aspartic acid	99
Threonine	30
Serine	20
Glutamic acid	60 (97[b])
Proline	73
Glycine	98[d]
Alanine	98[d]
Half-cystine	0
Valine	96
Methionine	99[e]
Isoleucine	93
Leucine	97
Tyrosine	96
Phenylalanine	99

[a] All hydrolyses were performed for 20 hours in 0.2 *M* NaOH at 110°, except as otherwise indicated.

[b] Hydrolysis for 96 hours in 6 *M* HCl at 110°.

[c] Determination was performed in the presence of serotonin (see text).

[d] Glycine and alanine were determined separately, not in the presence of other amino acids.

[e] Methionine plus methionine sulfoxides.

amounts of *O*-phosphate, are to be determined, the special procedure for use with oxidized proteins, described in detail below, should be used.

Correction Factors and Calculation of Results. The recoveries obtained from carbamylamino acids carried through the complete procedure are summarized in Table I. Since glycine and possibly alanine result from the decomposition of other hydantoins, the recoveries given for these two amino acids were determined separately. The recovery for each amino acid reflects losses inherent in the procedure; therefore, the amount of an NH_2-terminal residue obtained after subtraction of the blank should be corrected by dividing by the appropriate recovery given in Table I.

Hydantoins are formed from carbamylamino acids or peptides in high yield; the corrections in Table I are mainly for losses that occur during isolation and hydrolysis. Therefore, these corrections are highly reproducible, since they do not depend upon the particular environment of the NH_2-terminal residue in a peptide. The yields of amino acids obtained upon hydrolysis of the purified hydantoins are shown in Tables II and III.

Procedure for NH_2-Terminal Cystine, Cysteine, or Half-Cystine. Oxidize with performic acid[8] the residue obtained after the cyclization mixture has been evaporated to dryness. (Alternatively, the carbamyl protein can be oxidized before cyclization.) Obtain fraction A and heat to 100° in 3 *M* HCl as described above, but *omit* the step using a 5×0.4 cm column of Dowex 50-X2.[9] To separate cysteic acid hydantoin from free cysteic acid, a column of Dowex 1-X8 (10×0.9 cm) is required (Fig. 3). Wash the column at room temperature with 0.2 *M* HCl at 60–70 ml per hour. Collect the fraction from 35 to 70 ml of effluent. Remove the HCl and water on a rotary evaporator. Dissolve the residue in 2.0 ml of freshly prepared 0.2 *M* NaOH (be sure that the resulting solution is strongly basic). Carry out the hydrolysis as described for fraction A. Analyze the hydrolyzate for cysteic acid on the amino acid analyzer or on Dowex 1-X8.

If cysteic acid is found, NH_2-terminal cysteine can be distinguished from NH_2-terminal cystine by treating the protein with iodoacetamide (*not* iodoacetic acid) before it is carbamylated, preferably in a denaturing solvent to ensure complete reaction with SH groups.[10] In two experiments in which *S*-carboxamidomethylcysteine was carbamylated and cyclized in company with a standard mixture of amino acids, *S*-carboxy-

[8] See performic acid oxidation, Vol. IV [1].

[9] A second treatment with hot 3 *M* HCl reduces the blank due to pyrrolidonecarboxylyl peptides of cysteic acid, as suggested by Gaetjens *et al.*[6]

[10] G. R. Stark, *J. Biol. Chem.* **239**, 1411 (1964).

TABLE II
HYDROLYSIS OF HYDANTOINS[a]

Hydantoin	Hydantoin recovered (%) as amino acid	
	HCl 6 *M*, 110°, 96 hr	NaOH 0.1 *M*, 110°, 24 hr
Histidine	94	53
Arginine	98	97[b]
Tryptophan	0	78[c]
Aspartic acid	100	103
Threonine[d]	9	36
Serine	4	45[e]
Glutamic acid	98	67
Proline	107	109
Glycine	101	133[f]
Alanine	97	102
Half-cystine	70	0
Homocitrulline	89[g]	97[b]
Valine	97	97
Methionine	94	102[e]
Isoleucine[d]	94	101[e]
Leucine	98	101[e]
Tyrosine	90	102[e]
Phenylalanine	99	101[e]

[a] The hydrolyses were performed in evacuated sealed tubes with equal portions of a single stock solution of the hydantoins. The amino acids released were determined chromatographically.

[b] Arginine hydantoin forms ornithine quantitatively upon hydrolysis in alkali. Ornithine and lysine are not resolved in the chromatographic system used; the recoveries in this instance were obtained by dividing the ornithine-lysine recovery by 2.

[c] Tryptophan hydantoin, hydrolyzed 20 hours at 110° in 0.2 *M* NaOH in the absence of other hydantoins, gave an 88% recovery of tryptophan.

[d] Large amounts of the allo forms of threonine and isoleucine are formed upon acid or base hydrolysis of the hydantoins. The recoveries in the table include these forms.

[e] These are data from a separate experiment in which hydrolysis was catalyzed by 0.2 *M* NaOH at 110° for 24 hours. The recovery of valine in this experiment was 103%.

[f] Glycine is formed upon alkaline hydrolysis of serine and threonine hydantoins.

[g] This recovery is the sum of homocitrulline and lysine, since hydrolysis at the ϵ-carbamyl group is not complete in 96 hours at 110° in 6 *M* HCl.

methylcysteine was recovered in 71% and 65% yield. The hydantoins were isolated as described above and hydrolyzed in 6 *M* HCl at 110° for 48 hours.[10] Hydrolysis in alkali gives *no* *S*-carboxymethylcysteine. Since 56% or 67% of the expected amount of this amino acid is formed upon hydrolysis of the hydantoin in acid for 24 or 96 hours,[1] the hy-

TABLE III
ALKALINE HYDROLYSIS OF HYDANTOINS OF SERINE AND THREONINE[a]

	Hydantoin recovered (%) as amino acid			
	Serine hydantoin		Threonine hydantoin	
Amino acid found	NaOH 0.1 *M*	NaOH 0.2 *M*	NaOH 0.1 *M*	NaOH 0.2 *M*
Serine	36	38	—	—
Threonine	—	—	34	43
Glycine	4	3	34	20

[a] The hydantoins were hydrolyzed separately in evacuated sealed tubes at 110° for 24 hours. The amino acids released were determined chromatographically.

dantoin must be formed from the carbamylamino acid in nearly quantitative yield.

Carbamylation of Amino Acids or Small Peptides. Prepare a mixture of 1 ml of water, 0.3 ml of *N*-ethylmorpholine, and 100 mg of KNCO. Bring the solution to pH 8 with approximately 0.1 ml of glacial acetic acid. Introduce 0.5 μmole of NH_2 compound into a heavy-walled Pyrex tube (18 × 150 mm) and remove any solvent. Add 0.05 ml of the above solution and keep the tube at 50° for 4 hours.[11] Compounds such as

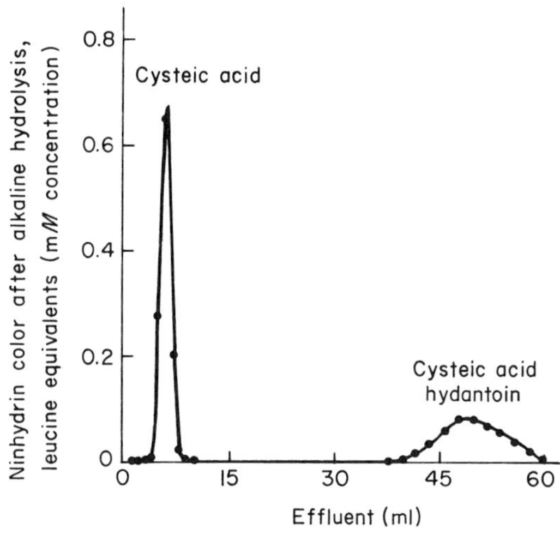

FIG. 3. Separation of cysteic acid hydantoin from free cysteic acid.

[11] This is 5 times less KNCO than recommended[1] in order to reduce further the amounts of glycine and alanine that arise from the cyanate itself. If at all possible, the carbamyl compound should be isolated before cyclization; for example, peptides of five or more residues can be separated from salts on columns of Sephadex G-25. Markus *et al.*[3] have found that the level of spurious glycine is lower if the carbamylation is carried out at 40°.

cystine and tyrosine, initially only very slightly soluble in water, dissolve as carbamylation proceeds. Evaporate the contents of the tube to dryness and proceed with the cyclization as above.[12]

Application of Method to Performic Acid-Oxidized Proteins. Modification of the usual procedure is required if the cyanate method is to be applied to oxidized proteins, because cysteic acid and cysteic acid peptides are not separated from neutral and acidic hydantoins on Dowex 50-X2. However, they may be eliminated by further chromatography on a column of Dowex 1-X8 (6×0.9 cm), which retains cysteic acid and cysteic acid peptides very strongly. The effluent curve obtained upon chromatography of a mixture of hydantoins on this column is shown in Fig. 4. The effluent from the column is divided into two fractions: the first 55 ml containing neutral hydantoins and glutamic acid hydantoin,

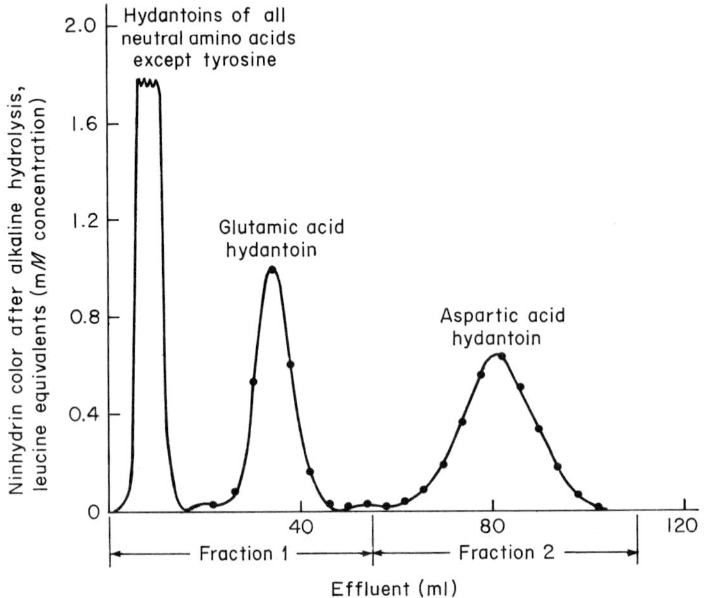

FIG. 4. Chromatography of a mixture of hydantoins on Dowex 1-X8. Eluent: 0.5 acetic acid, 25°. Flow rate: 60 ml/hour.

[12] In a study of the peptides produced by digestion of *S*-aminoethyl ribonuclease, B. V. Plapp, M. A. Raferty, and R. D. Cole [*J. Biol. Chem.* **242**, 265 (1967)] have obtained good agreement between the end groups expected and those actually found. In determining the end groups of some pure peptides, these authors confirmed the results obtained by analysis of the hydantoin by a subtractive method in which the amino acids retained on the Dowex 50 column (0.8×12 cm) were washed off with 40 ml of 1 *M* NH_4OH, then analyzed directly.

the second 55 ml containing aspartic acid hydantoin. Tyrosine hydantoin appears in both fractions and must be estimated by summation. When Dowex 1-X8 is used, carbamylamino acids (see the discussion below) are lost, and the recoveries of end groups are consequently somewhat low.

This procedure with Dowex 1-X8 should also be used in end-group determinations of proteins, such as phosvitin, which contain large amounts of serine *O*-phosphate or threonine *O*-phosphate.

Comments

Preparation of Carbamyl Proteins. The carbamylation procedure described is scaled for 50 mg of protein, but the amount taken in a specific case will depend on the molecular weight of the protein and on the sensitivity of the final amino acid analysis. If substantially less material is used, the quantities and volumes of reagents should be reduced accordingly; columns more narrow than those described above should be used when there is less material to be separated.

Carbamyl proteins are stable; for example, solutions of carbamyl ribonuclease have been stored in 50% acetic acid at —20° for several months without significant change in amino acid composition or decrease in yield of end group. If possible, more carbamyl protein than is necessary for a single determination should be prepared, since a separate procedure is required to determine NH_2-terminal cystine, cysteine, and half-cystine.

Since it is difficult to avoid small losses during the separation of the carbamyl protein from reagents, it is best to base the calculations on the amount of carbamyl protein taken for cyclization. A quantitative amino acid analysis of the carbamyl protein is extremely useful and should be performed whenever possible. Such an analysis gives the relative amounts of homocitrulline and lysine, showing that the carbamylation has been satisfactory; it also provides an accurate measure of the protein concentration of a stock solution or of the proportion of a solid that is protein. A comparison of this analysis with that of the starting material indicates whether any fragments, such as bound peptides, have been lost.

The amounts of lysine and homocitrulline obtained upon acid hydrolysis of a carbamyl protein are a measure of the extent of carbamylation. Lysine is formed in 24% yield when free homocitrulline is hydrolyzed in 6 *M* HCl at 110° for 22 hours. For seven different carbamyl proteins, the recovery of lysine ranges from 17% to 30% of the total of lysine plus homocitrulline.[1] The figures for a given protein are reproducible, however, and the variations probably reflect differences in rate of regeneration of lysine from homocitrulline in various peptide linkages rather than incomplete carbamylation; i.e., the nature of the adjacent amino acid residues may influence the rate of regeneration.

The remainder of the procedure is described for 0.5 μmole of carbamyl compound. This is not a minimal amount, for we have carried out successful determinations on less than 0.2 μmole. Improvements in the sensitivity of amino acid analyses[7] enable this amount to be reduced even further.

Cyclization. Cyclization is carried out in a solution of equal volumes of 50% acetic acid and concentrated HCl, partly because 50% acetic acid is often a better solvent for carbamyl proteins than water. Most hydantoins are quite stable under these conditions; after 1 hour at 100°, less than 3% of any amino acid is formed from the corresponding hydantoin with the exception of proline, which is formed in 21% yield. If the HCl contains no acetic acid, the amounts of the other amino acids are not appreciably altered, but approximately twice as much proline is released. Thus a maximal yield of proline hydantoin is obtained in a solvent that contains acetic acid.

Serine and threonine hydantoins decompose somewhat during the cyclization. The extent of decomposition is variable and probably depends upon the extent of contamination of the HCl and acetic acid by metal ions; therefore, analytical grade reagents as free as possible from metals should be used. If either serine or threonine is found as an end group, it is probably preferable for an investigator to determine the recoveries under his own conditions by performing a control experiment with the two amino acids, rather than to use the recoveries listed in Table I.

When the cyclization is carried out in the presence of air, a small amount of the cystine of a carbamyl protein becomes oxidized to cysteic acid. Since cysteic acid peptides are not held by Dowex 50 and contribute amino acids to the analysis, it is important that their formation be minimized by performing the cyclization in an evacuated sealed tube.

Spurious Glutamic Acid End Groups. If fraction A is not heated with 3 *M* HCl after the initial separation, substantial amounts of glutamic acid will appear in the final amino acid analysis. Treatment of carbamyl proteins with 6 *M* HCl at 100° results in the formation of small amounts of pyrrolidonecarboxylic acid and pyrrolidonecarboxylyl peptides, which are not held by Dowex 50. Of course, the amino acids formed upon hydrolysis of such compounds cannot be distinguished from those formed upon hydrolysis of hydantoins. This difficulty is overcome by exposing the mixture of hydantoins, pyrrolidonecarboxylic acid, and pyrrolidonecarboxyl peptides to HCl for a second time, followed by filtration through a very short column of Dowex 50-X2. The pyrrolidonyl compounds are (98%) hydrolyzed to glutamic acid plus glutamyl peptides, and these compounds are retained by the second column. With this procedure, only very small amounts of any amino acid not derived from the end group are found. For example, 0.16 residue of glutamic acid and slightly larger amounts of other "end groups" were found when the second treatment with HCl was omitted from the procedure employed to determine the end groups of α-chymotrypsinogen. When this treatment was included, no glutamic acid was found.[1]

Equilibrium between Carbamylamino Acids and Hydantoins. Only 80% of the mixture formed upon heating carbamylglycine to 100° for 1 hour in 6 *M* HCl passes through a column of Dowex 1 when eluted with 0.5 *M* acetic acid. The material bound to the resin is carbamylglycine. The small amount of free glycine that is present is not held. Significant but smaller amounts of several other carbamylamino acids remain after similar treatment. Since carbamylamino acids emerge with hydantoins in fraction A, they are not lost. However, carbamylhomocitrulline, in contrast to homocitrulline hydantoin, does not chromatograph as a sharp peak on Dowex 50 when eluted with water; approximately 80% emerges in fraction A. The empirical correction factor (Table I) has given good results when applied to the lysine end group of ribonuclease.[1]

Isolation of Hydantoins. Neutral and acidic hydantoins are readily eluted from Dowex 50-X2 columns with water. (A 10 × 0.9 cm column of Dowex 50-X2 contains approximately 2 g of resin; the amino acids formed upon hydrolysis of 100 mg of protein saturate only about 10% of its capacity of 10 meq.) However, the hydantoins of homocitrulline and tryptophan (fraction B) are retarded more than any of the others (fraction A) and can be completely separated from them (see Fig. 2a). This separation is useful in the later analyses.

Since the ε-NH_2 groups of lysine are completely carbamylated, the derivatives of histidine and arginine are the only basic hydantoins present. During elution with water they remain tightly bound to the resin, but can be eluted by 1 *M* NH_4OH, which is then easily removed by evaporation. The residue, which includes the amino acids and peptides formed from the protein during the cyclization procedure, is dissolved in 0.8 *M* HCl. The hydantoins of histidine and arginine have one positive charge in this solvent, as do any peptides that do *not* include a histidine or arginine residue. Histidine, arginine, or any peptide that includes one of these basic residues has two positive charges and can therefore be separated from the hydantoins on Dowex 50-X2. Arginine is eluted between 180 and 220 ml of effluent under the conditions described in Fig. 2b.

Small amounts of histidine-containing compounds, probably diketopiperazines and pyrrolidonecarboxylyl peptides, chromatograph in the region of arginine hydantoin, but, since the two hydantoins are well separated, they can be isolated in separate fractions (Fig. 2b). Fraction C-1 usually contains histidine hydantoin and only traces of other histidine-containing compounds, whereas fraction C-2 contains arginine hydantoin and only traces of other arginine-containing compounds.

Hydrolysis of Hydantoins. The data in Table I and in column 3 of Table II show that nearly quantitative yields of amino acids are obtained from most hydantoins by alkaline hydrolysis, but serine and threonine are important exceptions. Alkaline hydrolysis of fraction A will not reveal NH_2-terminal cystine, cysteine, or half-cystine. These should be determined separately with the aid of performic acid oxidation as described above. Glutamic acid hydantoin is partially destroyed by alkaline hydrolysis; if desired, NH_2-terminal glutamic acid can be determined in excellent yield by acid rather than alkaline hydrolysis of fraction A. Hydrolyzates of fraction A need not be analyzed on the short column, since basic amino acids are not present.

During alkaline hydrolysis of methionine hydantoin, oxidation occurs to a variable extent, resulting in a mixture of methionine and the diastereomeric sulfoxides. If the 150-cm column of the amino acid analyzer[7] is used, the sulfoxides are well resolved from aspartic acid and should be added to methionine. If columns of 55–60 cm are used, the more retarded sulfoxide is not separated from aspartic acid. Oxidation of the hydantoin with performic acid[8] before hydrolysis, as suggested by Lederer *et al.*,[13] converts methionine and the sulfoxides to methionine sulfone, which is resolved on the shorter columns.

Approximately equal amounts of isoleucine and alloisoleucine are found in both acid and alkaline hydrolyzates of isoleucine hydantoin (prepared from L-isoleucine); this shows that racemization at the α-carbon is essentially complete in either medium. Both threonine and allothreonine are formed by alkaline hydrolysis of threonine hydantoin, but are not sufficiently resolved on the amino acid analyzer to allow separate determination.

The hydantoins of serine and threonine are markedly unstable to acid hydrolysis, but a small amount of the free amino acid, which is much more stable, does form. Ninhydrin-positive products from dilute alkaline hydrolysis of these hydantoins are recorded in Table III. Both yield only the parent amino acid, glycine, and NH_3. The recoveries indicate that extensive decomposition occurs, with formation of products that are not amino acids. Free threonine and serine are stable to the

[13] F. Lederer, S. M. Coutts, R. A. Laursen, and F. H. Westheimer, *Biochemistry* **5**, 823 (1966).

conditions specified in Table III. The formation of an appreciable amount of glycine upon hydrolysis of threonine hydantoin should be taken into account if NH_2-terminal threonine is found. In the rather unlikely instance that NH_2-terminal threonine and glycine are both present, it may be desirable to perform an acid hydrolysis of fraction A, since threonine hydantoin does not give rise to glycine under these conditions.

Upon alkaline hydrolysis, the homocitrulline hydantoin of fraction B is converted quantitatively to lysine, and tryptophan is formed from tryptophan hydantoin in good yield. These amino acids may be determined by using the short column of the amino acid analyzer. Tryptophan hydantoin decomposes on acid hydrolysis, and homocitrulline hydantoin gives rise to a mixture of lysine and homocitrulline.

Fraction C-1 should be hydrolyzed in acid for the best yield of histidine from the hydantoin. Again, only analysis on the short column is required. Fraction C-2 *must* be hydrolyzed in acid, since alkaline hydrolysis would yield both ornithine from arginine hydantoin and lysine from the homocitrulline residues of neutral peptides; ornithine and lysine are not resolved easily.

Direct Analysis of Hydantoins. Hagel and Gerding[2] have developed a method for direct chromatographic analysis of hydantoins on columns of AG 1-X8 anion exchange resin, thereby eliminating the need for hydrolysis to amino acids, with attendant poor yields from threonine and serine, and complications from spurious end groups. The chromatographic separation obtained is shown in Fig. 5. The absorbance at 225 nm is shown in Table IV for each peak of Fig. 5, after normalization to a total of 1 μmole of hydantoin. The pH of each chromatographic fraction was adjusted to pH about 11 with NaOH, in order to obtain maximum absorbance. The method has been automated, with the use of a flow-through cuvette.[14] Within an experimental error of about 3%, the recovery of all hydantoins from the AG 1-X8 column is quantitative. Upon standing at pH about 9, threonine hydantoin becomes racemized to allothreonine hydantoin, which runs in the approximate position of serine hydantoin.[2] Similarly, isoleucine and alloisoleucine hydantoins may both be present; the allo isomer runs ahead of isoleucine hydantoin. See Hagel and Gerding[2] for more details.

Pyrrolidone carboxylyl peptides and pyrrolidone carboxylate do not interfere with the analysis, since these are present in small amount and have a much lower extinction than hydantoins. The same is true of diketopiperazines.[14] Even though the hydrolylic step has been eliminated,

[14] P. Hagel, personal communication.

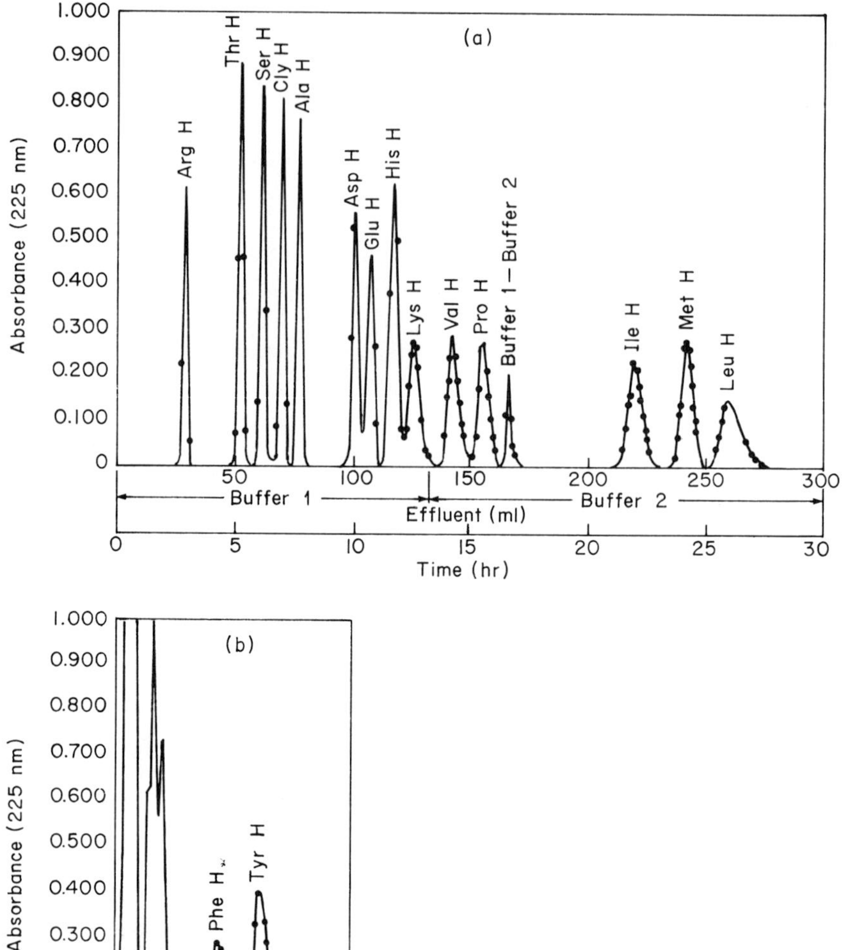

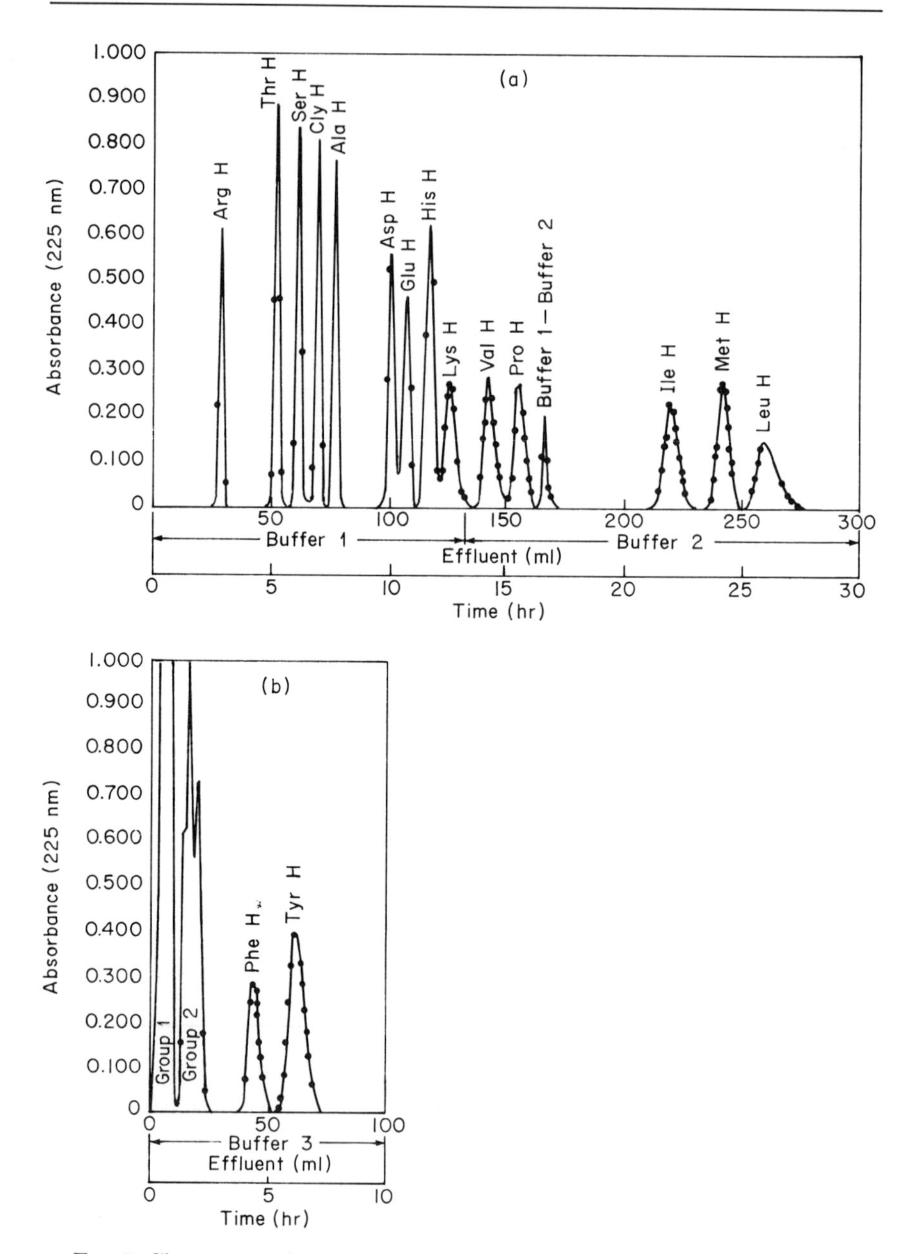

FIG. 5. Chromatographic fractionation of a synthetic mixture of hydantoins on columns of AG 1-X8 spherical particles, 5–45 μ diameter; (a) obtained by elution with buffer I and buffer II from a 0.900 × 95 cm column; (b) obtained by elution with buffer III from a 0.900 × 15 cm column. Load on both columns was about 1 μmole of each hydantoin (Table IV). Effluent was collected in 0.947-ml fractions at 20°C at a flow rate of 9.48 ml per hour. The hydantoin concentration in each

TABLE IV
ABSORBANCIES OF SYNTHETIC HYDANTOINS[a]

Hydantoin of	Net total absorbance of 1 μmole at 225 nm
Arginine	1.79
Threonine	1.76
Serine	1.88
Glycine	1.98
Alanine	1.99
Aspartic acid	1.86
Glutamic acid	1.78
Histidine	2.28
Lysine	1.77
Valine	1.75
Proline	1.53
Isoleucine	1.81
Methionine	1.91
Leucine	1.68
Phenylalanine	1.74
Tyrosine	3.39

[a] From P. Hagel and J. J. T. Gerding, *Anal. Biochem.* **28,** 47 (1969).

separation of hydantoins from the amino acids and peptides present in the cyclization mixture is still recommended if direct analysis is employed, in order to avoid overloading the AG 1 column.

Direct analysis of hydantoins seems to be a most useful adjunct to the cyanate procedure, especially if threonine or serine are suspected to be NH_2-terminal. However, no degradations of known proteins have been accomplished with this procedure so far.[14] One minor complication that can be anticipated is that the small amounts of carbamylamino acids in equilibrium with hydantoins under cyclization conditions (see above) will not be detected.

Discussion of Some Results. The cyanate method has often yielded quantitative results in excellent agreement with estimates of molecular weights of polypeptide chains obtained by physical methods. In analyses of papaya lysozyme[15] and swine pepsin,[16] several end groups, each in low

[15] J. B. Howard and A. N. Glazer, *J. Biol. Chem.* **242,** 5715 (1967).
[16] T. G. Rajagopalan, S. Moore, and W. H. Stein, *J. Biol. Chem.* **241,** 4940 (1966).

fraction was determined at 225 nm after addition of 2 ml of a solution containing a sufficient amount of NaOH to exclude NH_4Cl from the buffer. Buffer I: 0.3 *M* NH_4Cl + 5.5 ml of 25% NH_3 per liter (pH 8.78); buffer II: 1.5 *M* NH_4Cl + 230 ml of 25% NH_3 per liter (pH 9.88); buffer III: 2.5 *M* NH_4Cl + 260 ml of 25% NH_3 per liter (pH 9.68). From P. Hagel and J. J. T. Gerding, *Anal. Biochem.* **28,** 47 (1969).

yield, were obtained for impure preparations in which autodigestion had occurred, but a good yield of a single end group was found when a pure preparation of each protein was analyzed, demonstrating the utility of the method in revealing heterogeneity in some cases.

However, low values have been obtained in NH_2-terminal analyses of several bacterial proteins, such as arginine decarboxylase,[17] thioredoxin,[18] thiogalactoside transacetylase,[19] and aspartate transcarbamylase.[20] In view of the success of the cyanate method in other situations, it seems likely that these results reflect the actual situation, i.e., that some of the NH_2 termini are blocked. Brown *et al.*[19] found only 0.25 mole of aspartic acid and 0.2 mole of methionine per chain of thiogalactoside transacetylase by both the cyanate and the dinitrofluorobenzene procedures. They discussed several possibilities for their results, including *N*-acyl and pyrrolidonecarboxylyl termini, the latter derived from NH_2-terminal glutamine. They specifically ruled out the possibility that the NH_2 terminus was formylated in this case. However, formylmethionine, the initiating residue in bacterial protein synthesis, might not be removed completely from all bacterial proteins under all conditions of growth. Such a possibility might be especially probable under conditions of derepression, when a large amount of a single protein is being synthesized by the cell. This seems to be a reasonable but as yet unsubstantiated explanation in the case of aspartate transcarbamylase from *Escherichia coli,* where low values for the alanine of the catalytic subunit and the threonine of the regulatory subunit were obtained independently in two laboratories.[20] Later work by Weber[21] revealed that variable amounts of threonine and methionine were found at the NH_2 terminus of the regulatory subunit, strongly implying that proteolytic removal of the methionine was incomplete and suggesting that removal of the formyl group might have been incomplete as well.

[17] E. A. Boeker, E. H. Fischer, and E. E. Snell, *J. Biol. Chem.* **244**, 5239 (1969).
[18] L. Thelander, *Eur. J. Biochem.* **4**, 407 (1968).
[19] J. L. Brown, S. Koorajian, and I. Zabin, *J. Biol. Chem.* **242**, 4259 (1967).
[20] G. R. Stark and G. L. Hervé, *Biochemistry* **6**, 3743 (1967); K. Weber, *J. Biol. Chem.* **243**, 543 (1968).
[21] K. Weber, *Nature (London)* **218**, 1116 (1968).

[8] End-Group Analysis Using Dansyl Chloride

By WILLIAM R. GRAY

The methods of amino-terminal analysis and sequence determination based on dansyl chloride[1-3] have found wide application in the past several years, largely because of the ease with which one can study minute amounts of peptides and proteins. The most important developments since the earlier edition of this volume have been the thin-layer chromatographic systems based on those of Woods and Wang,[4] methods for rapid handling of many samples,[5] and simple methods for analyzing proteins.[6,7]

Reactions of Dansyl Chloride

Dansyl chloride is reactive toward a variety of bases. In sequence analysis, the most important of these are the primary and secondary amines present as the terminal amino groups of peptides and proteins. Reactive side-chain functions include thiol (cysteine), phenolic hydroxyl (tyrosine), amino (lysine), and imidazole (histidine) in decreasing order of reactivity. There is essentially no reactivity toward guanidinium (arginine), aliphatic hydroxyl (serine and threonine), amides (asparagine and glutamine), or indole (tryptophan) under normal conditions of pH and reagent concentration.

Compared with most other end-group reagents, dansyl chloride is rather susceptible to hydrolysis by water and hydroxyl ions. Amino groups react only as the free base ($R\text{-}NH_2$), not as the conjugate acid ($R\text{-}NH_3^+$), so that labeling must be carried out at alkaline pH. The labeling is always in competition with hydrolysis; when labeling is carried out with at least a severalfold excess of reagent, it can be shown that when the overall reaction is complete:

$$\% \text{ labeled} = 100 \left[1 - e \frac{-D \cdot \alpha k_a}{k_w} \right]$$

where D is the initial concentration of dansyl chloride; α is the proportion

[1] W. R. Gray and B. S. Hartley, *Biochem. J.* **89**, 59 p (1963); *Biochem. J.* **89**, 379 (1963).

[2] W. R. Gray, Vol. 11, p. 139 (1967).

[3] W. R. Gray, Vol. 11, p. 469 (1967).

[4] K. R. Woods and K.-T. Wang, *Biochim. Biophys. Acta* **133**, 369 (1967).

[5] W. R. Gray and J. F. Smith, *Anal. Biochem.* **33**, 36 (1970).

[6] W. R. Gray, unpublished results.

[7] C. Gros and B. Labouesse, *Eur. J. Biochem.* **7**, 463 (1969).

of amino group present as the free base; k_a is the second-order rate constant for the reaction between free base and dansyl chloride; k_w is the apparent first-order rate constant for hydrolysis of dansyl chloride. This last term is the sum of two components, hydrolysis by H_2O (independent of pH above pH 5) and hydrolysis by OH^- (increasingly dominant above pH 10).

Some important facts emerge from this formulation of the reaction process. First, successful labeling is dependent on *absolute concentration*

TABLE I
REACTIVITY OF VARIOUS BASES TOWARD DANSYL CHLORIDE

Base	pK_a	k[a]	% Labeled[b]
H_2O	—	4.1×10^{-5}	—
OH^-	—	15	—
NH_3	9.26	0.5	39
Alanine	9.87	10	>99
Arginine (α-)	9.04	6.4	>99
Asparagine	8.80	3.1	97
Aspartic acid	9.82	4.4	93
Cysteic acid	9.0	5.7	>99
Glutamine	9.13	4.8	99
Glutamic acid	9.67	5.0	97
Glycine	9.78	13	>99
Histidine[c] (α-)	9.18	12	>99
Histidine[c] (imid.)	6.1	0.5	53
Isoleucine	9.76	14.5	>99
Leucine	9.74	13.5	>99
Lysine[c] (α-)	8.95	3.6	98
Lysine[c] (ε-)	10.53	42	>99
Methionine sulfone	8.6	3.2	98
Phenylalanine	9.24	10	>99
Proline	10.60	360	>99
Serine	9.15	4.6	>99
Threonine	9.12	4.9	>99
Tryptophan	9.39	35	>99
Tyrosine[c,d] (α-)	9.11	(10)	>99
Tyrosine[c] (ionized phenolic group)	10.07	280	>99
Valine	9.72	13	>99
Diglycine	8.17	3.5	99
Triglycine	7.91	3.0	97
Leucylglycine	8.2	2.4	94

[a] k is the second-order rate constant for the reaction between DNS-Cl and the appropriate base. Measurements were made in water at 22°.

[b] Calculated percentage labeling of amino acids at pH 9.8, using 5 mM DNS-Cl (see text).

[c] Value of k is that for the corresponding group of the free amino acid.

[d] The value for k is only approximate.

of dansyl chloride, not on the ratio of dansyl chloride to amino group. Terms such as "10-fold excess of reagent" have no meaning: this could give rise to 1% or 99% labeling, depending on the volume of reactant solutions.

Second, both α and k_w vary with pH. At a given pH, one can calculate α from a knowledge of the pK_a of the amino group, and k_w from the second-order rate constants for hydrolysis by H_2O and OH^- (Table I). The overall effect of pH upon the labeling process is very pronounced (Fig. 1), showing a relatively narrow pH range in which successful labeling can be carried out without using a very high concentration of dansyl chloride. At low pH, too little of the peptide is present in the free-base form to compete with hydrolysis by H_2O; at high pH, the hydrolysis by OH^- rapidly swamps out the labeling process.

Second-order rate constants for reactions between dansyl chloride and

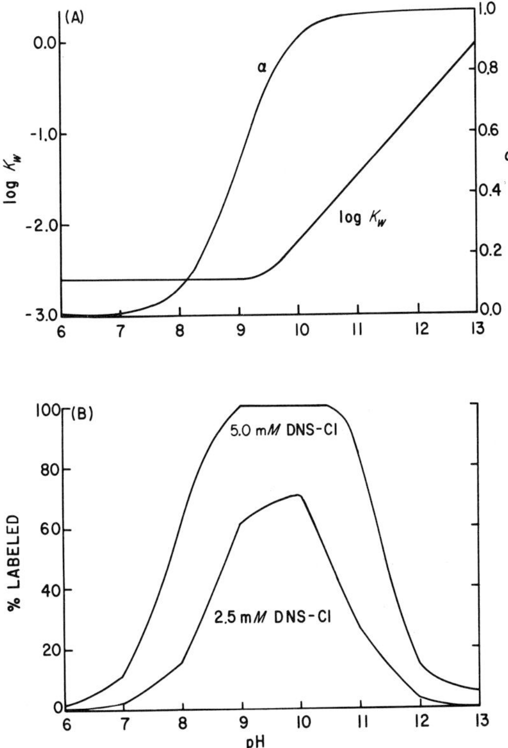

FIG. 1. (a) Variation of degree of dissociation (α) of an amino group (pK_a = 9.0) with pH. Variation of rate of hydrolysis of dansyl chloride with pH: k_w is the apparent first-order rate constant (sec^{-1}) in 0.5% acetone, 22°. (b) Labeling efficiency as a function of pH and dansyl chloride concentration for an amino group, pK_a = 9.0, K_s = 4.5 l/mole per second.

various bases have been measured[8] and are given in Table I. Using these figures, the extent of reaction was calculated for a standard set of conditions (5 mM DNS-Cl at pH 9.8). Most amino acids are almost completely labeled, though it can be seen that some peptides are incompletely dansylated. The conditions were originally chosen as a compromise to avoid overloading the separation system with DNS-OH; this is no longer considered necessary, so that we routinely use twice this concentration of reagent to achieve better labeling.

The rate constants listed in Table I refer to 0.5% acetone at 22°. Other solvents and catalysts can markedly influence the reaction rates. In 50% acetone (v/v), rates are about 8-fold lower than in water; there is little effect upon the extent of labeling, however, since k_a and k_w are affected equally and α is not greatly changed for amino groups. The extent of labeling of phenolic groups would probably be significantly affected because of a decrease in α, perhaps partly offset by an increase in k_a. Trimethylamine is a highly effective catalyst for both hydrolysis and labeling: a concentration of the free base of 10^{-4} M is sufficient to increase the rate of both processes about 7-fold. In general, there is little point in catalyzing the labeling process since it saves comparatively little working time and risks hydrolyzing the reagent before it is fully mixed with the peptide.

To achieve adequate concentrations of DNS-Cl, it is necessary to use a mixed organic-aqueous system for labeling. Acetone is the most convenient solvent for the DNS-Cl in work with peptides, because the solutions can be stored for long periods without deterioration. It is not very suitable for work with proteins, however, because of solubility problems, and for this it is best to use freshly prepared solutions in dimethyl formamide. Labeling has generally been poor in solvents containing large amounts of pyridine.

Stability of Dansyl Derivatives

Thiol and imidazole derivatives are unstable in acid, and are completely broken down under the conditions used for hydrolyzing labeled peptides. The degradation products are DNS-OH and the original group (e.g., bis-DNS-His yields α-DNS-His and DNS-OH). In mild alkali, they readily transfer the dansyl group to other bases, and this reaction must be borne in mind when hydrolysis methods other than acid are used. Alkyl esters are also unstable to acid and alkali, although they would not normally be formed by reactions with peptides and proteins.

Primary sulfonamide derivatives are generally very stable to acid

[8] W. R. Gray, Ph.D. Thesis, University of Cambridge, Cambridge, England, 1964.

TABLE II
STABILITY OF DANSYL AMINO ACIDS TO ACID HYDROLYSIS

Derivative	% Remaining[a]	Fluorescent by-products
DNS-Pro	23	DNS-OH[b]
DNS-Ser	65	$DNS-NH_2$[c] (25%), DNS-OH (10%)
DNS-Gly	82	DNS-OH
DNS-Ala	93	DMS-OH
DNS-Val, DNS-Leu, DNS-Ile	100	No detectable breakdown
DNS-Trp	—	$DNS-NH_2$, DNS-OH (traces of DNS-Gly, -Ser, -Ala)

[a] Percentage remaining as unchanged DNS-amino acid after 18 hours of hydrolysis (6.1 *N* HCl, 105°).
[b] DNS-OH: 1-dimethylaminonaphthalene-5-sulfonic acid.
[c] $DNS-NH_2$: 1-dimethylaminonaphthalene-5-sulfonamide.

hydrolysis, so that good recoveries of dansyl amino acids are obtained after hydrolyzing labeled peptides. DNS-Pro, having a secondary sulfonamide link is more susceptible to acid cleavage, although much less so than the corresponding DNP derivative.[9] Elimination of $DNS-NH_2$ is the major route of breakdown of DNS-Ser, rather than hydrolysis of the sulfonamide link; the same is likely to be true of DNS-Thr and DNS-Cys. DNS-Trp is usually destroyed completely by the hydrolysis conditions used; it can sometimes be recovered by incubating the labeled peptide with chymotrypsin or carboxypeptidase.

Table II lists the recoveries of several DNS-amino acids after hydrolysis by constant boiling HCl (approximately 6.1 *N*) at 105° for 18 hours. These figures were obtained from a series of timed hydrolyses, using free DNS-amino acids. Gros and Labouesse[7] gave markedly higher rates of breakdown for derivatives hydrolyzed at 110°.

The derivatives of *S*-aminoethylcysteine are especially prone to breakdown during hydrolysis. The principal unique product is DNS-taurine, formed from the side-chain label. Breakdown occurs whether the aminoethylcysteine is terminal or internal in the peptide, so its presence is not diagnostic. Since the breakdown probably proceeds via elimination of the side chain, it would be expected that the group attached to the α-amino group would end up as $DNS-NH_2$. It is unclear whether exclusion of oxygen from the hydrolysis tube would prevent the breakdown, or merely block the subsequent oxidation to DNS-taurine.

The derivatives of methionine, cystine, and *S*-carboxymethylcysteine become partly oxidized during hydrolysis, leading to variable amounts

[9] R. R. Porter and F. Sanger, *Biochem. J.* **42**, 287 (1948).

of methionine sulfoxide and sulfone and of cysteic acid. DNS-Methionine sulfone and DNS-cysteic acid are the most stable forms of the derivatives of these amino acids, and performic acid-oxidized peptides are the most convenient form to work with. Alternatively, the oxidation can be carried out on the DNS-amino acids, provided no chloride is present.

Recommended Procedure for Peptides

Reagents, Buffers, and Glassware

Dansyl chloride solution, 5 mg/ml in acetone (analytical grade used without further purification). The solution is stable for several weeks if kept in a vial with a tight-fitting polyethylene closure. Keeping the solution at room temperature, rather than in a refrigerator, minimizes absorption of water from the atmosphere upon opening. Most preparations of DNS-Cl contain small amounts of DNS-OH, which is very insoluble in acetone; it is of no consequence and need not be removed. If it is felt desirable to check the actual concentration of DNS-Cl, this may be done by measuring the extinction at 369 nm ($E = 3690$), after dilution.

Sodium bicarbonate solution, 0.2 *M* in ammonia-free water; it may be necessary to use distilled water which has been passed through a mixed-bed ion-exchange resin.

Hydrochloric acid, approximately 6.1 *N*. Analytical grade concentrated HCl, diluted with water (1:1 by volume) and distilled under nitrogen at atmospheric pressure: "constant-boiling HCl."

Glassware. Pyrex glass test tubes, 4 mm i.d. × 50 mm long, are cleaned by heating overnight in a glassblower's oven at 500°. This is much simpler and more reliable than washing. They are stored in a clean, dust-free container.

Labeling of Peptide. A solution of the peptide (0.5–5 nmoles) is transferred to a small glass test tube with a fine-tipped micropipette. After drying *in vacuo,* the peptide is redissolved in 15 μl of 0.2 *M* sodium bicarbonate solution, and dried a second time. This step helps to ensure removal of traces of ammonia that may have been present in the sample. The peptide is then dissolved in 15 μl of deionized water, and approximately 0.1 μl of the solution is spotted onto pH indicator paper. The pH at this point should be 8.5–9; if it is below 7.5, more base must be added, or labeling will be inefficient. It is important that the pH be checked *before* addition of the dansyl chloride solution, since the acetone represses ionization of the bicarbonate and of some indicator dyes. When the pH is correct, 15 μl of dansyl chloride solution is added; an effect of repressing ionization of the bicarbonate is to raise the apparent pH (measured with a glass electrode) from about 8.5 to nearly 10. When working with small volumes of reagents, it is essential to make sure that liquids are kept in the bottom of the test tubes and that thorough mixing takes place. Final concentration of DNS-Cl is approximately 10 m*M*. The tubes are then covered with Parafilm, and the reaction is allowed to proceed for 1 hour at 37° or 2 hours at room temperature. During this time, most of the excess reagent is hydrolyzed to DNS-OH, and the solution becomes colorless or pale yellow. Any dansyl chloride that remains is subsequently broken down during the acid hydrolysis; no labeling of newly exposed amino groups occurs under these conditions.

Hydrolysis. When the reaction is complete, the solutions are dried *in vacuo,* and 100 μl of 6.1 *N* HCl is added to each tube. The tubes are sealed as follows. A small glass rod is fused gently to the rim at the open end of the tube; the neck of the tube is then drawn out after heating in a fine oxygen-gas flame at a point as close as convenient to the open end of the tube. The drawn neck should not be sealed immediately while the glass is very hot, or losses due to oxidation may occur. Instead, it should be cooled by gently blowing on it, and then sealed by quickly heating the narrow neck of the tube. The presence of air in the tube is of little consequence during the hydrolysis, and there is no need to degas, or to seal under nitrogen or vacuum.

Hydrolysis is carried out for 6–18 hours at 105°, using any convenient method of heating, such as an oven, metal block, etc. As a routine, we use an overnight hydrolysis of about 16 hours. In this time, almost all peptide bonds are broken, and there is an adequate recovery of DNS-amino acids. When there is reason to suspect DNS-Pro (23% recovery), a shorter time is used, but there is less complete hydrolysis of peptide bonds.

Peptides containing amino-terminal valine or isoleucine are much more resistant to acid hydrolysis than others and give rise to a mixture of DNS-amino acid plus DNS-dipeptide. The amount of dipeptide is negligible when serine or threonine are the penultimate residue, but increases through the series glycine, alanine, aspartic acid, glutamic acid, leucine, and lysine. With valine and isoleucine, the dipeptide is the dominant product, and little free DNS-amino acid is released, except after prolonged hydrolysis. The DNS-dipeptides are readily distinguished from DNS-amino acids by their lower electrophoretic mobilities at pH 1.6 (see later). It is not possible to distinguish DNS-Ile·Val from DNS-Val·Ile in this way, however, and one may have to carry out one round of sequential degradation, plus dansylation, to do so (see this volume [26a]).

At the end of the hydrolysis period, the tubes are opened, and the HCl is removed *in vacuo* over NaOH pellets.

Recommended Procedure for Proteins

The very simple procedure recommended for peptides is usually unsuitable for proteins, mainly because of a need to use much more dansyl chloride, and larger volumes. This arises from several causes: reaction with the terminal amino group may be much slower because of steric effects; there are many lysine and tyrosine side chains to react simultaneously; it is often difficult or impossible to dissolve the required amount of protein in 15 μl of sodium bicarbonate solution (5 nmoles of protein per 15 μl may be equivalent to 15 mg/ml). In addition, many proteins tend to precipitate in the acetone–bicarbonate mixture, leading to low labeling efficiency. The earlier procedure[2,7] using 0.5 ml of 8 M urea–sodium bicarbonate was rather cumbersome and often ineffective. The following procedure has given good results with a wide variety of proteins, even some that remained insoluble during the labeling process.

Reagents, Buffers, Glassware

Sodium dodecyl sulfate (SDS), 1% aqueous solution (w/v)

N-Ethylmorpholine, Sequanal grade, Pierce Chemical Co.

Dansyl chloride solution: 25 mg/ml in anhydrous dimethyl formamide (Pierce Chemical Co., listed as specially purified for silylation reactions). This solution must be made up freshly for each use, since it rapidly absorbs moisture and the reagent hydrolyzes.

Glassware. Pyrex glass test tubes, 4 mm i.d. × 50 mm long, are cleaned by heating in a glassblower's oven at 500° overnight. They are then stored in a clean, dust-free container.

Labeling of Protein. Approximately 50–250 μg of performic acid-oxidized protein is placed in a small test tube, and 50 μl of 1% SDS is added. The protein is then dissolved by heating the mixture in a boiling water bath for 2–5 minutes; this ensures thorough unfolding of the peptide chains. After the solution has cooled, 50 μl of *N*-ethylmorpholine is added, and mixed thoroughly. *N*-Ethylmorpholine acts both as a base and a detergent, and seems to assist reagent to penetrate into insoluble protein particles. Dansyl chloride is then added and mixed thoroughly (75 μl; 25 mg/ml). We have never observed a protein to precipitate during this procedure, despite the fact that the final mixture contains only about 30% water. Reaction is allowed to proceed for 1 hour at room temperature, or longer, if convenient.

Hydrolysis. The labeled protein is precipitated by the addition of approximately 0.5 ml of acetone to the reaction mixture. Because of viscosity and density differences, it is best to mix the solutions by inverting the tube, several times, after covering with Parafilm. The protein usually precipitates as fine floccules that can be formed into a small pellet by a few minutes' centrifugation in a bench-top centrifuge. It sometimes takes several hours for a suitable precipitate to form, especially when very small amounts of protein are used. If larger reaction volumes have been used for any reason, it is best to evaporate almost to dryness, add 50 μl of water, and then precipitate with acetone. When the pellet has been compressed, it is washed once with 500 μl of 80% acetone, centrifuged down again, and dried. It is then hydrolyzed as described for peptides, and the dansyl amino acids are identified by electrophoresis or chromatography.

Identification of Dansyl Amino Acids

High-voltage electrophoresis[1,2] and chromatography on polyamide layers,[4] are the most widely used methods for analyzing DNS-amino acids. Despite a high capital outlay, the electrophoretic system is remarkably cheap to operate. Chromatography on polyamide layers is especially useful when working with very small amounts of material, and when only occasional analyses are needed. Because the apparatus required is so simple and widely available, it has become the method of choice in many laboratories.

Marker Mixtures of DNS-Amino Acids. These can be prepared either from the solid compounds or by labeling mixtures of amino acids as follows. Dissolve or suspend the required amino acids in 0.5 *M* $NaHCO_3$ to give a final concentration of 10 m*M* of each. To 0.1 ml of this solution add 0.1 ml of a solution of DNS-Cl in acetone, strong enough to give an excess of 2 μmoles over all reactive groups (e.g., for a mixture of 10 amino acids, concentration would be 12 μmoles/0.1 ml, or approximately 34 mg/ml). When reaction is complete, dilute to 1 ml with 10% formic acid, and store in a stoppered bottle. Samples of the single-labeled derivatives of lysine and tyrosine must be prepared separately by labeling 2-fold excess of amino acid with dansyl chloride. Formic acid (10%, v/v) is a convenient solvent for DNS-amino acids, since there is little tendency for them to oxidize; when marker mixtures are made from the solid derivatives, however, some are slow in dissolving. Final concentration of DNS-amino acids should be about 0.5 m*M*, so that 2 μl is equivalent to 1 nmole.

High-Voltage Electrophoresis at pH 4.4. Cleanest results are obtained by first extracting the DNS-amino acids from the hydrolyzate. For this purpose, we have found ethyl acetate to be very satisfactory, provided it is saturated with water. The hydrolyzate, after removal of HCl, is dried as a film covering about the bottom 5–10 mm of the inside of the glass tube. Most of the material present in peptide hydrolyzates is a mixture of sodium chloride and DNS-OH, with small amounts of amino acids, DNS-amino acids, and DNS-NH_2. Dry ethyl acetate does not penetrate the salt sufficiently to extract out the DNS-amino acids. The top phase of an ethyl acetate/water mixture, however, is wet enough to convert the salt into a thin film of solution from which the DNS-amino acids extract readily. Two extractions with 100 μl of this wet ethyl acetate removes the neutral DNS-amino acids almost completely, DNS-Asp and DNS-Glu approximately 80%, moderate amounts of α or ϵ-DNS-Lys and *O*-DNS-Tyr, and only traces of DNS-His, DNS-Arg, DNS-$CySO_3H$, and DNS-OH. It is the relative absence of salt and DNS-OH which is important in improving the appearance of the electrophoretic separation.

The extracts can be dried either by blowing a gentle stream of N_2 into the tubes or by immersing the tubes in water maintained at about 80°. Evaporation is quite rapid by this latter procedure, and we have never had problems with the liquid bumping out of the tubes; 10–15 minutes is normally adequate for drying. For spotting onto the electrophoresis paper, the DNS-amino acids are redissolved in 20 μl of 50% pyridine (v/v), in which they are all readily soluble.

Paper dimensions will naturally vary with the size of the available apparatus. On standard sheets of chromatography paper (46 cm $\times$ 57 cm), the origin line should be placed about 24 cm from the cathode end. It is best to place spots no closer together than 1.5 cm, to have at least 2.5 cm at the edges of the paper, and to have at least one standard mixture next to each unknown. A suitable arrangement would thus be to have a spot of marker mixture A (2 nmoles) 2.5 cm from the edge, followed by two unknowns, marker mixture B, two more unknowns, marker mixture A, etc., across the width of the paper. Best results are obtained with a fairly thick paper, such as Whatman No. 3 MM or No. 3.

Papers are wetted by spreading buffer from a pipette and allowing it to flow in evenly to the origin from both sides. Some people find it advantageous to support the origin line on a glass rod. As the buffer flows through the paper by capillary movement, it concentrates the DNS-amino acids into a sharp band at the origin. Patience is needed to ensure a good wetting. A buffer loading of approximately 25–30 ml/1000 cm^2 gives best results; the paper should be just wet enough to stick slightly to the polyethylene insulation. Any excess buffer is removed by blotting with a clean sheet of absorbent paper.

The exact pH of the buffer to be used should be determined by trial and error around pH 4.40; it varies somewhat with the electrophoresis apparatus and running temperature. The buffer stock for electrode vessels is made from pyridine/acetic acid/water (10:20:2500 by volume); a separate stock is adjusted carefully for pH and kept for wetting papers. Electrophoresis is carried out for 2.5 hours at 80 V/cm, or the equivalent in a flat-plate apparatus; machines vary in the amount of voltage drop across the wicks and electrode vessels, and suitable running times must be found for each. The current taken at 80 V/cm is approximately 4–4.5 mA/cm width of paper. A number of machines incorporate an inflatable bag to press the paper tightly against the cooling plate. This is a real advantage. A pressure of 4 psi serves both to ensure good thermal contact, and to prevent inflow of buffer from the reservoirs to the paper. When conditions are properly established, the DNS-amino acids migrate as circular spots about 1 cm in diameter, and the 50-cm migration path allows for excellent resolution (Fig. 2). Factors that adversely affect the running and resolution include using too weak a buffer or having the paper too wet or too dry.

There is sometimes a slight unevenness in electrophoretic mobility in different parts of the paper, due to uneven wetting at the origin or to slight drying of the paper near the edges. This latter effect is particularly noticeable on machines which lack an adequate pressurizing system. Each unknown sample contains a small amount of DNS-OH, and this, having a distinct blue fluorescence, acts as an excellent internal mobility marker. For this reason, it is suggested that DNS-OH be included in both standard mixtures A and B (see Fig. 2).

Many individual DNS-amino acids can be identified by this single electrophoresis at pH 4.4. Under optimum conditions, it is possible to make tentative assignments, even among DNS-Ser, DNS-Pro, and DNS-Ala, by measuring back from the DNS-OH reference point. However, it is always preferable to confirm assignments in this group and in the group DNS-Val, DNS-Ile, and DNS-Phe by rerunning at a second pH (see later). With the latter group, the existence of stable DNS-dipeptides may make assignments easier or harder, depending where the "extra" derivatives run. A number of these are shown in Fig. 2; the blue-fluorescing by-products are not present in the extracts, but are sometimes found when the entire hydrolyzate is electrophoresed.

DNS-Cysteic acid and DNS-carboxymethylcysteine would migrate too far under the electrophoresis conditions just described. The bulk of these derivatives, however, is not present in the extracted material, but is in the residues. These are best examined by electrophoresis at low pH.

High-Voltage Electrophoresis at Low pH. Below pH 2, there is little

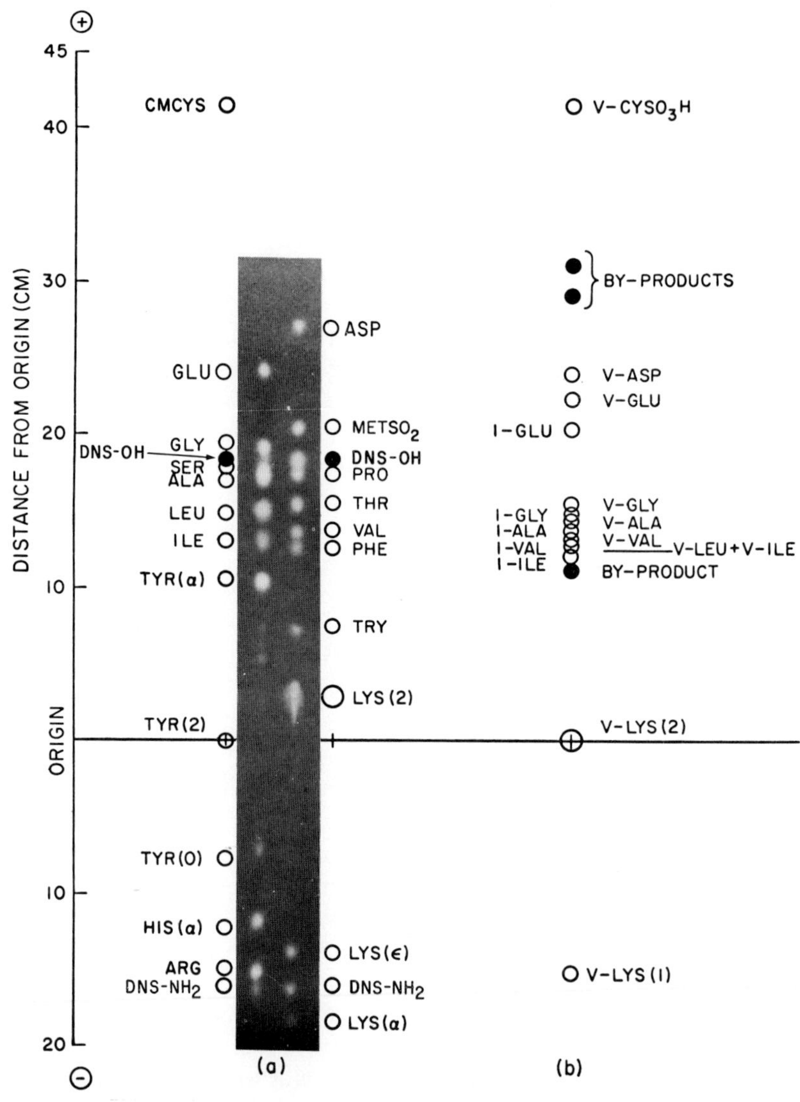

FIG. 2. Electrophoretic mobility of various dansyl derivatives at pH 4.38 (80 V/cm, 2.5 hours, 15°). Abbreviation system used: TYR (α) = α-DNS-Tyr; V-CYSO$_3$H = DNS-Val-CySO$_3$H; I-GLU = DNS-Ile-Glu, etc. Filled circles represent compounds having a blue fluorescence. (a) A suggested complementary pair of DNS-amino acid marker mixtures. DNS-CySO$_3$H moves about 55 cm in 2.5 hours. (b) Some stable DNS-dipeptides of valine and isoleucine which may be encountered in hydrolyzates.

change in electrophoretic mobility with change in pH. All dansyl compounds carry a full positive charge due to protonation of the dimethylamino group, while carboxyl groups are almost all electrically neutral. Only the carboxyl groups of ε-DNS-Lys and *O*-DNS-Tyr are significantly ionized. Full negative charges are retained only by the sulfonic acids DNS-OH and DNS-cysteic acid. A few groups contribute extra positive charges in certain derivatives. A suitable buffer for use in this system is 8% (v/v) formic acid (pH 1.6). The separation at this pH is based largely on mass differences, since most derivatives have a net charge of plus one.

Protonation of the dimethylamino groups leads to a marked increase in solubility of the DNS-amino acids, so that sharper bands are obtained than at pH 4.4. Even bis-DNS-Tyr runs as a compact spot.

Electrophoresis at this pH is useful in several different situations; in all these, the origin is placed as close to the anode end of the paper as is convenient.

Hydrolysis residues. After the hydrolyzate has been extracted with ethyl acetate, certain DNS-amino acids may still be in the film of aqueous phase, along with salt, free amino acids, and DNS-OH. Examination of these residues is made by electrophoresis at pH 1.6 for about 1 hour at 50 V/cm. Mobilities of the basic DNS-amino acids and DNS-OH are shown in Fig. 3. Being neutral, DNS-OH remains as a compact band despite the large amounts present, and the basic DNS-amino acids run well clear. DNS-Cysteic acid has a slightly negative mobility relative to DNS-OH and can usually be identified as a yellow-fluorescent protrusion toward the anode side of the DNS-OH spot.

Rerunning of selected regions of pH 4.4 electropherogram. Clarification of uncertain assignments from the pH 4.4 electrophoresis can be obtained by cutting out the appropriate areas of the paper, stitching to a second sheet, and rerunning at pH 1.6. It is best to include only the marker DNS-amino acids pertinent to the rerunning of the particular part of the pH 4.4 separation; e.g., DNS-Gly, DNS-Ala, DNS-Pro, and DNS-Ser. This is done by outlining the area to be rerun, while visualizing the spots with an ultraviolet (UV) lamp (see later): in this way, one can easily be sure to include the appropriate spots from the marker mixtures used at pH 4.4. It is helpful to add a small amount (1 nmole) of DNS-NH_2 to each spot to act as an internal mobility marker, much as DNS-OH is useful at pH 4.4. The paper to be rerun is then stitched to another sheet of paper of the appropriate width, using a sewing machine. Best results are obtained if this second sheet is of the same thickness as the paper to be rerun (e.g., Whatman No. 3 MM), and extends 2 cm beyond the edges of the small piece. The area of paper behind the stitched-on piece is then cut out carefully with fine scissors. Failure to

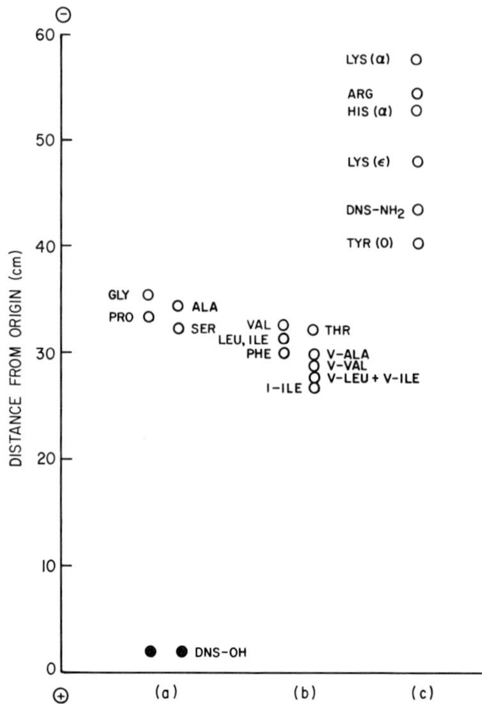

FIG. 3. Electrophoretic mobilities of dansyl derivatives at pH 1.9 (50 V/cm, 2 hours, 20°). Same scale and abbreviations as in Fig. 1. (a) DNS-amino acids which run near DNS-OH at pH 4.4. (b) Certain other DNS-amino acids and peptides. (c) Basic DNS-amino acids.

do this results in excessive broadening of the zones as they move from a low voltage gradient (double thickness of paper) to a higher gradient (single thickness). The paper is wetted with buffer at a loading of about 30 ml/1000 cm^2, care being taken to ensure a slow steady movement of the buffer from the edges of the stitched-on piece. This ensures that all DNS-amino acids become washed into a sharp zone in the center of the strip. Electrophoresis is carried out for about 2 hours at 40 V/cm, or the equivalent; longer times can be used to better separate some of the slower-moving compounds. Samples migrate smoothly across the stitched area if the paper is pressed firmly against the cooling plate by a flexible bag inflated to 4 or 5 psi.

Under these conditions, the DNS-amino acids which run near DNS-OH at pH 4.4 (DNS-Gly, DNS-Ser, DNS-Pro, DNS-Ala) are better resolved, and by using the internal marker of DNS-NH_2, it is usually possible to make a clear-cut assignment. The hydrophobic group,

DNS-Val, DNS-Ile, and DNS-Phe are completely resolved, running in that order; DNS-Leu has the same mobility as DNS-Ile, but is easily identified at pH 4.4. DNS-dipeptides, having higher masses than the DNS-amino acids, run more slowly, and several can be tentatively identified when the hydrophobic DNS-amino acids are rerun. Bis-DNS-Tyr is very insoluble at pH 4.4 and remains at the point of application. Because there are often small amounts of fluorescent materials at the origin (usually orange, rather than the whitish green of bis-DNS-Tyr) it is best to rerun this area at pH 1.6 if tyrosine is suspected.

Thin-Layer Chromatography on Polyamide Sheets. Polyamide layers, first exploited for DNS-amino acids by Woods and Wang,[4] have proved to be a very popular medium for thin-layer chromatography of these derivatives. The apparatus required is inexpensive and compact, and many laboratories are equipped for this procedure. Prior to the use of polyamide layers, several systems were described for thin-layer chromatography of DNS-amino acids on silica gel or alumina.[10] A problem with these media was a rapid destruction of the derivatives, so that plates had to be examined and recorded immediately; on paper, by contrast, the derivatives are stable for many weeks. They are also much more stable on the polyamide layer than on silica, and this is a significant advantage of this medium.

Although the apparatus required is inexpensive, the polyamide thin-layer sheets are not (approximately $1.50 for a 20 × 20-cm sheet). High running costs would be encountered in sequencing work if each end-group analysis required exclusive use of a single sheet, since many hundreds of samples are analyzed in the course of a single protein sequence determination. Running costs can be greatly reduced, however, by using smaller sheets (e.g., 10 × 10-cm), and by reusing the sheets after they have been washed in a suitable solvent.[11] Dansyl amino acids can be removed by soaking the used layer in acetone–90% formic acid (9:1, v/v) for 6 hours, followed by three washes in methanol, the polyamide sheets then being hung up to dry in air. The layers should not be exposed to high temperatures (more than 70°) for more than a few seconds, or damage will result. With adequate care in handling, the layers may be used six or more times. For best results, the washing should be carried out as soon as possible after development, or the compounds may become

[10] N. Seiler and J. Wiechmann, *Experientia* **20,** 559 (1964); Z. Deyl and J. Rosmus, *J. Chromatogr.* **20,** 514 (1965); M. Cole, J. C. Fletcher, and A. Robson, *J. Chromatogr.* **20,** 616 (1965); D. Morse and B. L. Horecker, *Anal. Biochem.* **14,** 429 (1966); B. Mesrob and V. Holeyšovský, *J. Chromatogr.* **21,** 130 (1966); G. Pataki, *Chromatogr. Rev.* **9,** 23 (1967).

[11] K.-T. Wang and P.-H. Wu, *J. Chromatogr.* **37,** 353 (1968).

irreversibly sorbed to the layer. Quite frequently it is possible to identify end groups on the basis of their mobilities on two one-dimensional chromatograms, rather than one two-dimensional one. Where this can be done, the samples can be run next to markers, with many samples on one sheet, as is done with the electrophoretic systems.

The solvent system most commonly used is based on that described by Woods and Wang,[4] with a rerun in the second dimension using a third solvent system. Samples are spotted about 2 cm from each edge in one corner of the sheet, 10 μl of either acetone–acetic acid (3:2, v/v)[4] or pyridine-water (1:1, v/v),[12] being used.

FIRST SOLVENT. Water–90% formic acid (200:3, v/v).[4] Development is allowed to proceed for 50 minutes, the solvent front moving approximately 15 cm. The sheet is dried, turned through 90°, and run in the second solvent.

SECOND SOLVENT. Benzene–acetic acid (9:1, v/v).[4] Develop for 1 hour. Dry.

At this stage, the separation is as shown in Fig. 4a. The sheet should be examined under a UV lamp (see later), and DNS-amino acids identified where possible. Particular attention should be paid to DNS-Leu/DNS-Ile, for the separation achieved at this stage may later be obscured by rerunning. The main ambiguities in the separation are: DNS-Thr/DNS-Ser; DNS-Asp/DNS-Glu, partly overlapping DNS-OH; DNS-cysteic acid/DNS-OH; and the basic DNS-amino acids. These can be resolved by additional runs with other solvents in the same direction as solvent 2.

THIRD SOLVENT. Ethyl acetate–methanol–acetic acid (20:1:1, by volume).[13,14] Run for 1 hour. Separation now appears as in Fig. 4b. DNS-Thr and DNS-Ser are resolved, as are DNS-Glu and DNS-Asp. There is no significant improvement in DNS-cysteic acid/DNS-OH, or the basic DNS-amino acids.

FOURTH SOLVENT. 1 *M* Ammonia–ethanol (1:1, v/v).[12] Resolves α-DNS-His, DNS-Arg, and ϵ-DNS-Lys.

When smaller polyamide sheets are used, the initial spot size and development times should be reduced.

Visualization of Derivatives

Many substances greatly quench the fluorescence of dansyl amino acids, including water, acids, and pyridine. Quantum yields of fluores-

[12] B. S. Hartley, *Biochem. J.* **119,** 805 (1970).

[13] K. Crowshaw, S. J. Jessup, and P. W. Ramwell, *Biochem. J.* **103,** 79 (1967).

[14] S. Magnusson, quoted by C P. Milstein and A. Feinstein, *Biochem. J.* **107,** 559 (1968).

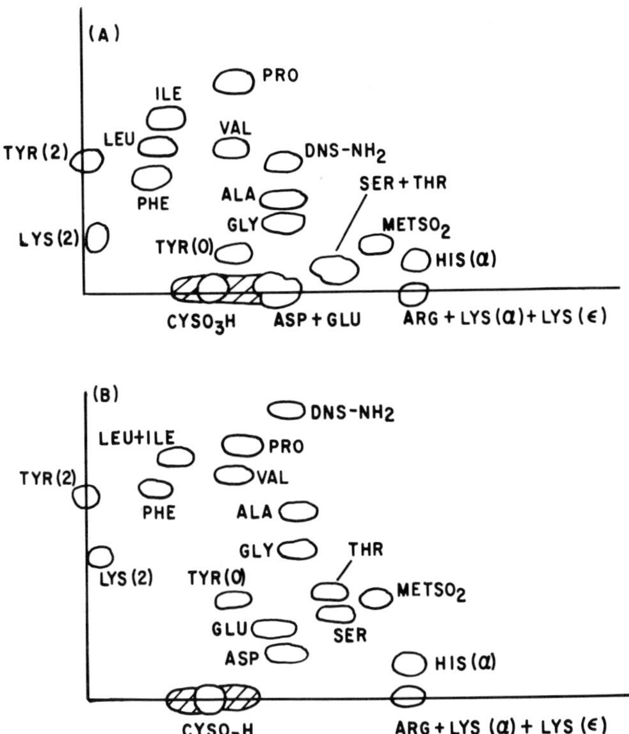

FIG. 4. Thin-layer chromatography of DNS-amino acids on polyamide sheets. (A) Chromatography in horizontal direction for 50 minutes, first solvent; followed by 1 hour in vertical direction, second solvent. (B) Same as (A), followed by re-running in vertical direction, third solvent. Abbreviations as in Fig. 1, solvent compositions in text. Hatched area is DNS-OH.

cence are thus highly dependent upon the solvent. In acids, where the dimethylamino group is protonated, the yield is essentially zero (<0.0001); in water, approximately 0.05; in 80% acetone, approximately 0.3; thoroughly dried on paper chromatograms, approximately 0.4. Correlated with changes in quantum yield are changes in fluorescence color from orange-yellow ($Q < 0.2$) to bright blue ($Q > 0.7$ approximately).

Since the buffers used for electrophoresis are volatile, they can be removed completely by heating the papers in an oven at 80–110°. After running at pH 1.6, some enhancement of fluorescence may be achieved by exposing the dried paper to ammonia fumes and then drying again. Immediately after drying, the spots have an intense yellow-green fluorescence which changes to a yellow-orange color as the paper picks up moisture from the air. Chromatograms may be stored for many weeks, preferably in large polyethylene bags or in aluminum foil; redrying of the paper is necessary for bringing up the intensity of many weaker spots.

The polyamide layers cannot be dried at high temperatures without damage. They should be left to air dry, or be dried in a stream of cold air from a blower. Since the solvents used are acidic, it may be helpful to expose layers to ammonia fumes. Although the dansyl amino acids are quite stable on the polyamide layers, they are usually washed off soon after development so that the layer can be reused.

Fluorescence can be excited by either longwave (366 nm) or shortwave (254 nm) UV light. The most convenient source is a "black-ray" fluorescent lamp, as commonly used for decorative lighting effects. These lamps have a long light source, giving even illumination of large areas, and the longwave light is much less harmful to the eyes than the shortwave. Even so, it is best to wear protective glasses or goggles; cheap ski goggles with yellow lenses are excellent. A quick test should be made to see whether they are suitable, by placing them immediately over a dansyl spot on a chromatogram. Fluorescence of the spot should be largely abolished as a result of absorption of the exciting UV light by the lens.

The limits of detection by eye depend greatly upon the compactness of the spot on the chromatogram. With the electrophoresis systems, this limit is between 10^{-5} and 10^{-4} μmole of DNS-amino acid; the polyamide system is more sensitive even than this. In practice, it is difficult to carry out end-group analysis on less than 0.1 nmole, and the usual working range is 1–5 nmoles. Bruton and Hartley,[15] by scaling down the volumes of peptide and reagent 10-fold, were able to carry out sequential degradation plus dansylation on less than 1 nmole of peptide.

[15] C. J. Bruton and B. S. Hartley, *J. Mol. Biol.* **52,** 165 (1970).

[9] Hydrazinolysis

By WALTER A. SCHROEDER

Although the hydrazinolytic procedure of Akabori *et al.*[1] for the determination of COOH-terminal amino acids is attractively simple in principle, its application often has been somewhat unsatisfactory. Quantitative use of the method has had limited success because of large and uncertain correction factors. The quantitativeness of the method has no

[1] S. Akabori, K. Ohno, and K. Narita, *Bull. Chem. Soc. Jap.* **25,** 214 (1952).

doubt suffered because of rather complicated methods for isolating and estimating the free amino acid. This fact, in turn, has made it difficult to study what conditions are necessary for achieving quantitative yields. The procedure to be described has given yields that in several cases approach the theoretical without the use of correction factors.[2]

The procedure carries out the hydrazinolysis in the presence of a catalyst to improve the yield. The free COOH-terminal amino acid(s) is then separated from the hydrazides by chromatography and is determined qualitatively and quantitatively with an amino acid analyzer.

Materials

Hydrazine. The quality of hydrazine is the critical factor in this procedure. Unfortunately, no criterion of quality is known except the satisfactory outcome of its use.

The following is suggested as a generally successful means of obtaining satisfactory hydrazine. The best available commercial grade of "anhydrous" hydrazine (95%+) is distilled from sodium hydroxide under nitrogen. Thus, 100 ml of hydrazine and 40 g of sodium hydroxide pellets are placed in a 250-ml round-bottom flask of a conventional glass distilling apparatus, and the whole apparatus is flushed with dry N_2 that has been passed through concentrated sulfuric acid. Under nitrogen at atmospheric pressure, the hydrazine is heated to the boiling point and then cooled to room temperature. A sturdy nonflexible capillary with a very small outlet is then inserted into the distilling flask. Dry nitrogen is passed through the capillary during distillation under oil pump vacuum. The hydrazine distills at 16–18° and is condensed at the temperature of Dry Ice. The product may be preserved in a stoppered flask in a desiccator over P_2O_5 at 2°. It can probably be preserved more satisfactorily by sealing in ampoules (5 ml per ampoule).

Many investigators are reluctant to distill hydrazine. Successful results have been obtained by using hydrazine directly from *previously unopened* bottles of hydrazine (anhydrous, 97+%; Matheson, Coleman, and Bell).[3] Unsuccessful results have been reported with hydrazine taken from bottles of old and uncertain history.

Other Materials

Amberlite CG-50. This resin of size less than 200 mesh (Fisher Scientific Co., Fair Lawn, New Jersey) is washed successively on a Büchner funnel with 1 *N* sodium hydroxide, water, 2 *N* hydrochloric acid, and then water until free of chloride. After drying at

[2] V. Braun and W. A. Schroeder, *Arch. Biochem. Biophys.* **118**, 241 (1967).

[3] Hydrazine from other suppliers has not been tested: it may be equally satisfactory. Hydrazine from a given supplier does not always have the same quality.

80° in an oven, it is stored under vacuum in a desiccator over P_2O_5.

8% Cross-linked Sulfonated Polystyrene Resin. Amberlite IR-120 (Type 150A, Beckman Instruments, Inc., Palo Alto, California) or equivalent is purified, if necessary, and then prepared as described for Dowex 50-X2 resin in article [15] by recycling with acid and base and finally suspending in pH 3.1 pyridine–acetic acid buffer as there described. After chromatographic use of this resin as described below, it is removed from the column and regenerated by washing with water, boiling with 0.5 *N* sodium hydroxide for 1 hour, filtering, washing with water and acid, and finally equilibrating with pH 3.1 buffer.

Phosphocellulose. Whatman phosphocellulose powder P-70 (unsifted) is washed successively on a Büchner funnel with 0.5 *N* sodium hydroxide, water, 1 *N* hydrochloric acid, water, and finally 0.4 *M* pyridine–formic acid buffer at pH 3.2 before it is suspended in the latter. This material is not regenerated.

Solutions

Buffer pH 3.1 (0.2 *M* in pyridine). The volumes of reagents are given in article [15]

Buffer pH 5.2 (0.2 *M* in pyridine): 16.1 ml of pyridine and 8.5 ml of glacial acetic acid diluted to 1 liter

Buffer pH 3.2 (0.4 *M* in pyridine): 32.2 ml of pyridine and 47 ml 98% formic acid diluted to 1 liter

Apparatus

Chromatographic tubes: A jacketed tube in which a 1 × 18-cm column may be poured and an unjacketed tube for a 1 × 30-cm column.

Pump: A constant-volume pump, such as a Milton Roy chromatographic minipump (Milton Roy Co., Philadelphia, Pennsylvania) or a Beckman Accu-Flo pump (Beckman Instruments, Inc., Spinco Division, Palo Alto, California), is necessary.

Fraction collector and test tubes

Constant-temperature bath

Oven

Amino acid analyzer

Hydrazinolysis

A weighed amount of protein (about 0.25 μmole usually) and about 50 mg of dry Amberlite CG-50 are placed in a 15 × 125-mm Pyrex tube, which is then drawn down to a tiny neck.[4] A 2-ml portion of anhydrous hydrazine is placed in the tube and drawn through the neck onto the sample by cooling the lower part in a bath of alcohol and solid CO_2.[5] The reaction mixture is frozen; then the tube is sealed under water aspirator vacuum.

The reaction mixture is heated in an oven at 80° for 10–100 hours. The sample should be shaken gently three or four times each day during the heating. The sample may be agitated gently throughout the heating by taping the tube with axis parallel to a shaft that is inclined at 45 degrees and rotated at 40–60 rpm. If the reaction mixture is not examined immediately after heating, it may be stored in the unopened tube in a freezer.

Finally, the reaction mixture is transferred to a 25-ml round-bottom flask, and the tube is rinsed twice with 1 ml of undistilled hydrazine. Absorption of moisture must be avoided during this transfer and rinsing. The reaction mixture is then frozen in an alcohol–solid CO_2 bath, and hydrazine is removed by lyophilization.[6]

Chromatographic Separation of Free Amino Acids and Hydrazides

In preparation for the separation, the residue of CG-50, the hydrazides, and the free amino acids is suspended in 3 ml of water and the resin is centrifuged and washed twice with 1 ml of water. The pH of the solution is then lowered to 1.5–2 with 2 N hydrochloric acid.

Chromatography on Sulfonated Polystyrene Resin. A 1×18-cm column of sulfonated polystyrene resin is poured and equilibrated with pH 3.1 buffer at 30°. The sample, prepared as above, is applied, and the chromatogram is developed with pH 3.1 buffer at 30 ml per hour. Fractions of appropriate size are collected. After 120 ml of this buffer has been used, a change is made to pH 5.2 buffer. Under these conditions, all neutral and acidic amino acids except tyrosine emerge between 15 and 60 ml of effluent volume, tyrosine between 108 and 120 ml, and lysine between 220 and 240 ml. Other amino acids and the hydrazides are strongly fixed.

Chromatography on Phosphocellulose. A 1×30-cm column of phos-

[4] In order to ensure completely anhydrous conditions, it is probably desirable at this point to redry resin and protein under vacuum at 50° for 24 hours.

[5] Absorption of CO_2 by the hydrazine must be prevented during this operation. If it is not, much ninhydrin-positive material will interfere with subsequent chromatography.

[6] Attempted removal of hydrazine in an evacuated desiccator over H_2SO_4 leads to bumping during the evacuation.

phocellulose is poured and equilibrated at room temperature with pH 3.2 buffer. After the sample has been applied, the chromatogram is developed with the pH 3.2 buffer at 30 ml per hour. Neutral and acidic amino acids appear between 20 and 30 ml of effluent volume; lysine, ammonia, and aspartic and glutamic acid monohydrazides between 42 and 75 ml; histidine between 80 and 90 ml; and arginine between 93 and 112 ml. Hydrazides are strongly fixed.

Assessment of Results. The volumes of effluent in which the amino acids emerge, as listed above, were determined by collecting small fractions and applying the ninhydrin procedure to detect those that contained the amino acids. Whether this information would be valid in another laboratory would have to be determined by the individual investigator. Once the correct volumes of emergence have been decided, appropriate portions of the effluent can be collected in large fractions.

Qualitative and Quantitative Determination of Free Amino Acids

After the appropriate portion of effluent has been evaporated to dryness, the type and quantity of the amino acids in the residue are determined by amino acid analysis on one of the commercially available machines. The monohydrazides of aspartic and glutamic acids must be hydrolyzed prior to determination as aspartic and glutamic acids.

Comments

Applicability. The rationale and supporting evidence behind this procedure may be found in the original paper.[2] Although proteins with all types of COOH-terminal amino acids were not available for testing, it seems probable that suitable conditions can be found for the determination of neutral and acidic amino acids (including cysteic acid) as well as histidine. Arginine is partly destroyed and also converted to ornithine. It is uncertain whether the procedure is satisfactory for lysine. It may not be applied with confidence to asparagine or glutamine. Because the COOH-terminal amino acids vary in stability as well as rate of release, the best time and temperature of hydrolysis must be determined individually for each specific case. Although results may be negative or yields somewhat less than theoretical, the procedure apparently does not give false-positive results.

Temperature. Although the effect of temperature is not known in detail, 80° appears to be a reasonable compromise. In a few tests at 100°, decreased yields resulted. A lower temperature probably would require exorbitantly long heating. Alteration of the temperature should be considered for each individual protein.

Reaction Time. The optimal reaction time depends upon the nature of the COOH-terminal amino acid and possibly also on the penultimate amino acid. Optimal times have varied between about 15 and 95 hours. In application to any protein, it is necessary, therefore, to vary the time over this range with a series of samples. On the basis of the results, it may be desirable to alter the temperature also.

Separation of Hydrazides and Free Amino Acids. If the procedure is being applied to a peptide or only occasionally to a protein, the solution of hydrazides and amino acids may be applied directly to the column of an amino acid analyzer. If the sample of protein did not exceed 4–5 mg, neutral and acidic amino acids may be determined with the "long" column, but the analysis for basic amino acids on the "short" column is unsatisfactory. Direct application of successive samples to the amino acid analyzer is not recommended because hydrazides are incompletely removed even if the amount of sodium hydroxide for regeneration is 5 times the usual amount. Incomplete removal becomes apparent in a raised baseline after the change to pH 4.25 buffer in the next analysis.

Either chromatographic procedure may be used equally well for separation. Phosphocellulose, however, has lesser capacity and the quantity should not exceed that from 10 mg of protein.

Acknowledgments

This procedure was devised by Dr. Volkmar Braun during the tenure of a NATO Fellowship. The investigation was partly supported by a grant (HE-02558) from the National Institutes of Health, U.S. Public Health Service. I appreciate discussions with Dr. Alex Glazer and Dr. John Pierce about their applications of the method with previously unopened bottles of hydrazine.

Contribution No. 4137 from the Division of Chemistry and Chemical Engineering, California Institute of Technology.

[10] Enzymatic Hydrolysis with Carboxypeptidases

By RICHARD P. AMBLER

Carboxypeptidases are enzymes that remove amino acids one residue at a time from the C termini of polypeptide chains. Three types of enzyme have been characterized which differ in the rate at which they release particular amino acids. Carboxypeptidases A (CPA) (from pancreas) release most rapidly amino acids with an aromatic or large aliphatic side chain (see the table), while carboxypeptidases B (CPB) (also from pancreas) release the basic amino acids lysine and arginine very much faster than any of the other common protein amino acids.

Approximate Relative Rates of Release[a] of Protein Amino Acids by Carboxypeptidase A

Rapid release	Tyr, Phe, Trp, Leu, Ile, Met, Thr, Gln, His, Ala, Val, homoserine, *S*-methyl cysteine[b]
Slow release	Asn, Ser, Lys,[c] $MetSO_2$
Very slow release	Gly, Asp, Glu, $CySO_3H$, *S*-carboxymethylcysteine
Not released	Pro, Arg, Hypro

[a] The presence of a "very slow" or "not released" amino acid as penultimate residue will generally decrease the rate of release of the C-terminal amino acid.

[b] H. Rochat, C. Rochat, F. Miranda, S. Lissitzky, and P. Edman, *Eur. J. Biochem.* **17,** 262 (1970).

[c] The rate of lysine release may be modified by changing the pH of digestion (see p. 147).

Carboxypeptidase C (CPC)[1] has been isolated from citrus fruit[2] and French beans,[3] and its activity was shown to be present in a wide range of plant material.[4] It can release proline as well as all the protein amino acids, but the release of glycine is very slow. The peptide bonds Gly-Pro, Pro-Gly, and Pro-Pro are resistant to hydrolysis.

When a carboxypeptidase acts on a protein, amino acids are removed until the residue exposed in the C-terminal position is one that is released at so slow a rate that hydrolysis is effectively ended. CPA will not release any proline or arginine at all, and glycine and the acidic amino acids are generally released only very slowly. In some proteins such a resistant residue may be the natural C terminus, while in others many amino acids will be released before the reaction is halted; since the frequency of occurrence of resistant residues is high, generally fewer than ten residues are released before reaction is at an end. In favorable circumstances the rate of appearance of the released amino acids during the digestion will give sufficient evidence for the C-terminal sequence to be deduced.

The most extensive review of the use of carboxypeptidase for end-group analysis of proteins and peptides is by Harris[5]; although now seventeen years old, it is not obsolete. The preparation and properties of CPA and CPB have been reviewed by Neurath.[6]

[1] The C of CPC originally referred to its occurrence in *c*itrus fruit.[2] The enzyme from French bean has been called phaseolain,[3] but carboxypeptidase C seems to be the most suitable trivial name for the whole group of enzymes.

[2] H. Zuber, *Hoppe-Seyler's Z. Physiol. Chem.* **349,** 1337 (1968).

[3] J. R. E. Wells, *Biochem. J.* **97,** 228 (1965).

[4] H. Zuber and P. Matile, *Z. Naturforsch. B* **23,** 663 (1968).

[5] H. Fraenkel-Conrat, J. I. Harris, and A. L. Levy, *Methods Biochem. Anal.* **2,** 339 (1955).

[6] H. Neurath, *in* "The Enzymes" (P. D. Boyer, H. Lardy, and K. Myrbäck, eds.), Vol. IV. p. 11. Academic Press, New York, 1960.

In a typical end-group determination, a sample of protein is incubated with CPA, CPB, or a mixture of CPA and CPB, and samples corresponding to a known amount of protein are withdrawn at known times. The different free amino acids in these portions are identified and quantitated, and the results are plotted against time as moles of each amino acid released per mole of protein (Figs. 1–5).

Preparation of Enzyme Solutions

CPA is generally prepared from ox pancreas by Anson's method.[7,8] Even after recrystallization the material is contaminated with other pancreatic proteases (especially chymotrypsin), and must be treated with diisopropyl phosphorofluoridate (DFP)[5] before being usable. Completely adequate DFP-treated enzyme is commercially available.

CPA is a euglobulin, insoluble at low ionic strengths and stable between pH 7 and 10. It is normally stored (and bought) as an aqueous suspension containing about 50 mg (w/v) of protein per milliliter, with toluene present as a preservative. In this state it is stable for years at 4°.

The protein crystals are dissolved immediately before use, after being washed free of contaminating amino acids.

Method (i) (adapted from Harris[5]). A suspension containing 1.25 mg of protein (normally 0.025 ml) is diluted with 1 ml of water in a 3-ml conical centrifuge tube, and centrifuged for 5 minutes at 2000 *g*. The supernatant is discarded, and the crystals are suspended in 0.1 ml of 1% (w/v) sodium bicarbonate and cooled in an ice bath. Sodium hydroxide 0.1 *N*[9] is added drop by drop with thorough mixing (e.g., on a Vortex-type mixer) until the protein has dissolved. Then the pH is brought back to 8–9 with 0.1 *N* HCl, and the volume is made up to 1.25 ml (i.e., 1 mg of protein per milliliter) with *N*-ethylmorpholine acetate (0.2 *M*, pH 8.5). If too much HCl is added and the pH drops below 8, the protein at once precipitates and should be discarded. The solution should be kept in ice and used as soon as possible.

Method (ii) (after Potts, Berger, Cooke, and Anfinsen[10]). The CPA crystals (1.25 mg) are washed as in method (*i*) and then stirred with 0.2 ml of 2 *M* ammonium bicarbonate. Most of the protein dissolves

[7] M. L. Anson, *J. Gen. Physiol.* **20,** 663 (1937).

[8] H. Neurath (1955), see Vol. 2, p. 77.

[9] Fresh sodium hydroxide solution must be used, because, if much carbon dioxide has been absorbed, the pH of the mixture will not be raised high enough to obtain rapid dissolution. About 3 drops (0.15 ml) of sodium hydroxide should be enough. With some recent batches, the protein has dissolved at once in the sodium bicarbonate.

[10] J. T. Potts, A. Berger, J. P. Cooke, and C. B. Anfinsen, *J. Biol. Chem.* **237,** 1851 (1962).

within 5 minutes. The solution is diluted with water to 1.25 ml; any undissolved protein is removed by centrifugation at 2000 *g* for 5 minutes.

There is little difference in activity between enzyme solubilized by methods (*i*) and (*ii*). Method (*ii*) uses a volatile buffer, but introduces ammonium ions, which is not desirable in samples that will be examined directly on an amino acid analyzer. Method (*i*) introduces very little nonvolatile salt—not enough to affect paper electrophoresis of peptide samples treated with the CPA.

CPA can also be brought into solution by a combination of high salt concentration and moderately alkaline pH[11,12] or by 10% (w/v) lithium chloride.[5]

The concentration of CPA solutions may be measured spectrophotometrically, assuming that the enzyme has a molar absorptivity of 8.6×10^4 at 278 nm[8] (MW 35,300).[13] Enzymatic activity may be measured with the synthetic substrate benzoylglycyl-L-phenylalanine.[14] The pH optimum of the enzyme, in the range 7–9, is broad.

CPB can be made from pig pancreas by the method of Folk, Piez, Carroll, and Gladner.[15] The enzyme made in this way is rather unstable and must be kept in frozen solution at —10°. It loses much activity if freeze-dried. The enzyme is generally contaminated with endopeptidases, and DFP treatment is essential. The free amino acid content of preparations may be large and should be determined by paper electrophoresis before use. If necessary the protein should be purified by gel filtration (e.g., through Sephadex G-25 in 0.01 *M* Tris·hydrochloric acid, pH 8.0, at 0°). CPB made by this method is commercially available in a DFP-treated form.

CPB can also be made, in crystalline form, from ox pancreas[16] or pancreatic juice[17] by activation of the purified zymogen. The crystalline ox enzyme is indefinitely stable when stored as an aqueous suspension at 4°, and can be solubilized before use with 1 *M* sodium chloride.

Both pig and ox CPB have molar absorptivities of 7.4×10^4 at 278 nm (MW 34,300).[13,14] Activity can be measured with the synthetic

[11] C. H. W. Hirs, S. Moore, and W. H. Stein, *J. Biol. Chem.* **235,** 633 (1960).

[12] G. Guidotti, R. J. Hill, and W. Konigsberg, *J. Biol. Chem.* **237,** 2184 (1962).

[13] If the concentration of enzyme made up by method (*i*) is to be measured, a buffer that does not absorb in the ultraviolet should be substituted for *N*-ethylmorpholine acetate (e.g., Tris·hydrochloric acid).

[14] J. E. Folk and E. W. Schirmer, *J. Biol. Chem.* **238,** 3884 (1963).

[15] J. E. Folk, K. A. Piez, W. R. Carroll, and J. A. Gladner, *J. Biol. Chem.* **235,** 2272 (1960).

[16] E. Wintersberger, D. J. Cox, and H. Neurath, *Biochemistry* **1,** 1069 (1962).

[17] J. H. Kycia, M. Elzinga, N. Alonzo, and C. H. W. Hirs, *Arch. Biochem. Biophys.* **123,** 336 (1968).

substrate benzoylglycyl-L-arginine.[15,18,19] The pH optimum is broad and in the range 7–9.

CPC has a molecular weight of about 150,000[2] and a pH optimum of about 5.3.[2,3] It can be assayed using a synthetic substrate such as benzyloxycarbonyl-L-leucyl-L-phenylalanine. It will soon be available commercially.[20]

Preparation of Protein Substrate

Although some proteins are readily digested by carboxypeptidases when native (e.g., intact tobacco mosaic virus[21]), it is generally necessary that the protein be denatured and in solution. Performic acid oxidation (Vol. XI [19]) is often a convenient method of denaturation. Digestions with carboxypeptidases have been performed in 6 *M* urea solution[22] and in 0.056 *M* sodium lauryl sulfate solution.[23,24]

CPA requires for optimal activity an ionic strength of at least 0.2,[25,26] but as too much salt present in the digestion mixture can affect adsorption onto ion-exchange resins, the volume should be kept small. A convenient buffer for digestion is 0.2 *M* *N*-ethylmorpholine acetate, which is volatile, but does not contain ammonia. If the reaction mixture is to be desalted and deproteinized by the resin bead method (see below), nonvolatile or ammonia-containing buffers can be used.

The normal pH for CPA or CPB digestion is 7.5–8.5. If the protein to be digested contains histidine or lysine near the C terminus, the rate of release may be markedly increased by higher pH (e.g., pH 9.2[27]). Glutamic or aspartic acid may be released more rapidly at low pH values (down to pH 5[28,29]).

[18] Worthington Biochemical Corp., Freehold, New Jersey. Data sheet *Carboxypeptidase B (porcine)*, 4–67.

[19] Folk *et al.*[15] defined a unit as percent of hydrolysis per minutes at 0.001 *M* substrate concentration. The Worthington Biochemical Corp.[18] use the unit recommended by the Enzyme Commission, the amount that will catalyze the hydrolysis of 1 μmole per minute. In many papers it is not clear which units are meant.

[20] From Röhm and Haas GmbH, Darmstadt, Germany.

[21] J. I. Harris and C. A. Knight, *J. Biol. Chem.* **214,** 215 (1955).

[22] Y. D. Halsey and H. Neurath, *J. Biol. Chem.* **217,** 247 (1955).

[23] G. Guidotti, *Biochim. Biophys. Acta* **42,** 177 (1960).

[24] D. R. Buhler, *J. Biol. Chem.* **238,** 1665 (1963).

[25] R. Lumry, E. L. Smith, and R. Glantz, *J. Amer. Chem. Soc.* **73,** 4330 (1951).

[26] Some proteins (e.g., β-lactoglobulin)[27] are not digested at high ionic strengths; this may be due to a conformational change in the substrate.

[27] E. W. Davie, C. R. Newman, and P. E. Wilcox, *J. Biol. Chem.* **234,** 2635 (1959).

[28] M. Green and M. A. Stahmann, *J. Biol. Chem.* **197,** 771 (1952).

[29] K. Titani, K. Narita, and K. Okunuki, *J. Biochem. (Tokyo)* **51,** 350 (1962).

Reaction of Carboxypeptidases with Protein, and Separation of Products of Digestion

Before a quantitative determination of the rate of release of amino acids from a protein is attempted, a preliminary experiment must be carried out to determine whether reaction takes place at all, and, if so, the optimal conditions (in particular, enzyme:substrate ratio) for a kinetic study.

Preliminary Experiment. Denatured protein (0.05 μmole) is dissolved in 0.2 *M* *N*-ethylmorpholine acetate, pH 8.5, at a concentration of 5 mg/ml in a 13 × 60 mm screw-cap culture tube. If the protein is not soluble at this pH, solubilization should be attempted with detergent (e.g., 0.056 *M* sodium lauryl sulfate) or 6 *M* (final concentration) urea. DFP-treated CPA (0.04 mg) prepared by either method (*i*) or (*ii*) (above) is added (equivalent to an enzyme:substrate ratio of 1:40). A blank should be run, containing all the constituents except substrate. The mixtures are incubated at 37° for 5 hours, and the reaction is ended by adding sulfonated polystyrene beads (H^+ form)[30] until the pH of the supernatant after shaking is 2.5–3 (indicator paper). The mixture is shaken for 20 minutes, the supernatant is removed, and the beads are washed twice with 2 (resin) volumes of water. The adsorbed amino acids are then eluted with three 2-volume portions of 5 *M* ammonia solution, and the combined eluates are evaporated to dryness (vacuum desiccator). The amino acids present are identified by high-voltage paper electrophoresis in a pH 2 system (Vol. XI [4]) as a 1-cm band on Whatman No. 1 paper, by paper chromatography, or by thin-layer chromatography (see this volume [6]).

If significant amounts of amino acid are not released from the protein under these conditions, further trials, using modified digestion conditions, should be made with small quantities of protein. Alternatives to try are: the use of CPB, instead of and as well as CPA; digestion in urea or detergent; and treatment with CPB after conversion of cyst(e)ine to *S*-β-aminoethylcysteine.[31] Hydrazinolysis (see this volume [9]) may indicate why the protein is not susceptible to digestion with carboxypeptidases.

A suitable amount of CPB to use in a trial experiment is 0.2 EC unit (about 0.02 mg)/0.05 μmole, digesting at 37° for 5 hours in 0.2 *M* pH 8.5 *N*-ethylmorpholine acetate. An enzyme blank must always be run.

[30] The resin used should be highly cross-linked (8–12%) and in large beads (20–60 mesh) to minimize adsorption of protein. Suitable resins are Zeo-Karb 225, S.R.C. 13 (Permutit Co., London), and Dowex 50-X8 (Dow Chemical Corp., Midland, Michigan).

[31] F. Tietze, J. A. Gladner, and J. E. Folk, *Biochim. Biophys. Acta* **26**, 659 (1957).

With CPA, the upper limit of enzyme:substrate ratio that can be used is determined by the amino acid blank derived, during digestion, from the enzyme itself. Leucine, threonine, and "serine + amides" are released in largest amount.[32] After 12 hours at 37° and pH 8.5, about 1 mole of each of these amino acids may be present for each mole of CPA used.

If the preliminary qualitative experiments have shown that some degradation with CPA can be achieved, suitable conditions must be chosen for a quantitative experiment in which the amounts of amino acid released after different times of digestion will be measured. If the preliminary experiment has shown that many amino acids are released, or that the terminal residues are aromatic amino acids, leucines, or valine (and as such will be released very rapidly), a very low enzyme:substrate ratio (e.g., 1:100 or 1:200) should be used in order to get conveniently slow amino acid release. Alternatively, digestion can be carried out at room temperature (25°) instead of 37°. If amino acids are released only slowly, an enzyme:substrate ratio of up to 1:20 may be used. CPB [1–10 Enzyme Commission (EC) units per micromole of substrate] may be included in the reaction mixture if the preliminary experiments have shown an arginine residue, or a lysine residue resistant to CPA, to be present near the C terminus; alternatively, a separate kinetic experiment, with CPB alone present, may be performed.

The amino acids released by carboxypeptidases should if possible be quantitated on an automatic amino acid analyzer; if none is available, the FDNB method[5] can be used for quantitation.

Quantitative Experiment. Protein (0.02–2 μmoles)[33] is dissolved in 0.2 *M* pH 8.5 *N*-ethylmorpholine acetate at a concentration of 5 mg/ml in a screw-cap culture tube.[34] CPA solution (made up by method (*i*) above) (1:20–1:200 micromole:micromole) and/or CPB solution (2–10 EC units/μmole) are added, and the mixture is incubated at 25° or 37°. (For choice of these variables, see above.) At convenient time intervals (e.g., 0.25, 0.5, 1, 2, 4, 8, and 16 hours) 0.02–0.25-μmole portions of the

[32] J.-P. Bargetzi, E. O. P. Thompson, K. S. V. Sampath Kumar, K. A. Walsh, and H. Neurath, *J. Biol. Chem.* **239**, 3767 (1964).

[33] The lower amount will be sufficient for only one quantitative determination at one time of digestion, and then only if a high sensitivity attachment is available on the amino acid analyzer.

[34] Internal standards[35] (e.g., norleucine or β-2-thienylalanine and α-amino-γ-guanidopropionic acid) may be added to the protein in mole/mole amounts. A convenient method of measuring protein concentration is to take a portion of the mixture (1 mg of protein) at this stage and analyze for amino acid content after acid hydrolysis.

[35] K. A. Walsh and J. R. Brown, *Biochim. Biophys. Acta* **58**, 596 (1962).

mixture are removed, and the reaction is ended by adding sufficient acetic acid to the portion to lower the pH to 2.5–3. Any protein that precipitates is removed by centrifugation, and the supernatant is dried (vacuum desiccator). The samples are then dissolved in 0.2 *M* pH 2.2 sodium citrate for application to the amino acid analyzer. In favorable conditions release can be very rapid. For instance, 100% of the C-terminal phenylalanine is released from dogfish M_4-lactic dehydrogenase in 1 minute when incubated in 6 *M* urea with 1:40 micromole:micromole of CPA.[36]

An alternative method of ending the reaction and of desalting and deproteinizing the mixture is to add each portion as soon as it is removed to sufficient sulfonated polystyrene beads (H^+ form) to lower the pH to 2.5–3; the beads are then washed, and the amino acids are eluted as described for the preliminary experiment. Internal standards should be used, and great care must be taken not to carry resin beads through with the eluting ammonia. Quantitation of released asparagine or glutamine is much easier, as the amino acid portions can be divided into three, one being hydrolyzed with acid and analyzed for neutral and acidic amino acids, while the other two are analyzed without hydrolysis; the increase in aspartic and glutamic acid values after hydrolysis gives the amide contents.

Interpretation of Results

The rate of hydrolysis of a peptide bond by a carboxypeptidase is affected by the nature of both the residues forming it, but the effect of the amino acid that is released predominates. The table is a guide to the relative release rates of the protein amino acids.

The results from quantitative experiments such as that described above are best interpreted from graphs of amino acid released (as moles per mole of protein) against time (Figs. 1–5). Good examples of such

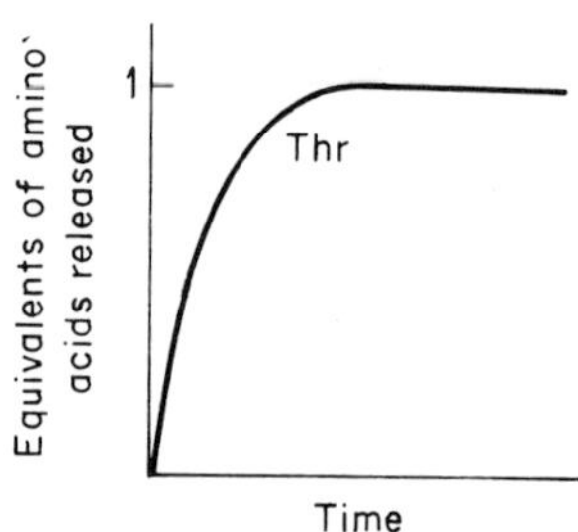

FIG. 1. Tobacco mosaic virus [J. I. Harris and C. A. Knight, *J. Biol. Chem.* **214**, 215 (1955)]: sequence ···Pro-Ala-Thr.

[36] W. S. Allison, J. Admiraal, and N. O. Kaplan, *J. Biol. Chem.* **244**, 4743 (1969).

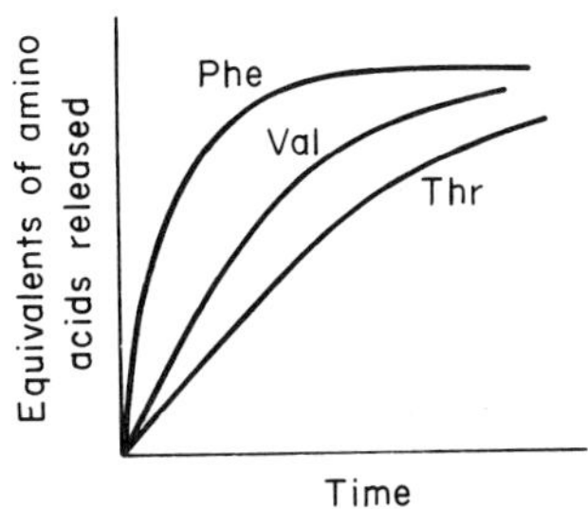

FIG. 2. Hypothetical protein: sequence ···Ser-Asp-Thr-Val-Phe.

graphs are given by Bargetzi *et al.*[32] and Winstead and Wold.[37] It is seldom possible to deduce more than the last three or four residues of a sequence, although there is a great temptation to overinterpret release data. If the molecular weight of a protein containing only one type of peptide chain is not known, a good value can often be obtained from the carboxypeptidase results, especially if the amounts of several of the released amino acids eventually reach the same limiting value (Figs. 2–5).

In the example shown in Fig. 1, carboxypeptidase releases only one amino acid from the peptide chain, and a stoichiometric value is attained. The classic example is the release by CPA of threonine from tobacco mosaic virus.[21,38]

In Fig. 2, several amino acids are released, but each at an appreciably different rate (as shown by the slopes at an early stage of reaction).

In Fig. 3, the rates of release of two or more amino acids are indistinguishable. This type of result is obtained if, in the sequence, an amino

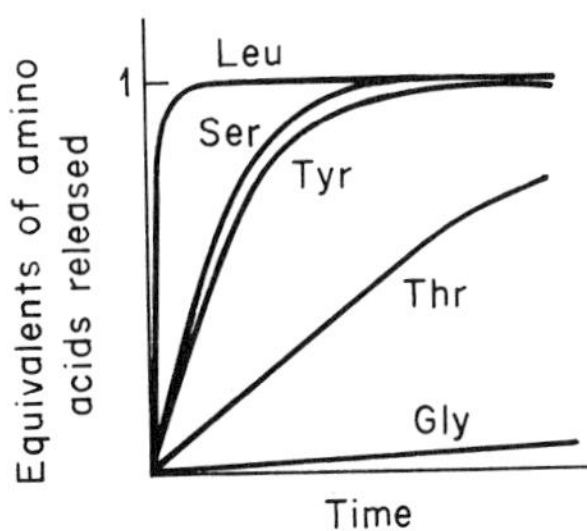

FIG. 3. Hypothetical protein: sequence ···Asp-Gly-Thr-Tyr-Ser-Leu.

[37] J. A. Winstead and F. Wold, *J. Biol. Chem.* **239**, 4212 (1964).

[38] It is of interest that a tobacco mosaic virus mutant has been isolated from which CPA releases many amino acids, to give a graph similar to Fig. 5. Substitution of leucine for proline as the residue next but one to the C terminus caused this change in reaction.[39]

[39] A. Tsugita and H. Fraenkel-Conrat, *Proc. Nat. Acad. Sci. U.S.* **46,** 636 (1960).

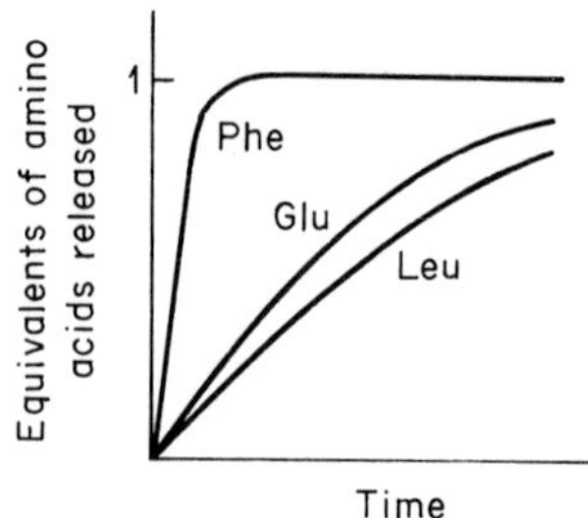

FIG. 4. ACTH [J. I. Harris and C. H. Li, *J. Biol. Chem.* **213**, 499 (1955)]: sequence ···Pro-Leu-Glu-Phe.

acid that is released slowly by carboxypeptidase is followed by one that is released very rapidly (see the table). A result approximating to this type was found with ACTH[40] (Fig. 4); the presence of the proline residue slowed the release of the leucine slightly; by very careful quantitation, the rate of glutamic acid release was shown to be greater.

With some proteins, the sequence degraded by carboxypeptidases will include more than one residue of a particular amino acid (Fig. 5); it then becomes difficult to place the second residue in sequence. Examples of such proteins are CPA itself (sequence ···His-Thr-$\frac{\text{Leu}}{\text{Val}}$-Asn-Asn),[32] turnip yellow mosaic virus (sequence ···Val-Thr-Ser-Thr[41]), and *Pseudomonas* azurin (sequence ···Lys-Gly-Thr-Leu-Thr-Leu-Lys[42]).

If the protein under investigation has more than one type of peptide chain, and carboxypeptidase releases amino acids from both, interpretation is very difficult. In some special cases the chains can be digested independently: thus one chain of human hemoglobin has C-terminal arginine, while the other has the sequence ···Tyr-His; with low concen-

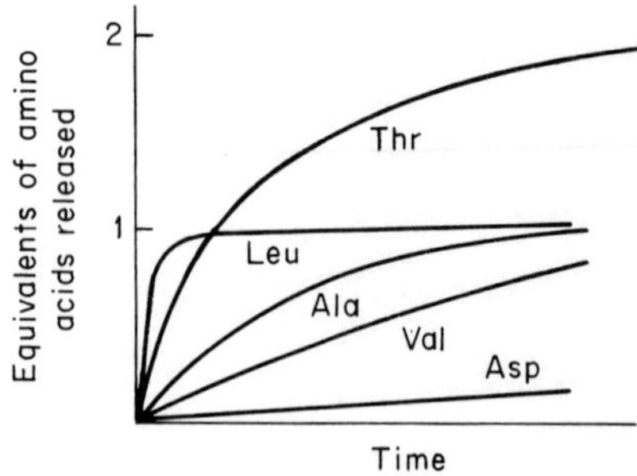

FIG. 5. Hypothetical protein: sequence ···Asp(Val,Thr)Ala-Thr-Leu.

[40] J. I. Harris and C. H. Li, *J. Biol. Chem.* **213**, 499 (1955).
[41] J. I. Harris and J. Hindley, *J. Mol. Biol.* **13**, 894 (1966).
[42] R. P. Ambler and L. H. Brown, *Biochem. J.* **104**, 784 (1967).

trations of CPB, only arginine is released, and with CPA, only tyrosine and histidine.[43] Peptide chains must be separated whenever possible before investigation with carboxypeptidases.

The most common cause of spurious results with carboxypeptidases is contamination of the enzyme preparation by endopeptidases. This is the probable explanation for the slow release of a large amount of leucine from β-lactoglobulin by CPA,[44] although the C-terminal sequence is -Cys-His-Ile.[45] Stoichiometric amounts of histidine and isoleucine were released rapidly, but no cysteine or cystine.

The results of digestion with CPB alone, or with a mixture of CPA and CPB, are interpreted in the same way. CPB probably has intrinsic CPA-type activity.[16,46] CPA will release lysine (but not arginine) from many[48] sequences at rates comparable to the rate of release of neutral amino acids.

Confirmation of C-Terminal Sequence by Isolation of C-Terminal Peptide

Kinetic experiments with carboxypeptidases can give in favorable circumstances a fairly reliable indication of the C-terminal sequence of a protein. However, the chances of misinterpretation are large, and whenever possible the results should be confirmed and extended by the recognition, isolation, and determination of the sequence of peptides containing the C-terminus of the protein.

Methods that have been used for finding C-terminal peptides are:

(*i*) *"Diagonal" Methods* (Vol. XI [36]). For instance, in a tryptic digest the C-terminal peptide is the only one not containing C-terminal lysine or arginine, and is therefore the only peptide[49] not altered in mobility if the electrophoresis paper is treated with CPB before the second dimension of the "diagonal" is run.[50]

[43] R. Zito, E. Antonini, and J. Wyman, *J. Biol. Chem.* **239**, 1804 (1964).

[44] E. B. Kalan, R. Greenberg, and M. Walter, *Biochemistry* **4**, 991 (1965).

[45] G. Frank and G. Braunitzer, *Hoppe-Seyler's Z. Physiol. Chem.* **348**, 1691 (1967).

[46] Both ox[16] and pig[15] CPB seem able to release neutral amino acids from a few very specific sequences.[47]

[47] R. P. Ambler, unpublished observations.

[48] CPA appears to be unable to release lysine to any measurable extent from the sequences ···Asp-Lys ···Glu-Lys or ···Pro-Lys.[47]

[49] Dipeptides of the type X-Lys or X-Arg are not rapidly split by CPB, but are easily recognized from their electrophoretic mobilities. The method fails if the C terminus of the protein is lysine or arginine, although in these cases the C-terminal peptide should be the only chymotryptic peptide to move off the diagonal after CPB treatment.

[50] M. A. Naughton and H. Hagopian, *Anal. Biochem.* **3**, 276 (1962).

(*ii*) *Subtractive Peptide Maps* (see Vol. XI [36], p. 337). The peptides both from intact protein and from the limit protein after exhaustive carboxypeptidase treatment are mapped. The peptides missing or in altered positions in the second map should be derived from the C-terminus of the molecule.[51]

(*iii*) *Cyanogen Bromide Cleavage* (Vol. XI [27]). After cyanogen bromide cleavage the only peptide that does not contain homoserine or homoserine lactone must (in the absence of nonspecific cleavage) be from the C terminus of the molecule.[32,42] This method should certainly be tried if there is any indication from carboxypeptidase experiments that a methionine residue is near the C terminus of the molecule.

Miscellaneous Methods Connected with Use of Carboxypeptidases

(*i*) *Hydrazinolysis*. The nature of the residue limiting further digestion of the protein with carboxypeptidases can often be obtained by hydrazinolysis (this volume [9]) of the carboxypeptidase-limit protein.[37,52,53]

(*ii*) *Digestion in* $H_2{}^{18}O$. Kowalsky and Boyer[54] suggested that, if carboxypeptidase acted on a protein while in buffer enriched with $H_2{}^{18}O$, the C-terminal amino acids could be distinguished from amino acids released either by carboxypeptidase or by contaminating proteases from other positions in the molecule. Amino acids from the C-terminal position would contain no excess ^{18}O in the carboxyl group, whereas amino acids released from internal positions in the molecule (including the penultimate residue) would contain, in the carboxyl group, half the excess ^{18}O present in the water of the reaction mixture. They used the method to show that all the tyrosine (and only the tyrosine) released from aldolase by CPA was C-terminal.

(*iii*) *Modification of Proteins by Treatment with Carboxypeptidases.* See (Vol. XI [76b]).

(*iv*) *Protein Synthesis in Cell-Free Systems.*[55] Carboxypeptidase has been used to investigate the nature of amino acid incorporation into cell-free systems.

(*v*) *Classification of Bence-Jones Proteins.*[56]

[51] B. C. Carlton and C. Yanofsky, *J. Biol. Chem.* **238,** 636 (1963).

[52] L. Campbell and P. D. Cleveland, *J. Biol. Chem.* **236,** 2966 (1961).

[53] R. G. Spiro, *J. Biol. Chem.* **238,** 644 (1963).

[54] A. Kowalsky and P. D. Boyer, *J. Biol. Chem.* **235,** 604 (1960).

[55] J. W. Davies, *Biochim. Biophys. Acta* **174,** 686 (1969).

[56] A. B. Edmundson, N. B. Simonds, F. A. Sheber, T. Johnson, and B. Bangs, *Arch. Biochem. Biophys.* **132,** 502 (1969).

Section III

Chain or Subunit Separation

[11] Significance of Interactions in Electrophoretic and Chromatographic Methods[1,2]

By JOHN R. CANN

When considering the significance of macromolecular interactions for the several methods of mass transport, one classically associates macromolecule–small ion interactions with electrophoresis.[3] This is so because the net electrical charge on a protein molecule at a given pH (and thus its electrophoretic mobility) is determined not only by the state of ionization of its acidic and basic groups but also by the binding of small ions other than H^+ by the protein. During the past decade or so it has become clear, however, that other types of interaction can also have a major effect on electrophoretic mobility and on the nature of the electrophoretic patterns of highly purified proteins. In fact, these new insights are providing the understanding required for unambiguous interpretation of both moving-boundary and zone patterns. The most dramatic of these is that the electrophoretic and chromatographic patterns of homogeneous but interacting macromolecules can and do show multiple peaks and zones despite instantaneous establishment of equilibrium. This is of great practical significance since many such patterns could easily be misinterpreted as indicating inherent heterogeneity. It cannot be overemphasized that unequivocal proof of inherent heterogeneity is afforded only by isolation of the various components. The interactions in question include macromolecular association–dissociation; kinetically controlled isomerization; complex formation between different macromolecules; and interaction of proteins with electrically neutral molecules such as undissociated buffer acid. Each will be considered in turn with respect to their detection, characterization, and implications for conventional analytical applications of electrophoresis and chromatography.

Macromolecular Association–Dissociation

One's first thought concerning the relationship of electrophoretic mobility to the state of molecular association of a protein is likely to be

[1] Supported in part by Research Grant 5R01 AI01482 from the National Institutes of Health, United States Public Health Service.

[2] Contribution No. 410 from the Department of Biophysics, Florence R. Sabin Laboratories, University of Colorado Medical Center, Denver, Colorado 80220.

[3] J. R. Cann, *in* "Physical Principles and Techniques of Protein Chemistry" (S. J. Leach, ed.), Part A, p. 411. Academic Press, New York, 1969.

that the greater the extent of association, the lower the mobility. Upon reflection, however, it becomes apparent that the situation is not nearly so simple. Thus, it is possible that the increase in size and, perhaps, asymmetry may be offset by an increase in zeta potential for one reason or another. Although it is not generally possible to predict *a priori* the relationship of the electrophoretic mobility of an aggregated protein molecule to that of the monomer, it is clear that changes in the state of aggregation sometimes do not influence the mobility significantly[4]—for example, acid-modified conalbumin.[5] In other cases, such as the low-temperature tetramerization of β-lactoglobulin A in acidic media, aggregation causes the mobility to increase.[6] In fact, the tetramerization of β-lactoglobulin was first detected electrophoretically. Comparison of moving-boundary electrophoretic patterns with sedimentation measurements led Ogston and Tilley[7] to conclude that the two peaks shown by the descending pattern at pH 4.65 are due to polymerization of the protein. Subsequent elucidation of this highly interesting reaction by Timasheff and his co-workers[6,8–13] is a model for studies on interacting systems, being noteworthy for its logical development and for its combined use of a variety of biophysical methods including electrophoresis. These investigations illustrate that the greater the diversity of methods brought to bear on the problem, the more precisely the interaction can be described; and their success stemmed in part from an appreciation of recent developments in the theory and practice of mass transport of interacting systems.

Moving-boundary electrophoretic experiments on reversibly associating-dissociating macromolecules present no interpretative difficulties if the rates of reaction are sufficiently slow. In that event, the time required

[4] These considerations are for electrophoresis in free solution and on supports, such as cellulose acetate, for which molecular sieving is of negligible importance. In the case of starch-gel or polyacrylamide-gel electrophoresis, whose extremely high resolving power resides in the coupling of electrophoretic migration with molecular sieving, migration velocities will in general decrease with increasing degree of molecular association.

[5] J. R. Cann and R. A. Phelps, *Arch. Biochem. Biophys.* **52,** 48 (1954).

[6] R. Townend, R. J. Winterbottom, and S. N. Timasheff, *J. Amer. Chem. Soc.* **82,** 3161 (1960).

[7] A. G. Ogston and J. M. A. Tilley, *Biochem. J.* **59,** 644 (1955).

[8] S. N. Timasheff and R. Townend, *J. Amer. Chem. Soc.* **82,** 3157 (1960).

[9] R. Townend and S. N. Timasheff, *J. Amer. Chem. Soc.* **82,** 3168 (1960).

[10] R. Townend, L. Weinberger, and S. N. Timasheff, *J. Amer. Chem. Soc.* **82,** 3175 (1960).

[11] S. N. Timasheff and R. Townend, *J. Amer. Chem. Soc.* **83,** 464 (1961).

[12] T. F. Kumosinski and S. N. Timasheff, *J. Amer. Chem. Soc.* **88,** 5635 (1966).

[13] J. J. Basch and S. N. Timasheff, *Arch. Biochem. Biophys.* **118,** 37 (1967).

for separation of the various species is short compared to the time for significant reequilibration to occur. Accordingly the electrophoretic patterns will resolve into two or more boundaries corresponding to reactants and products; and conventional analysis of the patterns yields their mobilities and concentrations in the initial equilibrium mixture. An example is the reversible dissociation of the arachins.[14,15] The ascending and descending patterns shown by such systems may be quite enantiographic[14] and give little, if any, hint of slow association–dissociation. Even a companion ultracentrifugal analysis, while revealing species with different molecular weights, probably would not be indicative of their interconversion.[15] This underscores the importance of fractionation for distinguishing between inherent heterogeneity and interaction. Fractionation may be achieved by removal from the electrophoresis cell of the protein disappearing across the slowest-migrating descending peak and the fastest ascending one.[16] For sufficiently slow interconversion, fractions may also be obtained by Sephadex chromatography or salt precipitation.[15] The aged and dialyzed fractions are examined electrophoretically and ultracentrifugally under the same environmental conditions and at the same protein concentration as used with the unfractionated material. For an interaction, the fraction will behave like the unfractionated material and show multiple electrophoretic and ultracentrifugal boundaries, while for inherent heterogeneity a single boundary will be obtained. In the case of interaction, such an experiment can also yield information on the rate of interconversion of species. Thus, depending upon the particular fraction, either the rapidly migrating or slowly migrating boundary should be observed to grow at the expense of the other with time lapsed between fractionation and analysis.[14,15]

The situation is considerably more complex for rapid association–dissociation reactions as, for example, in the case of the tetramerization of β-lactoglobulin A. When rapid reequilibration occurs between species during their differential transport, peaks in the electrophoretic patterns cannot be placed into correspondence with individual reactants and products, and an entirely different method of analysis of the patterns is required. When monomer is in rapid equilibrium with a single polymer of greater mobility, $m\mathrm{M} \rightleftharpoons \mathrm{M}_m$, the Gilbert theory of sedimentation[17-19]

[14] P. Johnson, E. M. Shooter, and E. K. Rideal, *Biochim. Biophys. Acta* **5,** 376 (1950).

[15] P. Johnson and E. M. Shooter, *Biochim. Biophys. Acta* **5,** 361 (1950).

[16] J. R. Cann, *in* "Physical Principles and Techniques of Protein Chemistry" (S. J. Leach, ed.), Part A, p. 384. Academic Press, New York, 1969.

[17] G. A. Gilbert, *Discuss. Faraday Soc.* **20,** 68 (1955).

[18] G. A. Gilbert, *Proc. Roy. Soc. Ser. A* **250,** 377 (1959).

[19] This volume [Part C].

is equally applicable to descending moving-boundary electrophoresis. This is so since his mathematical equations make the rectilinear approximation and can be applied directly to electrophoresis simply by substitution of electrophoretic velocities for sedimentation velocities. Accordingly, it is anticipated that the descending pattern will show a single migrating peak for dimerization ($m = 2$) and a bimodal reaction boundary for higher-order association ($m \geq 3$) provided, of course, that the total macromolecule concentration exceeds the characteristic value corresponding to the minimum in the bimodal reaction boundary. While Gilbert's theory applies to descending electrophoresis of systems for which the mobility of the polymer is greater than that of the monomer, for ascending electrophoresis of such systems it describes a physically impossible situation. Physical reasoning tells us, however, that the ascending pattern will show a hypersharp boundary migrating with the weight average mobility in the initial equilibrium mixture. As the initial boundary tends to spread due to diffusion and differential transport of monomer and polymer, polymer molecules in its dilute leading edge dissociate into monomer molecules which lag behind, thereby producing and maintaining a hypersharp front.

The predicted behavior is illustrated by the electrophoretic patterns of β-lactoglobulin A displayed in Fig. 1.[20] Whereas the descending pattern shows a bimodal reaction boundary, the ascending boundary is hypersharp. In cases where the mobility of the polymer is less than that

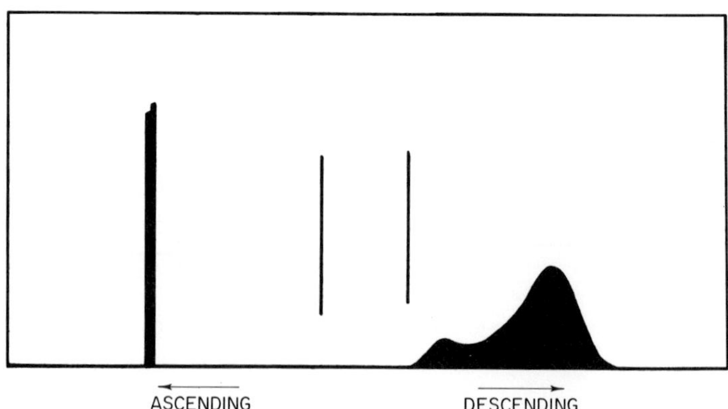

FIG. 1. Moving-boundary electrophoretic patterns of β-lactoglobulin A under optimal conditions for tetramerization, 0.1 ionic strength acetate buffer, pH 4.66, at 1°. Protein concentration, 1.58%; position of initial boundary indicated by vertical lines. Adapted from M. P. Tombs, *Biochem. J.* **67**, 517 (1957).

[20] M. P. Tombs, *Biochem. J.* **67**, 517 (1957).

of the monomer, the situation is exactly reversed; now it is the descending boundary that is hypersharp and the ascending boundary that is bimodal. In either event, the described nonenantiography is diagnostic of interaction and distinguishable from the Dole nonideal electrophoretic effects.[21] Thus, hypersharpening of ascending boundaries associated with the generation of pH and conductance gradients by the electrophoretic process per se generally tends to enhance rather than obliterate resolution of noninteracting macromolecules. After the initial observation has been made, supporting evidence for reversible polymerization is sought by electrophoretic and companion ultracentrifugal analyses over a wide range of protein concentration. Experiments at a single concentration or over a limited range of low concentrations could be quite misleading. Thus, at sufficiently low concentration the descending electrophoretic pattern (and the sedimentation pattern) of the system, $m\text{M} \rightleftharpoons \text{M}_m$, $m \geq 3$, will show a single peak whose electrophoretic mobility (and sedimentation coefficient) decreases with decreasing concentration and extrapolates to that of the monomer. A limited set of observations of this sort could easily be misinterpreted in terms of dimerization reaction. Indeed, the association–dissociation reaction might even go unsuspected if only an isolated observation at low concentration were to be made. On the other hand, if the concentration were to be progressively increased, a second more rapidly migrating peak would eventually appear and grow in area while that of the slower peak would now remain constant. Constancy of the area of the slower peak with increasing concentration is diagnostic for rapidly equilibrating higher-order polymerization reactions. Moreover, the mobility of the slower peak will be independent of concentration, while that of the faster one will at first increase with increasing concentration and then pass through a maximum; but it will always be less than that of the polymer. Extrapolation to infinite dilution of the mobility of the faster peak at low concentration gives the mobility of the monomer; the mobility of the polymer can be obtained from the patterns by extrapolation to infinite dilution from the very high concentration range, using a process of successive approximation.[22]

Another electrophoretic test for interactions of the type under consideration is provided by field-reversal experiments. Suppose that an electrophoretic experiment of duration t (seconds) yields electrophoretic patterns such as those displayed in Fig. 1. If at that time the polarity of the electric field is reversed, the hypersharp ascending boundary will generate a bimodal reaction boundary as it migrates back to its original

[21] J. R. Cann, *in* "Physical Principles and Techniques of Protein Chemistry" (S. J. Leach, ed.), Part A, p. 386. Academic Press, New York, 1969.

[22] L. M. Gilbert and G. A. Gilbert, *Nature* (*London*) **194**, 1173 (1962).

starting position. After t (seconds) of reverse electrophoresis, the first moment of the newly generated bimodal boundary will be located at the original starting position. Simultaneously, the bimodal descending boundary will sharpen as it migrates back to its original starting position; and after t (seconds) the newly generated hypersharp boundary will be located at the original starting position. This theoretically predicted behavior has been verified experimentally[23] with β-lactoglobulin A and is in contradistinction to that expected of a simple mixture of noninteracting macromolecules. In the latter case, coalescence of resolved boundaries in either limb of the Tiselius cell upon reversal of the field should yield a single boundary which is unimodal and symmetrical, although somewhat broader than the original starting boundary because of diffusion. Moreover, if for one reason or another resolution does not occur in one of the limbs during forward electrophoresis, it will not occur when the field is reversed. While the field-reversal test is unambiguous for interactions and is described above for rapid association–dissociation reactions, it is not qualitatively unique for the latter type of interaction. In fact, such experiments provided[24] one of the first pieces of evidence that the two or three peaks in the nonenantiographic electrophoretic patterns of highly purified proteins in acetate buffer (see below) constitute a single reaction boundary arising from interaction of the protein with undissociated acetic acid.

Finally, the degree of polymerization m must be ascertained from independent measurements made by use of a method such as the Archibald approach to equilibrium, equilibrium sedimentation, or light scattering. The latter method also provides crucial information concerning the rates of reaction. The equilibrium constant for the polymerization reaction is then estimated from the value of m and the area under the slower migrating peak of the electrophoretic reaction boundary.[25] The value thus obtained is to be compared with those derived from sedimentation patterns and light scattering.

Above we have considered the two limiting cases in which the rate of equilibration of the associating–dissociating reaction is either very slow or very rapid; but intermediate situations are also of interest. Belford and Belford[26] and Oberhauser *et al.*[27] have formulated the theory of kinetically controlled dimerization,

[23] T. Stamato and J. R. Cann, unpublished results quoted in J. R. Cann *in* "Interacting Macromolecules." Chapter III, Academic Press, New York, 1970.

[24] R. A. Phelps and J. R. Cann, *J. Amer. Chem. Soc.* **78,** 3539 (1956).

[25] L. W. Nichol, J. L. Bethune, G. Kegeles, and E. L. Hess, *in* "The Proteins" (H. Neurath, ed.), 2nd ed., Vol. II, Chapter 9. Academic Press, New York, 1964.

[26] G. G. Belford and R. L. Belford, *J. Chem. Phys.* **37,** 1926 (1962).

[27] D. R. Oberhauser, J. L. Bethune, and G. Kegeles, *Biochemistry* **4,** 1878 (1965).

$$2M \underset{k_2}{\overset{k_1}{\rightleftharpoons}} \mathrm{M}_2$$

where k_1 and k_2 are the specific rates of association and dissociation, respectively. A unique feature of this system is that the theoretical electrophoretic or sedimentation patterns may show three peaks when the half-time of reactions are of the order of the time of electrophoresis or sedimentation. Moreover, changes in total macromolecule concentration can induce remarkable changes in the patterns.[27] As in the case of rapidly reversible polymerization, the latter provides perhaps the most sensitive diagnostic test for interaction. Another kinetically controlled interaction of considerable interest is macromolecular isomerization,

$$\mathrm{A} \underset{k_2}{\overset{k_1}{\rightleftharpoons}} \mathrm{B}$$

Here A and B represent two interconvertible states of a macromolecule in which it migrates with different electrophoretic mobilities. The moving-boundary electrophoretic patterns of such systems can show one, two, or three peaks, depending upon the rates of reaction and time of electrophoresis.[28,29] The same predictions apply to zone electrophoresis and chromatography,[30,31] which can show one, two, or three zones or spots, depending upon the rates of interconversion. In fact, isomerization is one of the causes of multiple spotting in the chromatography of pure substances. An example[32] is the interconversion of glucuronic acid and its lactone, glucurone, during the course of development on paper. Kinetically controlled isomerization may be detected by (a) the distinctive manner in which the shape of the reaction boundary or zone changes with time of electrophoresis or length of the chromatographic column; (b) companion ultracentrifugal analysis to eliminate possible association-dissociation reaction; and (c) electrophoresis or chromatography of fractions to see whether they run true or give the same transport pattern as the unfractionated material.

Macromolecular Complex Formation

Reversible complex formation between different macromolecules is one of the most important of biological interactions, and the several methods of mass transport are particularly well suited for their detection and elucidation. Thus, for example, the significance of electrophoresis and ultracentrifugation for immunochemistry was recognized early. In fact,

[28] J. R. Cann and H. R. Bailey, *Arch. Biochem. Biophys.* **93**, 576 (1961).

[29] P. C. Scholten, *Arch. Biochem. Biophys.* **93**, 568 (1961).

[30] J. C. Giddings, *J. Chromatogr.* **3**, 443 (1960).

[31] R. A. Keeler and J. C. Giddings, *J. Chromatogr.* **3**, 205 (1960).

[32] S. M. Partridge and R. G. Westall, *Biochem. J.* **42**, 238 (1948).

one of Tiselius' first applications of his moving-boundary electrophoresis apparatus was the demonstration[33] that rabbit antibodies to ovalbumin are associated with the γ-globulin fraction of the serum proteins. Contemporaneously, Heidelberger and Pedersen[34] used the ultracentrifuge to demonstrate soluble antigen–antibody complexes in the antigen-excess zones of the precipitin reaction between ovalbumin and its rabbit antibodies. More recently, Singer and his co-workers[35] used both electrophoresis and ultracentrifugation to establish that the composition of the single soluble complex formed in heavy antigen excess is Ag_2Ab; that precipitating antibodies are bivalent; and that the framework theory of antigen–antibody precipitation is essentially correct. In addition, equilibrium constants and other thermodynamic parameters of the antigen–antibody reactions have been evaluated and a partial assessment of the forces responsible for the specific antigen–antibody bond has been possible.

Over the years a number of other such systems have been characterized quantitatively by means of moving-boundary electrophoresis. Examples include the electrophoretic demonstration of the specific Michaelis-Menten complex between pepsin and serum albumin[36,37]; complex formation between ovalbumin and nucleic acid[38]; between conalbumin and lysozyme,[39] and insulin-protamine.[40] In each case the electrophoretic mobility of the complex is intermediate in value between those of the two reactants. This is the situation most frequently encountered in practice, and such interactions are particularly amenable to analysis by moving-boundary electrophoresis. Experimental design is straightforward, and the weak electrolyte moving-boundary theory[41] can be used for quantitative interpretation of the electrophoretic patterns. Specifically, the theory is applicable to simple systems of the type, $n\text{A} + m\text{B} \rightleftharpoons \text{A}_n\text{B}_m$, in which only a *single* complex is formed in significant concentration and equilibration is rapid. When two or more complexes are formed, rigorous analysis in terms of equilibrium constants becomes impossible.[39] On the

[33] A. Tiselius, *Biochem. J.* **31,** 1464 (1937).

[34] M. Heidelberger and K. O. Pedersen, *J. Exp. Med.* **65,** 393 (1937).

[35] S. J. Singer, *in* "The Proteins" (H. Neurath, ed.), 2nd ed., Vol. III, Chapter 15. Academic Press, New York, 1965.

[36] J. R. Cann and J. A. Klapper, Jr., *J. Biol. Chem.* **236,** 2446 (1961).

[37] J. R. Cann, *J. Biol. Chem.* **237,** 707 (1962).

[38] L. G. Longsworth, *in* "Electrophoresis" (M. Bier, ed.), pp. 125–126. Academic Press, New York, 1959.

[39] L. Ehrenpreis and R. E. Warner, *Arch. Biochem. Biophys.* **61,** 38 (1956).

[40] S. N. Timasheff and J. G. Kirkwood, *J. Amer. Chem. Soc.* **75,** 3124 (1953).

[41] J. R. Cann and W. B. Goad, *in* "Interacting Macromolecules," Chapter II. Academic Press, New York, 1970.

other hand, it should be noted that, even though rigorous analysis hinges upon formation of a single complex, it is not necessary that the stoichiometry of the complex be 1:1 on a molar basis.[38] In the case of a single complex, the weak electrolyte moving-boundary theory predicts the gross features of the electrophoretic patterns of reaction mixtures and admits an explicit relationship between the equilibrium constant and quantities readily obtained from the descending pattern.

Although the weak electrolyte moving-boundary theory permits quantitative interpretation of the electrophoretic patterns, it does not describe the effect of the interaction on the shape of the boundaries. It is particularly significant, therefore, that Gilbert and Jenkins[42] have solved the transport equations for the interaction, $A + B \rightleftharpoons C$, where A and B are different macromolecules in rapid equilibrium with their complex C. Their solution permits computation of theoretical electrophoretic patterns. Several cases can be distinguished according to the relative mobilities μ of A, B, and C. The case pertinent to this discussion is the one in which $\mu_A > \mu_C > \mu_B$. Consider, for example, specific complex formation between pepsin (E) and serum albumin (S) under conditions such that proteolysis has negligible effect on electrophoretic patterns obtained shortly after mixing E and S ($E + S \rightleftharpoons ES$ with $\mu_E > \mu_{ES} > \mu_S$). Both theories predict that the descending and ascending patterns will each show two moving boundaries. The slower migrating boundary in the descending pattern corresponds to unbound S, while the more rapid one is a reaction boundary within which the equilibrium, $E + S \rightleftharpoons ES$, is being continually readjusted during differential transport of E and ES. The reaction boundary migrates with the constituent mobility of E in the underlying equilibrium mixture. In the ascending pattern the more rapidly migrating boundary corresponds to E, while the slower one is a reaction boundary migrating with the constituent mobility of S in the underlying solution. In addition, the Gilbert-Jenkins theory tells us that the descending reaction boundary will be unimodal and the ascending one bimodal, provided that the equilibrium concentrations of E and S are not the same. If they are the same, both reaction boundaries will be bimodal. (By its very nature the weak electrolyte moving-boundary theory cannot make such detailed predictions.) As illustrated in Fig. 2, the theory accurately predicts the experimentally observed patterns.

A few words of caution concerning the ascending reaction boundary are in order. One's first inclination might be to equate the fast migrating peak (apparent mobility 4.96) in this boundary with the complex. Actually, however, the areas and migration velocities of the peaks in a

[42] G. A. Gilbert and R. C. Ll. Jenkins, *Proc. Roy. Soc. Ser. A* **253**, 420 (1959).

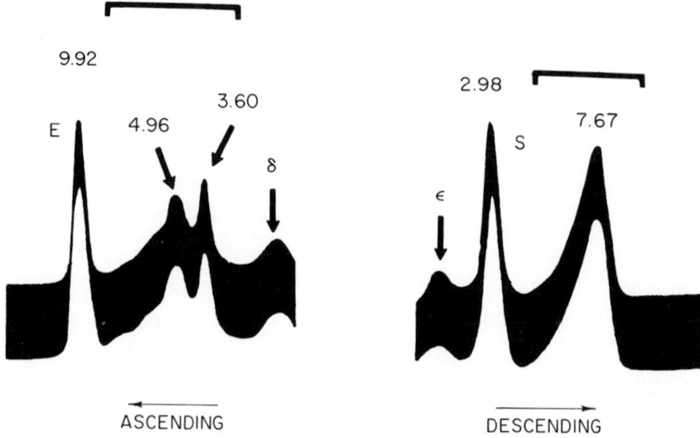

FIG. 2. Moving-boundary electrophoretic patterns for specific complex formation between pepsin and bovine serum albumin (BSA). Pepsin (1%)–BSA (1%) 0.1 ionic strength phosphate buffer, pH 5.35. Field was applied 25 minutes after mixing at 0°; time of electrophoresis, 90 minutes at about 5 V cm^{-1}. Reaction boundaries are indicated by brackets. Mobilities and apparent mobilities ($-10^5 \times cm^2 sec^{-1} V^{-1}$) are shown above corresponding peaks. The descending mobilities of pepsin and BSA in solutions of pure constituents are $-(9.42 \pm 0.05) \times 10^{-5}$ and $-(2.85 \pm 0.03) \times 10^{-5}$ $cm^2 sec^{-1} V^{-1}$, respectively; and the apparent ascending mobilities, $-(9.68 \pm 0.02) \times 10^{-5}$ and $-(2.99 \pm 0.02) \times 10^{-5}$ $cm^2 sec^{-1} V^{-1}$, respectively. Adapted from J. R. Cann and J. A. Klapper, Jr., *J. Biol. Chem.* **236,** 2446 (1961).

reaction boundary cannot be identified in the conventional manner with the concentrations of species in solution. Nevertheless, the relative area of the peak in question proved in practice[36] to be an empirical index of the extent of binding of albumin by pepsin. Analysis of mixtures containing varying ratios of pepsin to albumin indicated a single complex with a mobility intermediate between those of the reactants. (The two minor peaks with apparent mobilities greater than 4.92 were observed only at high ratios and have been attributed to higher-order complexes containing a greater amount of pepsin than albumin.) Several independent lines of evidence were advanced that the complex is the specific Michaelis-Menten complex, capable of being activated to give rise to reaction products. Accordingly, the electrophoretic patterns were interpreted quantitatively in terms of the dissociation constant K of the enzyme–substrate complex ES. The weak electrolyte moving-boundary theory admits[36,38,41] the relationship

$$K = m_E m_S / m_{ES} = m_E(\bar{m}_S - \bar{m}_E + m_E)/(\bar{m}_E - m_E) \tag{1}$$

$$m_E = (\bar{m}_E + m'_S - \bar{m}_S)(\bar{\mu}_E - \mu_S)/(\mu_E - \mu_S) \tag{2}$$

in which m_E and m_S are the molar concentrations of uncombined enzyme and uncombined substrate in the equilibrium mixture; m_{ES} is the concentration of ES; $\bar{m}_E$ is the constituent concentration of the enzyme, that is, the total concentration of this constituent combined and uncombined; and $\bar{m}_S$ is the constituent concentration of the substrate. The constituent concentrations are fixed by the amounts of E and S used to prepare the reaction mixture. The quantity m'_S is the concentration of S in the region between the two descending boundaries. Its value is determined from the area under the slower-migrating boundary. The constituent mobility of the enzyme $\bar{\mu}_E$ is the mean mobility of the descending reaction boundary. The mobilities of uncombined enzyme and uncombined substrate, μ_E and μ_S, are taken as the mobilities of E and S in solutions of the pure constituents. (Note that by using quantities derived from the descending pattern one avoids serious errors[21] due to Dole non-ideal electrophoretic effects.) The value of K determined in this manner is of the same order of magnitude as the value of the Michaelis-Menten constant. These experiments illustrate how the ability to interpret the electrophoretic patterns of interacting systems in a quantitative fashion can help to elucidate the nature of important biological reactions. Thus, they constitute the first unambiguous physical demonstration of a Michaelis-Menten complex between a proteolytic enzyme and its macromolecular substrate.

It is possible that patterns similar to those displayed in Fig. 2 for a known mixture of complexing macromolecules may be shown by a highly purified protein thought for one reason or another to be a single component but actually a complex between two different proteins.[43] The patterns will be recognized immediately as indicative of interaction by their nonenantiography, but unambiguous interpretation depends upon analyses at different concentrations and upon fractionation experiments. As the concentration is decreased, the complex will dissociate by mass action. Consequently, the constituent mobility of the descending reaction boundary will increase; and the relative area of the faster migrating peak in the ascending reaction boundary will decrease. Fractions can be obtained by removal from the Tiselius cell of the protein disappearing across the slow-moving descending boundary and the fast-moving ascending one.[16] After reconstituting the concentration and equilibration by dialysis against buffer, the fractions are analyzed electrophoretically to see whether they run true. Likewise, mixtures of the two fractions are analyzed. If the patterns of the mixtures are the same as those of the unfractionated material, it is certain that one is dealing with complex

[43] Although the most frequently observed type of complex is of intermediate mobility, it is conceivable that other cases may sometimes be encountered, e.g., $\mu_C > \mu_A = \mu_B$ or $\mu_C < \mu_A < \mu_B$. The weak electrolyte moving-boundary theory provides little insight here, and one must appeal to the Gilbert-Jenkins theory.

formation. The fractions and their admixtures should then be analyzed for biological activity.

Essentially the same considerations apply to zone electrophoresis and chromatography. The theoretical calculations of Bethune and Kegeles[44] predict that the zone electrophoretic pattern and chromatogram for rapidly equilibrating, bimolecular complex formation ($A + B \rightleftharpoons C$ with $\mu_A > \mu_C > \mu_B$) can show one, two or three peaks depending upon the value of the equilibrium constant. In the case of a pattern showing three peaks, the central one contains the complex C along with equilibrium concentrations of A and B, and the peripheral peaks correspond to A and B. A striking example is provided by tryptophan synthetase from *E. coli*.[45] In this system, two separable proteins A and B must be combined to give an enzyme which catalyzes the transformation of indole and serine to tryptophan. Chromatography of the original extract of tryptophan synthetase on DEAE-cellulose (Fig. 3a and b) gives two major peaks of activity (and an additional, fast-running, minor one) when protein B is added to the assay system; two, when protein A is added, and one, with no addition of A and B. The latter peak, which contains the complex, is intermediate in position between the A and B peaks of the chromatogram. When rechromatographed (Fig. 3c) it produces the pattern of the original extract (except for the small peak of A activity noted above). Moreover, when pure A and B are isolated, each chromatographs as a single peak; but when admixed they give a pattern resembling that of the unfractionated system.

Macromolecule-Small Molecule Interactions

Longsworth and Jacobsen[46] were the first to note that the moving-boundary electrophoretic patterns of some proteins at pH values acid to their isoelectric points are suggestive of interactions. Cann and his coworkers[47] have pursued this matter and observed that a variety of proteins display nonenantiographic electrophoretic patterns in acidic media containing acetate, formate, or other carboxylic acid buffers. The nature of the patterns is critically dependent in a reversible fashion upon the concentration of buffer in the solvent medium but insensitive to ionic strength. The behavior typically shown by proteins in media containing varying concentration of buffer is illustrated in Figs. 4 and 5. The characteristic nonenantiography is quite bizarre and not at all like that shown by mixtures of noninteracting proteins. Moreover, increasing the

[44] J. L. Bethune and G. Kegeles, *J. Phys. Chem.* **65,** 1755 (1961).

[45] I. P. Crawford and C. Yanofsky, *Proc. Nat. Acad. Sci. U.S.* **44,** 1161 (1958).

[46] L. G. Longsworth and C. F. Jacobsen, *J. Phys. Colloid Chem.* **53,** 126 (1949).

[47] J. R. Cann and W. B. Goad, *Advan. Enzymol.* **30,** 139 (1968).

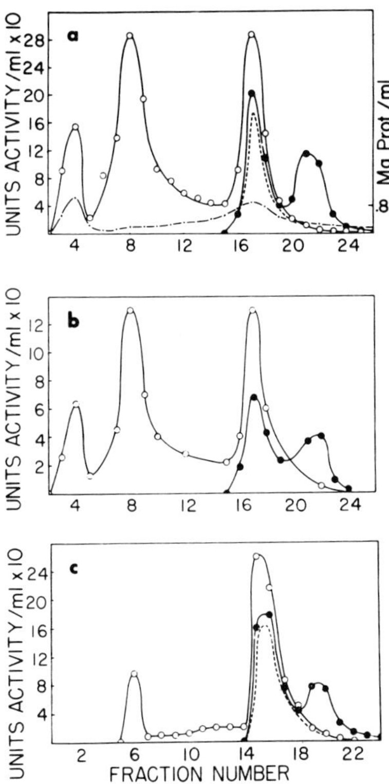

Fig. 3. Chromatography on DEAE-cellulose of partially purified *Escherichia coli* tryptophan synthetase, an interacting system of the type $A + B \rightleftharpoons C$. ○, protein A enzymatic activity, with protein B added to assay mixture; ●, protein B enzymatic activity, with protein A added to assay mixture; - - -, enzymatic activity with no addition of proteins A or B; – · –, milligrams of protein ml^{-1}. Panels a and b are the same chromatogram of the original extract assayed by two different methods; panel c is a chromatogram of the material from the central zone having activity without the addition of A or B. In panels a and c, the assay reaction is indole + serine → tryptophan; in b, the assay reaction is indoleglycerophosphate + serine → tryptophan + triose phosphate. From I. P. Crawford and C. Yanofsky, *Proc. Nat. Acad. Sci. U.S.* **44,** 1161 (1958).

buffer concentration at constant pH and ionic strength results in progressive and characteristic changes in the patterns, notably in the appearance and growth of fast-moving peaks at the expense of slow ones. The patterns are also dependent upon protein concentration. Consider, for example, ascending patterns of the sort shown in Fig. 4 for 1.3% protein in 0.04 *M* formate buffer. As the protein concentration is lowered, the relative area of the broad slow-moving ascending peak decreases progressively (with concomitant decrease in the apparent mobility of the

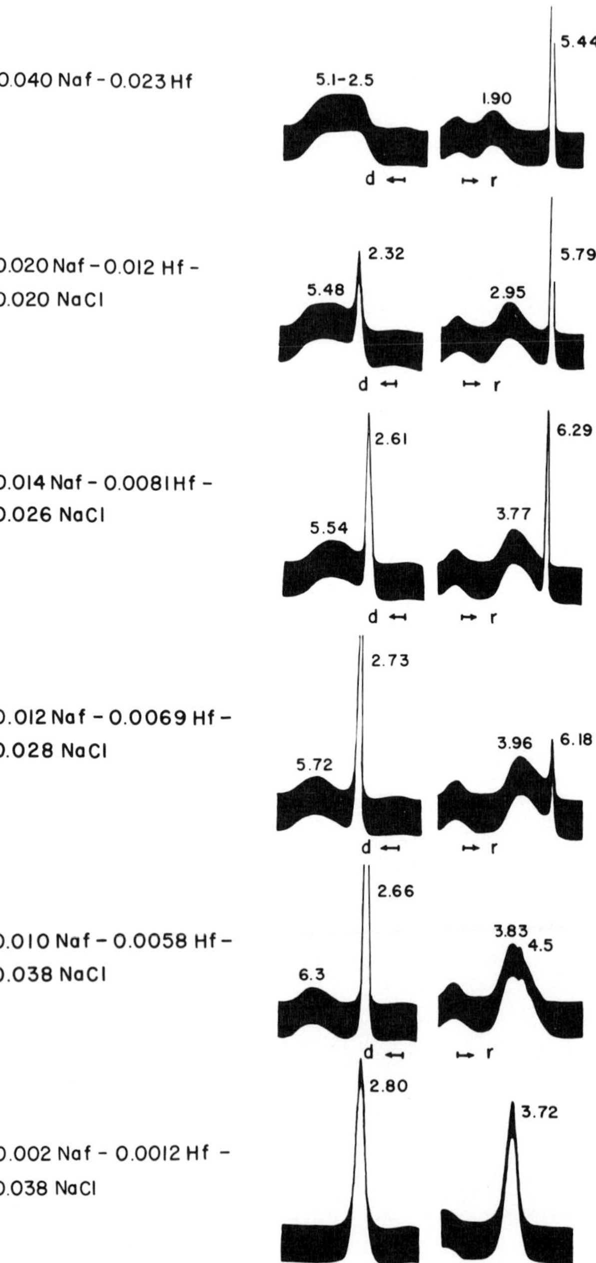
BUFFER COMPOSITION
(MOLAR CONCENTRATION)
0.040 Naf - 0.023 Hf
5.1-2.5
1.90
5.44
d
r
0.020 Naf - 0.012 Hf -
0.020 NaCl
5.48
2.32
2.95
5.79
d
r
0.014 Naf - 0.0081 Hf -
0.026 NaCl
5.54
2.61
3.77
6.29
d
r
0.012 Naf - 0.0069 Hf -
0.028 NaCl
5.72
2.73
3.96
6.18
d
r
0.010 Naf - 0.0058 Hf -
0.038 NaCl
6.3
2.66
3.83
4.5
d
r
0.002 Naf - 0.0012 Hf -
0.038 NaCl
2.80
3.72
d
r

sharp fast-moving peak) until at about 0.2% it has disappeared entirely; and the patterns have become reasonably enantiographic. Fractionation experiments and other measurements have established that the peaks in a given pattern constitute a reaction boundary, modified in some instances by mild convective disturbances, and are in no sense indicative of inherent heterogeneity. Evidence has been advanced to support interpretation of these observations in terms of reversible complexing of the protein molecules with undissociated buffer acid, with concomitant subtle structural changes which increase the net positive charge on the protein but do not change its frictional coefficient significantly.

A theoretical basis for this interpretation has been provided by a theory of electrophoresis of interacting systems of the type, $P + nHA \rightleftharpoons P(HA)_n$, where P represents a protein molecule or other macromolecular ion in solution and $P(HA)_n$ its complex with n moles of a small uncharged constituent, HA, of the solvent medium—for example, undissociated buffer acid. It is assumed that the electrophoretic mobility of the complex differs from that of the uncomplexed protein and that equilibrium is reestablished instantaneously during differential transport of P and $P(HA)_n$. Theoretical electrophoretic patterns were computed by numerical solution of the transport equations on a high speed electronic computer. These computations account for the essential features of the electrophoretic behavior described above, including the nonenantiography, dependence of the patterns on buffer and protein concentrations, and the mild convective disturbances sometimes observed in the ascending patterns. The results show that resolution occurs because of changes in the concentration of HA along the electrophoresis column accompanying reequilibration during differential transport of P and $P(HA)_n$ and maintenance of the resulting concentration gradients of the electrically neutral molecule. The peaks in the patterns correspond to different equilibrium compositions, not simply to P or $P(HA)_n$. Accordingly, the areas and mobilities of the peaks, which constitute a single reaction boundary, cannot be placed into correspondence with the concentrations and mobilities of P and $P(HA)_n$. The changes in the patterns brought about by lowering the protein concentrations are explained in terms of the progressively decreasing strength of the gradients of HA which deter-

---

FIG. 4. Moving-boundary electrophoretic patterns of ovalbumin in media containing varying concentration of formate buffer (Naf-Hf), pH 4.0; protein concentration, 1.3%. Essentially the same behavior is shown in media containing acetate or other carboxylic acid buffers. Thus, the particular type of nonenantiography displayed by the patterns in 0.04 M sodium formate buffer is shown typically by a variety of proteins in media containing 0.01 M sodium acetate buffer. From J. R. Cann and R. A. Phelps, *J. Amer. Chem. Soc.* **79**, 4672 (1957).

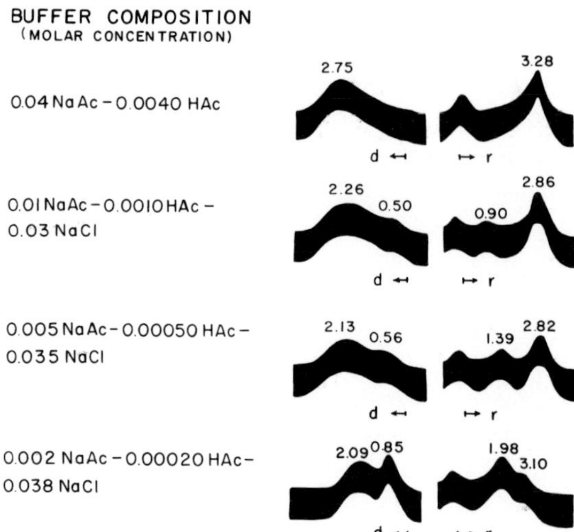

Fig. 5. Moving-boundary electrophoretic patterns of bovine γ-pseudoglobulin in media containing varying concentration of acetate buffer (NaAc-HAc), pH 5.7. Protein concentration 1.3%. From J. R. Cann and R. A. Phelps, *J. Amer. Chem. Soc.* **79**, 4672 (1957).

mine resolution. At sufficiently low protein concentration, the concentration of HA would be in such large excess as to be considered constant. Under such conditions the initial equilibrium composition would be maintained during electrophoretic transport, and the patterns would show a single peak migrating with the weight average mobility.

Whenever nonenantiographic patterns such as those presented in Figs. 4 and 5 are encountered, they must be viewed as indicative of interaction. Cognizance must also be taken of the fact that the patterns may sometimes be reasonably enantiographic, e.g., the patterns in Fig. 5 for 0.002 *M* NaAc–0.00020 *M* HAc–0.038 *M* NaCl. Nor are these considerations restricted to pH values acid to the isoelectric point of the protein. There are at least two known cases of protein–buffer interaction at pH values alkaline to the isoelectric point: (a) bovine serum albumin in phosphate–borate buffer, pH 6.2, or borate buffer, pH 8.9[48]; and (b) β-lactoglobulin in media containing acetate buffer.[49] β-Lactoglobulin is a particularly interesting case since it differs from most other proteins in that its electrophoretic behavior reveals interaction with acetic acid at pH values both acid and alkaline to its isoelectric point. In general, the patterns for protein–buffer interactions alkaline to the isoelectric point

[48] J. R. Cann, *Biochemistry* **5,** 1108 (1966).
[49] S. N. Timasheff, personal communication (1967).

exhibit the same nonenantiography as on the acid side. Clearly, conclusions as to inherent heterogeneity depend upon fractionation experiments.

If certain precautions are taken, fractionation provides an unambiguous method for distinguishing between interaction and inherent heterogeneity. The protein disappearing across the fast-moving ascending and slow-moving descending peaks is isolated. The resulting fractions are then analyzed electrophoretically under conditions identical with those used in the original separation. For interactions the fractions will behave like the unfractionated material and show multiple peaks, while for heterogeneity a single peak will be obtained. It must be stressed that validity of the fractionation test rests upon fulfillment of two requirements. First, the fractions should be analyzed at about the same protein concentration as the unfractionated control. If the concentration of the fraction is too low, the limit may be approached in which the system behaves like a single component with average transport properties. In that event, the patterns of the fraction will show a single peak thereby leading to an incorrect conclusion of heterogeneity. It is imperative, therefore, either to (a) carry out the fractionation at a sufficiently high protein concentration (perhaps 3%, depending on circumstances) so that the concentration of the fractions falls within the range (about 0.5–1.5%) usually used for analysis; (b) concentrate the fractions before dialysis; or (c) calibrate the method by analyzing the unfractionated material at several different concentrations. The second requirement is that prior to analysis the fraction be reequilibrated by dialysis against the buffer or readjusted to the same conductance and pH as the buffer by addition of a small quantity of a buffer of appropriate concentration and pH. This is an absolute requirement for, if it is not met, a fraction from an interacting system will show a single peak, once again leading to an incorrect conclusion of heterogeneity.

Occasionally fractionation is impractical and appeal must be made to other kinds of experiments.[48] Thus, prolonged electrophoresis with back compensation to increase the effective length of the electrophoresis column permits examination of the resolution of peaks in greater detail. In general, prolonged electrophoresis does not improve the resolution of poorly resolved reaction boundaries; and the boundary tends to change its shape. In the case of serum albumin in pH 8.9 borate buffer, for example, the two ascending peaks retained their integrity; but the gradient curve never returned to the base line between them. Rather, the peaks became separated by a high plateau. In the descending pattern, the faster-moving peak retained its sharpness while the slow one spread excessively to form a high and long plateau. The effect of the electric field strength on the patterns can also be examined. At sufficiently low field strength and correspondingly longer time of electrophoresis, the gradients of small molecule upon which resolution depends cannot be maintained against diffusion. Consequently, resolution cannot occur. Nor will the symmetrical, unimodal boundaries show excessive spreading. This test has also been applied to the serum albumin-borate interaction with results described.

Protein-buffer interactions may be distinguished from other types of interaction by the combined use of ultracentrifugation and electrophoresis. After eliminating association-dissociation reactions by sedimentation analyses, protein-buffer interaction can be identified electrophoretically by systematic variation of buffer concentration at constant pH and ionic strength (Figs. 4 and 5).

The question naturally arises as to how one can characterize highly purified proteins with respect to parameters such as their isoelectric pH when strong interactions with the solvent is indicated and noninteracting buffers cannot be found. The method suggested by both theory and experimentation would be to lower the protein concentration to the point where the concentration of the interacting constituent of the buffer is in sufficiently large excess as to be effectively constant. Under such conditions the electrophoretic patterns of a truly homogeneous but interacting protein will show a single peak. Alternatively, it may be possible to increase the buffer concentration sufficiently that all the protein is complexed and the interacting small molecule is in overwhelming excess. While this sometimes yields a single peak, enantiography may not be restored, i.e., the descending peak remains broad and the ascending one hypersharp.

Above we have been concerned with the implications of protein–buffer interactions for moving-boundary electrophoresis, but such interactions are also of considerable importance for zone electrophoresis.[47] An extension of the theory of electrophoresis of interacting systems mentioned previously predicts that a single macromolecule, which interacts cooperatively and reversibly with an uncharged constituent of the solvent with concomitant change in electrophoretic mobility, can give two well resolved and intense zones despite instantaneous establishment of equilibrium.[50] Examples of multiple zones due to protein-buffer interaction include bovine serum albumin in phosphate-borate buffer, pH 6.1, on

[50] Less cooperative interactions (i.e., lower values of n) can give single trailing zones of the type usually attributed either to adsorption of the protein on the solid support or to known inherent heterogeneity. In the absence of demonstrated heterogeneity, protein–buffer interaction must also be entertained as a possible explanation of trailing zones.

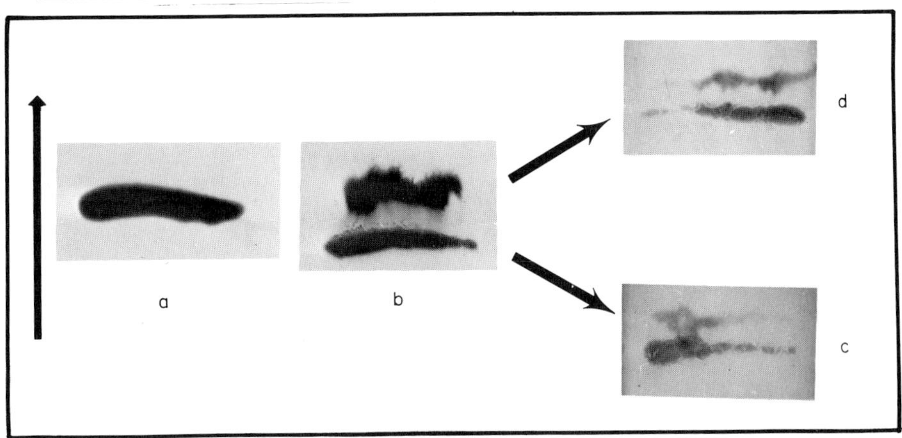

FIG. 6. Demonstration that the two zones shown by bovine serum albumin (BSA) in phosphate–borate buffer on cellulose acetate arise from reversible protein–buffer interaction and are not due to inherent heterogeneity. (a) Zone pattern of BSA in 0.0043 *M* Na_2HPO_4–0.03 *M* NaH_2PO_4, pH 6.07, 5% protein; (b) pattern in 0.012 *M* Na_2HPO_4–0.0081 *M* NaH_2PO_4–0.4 *M* H_3BO_3, pH 6.15, 5%; (c) zone pattern in phosphate–borate buffer of material eluted from the slower migrating zone of pattern b; (d) pattern of material from faster zone. Vertical arrow indicates direction of migration. From J. R. Cann, *Biochemistry* **5,** 1108 (1966).

cellulose acetate (Fig. 6b); bovine serum albumin in borate buffer, pH 7.2, on cellulose acetate (Fig. 7); and conalbumin in borate buffer, pH 8.9, on starch gel.[51] Such patterns could easily be misinterpreted as indicating inherent heterogeneity; but with certain precautions, fractionation provides an unambiguous method for distinguishing between interaction and heterogeneity. The protein eluted from each unstained zone is subjected to zone electrophoresis in the same buffer that was used for the original separation. For interactions the fractions will behave like the unfractionated material and show multiple zones, while for heterogeneity a single zone will be obtained. A case in point is presented in Fig. 6. Whereas bovine serum albumin gives a single zone in 0.043 ionic strength phosphate buffer at pH 6.1 (Fig. 6a), it shows two zones in a phosphate–borate buffer of about the same pH and ionic strength (Fig. 6b). Each of the fractions from the latter pattern also shows two zones (Figs. 6c and d), which demonstrates that the duplicity of zones arises from reversible interaction of albumin with boric acid or borate anion and is not due to inherent heterogeneity.

Validity of the fractionation test is critically dependent upon careful scrutiny of the concentration dependence of the electrophoretic patterns.

[51] W. C. Parker and A. G. Bearn, *Nature* (*London*) **199,** 1184 (1963).

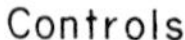

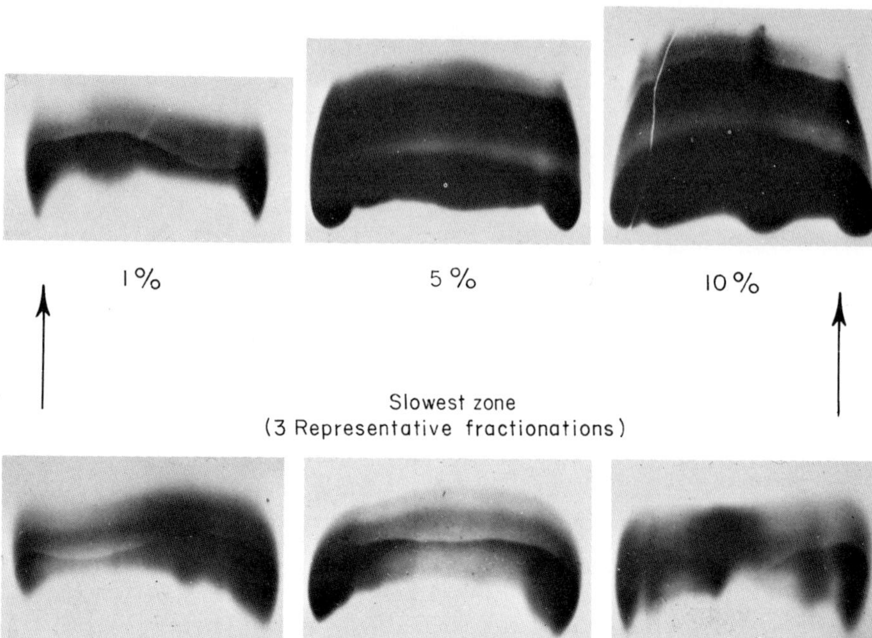

FIG. 7. Demonstration that the two or three zones shown by bovine serum albumin in borate buffer (0.0108 M NaOH–0.260 M H_3BO_3 originally added to prepare buffer, pH 7.15) on cellulose acetate arise from interaction of the protein with the buffer and are not due to an inherent heterogeneity. Zone patterns of unfractionated controls at three different protein concentrations (0.5% protein showed only a single zone; the fastest zone shown by 10% protein shaded off in the direction of migration, but the shaded portion could not be captured photographically) and of three representative fractions obtained from 10% protein by cutting the slowest zone from unstained strips and removing the protein by centrifugation. Vertical arrow indicates direction of migration. From J. R. Cann, *Biochemistry* **5**, 1108 (1966).

Although in certain instances the generation of multiple zones due to interaction may be relatively insensitive to protein concentration, in general this will not be so. An example is the interaction of conalbumin with borate buffer at pH 8.9.[51] This system gives a single zone at low protein concentration, two at intermediate, and three at high concentration. Bovine serum albumin exhibits similar properties in borate buffer, pH 7.2. Although resolution into two or three zones occurs at protein concentrations in the range 1–10% (Fig. 7), patterns obtained at lower concentrations show only a single zone. The possibility of macromolecular association was eliminated in both cases by ultracentrifugal analyses. Experiments on the albumin system designed to distinguish between interaction with the buffer and heterogeneity illustrate the sole pitfall of the fractionation test alluded to above. Thus, a preliminary experiment gave a fraction whose concentration was inadvertently so low that its electrophoretic pattern showed only a single zone as in a control analysis of unfractionated protein at low concentration. In subsequent experiments, therefore, care was taken to obtain fractions of sufficiently high concentration to ensure against possible misinterpretation should they too give a single zone. These fractions consistently showed not one but two or three zones (Fig. 7), thereby demonstrating that the multiple zones arise from interaction of albumin with borate buffer and are not indicative of inherent heterogeneity. The point to be made is that an isolated observation of the sort made in the preliminary experiment, but without the appropriate control, most likely would be misinterpreted to mean heterogeneity.

The recommended procedure for dealing with this problem involves preliminary examination of the concentration dependence of the zone patterns of the unfractionated protein; fractionation at a sufficiently high protein concentration to ensure against possible misinterpretation in case the fractions show a single zone on reanalysis; or restoration of the concentration of the fractions to that used in the original separation followed by dialysis. The requirement that unconcentrated fractions be reequilibrated by dialysis against buffer prior to analysis while an absolute one in moving boundary electrophoresis, will in general be met by the elution process itself.

Aside from fundamental investigations on protein interactions per se, it is obviously desirable to avoid conditions conductive to the production of electrophoretic patterns which do not faithfully reflect the inherent state of homogeneity of the protein. Multizoning due to protein–buffer interactions have been encountered with carboxylate, phosphate–borate, borate, Tris-borate, and amino acid buffers. Choice of a buffer in which a particular protein at specified pH and ionic strength will not show multiple zones due to interaction is largely empirical, although there are some guide lines. For example, electrophoresis in phosphate and barbital buffers appears to be free from these complications, at least for bovine serum albumin at pH 8.9, 7.2, 6.1, and 2.3. If strong interaction with the solvent is unavoidable, one can attempt to minimize interpretative difficulties by lowering the protein concentration and/or increasing the buffer concentration. At a sufficiently low protein to buffer concentration ratio a truly homogeneous but interacting protein will in principle show a single zone. But these conditions may be difficult to realize in practice, and failure to do so must not be accepted in lieu of fractionation as indicative of inherent heterogeneity.

Finally, a word of caution concerning interpretation of isoelectric focusing patterns is in order. It is conceivable that a given protein may interact with the supporting ampholyte to give multiple banding. Also, the considerations presented above for zone electrophoresis are equally applicable to chromatography. Possible complications of this sort will become apparent, however, when the fractions themselves are analyzed to see whether they run true.

[12] The Applicability of Acrylamide Gel Electrophoresis to Determination of Protein Purity

By ROBERT F. PETERSON

Acrylamide gel electrophoresis has been covered extensively by several authors in this series of volumes in articles on analytical disc electrophoresis, slab vertical cells, and preparative disc electrophoresis.[1-4] At this time it would seem appropriate to summarize the experience of the past several years and detail the precautions to be taken with this widely used procedure. The two basic questions we ask of analytical gel electrophoresis are: Is a protein pure if it migrates as a single band? and: Is it necessarily heterogeneous if it shows more than one band? At each step of the procedures we shall discuss the precautions necessary to avoid misleading results.

All the chemicals used for acrylamide gel electrophoresis, excepting buffer components, should be stored at refrigerator temperatures, 0–5° to prevent deterioration. Acrylamide monomer and the cross-linking agent, methylenebisacrylamide, as well as the commercial mixture of these, Cyanogum-41,[5] slowly polymerize, resulting in decreased solubility in buffer solutions. Ammonium persulfate should be tested by measuring the pH of a 1% solution in water; if the pH is below 2, a new batch should be used. The amines which are used for the other half of the

[1] See Vol. 5 [3], pp. 48–49.

[2] See Vol. 10 [103], pp. 676–680.

[3] See Vol. 11 [17], pp. 179–186.

[4] See Vol. 12, part B [96], pp. 32–37.

[5] Reference to a company or product name does not imply approval or recommendation of the product by the U.S. Department of Agriculture to the exclusion of others that may be suitable.

catalyst pair discolor with time. If a catalyst, such as 0.2% ammonium persulfate and 0.1% *N,N,N′,N′*-tetramethylethylenediamine (TEMED), is not active enough to gel 5% acrylamide solution within 20 minutes, the gel produced in a slab cell may be heterogeneous, resulting in uneven mobilities across its width for identical samples.

The high resolution of gel electrophoresis methods is due to the production of very thin starting zones, followed by imposition of the constraint of molecular sieving on movement due to the electrical force field. The species to be examined must enter the gel. If an examination of the dyed gel shows a heavy residue where the sample should have entered the gel, the pore size is too small. Blattler and Reithel[6] have published a table correlating molecular weights of proteins which are totally excluded by gels to the acrylamide concentration. A series of experiments with decreasing concentrations of acrylamide will provide data for choosing a gel which will give the optimum separation of components.

When the proteins are highly aggregated, urea and often 2-mercaptoethanol are used to obtain monomers of the proteins. If analytical grade urea, low in cyanate, is used and the buffer solutions are acidic, carbamylation of the lysine groups of the protein will not occur. However, at alkaline pH there is always a possibility that artifactual bands may be produced. Urea solutions may be passed through mixed-bed ion exchangers (Amberlite MB-1 with 5% IR-128 added) to remove cyanate.[7] A protein solution, once dissolved in urea at alkaline pH, should be used immediately. If carbamylation is suspected, cyanate may be added to protein samples to see whether the suspected bands are intensified. Guanidine hydrochloride cannot be used as a dissociating agent because of the high conductivity of its solutions. Sodium dodecyl sulfate (SDS) is an excellent dispersing agent, but completely alters the determination, since small charge differences are equalized owing to the high adsorption of SDS.

Electrophoresis in buffers containing 0.1% SDS and mercaptoethanol at pH 7–9 is an excellent method for determining monomer chain molecular weights[8] (cf. Weber, Part C). Because of the high adsorption of SDS, averaging 1.4 g per gram of protein, differences in net charge due to amino acid replacement are imperceptible, and mobility is strictly a function of molecular weight. The logarithm of molecular weight plotted against distance moved in the gel is a straight-line function. Hetero-

[6] D. P. Blattler and F. J. Reithel, *J. Chromatogr.* **46**, 286 (1970).

[7] M. P. Thompson, personal communication.

[8] A. L. Shapiro, E. Viñuela, and J. V. Maizel, Jr., *Biochem. Biophys. Res. Commun.* **28**, 815 (1967).

geneity, due to higher or lower molecular weight contaminants, is thus easily detected.

The choice of buffers may be illustrated with protein polymorphs which differ by replacement of a single charged amino acid. The series of β-caseins A^1, A^2, and A^3, which contain respectively 6, 5, and 4 histidine residues per molecule of 24,000 molecular weight, can be resolved in 10% Cyanogum-41 gels at pH 3 in formic–acetic acid buffers containing 4.5 *M* urea. The same β-caseins are not resolved at pH 9 because the histidine side chains are not charged. Analysis at a single pH is not a sufficient test for heterogeneity.[9]

Oxygen inhibits polymerization. For this reason deaerated water is used to cover the acrylamide solutions being polymerized in tubes. In flat slab cells, such as those of Raymond's design,[10] air is excluded by rubber gaskets. The area exposed around the slot formers will not gel completely unless the assembled cell is put in a polyethylene bag and nitrogen is used to flush out the air. When acrylamide concentrations are materially below 5%, this precaution may be essential.

Catalyzing the polymerization of acid solutions of acrylamide can be accomplished by increasing the proportion of amine to persulfate over the amount used in alkaline solutions. However, the amine increases the pH of the buffer and necessitates a lengthy prerun to remove the catalyst by-products from the gel. In this laboratory we use 0.5 ml of 30% hydrogen peroxide and 0.35 g of thiourea per 100 ml of acrylamide solution. The hydrogen peroxide cannot be detected with tolidine–KI reagent a few minutes after mixing and makes no contribution to the ionic strength of the buffers. Therefore, the necessity of lengthy preruns to reduce the conductivity of the gels is eliminated.

In water-cooled cells higher voltages may be applied, with consequent improvement in resolving power. If identical samples produce a curved front across a gel slab, the voltage is too high. In analytical gels the proteins are usually rendered visible by dyeing the gel in acetic acid solutions of an acid dye. If the protein is not precipitated by acetic acid or is soluble in acetic acid, this procedure will not work. Amido Black 10B (Color Index 20470) has been used by many workers. Insufficient dyeing time will cause a concentrated protein band to appear as two bands. A recent method of Fazekas de St. Groth consists of fixing and dyeing simultaneously in 0.025% solution of Coomassie Brilliant Blue which contains 10% trichloroacetic acid in acetic acid–methanol–water (14:40:160, v/v/v).[11] The acrylamide gels are dyed for 24 hours

[9] R. F. Peterson, Jr. and F. C. Kopfler, *Biochem. Biophys. Res. Commun.* **22**, 388 (1966).

[10] S. Raymond, *Clin. Chem.* **8**, 455 (1962).

[11] S. Fazekas de St. Groth, R. G. Webster, and A. Datyner, *Biochim. Biophys. Acta* **71**, 377 (1963).

and then washed free of excess dye with acetic acid–methanol–water (14:40:160, v/v/v). In our laboratory this is very effective for the detection of protein at low concentration and is successful in dyeing proteins complexed with SDS in molecular weight determination gels.

Until the advent of scaled-up disc electrophoresis as a preparative method, proteins had been detected by dye binding and had not been subjected to chemical analysis. Work in this laboratory showed that casein components, isolated from urea-containing gels, were being extensively degraded. A prerun removed persulfate ions from the gel. However, it did not stop the degradation; nor did photopolymerizing the acrylamide preparative columns. However, a prerun with hydroquinone would scavenge the gels and proteins could be recovered without damage.[12]

Because the termination step in free-radical polymerization involves the combination of two free radicals to form a homopolar bond (and this might not be possible in a rigidly cross-bonded gel), free radicals may exist for a long time after gel polymerization. When acrylamide monomer was polymerized in aqueous solution using the γ-ray flux from the Yale electron accelerator, a strong electron spin resonance (ESR) signal was obtained at 3400 gauss. Persulfate-amine polymerized gels had a weaker resonance at the same point, and this was not decreased by lengthy electrophoresis to remove the persulfate. When caseins were analyzed for tryptophan, this amino acid was found to be totally destroyed after preparative gel electrophoresis in 10% acrylamide gels. β-Lactoglobulin was eluted as a symmetrical peak from 10% acrylamide gels which did not contain urea. Nevertheless, the tryptophan content was considerably reduced. Histidine and tyrosine were essentially unaffected by preparative gel electrophoresis. Again, a preliminary run with a scavenging agent such as cysteine, thioglycolate, or hydroquinone would prevent damage to proteins during preparative electrophoresis on acrylamide gels.[13]

A prerun interferes with the stacking gel-running gel systems.[2] However, Hjertén and co-authors[14] pointed out in 1965 that equal resolution could be obtained in single running gel systems, if the sample of protein had a lower conductivity than the following buffer. The practice of embedding the sample in a separate gel, which is photopolymerized with a riboflavin catalyst, has been discarded in most preparative work, since addition of sucrose to the sample will produce good layering on the

[12] R. F. Peterson and E. Bradshaw, *Abstr. 154th Meeting ACS,* C-135, Chicago, Illinois, Sept. 1967.

[13] R. F. Peterson, Sr. and R. F. Peterson, Jr., *Abstr. 158th Meeting ACS, B.*-151, New York, Sept. 1969.

[14] S. Hjertén, S. Jerstedt, and A. Tiselius, *Anal. Biochem.* **11,** 219 (1965).

running gel. Pastewka and co-workers[15] have recently noted that exposure of hemoglobin solutions to the fluorescent light used for riboflavin-catalyzed polymerization of stacking gels resulted in the formation of artifacts.

Cann (cf. [11]) has shown that artifacts can be produced by protein–buffer interactions in gel electrophoresis.[16] Bovine serum albumin may show two or three zones due to reversible protein–borate buffer interactions. Conalbumin also exhibits this behavior. Myoglobin, however, has been separated by pH-gradient preparative electrophoresis into three components which may be identical with the species previously suspected to be buffer interaction artifacts.[17]

With all the possible causes mentioned for artifacts, no infallible method exists for determining heterogeneity. Analytical gel electrophoresis at acid and basic pH are a minimum for asserting purity. Analytical molecular weight determination in SDS gels is sensitive to small amounts of impurities which differ in molecular weight.

We still have no detection method for genetic variants which involve nonpolar amino acids. The mobility differences observed by Kalan *et al.*[18] for two genetic forms of swine whey protein, which differ only in alanine and valine content, can perhaps be attributed to changes in the environment of charged groups.

[15] J. V. Pastewka, A. T. Ness, and A. C. Peacock, *Anal. Biochem.* **35**, 160 (1970).
[16] J. R. Cann, *Biochemistry* **5**, 1108 (1966).
[17] R. F. Peterson, unpublished observations (1970).
[18] E. B. Kalan, R. R. Kraeling, and R. J. Gerrits, *Int. J. Biochem.* **2**, 232–44 (1971).

Section IV

Cleavage of Disulfide Bonds

[13] Reduction of Disulfide Bonds in Proteins with Dithiothreitol

By WILLIAM KONIGSBERG

The cleavage of disulfide bridges in proteins can be accomplished in several ways. The simplest way of converting the half-cystine residues to a chemically stable form after disulfide bond cleavage is by performic acid oxidation.[1] Because of the excellent solvent and denaturing properties of performic acid, the use of urea and guanidine-HCl as denaturants can be avoided, thus permitting the oxidized protein to be recovered simply by lyophilization. The main disadvantages of performic acid oxidation are its lack of reversibility and the fact that it destroys tryptophan, giving rise to a series of products that are unstable. For these reasons, reductive disulfide bond cleavage is the most generally used procedure. The reagent of choice for the reduction was β-mercaptoethanol[2] until recently, when dithiothreitol (DTT) became commercially available. The use of DTT as a protective reagent for sulfhydryl groups was first described by Cleland,[3] and the reagent bears his name. DTT has now been widely adopted for reducing disulfide bridges, regenerating protein sulfhydryl groups from other reversible SH blocking reagents, and protecting protein SH groups from air oxidation.

Principle

The reaction of DTT with disulfides occurs as follows:

$$\begin{array}{c} R{-}S{-}S{-}R \\ + \\ HS{-}CH_2{-}\underset{}{\overset{OH}{\overset{|}{C}}}H{-}\overset{OH}{\overset{|}{C}}H{-}CH_2{-}SH \end{array} \rightleftharpoons \begin{array}{c} RSH \\ + \\ R{-}S{-}S{-}CH_2{-}\underset{OH}{\underset{|}{C}}H{-}\underset{OH}{\underset{|}{C}}H{-}CH_2{-}SH \end{array} \qquad (1)$$

$$R{-}S{-}S{-}CH_2{-}\overset{OH}{\overset{|}{C}}H{-}\overset{OH}{\overset{|}{C}}H{-}CH_2{-}SH \rightleftharpoons \begin{array}{c} \text{(cyclic: } CH_2{-}S{-}S{-}CH_2{-}C(H)(OH){-}C(H)(OH){-}) \end{array} + RSH \qquad (2)$$

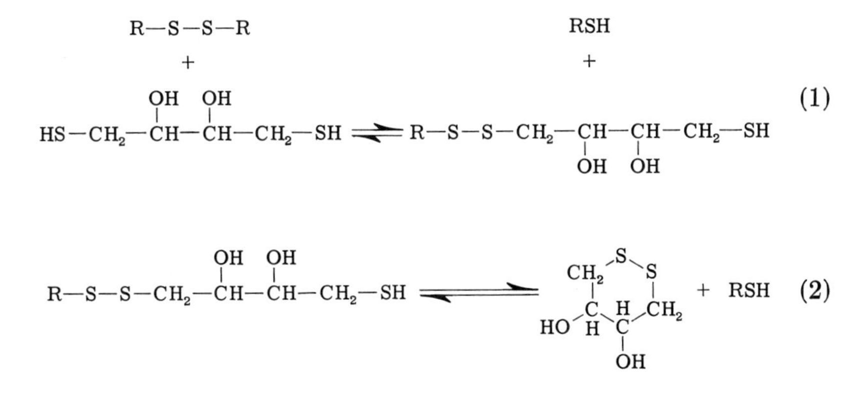

[1] C. H. W. Hirs, Vol. 11, p. 197.

[2] C. H. W. Hirs, Vol. 11, p. 199.

[3] W. W. Cleland, *Biochemistry* **4**, 480 (1964).

The overall reaction proceeds nearly to completion because the formation of a 6-membered ring containing a disulfide bridge in reaction (2) is energetically favored over the mixed disulfide formed in reaction (1). For instance, attempts to measure the overall equilibrium constant of reactions (1) and (2) where R—S—S—R was cystine failed even though a 10-fold excess of the oxidized, cyclized form of DTT was used,[3] since reaction (1) proceeded almost exclusively to the right. The oxidation reduction potential of DTT at pH 7 and 25° is —0.33 V[3] compared to —0.22 V for cysteine.[4] These two values allow one to calculate an overall equilibrium constant of 10^4 for the reduction of cystine by DTT. Since the oxidation-reduction potential of β-mercaptoethanol is close to that of cysteine, the apparent overall equilibrium constant for the reaction

$$2HOCH_2CH_2SH + R—SS—R \rightleftharpoons HOCH_2CH_2S—SCH_2CH_2OH + 2RSH$$

is close to unity; thus a large excess of β-mercaptoethanol must be used to attain complete reduction.

Procedures

Limited Reduction. It is often desirable to limit the reduction of disulfide bonds to those in exposed positions in the native molecule. Disulfide bridges holding together two or more polypeptide chains are often accessible to solvents and reagents without unfolding of the polypeptide chain whereas intrachain disulfide bridges usually are not. This difference makes it possible to selectively reduce the exposed disulfide bonds without affecting those in the interior. A typical procedure is the one described for the reduction of the interchain disulfide bridges in a γ_G-immunoglobulin.[5]

The protein at a concentration of 2% was dissolved in 0.15 *M* Tris buffer, pH 8.0. The solution was also 0.15 *M* in NaCl, which maintained the high solubility of the immunoglobulin. The solution also contained 0.01 *M* dithiothreitol (~60-fold molar excess compared to the moles of protein) and 0.002 *M* EDTA to chelate any metals that might catalyze the oxidation of SH groups. The reaction mixture was allowed to stand at 25° for 2 hours. After this time an alkylating agent (in this case iodoacetamide) was added, and the reaction was allowed to stand for 20 minutes before removal of the excess reagents by dialysis. The effect of varying DTT concentration, time, and the concentration of alkylating agent in this system was studied by Gall *et al.*[6] Between the limits of a

[4] J. S. Fruton and H. T. Clarke, *J. Biol. Chem.* **106,** 667 (1934).

[5] G. M. Edelman, W. E. Gall, M. J. Waxdal, and W. H. Konigsberg, *Biochemistry* **7,** 1950 (1968).

[6] W. E. Gall, B. A. Cunningham, M. J. Waxdal, W. H. Konigsberg, and G. M. Edelman, *Biochemistry* **7,** 1973 (1968).

30- to 100-fold molar excess of DTT and reduction times of 1–2 hours, the reaction with interchain disulfide bonds went to completion without reduction of any intrachain disulfide bridges. When separating light and heavy chains, after partial reduction but without prior alkylation, 0.002 *M* DTT was added to the 1 *M* propionic acid used in separating the chains on Sephadex G-100.

Complete Reduction. The requirement for complete reduction is that the polypeptide chain be unfolded, exposing all the disulfide bridges, and that excess reducing agent be present. A typical general procedure is as follows: A 2% solution of protein in 6 *M* guanidine hydrochloride (recrystallized) containing 0.5 *M* Tris buffer, pH 8.1, and 0.002 *M* EDTA was placed in a flask. After flushing with N_2 the flask was stoppered and heated at 50° for 30 minutes to completely denature the protein. A 50-fold molar excess of DTT (compared to moles of disulfide in the protein) was added. The flask was flushed once again with N_2, then the reaction was allowed to proceed at 50° for 4 hours. After cooling, the appropriate alkylating agent was added. This method was applied to the CNBr fragments of a γ_G myeloma protein,[7] where complete reduction of all the inter- and intrachain disulfide bridges was achieved.

Comments

The cleavage of disulfide bridges is of importance in both functional and structural studies of proteins. Of the three available methods for accomplishing this, oxidation, sulfitolysis and reduction, the third procedure is often the method of choice since the first method suffers from lack of reversibility as well as undesirable side reactions, and the second from the difficulty in achieving complete reaction. While β-mercaptoethanol has been extensively used for both selective and complete reduction of disulfide bridges, it can be replaced by DTT. This reagent can be used at a much lower concentration than β-mercaptoethanol by virtue of its lower oxidation-reduction potential and its resistance to air oxidation compared to β-mercaptoethanol. This can be of particular importance when radioactive alkylating agents are to be employed, since only a very small molar excess of DTT is required for complete reduction and therefore only minimal quantities of radioactive material need be wasted in alkylating excess reducing agent. When complete reduction is undertaken in a denaturing solvent such as 8 *M* urea of 6 *M* guanidine HCl, care must be exercised in purifying these chemicals since the presence of cyanate in the urea will cause modification of protein at NH_2 groups and polymeric UV-absorbing materials in crude guanidine HCl

[7] M. J. Waxdal, W. H. Konigsberg, W. L. Henley, and G. M. Edelman, *Biochemistry* **7**, 1959 (1968).

can be difficult to remove at a later stage. An additional advantage of DTT is its relative lack of the characteristic unpleasant thiol odor. A disadvantage of DTT is its relatively high cost relative to β-mercaptoethanol. Nevertheless the reagent has come into wide use for both reduction of disulfide bridges and protection of sulfhydryl groups against air oxidation.

Section V

Cleavage of Peptide Chains

[14] Maleylation of Amino Groups

By P. J. G. BUTLER and B. S. HARTLEY

Maleic anhydride readily acylates α- and ε-amino groups in proteins at mild alkaline pH.[1,2] The extent of reaction can be conveniently measured by the absorption at 250 nm. The maleyl proteins have a large excess of negative charge at or above neutral pH, and the resulting electrostatic repulsion has a dissociating and solubilizing effect, as already described for succinyl derivatives.[3] This effect is particularly useful for determining molecular weights of subunits of multimeric proteins, since the subunits can be examined by ultracentrifugation or gel filtration in the same buffers as the native protein.[4,5] However, unlike the succinyl group, the maleyl group is readily removed by hydrolysis of the amide bond at mildly acid pH (Fig. 1), as a result of an intramolecular catalysis by the adjacent, locked carboxylic acid group. The unsubstituted protein can thus be recovered.

Maleylation of the lysyl residues allows specific cleavage by trypsin at the arginyl bonds. The maleyl amino group is very stable at alkaline pH and is therefore unlikely to hydrolyze during prolonged enzymatic digestions at alkaline pH, a hazard inherent in alternative reversible blocking groups such as trifluoroacetyl-,[6] or tetrafluorosuccinyl-,[7] which are alkali labile. Cleavage at some arginyl residues in maleylated proteins seems to occur much more slowly than at others, probably because of inhibition by adjacent negative charges, and this kinetic specificity can be an advantage in obtaining large peptides for sequence studies. The maleyl group can, of course, be removed from each peptide by acid incubation to allow further tryptic cleavage at lysine residues. The charge change of +2 consequent on such unblocking is the basis of a "diagonal" electrophoretic separation[7a] which allows N-terminal and lysine-contain-

[1] P. J. G. Butler, J. I. Harris, B. S. Hartley, and R. Leberman, *Biochem. J.* **103,** 78P (1967).

[2] P. J. G. Butler, J. I. Harris, B. S. Hartley, and R. Leberman, *Biochem. J.* **112,** 679 (1969).

[3] I. M. Klotz, Vol. 11, p. 576.

[4] P. J. G. Butler, H. Jörnvall, and J. I. Harris, *FEBS (Fed. Eur. Biochem. Soc.) Lett.* **2,** 239 (1969).

[5] C. J. Bruton and B. S. Hartley, *Biochem. J.* **108,** 281 (1968).

[6] R. F. Goldberger, Vol. 11, p. 317.

[7] G. Braunitzer, K. Beyreuther, H. Fujiki, and B. Schrank, *Hoppe-Seyler's Z. Physiol. Chem.* **349,** 265 (1968).

[7a] J. R. Brown and B. S. Hartley, *Biochem. J.* **101,** 214 (1966).

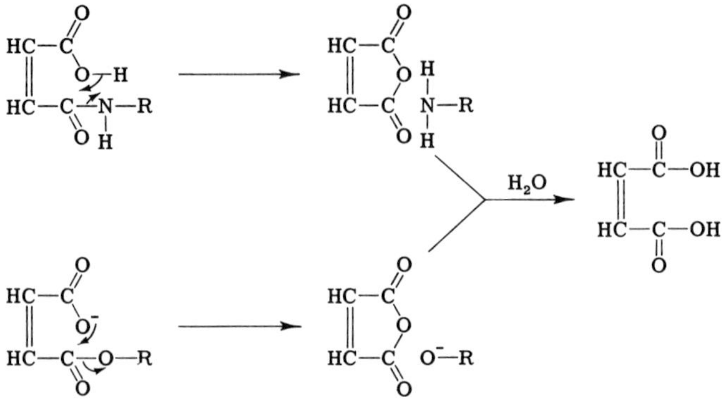

FIG. 1. Possible mechanisms for unblocking of maleyl derivatives.

ing peptides to be specifically purified from enzymatic digests. An alternative procedure, using ^{14}C-labeled maleic anhydride, has allowed specific purification of maleylated N-terminal peptides from as little as 1 mg of protein, which can then be unblocked and sequenced by a micromodification of the DNS-Edman procedure.[8]

Reaction Conditions

As with most acylating reagents, the reaction of maleic anhydride with protein amino groups is competitive with hydrolysis of the reagent by water or hydroxyl ions. Optimal conditions will therefore employ a high concentration of reagent in excess of the total concentration of amino groups, at a pH which maximizes the concentration of unprotonated amino groups but minimizes hydrolysis of the reagent. The effect of this competition is demonstrated in Fig. 2, which shows the fraction of amino groups reacting at different pH values.

In practice, complete blocking of amino groups is found after a protein has been allowed to react at concentrations of 5–10 mg/ml with 0.03–0.10 *M* maleic anhydride[9] at pH 8.5–9.0 between 0° and 30°.[2] Pyrophosphate or carbonate buffers are convenient; amine buffers should obviously be avoided. Urea or guanidine hydrochloride do not affect the reaction and can therefore be added to solubilize an insoluble protein or to unfold the protein and expose "buried" amino groups. The fast liberation of acid can be followed by titrating with sodium hydroxide in a pH-stat after the addition of maleic anhydride in a small quantity of a dry, miscible organic solvent such as dioxane or acetonitrile. It is often

[8] C. J. Bruton and B. S. Hartley, *J. Mol. Biol.* **52**, 165 (1970).

[9] An error in Table 1 of Butler *et al.*[2] indicates protein and reagent concentrations of 0.1–30 μM instead of 0.1–30 m*M*.

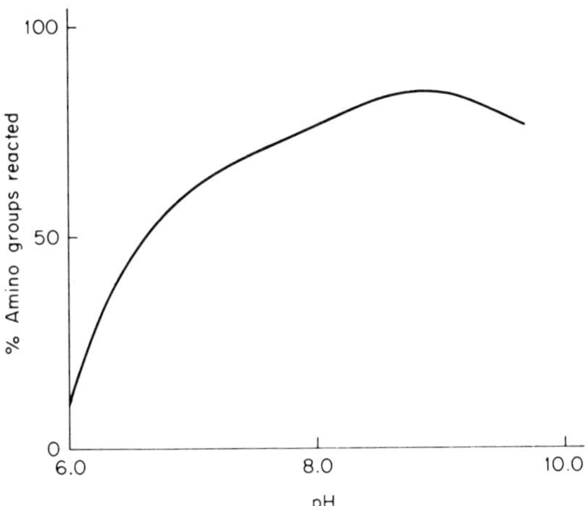

FIG. 2. Effect of pH on the extent of reaction of maleic anhydride with protein amino groups. Maleic anhydride (5.0 mM) reacting with chymotrypsinogen (0.1 mM, 1.5 mM in amino groups) at 2°.

convenient to conduct the reaction at 2° to avoid pH fluctuations, and to add the reagent in aliquots to a rapidly stirred solution, since the reaction is complete within 5 minutes. Alternatively, solid maleic anhydride can be added to a well-stirred, buffered protein solution, the pH of which is monitored with a pH-meter, with the addition of solid sodium carbonate to maintain the pH. Dilution is thereby avoided, and the relatively slow rate of solution of the reagent avoids pH fluctuation. For small samples, addition of a trace of phenolphthalein indicator allows one to carry out a similar operation without a pH meter.

Reaction is complete when no further acid is liberated, and the maleylated protein can be recovered after dialysis or gel filtration at alkaline pH.

Side Reactions

In theory the reaction should be highly specific for amino groups, since the phenolic group of tyrosyl residues or the imidazole group of histidyl residues yield products which are extremely labile to hydrolysis at pH 9 (e.g., as shown in Fig. 1). Serine and threonine hydroxyl groups are much less nucleophilic and should react to a negligible extent in the above reaction conditions. There have nevertheless been reports that some seryl or threonyl hydroxyl groups may be maleylated,[10,11] but

[10] M. H. Freedman, A. L. Grossberg, and D. Pressman, *Biochemistry* **7,** 1941 (1968).
[11] H. Jörnvall, *Eur. J. Biochem.* **16,** 25 (1970).

reaction with 0.8 *M* hydroxylamine at pH 9, for 18 hours at 4° appeared to unblock these without affecting the maleylated amino groups.[10]

Thiol groups present a special problem, because these can be *alkylated* by maleic anhydride or, more seriously, by the hydrolysis product maleic acid. Such reaction will be quantitative under most maleylation conditions, to yield stable *S*-(2-succinic acid)-cysteinyl residues.[12,13] Hence, if it is necessary to recover the unmodified peptide after unblocking, thiol groups should first be reversibly protected by reaction with sulfite or by disulfide exchange (e.g., with cystine). Possible dangers arise from the use of impure maleic anhydride, since the reagent might form polymers, e.g.,

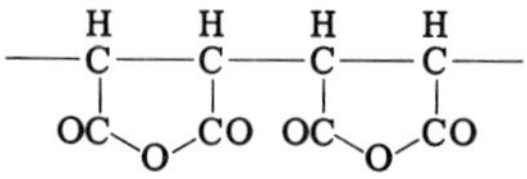

```
    H     H     H     H
 ---C-----C-----C-----C---
    |     |     |     |
   OC     CO   OC     CO
     \   /       \   /
       O           O
```

which could cross-link peptides and whose product would be less easily hydrolyzed at pH 3.5 than the maleyl group. Sublimation of the reagent or recrystallization from chloroform is therefore recommended.

Estimation of Maleyl Groups

The maleyl groups bound to the amino groups of a protein are readily estimated spectrophotometrically in 0.1 *M* sodium hydroxide. Although the spectrum of the maleyl groups will not change about pH 7, it is necessary to use such a denaturing solvent in order to obtain a constant absorption from the protein both before and after the reaction with maleic anhydride. The absorption spectrum of ϵ-maleyllysine (and of lysine plus maleic acid) is shown in Fig. 3, giving $\epsilon_{250} = 3360$ and $\epsilon_{280} = 308$ for the maleyl amino group and $\epsilon_{250} = 939$ and $\epsilon_{280} = 59$ for maleic acid and amino groups. From these both the extent of blocking and also the extent of subsequent unblocking of the amino groups can be estimated.

Removal of Maleyl Blocking Groups

The maleyl blocking groups are readily removed by an intramolecularly catalyzed hydrolysis at acid pH (Fig. 1). The half-life for hydrolysis at different pH values is shown in Fig. 4. From this the optimum pH for general use is 3.5, giving the maximum rate of unblocking with the minimum of side reaction. At 37° this pH gives a half-life for the unblocking reaction of between 11 and 12 hours.

[12] E. J. Morgan and E. Friedmann, *Biochem. J.* **32**, 733 (1938).

[13] D. G. Smyth, O. O. Blumenfeld, and W. Konigsberg, *Biochem. J.* **91**, 589 (1964).

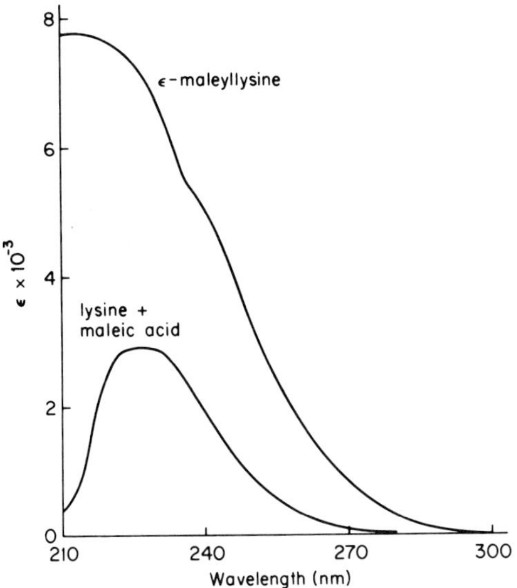

FIG. 3. Spectra of ε-maleyllysine and of lysine plus maleic acid in 0.1 *M* sodium hydroxide.

A convenient buffer for the reaction is 1% (v/v) pyridine, 5% (v/v) acetic acid. This has the advantage of volatility and also that the protein or peptide can be dissolved in 1% pyridine and then an appropriate amount of acetic acid added to give a final concentration of 5%. Even

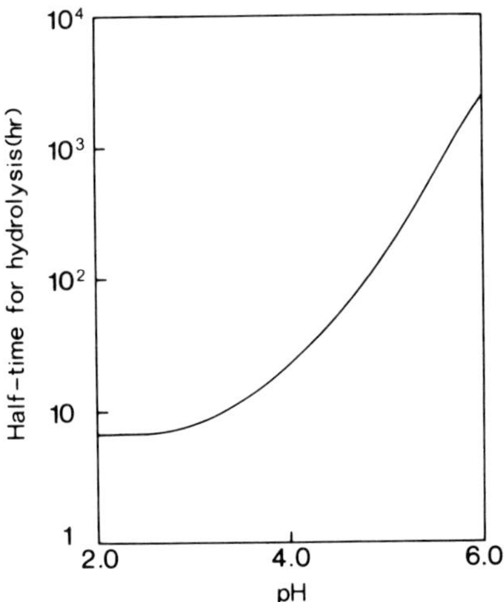

FIG. 4. Half-life of ε-maleyllysine at 37°.

though the material may precipitate under these conditions, the precipitate is wet and the reaction proceeds at the same rate as that with a soluble material. No significant difference was found in the rate of unblocking of maleylchymotrypsinogen at pH 3.5, in aqueous solution, where it is insoluble, or in 5 *M* guanidine hydrochloride, where it is soluble.[2]

Possible reactions which might interfere with unblocking at pH 3.5 are the geometrical inversion of maleyl groups to fumaryl groups or condensation, with the elimination of water, to yield a maleoyl derivative:

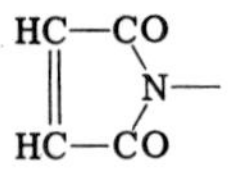

$$\begin{array}{l} HC{-}CO \\ \| \qquad \backslash \\ \| \qquad\; N{-} \\ \| \qquad / \\ HC{-}CO \end{array}$$

This latter derivative would be unstable at alkaline pH (cf. *N*-ethylmaleimide) and the ring would reopen to regenerate the maleyl derivative. No evidence for either of these side reactions has yet been reported.

Purification of N-Terminal Peptides or Peptides Containing Lysine

The maleylated protein or polypeptide can be digested with trypsin, whereupon specific cleavages occur at arginyl bonds as outlined above. After separation, these peptides can be unblocked and redigested with trypsin, to cleave the lysyl bonds. The technique is useful even without separation of the arginyl peptides, since the denatured proteins are frequently insoluble at pH 7–9 and therefore fail to digest cleanly with trypsin, whereas denatured maleyl proteins are generally soluble at pH 7–9 and so digest more easily. The digest can then be incubated at pH 3.5 to remove the maleyl groups and then readjusted to pH 7–9 when the mixture of unblocked peptides with C-terminal arginine is frequently soluble and fully digestible with trypsin.

Digests of the maleylated protein with other enzymes, such as chymotrypsin, thermolysin, or subtilisin, will yield fragments with internal maleylated lysine residues. After paper electrophoresis at pH 6.5, a strip of the paper can be incubated in a desiccator over the vapor of 1% pyridine, 5% acetic acid at 60° overnight to remove the maleyl groups. Subsequent electrophoresis at right angles to the initial direction produces a "diagonal electrophoretic map"[7a] in which the lysine or N-terminal peptides lie away from the diagonal position by virtue of the charge change of +2.[2] The mobilities of these peptides indicate their charge pattern (Fig. 5), and their approximate molecular weights can be

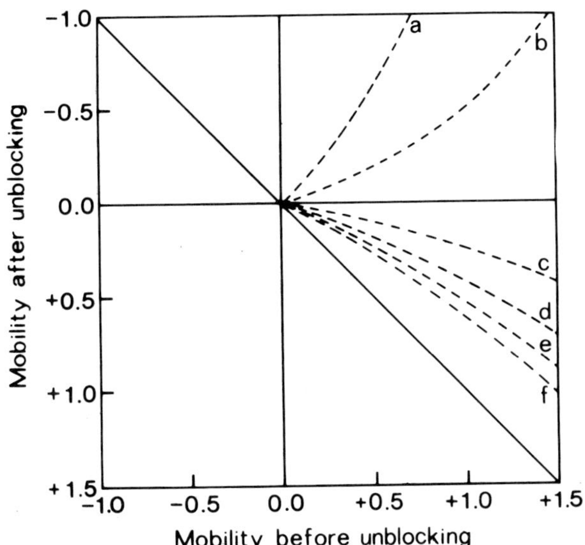

FIG. 5. Predicted positions for peptides on diagonal electrophoretic separation at pH 6.5 for lysine- or α-amino-containing peptides. Unchanged peptides (not containing maleyl groups) will lie on the solid diagonal line. Peptides lying on the broken lines will have undergone charge changes: a, from -1 to $+1$ or from -2 to $+2$; b, from -3 to $+1$; c, from -5 to -1; d, from -3 to -1; e, from -4 to -2; f, from -5 to -3. Mobilities are calculated with respect to aspartic acid ($= +1.0$). See R. E. Offord, *Nature* (*London*) **211,** 591 (1966).

calculated from the mobility data of Offord.[14] The diagonal electrophoretic map can then be used as a guide to the selective purification of each lysine or N-terminal peptide, as described by Brown and Hartley.[7a] If cysteine residues have first been disulfide-exchanged with cystamine, or aminoethylated, peptides containing these groups can be similarly purified.[15]

The use of ^{14}C-labeled maleic anhydride allows such operations to be carried out on very small quantities of protein. For example, Bruton and Hartley[8] used 0.2 mg of methionyl-tRNA synthetase from *Escherichia coli* to show that maleylation produces a monomer of molecular weight 51,000 from the dimer of molecular weight 96,000 or the tetramer of molecular weight 187,000. When a thermolysin digest of 1 mg (20 nmoles) of the maleylated enzyme was passed through Zeocarb 225 cation-exchange resin (H^+ form) in 0.1 *M* acetic acid, only the maleylated N-terminal peptide emerged in the breakthrough, because all the

[14] R. E. Offord, *Nature* (*London*) **211,** 591 (1966).

[15] R. N. Perham, *Biochem. J.* **105,** 1203 (1967).

other fragments were cationic. This tetrapeptide was purified in 6 nmole yield and unblocked, whereupon 1 nmole sufficed for determination of its sequence by a micromodification of the DNS-Edman procedure.

Analogous Reagents

The concept of intramolecular catalysis as a means to introduce specificity and lability into chemical modifications of proteins is obviously not restricted to the maleyl group. The table shows the structures of derivatives of 2,3-dimethylmaleic anhydride (II) and 2-methylmaleic (citraconic) anhydride (III) which have been used instead of maleic anhydride.[16] These react in a similar way, to give somewhat less stable derivatives, which can thus be unblocked more readily. The 2,3-dimethylmaleyl-amino bond is labile below pH 8 and therefore can only

STRUCTURES AND PROPERTIES OF MALEYL GROUPS AND ANALOGOUS SUBSTITUENTS

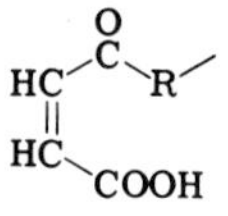

(I) HC(=HC–COOH)–C(=O)–R–

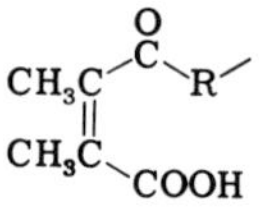

(II) CH_3C(=CH_3C–COOH)–C(=O)–R–

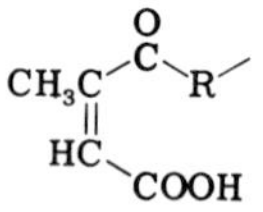

(III) CH_3C(=HC–COOH)–C(=O)–R–

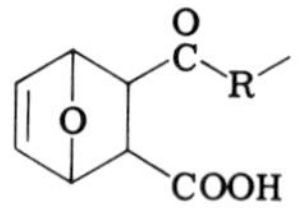

(IV) [bicyclic O-bridged ring]–C(=O)–R–, COOH

Group[a]	R = —OH ϵ_{250}	ϵ_{260}	ϵ_{280}	R = —NH— ϵ_{250}	ϵ_{260}	ϵ_{280}	$t_{1/2}$ at pH 3.5
I[b]	939	365	59	3360	1846	308	11 hours (37°)[b]
II	—	—	—	—	—	—	2 minutes (20°)[c]
III[d]	942	303	20	2180	775	60	2 hours (20°)[c]
IV	—	—	—	—	—	—	5 hours (25°)[e]

[a] I, maleyl; II, 2,3-dimethylmaleyl; III, 2-methylmaleyl (citraconyl); IV, exo-*cis*-3,6-endoxo-Δ^4-tetrahydrophthalyl.

[b] P. J. G. Butler, J. I. Harris, B. S. Hartley, and R. Leberman, *Biochem. J.* **112,** 679 (1969).

[c] H. B. F. Dixon and R. N. Perham, *Biochem. J.* **109,** 312 (1968).

[d] I. Gibbons, personal communication (1970).

[e] M. Riley and R. N. Perham, *Biochem. J.* **118,** 733 (1970).

[16] H. B. F. Dixon and R. N. Perham, *Biochem. J.* **109,** 312 (1968).

be used in the pH range above about pH 8.5. The 2-methylmaleyl-amino bond has a stability between that of the dimethylmaleyl and the maleyl derivatives. The two isomeric derivatives that it forms with each reacting amino group have a similar stability, with an average half-life between 1 and 2 hours in 40 m*M* hydrochloric acid at 25°. The extinction coefficients of 2-methylmaleylbutylamine and of disodium 2-methylmaleate at pH 8 have been determined[17] and are shown in the table. No values are available for the 2,3-dimethylmaleyl derivative. The mechanism for the increased catalysis with the methylmaleyl derivatives is being investigated by Kirby and Lancaster.[18]

Recently exo-*cis*-3,6-endoxo-Δ^4-tetrahydrophthalic anhydride (see the table, group IV) has been used to block reversibly the amino groups of lysozyme.[19] The derivatives are intermediate in lability between maleyl and 2-methylmaleyl (see the table), and the authors suggested that the absence of an activated olefinic bond may prevent alkylation of thiol groups.

The choice of reagent will be governed by the degree of stability which is desired in the product. Maleyl groups are most stable and can therefore be safely manipulated over the widest range of pH and temperature. The unblocking conditions are generally safe but may cause deamidation of particularly labile amide groups (e.g., in -Asn-Gly-). 2-Methylmaleyl groups are more labile and may be advantageous in such cases, but exist as diastereomeric products with slight differences in pK_a[16] that may produce problems in ion-exchange or electrophoretic separations. 2,3-Dimethylmaleyl groups are extremely labile and can be manipulated only above pH 8.5, but may be advantageous in cases where unblocking at neutral pH and low temperature is a necessity. Endoxotetrahydrophthalyl groups are intermediate in stability between maleyl and 2-methylmaleyl groups and may prove to be a satisfactory compromise for some problems.[19]

[17] I. Gibbons, personal communication (1970).

[18] A. J. Kirby and P. W. Lancaster, *Biochem. Soc. Symp.* p. 99 (1970).

[19] M. Riley and R. N. Perham, *Biochem. J.* **118**, 733 (1970).

Section VI

Separation of Peptides

[15] Separation of Peptides by Chromatography on Columns of Dowex 50 with Volatile Developers

By WALTER A. SCHROEDER

Soon after Moore and Stein[1] had devised their successful procedure for the chromatographic determination of amino acids in peptide or protein hydrolyzates on ion-exchange columns, the method was applied to the separation of the peptides in the partial hydrolyzate of a protein.[2] Unfortunately a peptide so separated is a rather minor contaminant of the salts that compose the developing solvents. Although the problem can be circumvented by conversion to the DNP-peptides and extraction with organic solvents, by the use of volatile salts that can be sublimed away, or by a desalting procedure of some type, each method has its own disadvantages; no truly satisfactory solution to the problem was achieved until readily volatile organic solvents began to be used in the developers. The isolation of a peptide after chromatographing then became a simple matter: the portion of the effluent that contained a peptide need only be evaporated to yield a residue of peptide.

The partial hydrolyzate of a protein is likely to be a far more complex mixture than that of the mixture of amino acids in a total hydrolyzate. It is improbable, then, that a single chromatogram will separate the peptide mixture completely into its components or that one procedure will be equally satisfactory for all proteins. Nevertheless the simpler mixtures thus obtained should be amenable to separation by another method that preferably differs in kind from the first. The dipolar ionic character of peptides allows the use of both cation and anion exchangers in one or the other sequence, and thus provides very different conditions of chromatography with a consequent increase in capacity to achieve difficult separations. This article describes a method for the separation of peptides on the cation exchanger Dowex 50, and article [16] for the separation on the anion exchanger Dowex 1.

Although the procedure describes the use of Dowex 50-X2, Dowex 50-X8 is also satisfactory with some modification in conditions of development and may offer advantages.

[1] S. Moore and W. H. Stein, *J. Biol. Chem.* **192**, 663 (1951).

[2] W. A. Schroeder, L. Honnen, and F. C. Green, *Proc. Nat. Acad. Sci. U.S.* **39**, 23 (1953).

Procedures

Materials

The cation exchange resin, Dowex 50-X2. The experimental procedure described below has always used (and reused) a large lot of Dowex 50-X2 (200–400 mesh) obtained in 1953 (Dow Chemical Co.). This material required thorough purification before use as Moore and Stein[1] described. An equivalent analytical grade resin is AG50W-X2 (200–400 mesh) (Bio-Rad Laboratories, Richmond, California). This purified and sized material (actual wet mesh size is 80–200) can be used directly after the regeneration procedure to be described below.[3]

Pyridine,[4] reagent grade, distilled. Reagent grade pyridine may vary considerably in quality from lot to lot of the same supplier. In certain instances it may be desirable to distill from ninhydrin.[5]

Glacial acetic acid, reagent grade undistilled

Concentrated hydrochloric acid, reagent grade

Sodium hydroxide, reagent grade

Solutions

Sodium hydroxide, 1 *N*

Hydrochloric acid, 2 *N*

Pyridine, 2 *M*

Buffer pH 3.1 (0.2 *M* in pyridine); 64.5 ml of pyridine and 1114 ml of glacial acetic acid diluted to a volume of 4 liters; this solution may discolor on standing at room temperature; the discoloration presumably is due to some minor contaminant that varies from lot to lot of pyridine even from the same supplier; discoloration does not occur if the solution is refrigerated.

[3] R. T. Jones (personal communication) has used AG50W-X2 of 270–325 wet mesh size (44–55 μ) with success. The resolution on this size is better than on Aminex 50-X2, 120–325 mesh (Bio-Rad Laboratories), and therefore is also probably better than on AG50W-X2 (200–400 mesh). It may be that in some applications it will be desirable to use the resin of smaller and more uniform size (270–325 mesh).

[4] Although pyridine has an unpleasant odor, it is rated as less toxic than benzene, carbon tetrachloride, and other common solvents (M. N. Gleason, R. E. Gosselin, and H. C. Hodge, "Clinical Toxicology of Commercial Products," 2nd ed. Williams and Wilkins, Baltimore, Maryland, 1963). In a well-ventilated room, pyridine will not often be evident by odor if solutions are poured from container to container in a hood and if solutions are discarded in a hood rather than into an open sink.

[5] R. L. Hill, cited by R. T. Jones, *Cold Spring Harbor Symp. Quant. Biol.* **29**, 297 (1964).

Buffer pH 5.0 (2 *M* in pyridine); 645 ml of pyridine and 573 ml of glacial acetic acid diluted to a volume of 4 liters; discoloration of this solution has not been observed.

Buffer pH 5.6 (8.5 *M* in pyridine)[6]; 684 ml of pyridine and 180 ml of glacial acetic acid diluted to a volume of 1 liter.

Apparatus

Chromatographic tubes. Jacketed tubes of appropriate size with a coarse sintered-glass support[7] for the chromatographic column are necessary. A 12/1 ball joint at the bottom of the tube and an 18/9 socket joint at the top are convenient and may be used with a 12-mm socket connector (Part No. 312561, Beckman Instruments, Inc., Spinco Division, Palo Alto, California) and filter connector (Part No. 313347, Beckman Instruments, Inc.) and their appropriate fittings (Parts No. 313336 tube swivel fitting and No. 313285 tube fitting). The procedure to be described has been used with columns of 0.6×60 cm to 3.5×100 cm.

Gradient vessels: The gradient vessels are patterned after those of Bock and Ling[8] (as shown in their Fig. 6). The most convenient arrangement will depend somewhat upon the size of the chromatographic column and thus the total volume of gradient. When the total volume of gradient is 2 liters or less, the apparatus in Fig. 1 is convenient. It consists of two glass cylinders such that the mixer has a diameter D and the reservoir $1.4\,D$: the cross-sectional areas are in the ratio 1:2. The bottom of each cylinder is flat, and the capillary inlets and outlets are positioned as shown, so that the entire volume of each cylinder may be used. A convenient length for the cylinders is about 30 cm. If a series of diameters is chosen so that the smallest is about 20 mm and each succeeding size is greater by a factor of 1.4, then a particular cylinder may be used as either mixer or reservoir depending upon the size of column and the volume of gradient. The use of these vessels will be described below. If the volume of gradient is more than 2 liters, it is recommended that three bottles of equal diameter and appropriate volume be connected at the bottom, much as in Fig. 1. The ratio 1:2 is then maintained by using one as the mixer and the other two as the "reservoir." Tubing from

[6] W. Konigsberg and R. J. Hill, *J. Biol. Chem.* **237**, 2547 (1962).

[7] If very finely divided resins are used, medium or fine-sintered supports are necessary.

[8] R. M. Bock and N.-S. Ling, *Anal. Chem.* **26**, 1543 (1954).

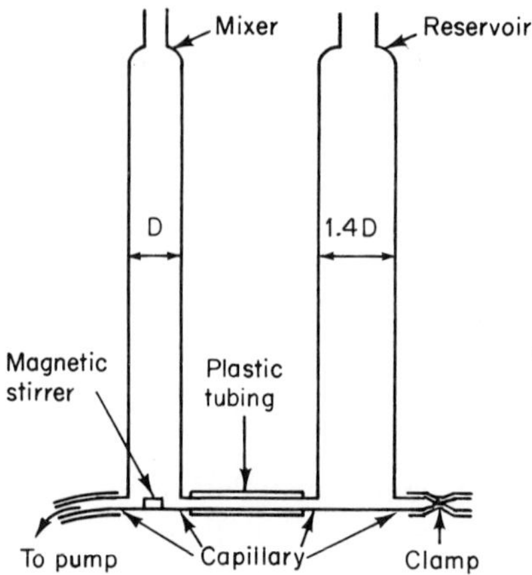

FIG. 1. Gradient vessels.

gradient vessels to the pump and from the pump to the column should be of polyethylene or Teflon.

Pump: A constant-volume pump, such as a Milton Roy chromatographic Minipump (Milton Roy Co., Philadelphia, Pennsylvania) or a Beckman Accu-Flo Pump (Beckman Instruments, Inc., Spinco Division, Palo Alto, California), is necessary to pass developer through the column.

Fraction collector and test tubes

Constant-temperature bath

Preparation of Dowex 50-X2 Resin

If columns as large as 3.5 × 100 cm are used, it is necessary to have about 5 lb of resin available. However, regardless of the size of column used, it is convenient to prepare 5 lb of resin for use and to store it in buffer so that a column may be poured at any time.

Purified and sized[9] resin (5 lb) is placed in water equal to 5 times its settled volume, stirred thoroughly, and allowed to settle. When the bulk of the resin has settled, the finely divided material in the supernatant fluid is removed by decantation. This settling may have to be repeated.[10] The resin is then transferred to a Büchner funnel (25 cm in diameter) and washed successively with 5 liters of water, 5 liters of 1 *N* sodium hydroxide, 5 liters of water, 3 liters of 3 *N* hydrochloric acid, 5 liters of

[9] A detailed description of sizing by a hydraulic method is given by S. Moore, D. H. Spackman, and W. H. Stein, *Anal. Chem.* **30**, 1185 (1958).

[10] Analytical grade resins usually have little finely divided material that can be removed in this way.

water, 4 liters of 2 *M* pyridine, and 3 liters of pH 3.1 buffer. The resin is then transferred to a flask with sufficient pH 3.1 buffer so that the ratio of supernatant buffer to settled resin is 2:1.

After the chromatographic procedure the resin has not been regenerated in the chromatographic tube,[11] but has been removed and stored until sufficient used resin has accumulated to warrant regeneration. The above method of washing on the Büchner funnel should be applied to the resin before it is used again; smaller amounts of resin should be washed with proportionately smaller volumes of liquid.

Preparation and Equilibration of the Chromatographic Column

The procedure below is described in general terms; for quantities appropriate to a given size of column, Table I should be consulted. The bottom of the tube is closed, and the thoroughly suspended resin is poured down the side of the chromatographic tube. No particular care need be taken to avoid the entrainment of air bubbles during the pouring. The volume of suspension should be such that the packed resin will form a column 15–20 cm long. (If the ratio of supernatant fluid to settled resin is 2:1, a 45- or 60-cm depth of suspension will give about 15 to 20

TABLE I

CONVENIENT PARAMETERS FOR VARIOUS SIZES OF COLUMNS OF DOWEX 50-X2

Column size (cm)	Approximate ratio of column volumes	Equilibrating volume (ml)	Sample volume (ml)	Developer volume (ml)		Fraction size (ml)	Rate of developer flow (ml/hr)
				pH 3.1 buffer in mixer	pH 5.0 buffer in reservoir		
0.6 × 60		50	1	83	166	1	10
	1:4						
1 × 100		100	2	333	666	1.5	20
	1:2						
1.4 × 100		250	5	666	1332	3	30
	1:2.5						
2.2 × 100		500	15	1500	3000	5	60
	1:2.5						
3.5 × 100		1000	30	4300	8600	10	120

[11] R. T. Jones (personal communication) reequilibrates directly in the column by passing through pH 3.1 buffer. After approximately 30 minutes (0.9 × 60 cm column; flow rate, 30 ml per hour), a rapid rise in pressure (80–90 psi) occurs. Passage of fluid is then stopped, the top fitting is removed from the column, and the resin is allowed to expand for 30 minutes. Passage of developer then is begun again: flow must sometimes be stopped several times before equilibration is complete.

cm of packed resin.) After the resin has settled for a few minutes, the bottom of the tube is opened, and the packing is carried out under air pressure equal to 100–150 Torr. After the resin has settled to a constant height, all except a 5–10-cm layer of supernatant fluid is removed, and more suspension of resin is added to produce another section. This procedure is continued until the column is the desired length.

After the packing of the column has been completed, water at 38° is circulated through the jacket and an appropriate volume of pH 3.1 buffer is pumped through the column (see Table I).

Application of the Sample

The sample is dissolved in a volume of water[12] appropriate to the size of the column (Table I), and the pH is reduced to about 2 with hydrochloric acid. After the buffer above the column has been removed, the sample is carefully layered on the column with a bent-tip dropper. When the sample has been completely forced into the column by air pressure, the sides of the tube are rinsed several times with small volumes of pH 3.1 buffer. Finally, the tube above the column is filled with pH 3.1 buffer.

Development of the Chromatogram

The developing system is prepared in the following way. The tubing from the mixer through the pump to the top of the column is filled with pH 3.1 buffer, and then closed to prevent outflow of liquid. Likewise, the connection between the two gradient vessels is filled with this buffer and then clamped shut. Any excess buffer is removed from both vessels. A given volume of pH 3.1 buffer is placed in the mixer, and twice that volume of pH 5.0 buffer in the reservoir. The two buffers are of approximately equal density and therefore the system can be placed in approximate hydrostatic equilibrium by adjusting the menisci to the same height.[13] When the clamp between mixer and reservoir is opened, a final slight adjustment may be necessary to achieve hydrostatic equilibrium, which is evidenced by the absence of schlieren in either cylinder. Finally, magnetic stirring is begun in the mixer, the connection is made to the top of the chromatographic tube, and pumping is begun.

[12] In an earlier description of this procedure [W. A. Schroeder, R. T. Jones, J. Cormick, and K. McCalla, *Anal. Chem.* **34**, 1570 (1962)], the sample in water was diluted with sufficient 1 *M* pyridine–acetic buffer at pH 3.5 to give 0.2 *M* pyridine in the final solution, and the pH was then adjusted with hydrochloric acid. From the experience of Jones,[3] the procedure described is satisfactory if the sample volume is small.

[13] The menisci themselves are a convenient leveling device. If one sights along one meniscus to the other, the levels may easily be adjusted to the same height.

All the developer in the gradient vessels is then pumped through the column at an appropriate speed (Table I). The effluent is collected in fractions of suitable size.

During the chromatogram, the column shrinks appreciably in length.

Assessment of Results

After the chromatogram has been completed, the success of the separation is assessed by an examination of the fractions of effluent. The color reaction with ninhydrin is most frequently applied to determine in which fractions the peptides have emerged. The presence of pyridine does not interfere with the reaction. Many procedures have been described and may be used.

The details of a sensitive method for small samples have been given (see Vol. XI [35]). An earlier readily applicable manual method with prior alkaline hydrolysis of the sample was described by Hirs, Moore, and Stein.[14] An automated method with the Technicon Autoanalyzer (Technicon Corp., Tarrytown, New York) has been used by Schroeder *et al.*[12,15] It is also possible to divert a portion of the effluent for automatic examination and recording as is done in automatic amino acid analysis (see Vol. XI [1]) or peptide mapping (see Vol. XI [37]; see also Jones[5]).

In certain instances special tests may be applied to the fractions. Thus Spackman, Stein, and Moore[16] have described a test for cystine-containing peptides, and Nolan and Smith[17] have used a well-known method to detect carbohydrate-containing peptides.

Concluding Operations

If a chromatographic separation is successful, a plot of color produced by one of the above methods vs. the fraction number of the effluent will show a series of more or less well-separated peaks. Those fractions that contain a given peak are therefore combined and the solvent is removed. When large columns have been used, the large volume of solvent is most conveniently removed in a rotary evaporator. However, the combined fractions of a peak from a small column may be contained in only 5 or 10 ml; it is simpler to remove the solvents in a stream of air while the containing vessel is warmed to 40° in a water bath. The residues so obtained are contaminated to a degree by impurities from resin and solvents, and frequently are oily unless the amount is large. They can be dissolved in water and freeze-dried to give solid materials.

[14] C. H. W. Hirs, S. Moore, and W. H. Stein, *J. Biol. Chem.* **219**, 623 (1956).
[15] W. A. Schroeder and B. Robberson, *Anal. Chem.* **37**, 1583 (1965).
[16] D. H. Spackman, W. H. Stein, and S. Moore, *J. Biol. Chem.* **235**, 648 (1960).
[17] C. Nolan and E. L. Smith, *J. Biol. Chem.* **237**, 446 (1962).

The peptides so isolated are ready for further examination by rechromatography, analysis, etc.

Practical Considerations

Column Size and Quantity of Sample. Experience with this procedure since an earlier description was published[12] has led to the realization that the columns can handle a larger quantity of sample than was initially used. For example, hydrolyzates from 0.4–1.0 g of protein (hemoglobin in this instance) were chromatographed on a 3.5 × 100-cm column.[12] Clearly the weight of sample that can be successfully chromatographed will depend somewhat on the size of the protein (therefore the number of micromoles per gram) and the complexity of the mixture. However, hydrolyzate from 200–250 mg of a hemoglobin chain may be chromatographed successfully on a 1 × 100-cm column. This loading is about 5 times as great as that mentioned above and probably is near the maximum. In another instance, the tryptic hydrolyzate of 1300 mg of catalase protein was chromatographed on a 2.2 × 100-cm column. The mixture in this hydrolyzate was more complex than that from a hemoglobin chain, and the separation of peptides was somewhat less satisfactory than in another chromatogram at lesser load.

The 0.6 × 60-cm columns have been most effectively used in operations with small amounts of peptides. Thus, if several micromoles of a peptide are hydrolyzed by an enzymatic or chemical method, the resulting peptides can be chromatographed on such a column. These chromatograms are easily completed in a day.

Use of Resin with Different Cross-Linking. The above chromatographic procedure on Dowex 50 ordinarily yields simpler mixtures that need to be separated further. Although such separation often can be effected by chromatography on Dowex 1, another type of Dowex 50 may also be used. Thus, Jones[5] has substituted a finely divided, more highly cross-linked cation-exchanger resin equivalent to Dowex 50-X8 in the procedure already described. The behavior of peptides is definitely altered by such a substitution, and separations of given peptides take place on one, but not on the other. There are, therefore, some advantages in using a more highly cross-linked resin, and it should perhaps be the resin of choice.

The same basic procedures described above apply also to the use of a more highly cross-linked resin equivalent to Dowex 50-X8. A suitable product is fraction C from a hydraulic sizing procedure[9] or its equivalent such as Aminex MS Fraction C (Bio-Rad Laboratories). Purification, preparation for column packing, and equilibration are the same as described above for Dowex 50-X2.

Because, under the same conditions of elution, peptides emerge more rapidly from the more highly cross-linked resin, a modified gradient is necessary. The following gradient has been satisfactory; it is first concave (and is formed by interchanging mixer and reservoir of Fig. 1) and then convex (by reverting to the configuration in Fig. 1). For a 1×100-cm column, the mixer contains 666 ml of buffer pH 3.1 (0.1 M in pyridine, *not* 0.2 M) and the reservoir has 333 ml of buffer pH 5.0 (M in pyridine, *not* 2 M). At the end of this gradient, a second one with 166 ml of buffer pH 5.0 (1 M in pyridine) in the mixer and 332 ml of buffer pH 5.0 (2 M in pyridine) in the reservoir is used. It may be desirable to use a third gradient with 166 ml of buffer pH 5.0 (2 M) in the mixer and 332 ml of buffer pH 5.0 (5 M) in the reservoir. This program of elution has been effective in retarding and separating more rapidly moving peptides and then eluting the more slowly moving peptides with reasonable speed. Flow rates of 30–35 ml per hour have been used; a faster rate may well be satisfactory.

Yield of peptides appears to be better from columns of Dowex 50-X8 than from those of Dowex 50-X2, although very basic or hydrophobic peptides do not always elute in discrete peaks.

Columns of Dowex 50-X8 do not shrink under the above conditions of elution and may, therefore, be regenerated and reused without repacking. Some caution must be observed in frequent reuse of a column. Peptides may be so strongly fixed that they are not removed under the above conditions of elution. However, with successive reuse of the column, such peptides seem to move slowly through the column and appear as a background under the normal peaks. This background becomes evident in poor quality amino acid analyses of the peptides. If this occurs, the resin should be removed, suspended in five volumes of 2 N sodium hydroxide per volume of settled resin, and heated on a boiling water bath for 2 hours with stirring. After this has been repeated, the resin is prepared in the normal way, and the column is repoured. The resin is effectively cleaned of any contaminating peptides by this alkaline hydrolysis.

Type of Gradient. The type of gradient described above for Dowex 50-X2 has been effective in the separation of peptides in hydrolyzates of hemoglobin and catalase. A rather similar gradient has been used by Kimmel, Rogers, and Smith[18] to separate peptides from carboxymethylated papain. In the case of hemoglobin and catalase, most of the peptides emerge from the column by the time that about 0.6 of the gradient has been used. With papain, this clearly is not the case. Because of the individuality of proteins, the type of development or gradient

[18] J. R. Kimmel, H. J. Rogers, and E. L. Smith, *J. Biol. Chem.* **240**, 266 (1965).

may need to be altered regardless of the cross-linking of the resin by using only concave or linear or shallower or steeper gradients.

With the convex gradient that has been described, relatively little change in pH and concentration occurs in the last 0.3–0.4 of the gradient volume. It may consequently be advantageous to alter the gradient after about 0.6 of the gradient volume has been used. For example, in some chromatograms the development has been stopped at this point. After the connection between mixer and reservoir vessels was closed, the pH 5.0 buffer (2 *M* in pyridine) was removed from the reservoir and replaced with an equal volume of pH 5.6 buffer (8.5 *M* in pyridine), while the solution in the mixer (now about 1.7 *M* in pyridine) was not changed. Further development was then made with this gradient, and on occasion the development was finally completed with pH 5.6 buffer. With this modified gradient and final developer, the column of Dowex 50-X2 will shrink to less than half its original volume, and at some stages the rate of flow of developer through the column may have to be decreased because of transient high pressures.

A rather different type of gradient may also be useful in certain instances. For example, chymotryptic hydrolyzates are usually rather complex, and chromatography on Dowex 50 and then Dowex 1 may not always yield an entirely homogeneous peptide. The gradient to be described has been very useful for a third stage of purification. The composition of the developers is given in Table II. The column of Dowex 50-X2 is equilibrated with Developer B, and the sample is applied in Developer A and rinsed in with Developer B. A concave gradient is used. After the sample has been applied, the volume above the column as well as the line to the gradient vessels is filled with Developer B. For a 0.6 × 60 cm column, 66 ml of Developer B is placed in the mixer and 33 ml of Developer C in the reservoir. This gradient, which is concave in concentration at constant pH, may be completed by midafternoon at a flow rate of 15 ml per hour. A second gradient (concave in both pH and con-

TABLE II
DEVELOPERS FOR MODIFIED GRADIENT

Developer	pH	Pyridine (ml/l)	Glacial acetic acid (ml/l)
A	3.98	2.0	8.4
B	3.98	4.05	15.0
C	3.98	8.1	30.0
D	6.12	83.0	5.6
E	6.12	166.0	11.2

centration) is produced with 100 ml of Developer C in the mixer and 50 ml of Developer D in the reservoir. At a slower flow rate, this gradient is complete by the next morning. Experience with these gradients has been limited, but the results have been gratifying. Sometimes a third gradient of Developers D and E (100 and 50 ml, respectively) is necessary. After the chromatogram is complete, the column may be reequilibrated with Developer B. Application of these gradients to chromatograms on Dowex 50-X8 may be useful.

The addition of 10% *n*-propanol to the buffers may aid in the elution of very strongly fixed peptides. *n*-Propanol alters the apparent pH of the buffers which, accordingly, should be prepared in the following way. Full amounts of pyridine and *n*-propanol (in place of an equal volume of water) are mixed with about three-fourths of the water that will finally be used. Acetic acid is then added until the appropriate pH has been attained, and then water almost to the final dilution. After the pH has been checked and adjusted if necessary, final dilution with water is made.

Various Parameters. Examination of Table I will show that the various parameters are not always in proportion to the ratio of column volumes. Thus, on the basis of cross-sectional area, the flow rate of developer through a 3.5 × 100-cm column should be about 350 ml per hour if the 0.6 × 60 cm column is taken as standard. The slower rate is somewhat of a convenient compromise; examination of fractions, combining fractions, evaporation, etc., may be carried out at a rate that does not lag too far behind the actual progress of the chromatogram. The faster rates of flow with the smaller columns, on the other hand, permit these to be finished quickly and conveniently. Although this factor has not been examined in detail, the fast rate of flow for the 0.6 × 60 cm column has not had an obviously deleterious effect on separations; a still faster rate may be satisfactory.

Sample volume probably is unimportant as long as the pH is 2–2.5. The choice will frequently be dependent on other factors. For example, if 1 g of protein were hydrolyzed at a substrate concentration of 1%, the 100 ml of hydrolyzate would be inconveniently large and its application too time-consuming even for the largest column. Lyophilization and solution in a smaller volume are to be preferred.

Acknowledgments

The above procedure was devised in the course of investigations under a grant (HE-02558) from the National Institutes of Health, U.S. Public Health Service. The assistance and suggestions of Jean Cormick, Richard T. Jones, Barbara Robberson, and J. Roger Shelton are gratefully acknowledged.

Contribution No. 4139 from the Division of Chemistry and Chemical Engineering, California Institute of Technology.

[16] Separation of Peptides by Chromatography on Columns of Dowex 1 with Volatile Developers

By WALTER A. SCHROEDER

As described in article [15] chromatography on the cation-exchange resin Dowex 50 separates a complex mixture of peptides into a series of simpler mixtures. Usually these simpler mixtures yield to rechromatography on the anion-exchange resin Dowex 1.

One failing of many published procedures that use Dowex 1 is the rather irregular gradient, which at some point has an abrupt change in pH. This change usually results in the rapid elution of some peptides, inadequate separation, and the consequent necessity for rechromatography. Although a peptide mixture can be examined, e.g., by paper chromatography and electrophoresis, in order to determine the number and charge of the components and conditions designed[1] for a separation on Dowex 1, the procedure below was devised for general application. It is a modification of earlier methods[1] that were similar to those of Rudloff and Braunitzer.[2] A high initial pH has been chosen to retard very basic peptides that normally are only slightly retarded on Dowex 1, and a gradual gradient has been devised in order to eliminate the poor separations caused by abrupt changes in pH.

Whether peptides are first separated on Dowex 50 and then on Dowex 1 or vice versa is immaterial. The consecutive use of Dowex 50 and then Dowex 1 is suggested in this and in article [15] because the conditions of chromatography cause much shrinkage of Dowex 50-X2 columns, which must be repacked instead of reequilibrated. Dowex 1 columns, however, have simply been reequilibrated. They may therefore be used repeatedly with little effort for the many chromatograms required for the second step of purification, whereas each Dowex 50-X2 column would have to be repoured. The advantage of such a sequence of chromatograms is nil if Dowex 50-X8 is used because Dowex 50-X8 columns can be regenerated without repouring.

Procedure

Materials

The anion-exchange resin, Dowex 1-X2. As the analytical grade resin, this material has the designation AG 1-X2 (200–400 mesh)

[1] W. A. Schroeder, R. T. Jones, J. Cormick, and K. McCalla, *Anal. Chem.* **34,** 1570 (1962).

[2] V. Rudloff and G. Braunitzer, *Hoppe-Seyler's Z. Physiol. Chem.* **323,** 129 (1961).

(Bio-Rad Laboratories, Richmond, California), the actual wet size is 80–200 mesh.

Pyridine,[3] reagent grade, distilled

N-Ethylmorpholine (Eastman, practical), distilled and stored in a freezer

α-Picoline (Matheson, Coleman, and Bell, practical), distilled

Glacial acetic acid, reagent grade

Concentrated hydrochloric acid, reagent grade

Sodium hydroxide, reagent grade

Solutions

Sodium hydroxide, 0.5 *N*

Hydrochloric acid, 1 *N*

Acetic acid, 0.5 *N*, 1 *N*, and 2 *N*

Buffer, pH 9.4: 60 ml of *N*-ethylmorpholine, 80 ml of α-picoline, 40 ml of pyridine, and approximately 0.5 ml of glacial acetic acid to give a pH of 9.4 when diluted to 4 liters with water. The water for this and the pH 8.4 buffer should be bubbled with nitrogen to remove carbon dioxide. Both buffers should be stored in a type of dispensing bottle such that the entering air passes through a carbon dioxide absorbent such as soda lime or Mallcosorb (Mallinckrodt)

Buffer, pH 8.4: the same quantities as for the pH 9.4 buffer except that the quantity of acetic acid must be increased to about 3 ml

Buffer, pH 6.5: the same quantities as above, but about 37 ml acetic acid is required

Apparatus

Chromatographic tubes: The same type as that described in article [15] is needed. Only 0.6 × 60-cm and 1 × 100-cm columns have been used in the procedures to be described. The procedure could no doubt be sealed-up.

Gradient vessels and solvent changers: The gradient is produced with a constant volume mixer and with a series of solvents successively introduced into the changing solution in the mixer. These changes may be made manually but can be made automatically by the device shown in Fig. 1. The apparatus in this figure is designed for use with 0.6 × 60-cm columns. The mixing vessel is conveniently made from a 34/28 inner joint which has a flat bottom and a volume of 40 ml. The outer joint that forms the top of the mixing vessel is fitted with spherical joints on inlet

[3] See comments in footnote 4 of article [15].

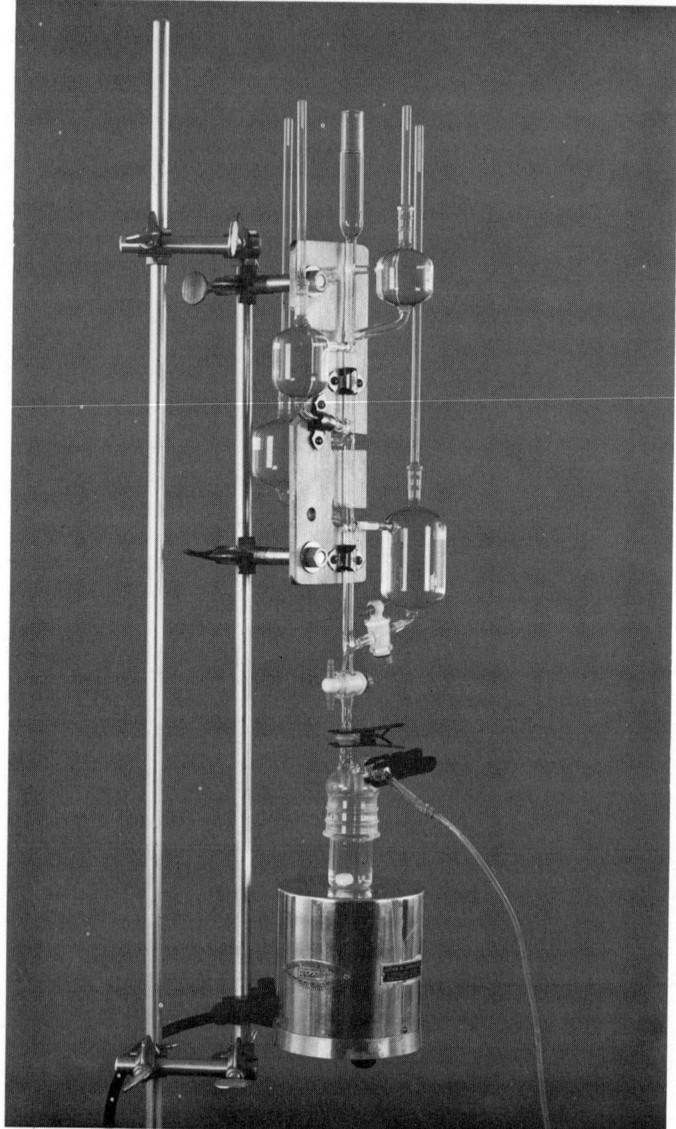

FIG. 1. Gradient vessel with automatic solvent changer.

and outlet for ease of connection, and the outlet reaches to the bottom of the vessel. To the inlet is connected (Fig. 1) a device (patterned after Anderson, Bond, and Canning[4] and Brusca and

[4] N. G. Anderson, H. E. Bond, and R. E. Canning, *Anal. Biochem.* 3, 472 (1962).

Gawienowski[5]) by means of which five solvents are automatically and successively allowed to flow into the mixing chamber. The upper tube connected to the central stem is calibrated to contain 10 ml, and the bulbs in order down the stem contain 30, 40, 60, and 100 ml, respectively.[6] The original references[4,5] should be consulted for the operating principle of this changer. The mixer is magnetically stirred. These solvent changes may also be made by a series of time-controlled solenoids, as described by Rombauts and Raftery.[7] Their device can readily be altered to change the volume of any solvent and in this respect is more adaptable than the changer in Fig. 1. Because it is time-controlled, it should be used with a constant-volume pump. If a 1 × 100-cm column is to be used, a 135-ml mixer is needed. If automatic changers are not used, the reservoir should be attached directly to the top of the mixer. All tubing between mixer and column must be of polyethylene or Teflon.

Fraction collector and test tubes

Constant-temperature bath

Constant-volume pump

Preparation of Dowex 1-X2 Resin

Unless very large columns are used, 1 lb of resin is adequate for several 0.6 × 60-cm and 1 × 100-cm columns.

Any finely divided material in the resin is first removed by settling and decantation as described for Dowex 50 (article [15]). After removal of the finely divided material, the remainder of the resin is washed on a Büchner funnel with 2 liters of water at 60°, and at room temperature with a liter of 0.5 *N* sodium hydroxide, 3 liters of water, 0.5 liter of 1 *N* hydrochloric acid, and 3 liters of water. It is stored wet as the chloride until needed for columns.

Inasmuch as columns of Dowex 1 can be reequilibrated easily and seldom need be repoured, only resin for packing the desired columns need be prepared. A 0.6 × 60-cm column requires less than 20 ml of packed resin, and a 1 × 100-cm column about 80 ml. Before a column is packed, the following solvents are passed through a 100-ml portion of wet resin

[5] D. R. Brusca and A. M. Gawienowski, *J. Chromatogr.* **14**, 502 (1964).

[6] The stopcock between the central stem and the largest bulb is not essential. It was originally installed because of the possibility that even a small amount of this developer might have an untoward effect on the early portion of the gradient. The stopcock permits this bulb to be isolated from the system until part of the gradient has been completed.

[7] W. A. Rombauts and M. A. Raftery, *Anal. Chem.* **37**, 1611 (1965).

under gravity on a Büchner funnel: 200 ml of water, 200 ml of 0.5 N sodium hydroxide, 500 ml of water, 200 ml of 1 N acetic acid, 200 ml of water, and 500 ml of pH 9.4 buffer. Finally, the resin is mixed with sufficient pH 9.4 buffer so that the ratio of supernatant buffer to settled resin is 2:1.

Preparation and Equilibration of the Chromatographic Column

Satisfactory columns of Dowex 1 are somewhat difficult to pour because of the tendency to trap bubbles of air, which then enlarge during the chromatographic procedure. Actually a few scattered bubbles do not impair the separations, and bubbles commonly appear in the upper few centimeters of the column during the chromatogram. The column may be poured at 38° in same manner as described for Dowex 50 [15] with some care to reduce the amount of entrained air. During equilibration with pH 9.4 buffer at as fast a rate as convenient, the flow at the exit of the column is restricted to increase the back pressure on the column. Any bubbles that are present will dissolve. A volume of 60 ml is used to equilibrate the 0.6 × 60-cm column and of 100 ml for the 1 × 100-cm column.

Application of the Sample

The sample is dissolved in 1–2 ml pH 9.4 buffer, and the pH is raised to about 10.5 with sodium hydroxide. After the buffer above the column, as well as a small amount of yellow material that may collect at the top of the column, has been removed, the sample is carefully added to the top of the column and allowed to flow in under gravity or slight air pressure. After the sides of the tube have been rinsed several times with pH 9.4 buffer, which is allowed to flow in, the tube above the column is filled with pH 9.4 buffer.

Development of the Chromatogram

The following description is based upon a manual changing of the developers. The tubing from the top of the column through the pump to the gradient mixer, the mixer itself, and the reservoir are filled with pH 9.4 buffer. The mixing chamber is magnetically stirred, and the flow of pH 9.4 buffer is started through the column. After the appropriate volume (see the table) has flowed through, the pH 9.4 buffer is removed from the *reservoir* and replaced with pH 8.4 buffer, it in turn by pH 6.5 buffer, then 0.5 N acetic acid, and finally 2 N acetic acid. The volumes of developer for two sizes of column are given in the table.

If the automatic device in Fig. 1 is used, the upper tube attached to the central stem contains the pH 9.4 buffer, and the bulbs contain pH

VOLUME OF DEVELOPERS FOR DOWEX 1 CHROMATOGRAMS

Developer	Column 0.6 × 60 cm (ml)	Column 1 × 100 cm (ml)
Buffer pH 9.4	10	40
Buffer pH 8.4	30	120
Buffer pH 6.5	40	160
Acetic acid 0.5 *N*	60	240
Acetic acid 2 *N*	100	400

8.4 buffer, pH 6.5 buffer, 0.5 *N* acetic acid, and 2 *N* acetic acid, respectively.[8]

If manual changes are made, it is most convenient to place a little more than the required amount of a given developer in the reservoir at each replacement, and then remove excess and replace with the next solvent as a given fraction is collected. Thus, if 1-ml fractions are collected from a 0.6 × 60 cm column, pH 8.4 buffer would be placed in the reservoir at the end of the 10th fraction, pH 6.5 buffer at the end of the 40th fraction, 0.5 *N* acetic acid at the end of the 80th fraction, and 2 *N* acetic acid at the end of the 140th fraction.

The development is carried out at 60 ml per hour with either size column. The chromatogram in the smaller size is completed in 4–5 hours, and the column is then reequilibrated overnight. A chromatogram on the larger column is most conveniently started during the afternoon. It is then completed by morning and may be equilibrated during the day. If bubbles develop, they may be dissolved by equilibrating as above with restricted flow at the exit in order to increase the back pressure; bubbles in the upper few centimeters can usually be removed by gentle stirring.

Assessment of Results and Concluding Operations

The remarks under these headings in the preceding article [15] on chromatographic procedures with Dowex 50 are equally applicable here.

Practical Considerations

The pH Curve. The recommended sequence of developers results in a pH curve as a function of effluent volume such as that shown in Fig. 2. A

[8] The filling is carried out as follows. The lower stopcock is closed. The fluid for the lowest bulb is poured in carefully from a beaker until the glass joint of the bulb is partly full. A rubber bulb is attached to the appropriate capillary standpipe, which is then filled with fluid and placed in position. After all bulbs and the upper tube have been filled, the rubber bulbs are removed from the standpipes, the stopcock is opened, and the central stem is flushed out with a few milliliters of the fluid in the upper tube.

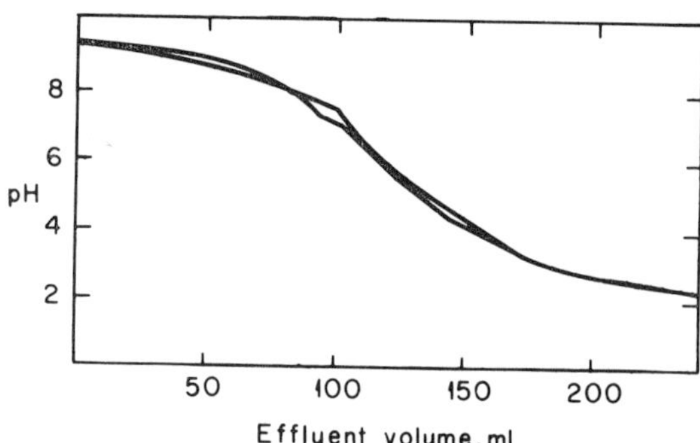

FIG. 2. Composite of five successive pH curves as a function of effluent volume.

0.6 × 60 cm column with the automatic changer was used. The curve is a composite of five curves and shows the extremes that were observed. The variations have had little obvious effect on separations and probably result from the rapid rate of development, individual differences in mixing in the gradient vessel, and variation from column to column.

The pH curve can easily be altered. It may be especially desirable to do so if several basic peptides are present. Then the gradient may be modified as follows for a 0.6 × 60-cm column: 10 ml of buffer pH 9.4; 30 ml of buffer pH 9.0 (prepared as described under *Solutions*, but with about 1 ml of acetic acid); 40 ml of buffer pH 8.4; 60 ml of buffer pH 6.5; 100 ml of 0.5 *N* acetic acid; and 100 ml of 2 *N* acetic acid. On the other hand, such changes as omission of pH 6.5 buffer and substitution of 1 *N* for 0.5 *N* acetic acid cause a quick decrease in pH—perhaps desirable if mainly acidic peptides are present.

For a 0.6 × 60-cm column with the recommended gradient, it is expected that basic peptides will emerge between 10 and 90 ml, neutral peptides between 90 and 130 ml, and acidic peptides thereafter. Histidine behaves as a neutral aromatic amino acid, and its presence or that of phenylalanine or tyrosine in the peptide retards the point of emergence over that to be expected on the basis of charge alone.

Column Size and Quantity of Sample. The 0.6 × 60-cm columns have proved to be very satisfactory. The simplicity of operation and speed recommend them for all except the separation of complex mixtures and very large amounts. The quantity that can be successfully chromatographed is mainly a function of the ease of separation; unfortunately, this is usually unknown until after the chromatogram is completed. Thus, a column of this size probably would separate easily a mixture that contained as much as 10 μmoles each of a basic, neutral, and acidic peptide. On the other hand, perhaps the quantity would have to be reduced to as little as 1 μmole each of two neutral peptides. In general, a mixture that contains several micromoles of each peptide is satisfactory.

Acknowledgments

This procedure was developed during investigations under a grant (HE-02558) from the National Institutes of Health, U.S. Public Health Service. Donald Babin, Jean Cormick, Philip Lieberman, and Barbara Robberson have devoted much time and thought to the development of the method.

Contribution No. 4138 from the Division of Chemistry and Chemical Engineering, California Institute of Technology.

[17] Chromatography of Proteins and Peptides on Diethylaminoethyl Cellulose

By DEBDUTTA ROY and WILLIAM KONIGSBERG

Diethylaminoethyl (or DEAE-) cellulose is an anion exchanger with residues linked to the hydroxyl groups of the cellulose. This type of

$$-C_2H_5-\overset{H}{\underset{+}{N}}\begin{matrix} C_2H_5 \\ C_2H_5 \end{matrix}$$

chemically modified cellulose was first introduced in 1956 by Peterson and Sober[1] and has been employed widely ever since as a powerful tool for the separation of proteins and nucleic acids. DEAE-cellulose is particularly useful for acidic polyelectrolytes; it offers the advantages of high capacity, good resolving power, and a minimum of surface effects which tend to denature proteins.

Principle

The separating potential of ion-exchange cellulose depends on the formation of multiple electrostatic bonds between charged sites on the surface of the exchanger and sites with opposite charge on the surface of the protein. The number of these electrostatic interactions determines the concentration of eluent ions required to release the adsorbed protein. Thus a mixture of proteins differing in net charge will have different binding properties. However, the binding strength of a protein to an ion exchanger is not dependent solely on the interaction between charged sites but also on nonionic interaction between the hydroxyl groups of

[1] E. A. Peterson and H. A. Sober, *J. Amer. Chem. Soc.* **78,** 756 (1956).

the cellulose and polar residues on the surface of the protein.[2] These nonionic interactions (mainly hydrogen bonds) markedly influence the elution characteristic of the protein under investigation, and for these reasons urea is often used in the elution buffers to minimize nonionic absorption.

Elution of the protein from DEAE-cellulose can be achieved in three ways: (1) reduction in the pH of the eluent, which lowers the net negative charge on the protein, thus weakening the binding to the exchanger; (2) increasing the pH of the eluent, thus suppressing the ionization of the basic groups of the absorbent and reducing its anion-binding capacity; and (3) increasing the ionic strength of the eluent, which decreases the strength of the ionic interactions between the protein and the exchanger.

The effective operating pH range for DEAE-cellulose is between 9 and 4.5, above the average pK of carboxyls on the peptide or protein. The presence of a polyelectrolyte may shift the operating range slightly to the right as illustrated by the titration curve of DEAE-cellulose in 0.1 M sodium chloride with and without the presence of a polyelectrolyte, as shown in Fig. 1. Another important consideration in the use of DEAE-cellulose is the time required to reach equilibrium. It can be seen from the curve in Fig. 2, that the kinetics of adsorption are nonlinear. A possible interpretation of these results may be that rapid adsorption occurs

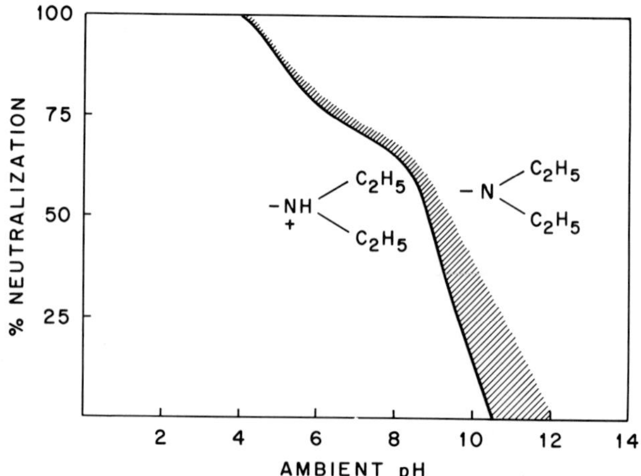

FIG. 1. Ionization characteristics of diethylaminoethyl cellulose. Redrawn from the Whatman laboratory manual on advanced ion-exchange cellulose (C. M. Thompson, W & R Balston Ltd.).

[2] R. V. Tomlinson and A. M. Tener, *Biochemistry* **2**, 697 (1963).

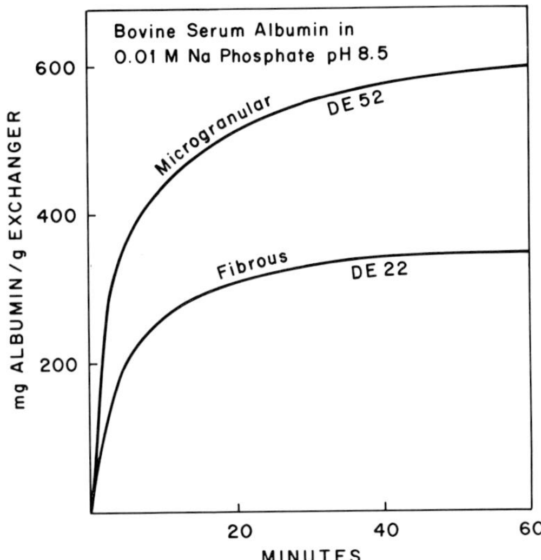

FIG. 2. Protein capacity and kinetics of absorption of microgranular and fibrous diethylaminoethyl cellulose. Redrawn from the Whatman laboratory manual on advanced ion-exchange cellulose (C. M. Thompson, W & R Balston Ltd.).

on the surface, accounting for the fast step, and that the rate-limiting factor in the slow phase may be the diffusion of the protein to charged sites in the interior of the exchanger through pores in the cellulose matrix which may be only slightly larger than the effective diameter of the protein being adsorbed. The principal effect of the failure to reach equilibrium is the trailing of charged solutes as they are eluted from the column.

Practical Considerations

Types of DEAE-Cellulose Available

DEAE-cellulose can be purchased from a number of companies. The DEAE-cellulose supplied under the Whatman trade name (W and R Balston Ltd.) is available in microgranular (DE52) and fibrous (DE22) forms. The main differences between the two types are the capacity, kinetics of equilibration (see Fig. 2), and the flow rates. The fibrous form has a lower capacity, takes longer to reach equilibrium but has flow rates that are much faster than those of the microgranular form. The microgranular form, compared to the fibrous form, has superior resolving power per volume of resin bed and is recommended for routine separation work where maximum resolving power is needed. The fibrous form is useful for batch separations where larger volumes have to be handled, as in the initial fractionation of a crude protein extract. Both the fibrous and microgranular types are supplied in the dry and wet forms, but the wet (preswollen) types DE52 (microgranular) and DE22 (fibrous) are advantageous since they contain fewer fines, which are produced by fragmentation of the dry material, and do not need vigorous precycling.

Precycling

The backbone of DEAE-cellulose is made up of polysaccharide chains held together by hydrogen bonds. The proportion of hydroxyl groups participating in these bonds is maximal in the fully dried state. If left intact, these interactions will limit the accessibility of polyelectrolytes during chromatography. To have the full capacity of a dried exchanger, it is essential to convert it to a stable optimum state of swelling by breaking the hydrogen bonds. Since simple immersion in water or buffer will not accomplish this, it is necessary to precycle the exchanger.

The most convenient way to precycle is to stir a weighed amount of exchanger into 15 volumes of 0.5 hydrochloric acid and leave for at least 30 minutes (not more than 2 hours). The resin is then washed with water on a Büchner funnel until the pH of the effluent is 4. The resin is stirred with 15 volumes of 0.5 *N* sodium hydroxide and allowed to stand for 30 minutes. It is finally washed until the pH of the wash is neutral. Treatment with 0.5 *N* sodium hydroxide should be repeated if the pH of the effluent does not reach neutrality in a reasonable time. Occasionally it is difficult to filter the exchanger in the presence of alkali, but this problem can be overcome by the addition of sodium chloride.

Removal of Fines

Small particles (less than 10 μ in diameter) are formed during manufacture of the DEAE-cellulose. These fines slow down the flow rate of the column, particularly in columns made with fibrous DEAE-cellulose. To remove these, the product is suspended in the initial buffer (total volume should be 6 ml per gram of wet filtered exchanger or 30 ml per gram of dry exchanger) in a measuring cylinder and allowed to settle. The time of settling can be calculated from the equation $t = nh$, where t = time of settling, n = a factor between 1.3 and 2.4, and h = the total height of the suspension in the cylinder (cm); $n = 2.4$ for DE52 and 1.3 for DE22. After this time, the supernatant liquid is decanted or drawn off with a water pump. For most purposes this procedure should be repeated at least once, but to get the maximum flow rate with fibrous grade exchanger a more complete removal of fines is needed.

Degassing

DEAE-cellulose will bind carbonate and bicarbonate anions that are introduced by prolonged exposure of a suspension of the resin at alkaline pH. Unless bicarbonate buffers are being used, the presence of this anion will interfere with the resolving power and prevent reproducible results with DE52. Removal of carbonate and bicarbonate can be accomplished in two ways: (1) By direct ion exchange with the anions of the working buffer using the equilibration procedure described later. This method is time consuming because of the high affinity of DE52 for carbonate and bicarbonate. (2) Conversion to free carbon dioxide followed by degassing. This is the recommended procedure and should be done by dispersing the fully swollen exchanger in the acidic component of the buffer (pH should be below 4.5). A good vacuum (water aspirator) is applied while stirring. When bubbling has ceased, the vacuum is released and the slurry is titrated with the basic component of the buffer to give the desired pH. The exchanger can then be equilibrated with the elution buffer.

Equilibration

Inadequate and nonreproducible resolution is often due to incomplete equilibration of the exchanger with the starting buffer. This applies equally to single-step elution where the exchanger remains in equilibrium and to gradient elution where the column may not be completely in equilibrium with the eluent except at the beginning of the gradient. There are two commonly used methods for equilibration of exchanger before the column is packed. In the first method the exchanger is stirred in the initial buffer (15–30 ml per gram of dry exchanger) with frequent changes of buffer until the correct pH and conductivity are reached. The disadvantage of the method is that it requires large volumes of buffer for complete equilibration. The second method involves the suspension of the exchanger in a solution of a neutral salt (e.g., sodium chloride) at the correct concentration of the eluting counterion, followed by adjustment of the pH with acid or alkali at the same counterion concentration. When the pH has been adjusted to the correct value, the exchanger is finally washed with the starting buffer. Careful consideration must be given in choosing buffers, particularly for enzyme separation. The buffer ions should neither inhibit activity of the enzyme nor interfere in the enzyme assay. The eluting ions should not interfere with the method used to determine protein concentration.

The ionic strength of the starting buffer should be kept low in order to permit binding of the protein to the cellulose. Proteins will be eluted more rapidly with di- or trivalent anions as compared to monovalent anions; thus the nature of the eluting anion should be chosen according to the exchanger–polyelectrolyte affinity. Buffers that are commonly used are: (a) phosphate in the pH range 7.0; (b) Tris·HCl or sodium carbonate/bicarbonate in the pH range 8.0. Combinations of Tris·HCl and phosphate also can be used. If a volatile buffer is desired, ammonium acetate, formate, or bicarbonate can be used. Increasing concentrations of neutral salts, e.g., sodium chloride, have been used in phosphate buffers of low concentration to make ionic strength gradients. EDTA, magnesium, and thiol compounds may be added to the buffers to protect proteins from inactivation during chromatography. Proteins that are insoluble in these buffers have been chromatographed in buffers containing 6–8 M urea.

Column Packing

Optimum performance is dependent on evenly packed columns. Since all ion exchangers have a wide spectrum of particle sizes, the aim of all column pouring techniques is to obtain the most uniform distribution of exchanger particles throughout the column. This will ensure uniform flow of the mobile phase and will favor equilibration of the solute with the adsorbing sites.

The most suitable column-pouring techniques are: (1) pouring in a single pass with or without an extension tube fitted at the top of the column; (2) pouring in several passes; and (3) pouring from a stirred reservoir above the column. With fibrous exchanger, single pass or stirred reservoir should be used because of the possibility of accumulation of fines at one position in the column, causing the formation of bands of differing properties. With microgranular exchangers, any of the methods can be used.

The consistency of the slurry before the column is poured has a marked influence on the flow rate of the column due to the uneven settling of the exchanger in the column. When a very dilute slurry is used to pack columns, localized fractionation of the exchanger on the basis of size occurs, leading to a considerable reduction of flow rate. Microgranular exchangers have high packing densities and will behave this way, so it is advisable to use the thickest slurry (10% dry weight of total volume) that can be poured without trapping air bubbles.

In practice the column should be set up in a completely vertical position, using a plumb line if necessary, and the initial buffer should be poured to about 5 cm from the bottom. The resin slurry should be poured in with stirring, and the effluent allowed to run to waste. When the column is filled, a pump is connected and the initial buffer is pumped at the flow rate equal to that of the actual rate that will be used during elution. When the buffer level reaches within 1 cm of the settled exchanger, more slurry is added and the process is repeated until the desired bed height is reached. Finally, the column is settled by passing through at least one bed volume of starting buffer.

Column Design

Columns of varying dimensions may be used, and the total volume of the column depends on the amount of sample to be applied. Tall, narrow columns in general give better separation, but sometimes the flow rates of these columns are very low, and it is usually necessary to compromise, even when the resolving power is important. Generally the ratio of length to diameter of columns used varies from 5 to 15. To avoid unnecessary loss of resolution the dead space above and below the column bed should be kept to a minimum. The column should be fitted with a porous glass or Teflon disk at the bottom. To keep the top surface of the bed undisturbed, a piece of filter paper should be carefully placed on top of the resin bed.

Application of the Sample

The material to be chromatographed is first equilibrated in the starting buffer. This can be done either by dialyzing the sample in the starting buffer or by dissolving the dry sample into starting buffer and adjusting the pH to that of the initial buffer. The volume of the sample to be applied is important. Sharper bands can be obtained if the sample is applied in a small volume. In general, the initial band of the protein should not occupy more than 10% of the total column volume. If the protein is bound very firmly, larger sample volumes can be applied, because they will be adsorbed in a concentrated band at the top of the column and their behavior during subsequent elution will be independent of the volume applied.

The general procedure for the application of the sample requires prior removal of the buffer at the top of the column. The sample is then carefully layered over the surface of the resin with a Pasteur pipette. Care must be taken so that the surface of the resin is not disturbed since this will result in an assymetric elution profile. (Very gentle air pressure can be applied to push the sample into the column if the sample moves into the resin very slowly.) After the sample has been applied, the inside of the column is rinsed with the buffer and allowed to flow into the resin. The top of the column is then filled with the starting buffer and the elution line is connected.

Elution Conditions

Temperature. The choice of temperature is dependent mainly on the thermostability of the sample. Cooling is frequently used to avoid the denaturation of labile enzymes and to retard microbial action. Lower temperature increases capacity of the exchanger for polyelectrolytes, but this increase is often accompanied by a decrease in resolving power. Increased temperature has the opposite effect and also decreases nonionic interactions. For reason of convenience the columns are usually run at room temperature. Temperature control is then not required except in very special cases.

Flow Rate. Reproducibility of a chromatogram depends largely on the temperature, packing, removal of fines, and flow rate. It is therefore advisable that a pump be used to maintain a constant flow rate. Pumps with positive or semipositive pumping action which will deliver the required flow rate at back pressures up to 20 psi should be used. Suitable peristaltic pumps are those where the plastic tubing is squashed between rollers or bed plates. Columns which are run by gravity flow tend to slow down as the ionic strength of the buffer increases. High flow rates can be used if the enzyme is unstable due to contact with the resin, but too high a flow rate can lead to loss of resolution. Conversely, a low flow rate improves resolution, but when it is too low, zone diffusion and broadening of the peak occurs. Flow rate can be calculated from the relationship: $F = (K\Delta PA)/L$ where $F = \text{ml/hr}$, K is a constant dependent on the nature of the exchanger and the conditions, ΔP is the pressure drop across the column in grams per square centimeter (psi $\times$ 70.3), A is the area in square centimeters, and L is the length of the column bed in centimeters. Typical values of K are 5 for DE52 and 20 for DE23. The usual range of flow rates lies between 5 and 90 ml/cm^2 per hour.

Elution Methods

Before chromatography of a protein mixture on DEAE-cellulose one should check the solubility, effect of salts, and pH stability of the proteins of interest. If the material is stable or soluble only over a narrow range of pH, then the elution conditions will be mainly limited to variation in the ionic strength of the buffer. If, on the other hand, it is stable or soluble over a wide range of pH, but a narrow range of ionic strength, then a pH gradient should be used.

In general, the elution of the protein from the column can be by the following procedures. If the protein of interest is very slightly bound and the impurities very firmly, washing the column with the starting buffer elutes the desired protein. If, on the other hand, it is very tightly bound and the impurities very weakly bound, washing the column with the starting buffer removes the impurities and the protein may be eluted with a high ionic strength buffer. Sometimes stepwise elution is used to get "cuts" that can be further purified by other methods; this is useful for very unstable enzymes. The biggest drawback of this type of elution is that it does not give the kind of resolution obtained with gradient elution.

In gradient elution the pH or the molarity of the buffer is changed very gradually and smoothly to distribute the protein into as many chromatographically distinct components as possible. This is usually the method of choice for the examination of proteins or peptides. The most commonly used gradient is the linear gradient, which can be easily made with two identical vessels connected with each other through a tube filled with the buffer. Two aspirator bottles connected to each other serve the purpose. For the construction of other kinds of gradient the original literature should be consulted.[3]

An example of gradient elution of peptides from a DE52 column is the separation of tryptic peptides of maleylated and citraconylated catalytic chain of aspartate transcarbamylase. A linear gradient of sodium bicarbonate ($0.02\,M$ to $0.5\,M$) has been used to obtain resolution of the peptides[4] (Fig. 3).

A recent advance in ion-exchange chromatography of enzymes takes advantage of their binding specificity. Pogell[5] eluted fructose-1,6-diphosphatase from CM-cellulose columns using dilute solutions of fructose 1,6-diphosphate and obtained a high degree of purification. Cuatrecasas *et al.*[6] purified staphylococcal nuclease, α-chymotrypsin, and carboxypeptidase A by affinity absorption to inhibitor-Sepharose columns and elution of the bound enzyme by changing salt concentration or pH, or by addition of a competitive inhibitor in the eluent. Such methods may prove useful in the field of enzyme chromatography on DEAE-cellulose.

Examination of the Effluent

The most commonly used procedure for protein determination is to read the optical density of the fractions at 230 or 280 mμ. Chemical measurements of protein (e.g., Folin-Lowry, nitrogen, and ninhydrin) can also be used. In enzymatic work, enzyme activity of each fraction can be determined.

[3] R. M. Block and N.-S. Ling, *Anal. Chem.* **26**, 1543 (1954).
[4] D. D. Roy and W. Konigsberg, unpublished work.
[5] B. M. Pogell, *Biochem. Biophys. Res. Commun.* **7**, 225 (1962).
[6] P. Cuatrecasas, M. Wilchek, and C. B. Anfinsen, *Proc. Nat. Acad. Sci.* **61**, 636 (1968).

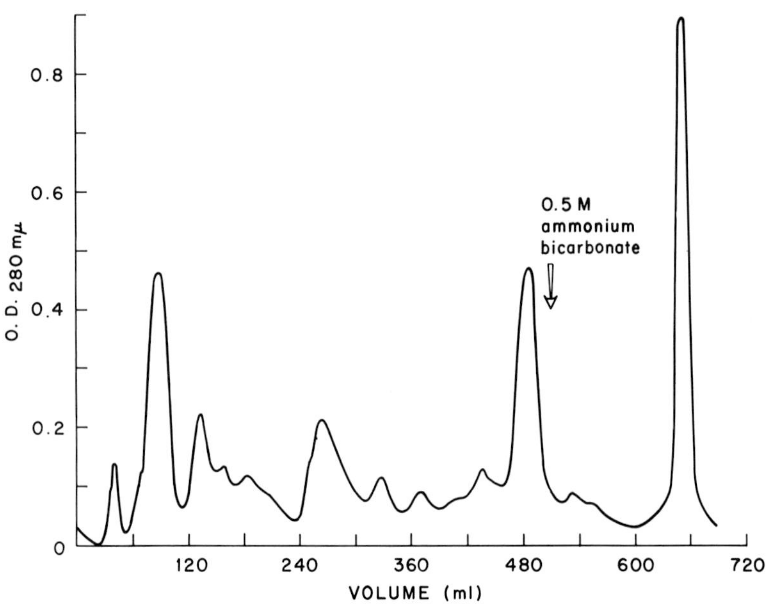

FIG. 3. Separation of tryptic peptides of maleylated aspartate transcarbamylase on DE52. 100 mg of the digest applied to 0.9 × 60 cm column equilibrated with 0.02 *M* $NaHCO_3$, pH 8.3. Eluted with linear gradient of pH 8.3 carbonate-bicarbonate buffers, 0.02–0.5 *M* in Na^+, 500 ml total volume. Final wash with 0.5 *M* ammonium bicarbonate.

Reuse and Storage

Ion-exchange celluloses may be reused many times. If complete elution of the protein has been achieved, the exchanger can be reused after washing with concentrated buffer or strong salt (e.g., 1.0 *M* NaCl) solution, followed by reequilibration. It is essential to give the exchanger the full pretreatment, including acid-alkali pretreatment, *if* (1) some material, such as pigment or protein, has been left in the column; or (2) the exchanger has been left for a long time before reuse. Sometimes denatured protein remains firmly bound to the exchanger. In such a case the contaminated portion of the column bed should be discarded before pretreatment.

Ion-exchange cellulose should be stored in wet form in a tightly stoppered bottle. If the exchanger has been used once, storage in saturated NaCl solution is recommended to inhibit microbial growth. Traces of chloroform, carbon tetrachloride, butanol, or toluene are recommended for storage of fresh exchangers. They can also be used during prolonged chromatography to prevent bacterial contamination.

Discussion

Purification of proteins and peptides on DEAE-cellulose is an inexpensive and conservative method. For highly complex mixtures of proteins, it can be used profitably in the early stages of purification because of its high capacity. The zones containing the proteins of interest can be further purified by other methods or by rechromatography on DEAE-cellulose using a shallower gradient or a different pH. Reproducibility of the chromatogram requires particular care in the packing, ratio of the amount of the sample to the amount of the absorbent, flow rate, and the establishment of the same gradient.

A single peak in the chromatogram does not necessarily indicate homogeneity, nor do multiple peaks signify heterogeneity. pH or ionic strength artifacts can result in partitioning of a pure sample causing premature elution of part of the material at the front while the remaining portion is eluted at its normal position. In addition to pH and ionic strength artifacts, double zoning of a homogeneous substance may be caused by overloading of the column, by excessively high protein concentration, by denaturation, by reversible dissociation–association phenomena,[7] by reaction with gases dissolved in the buffer,[8] or by complex formation.[9]

Suspected artifacts can be checked by rechromatography of each zone using the same gradient employed for the first separation. Homogeneity of material in each zone is best established by other methods, such as acrylamide gel electrophoresis, ultracentrifugation, and NH_2-terminal sequence determinations.

[7] F. J. Gutter, E. A. Peterson, and H. A. Sober, *Arch. Biochem. Biophys.* **80**, 353 (1959).

[8] W. Bjork and H. G. Boman, *Biochim. Biophys. Acta* **34**, 503, (1959); and W. Bjork, *J. Chromatogr.* **2**, 536 (1959).

[9] S. E. G. Åqvist and C. B. Anfinsen, *J. Biol. Chem.* **234**, 1112 (1959).

[18] Terminal Pyrrolidonecarboxylic Acid: Cleavage with Enzymes

By R. F. Doolittle

One of the most vexing situations in amino acid sequence work arises when the terminal α-amino group is blocked and unable to react with coupling agents used in stepwise degradation procedures or even simple detection reagents like ninhydrin. One very common type of blockage occurs when a terminal glutaminyl (I) or glutamyl (II) residue cyclizes, either spontaneously or enzymatically, to yield a pyrrolidonecarboxylyl residue (III).[1]

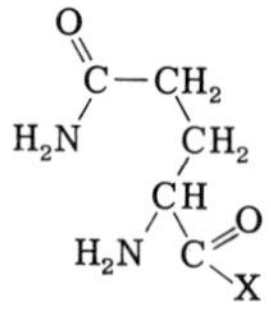

$$\underset{\text{(I)}}{H_2N{-}C({=}O){-}CH_2{-}CH_2{-}CH(NH_2){-}C({=}O){-}X} \qquad \underset{\text{(II)}}{HO{-}C({=}O){-}CH_2{-}CH_2{-}CH(NH_2){-}C({=}O){-}X}$$

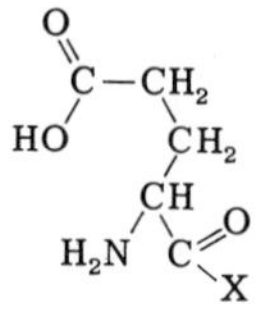

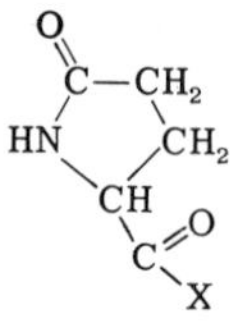

$$\underset{\text{(III)}}{\text{HN}{-}C({=}O){-}CH_2{-}CH_2{-}CH(\text{ring closed to HN}){-}C({=}O){-}X}$$

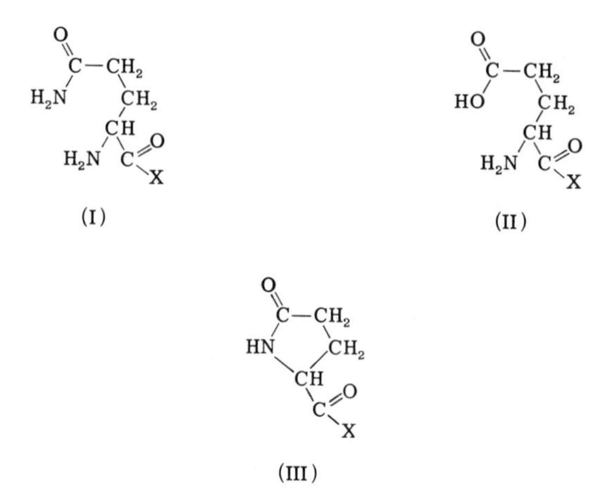

The difficulties and frustrations of determining the structures of peptides with blocked amino-terminal groups[2] led the author to search for an enzyme that could open terminal pyrrolidone rings. A fluorescent pseudomonad which could grow on pyrrolidonecarboxylic acid (PCA)[3] as its sole source of carbon and nitrogen was subsequently isolated from the soil. Although crude extracts from this organism did not exhibit hydrolytic activity toward the amide linkage of free PCA when tested *in vitro,* surprisingly enough they were found to contain an enzyme which removed the entire terminal pyrrolidone residue from PCA-peptides.

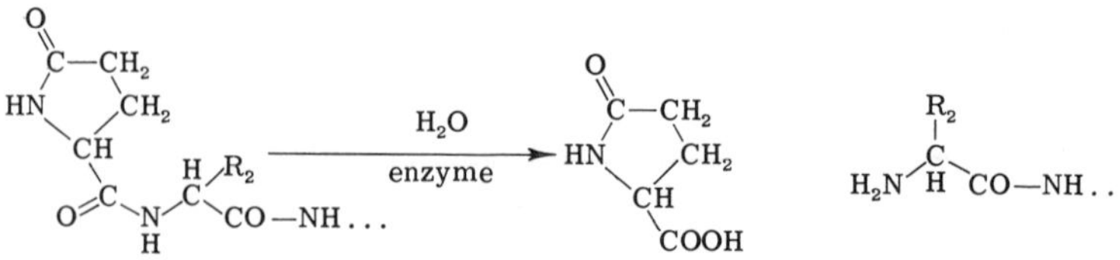

$$\text{PCA}{-}C({=}O){-}NH{-}CH(R_2){-}CO{-}NH\ldots \xrightarrow[\text{enzyme}]{H_2O} \text{PCA}{-}COOH \quad H_2N{-}CH(R_2){-}CO{-}NH\ldots$$

In earlier publications[4,5] the enzyme was called pyrrolidonyl peptidase,

[1] B. Blombäck, Vol. 11, pp. 398–411.

[2] B. Blombäck and R. F. Doolittle, *Acta Chem. Scand.* **17,** 1816 (1963).

[3] In this article the abbreviation PCA will be used for both free pyrrolidonecarboxylic acid and the residue in peptide linkage. In the past the designation Pyr has been widely used for the latter, as well as the abbreviation Glp.

[4] R. F. Doolittle and R. W. Armentrout, *Abstr. Int. Congr. Biochem. 7th,* p. 608 (1967).

[5] R. F. Doolittle and R. W. Armentrout, *Biochemistry* **7,** 516 (1968).

a name which does not precisely describe the reaction catalyzed. Accordingly, we now refer to the enzyme as pyrrolidonecarboxylyl peptidase. Although the enzyme was first found in the pseudomonad strain isolated by soil enrichment, the same activity has now been found in a variety of microorganisms, including *Bacillus subtilis*[5,6] and a wide assortment of both gram-negative and gram-positive bacteria.[6] It is altogether absent from many bacterial species, including *Escherichia coli* and many strains of *Pseudomonas*.[5,6] A similar enzymatic activity has been detected also in certain animal tissues.[7,8] The present article is based on studies on enzyme isolated from the original pseudomonad strain, the emphasis being on application of the enzyme to protein and peptide structural studies rather than on a characterization of the enzyme itself. A related article has appeared in Volume 19 of this series.[9]

Assay Method

The standard assay used in our laboratory has depended on the enzyme-catalyzed hydrolysis of L-pyrrolidonecarboxylyl-L-alanine (PCA-Ala). The progress of the reaction is followed by monitoring the exposure of the alanine α-amino groups by use of the ninhydrin method.

A stock solution of PCA-Ala (MW 200) containing 25 μmoles/ml (5.0 mg/ml) is prepared in glass-distilled water. Ordinarily, 50 μl of buffered (pH 7.3) enzyme solution is added to 25 μl of the stock substrate solution, the final PCA-Ala concentration being 8.3×10^{-3} *M*. Digestions are carried out at 30° in stoppered tubes. At appropriate times, 1.0 ml of absolute ethanol is added to the mixture to poison the enzyme and precipitate the bulk of the protein. The stoppered tubes are placed in the cold for at least 15 minutes to aid precipitation and then centrifuged in a table-top centrifuge. The supernatants are decanted into clean tubes, and 0.5-ml aliquots are removed for ninhydrin assay. In the latter case we have followed the procedure of the Rockefeller group rather literally, adding 0.5 ml of the ninhydrin reagent, capping loosely, and placing the tubes in a boiling water bath for exactly 15 minutes. After cooling, the solutions are diluted with 5 ml of 50% isopropanol, and the absorbance is read at 570 nm. All these operations, including the original digestions, can be carried out in disposable flint glass tubes (13 × 100 mm). These tubes fit well into a Bausch & Lomb Spectronic 20 spectrophotometer and are matched well enough to assure adequate precision for most routine assay situations. The ninhydrin reagent is

[6] A. Szewczuk and M. Mulczyk, *Eur. J. Biochem.* **8**, 63 (1969).

[7] R. W. Armentrout, *Biochim. Biophys. Acta* **191**, 756 (1969).

[8] A. Szewczuk and J. Kwiatkowska, *Eur. J. Biochem.* **15**, 92 (1970).

[9] R. F. Doolittle, Vol. 19, pp. 555–569.

made up according to Hirs *et al.*,[10] 2.0 g of ninhydrin and 0.3 g of hydrindantin being dissolved in 75 ml of ethylene glycol monomethylether (Methyl Cellosolve) followed by addition of 25 ml of 4 *M* sodium acetate buffer, pH 5.5. Under these conditions a leucine standard of 100 nmoles corresponds to about 0.40 absorbance unit. If all the PCA-Ala in a digestion tube is hydrolyzed, a nearly full-scale deflection of 1.2 absorbance units above background is obtained. In crude extracts of the enzyme there is frequently a significant exposure of amino groups which is independent of the presence of added substrate and presumably results from endogenous proteolysis. As a control, then, a separate set of incubation mixtures is set up without substrate (PCA-Ala), but containing instead an equivalent amount of pyrrolidonecarboxylic acid (PCA).

A unit of enzyme activity is defined as the amount of enzyme which releases 1 nmole of alanine per minute under the conditions described above, the pH being maintained at 7.3 with phosphate buffers. We have found it convenient to calculate relative specific activities using the absorbance at 280 nm (A_{280}) of the enzyme solutions as an index of protein concentration.

Szewczuk and Mulczyk[6] use a colorimetric assay employing L-pyrrolidonyl-β-naphthylamide (now available commercially from Cyclo Chemical Corp.) as a substrate and measure the released β-naphthylamine by the Bratton-Marshall reaction.[11] We have also tried to effect a more direct assay by using L-pyrrolidonyl nitroanilide and measuring the yellow color of freed nitroaniline at 390 nm.[12] Although the enzyme is active toward this substrate, the poor solubility of the material in water at neutral pH (approximately 1 mg/ml) limits its usefulness. We still find the ninhydrin assay and PCA-Ala to be the most sensitive and convenient. Several PCA-dipeptides, including PCA-Ala, are available commercially (Cyclo Chemical Corp.). In addition, we have published a complete description of the synthesis and properties of several others.[13]

Culture of the Pseudomonad Strain

Slants of the original pseudomonad strain have been sent to numerous laboratories around the world, several of which are now carrying the strain in culture. The strain can also be obtained from the American Type Culture Collection, Rockville, Maryland (No. 25289). It has been our custom to use a recipe for slants which contains PCA as the sole source of carbon and nitrogen. Care must be taken in maintaining the

[10] C. H. W. Hirs, S. Moore, and W. H. Stein, *J. Biol. Chem.* **219**, 623 (1956).
[11] A. C. Bratton and E. K. Marshall, *J. Biol. Chem.* **128**, 537 (1939).
[12] R. F. Doolittle and V. Woods, unpublished experiments (1969).
[13] J. A. Uliana and R. F. Doolittle, *Arch. Biochem. Biophys.* **131**, 561 (1969).

strain, however, since Stanier has subsequently found that 166 of 175 strains of *Pseudomonas* in his collection will grow on PCA, but when one of these was chosen at random and sent to us, it failed to exhibit pyrrolidonecarboxylyl peptidase activity.

We have used two different liquid media in tandem for large-scale culturing of the organism. First, slants of the strain are rinsed into a minimal medium containing 0.5% PCA and the usual salts (but not ammonium salts) and trace metals. When this inoculum is grown up, it is transferred to a large volume (5 liters) of a glucose-citrate minimal medium (including ammonium salts). When the latter culture is grown up it can be harvested by centrifugation or used as an inoculum for a 100-liter batch in a fermentor, the glucose-citrate medium being used in the latter case also. The substitution of the glucose-citrate–ammonium medium for the PCA type is primarily a matter of economy, although growth is somewhat more vigorous on the glucose type. Each of the steps involves 24–48 hours' growth at 30°. The recipes used for the preparation of these culture media are available in an earlier volume.[9]

The bacteria are readily harvested after a fermentor run using a Sharples-type centrifuge. Our average yield from 100 liters of culture is about 500 g wet weight. The bacteria are then freeze-dried, the dry weight yield ranging from 80 to 120 g. After pulverization the material is stored at −20° until needed. The dried cells readily withstand a few days at room temperature for mailing purposes.

Preparation of the Bacterial Enzyme

The bacterial enzyme can be prepared either from bacterial pellets frozen directly after harvesting or from pulverized freeze-dried suspensions. The latter material is especially useful when shipment at room temperature is necessary. In either case, cell disruption is readily accomplished by sonication. The cells are suspended in a 0.05 *M* phosphate buffer, pH 7.3, containing 0.01 *M* mercaptoethanol and 0.001 *M* EDTA. This buffer, which is used throughout most of the early purification steps, should also contain 0.1 *M* 2-pyrrolidone as an enzyme stabilizer.[14] The actual sonication conditions will vary with the type of sonifier employed; we use a glass "flow-around" vessel packed in ice and submit the suspension to a series of intermittent blasts with a Branson sonifier. Care must be taken to maintain the temperature of the suspension in the range 0–10°.

After sonication the material is centrifuged at 39,000 *g* at 0° for 30–60 minutes. The supernatant liquid containing the enzyme activity is

[14] R. W. Armentrout and R. F. Doolittle, *Arch. Biochem. Biophys.* **132,** 80 (1969).

carefully decanted; if necessary, it should be centrifuged a second time. The A_{280} of the supernatant should be at least 50; if it is higher, the liquid can be diluted with the phosphate buffer described above.

Nucleic acids may be precipitated by adding 1 volume of 1% protamine sulfate to 6 volumes of cold supernatant containing the enzyme. Although this step is not absolutely necessary and was not used in our early preparations, the removal of nucleic acids does seem to facilitate the subsequent gel filtration step on Sephadex G-200. After centrifugation (15,000 *g* for 20 minutes), the pellet is discarded and the supernatant is used immediately for ammonium sulfate fractionation.

A saturated ammonium sulfate solution is added directly, with stirring, to the supernatant fluid from the protamine sulfate step until 42% saturation is achieved. After standing at 0° for 1 hour, the preparation is centrifuged (39,000 *g* for 30 minutes). The supernatant fluid is carefully decanted and discarded, the walls of the tubes being wiped dry with tissues. The pellets are dissolved in a minimal volume of phosphate buffer and diluted to an A_{280} of approximately 250. The material is stored frozen.

The next step in the purification entails gel filtration on Sephadex G-200. In our early work it was necessary that this step be carried out swiftly to prevent activity losses during the chromatography run. For this reason we employed moderate-sized columns (2.5 × 30–40 cm),

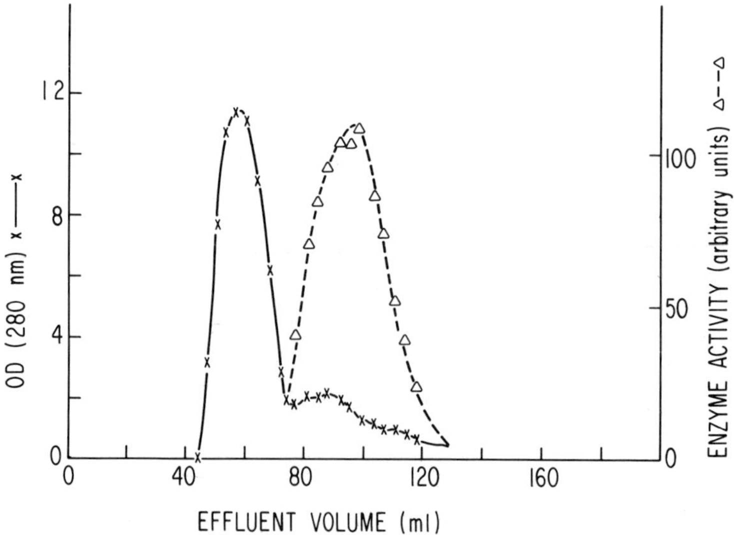

FIG. 1. Gel filtration on Sephadex G-200 (2.5 × 32 cm) of 42% ammonium sulfate cut.

preferring to run several of these rather than a single larger column with a longer run time. The discovery that 2-pyrrolidone added to the buffer stabilizes the enzyme obviates the need for very fast runs, and a 5 × 60-cm column is suitable for large-scale purifications. The enzyme activity comes behind the main protein peak on G-200 (Fig. 1). After removal of aliquots for assay and absorbance measurement, the enzyme activity is precipitated from the pool by the addition of solid ammonium sulfate (0.45 g/ml). After at least 1 hour at 0°, the material is centrifuged and the pellets are stored frozen. If no further purification is to be undertaken, the slurried pool is divided into several portions before centrifuging, and the subpellets are frozen separately (Table I). Enzyme preparations at this stage are generally quite suitable for digestion of PCA-peptides (see below), having a specific activity 20 to 40-fold that of the original crude extract (sonicate supernatant).

Further purification of the enzyme is readily accomplished by ion-exchange chromatography on DEAE-Sephadex (A-25). The pooled pellets from Sephadex G-200 columns (above) are dissolved in a minimal volume of 0.05 *M* phosphate buffer, pH 8.0, and desalted either by rapid dialysis or gel filtration on Sephadex G-25. The enzyme activity is readily eluted at pH 8.0 (0.05 *M* phosphate) using a 0–3% NaCl gradient. The active tubes, which come early in the gradient, are pooled and solid ammonium sulfate is added (0.6 g/ml) to precipitate the enzyme. Once again, if no further purification steps are planned, it is convenient to divide the slurry into a half-dozen portions before centrifuging. The pellet or subpellets are stored frozen at −20°. Enzyme preparations at this stage of purification have been employed successfully in the removal of PCA residues from a large molecular weight protein without detecting any other proteolysis.[14] The material ranges from 60 to 100-fold purified over the crude bacterial extract.

Additional purification is not necessary if the enzyme is only to be used as a reagent in protein structure work. In fact, because the enzyme activity tends to become less stable with increasing purification, it is usually undesirable to carry the purification further. On the other hand, complete characterization of the enzyme necessitates purer material, and a Buchler Polyprep gel electrophoresis can be usefully employed at this stage. This approach takes advantage of the great mobility of the enzyme in acrylamide gels in Tris·borate buffer systems at pH 8.4. Suitable stabilizing agents must be added to the system; in particular, thioglycolate is substituted for mercaptoethanol. The active pools from these electrophoresis runs exhibit two bands when examined on analytical gel electrophoresis and are about 200-fold purified relative to starting crude extracts.[14]

TABLE I
SUGGESTED ISOLATION SCHEME FOR PYRROLIDONECARBOXYLYL PEPTIDASE

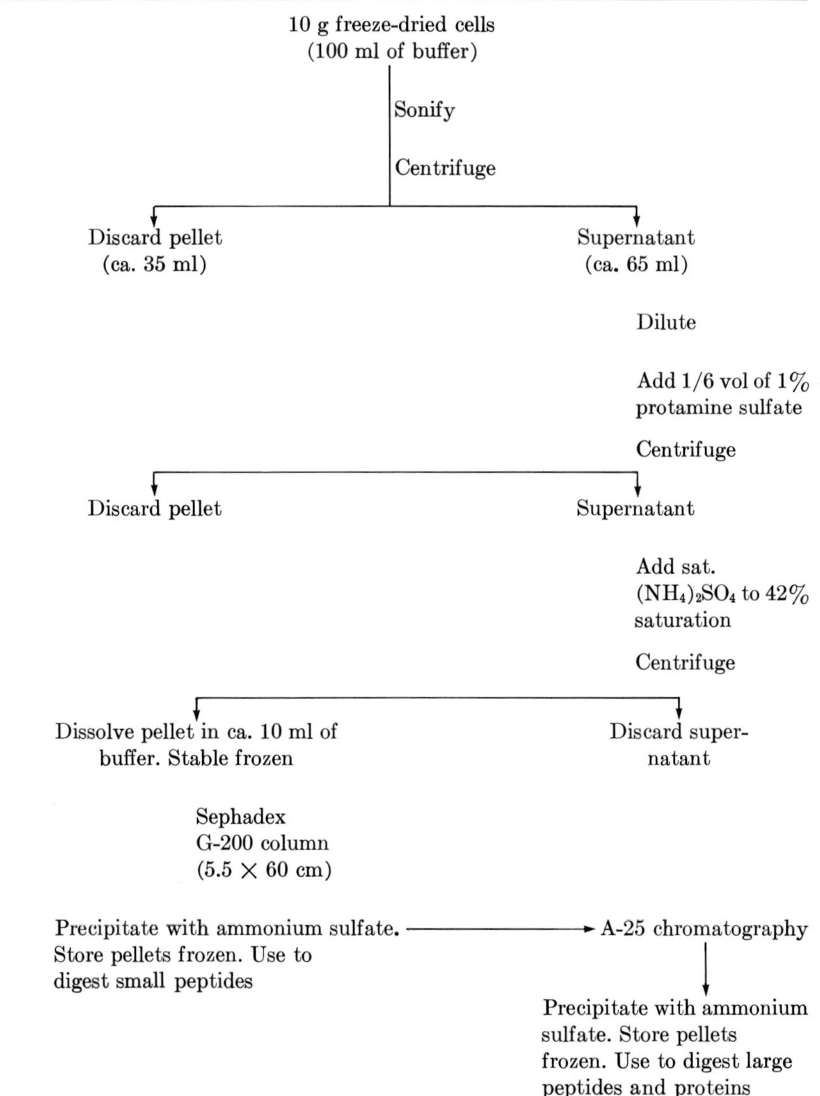

Digestion of PCA-Peptides and PCA-Proteins

PCA-Peptides. Peptides suspected of having terminal PCA can be conveniently studied with "G-200 enzyme." The designation "peptide" here includes any material which has a significant mobility on paper electrophoresis at pH 2.0, since this system offers a very convenient method of scanning a preliminary digest on a small amount of unknown material. At pH 2.0, peptides which lose a terminal PCA residue have their cathodic mobility significantly increased by the exposure of the (penultimate) α-amino group. Ninhydrin-negative peptides also become ninhydrin-positive.

Typically, a frozen G-200 (or A-25) subpellet is thawed and dissolved in a suitable volume of 0.05 *M* phosphate buffer containing 0.01 *M* mercaptoethanol and 0.001 *M* EDTA to yield an A_{280} between 2 and 4 (or about 1.0 for an A-25 preparation). The material is then dialyzed against the same buffer to remove the residual ammonium sulfate and 2-pyrrolidone (the stabilizer is also an inhibitor). Usually three successive 1-hour dialyses against 250 ml of buffer each are sufficient. Alternatively, the subpellet can be dissolved in a smaller volume of the same buffer. Passage over a small (1.0×10-cm) Sephadex G-25 column removes the ammonium sulfate and the 2-pyrrolidone. The optimum concentrations of enzyme and substrate will vary with the particular peptide under study, and it is useful to make a preliminary examination of the reaction progress on a small amount of the peptide before attempting to prepare a large amount for subsequent sequential degradation or other follow-up characterization. To this end, about 0.1 μmole of peptide is dissolved in 100 μl of the enzyme solution and incubated at 30°. Aliquots (20 μl each) are removed at 0, 2, 4, 8, and 20 hours and applied directly to electrophoresis paper to stop the reaction. The papers are then electrophoresed at pH 2.0 (8% acetic acid, 2% formic acid). We are pleased with our results on an LKB low voltage apparatus and use 300 V for 3 hours. On the other hand, the kind of rig is not critical and we have used 2000 V for an hour with the same buffer in a Savant water-cooled plate apparatus.

If the original peptide contains lysine, ninhydrin staining can be used to follow the progress of the reaction. If the peptide contains arginine, the Yamada-Itano stain[15] is recommended because of its great sensitivity and the subsequent saving of material during the initial scan. If the peptide contains neither lysine nor arginine, then the chlorine gas method[16] is useful. One cannot depend only on the appearance of ninhydrin-positive material and still know when all the material has been digested.

Besides the change in mobility of the substrate peptide, successful removal of terminal PCA residues can be observed by following the appearance of the PCA itself. At pH 4.1 PCA has a mobility (anodic) twice that of free aspartic acid. We use a 0.1 *M* pyridine–acetic acid

[15] S. Yamada and H. A. Itano, *Biochim. Biophys. Acta* **130,** 538 (1966).

[16] F. Reindel and W. Hoppe, *Chem. Ber.* **87,** 1103 (1954).

buffer, pH 4.1, 300 V, 2 hours and locate the PCA with the chlorine gas method.[16]

If the digestion appears satisfactory and complete in the pilot study, another subpellet is thawed and dialyzed; proportionately more enzyme solution is used to digest a substantial amount of the peptide for an appropriate length of time. The new species, without the terminal PCA, can be recovered by preparative electrophoresis or ion-exchange chromatography on Dowex 50-X2. Alternatively, the digestion can be terminated by the addition of pyridine so that stepwise degradation using the phenylisothiocyanate method may be undertaken directly.

PCA-Proteins. Removal of terminal PCA moieties from proteins has been less well studied. In general we have favored sustained digestions with "A-25" preparations of the enzyme. Removal of the PCA can be determined by various quantitative end-group procedures directed toward the newly exposed penultimate amino acid. Enzyme incubations of 0, 8, and 24 hours at 30° are recommended. The digestions are stopped by the addition of the appropriate organic solvent, depending on which amino-terminal procedure will be attempted (phenyl isothiocyanate, cyanate, fluorodinitrobenzene, dansyl, etc.). If availability of material permits, control incubations of enzyme solution alone and protein alone should be run simultaneously. In our hands, "A-25 enzyme" solutions of A_{280} = 1.0–3.0 do not reveal significant quantities of endogenous end groups when incubated as controls. The same preparations readily remove the terminal PCA group from the β chain of native bovine fibrinogen, however, exposing the penultimate phenylalanine residue. No other amino-terminal groups are exposed during the digestion.[14]

Although the terminal PCA in native bovine fibrinogen is readily removed by pyrrolidonecarboxylyl peptidase, other proteins may present more difficult situations. It is generally accepted that the fibrinopeptide portions of fibrinogen molecules are on the surface of the protein. It might have been anticipated, then, that the terminal PCA would be accessible. In other proteins, however, the terminal PCA may be tucked more inside, and it may be necessary to unfold the target protein by suitable means before digestion with the enzyme. Szewczuk and Mulczyk[6] have reported the release of PCA from human seromucoid using an enzyme prepared from *B. subtilis,* but they did not attempt to establish any newly exposed amino-terminal residues.

Some Pertinent Properties of Bacterial Pyrrolidonecarboxylyl Peptidase

Stability. Pyrrolidonecarboxylyl peptidase is a sulfhydryl-dependent enzyme and is easily poisoned by iodoacetamide and organomercurials.

It is necessary to keep a reducing agent in solutions employed for isolation steps as well as the final mixture. The active enzyme sulfhydryl group can be reversibly blocked by reaction with tetrathionate under conditions similar to those used to block other sulfhydryl proteases.[17] Subsequent short-term incubations with mercaptoethanol restore full activity.[14]

The bacterial enzyme is remarkably stabilized by the presence of 0.1 *M* 2-pyrrolidone (IV).

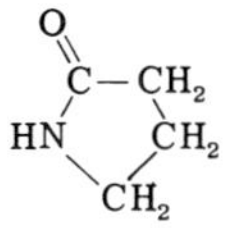

```
 O
  \\
   C—CH2
  /    \
HN      CH2
  \    /
   CH2
```

(IV)

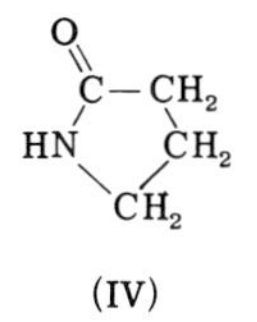

The substance, which was selected as a potential substrate analog, behaves as a noncompetitive, fully reversible inhibitor of the enzyme.[14] It must be dialyzed away before the enzyme is used for digestion purposes.

Generally speaking, the purer the preparation of pyrrolidonecarboxylyl peptidase, the more liable it is to decay during storage. Hence, "G-200 enzyme," frozen as pellets after precipitation with ammonium sulfate, is stable at —20° for several months. "A-25 enzyme" stored the same way loses about half its activity in 6 weeks,[14] although recently we have found that storage at —70° preserves these activities for considerably longer periods. Attempts to stabilize the activity by attachment to inert supports have been hindered by the great sensitivity of the enzyme to diazotization.[18]

Specificity. A consideration of specificity for pyrrolidonecarboxylyl peptidase has two aspects. On the one hand, we want to know whether it will attack all or most PCA-peptides, no matter what the nature of the adjacent amino acid. Second, we have to know whether the enzyme is specific for the peptide bond linking PCA to other amino acids or whether this activity is only part of a much broader range of specificity.

With regard to the first aspect, to date the enzyme has successfully removed PCA from 11 different penultimate amino acids (Table II). It does not cleave L-pyrrolidonecarboxylyl-L-proline (PCA-Pro) peptide bonds.[13]

The rate at which the enzyme hydrolyzes different PCA-dipeptides varies considerably. Of all the substrates tried so far, L-pyrrolidone-

[17] P. T. Englund, T. P. King, L. C. Craig, and A. Walti, *Biochemistry* **7**, 163 (1968).
[18] R. F. Doolittle and F. Lau, unpublished observations (1970).

TABLE II
TABULATION OF AMINO ACIDS PENULTIMATE TO TERMINAL PYRROLIDONECARBOXYLIC ACID RESIDUES AND RESULT OF ENZYME ACTION

Penultimate amino acid	PCA removed by enzyme	Reference
Alanine	Yes	*a*
Arginine	—	
Asparagine	—	
Aspartic acid	Yes	*b*
Cysteine	—	
Glutamine	—	
Glutamic acid	—	
Glycine	Yes	*c*
Histidine	Yes	*a, d, j*
Isoleucine	Yes	*e*
Leucine	Yes	*e*
Lysine	Yes	*f*
Methionine	—	
Phenylalanine	Yes	*a, e*
Proline	No	*e*
Serine	—	
Threonine	Yes	*g, h*
Tryptophan	—	
Tyrosine	Yes	*e, i*
Valine	Yes	*a*

[a] R. F. Doolittle and R. W. Armentrout, *Biochemistry* **7**, 516 (1968).
[b] H. Itano, personal communication, 1969.
[c] R. F. Doolittle, G. L. Wooding, Y. Lin, and M. Riley, *J. Mol. Evol.* **1** (in press, 1971).
[d] R. L. Jackson and C. H. W. Hirs, *J. Biol. Chem.* **245**, 624 (1970).
[e] J. A. Uliana and R. F. Doolittle, *Arch. Biochem. Biophys.* **131**, 561 (1969).
[f] S. H. Ferreira, D. C. Bartelt, and L. J. Greene, *Biochemistry* **9**, 2583 (1970).
[g] E. Appella, R. G. Maye, S. Dubiski, and R. A. Reisfield, *Proc. Nat. Acad. Sci. U.S.* **60**, 975 (1968).
[h] R. L. Jackson and C. H. W. Hirs, *J. Biol. Chem.* **245**, 637 (1970).
[i] A. H. Kang and J. Gross, *Biochemistry* **9**, 796 (1970).
[j] Y. Baba, H. Matsuo, and A. V. Schally, *Biochem. Biophys. Res. Comm.* **44**, 459 (1971).

carboxylyl-L-alanine (PCA-Ala) is the most readily attacked (Table III). The second most easily hydrolyzed of the PCA dipeptides is PCA-Ile, which is attacked at 50% the rate of PCA-Ala. Next come PCA-Val, PCA-Leu, and PCA-Phe, all of which are hydrolyzed at about 20% the rate of PCA-Ala. Crude extracts exhibit the same order of relative activity, negating any notion that we have inadvertently purified one enzyme of a family by using PCA-Ala as our test substrate during the isolation procedure.

With regard to the strictness of specificity toward PCA-peptide

TABLE III
RELATIVE INITIAL ACTIVITY OF PYRROLIDONECARBOXYLYL PEPTIDASE DIRECTED TOWARD A VARIETY OF PCA-DIPEPTIDES[a]

PCA-Dipeptide[b]	Percent of activity found against PCA-Ala		
	Expt. I	Expt. II	Expt. III
PCA-Ala	100	100	100
PCA-Ile	57	50	49
PCA-Val	20	22	24
PCA-Phe	NR[c]	14	14
PCA-Leu	17	19	19
PCA-Tyr	5	9	10
PCA-Pro	0	NR[c]	0

[a] Adapted from J. A. Uliana and R. F. Doolittle, *Arch. Biochem. Biophys.* **131,** 561 (1969).

[b] In all cases the final substrate concentration was approximately $8 \times 10^{-3}\ M$, except in the case of PCA-Pro, where it was $5 \times 10^{-3}\ M$.

[c] NR = not run in that particular experiment.

linkages, it is perhaps premature to state rigidly that this is the only bond which the enzyme hydrolyzes. We know that with "G-200 enzyme" preparations (20 to 40-fold purified) no other peptide bonds are detectably cleaved in bovine fibrinopeptide B (21 residues),[5] and "A-25" preparations (60- to 100-fold purified) do not cleave other peptide bonds in native bovine fibrinogen (MW 330,000).[14] On the other hand, it is possible that the enzyme may cleave other terminal residues which lack α-amino groups. The *Pseudomonas* enzyme does not remove acetyl groups from *N*-acetyl tyrosine ethyl ester, but it has not been tested on *N*-acetylated or *N*-formylated peptides. One unnatural situation which may be attacked by the enzyme is that in which an *S*-carboxamidomethylcysteinyl residue exists at the amino-terminal position as a result of enzymatic or other cleavage after routine reduction and alkylation of a protein. We have demonstrated that at neutral pH *S*-carboxamidomethyl-L-cysteine (V) cyclizes at a rate comparable to that of glutamine, forming a 6-membered ring (VI) equivalent to the 5-membered ring of PCA.

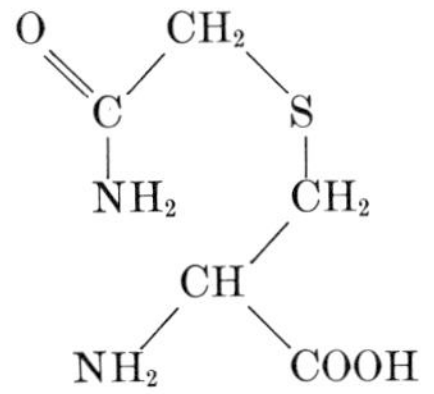

```
O     CH2                    O     CH2
 \\  /   \                    \\  /   \
   C      S                     C      S
   |      |                     |      |
  NH2    CH2                   NH      CH2
         /                        \   /
      CH                            CH
     /  \                             \
  NH2    COOH                          COOH
     (V)                           (VI)
```

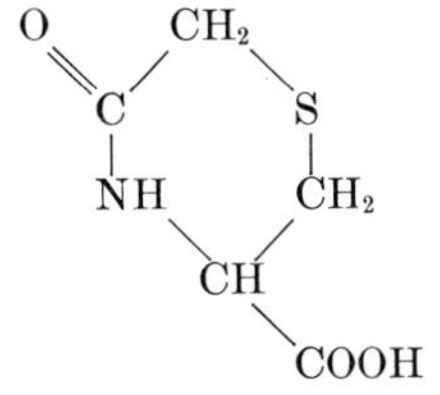

In summary, an enzyme preparation is available that can be used to advantage in structural studies on PCA-peptides and PCA-proteins. The enzyme not only allows the possibility of positively identifying the nature of many blocked α-amino groups, but also permits stepwise degradation techniques to be employed on peptides formerly inaccessible to such methods.

[19] Gas–Liquid Chromatographic Analysis of Constituent Carbohydrates in Glycoproteins

By Vernon N. Reinhold

Procedures applicable to the analysis and structural study of glycoproteins by more conventional techniques have been reviewed in a previous volume.[1] Application of gas–liquid chromatography (GLC) to the analysis of steroids,[2] fatty acids,[3,4] Krebs cycle and allied acids,[5] amino acids,[6] and carbohydrates[7] have been reported in this and in earlier volumes. These latter articles will greatly assist those uninitiated in the problems of GLC.

The conditions of cleavage, derivatization, and GLC described in this article provide a quantitative determination of the commonly found glycoprotein sugars (i.e., L-arabinose, L-fucose, D-xylose, D-mannose, D-galactose, D-glucose, *N*-acetyl-D-galactosamine, *N*-acetyl-D-glucosamine, and the neuraminic acids as the *N*-acetyl derivatives. A general procedure will be described, allowing quantitation of the above-mentioned sugar components, along with two modifications. One modification involves cleavage under mild conditions to determine the neuraminic acid and fucose values; the second, under more vigorous conditions, to assess total recovery of the amino sugars. For some glycoproteins the general procedure determines all the expected components quantitatively, but this cannot be assumed to be the case with all samples until verified by the modifications.

The GLC described here was carried out on a dual-column instrument

[1] R. G. Spiro, Vol. 8, p. 3.
[2] H. H. Wotiz and S. J. Clark, Vol. 15, p. 158.
[3] R. G. Ackman, Vol. 14, p. 329.
[4] S. Lipsky and R. Landowne, Vol. 6, p. 513.
[5] N. W. Alcock, Vol. 13, p. 397.
[6] J. J. Pisano, this volume [3].
[7] C. C. Sweeley, W. W. Wells, and R. Bentley, Vol. 8, p. 95.

with a flame-ionization detector and the *O*-trimethylsilyl (TMS) ether sugar derivatives.

Procedures

General Procedure

The glycoprotein (0.5–2.0 mg, depending on the amount and composition of carbohydrate) is placed in a Teflon-lined screw-cap culture tube (0.9×10 cm)[8] with mesoinositol (25–50 μg) as internal standard. The mesoinositol and the glycoprotein are heated together with 1 ml of 500 m*M* HCl in anhydrous methanol for 16 hours at 65°. The sample is dried under a stream of nitrogen and *N*-acetylation of the free amino groups is achieved by suspending the residue in 0.1 ml of pyridine and adding 0.1 ml of acetic anhydride. The sample is immediately evaporated to dryness and the simultaneous partial *O*-acetylation is reversed by refluxing in 1 ml of 500 m*M* HCl in anhydrous methanol for 1 hour. After solvent evaporation, the sample is ready for derivatization.

Modification I

The glycoprotein and internal standard are prepared as above, but hydrolysis is performed by treatment of the sample with aqueous 2 *N* trifluoroacetic acid at 121° for 2 hours.[9] The sample is now taken through the general procedure. The modification provides for analysis of the methyl 2-acetamido-2-deoxy hexosides.

Modification II

To be sure of maximum recoveries of neuraminic acid and fucose, the methanolysis is performed for 1.5 hours, instead of the usual 16 hours, followed directly by solvent evaporation and derivatization.

Derivatization

A trimethylsilylating reagent[10] containing pyridine, hexamethyldisilazane, and trimethylchlorosilane (1.0:0.2:0.1), used with the published precautions,[7] converts all the constituent carbohydrates to their *O*-TMS ethers in 15 minutes at room temperature. Tri-Sil,[11] a similar commercially prepared silylating reagent, is equally effective under the same conditions.

[8] Corning Glass, Corning, New York.

[9] P. Albersheim, P. Nerins, P. English, and A. Karn, *Carbohyd. Res.* **5**, 340 (1967).

[10] C. C. Sweeley, R. Bentley, M. Makita, and W. W. Wells, *J. Amer. Chem. Soc.* **85**, 2497 (1963).

[11] Pierce Chemical Co., Rockford, Illinois.

Bis(trimethylsilyl)trifluoroacetamide,[12] (mixed in equal amounts with acetonitrile), introduced as a silylating reagent for amino acids,[13] converts the sugars to their corresponding *O*-TMS ethers when heated for 30 minutes at 100°. A small amount of anomerization does occur for some sugars, but is no great problem. However, the reagent is unsatisfactory for neuraminic acid derivatization. The sample dissolved in the reagent can be injected directly without extraction.

Extraction

The silylation reagent is evaporated under a stream of nitrogen, and 50–75 μl of heptane is added to dissolve the *O*-TMS ethers. Extraction of the silylated glycosides avoids excessive tailing of the pyridine solvent peak and provides a clear solution for injection.

Gas–Liquid Chromatography

Stainless steel or glass columns packed with lightly loaded OV-11 or OV-17[14] on glass beads[15–17] have given the resolution demanded for isomer separation. The packing can be prepared or purchased commercially,[12] and the column packing procedures should involve only gentle tapping of the column.[18]

Quantitation

Known concentrations of mesoinositol added as internal standard provide relative retention time data and quantitation of the individual sugars. The integrated peak areas multiplied by the detector response factor and compared to the known amount of internal standard provide directly the concentration in the sample injected. It is recommended that the "solvent flush" method[19] be utilized when injecting samples; good technique in sample application cannot be overemphasized.

[12] This and other silane reagents are available from several suppliers including Regis Chemical Co., Chicago, Illinois; Supelco, Inc., Bellefonte, Pennsylvania; Applied Science Laboratories, State College, Pennsylvania; Analabs, Inc., North Haven, Connecticut.

[13] D. L. Stalling, C. W. Gehrke, and R. W. Zumwalt, *Biochem. Biophys. Res. Commun.* **31**, 616 (1968).

[14] Ohio Valley Specialty Chemical Co., Marietta, Ohio.

[15] H. L. MacDonell, *Anal. Chem.* **40**, 221 (1968).

[16] A. M. Filbert and M. L. Hair, *J. Gas Chromatogr.* **6**, 218 (1968).

[17] Corning GLC-110 available through most gas chromatographic supply houses; see footnote 12.

[18] F. A. Vandenheuvel, G. J. Hinderks, J. C. Nixon, and W. G. Layng, *J. Amer. Oil Chem. Soc.* **42**, 283 (1965).

[19] Solvent, 1 or 2 μl, is drawn into the syringe barrel followed by a small volume of air and then sample. On injection, this allows the sample to be washed out with a slug of solvent.

Results

The data in the table provide a summation of elution times and retention times relative to mesoinositol obtained with OV-11 and OV-17. Figure 1 is a chromatogram obtained from the analysis of human α_1-acid

GAS–LIQUID CHROMATOGRAPHIC RETENTION VALUES FOR SOME COMMONLY OBSERVED GLYCOPROTEIN *O*-TRIMETHYLSILYL SUGAR DERIVATIVES

	OV-11		OV-17	
Glycosides	t'_R(min)	$\frac{t'_{R(sample)}}{t'_{R(inositol)}} \times 100$	t'_R(min)	$\frac{t'_{R(sample)}}{t'_{R(inositol)}} \times 100$
Methyl α-L-arabinopyranoside	4.0	47.1	4.2	45.3
Methyl β-L-arabinopyranoside	4.1	50.0	4.5	48.7
Methyl n?-L-fucofuranoside	4.2	51.0	4.7	49.6
Methyl α-L-fucopyranoside	4.4	52.9	5.2	52.1
Methyl β-L-fucopyranoside	4.6	54.7	5.4	54.7
Methyl α-D-xylopyranoside	4.8	58.5	5.6	59.4
Methyl β-D-xylopyranoside	4.8	58.5	5.6	59.4
Methyl α-D-mannopyranoside	6.0	71.6	6.6	71.9
Methyl β-D-mannopyranoside	6.2	73.5	6.8	74.4
Methyl n?-D-galactofuranoside	6.4	76.4	7.0	77.0
Methyl α-D-galactopyranoside	6.6	79.2	7.4	79.4
Methyl β-D-galactopyranoside	6.8	82.0	7.6	82.9
Methyl α-D-glucopyranoside	7.5	88.5	8.0	86.7
Methyl β-D-glucopyranoside	7.5	88.5	8.0	86.7
Mesoinositol (internal standard)	8.2	100.	9.2	100.0
Methyl 2-acetamido-2-deoxy-α-D-galactopyranoside	9.4	112.	10.8	116.0
Methyl 2-acetamido-2-deoxy-β-D-galactopyranoside	10.2	122.	11.6	126.1
Methyl 2-amino-2-deoxy-α-D-glucopyranoside	7.6	92.5	8.0	86.7
Methyl 2-acetamido-2-deoxy-α-D-glucopyranoside	9.6	112.	11.1	121.
Methyl 2-acetamido-2-deoxy-β-D-glucopyranoside	9.6	112.	11.0	120.
Methyl (methyl 5-amino-3,5-dideoxy-D-glycero-D-galacto-β-nonulo-pyranosid) onate	10.4	125.	12.0	130.
Methyl (methyl 5-amino-3,5-dideoxy-D-glycero-D-galacto-α-nonulo-pyranosid) onate	12.0	143.	13.6	148.
Methyl (methyl 5-acetamido-3,5-dideoxy-D-glycero-D-galacto-α-nonulo-pyranosid) onate	12.4	148.	14.0	151.
Methyl (methyl 5-acetamido-3,5-dideoxy-D-glycero-D-galacto-β-nonulo-pyranosid) onate	13.6	163.	15.4	196.

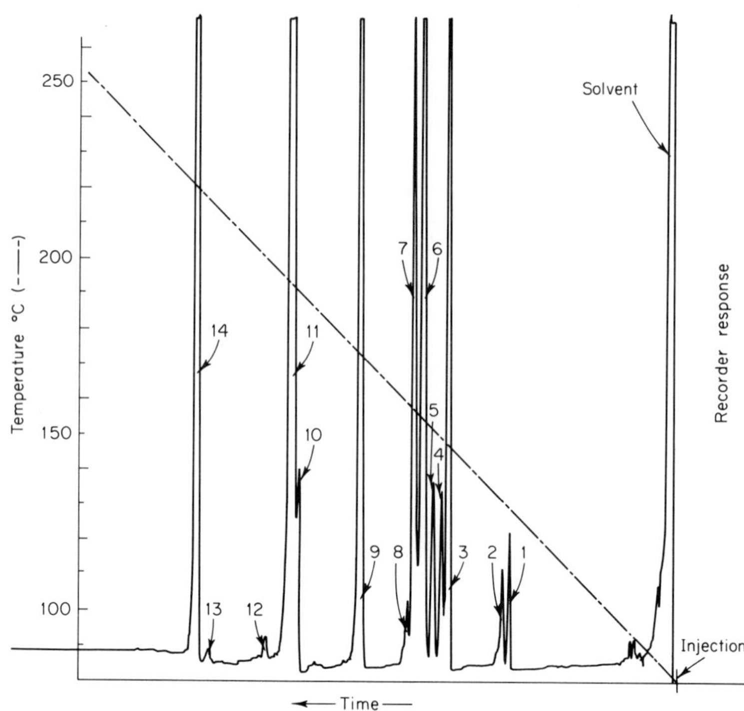

FIG. 1. Chromatogram of methyl glycosides obtained from α_1-acid glycoprotein following the general procedure. Stainless steel column (5 foot × 3 mm) packed with 0.05% OV-17 organosilicone on 120–140-mesh glass beads. Initial temperature 80°C and programmed at a rate of 10° per minute; helium carrier gas, 50 ml per minute; flame ionization detection; chart speed, 0.5 inch per minute. Components as numbered: (1) methyl-α-L-fucopyranoside; (2) methyl-β-L-fucopyranoside; (3) methyl-α-D-mannopyranoside; (4) methyl-β-D-mannopyranoside; (5) methyl-α-D-galactofuranoside; (6) methyl-α-D-galactopyranoside; (7) methyl-β-D-galactopyranoside; (8) methyl-α-amino-2-deoxy-α-D-glucopyranoside; (9) mesoinositol (internal standard); (10) methyl-2-acetamido-2-deoxy-β-D-glucopyranoside; (11) methyl-2-acetamido-2-deoxy-α-D-glucopyranoside; (12) methyl (methyl-5-amino-3,5-dideoxy-D-glycero-D-galacto-α-nonulopyranosid) onate; (13) methyl (methyl 5-acetamido-3,5-dideoxy-D-glycero-D-galacto-α-nonulopyranosid)onate; (14) methyl (methyl 5-acetamido-3,5-dideoxy-D-glycero-D-galacto-β-nonulopyranosid)onate.

glycoprotein following the general procedure. Peaks 8 and 12 are the free amines of acetamidohexoses and neuraminates, respectively.

Comments

Cleavage. Constituent analysis of carbohydrates in glycoproteins is hampered by the variable lability of the bonds linking each sugar moiety and the considerable susceptibility to destruction of some components under conditions yielding quantitative cleavage.

An approach to this problem has been to introduce a correction related to the rate of destruction for each of the susceptible sugars being studied. However, since carbohydrate cleavage may vary for different glycoproteins, the introduction of correction factors must be considered an approximation. Frequently, cleavage has been performed under conditions that are sufficiently mild as to avoid component destruction. The hope then has been that cleavage from the heteropolymer is quantitative. Thus, variation in linkage susceptibility would demand separate analyses for the neuraminic acid, amino sugar, and neutral sugar components. Methanolysis offers the best single cleavage procedure for analyzing the total carbohydrate content of glycoproteins.[20] The marked stability of the methyl glycoside derivatives offsets the GLC problem of anomer separation.

Treatment of the glycoprotein following Modification II allows the determination of *N*-glycolylneuraminic acid, which, during longer methanolysis, is converted to the free amine; see peak 12 in Fig. 1. The *O*-acetyl derivatives cannot be quantitated by a methanolysis procedure.

Reacetylation. Methanolysis causes partial de-*N*-acetylation, yielding a mixture of free amino and 2-acetamido hexosides. Reacetylation conditions lack the specificity to achieve only *N*-acetylation, and the simultaneous partial *O*-acetylation is reversed by a methanolysis step. This de-*O*-acetylation can also be done in ammonia–methanol solutions, but neuraminic acid shows about 3–5% conversion to the amide, which elutes soon after the methyl ester. Other components are not affected.

Acknowledgment

This article is based on work carried out in the laboratories of Professor Roger Jeanloz, Massachusetts General Hospital, Harvard Medical School, Boston, Massachusetts and Professor Klaus Biemann, Department of Chemistry, Massachusetts Institute of Technology, Cambridge, Massachusetts. The author is deeply indebted to them for support, interest, and encouragement in this work. The author is a Fellow of the Helen Hay Whitney Foundation.

[20] J. R. Clamp, G. Dawson, and L. Hough, *Biochim. Biophys. Acta* **148,** 342 (1967).

Section VII

Sequence Determination

[20] Leucine Aminopeptidase in Sequence Determination of Peptides

By Albert Light

The aminopeptidases of swine kidney consist of a group of intracellular enzymes with hydrolytic activity toward the amino-terminal residue of peptides and polypeptides.[1,1a] The enzymes are useful in sequence studies for amino acid analysis and to deduce a partial sequence of a peptide or protein. Complete hydrolysis of a peptide with the enzyme provides residue values for glutamine, asparagine, tryptophan, and other acid-labile amino acids. On partial hydrolysis, amino acids are released sequentially from the amino terminus, and the partial sequence complements the information obtained from hydrolysis with carboxypeptidase which releases amino acids from the carboxyl terminus.

The enzymes are found in the supernatant and microsomal fractions of swine kidney, and appropriate purification schemes are available for each. Smith and co-workers[2] were the first to purify the supernatant enzyme, leucine aminopeptidase (LAP). In a series of studies, the properties of the enzyme were described,[3,4] and Hill and Smith[5] showed that the enzyme is capable of degrading polypeptides in a sequential manner. Their efforts led to the widespread use of the enzyme in sequence studies.

More recently, Pfleiderer and co-workers[6–8,8a] discovered a particle-bound enzyme, aminopeptidase M, in the microsomal fraction and reported on the purification, properties, and mechanism of hydrolysis of the aminopeptidase. LAP and aminopeptidase M have several properties in common but differ in divalent cation requirements (Table I) and in their specificities (Tables II and III). Both enzymes are commercially available and suitable for complete and partial hydrolysis of peptides.

[1] E. L. Smith and R. L. Hill, *in* "The Enzymes" (P. D. Boyer, H. Lardy, and K. Myrbäck, eds.), 2nd ed., Vol. 4, p. 37. Academic Press, New York, 1960.

[1a] R. J. DeLange and E. L. Smith, *in* "The Enzymes" (P. D. Boyer, ed.), 3rd ed., Vol. 3, p. 81. Academic Press, New York, 1971.

[2] D. H. Spackman, E. L. Smith, and D. M. Brown, *J. Biol. Chem.* **212**, 255 (1955).

[3] E. L. Smith and D. H. Spackman, *J. Biol. Chem.* **212**, 271 (1955).

[4] R. L. Hill and E. L. Smith, *J. Biol. Chem.* **224**, 209 (1957).

[5] R. L. Hill and E. L. Smith, *J. Biol. Chem.* **228**, 577 (1957).

[6] G. Pfleiderer and P. G. Celliers, *Biochem. Z.* **339**, 186 (1963).

[7] E. D. Wachsmuth, I. Fritze, and G. Pfleiderer, *Biochemistry* **5**, 169 (1966).

[8] E. D. Wachsmuth, I. Fritze, and G. Pfleiderer, *Biochemistry* **5**, 175 (1966).

[8a] G. Pfleiderer, Vol. XIX, p. 514.

TABLE I
PROPERTIES OF AMINOPEPTIDASES

Property	Leucine aminopeptidase	Aminopeptidase M
Molecular weight	300,000	280,000
Temperature stability	70°[a]	65°
pH stability	8.0–8.5	3.5–11.0
Activators	Mn^{2+} or Mg^{2+}	None
pH optimum	9.1 (Mn^{2+}); 9.3 (Mg^{2+})[b]	8.8–9.0[c]; 7.0[d]
Inhibitors	EDTA; Hg^{2+}, Cd^{2+}, Cu^{+}, Pb^{2+}; *p*-chloromercuribenzoate	—

[a] The stability is reported for a partially purified preparation.
[b] Hydrolysis of peptides is performed at pH 8.5 because of greater enzyme stability.
[c] For leucine *p*-nitroanilide and alanine *p*-nitroanilide at saturating concentrations.
[d] For peptides.

Purification of the Enzymes

Leucine Aminopeptidase (*LAP*). The enzyme is obtained from swine kidneys by the procedure of Hill, Spackman, Brown, and Smith.[9] An acetone-dried powder of the gland is prepared, and it serves as a convenient starting material for extraction and subsequent fractionation. An extract of the powder is submitted to two ammonium sulfate precipitations, and the 0.4 to 0.5 fraction is then precipitated with 0.01 *M* magnesium chloride. Heat inactivation at 70° removes an inactive fraction. Further purification is accomplished with acetone precipitation, the active component being recovered at 30% acetone.[10] LAP is stabilized in the presence of 0.005 *M* $MgCl_2$, and the salt is included in all operations after the ammonium sulfate precipitation. The acetone-precipitated fraction is further purified by electrophoresis or chromatography. The early use of paper electrophoresis[2] was replaced by electrophoresis on starch columns, which increased the sample load, gave a better separation, and facilitated the recovery of the active fraction.[5] Folk *et al.*[11] developed a chromatographic procedure with DEAE-cellulose to further purify the acetone fraction and increased the specific activity to 160 min^{-1} per milligram of protein nitrogen for the peak fractions. LAP after the acetone precipitation or after electrophoresis is contaminated with endo-

[9] R. L. Hill, D. H. Spackman, D. M. Brown, and E. L. Smith, *Biochem. Prep.* **6**, 35 (1958).

[10] An increase in purification is achieved if acetone is first added to bring the concentration to 20%. The supernatant fraction is brought to 30% acetone, and the active precipitate is collected by centrifugation.

[11] J. E. Folk, J. A. Gladner, and T. Viswanatha, *Biochim. Biophys. Acta* **36**, 256 (1959).

TABLE II
RATES OF HYDROLYSIS OF AMINO ACID AMIDES BY LEUCINE AMINO PEPTIDASE[a,b]

Substrate	Mn^{2+} activation			Mg^{2+} activation		
	Enzyme concentration (μg/ml)	C_0	Relative rate	Enzyme concentration (μg/ml)	C_0	Relative rate
L-Leucinamide	0.082	14,000	100	0.163	6600	100
DL-Norleucinamide	0.082	14,200	101	0.163	7200	105
DL-Norvalinamide	0.082	11,800	84	0.163	7200	109
DL-α-Amino-*n*-butyramide	0.082	5,100	36	0.163	5100	77
L-Alaninamide	1.37	470	3.4	3.26	325	4.9
Glycinamide	1.64	18	0.13	3.26	7	0.10
L-Isoleucinamide	0.328	2,800	20	0.543	1120	17
L-Alloisoleucinamide	1.01	1,000	7.1	—	—	—
L-Valinamide	0.328	2,400	17	—	—	—
L-Tryptophanamide	0.328	3,400	24	—	—	—
L-Phenylalaninamide	0.541	3,600	26	0.543	1140	17
L-Tyrosinamide	0.328	2,200	16	—	—	—
L-Histidinamide	0.541	2,700	19	0.543	680	10
L-Lysinamide	1.37	1,000	7.1	3.26	800	12
L-Argininamide	1.01	1,000	7.1	—	—	—
L-Isoglutamine	1.37	310	2.2	—	—	—
L-Aspartic acid diamide	1.37	410	2.9	3.26	250	3.8
L-Serinamide	1.64	106	0.76	—	—	—
L-Prolinamide	1.64	100	0.71	—	—	—
Hydroxy-L-prolinamide	1.64	80	0.57	—	—	—
DL-α-Aminocaprylamide	0.082	10,500	75	—	—	—
DL-*tert*-Leucinamide	1.64	16	0.11	—	—	—
α-Aminoisobutyramide	1.64	0	0	—	—	—
D-Leucinamide	1.64	0	0	3.26	0	0

[a] E. L. Smith and D. H. Spackman, *J. Biol. Chem.* **212**, 271 (1965).
[b] The value for L-leucinamide is given as 100 and for the other amides as the relative rate. C_0 = zero-order proteolytic coefficient.

peptidases, as seen by the release of amino acids from cytochrome *c*.[12,13] The heme protein is *N*-acetylated[14] and should be refractory to hydrolysis. The liberation of amino acids is decreased, but not entirely eliminated, if the enzyme preparation is treated with diisopropyl phosphorofluoridate (DFP)[5,12] and iodoacetate.[12]

[12] R. Frater, A. Light, and E. L. Smith, *J. Biol. Chem.* **240**, 253 (1965).
[13] L. J. Deftos and J. T. Potts, Jr., *Biochim. Biophys. Acta* **171**, 121 (1969).
[14] E. Margoliash, E. L. Smith, G. Keil, and H. Tuppy, *Nature* (*London*) **192**, 1121 (1961).

TABLE III
MAXIMUM RATES OF HYDROLYSIS OF VARIOUS AMINO ACID DERIVATIVES BY AMINOPEPTIDASE M[a,b]

L-Amino acid	*p*-Nitroanilide Buffer	*p*-Nitroanilide 20% Dioxane	Amide Buffer	Amide 20% Dioxane
Alanine	100	84	1.15	0.32
Norvaline	—	—	0.44	—
Norleucine	—	—	0.35	0.12
Leucine	71	15	0.32	0.13
Phenylalanine	83	14	0.29	0.085
Serine	—	—	0.2	0.2
Valine	—	—	0.11	0.1
Glycine	22	42	0.085	0.21
α-Aminobutyric acid	—	—	0.6	0.21
Histidine	—	—	0.058	0.032
Proline	—	—	0.0042	0.0042
Sarcosine	—	—	0.0002	—

[a] E. D. Wachsmuth, I. Fritze, and G. Pfleiderer, *Biochemistry* **5**, 175 (1966).
[b] Values are expressed as percentage of V_{max} of alanine *p*-nitroanilide measured in 0.06 *M* phosphate buffer at 37° and in 20% (v/v) dioxane solution, pH 7.

Himmelhoch and Peterson[15,15a] described a new method for preparing LAP which is free of endopeptidase activity. The extraction of swine kidneys incorporates hexyldecyltrimethylammonium bromide (CETAB) in a Tris·succinate–$MgCl_2$–sucrose buffer. The extract of the tissue is easier to prepare than the acetone-dried powder. Furthermore, the endopeptidases are removed in this step, and they remain in the particulate fraction. Purification of the active fraction on columns of DEAE-cellulose and Sephadex G-200 increased the specific activity 500-fold. A final chromatographic separation on DEAE-cellulose with a linear salt gradient brought the specific activity to a range of 98–158 min^{-1} per milligram of protein N, the higher value being equal to the best value reported before. The enzyme is recovered in an overall yield of 50%. The purified enzyme is free of endopeptidase activity when tested on protein substrates known to be resistant to LAP hydrolysis.[12] Thus cytochrome *c* and mercuripain, which has a resistant Ile-Pro sequence, are refractory to the action of the purified enzyme. The complete absence of free amino acids eliminated the possibility that internal peptide bonds were hydrolyzed. It is interesting to note that performic acid-oxidized

[15] S. R. Himmelhoch and E. A. Peterson, *Biochemistry* **7**, 2085 (1968).
[15a] S. R. Himmelhoch, Vol. XIX, p. 508.

ribonuclease resisted hydrolysis with the purified enzyme but was susceptible to hydrolysis when a crude fraction was used.

Aminopeptidase M. The purification of the particle-bound aminopeptidase was described by Wachsmuth, Fritze, and Pfleiderer.[7] Swine kidneys are homogenized in 0.1 *M* Tris buffer, pH 7.3. The microsomal particles, which account for the bulk of the aminopeptidase activity of the tissue, are precipitated from the clarified extract at pH 5.0. In order to increase the solubility of the enzyme, the particles are treated with toluene, which causes swelling, and by tryptic hydrolysis at pH 7.3 and 37° for 1 hour, which releases the enzyme from the insoluble particles. Ammonium sulfate precipitation between 20% and 80% saturation and at 60% to 80% concentrates the enzyme and increases the specific activity. Further purification is gained on ion-exchange chromatography on columns of DEAE-Sephadex A-50. The specific activity reaches a constant value on repeated precipitation with ammonium sulfate (65–80%).

LAP has also been isolated and characterized from bovine crystalline lens[16-19] and from plasma.[20] Aminopeptidase purified from *E. coli* B shows a restricted specificity for imide bonds and releases the terminal residue, X, from the sequence X-Pro.[21] The aminopeptidases of thermophilic microorganisms[22] act at elevated temperatures, and they may be valuable for the release of amino acids from protein substrates.

Enzyme Assays

The protein concentration of a solution of LAP is estimated spectrophotometrically at 280 nm with an extinction of 0.84 for 1 mg of protein per milliliter.[2] Other methods of protein analysis may be used, such as the colorimetric procedure of Lowry *et al.*[23] and micro-Kjeldahl nitrogen determination. Aminopeptidase M is available as a solid, and a stock solution can be prepared of known concentration.

The activity of LAP is measured with leucine substrates, and of

[16] A. Spector, *J. Biol. Chem.* **238,** 1353 (1963).

[17] A. Spector and G. Mechanic, *J. Biol. Chem.* **238,** 2358 (1963).

[18] D. Glässer and H. Hanson, *Hoppe-Seyler's Z. Physiol. Chem.* **329,** 249 (1962).

[19] H. Hanson, G. Glässer, and H. Kirschke, *Hoppe-Seyler's Z. Physiol. Chem.* **340,** 107 (1965).

[20] A. B. Kurtz and E. D. Wachsmuth, *Nature (London)* **221,** 92 (1969).

[21] A. Yaron and D. Mlynar, *Biochem. Biophys. Res. Commun.* **32,** 658 (1968).

[22] H. Zuber, *in* "Structure-Function Relationships of Proteolytic Enzymes," (P. Desnuelle, H. Neurath, and M. Ottensen, eds.), p. 188. Academic Press, New York, 1970.

[23] O. H. Lowry, N. J. Rosebrough, A. L. Farr, and R. J. Randall, *J. Biol. Chem.* **193,** 265 (1951).

aminopeptidase M with alanine substrates. Hydrolysis of amides is followed titrimetrically with alcoholic potassium hydroxide,[24] spectrophotometrically at 238 nm,[25] or with the use of a pH-stat.[26] Alternatively, the ammonium ion liberated may be distilled and analyzed colorimetrically.[27] Hydrolysis of *p*-nitroanilides[28] or β-naphthylamides[29] is followed spectrophotometrically.

LAP is activated prior to assay with 0.002 *M* $MnCl_2$ or 0.005 *M* $MgCl_2$ in 0.1 *M* Tris buffer, pH 8.6, for 1 hour at 40°.

Enzyme Specificity

The relative rates of hydrolysis of amino acid derivatives are given in Table II for LAP and Table III for aminopeptidase M. The enzymes show a broad specificity, and all amino acids of the L configuration with a free α-amino group are hydrolyzed. However, the amino acid residues with hydrophobic side chains are most easily hydrolyzed, and leucyl and alanyl residues are the best substrates for LAP and aminopeptidase M, respectively. In general, the relative rates of hydrolysis of amino acid derivatives are also relevant to the rate of release of amino acids from peptides and proteins.

Hydrolysis of Peptides

The reaction conditions used for the partial or complete hydrolysis of peptides is described below for the two commercially available enzyme preparations.

Leucine aminopeptidase: Typically, about 0.1 μmole of peptide is dissolved in 0.1 ml of 0.1 *M* Tris buffer, pH 8.6, containing 0.0025 *M* $MgCl_2$. The enzyme in Tris·$MgCl_2$ buffer is added to a final concentration of 0.05–0.5 mg/ml. The mixture is maintained at 37°, and samples are removed as a function of time and adjusted to pH 2.2 with 0.2 *M* citrate buffer. The samples are stored in the frozen state until analyzed.

Aminopeptidase M: Approximately 0.1 μmole of peptide is dissolved in 0.1 ml of 0.1 *M* sodium phosphate buffer, pH 7.0. About 0.05 ml to 0.1 ml of a 0.1% solution of the enzyme is added, and the reaction mixture is incubated at 37° for varying periods of time. Aliquots of the reaction are added to 0.2 *M* citrate buffer, pH 2.2, and stored for analysis.

[24] W. Grassmann and W. Heyde, *Hoppe-Seyler's Z. Physiol. Chem.* **183,** 32 (1929).

[25] M. A. Mitz and R. J. Schlueter, *Biochim. Biophys. Acta* **27,** 168 (1958).

[26] G. F. Bryce and B. R. Rabin, *Biochem. J.* **90,** 509 (1964).

[27] G. Pfleiderer, P. G. Celliers, M. Stanulovic, E. D. Wachsmuth, H. Determann, and G. Braunitzer, *Biochem. Z.* **340,** 552 (1964).

[28] H. Tuppy, U. Wiesbauer, and E. Wintersberger, *Hoppe-Seyler's Z. Physiol. Chem.* **329,** 278 (1962).

[29] M. D. Green, K. C. Tsou, R. Bressler, and A. M. Seligman, *Arch. Biochem. Biophys.* **57,** 458 (1955).

The reaction conditions as described should be suitable for most peptides. However, preliminary experiments may be necessary to establish the optimum conditions of time and enzyme concentration required to release the amino acids in a sequential manner. Complete hydrolysis of a peptide may require an extended reaction time of 18–24 hours or the use of the higher levels of enzyme. If lysine is a component of the peptide, it is useful to substitute barbital buffer or ammonium bicarbonate for the Tris buffer, since lysine and Tris emerge as a single peak on the B column of amino acid analyzers.

Amino acid analyses of hydrolyzed peptides are performed on an amino acid analyzer with the lithium citrate system of Benson *et al.*[30] Asparagine and glutamine are resolved, and the amide content is directly evaluated. Alternatively, analyzers operating with sodium citrate buffers may be used, but the serine content will include asparagine and glutamine, which are not resolved with this system. A sample of the enzymatic digest must also be hydrolyzed with acid, and a second analysis will give an increase in the aspartic and glutamic acid values corresponding to the amide content.[31]

The complete hydrolysis of the oxidized A chain of insulin[5] and of glucagon[31] with LAP was shown by Hill and Smith to be in accord with their known compositions. Aminopeptidase is a suitable reagent for the analysis of peptides, but a combination of enzymes, such as papain, prolidase, and LAP, are required for the hydrolysis of proteins.[32]

The hydrolysis of peptides is conveniently performed with aminopeptidase M, which hydrolyzes all peptide bonds including the imide bond of prolyl residues. If large peptides are analyzed, a preliminary hydrolysis with Pronase[33] produces a mixture of small peptides which are completely hydrolyzed in the presence of aminopeptidases.

Occasionally, an incomplete hydrolysis of a peptide will result if proline is present because of diketopiperazine formation. It was noted that the hydrolysis of the peptide, Ala-Tyr-Glu-Pro-Val-Trp, gave the amino acids Ala, Tyr, Val, and Trp, and a diketopiperazine of Glu-Pro.[34] Proline dipeptides are known to cyclize readily,[35] and this may provide a sufficient driving force to release the cyclic dipeptide.[36] The diketo-

[30] J. V. Benson, Jr., M. J. Gordon, and J. A. Patterson, *Anal. Biochem.* **18**, 228 (1967).

[31] R. L. Hill and E. L. Smith, *Biochim. Biophys. Acta* **31**, 257 (1959).

[32] R. L. Hill and W. R. Schmidt, *J. Biol. Chem.* **237**, 389 (1962).

[33] C. Bennett, W. H. Konigsberg, and G. M. Edelman, *Biochemistry* **9**, 3181 (1970).

[34] S. G. Waley, J. C. Miller, I. A. Rose, and E. L. O'Connell, *Nature (London)* **227**, 181 (1970).

[35] E. L. Smith and M. Bergmann, *J. Biol. Chem.* **153**, 627 (1944).

[36] K. Takahashi, W. H. Stein, and S. Moore, *J. Biol. Chem.* **242**, 4682 (1967).

piperazine will not be hydrolyzed further, and the apparent composition of the peptide will be incorrect, with two amino acids missing.

Sequence Studies

Partial Hydrolysis. Typical examples of a time-dependent release of amino acids from peptides submitted to hydrolysis with LAP are shown below.

	Ala—	Ala—	Tyr—	Lys[37]
30 minutes	0.94		0.26	0.28
90 minutes	2.00		0.86	0.88
24 hours	2.00		1.00	1.00

	Ala—	Asn—	Gly—	Val—	Ala—	Glu—	Trp[38]
3 minutes	0.95	0.60	0.20	0.08			
10 minutes	1.40	1.00	0.62	0.40			
30 minutes	1.95	1.05	1.02	0.91	—	0.75	

Short reaction times and a quantitative analysis of the amino acids released are necessary in order to derive a partial sequence. The reliability of the sequence information decreases as the number of amino acids released becomes large, and useful information may be limited to the first two or three residues.

Prolyl and Aspartyl Residues. If proline is a constituent of a peptide, hydrolysis with LAP proceeds sequentially until the imide bond of a prolyl residue is encountered. Further hydrolysis is limited, and the amino acids released are assigned to positions amino terminal to X-Pro, where X may be any amino acid. A chymotryptic peptide derived from papain was digested with LAP, and the first four amino acids were released in equal amounts.[39]

$\overrightarrow{\text{Ser}}$-$\overrightarrow{\text{Ile}}$-$\overrightarrow{\text{Ala}}$-$\overrightarrow{\text{Asn}}$-Gln-Pro-Ser-Val-Val-Leu

The assignment of residues to positions on either side of the prolyl residue simplified further studies on the sequence of the peptide. Furthermore, the peptide fragment can be purified and its sequence established. It should be noted that LAP can hydrolyze the imide bond at high enzyme-to-substrate ratios and with extended periods of digestion.[40]

Hydrolysis with the aminopeptidases is hindered at aspartyl residues if the residue has undergone an interconversion of the α- and β-carboxyl

[37] H. Matsubara and R. M. Sasaki, *J. Biol. Chem.* **243**, 1732 (1968).

[38] K. Weber and W. Konigsberg, *J. Biol. Chem.* **242**, 3563 (1967).

[39] A. Light and J. Greenberg, *J. Biol. Chem.* **240**, 258 (1965).

[40] K. Titani, M. Wikler, and F. W. Putnam, *J. Biol. Chem.* **245**, 2142 (1970).

groups.[41,42] If either a β-peptide bond or an α,β-imide bond of the aspartyl residue is produced, hydrolysis will be limited or completely inhibited at this position of the sequence. A peptide obtained from fetal hemoglobin showed a quantitative release of six amino acids, a 50% yield of aspartic acid, and a lower yield of the three remaining residues.[43]

Val-Leu-Thr-Ser-Leu-Gly-Asp-Ala-Ile-Lys

The residues on the amino-terminal side of the aspartyl residue were completely hydrolyzed, and the limited hydrolysis of the aspartyl residue suggested that it was partly rearranged. The distribution of amino acids provided a partial sequence.

Edman Degradation. Enzymatic hydrolysis of a peptide[12] after each step of the Edman procedure provides a useful way to perform a difference analysis.[44] Asparagine, glutamine and other acid-labile residues are directly determined, and their positions in the sequence of a peptide is readily evaluated. For example, a tetrapeptide from egg white avidin was submitted to two steps of the Edman procedure.[45] The analysis of the peptide residue with acid hydrolysis after the first step and with aminopeptidase after the second step clearly showed that glutamine was removed in the second step.

Thr-Gln-Lys-Glu

Step 1: *Thr, 0.05 (0)*; Glu, 2.00 (2); Lys (not determined)
Step 2: Thr, 0.02 (0); *Gln, 0.12 (0)*; Glu, 1.00 (1); Lys

Specific Applications. The *S*-peptide of ribonuclease was converted to the carbamyl derivative and submitted to hydrolysis with trypsin, with the following result[46] (the sequence is shown in part):

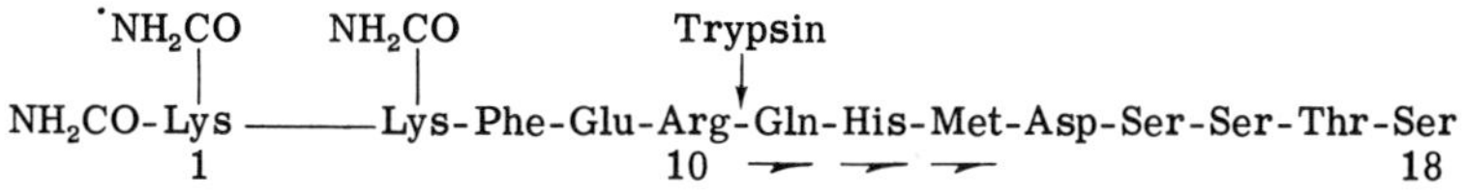

```
 ·NH2CO     NH2CO          Trypsin
   |          |               ↓
NH2CO-Lys ------Lys-Phe-Glu-Arg-Gln-His-Met-Asp-Ser-Ser-Thr-Ser
   1                        10  →    →    →                 18
```

Trypsin catalyzed the hydrolysis of the peptide bond at the arginine residue and released a peptide with amino-terminal glutamine. The tryptic digest, without further separation, was submitted to hydrolysis

[41] D. L. Swallow and E. P. Abraham, *Biochem. J.* **70**, 364 (1958).

[42] M. A. Naughton, F. Sanger, B. S. Hartley, and D. C. Shaw, *Biochem. J.* **77**, 149 (1960).

[43] W. A. Schroeder, J. R. Shelton, J. B. Shelton, J. Cormick, and R. T. Jones, *Biochemistry* **2**, 992 (1963).

[44] W. Konigsberg and R. J. Hill, *J. Biol. Chem.* **237**, 2547 (1962).

[45] R. J. DeLange, *J. Biol. Chem.* **245**, 907 (1970).

[46] D. G. Smyth, W. H. Stein, and S. Moore, *J. Biol. Chem.* **237**, 1845 (1962).

by LAP, which released amino acids exclusively from the newly exposed glutamine sequence. The carbamyl peptide itself was not attacked by LAP because the α-amino group was protected. This example illustrates a way to use a blocked peptide and restrictive proteolysis to obtain a partial sequence of the carboxyl-terminal region of a peptide.

Complete enzymatic hydrolysis of peptides is widely used in evaluating the optical purity of chemically synthesized peptides.[47] A problem associated with synthesis is racemization of amino acids on formation of peptide linkages. Since the aminopeptidases hydrolyze L-amino acids only, the stoichiometry of the amino acids released provides an estimate of the optical purity. At the same time, the recovery of amino acids eliminates the possibility that side reactions occurred. It is interesting to note that blocked peptides with *S*-benzylcysteinyl, *O*-methyltyrosyl, and α-carboxyl amides were completely hydrolyzed.[48] Ethyl esters[49] and hydrazides[50] of the α-carboxyl group were also susceptible.

[47] M. Bodanszky and M. A. Ondetti, "Peptide Synthesis." Wiley (Interscience), New York, 1966.

[48] Z. Beránková, I. Rychlík, and F. Šorm, *Coll. Czech. Chem. Commun.* **26,** 1708 (1961).

[49] H. Kienhuis, A. van de Linde, J. P. J. van der Holst, and A. Verweig, *Rec. Trav. Chim. Pays-Bas* **80,** 1278 (1961).

[50] A. Wergin, *Naturwissenschaften* **52,** 34 (1965).

[21] Carboxypeptidases A and B

By Richard P. Ambler

Carboxypeptidases are enzymes that remove amino acids one at a time from the C termini of tripeptides, higher peptides, and proteins. Although any of the normal protein (L) amino acids may be removed by one or another of the known enzymes, the differences in rate of hydrolysis are very great, and it is normally possible to find digestion conditions in which only a small number of residues are removed and so obtain information about the C-terminal sequence of the molecule under investigation.

The only carboxypeptidases that have been used extensively for structural studies are the pancreatic carboxypeptidases A and B.[1] The A enzyme (CPA) removes C-terminal aromatic and long side-chain

[1] H. Neurath, *in* "The Enzymes" (P. D. Boyer, H. Lardy, and K. Myrbäck, eds.), Vol. IV, p. 11. Academic Press, New York, 1960.

aliphatic residues most rapidly (Table II), and removes glycine and acidic amino acids only slowly. The B enzyme ("basic") has a narrower specificity, removing lysine and arginine rapidly, but it may also remove neutral amino acids in some cases. Plant sources contain a carboxypeptidase CPC[2,3] that will remove C-terminal proline as well as many of the protein amino acids.

In studying the action of a carboxypeptidase on a peptide substrate, two products can be investigated—the amino acids liberated, and the residual peptides formed. Ideally, the rates of disappearance of the original peptide, and of formation of the products, should all be followed quantitatively, but considerations of material, equipment, and labor make this impracticable. The methods described below represent different compromises with this counsel of perfection.

Preparation of Enzyme and Sample

The enzymes used should be free of endopeptidase activity, although contamination is not likely to produce such misleading results with a peptide as with a whole protein. Chymotrypsin is the most likely contaminant, and should be inactivated by treatment with diisopropyl phosphorofluoridate (DFP)[4] (this volume [10]).

The methods described (this volume [10]) for solubilizing CPA are applicable for peptide studies. Neither method introduces much nonvolatile salt into the digestion mixture, and either method can be used if the products are to be investigated by paper electrophoresis or chromatography.

CPB and CPC preparations suitable for investigation of peptide sequences have been discussed (this volume [10]).

The peptide to be treated should be stored frozen in solution in a volatile buffer, at a known concentration (1 mM is convenient) determined by amino acid analysis. Samples for treatment with a carboxypeptidase are dispensed by volume and then evaporated to dryness in a vacuum desiccator in the digestion container.

Choice of Reaction Conditions

The time of digestion and the amount of enzyme to be used are best determined by a trial experiment, using Method (*d*) below. Suitable trial conditions are 10 μg of CPA, or 0.1 Enzyme Commission unit, and 0.02 μmole of sample kept at pH 8.5 for 4 hours at 37°. This represents

[2] H. Zuber, *Hoppe-Seyler's Z. Physiol. Chem.* **349**, 1337 (1968).

[3] J. R. E. Wells, *Biochem. J.* **97**, 228 (1965).

[4] H. Fraenkel-Conrat, J. I. Harris, and A. L. Levy, *Methods Biochem. Anal.* **2**, 339 (1955).

0.5 mg of CPA per micromole, an enzyme: substrate ratio of about 1:70. If digestion under these conditions proceeds too far, milder conditions can be used, down to as low as 0.025 mg CPA per micromole for 10 minutes at 25°. The upper limit of the amount of enzyme that can be used is reached when the "enzyme blank" (the free amino acids released by autodigestion) becomes so high as to make interpretation of the results ambiguous. For this reason an enzyme blank should always be included in each batch of samples.

Portions of the incubation mixture may be removed at different times, although it is often easier to perform parallel digestions in separate tubes and to vary the degree of digestion by altering the amount of enzyme. If the time course of the reaction is to be determined, further reaction in the samples removed should be stopped by acidification to below pH 4 (with acetic acid) and freezing, followed by freeze-drying. CPC has a pH optimum of 5–6, but is inactive at pH 2.

Methods of Separating the Products of Reaction

Alternative methods of examining the products of reaction of peptides with CPA and CPB are summarized in Table I.

(*a*) The whole reaction mixture is examined on an automatic amino acid analyzer. (i) CPA and CPB. The peptide (0.005–0.5 μmole[5]) is dissolved in 0.2–0.5 ml 0.2 *M* *N*-ethylmorpholine acetate (pH 8.5) in a (13 × 60 mm) screw-cap culture tube, the enzyme solution is added, and the mixture is incubated. At the end of the reaction, the mixture is dried, and the residue is dissolved in suitable buffer and analyzed with an automatic amino acid analyzer.

The interpretation of the analyses is sometimes complicated by the presence of peptides. The positions of peptides on the effluent curves are generally different from those of amino acids and, when coincidence occurs, the peaks given by the peptides on highly cross-linked resins are not as sharp as those given by amino acids. If asparagine or glutamine is released, it is necessary to perform a special analysis for these constituents since they normally overlap serine; however, asparagine and glutamine are usually resolved only from serine, not from each other. If ambiguity arises in recognition of the effluent peaks, the carboxypeptidase experiment must be repeated by one of the other methods (e.g., Method *b* below).

Example: Peptide T-β1 from the β chain of human hemoglobin[6] (about 0.5 μmole of peptide at pH 7.6 and 25°):

[5] The lesser quantity of material should be used only if a high-sensitivity attachment is available on the amino acid analyzer.

[6] G. Guidotti, R. J. Hill, and W. Konigsberg, *J. Biol. Chem.* **237**, 2184 (1962).

. . . Gly-Gly-Glu-Ala-Leu-Gly-Arg	Ala	Leu	Gly	Arg
(1) CPB for 30 min				0.95
(2) CPB for 60 min, CPA (0.7 mg/μmole) for 30 min		0.05	0.30	0.98
(3) CPB for 120 min, CPA (0.7 mg/μmole) for 90 min	0.50	1.10	1.02	0.95

(ii) CPC. The peptide (0.005–1.0 μmole[5]) is dissolved in 0.1 *M* sodium citrate (pH 5.3). The volume used depends on the number of samples that will be taken in the course of the digestion. Samples are taken at suitable time intervals after the addition of the enzyme, and at once acidified by mixing with a 10-fold excess of 0.2 *M* sodium citrate (pH 2.2), and stored frozen until analyzed. Unambiguous sequences have been obtained for rather longer sequences than are usual with CPA or CPB.

Example: Peptide Tr_1 from α-thyrocalcitonin[7]

(0.63 μmole of peptide in 0.5 ml of buffer at 37° with 40 mU CPC)

```
—S—S
    |
---- Cys-Val-Leu-Ser-Ala-Tyr-Trp-Arg
```

0.08-ml samples taken after 10, 30, 60, 120, and 300 minutes

The sequence of the seven C-terminal amino acids was deduced from release curves (this volume [10]).

(*b*) The amino acids released are separated from peptides by gel filtration; fractions are then analyzed. The digestion conditions are as in Method (*a*). The dry mixture is dissolved in 1 *M* acetic acid and fractionated by gel filtration through a (150 × 1 cm) column of Sephadex G-25 (fine beads) equilibrated with 1 *M* acetic acid.[8] The peptide and amino acid peaks are detected by ninhydrin analysis before and after alkaline hydrolysis or by application of portions of each fraction onto filter paper, which is stained first with ninhydrin and then with the chlorine/*o*-tolidine reagent.[9] If the residual peptides are large it may be possible to use a shorter column. If the peptide contains tryptophan, or a high proportion of other aromatic amino acids, there may be no separation from free amino acids at all.

The amino acid and peptide fractions are evaporated to dryness, and samples of each are subjected to hydrolysis and analyzed. Another portion of the free amino acid fraction is analyzed without hydrolysis, and

[7] R. Neher, B. Riniker, H. Zuber, W. Rittel, and F. W. Kahnt, *Helv. Chim. Acta* **51,** 917 (1968).

[8] Gel filtration through highly cross-linked acrylamide gels (Bio-gel P-2, P-4, or P-6) or Sephadex G-10 will often be extremely effective for such separations.

[9] F. Reindel and W. Hoppe, *Chem. Ber.* **87,** 1103 (1954).

TABLE I

METHODS FOR SEPARATION OF THE PRODUCTS OF CARBOXYPEPTIDASE DIGESTION OF PEPTIDES

Method	Advantages	Disadvantages	Recommended occasions for use	References
(*a*) The whole reaction mixture separated by amino acid analyzer	(1) Quantitative (2) Unaffected by quite large amounts of salt in sample (3) Very sensitive if a high-sensitivity analyzer is available	(1) Ambiguity between serine, asparagine, and glutamine (2) Results confused by residual peptides (3) Very little information obtained about residual peptides (4) Requires amino acid analyzer	Kinetic studies of amino acid release if Method (*d*) below shows very extensive degradation of peptide	[a,b,c]
(*b*) Released amino acids separated from residual peptides by gel filtration; fractions then analyzed separately	(1) Quantitative (2) Unaffected by salt	(1) Very tedious for many samples (2) Residual peptides probably not resolved from each other (3) Conditions for gel filtration cannot be standardized		[a,d]
(*c*) Analysis by electrophoresis and densitometry at several times during reaction	(1) Sequence determined by order of amino acid release	(1) Identification of amino acids and discrimination from peptides not reliable		[e]

(*d*) Released amino acids separated from peptides by electrophoresis, then identified by an electrophoresis at a different pH	(1) Quick and sensitive (2) Good recognition of amides (3) Some information obtained (from mobilities) about residual peptides (4) Does not require amino acid analyzer	(1) Requires some knowledge of properties of the peptide (2) Amino acids released not quantitated	Initial experiments to determine conditions for a more thorough investigation; results may be adequately conclusive in many cases	f
(*e*) Purification and characterization of residual peptides	(1) A complete subtractive method (2) Amino acids released (including amides) can be quantitated	Tedious for many samples	Determination of C-terminal sequence of small (4–10 residues) peptides if Method (*d*) shows condition for obtaining a good series of residual peptides	f
(*f*) Analysis of released amino acids as DNP derivatives	Does not require specialized equipment	(1) Slow and tedious (2) Less accurate or sensitive than quantitative amino acid analyses (3) Amino acid identification ambiguous	Not recommended unless equipment for alternative methods is not available	g,h

[a] G. Guidotti, R. J. Hill, and W. Konigsberg, *J. Biol. Chem.* **237,** 2184 (1962).
[b] C. H. W. Hirs, S. Moore, and W. H. Stein, *J. Biol. Chem.* **235,** 633 (1960).
[c] B. C. Carlton and C. Yanofsky, *J. Biol. Chem.* **238,** 2390 (1963).
[d] D. G. Smyth, W. H. Stein, and S. Moore, *J. Biol. Chem.* **237,** 1845 (1962).
[e] R. E. Canfield, *J. Biol. Chem.* **238,** 2698 (1963).
[f] R. P. Ambler, *Biochem. J.* **89,** 349 (1963).
[g] H. Fraenkel-Conrat, J. I. Harris, and A. L. Levy, *Methods Biochem. Anal.* **2,** 339 (1955).
[h] D. T. Gish, *J. Amer. Chem. Soc.* **83,** 3303 (1961).

the amide content is determined by the decrease in "serine" (cf. above) and the increase in aspartic acid and glutamic acid content.

Example: Peptide (*O*-Tryp 4)Chy 1 from ribonuclease[10] (0.5 μmole of peptide kept at 37° for 18 hours with 0.5 mg of CPA per micromole):

$$\begin{pmatrix}\text{Pyr}\\ \text{Glu}\end{pmatrix}\text{-His-MetSO}_2\text{-Asp-Ser-Ser-Thr-Ser}$$

Residual peptide had composition:	Asp 1.02
(separated from amino acids on G-25 column,	Glu 1.04
100 × 0.9 cm, 4°, in 50% acetic acid)	$MetSO_2$ 1.00
	Ser 0.94
	His (not determined)
Amino acids released (estimated before gel filtration):	Ser 1.70
	Thr 0.99

(*c*) ANALYSIS OF MIXTURE BY ELECTROPHORESIS AND DENSITOMETRY AFTER VARIOUS TIMES OF REACTION. Peptide (0.2–0.3 μmole) is dissolved in 0.1 ml of 0.1 *M* ammonium carbonate (pH about 8.0), and carboxypeptidase (solubilized in 2 *M* ammonium carbonate—Method (ii), see [10]) added. Portions (0.01 ml) of the reaction mixture are taken out at timed intervals, and applied directly to filter paper for electrophoresis by the Dreyer-Bynum method[11] (see Vol. XI [4]). At the end of the reaction, when all the sample has been applied to the paper, electrophoresis is performed in 4% (v/v) formic acid, for 1.5 hours at 45 V/cm. The paper is stained with a cadmium/ninhydrin reagent, and the amount of amino acids and peptides present is estimated by densitometry. Great care must be taken in the interpretation of the results to avoid confusing released amino acids with residual or unchanged peptides. Ambiguity in identification of released amino acids may sometimes be resolved with a collidine/ninhydrin reagent, which gives characteristic colors for different amino acids, instead of the quantitative cadmium reagent, or by use of a two-dimensional separation system; in these cases, a duplicate sample should be taken at the end of the reaction.

Example: Peptide P2a from egg white lysozyme[12] [0.2 μmole of peptide treated at 23° with 0.5 mg of CPA per micromole (times not given)]:

Leu-Ser-Ser-Asp-Ile-Thr-Ala

Analyses of portions at (short-time) intervals showed that alanine was released first, followed by threonine and then isoleucine

(*d*) SEPARATION OF AMINO ACIDS FROM PEPTIDES BY ELECTROPHORESIS, FOLLOWED BY IDENTIFICATION OF RELEASED AMINO ACIDS BY ELECTROPHORESIS

[10] D. G. Smyth, W. H. Stein, and S. Moore, *J. Biol. Chem.* **237**, 1845 (1962).
[11] W. J. Dreyer, *Brookhaven Symp. Biol.* **13**, 243 (1960).
[12] R. E. Canfield, *J. Biol. Chem.* **238**, 2698 (1963).

AT A DIFFERENT pH. Peptide (0.01–0.05 μmole) is dissolved in 0.1 ml of 0.2 *M* *N*-ethylmorpholine acetate (pH 8.5)[13] in a (28 × 6 mm) Durham fermentation tube; a similar quantity is reserved for use as a marker, and two or more samples of the same peptide may be taken for digestion, so that different conditions (time, pH, temperature, and amount of enzyme) can be tested. After incubation with enzyme, the samples are evaporated to dryness in a vacuum desiccator, and then digests and markers are subjected to electrophoresis at pH 3.5 or 6.5 (pyridine acetate buffers, 1 hour at 60 V/cm; samples are applied as 1-cm bands on Whatman No. 1 paper). After electrophoresis, the band where the neutral (monoaminomonocarboxylic) amino acids will lie is cut out (unstained), and sewn by machine[14] onto another sheet of Whatman No. 1 paper. The amino acids are separated by electrophoresis at about pH 2 [e.g., in 2% formic acid, 8% acetic acid (v/v), 120 V/cm, 20 minutes]. Both ends of the first electrophoresis paper and all the second sheet are stained with ninhydrin/collidine. Basic and acidic amino acids released and residual and unchanged peptide are recognized on the first paper, and neutral amino acids on the second. The choice between pH 3.5 and pH 6.5 for the initial electrophoresis is determined by the electrophoretic mobility (at pH 6.5) and origin of the peptide digested. If the residual peptides are likely to be neutral at pH 6.5 (e.g., by removal of a lysine or arginine residue from a basic tryptic peptide) or by removal of neutral amino acids from a neutral chymotryptic or peptic peptide, the separation should be at pH 3.5. In other cases it should be at pH 6.5, under which condition most information about the size of residual peptides can be derived from their electrophoretic mobility. The results obtained from the nature and approximate amount of amino acids released should be completely reconcilable with the mobilities and quantities of the unchanged and residual peptides.

Example: Peptide C4b from *Pseudomonas* azurin[15] (0.03 μmole of peptide kept at pH 8.5 and 37° for 7 hours with 0.1 mg of CPA per micromole):

Gly-His-Asn-Trp $m = +0.36$

Amino acids released: Trp (++++) Asn (++)

Residual peptides: (a) $m = +0.60$ (+++), and (b) $m = +0.79$ (+) (m is the mobility relative to lysine (= 1) at pH 6.5, corrected for endosmosis)

(*e*) ISOLATION AND CHARACTERIZATION OF RESIDUAL PEPTIDES BY ELECTROPHORESIS. Peptide (0.05–0.3 μmole) is digested as in Method (*d*), and

[13] Other volatile buffers in the range pH 5–10 may be used for digestion to modify the specificity of CPA.

[14] M. A. Naughton and H. Hagopian, *Anal. Biochem.* **3,** 276 (1962).

[15] R. P. Ambler and L. H. Brown, *Biochem. J.* **104,** 784 (1967).

the products are separated as a wide band (2–10 cm) by paper electrophoresis at pH 3.5 or 6.5. Peptides and amino acids are located by staining side strips, and the bulk of the material is eluted. The neutral amino acids are identified by a one- or two-dimensional system, and absence of contaminating peptides is confirmed by parallel separation of hydrolyzed and unhydrolyzed portions. If necessary, the neutral amino acids may be hydrolyzed (to convert any amides present to amino acids) and relative amounts measured by automatic amino acid analysis. Samples of the residual peptides are hydrolyzed and analyzed for amino acid content qualitatively or quantitatively.

Example: Peptide C8aT3 from *Pseudomonas* azurin[15] (0.2 μmole of peptide kept at pH 8.5 and 37° for 6 hours with 0.025 mg of CPA per micromole):

Asp-Ser-Val-Thr-Phe $m = -0.40$

Amino acids released: Phe (+++) Thr (++) Val (++)

Residual peptides: (a) $m = -0.48$ (++) composition (Val, Ser, Thr, Asp), (b) $m = -0.57$ (+) composition (Val, Ser, Asp), and (c) $m = -0.73$ (++) composition (Ser, Asp) (m is the mobility relative to aspartic acid ($= -1$) at pH 6.5, corrected for endosmosis)

(*f*) Quantitation of the released amino acids by the FDNB method. This method, fully described by Fraenkel-Conrat, Harris, and Levy[4] should be regarded as obsolete and used only if the equipment needed for the alternative methods is not available.

Other Uses of Carboxypeptidases in Sequence Studies

It is convenient in some cases to remove the C-terminal lysine residue from tryptic peptides, as the presence of a partly phenylthiocarbamylated ϵ-amino group complicates the characterization of peptides formed by PTC degradation. The terminal lysine residue is readily removed by CPB, and the residual peptide may be isolated after electrophoresis (cf. Method (*e*) above).

If CP releases two amino acids from a tetrapeptide, it may be difficult to distinguish which was released first. If the N-terminal amino acid is removed (e.g., by leucine aminopeptidase or PTC degradation), it is probable that only one amino acid will be released from the tripeptide.

Example: Peptide C9S1b from *Pseudomonas* cytochrome 551[16]:

Ala-Gly-Gln-Ala

CPA released Gln and Ala, at rates that were not distinguished; leucine aminopeptidase treatment removed only Ala from the peptide, and the product, when treated with CPA, released only Ala

[16] R. P. Ambler, *Biochem. J.* **89**, 349 (1963).

Neither cystine, cysteine, nor cysteic acid is released rapidly by CPA, but after conversion of cysteine to *S*-β-aminoethylcysteine, the residue is readily removed by CPB.[17] *S*-methylcysteine[17a] is readily removed by CPA.[17b]

Peptides formed by cyanogen bromide cleavage of proteins contain C-terminal homoserine lactone. Under conditions of CP digestion (pH 8.5 and 37°) some opening of the lactone ring takes place, and homoserine will generally be rapidly released.

Interpretation of Results

The interpretation of results obtained from the time course of digestions has been discussed (see this volume [10]).

Specificity of Carboxypeptidases

Table II is a guide to the rates of release of different amino acids by CPA. An amino acid may drop to a lower class if the penultimate residue is proline or an acidic amino acid.

Conflicting reports have been published on the removal of acidic amino acids by CPA.[6,18] In most cases where acidic amino acids were released at appreciable rates the digestion was performed at about 40°, whereas in most studies in which acidic residues were not released the digestion temperature was about 25°. Residues adjacent to a proline residue are more likely to be released at the higher temperature.[19] In

TABLE II
APPROXIMATE RELATIVE RATES OF RELEASE[a] OF PROTEIN AMINO ACIDS BY CARBOXYPEPTIDASE A

Rapid release:	Tyr, Phe, Trp, Leu, Ile, Met, Thr, Gln, His, Ala, Val, homoserine
Slow release:	Asn, Ser, Lys,[b] $MetSO_2$
Very slow release:	Gly, Asp, Glu, $CySO_3H$, *S*-carboxymethylcysteine
Not released:	Proline, hydroxyproline, Arg

[a] The presence of a "very slow" or "not released" amino acid as penultimate residue will generally decrease the rate of release of the C-terminal amino acid.

[b] The rate of lysine release may be modified by changing the pH of digestion (this volume [10]).

[17] F. Tietze, J. A. Gladner, and J. E. Folk, *Biochim. Biophys. Acta* **26**, 659 (1957).

[17a] R. L. Heinrikson, *Biochem. Biophys. Res. Comm.* **41**, 967 (1970).

[17b] H. Rochat, C. Rochat, M. Miranda, S. Lissitzky, and P. Edman, *Eur. J. Biochem.* **17**, 262 (1970).

[18] D. T. Gish, *Amer. Chem. Soc.* **83**, 3303 (1961).

[19] E. L. Smith, *in* "The Chemical Structure of Proteins" (G. E. W. Wolstenholme and M. P. Cameron, eds.), p. 109. Churchill, London, 1954.

some cases lowering the pH of digestion may increase the rate of release of acidic amino acids, even at 25°.[20]

Bovine CPB will certainly remove some nonbasic amino acids,[6,15] and there is good evidence that this represents an inherent broadness in specificity.[21]

[20] K. Titani, K. Narita, and K. Okunuki, *J. Biochem.* (*Tokyo*) **51**, 350 (1962).

[21] E. Wintersberger, D. J. Cox, and H. Neurath, *Biochemistry* **1**, 1069 (1962).

[22] Preparation and Specificity of Dipeptidyl Aminopeptidase I

By J. Ken McDonald, Paul X. Callahan, and Stanley Ellis

The dipeptidyl aminopeptidases are enzymes that catalyze the consecutive removal of dipeptide moieties from the unsubstituted NH_2 termini of polypeptide chains. Four such enzymes have thus far been characterized and shown to have distinctive substrate specificities.[1] Within this group, dipeptidyl aminopeptidase I (DAP I) exhibits the broadest substrate specificity. Preparations of the enzyme suitable for sequencing purposes can be made quite simply from either bovine spleen[2] or rat liver,[3] and are stable in a freeze-dried state for an indefinite period.

When DAP I was first observed in bovine pituitary extracts, it was characterized as a sulfhydryl-dependent Ser-Tyr-β-naphthylamidase with an absolute or near-absolute chloride requirement and a pH 4 optimum.[4] Comparative studies later revealed that pituitary DAP I was probably similar or identical to beef spleen cathepsin C. A highly purified preparation of the latter, prepared according to Metrione *et al.*,[2] was likewise shown to have a halide requirement,[5] together with a much broader substrate specificity than was hitherto realized. Like pituitary DAP I, the beef spleen preparation catalyzed the consecutive release of

[1] J. K. McDonald, P. X. Callahan, R. E. Smith, and S. Ellis, *in* "Tissue Proteinases: Enzymology and Biology" (A. J. Barrett and J. T. Dingle, eds.), p. 69. North-Holland Publ., Amsterdam, 1971.

[2] R. M. Metrione, A. G. Neves, and J. S. Fruton, *Biochemistry* **5**, 1597 (1966).

[3] J. K. McDonald, B. B. Zeitman, T. J. Reilly, and S. Ellis, *J. Biol. Chem.* **244**, 2693 (1969).

[4] J. K. McDonald, S. Ellis, and T. J. Reilly, *J. Biol. Chem.* **241**, 1494 (1966).

[5] J. K. McDonald, T. J. Reilly, B. B. Zeitman, and S. Ellis, *Biochem. Biophys. Res. Commun.* **24**, 771 (1966).

dipeptides from large hormonal polypeptides without exhibiting any complicating endopeptidase activity.[3,6]

The beef spleen enzyme (EC 3.4.4.9) was originally called "cathepsin C."[7] However, it was more recently renamed "dipeptidyl transferase,"[2] a term which emphasizes its activity at alkaline pH. On the other hand, we have retained the name "dipeptidyl aminopeptidase I," which is more descriptive of its hydrolytic activity at acid pH, and apparent degradative function as suggested by its lysosomal localization.[1]

Assay Method

Perhaps the simplest and most precise method of assay is a fluorometric assay involving the use of a fluorogenic dipeptide-β-naphthylamide substrate. K_m values for most dipeptide β-naphthylamides fall in the range of 0.1–0.2 mM at pH 6.0 and 37°.[3] By comparison, the dipeptide amides have K_m values at least 100 times greater, and their use as substrates requires rather laborious methods of product detection. When a β-naphthylamide substrate is used, a fluorometer coupled to a strip chart recorder can be used as a rapid means of direct assay.[4] Such a system provides for the continuous measurement of liberated β-naphthylamine, which has a fluorescence emission at 410 nm when excited at 335 nm. The method is exceedingly sensitive, requiring the smallest possible amount of enzyme (10^{-5} to 10^{-4} unit), and is particularly appropriate for kinetic studies. For the purpose of routine assay, Gly-Phe-β-naphthylamide is used as the substrate of choice. Although other dipeptide β-naphthylamides, e.g., Gly-Arg-β-naphthylamide, are hydrolyzed much more rapidly (see Table II), Gly-Phe-β-naphthylamide is a specific assay substrate that offers ample sensitivity, and is commercially available from Fox Chemical Company (Los Angeles, California). A unit of activity is defined as the amount of DAP I that hydrolyzes 1 μmole of Gly-Phe-β-naphthylamide per minute at pH 6.0 and 37° under the conditions of the assay.

Assays are routinely performed in 12×75-mm Pyrex tubes containing 0.7 ml of 60 mM 2-mercaptoethylamine hydrochloride–50 mM cacodylic acid–NaOH buffer, pH 6.0, and 3.2 ml of 0.25 mM Gly-Phe-β-naphthylamide. The mixture is brought to 37° in a water bath, and the reaction is initiated by adding 0.1 ml of an appropriate dilution of DAP I. Usually, 2×10^{-4} unit of DAP I is appropriate. The reaction mixture is immediately placed in the water-jacketed cuvette holder (maintained at 37°) of a fluorometer that is coupled to a strip chart

[6] J. K. McDonald, P. X. Callahan, B. B. Zeitman, and S. Ellis, *J. Biol. Chem.* **244,** 6199 (1969).

[7] H. H. Tallan, M. E. Jones, and J. S. Fruton, *J. Biol. Chem.* **194,** 793 (1952).

recorder. We have routinely used a Turner fluorometer (Model 111) coupled to a 1-mA Rustrack recorder operated at a chart speed of 30 inches per hour. The reaction rate is given by the rate of free β-naphthylamine production in terms of fluorescence units per minute. A standard solution of β-naphthylamine hydrochloride in the assay buffer (pH 6.0) is used to calibrate the fluorometer. Although individual fluorometers vary in sensitivity, a neutral solution of β-naphthylamine hydrochloride containing 1 nmole per milliliter usually gives a full-scale reading (100 fluorescence units) at sensitivity 10. Corrections for quenching are usually not required except for the acid quenching that arises if the pH drops below 5.5.

The light source used in the Turner fluorometer is a 4-W germicidal lamp (GE F4T4/BL, major emission at 360 nm). The excitation wavelength (335 nm) is obtained by passing the light through a primary filter consisting of a Wratten 34A (gelatin) filter supported between two Corning 7-54 filters. The fluorescence emission (410 nm) is passed through a secondary filter consisting of a Wratten 2A that screens out light with a wavelength less than 405 nm.

If a fluorometric procedure is not possible, the rate of β-naphthylamine formation can be measured colorimetrically by removing timed aliquots for diazotization.[8]

DAP I activity has also been measured on Gly-Phe-NH_2. At pH 6.0 a deamidation reaction occurs in which the liberated ammonia can be measured by a diffusion method.[9] At pH 6.8 a transferase reaction occurs in which the dipeptide product is transferred to hydroxylamine, and the resulting dipeptide-hydroxamic acid is measured colorimetrically.[10] The esterase activity of DAP I has been measured on Gly-Phe-OMe by automatic titration at pH 5.0,[11] and a direct colorimetric method has been described that utilizes Gly-Phe-*p*-nitroanilide.[12]

Preparation of DAP I Suitable for Sequence Studies

The purification scheme devised by Metrione *et al.*,[2] serves as the basis for the preparation of DAP I. Changes in the procedure are intended to shorten the method and to eliminate contaminating activities. As will be seen, the practicability of using DAP I as a sequencing reagent is greatly enhanced by the relative ease with which a suitable preparation can be made. Metrione *et al.* reported that the specific

[8] J. A. Goldbarg and A. M. Rutenburg, *Cancer* **11**, 283 (1958).
[9] D. Seligson and H. Seligson, *J. Lab. Clin. Med.* **38**, 324 (1951).
[10] G. de la Haba, P. S. Cammarata, and J. S. Fruton, Vol. II, p 64.
[11] I. M. Voynick and J. S. Fruton, *Biochemistry* **7**, 40 (1968).
[12] R. J. Planta and M. Gruber, *Anal. Biochem.* **5**, 360 (1963).

activity of their beef spleen preparation was about 9 times greater than that achieved by de la Haba *et al.*[10] Subsequently, in this laboratory, the method was applied to the purification of DAP I from rat liver, yielding a preparation with exceptionally high specific activity.[3] In addition to liver being a particularly rich source of DAP I, the purification of the liver enzyme was further facilitated by using ultrafiltration to recover the activity from dilute pools. Such a procedure is relatively innocuous compared to reprecipitation with ammonium sulfate.

Steps were also taken to eliminate trace contamination by two peptidase activities that were found in the best preparations of DAP I. The use of such preparations in prolonged digestions of peptide hormones revealed low levels of a dipeptidase activity that caused a partial breakdown of certain liberated dipeptides.[3,6] In addition to the dipeptidase, low levels of a carboxypeptidase were detected that attacked COOH-terminal residues adjacent to proline. Both contaminants showed optimum activity around pH 5.5. The properties of these two contaminating enzymes were the subject of a recent communication[13] in which it was reported that the dipeptidase (termed "Ser-Met dipeptidase") was inhibited by EDTA, and the carboxypeptidase (termed "catheptic carboxypeptidase C") by diisopropyl phosphorofluoridate (DFP). These inhibitors can be effectively employed in preparing DAP I for sequencing applications and thereby obviate the necessity of separating the contaminating peptidases by laborious fractionation procedures.

The last two steps in the purification of DAP I, as described by Metrione *et al.*, involve gradient elution chromatography from DEAE-cellulose, and subsequently CM-cellulose. These are laborious, low-return steps that do not significantly improve the preparation of DAP I intended for sequence studies. They were therefore eliminated, leaving only one column step (gel filtration) to be performed. An abridged purification procedure is described below for the preparation of DAP I from a 2-kg quantity of beef spleen. A brief summary is shown in Table I. The method employs treatment with DFP and EDTA as a means of eliminating the activities of "Ser-Met dipeptidase" and "catheptic carboxypeptidase C." Except for the acid extraction, the entire process was conducted at 5°. (It is reported, however, that the gel filtration step can be conducted at room temperature without loss of activity.[2])

Acid Extraction. About 6 lb of freshly collected beef spleens were obtained from an abattoir and brought to the laboratory in crushed ice. The spleens were placed in plastic bags and stored at —20° until used. When the purification was commenced, they were allowed to undergo a

[13] J. K. McDonald, B. B. Zeitman, and S. Ellis, *Biochem. Biophys. Res. Commun.* In press.

TABLE I
PURIFICATION OF DIPEPTIDYL AMINOPEPTIDASE I FROM 2 KG OF BEEF SPLEEN

Fraction	Protein (mg)	Activity: Total (units[a])	Activity: Specific (units/mg)
Acid extract	90,600	4520	0.05
40–70 AS fraction	1,440	2100	1.45
Heated fraction	804	1990	2.48
Sephadex G-200	82	1520	18.6

[a] One unit of DAP I hydrolyzes 1 μmole of Gly-Phe-β-naphthylamide per minute per milligram of protein at pH 6.0 at 37°.

surface thaw that facilitated the removal of the outer membrane. The semifrozen tissue was cut into strips and passed twice through a power meat grinder. A 2-kg quantity of the ground spleen was dispersed in 4 liters of water containing disodium EDTA (2.4 g) at 1.5 mM. The stirred suspension was adjusted to pH 3.5 with about 65 ml of 6 N H_2SO_4. The mixture was stirred for an additional hour at room temperature before being readjusted to pH 3.5 with additional H_2SO_4. The suspension was then incubated at 37° for 22 hours, with slow stirring for the first 5 hours. A clear supernatant, referred to as the "acid extract," was obtained by centrifuging the mixture at 10,000 g for 30 minutes at 5°.

Ammonium Sulfate Fractionation. The acid extract was brought to 40% of saturation by adding 228.5 g of ammonium sulfate per liter of acid extract. The mixture was allowed to stand for 2 hours before the precipitate was removed by centrifugation at 10,000 g for 30 minutes. The supernatant was then adjusted to 70% of saturation by adding 190 g of ammonium sulfate per liter of supernatant. The suspension was allowed to stand overnight. The protein was sedimented at 10,000 g, recovered in water to a final volume of 90 ml, and dialyzed extensively against 1% NaCl. The dialyzed solution was adjusted to pH 5.0, if necessary, and clarified at 20,000 g for 15 minutes. This supernatant, referred to as the "40–70 AS fraction," had a (Lowry) protein concentration of about 1.5% in a final volume of about 130 ml.

Heat Treatment. The 40–70 AS fraction was dispensed into 13-ml, conical, glass centrifuge tubes, heated to 65° for 40 minutes, and chilled on ice. The precipitated protein was removed by centrifugation, and the supernatants were pooled. The total volume was concentrated to less than 10 ml by means of ultrafiltration under nitrogen at 50 psi with a 65-ml Diaflo ultrafiltration assembly available from Amicon Corporation (Cambridge, Massachusetts). A UM-10 Diaflo membrane (formerly

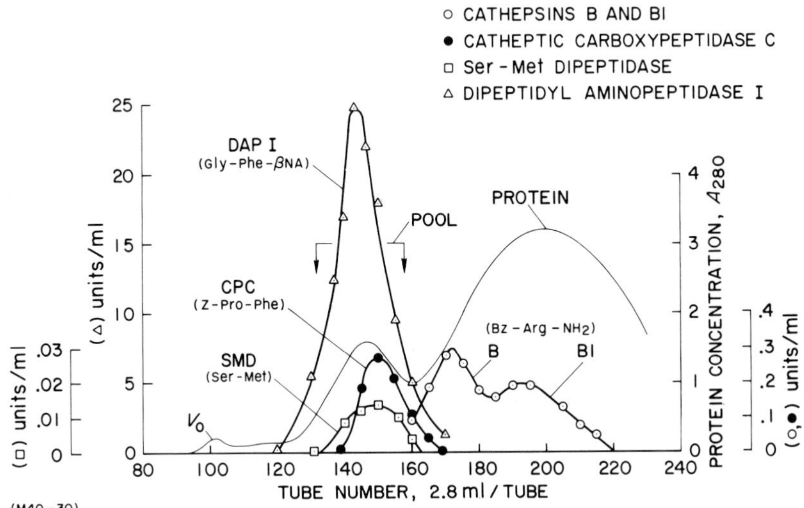

FIG. 1. Gel filtration of a heat-treated fraction of beef spleen dipeptidyl aminopeptidase I (DAP I) on a column (3.2 × 94 cm) of Sephadex G-200 equilibrated with 0.1 *M* NaCl–2 m*M* 2-mercaptoethanol–0.1 *M* acetate buffer, pH 4.5, at 5°. An 800-mg quantity of protein was applied in 8 ml of 1% NaCl, and the flow rate was maintained at 17 ml per hour. V_0 designates the void volume of the column. "Catheptic carboxypeptidase C" (CPC) was assayed on Z-Pro-Phe at pH 5.5. "Ser-Met dipeptidase" (SMD) was assayed on Ser-Met at pH 5.5. Cathepsins B and B1 were assayed on Bz-Arg-NH_2 at pH 5.5. For all activities, a unit of enzyme hydrolyzes 1 μmole of substrate per minute at 37°.

known as UM-1) was utilized. This membrane is impermeable to components having molecular weights in excess of 10,000. The concentrate was dialyzed against 0.1 *M* NaCl–2 m*M* 2-mercaptoethanol–0.1 *M* acetate buffer, pH 4.5. The dialyzed solution is referred to as the "heated fraction."

Gel Filtration. Because the heated fraction was in concentrated form (5–10 ml), it was possible to apply the entire amount (representing 2 kg of spleen) as a very discrete band on a column of Sephadex G-200 that was equilibrated with 0.1 *M* NaCl–2 m*M* 2-mercaptoethanol–0.1 *M* acetate buffer, pH 4.5. The effluent diagram is shown in Fig. 1. Whereas the heat treatment appeared to destroy all the cathepsin A and cathepsin D in the preparation, approximately 5–10% of cathepsins B and B1 survived. Activities that have been described as catheptic carboxypeptidases A and B[14] were located in the cathepsin B peak. As shown in Fig. 1, it was possible to exclude the foregoing activities when the DAP I fractions were pooled to form the "G-200 fraction."

[14] L. M. Greenbaum and R. Sherman, *J. Biol. Chem.* **237**, 1082 (1962).

DFP and EDTA Treatment. It was not possible, as is evident in Fig. 1, to exclude "Ser-Met dipeptidase" and "catheptic carboxypeptidase C" from the G-200 fraction. Although these activities are present at very low levels, and are plotted on greatly expanded scales in Fig. 1, their presence may be manifested when DAP I is used for extended periods of peptide digestion. Inhibitors were therefore incorporated to preclude this possibility. The G-200 fraction was first treated with DFP to inactivate the carboxypeptidase. This was carried out by combining 1 part of 80 m*M* DFP (in isopropanol) with 20 parts of the DAP I pool off Sephadex. The mixture was held on ice for 30 minutes, and then concentrated 10- to 15-fold (to give about 2.5% protein) by ultrafiltration as described under *Heat Treatment.* The concentrated enzyme was next dialyzed against 1% NaCl–4 m*M* 2-mercaptoethanol that contained 1 m*M* EDTA as a dipeptidase inhibitor.

Preservation. The inhibitor-treated G-200 fraction was dispensed into small glass ampoules at the rate of 0.2 ml (about 4 mg protein) per ampoule. The contents were then subjected to a rapid freeze in a Dry Ice–acetone bath, and finally dried and sealed under vacuum. Up to 10% of the activity may be lost as a consequence of the DFP treatment, dialysis, and freeze-drying, but the remaining DAP I activity was stable under refrigeration for at least two years. The contents of an ampoule can be reconstituted in 2 ml of 1% NaCl–1 m*M* EDTA and stored on ice as a working solution. Such a solution loses about 10% of its activity in a month.

Substrate Specificity

The broad substrate specificity of DAP I was first observed with β-corticotropin used as a substrate. At pH 5.5, preparations of DAP I (from beef spleen, beef pituitary glands, and rat liver) catalyzed the successive removal of dipeptides from the NH_2 terminus of the hormone. A time-course analysis of the products by paper chromatography provided an unambiguous illustration of the exact sequence of dipeptide release. The course and extent of degradation are illustrated in Fig. 2 for several hormonal polypeptides. It was necessary to satisfy both the sulfhydryl[15] and chloride[4,5] requirements of DAP I to achieve adequate rates of hydrolysis. Glucagon, for example, was virtually immune to attack in the absence of chloride.[6]

An analysis of the specific bonds cleaved in polypeptide substrates (Fig. 2) revealed that DAP I can remove dipeptides having penultimate residues with both hydrophilic and hydrophobic side chains. Dipeptides

[15] J. S. Fruton and M. J. Mycek, *Anal. Biochem. Biophys.* **65,** 11 (1956).

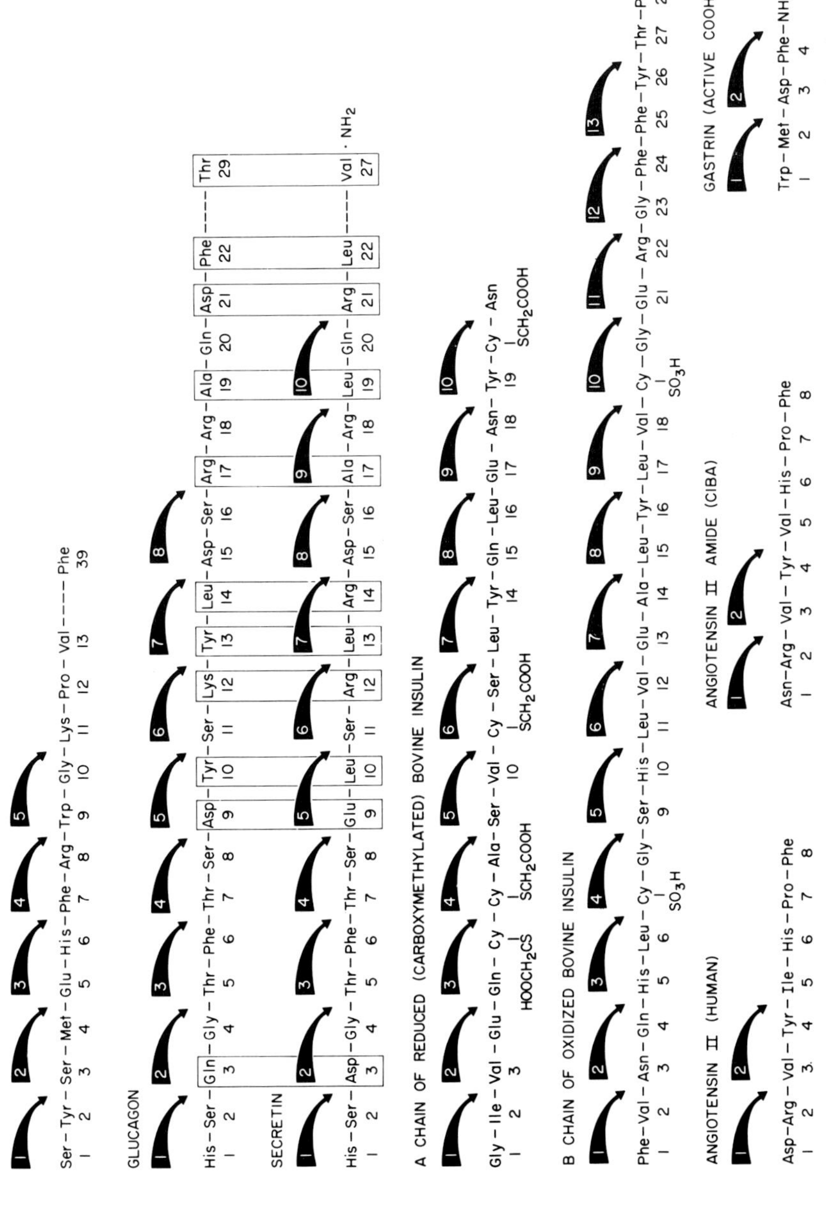

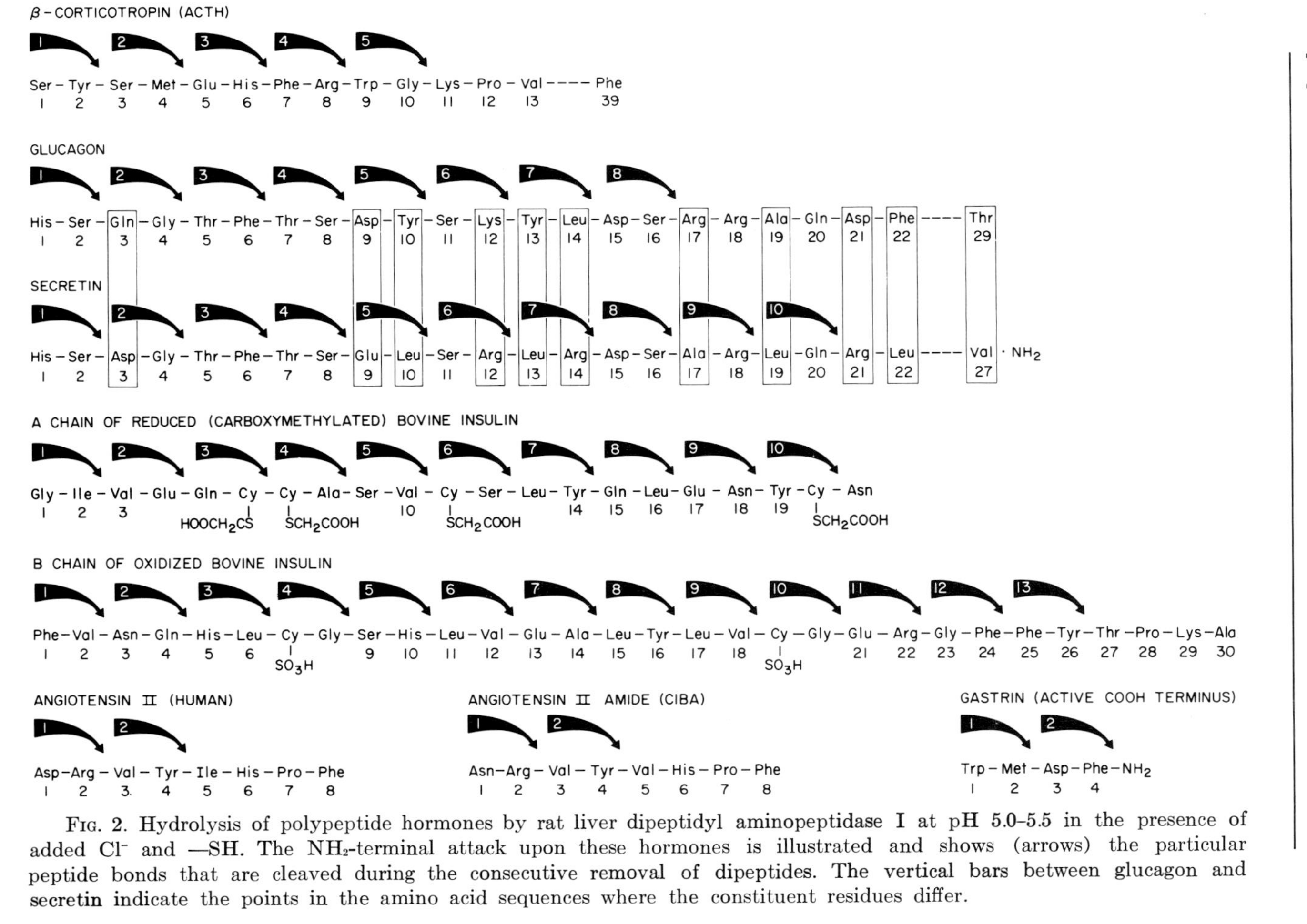

FIG. 2. Hydrolysis of polypeptide hormones by rat liver dipeptidyl aminopeptidase I at pH 5.0–5.5 in the presence of added Cl^- and —SH. The NH_2-terminal attack upon these hormones is illustrated and shows (arrows) the particular peptide bonds that are cleaved during the consecutive removal of dipeptides. The vertical bars between glucagon and secretin indicate the points in the amino acid sequences where the constituent residues differ.

with penultimate basic residues were particularly susceptible to attack. However, the properties of the NH_2-terminal residue can greatly modify these rates. In addition, a comparison of the rates of hydrolysis for a wide variety of dipeptides β-naphthylamides (Table II) and esters showed that penultimate arginyl and lysyl bonds were hydrolyzed more rapidly than any other.[3] On the other hand, as shown in Fig. 2, the degradation of β-corticotropin, glucagon, and secretin was terminated when either lysine or arginine emerged at an NH_2-terminal position. These results are in accord with those of Izumiya and Fruton,[16] who reported that Lys-Phe-NH_2 and Lys-Tyr-NH_2 were not attacked. Although dipeptides with a penultimate arginyl residue are favored substrates, dipeptides containing two basic residues were immune to attack by DAP I. (See Table II and residues 17 and 18 of glucagon in Fig. 2).

Groupings of acidic residues were found to retard the action of

TABLE II
RATES OF DIPEPTIDE β-NAPHTHYLAMIDE HYDROLYSIS BY HIGHLY PURIFIED (RAT LIVER) DIPEPTIDYL AMINOPEPTIDASE I[a]

Substrate (-β-naphthylamide)	Activity: Specific (μmoles/min/mg protein)	Activity: Relative (%)
Gly-Arg	300	100
Cbz-Gly-Arg	0	0
Ala-Arg	215	72
Pro-Arg	164	55
Gly-Trp	150	50
Ser-Met	100	33
Ala-Ala	46	15
Glu-His	33	11
Ser-Tyr	28	9
Phe-Arg	23	8
Gly-Phe[b]	17	5.6
His-Ser	16	5.3
Leu-Ala	10	3.4
Lys-Ala	0	0
Arg-Arg	0	0
Gly-Pro	0	0

[a] Reaction mixtures contained 0.2 m*M* substrate; 10 m*M* 2-mercaptoethylamine hydrochloride; 10 m*M* cacodylic acid–NaOH, pH 6.0. Temperature 37°. Reaction rates were measured by a direct fluorometric method.

[b] Assay substrate.

[16] N. Izumiya and J. S. Fruton, *J. Biol. Chem.* **218**, 59 (1956).

DAP I, as judged by its sluggish attack on Glu_4.[3] Experience gained from the digestion of the polypeptides shown in Fig. 2 revealed that acidic residues, including cysteic acid do not significantly retard the action of DAP I provided these residues are NH_2-terminal. Penultimate acidic residues do retard the action of DAP I, but to a lesser degree if the pH of the reaction is lowered to 4.0–4.5 to suppress the ionization of aspartyl and glutaminyl residues.

In the case of β-corticotropin, the prolyl residue (in position 12) also constituted an obstruction. The emergence of prolyl residues also terminated the degradation of the B chain of oxidized bovine insulin, angiotensin II amide (Ciba), and human angiotensin II. On the basis of these results, it was concluded that DAP I was unable to cleave the bond on either side of a prolyl residue, thereby halting the action of DAP I either one or two residues in advance of proline. A prolyl residue that resides initially at the NH_2 terminus has no effect. Pro-Phe-NH_2, for example, is known to be hydrolyzed about as well as Gly-Phe-NH_2 [16]; and, as seen in Table II, Pro-Arg-β-naphthylamide is an excellent substrate.

As regards size limitations for peptide substrates of DAP I, there appears to be no upper limit, provided the NH_2 terminus remains accessible to the enzyme. What appeared to be a problem arising from the inaccessibility of an NH_2 terminus is seen in the case of insulin. Although the separate chains were readily degraded (Fig. 2), the intact hormone was resistant to attack. In general, there does not appear to be a lower size limit for peptide substrates. However, some tripeptides are rather resistant to attack. For example, Phe_3 and Gly_3 were hydrolyzed at relatively slow rates, and Ala_3 was resistant. On the other hand, the corresponding tetrapeptides Phe_4, Gly_4, and Ala_4, were readily split to dipeptides.[3]

In summary, peptide degradation by DAP I is blocked by the emergence of an NH_2-terminal lysyl or arginyl residue, or by a prolyl residue. (Notably, the use of tryptic fragments obviates the former restriction.) Peptides with an odd number of residues will generate a COOH-terminal tripeptide fragment that may or may not be susceptible to the scission of one last dipeptide by DAP I.

[23] Sequencing of Peptides with Dipeptidyl Aminopeptidase I

By PAUL X. CALLAHAN, J. KEN McDONALD, and STANLEY ELLIS

The methodology contained in this article provides the means by which dipeptidyl aminopeptidase I (DAP I) can be utilized in a new approach to the sequencing of small amounts (0.5 μmole or less) of peptides.[1] The ability of DAP I to remove dipeptides, in sequence, from the NH_2 termini of oligopeptides can be followed, for at least several dipeptide products, by time-course chromatography.[2,3] Although often this is insufficient for establishing sequence, it is possible to modify the peptide substrate in such a way that DAP I will generate the alternate, overlapping dipeptides[1]—an advantage not possible with monoaminoacyl aminopeptidases. Certain advantages over chemical methods of sequencing are inherent in this approach. Asparagine and glutamine are recovered as such (not as aspartic or glutamic acid), and there is no tendency to form pyroglutamic acid, as occurs with chemical methods. Additionally, there is no tendency to "lose" a residue as a result of side reaction occurring during sequencing,[4] and the yield of the last dipeptide is equal to the yield of the first. Sequencing with DAP I is, therefore, limited not by the length of the peptide, but by the appearance of a noncleavable bond.

The inability of DAP I to remove dipeptides with an NH_2-terminal arginine or lysine is not a problem if the peptide substrates are derived from tryptic digests wherein most, if not all, lysines and arginines are COOH terminal. A limitation arises from the inability of DAP I to cleave the bond on either side of proline. It is possible, however, to take advantage of this limitation, and allow the digestion to come to a complete stop at this point. One or two residues can then be removed by the Edman method and sequencing with DAP I be continued.

The full potential of DAP I, in the sequencing of peptides, could not be realized until a methodology for identifying dipeptides had been

[1] P. X. Callahan, J. K. McDonald, and S. Ellis, *Fed. Proc., Fed. Amer. Soc. Exp. Biol.* **28**, 661 (1969); *ibid.* **30**, 1045 (1971).

[2] J. K. McDonald, B. B. Zeitman, T. J. Reilly, and S. Ellis, *J. Biol. Chem.* **244**, 2693 (1969).

[3] J. K. McDonald, P. X. Callahan, B. B. Zeitman, and S. Ellis, *J. Biol. Chem.* **244**, 6199 (1969).

[4] Discussed in this volume [24, 27, 28].

developed. This methodology has been published[5] and will not be discussed in detail in this article. Instead, emphasis will be placed on the separation of the dipeptide products that result from the action of DAP I on oligopeptides, their recognition in mixtures when separation of the dipeptides is unnecessary, and the approaches necessary for the reordering of those dipeptide products.

Digestion of the Peptide

Substrate. Peptide substrates should be as pure as possible since peptide contaminants are also potential substrates. Whereas some dipeptides have low color yields (10% or less of the value seen with most dipeptides), a 10% contaminant of a high color yield dipeptide cannot be distinguished by its intensity from a major dipeptide having a low color yield. This can lead to difficulties in interpreting time-course chromatograms and result in possible ambiguities in sequence information. Nonpeptide impurities, such as pyridine, acetate, salts, do not seem to interfere with the digestion.

Modification of Substrate. The reordering of peptide sequence is based partly on a knowledge of the overlapping alternate dipeptides. In order to obtain this information, the peptide substrate is modified by deleting or adding an amino acid residue at the NH_2 terminus. DAP I is then used to generate the alternate dipeptides. A modification of the Edman method can be used to shorten the peptide substrate or the *N*-carboxyanhydride method to lengthen it. The Edman procedure is the method of choice since yields of the modified peptide, at peptide quantities of 0.5 μmole or less, are markedly better.

a. Edman degradation. The NH_2 terminus of the peptide is removed by a procedure essentially that of Blombäck *et al.*[6] with the following modifications.

The buffer, chosen for its volatility and lack of residue, is composed of 3.3 ml of *N*-ethylmorpholine, 0.5 ml of glacial acetic acid, 25 ml of ethanol, and 21.65 ml of water. The peptide (generally 0.25 μmole) is dissolved in 500 μl of pyridine, and 100 μl of buffer and 100 μl of water are added. All reagents are stored under nitrogen, and all degradative steps are performed under nitrogen.

A 1500-fold excess of phenylisothiocyanate, 5%, in heptane, is used to drive the reaction to completion.

Cleavage with trifluoroacetic acid (0.1 ml for each 0.1 μmole of pep-

[5] P. X. Callahan, J. A. Shepard, T. J. Reilly, J. K. McDonald, and S. Ellis, *Anal. Biochem.* **38**, 330 (1970).

[6] B. Blombäck, M. Blombäck, P. Edman, and B. Hessel, *Biochim. Biophys. Acta* **115**, 371 (1966).

tide) is performed twice, with evaporation to dryness between treatments to reduce the possibility of subsequently extracting small PTC-coupled peptides into the 1-chlorobutane.

The phenylthiohydantoin is estimated spectrophotometrically, and NH_2-terminal analysis on the modified peptide is performed by the method of Gray and Hartley,[7] according to Gray,[8] to confirm complete removal of the NH_2-terminal amino acid without damage to the remainder of the molecule.

Yields are invariably better than 97%, and the des-amino acid peptide can generally be used as a substrate for DAP I without further purification.

b. *N*-CARBOXYANHYDRIDE CONDENSATION. The NH_2-terminus of the peptide is reacted with an anhydro-*N*-carboxyamino acid by the method of Denkewalter *et al.*,[9] except that volatile buffers are used to facilitate the isolation of the peptide. The peptide (generally 1 μmole) is dissolved in 2 ml of 2.0 *M* *N*-ethylmorpholine, and the pH is adjusted to 10.4 with 2.0 *M* triethylamine. The solution, maintained at 0°, is vigorously mixed during the addition of 1.5 μmoles of the anhydro-*N*-carboxyamino acid for each micromole of free amino group. The solution is then adjusted to pH 3 with about 20 drops of glacial acetic acid, extracted 3 times with 2-ml portions of ethyl acetate, and freeze-dried. Excess free amino acid is removed by passing the redissolved freeze-dried material through a 0.9×25-cm column of Biogel P-2. The material contained in the void volume of the column is freeze-dried to give the modified peptide (in yields approaching 95%) for digestion with DAP I. This approach is not as useful in generating a modified peptide as the Edman reaction since, at peptide concentrations lower than 0.5 μmole, an apparently constant amount of unmodified peptide remains. This is presumed to be due to carbonate in the reaction system which inhibits the condensation.

Enzyme. The freeze-dried enzyme[2,10] is reconstituted in 1% NaCl–0.5 m*M* EDTA to provide a working solution containing 25–30 units per milliliter. This solution is relatively stable when stored on ice, showing losses of less than 10% in a month. Generally, DAP I is used in digests at a ratio of 3 units per micromole of peptide substrate. Less enzyme is needed for small, easily hydrolyzed peptides.

Buffer-Activator and Substrate. In the following discussions, the

[7] W. R. Gray and B. S. Hartley, *Biochem. J.* **89**, 379 (1963).

[8] W. R. Gray, Vol. 11, p. 139.

[9] R. G. Denkewalter, H. Schwam, R. G. Strachan, T. E. Beesley, D. F. Veber, E. F. Schoenewaldt, H. Barkemeyer, W. J. Paleveda, Jr., T. A. Jacob, and R. Hirschmann, *J. Amer. Chem. Soc.* **88**, 3163 (1966).

[10] J. K. McDonald, P. X. Callahan, and S. Ellis. This volume [22].

"peptide substrate" refers to both the original peptide and the modified peptide. Digestion of these peptides produces the "original dipeptides" and the "alternate dipeptides," respectively.

The buffer-activator is 16 m*M* HCl–0.8% pyridine–52 m*M* acetic acid–15 m*M* mercaptoethanol–0.5 m*M* EDTA, pH 5.0. The peptide substrate is dissolved in this solution at a concentration of 0.5 m*M* in the following manner. The peptide sample is initially dissolved in 4% pyridine (0.4 ml/μmole) and diluted with water (0.79 ml/μmole). To this solution are added 0.5% acetic acid (0.4 ml/μmole), 0.1 *M* HCl (0.32 ml/μmole), 0.1 *M* EDTA (0.01 ml/μmole), and 0.375 *M* mercaptoethanol (0.08 ml/μmole). Final pH of 5.0 is usually obtained. Adjustment, if necessary, is made with 4% pyridine or 0.5% acetic acid.

Digestion Conditions for Time-Course Analysis. The buffer–activator–peptide solution is preincubated at 37° for 5 minutes, and the hydrolysis is initiated by adding 3 units (about 0.1 ml) of DAP I per micromole of substrate. At this enzyme concentration, the digestion is usually complete in less than 4 hours.

Aliquots (0.02 μmole) are taken at logarithmic time intervals (i.e., 0-, 1-, 2-, 4-, 8-minute, etc.), inactivated by the addition of 1 part of 12 *M* HCl per 5 parts of digest, and spotted directly on paper. Chromatography is performed on Whatman No. 1 chromatography paper in a 4-trough chromatography cabinet. The chromatograms are developed for 16–20 hours with 1-butanol–formic acid–water (70:15:15). Visualization of the dipeptide products is facilitated by the differences in color and rates of color development obtained with the polychromic ninhydrin spray.[11] Although it is usually not possible to establish peptide sequence from the time-course chromatogram alone, it is generally possible to recognize the first 7 or 8 ninhydrin-positive areas by intensity and to determine which other peptides are visible midway or late in the time-course reaction. The sequential appearance of dipeptides is often more apparent with a reduction of the ratio of enzyme to substrate, but at the expense of a longer digestion time. This time-course analysis not only demonstrates the extent to which the peptide substrate is degraded but, as will be seen in a later discussion, also aids in reordering the dipeptide products.

Separation of the Dipeptide Products

Principle. As reported,[5] identification of dipeptides is based on column chromatographic elution times used in combination with paper chromatographic R_f values. In this procedure for identifying dipeptides,

[11] E. D. Moffat and R. I. Lytle, *Anal. Chem.* **31**, 926 (1959).

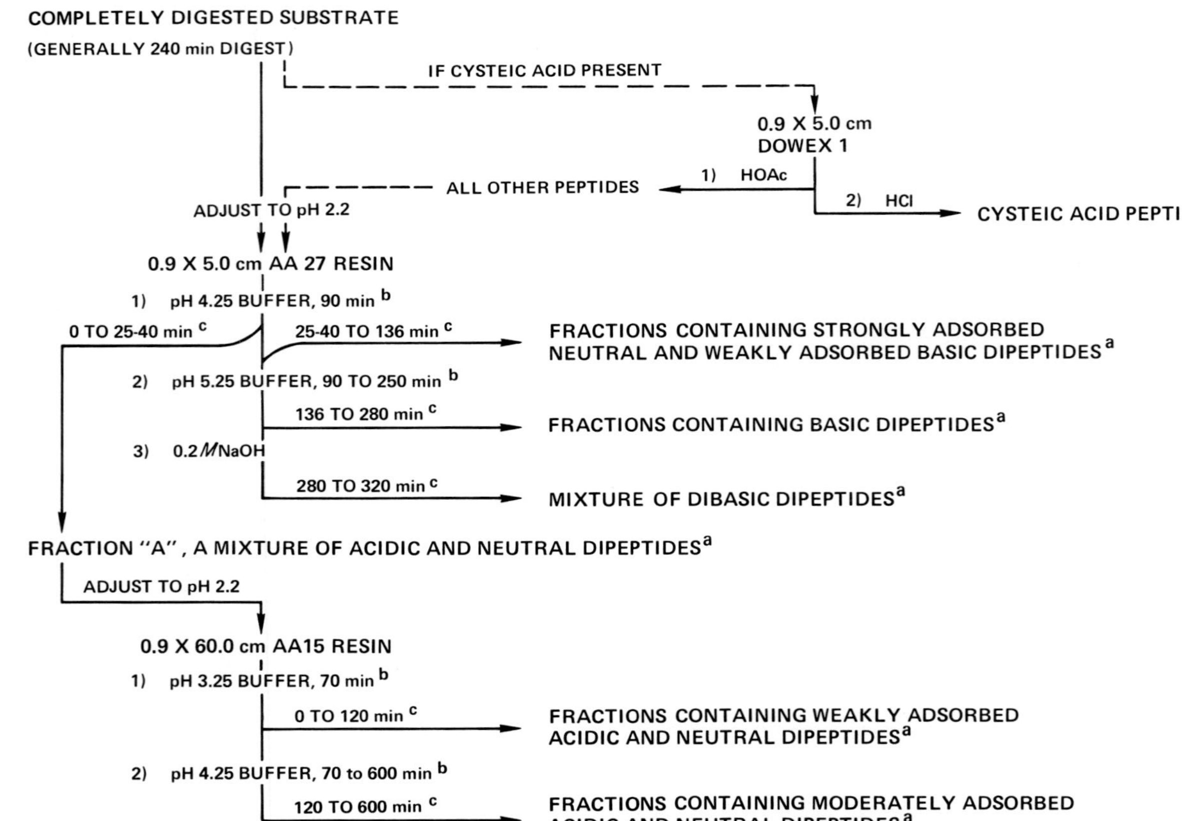
COMPLETELY DIGESTED SUBSTRATE
(GENERALLY 240 min DIGEST)
IF CYSTEIC ACID PRESENT
0.9 X 5.0 cm
DOWEX 1
1) HOAc
ALL OTHER PEPTIDES
2) HCl
CYSTEIC ACID PEPTIDES
ADJUST TO pH 2.2
0.9 X 5.0 cm AA 27 RESIN
1) pH 4.25 BUFFER, 90 min b
0 TO 25-40 min c
25-40 TO 136 min c
FRACTIONS CONTAINING STRONGLY ADSORBED NEUTRAL AND WEAKLY ADSORBED BASIC DIPEPTIDES a
2) pH 5.25 BUFFER, 90 TO 250 min b
136 TO 280 min c
FRACTIONS CONTAINING BASIC DIPEPTIDES a
3) 0.2 M NaOH
280 TO 320 min c
MIXTURE OF DIBASIC DIPEPTIDES a
FRACTION "A", A MIXTURE OF ACIDIC AND NEUTRAL DIPEPTIDES a
ADJUST TO pH 2.2
0.9 X 60.0 cm AA15 RESIN
1) pH 3.25 BUFFER, 70 min b
0 TO 120 min c
FRACTIONS CONTAINING WEAKLY ADSORBED ACIDIC AND NEUTRAL DIPEPTIDES a
2) pH 4.25 BUFFER, 70 to 600 min b
120 TO 600 min c
FRACTIONS CONTAINING MODERATELY ADSORBED ACIDIC AND NEUTRAL DIPEPTIDES a

elution times were determined by using separate samples on the 5-cm basic and 60-cm neutral and acidic hydrolyzate columns. In order to conserve peptide substrate, the identification procedure has been modified to provide for the collection of dipeptide fractions, and the determination of their elution times on a single aliquot of the reaction mixture. This scheme has a small disadvantage in that it requires minor adjustment of some elution times but provides, as a bonus, markedly decreased analysis times. The scheme is depicted on a flow diagram in Fig. 1.

Preparative Column Chromatography. If the peptide digest (equivalent to 0.1–0.5 μmole of substrate) contains cysteic acid moieties, it is adjusted to pH 2.2 and initially chromatographed on a 0.9×5-cm column of Dowex 1. The column is washed with 2 *M* acetic acid to elute all dipeptides except those containing cysteic acid. The cysteic acid dipeptides are subsequently eluted with 2 *M* HCl and retained for identification. An aliquot (1/5) is applied to the 60-cm column and developed with pH 3.25 buffer to quantitate the cysteic acid dipeptide(s). The initial dipeptide eluate and the acetic acid wash are combined and evaporated to dryness and handled in the same manner as a digest which contained no cysteic acid, as follows.

The peptide, digested to completion with DAP I, is adjusted to pH 2.2 and applied to a 0.9×5-cm column of Beckman/Spinco AA-27 resin (spherical sulfonated polystyrene, 15 ± 6 μ, 8% cross-linkage). The 5-cm column referred to herein is the 20-cm basic, hydrolyzate column filled to about 5.3 cm with resin. The 60-cm column is the 60-cm neutral and acidic, hydrolyzate column filled to about 55 cm with resin. The analyzer is equipped for stream division, 20% of the effluent being used for color development and the remainder for collection of fractions. The collection time, which is governed by the spacing of the separated components, is initially 1 minute and is increased to 2 or 4 minutes after about 2 hours of development time. Buffer flow rate is 68 ml per hour, and the ninhydrin flow rate is 34 ml per hour. The column is developed with 0.2 *M* sodium citrate buffer, initially pH 4.25, for 90 minutes, and then with pH 5.25

Fig. 1. Flow sheet for column separation of dipeptides resulting from DAP I digests. With the exception of the Dowex 1 column, the column separations illustrated here utilize the resins and buffers designed for amino acid analysis of hydrolyzates. (a) The column effluent is stream-split to provide for the collection of fractions that correspond to the recorded peaks. Published charts and tables [P. X. Callahan, J. A. Shepard, T. J. Reilly, J. K. McDonald, and S. Ellis, *Anal. Biochem.* **38**, 330 (1970)] are then used to relate the observed elution times to specific dipeptide possibilities. (b) Times refer to events as they are performed during the development of the column. (c) Times refer to events as they appear on the recording.

for an additional 160 minutes. The eluate from 0 to 25–40 minutes is retained as fraction "A" for later refractionation. The exact pooling of this fraction is determined by the position of the valley between the dipeptide peaks. Fraction "A" is composed of the neutral and acidic dipeptides which normally elute from the 60-cm column in less than 400 minutes. An example of this column fractionation procedure that complements the flow diagram is illustrated in Fig. 2.

The remaining pH 4.25 eluate from the 5-cm column (from 25–40 to 136 minutes) contains those neutral dipeptides that are strongly adsorbed on a 60-cm column and normally eluate from that column between 400 and 1200 minutes. These dipeptides are usually sufficiently resolved on the 5-cm column, and their elution times on this column, when multiplied by a factor of approximately 10 (depending on the lengths of the resin beds in the two columns), equal their elution times on the 60-cm column.

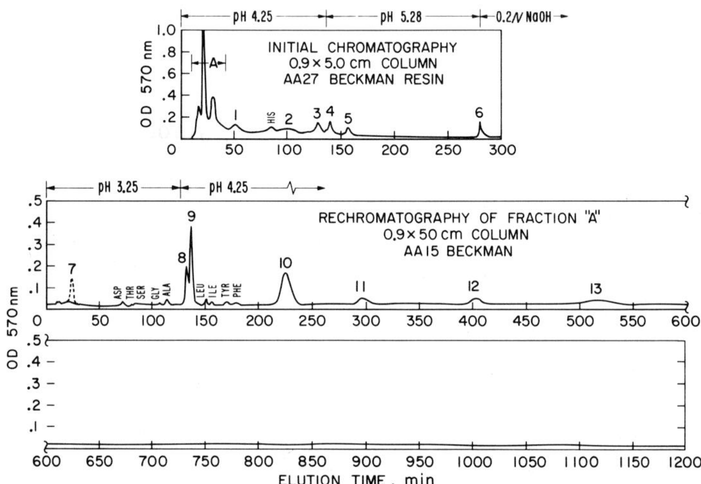

FIG. 2. Column chromatography of DAP I-digested (240 minutes) B chain of oxidized insulin. A 240-minute digest (to completion), containing 0.25 μmole of the B chain of oxidized insulin was applied to the 5-cm column after isolation of the cysteic acid dipeptides on Dowex 1 (peak 7). The 5-cm column was developed with 0.2 *M* sodium citrate buffer, initially pH 4.25 for 90 minutes, followed by pH 5.28 (the buffer change is apparent at 136 minutes). At 250 minutes, the column was stripped with 0.2 *M* NaOH. Fraction "A" was reapplied to the 60-cm column and developed with 0.2 *M* sodium citrate buffer, initially pH 3.25 for 70 minutes, and thence with pH 4.25 buffer (the buffer change is apparent at 120 minutes). Fractions were combined, desalted as described in the text, and identified by paper chromatography (Table II). Fraction 13 results from the inclusion of a small portion of fraction 1 in fraction "A."

TABLE I
ELUTION TIMES OF BASIC DIPEPTIDES[a]

NH_2-Terminal amino acid	COOH-Terminal amino acid									
	Lys	His	Asp	Ser	Glu	Ala	Val	Leu	Tyr	Phe
Lys			47.5		84			165	178	188
His			41	94.5	74	144	150	157	175	181
Arg			119		142		186.5	215.5	252	
Asp	32	22.5								
Ser		124								
Glu	88									

[a] The 5-cm column was developed with 0.2 *M* sodium citrate buffer, initially at pH 4.25 for 90 minutes, and then at pH 5.25. Buffer change appeared on the recording at 136 minutes.

It was possible to save 700 minutes of analysis time by adopting the 5-cm column to separate the strongly adsorbed, neutral dipeptides.

Two other groups of dipeptides are also eluted from the 5-cm column with the pH 4.25 buffer. The first group is comprised of the dipeptides of lysine or histidine that occur in combination with aspartic and glutamic acids. These dipeptides, which are unresolved on the 5-cm column when eluted with pH 5.28 buffer, have elution times as listed in Table I. The second group is comprised of some of the more easily eluted lysine and histidine dipeptides. The pH 4.25 buffer causes these dipeptides to be more easily eluted from the column. Examples of their elution times with pH 4.25 buffer are also listed in Table I.

Those dipeptides eluted with the pH 5.28 buffer on the 5-cm column (136–280 minutes) are monobasic dipeptides. Because of the prior development of this column with pH 4.25 buffer, however, their elution times differ slightly from the published values.[5] Those dipeptides containing histidine and lysine are eluted 10 minutes earlier, with respect to the pH 4.25 to pH 5.28 buffer change, than the published values, and those dipeptides containing arginine are eluted 1 minute earlier.

At 250 minutes, the 5-cm column is stripped with 0.2 *M* NaOH (recorded at 280 minutes) to remove dibasic dipeptides. It is not necessary to resolve mixtures of dibasic dipeptides since DAP I will not release dipeptides of this type except as a COOH-terminal fragment.

Fraction "A," retained from the 5-cm column, is now adjusted to pH 2.2 with 6 *M* HCl and refractionated on a column (0.9 $\times$ 60 cm) of Beckman/Spinco AA-15 resin (spherical sulfonated polystyrene, 22 $\pm$

$6\ \mu$, 8% cross-linkage). The column is developed with 0.2 *M* sodium citrate buffer, initially pH 3.25, for 70 minutes, and then with pH 4.25 buffer for an additional 530 minutes. The fractions eluted in the first 120 minutes contain weakly adsorbed acidic and neutral dipeptides whose elution times correspond to the published values of those dipeptides eluted prior to 120 minutes. At 120 minutes, the buffer change appears on the recording. If it occurs earlier or later than 120 minutes, all elution times occurring after the buffer change are adjusted to conform to a 120-minute buffer change time.

The fractions eluted after 120 minutes contain the moderately adsorbed acidic and neutral dipeptides, whose elution times conform to the published values.[5] Column development past 600 minutes is unnecessary since all dipeptides that would have been eluted after 500 minutes of development are retained on the 5-cm column. The contents of the tubes comprising each peak are combined with a water rinse and held for desalting.

Desalting of Dipeptides. Desalting of peptide fractions is accomplished on a 0.9 × 1-cm column of Beckman/Spinco type 150A resin which had previously been converted to the Na^+ form with 0.063 *M* NaOH and washed to neutrality with distilled water. The fraction to be desalted is diluted to 0.10 *M* Na^+, adjusted to pH 2.2 with HCl (initially 12 *M*, then 1 *M*), and passed through the column at a rate of 60 ml per hour or less. Under these conditions, the column accepts 200 ml of solution without noticeable loss of dipeptides. The column is next washed with 3 ml of distilled water to remove residual citrate buffer. The dipeptide is eluted from the column with 2 ml of 63 m*M* NaOH followed by 2 ml of distilled water, and the eluate is immediately acidified with 0.2 ml of 6 *M* HCl (to prevent racemization and loss of base-sensitive components) and taken to dryness in a high vacuum over NaOH pellets. A tendency to "bump" is minimized by slant-freezing the sample before applying the vacuum. The residue is extracted twice with 1-ml portions of isopropanol containing 5% 12 *M* HCl. The extracts are either centrifuged or filtered through glass wool, combined, dried over NaOH pellets, and used as such for identification.

Identification of Dipeptide Products

Principle. Elution times and R_f values are used to identify the dipeptides.[5] Since only a limited number of dipeptides can have a given elution time (plus or minus 4.1%), the number of possible dipeptides having this elution time is reduced from 400 to between 1 and 15. The elution time also defines which dipeptides these 1 to 15 may be and makes possible the choice of an appropriate solvent system for their final identification by paper chromatography. In general, the 1-butanol–formic acid–water (70:15:15) solvent is the best first choice since it not only provides the greatest resolving power, but also relates the column chromatographic fraction to the time-course chromatogram. Occasionally, final identification of a dipeptide is not possible by paper chromatography. It is then necessary to determine the NH_2-terminal amino acid. This is most easily accomplished using the dansylation method of Gray and Hartley[7] according to Gray.[8] The dansyl (DNS-) derivative is chromatographed on polyamide thin-layer sheets[12] by the method of Woods and Wang,[13] using water–formic acid (200:3) in the first dimension. Better resolution of *N*-(1-dimethylaminonaphthalene-5-sulfonyl)alanine (DNS-Ala) from DNS-NH_2 is obtained if benzene–acetic acid (14:1) is substituted in the second dimension for the recommended ratio of 9:1. The resolution of other DNS-amino acids is unaffected. In most instances, the dansyl derivative of the NH_2 terminus of a dipeptide can be identified using heptane–1-butanol–acetic acid (30:30:10) in one dimension. With this system, NH_2-terminal analysis can routinely be performed on all fractions as a method of confirming the assigned identity of the dipeptide.

The Dipeptide Products of the B Chain of Oxidized Insulin. As an example of this procedure, the experimental values obtained from a DAP I digest of the B chain of oxidized insulin are presented. The paper chromatographic $R_{Ala}1$ values[14] (1-butanol–formic acid–water system) of the isolated dipeptide products are tabulated in Table II along with the column chromatographic elution times. Often, as will be seen in Table II, the elution time in combination with the $R_{Ala}1$ value in this solvent is sufficient to identify the dipeptide in question. Reference should be made to the published charts and tables that relate these values to specific dipeptides.[5]

As listed in Table II, the elution time of 51 minutes for fraction 1 limits its identification to four possibilities: Phe-Pro, Tyr-Met, Ile-Phe, or Phe-Val. Paper chromatography ($R_{Ala}1 = 2.13$) demonstrated it to be Phe-Val.

[12] Polyamide sheets from Cheng Chin Trading Co., Ltd. (Taipei, Taiwan) were the only ones found to give satisfactory results. They were developed in an Eastman Chromogram apparatus, Distillation Products Industry (Rochester, New York).

[13] K. R. Woods and K.-T. Wang, *Biochem. Biophys. Acta* **133**, 369 (1967).

[14] $R_{Ala}1$ indicates that mobilities in this solvent system are determined with respect to alanine.

TABLE II
IDENTIFICATION OF FRACTIONS DERIVED FROM THE B CHAIN OF OXIDIZED INSULIN

Fraction[a]	Column (cm)	Buffer (pH)	Elution time (min)	$R_{Ala}1$[b]	R_f2[c]	Possible dipeptides[d]	Identified dipeptides
1	5	4.25	51	2.13		Phe-Val	Phe-Val
2	5	4.25	100	2.04	0.70	Phe-Tyr	Phe-Tyr
				0.32	0.21	+unknown	+Thr-(Pro, Lys, Ala)
3	5	4.25	128	0.15		Ser-His	Ser-His
4	5	5.28	140	0.23		Glu-Arg	Glu-Arg
5	5	5.28	158	0.98		His-Leu	His-Leu
6	5	(NaOH)	280	Streak		—	Residual chains
7	60	3.25	25	0.175		$CySO_3H$-dipeptides	$CySO_3H$-Gly
8	60	4.25	126	0.23		Asn-Gln or Gln-Asn	Asn-Gln
9	60	4.25	127	0.892	0.25	Glu-Ala or Ala-Glu or Ala-Thr	Glu-Ala
10	60	4.25	218	2.16		Leu-Val	Leu-Val
11	60	4.25	282	1.55		Gly-Phe	Gly-Phe
12	60	4.25	390	2.08		Leu-Tyr	Leu-Tyr
13	60	4.25	499	2.13 (trace)		Phe-Val	Phe-Val (trace)

[a] Fractions are numbered as obtained from preparative column chromatography of a DAP I digest of the B chain of oxidized insulin (Fig. 2).
[b] $R_{Ala}1$ refers to mobility in the 1-butanol–formic acid–water (70:15:15) system with respect to alanine.
[c] R_f2 refers to mobility in the 1-butanol–acetic acid–water (450:50:125) system with respect to the solvent front.
[d] Possible dipeptides are those dipeptides which meet both R_f and elution time constraints.

Fraction 2 (100 minutes) showed two ninhydrin-positive spots on paper chromatography. One of these ($R_{Ala}1 = 2.04$) allowed for a choice between Phe-Tyr and Tyr-Phe which was resolved by the use of the second solvent system, 1-butanol–acetic acid–water (450:50:125), which gave an R_f2 of 0.70 and demonstrated the peptide to be Phe-Tyr. The other spot ($R_{Ala}1 = 0.32$) was not compatible with any dipeptide or free amino acid having an elution time of 100 minutes. NH_2-terminal analysis of fraction 2 demonstrated threonine, in addition to phenylalanine (of Phe-Tyr). Hydrolysis resulted in free phenylalanine, tyrosine, proline, alanine, threonine, and lysine. Fraction 2, therefore, was a mixture of the dipeptide Phe-Tyr and an unhydrolyzed tetrapeptide, Thr-(Pro, Ala, Lys). Two cycles of Edman degradation followed by dansylation[15] demonstrated this to be Thr-Pro-Lys-Ala, the COOH-terminal fragment of the B chain.

Fraction 3 (at 130 minutes) was identified by paper chromatography and confirmed by NH_2-terminal analysis and hydrolysis to be Ser-His, and fraction 4 was similarly shown to be Glu-Arg.

Fraction 5 had an elution time (158 minutes) which allowed three dipeptides as choices (Lys-Val, His-Leu, and Leu-His). Paper chromatography in solvent 1 produced an $R_{Ala}1$ value of 0.98 unit which restricted the possibilities to His-Leu. This choice was confirmed by NH_2-terminal analysis.

Fraction 6 was eluted with the NaOH front (280 minutes) allowing a choice of 9 dibasic dipeptides (arginine, histidine, and lysine in combination with each other). Paper chromatography, however, demonstrated only a faint streak, indicating that none of these choices were correct. Hydrolysis and NH_2-terminal analysis demonstrated fraction 6 to consist of a small amount of a mixture of residual chain lengths.

The elution time of fraction 7 (25 minutes, determined on an aliquot of the Dowex 1 fraction) dictates that the dipeptide must be a cysteic acid dipeptide. Since an insufficient number of cysteic acid dipeptides have been evaluated by column and paper chromatography at this time, it is not possible to identify this dipeptide on the basis of these values. This dipeptide was therefore hydrolyzed to identify the component residues, cysteic acid and glycine. NH_2-terminal analysis by the dansyl technique[8] identified the fraction 7 dipeptide as $CySO_3H$-Gly.

Fraction 8 had an elution time of 126 minutes. This allowed 13 dipeptides as possibilities but paper chromatography in solvent 1 ($R_{Ala}1 = 0.23$) reduced the possibilities to Asn-Gln and perhaps Gln-Asn, since

[15] W. R. Gray and J. F. Smith, *Anal. Biochem.* **33**, 36 (1970).

the glutamine series had not yet been characterized by paper chromatography. NH_2-terminal analysis and amino acid analysis confirmed the choice of Asn-Gln.

Fraction 9, with an elution time of 127 minutes, had 11 possibilities. After paper chromatography in solvent 1, 4 possibilities remained, and after solvent 2, Glu-Ala, Ala-Glu, and Ala-Thr remained. Since Ala-Glu and Glu-Ala are one of the unresolvable dipeptide pairs, further chromatographic analysis was abandoned, and the dipeptide was identified by NH_2-terminal analysis and amino acid analysis to be Glu-Ala.

Fraction 10, with an elution time of 218 minutes, allowed 5 possible dipeptides, Val-Met, Met-Val, Leu-Val, Tyr-Glu, and Phe-Ser. Paper chromatography in solvent 1 established this peptide to be Leu-Val.

Fractions 11, 12, and 13 had, as a result of their column chromatographic elution times, 3, 7, and 3 possible dipeptides. They were distinguished by paper chromatography to be, respectively, Gly-Phe, Leu-Tyr, and a trace of Phe-Val. The latter resulted from the inclusion of a portion of fraction 1 in the unresolved fraction "A" from the 5-cm column.

Ordering of Dipeptide Fragments

Principle. For small peptides, time-course analysis alone will often establish the order of appearance of the product dipeptides. For longer peptides, where the order of dipeptide release becomes indistinguishable, it is necessary to generate the overlapping, alternate dipeptides from the modified peptide. The identity of these alternate dipeptides (without knowledge of their order of appearance) will establish the amino acid sequence of the unmodified peptide provided no duplicate amino acids are present. In the presence of duplicate amino acids, reliance must be placed on the order of appearance of the dipeptides in the time-course chromatogram. This is illustrated in the reordering of the B chain of oxidized insulin.

Ordering the Dipeptides Produced from the B Chain of Oxidized Insulin. The original set of dipeptides from the unmodified B chain of oxidized insulin consisted of 2 equivalents each of Cys-Gly and Leu-Val, 1 equivalent of Ser-His, Asn-Gln, Glu-Arg, Glu-Ala, His-Leu, Gly-Phe, Phe-Tyr, Leu-Tyr, and Phe-Val, and the COOH-terminal tetrapeptide Thr-Pro-Lys-Ala. The modified (des-Phe) B chain of oxidized insulin yielded the alternate dipeptides Gln-His, His-Leu, Tyr-Leu, Ala-Leu, Leu-Cys, Val-Cys, Gly-Glu, Gly-Ser, Val-Glu, Val-Asn, and the COOH-terminal nonapeptide. The latter was subjected to a single Edman degradation followed by NH_2-terminal analysis.[15] Glycine was thereby shown to be the penultimate residue. This information, in conjunction with knowledge of the original dipeptides demonstrated the sequence to be Arg-Gly-Phe-Phe-Tyr-Thr-Pro-Lys-Ala. The NH_2-terminal arginyl residue rendered this nonapeptide resistant to attack by DAP I. From this it is seen that tryptic peptides are more appropriate substrates. A knowledge of the first dipeptides generated in both the original and modified peptide establishes the initial sequence as:

residue position:	1 2 3 4 5 6 7 8
original dipeptides:	Phe-Val, Asn-Gln, His-Leu, Cys-Gly
alternate dipeptides:	Val-Asn, Gln-His, Leu-Cys

but leaves 6 possible configurations for the sequence of amino acids from position 9 to position 20. However, recognition of the order of appearance of the first 7 or 8 dipeptides in the time course analysis of the original peptide allowed the establishment of the correct sequence. The time course chromatogram of the alternate dipeptides offered little help in this instance establishing only the first two dipeptides. The strongly acidic dipeptide Leu-$CySO_3H$ with the cysteic acid in the penultimate position proved to be rate-limiting and rendered the time course unreadable beyond that point.

Sequencing with DAP I by Other Means

Sequencing without Dipeptide Isolation. For small tryptic peptides of 4–12 residues having known amino acid composition, the small number of possible dipeptide products makes sequencing feasible without the necessity of generating the alternate dipeptides or performing preparative column chromatography. A time-course chromatogram requiring only 3 nmoles per origin coupled with a determination of elution times of a complete digest on an expanded range analyzer (requiring 5 nmoles) is often sufficient to establish the sequence of a peptide with as little as 10–20 nmoles of that peptide.

a. Time-course analysis. The peptide is digested as described under "Digestion Conditions for Time-Course Analysis" in a tightly stoppered culture tube (6×5 mm) using half the units of enzyme to augment the interpretation of the order of appearance of the dipeptides. Samples (2–3 nmoles) are taken at 5-, 10-, 30-, 120-, and 240-minute intervals and applied to a thin layer of Avicel microcrystalline cellulose on 20×20-cm glass plates (No. 12200-7, Brinkman Instruments, Inc., Westbury, New York). Thin-layer plates are developed with 1-butanol–formic acid–water (70:15:15) and stained with polychromic ninhydrin spray (100 mg of ninhydrin, 50 ml of ethanol, 10 ml of acetic acid, and 2 ml of

2,4,6-collidine).[11] One nanomole of the 240-minute digest is retained for NH_2-terminal analysis by the method of Gray,[8] and 5 nmoles are retained for dipeptide analysis on an amino acid analyzer.

b. ANALYTICAL COLUMN CHROMATOGRAPHY. The 240-minute digest is applied to the 5-cm column of an amino acid analyzer equipped with high sensitivity cuvettes and an expanded range recorder (full scale $\cong$ 10 nmoles). The column is developed as described under "Separation of Dipeptide Products" with the exception that the first 25 minutes of the 5-cm column eluate is collected directly from the column (ninhydrin and buffer flow from another column are maintained through the reaction coil during this collection). The remainder of the 5-cm column is developed normally, and the 25-minute collection is diluted with one-half its volume of water and adjusted to pH 2.2 for resolution on the 60-cm column.

c. ORDERING OF PEPTIDE SEQUENCE. Steps a and b provide a knowledge of elution times, chromatographic R_f values, order of sequential appearance of the dipeptides, and N-terminal amino acid composition of the dipeptides. This information, when correlated, is usually sufficient to establish the sequence of the peptide. Some dipeptides have R_f values on microcrystalline-cellulose, thin-layer plates that differ from values found on paper. These differences appear to be correlated with differences in the R_f values of the free amino acids on thin-layer chromatography, but it is not known to what extent they are related. Caution must therefore be taken when interpreting thin-layer values.

In this laboratory, the technique has been used with certain tryptic peptides derived from prolactin as illustrated in the following two examples. On a time-course chromatogram of a peptide containing leucine, phenylalanine, and arginine in equal amounts, two spots appeared simultaneously, having $R_{Ala}1$ values of 2.21 and 0.32. Column chromatography demonstrated elution times of 70 minutes and 145 minutes on the 5-cm column and no peptides on the 60-cm column. NH_2-terminal analysis on the products demonstrated phenylalanine and arginine. The sequence of this peptide was thereby determined to be Phe-Leu-Arg.

The other prolactin peptide contained equal amounts of glutamic acid, alanine, isoleucine, phenylalanine, and lysine. Time-course chromatographic analysis of a DAP I digest showed the initial appearance of two components with $R_{Ala}1$ values of 2.30 and 0.58. This was followed by the disappearance of the 0.58 component and the simultaneous appearance of two additional components with $R_{Ala}1$ values of 1.80 and 0.24. The unresolved complete digest showed NH_2 termini of glutamic acid, isoleucine, and lysine. Column chromatographic analysis of the digest demonstrated elution times to be 80 minutes on the 5-cm column, and 205 minutes and 250 minutes on the 60-cm column. The 80-minute elution time coupled with the 0.24 $R_{Ala}1$ value identified this product as lysine. This residue was designated as the COOH terminus since it was a late product in the time course and since DAP I could only release a basic residue if it appears at the COOH terminus, as in a tryptic peptide. This leaves as possible dipeptides Glu-Ala (128 minutes and 0.892, elution time and $R_{Ala}1$ value, respectively), Glu-Phe (179 minutes and 1.68), Ile-Ala (205 minutes and 1.79) and Ile-Phe (490 minutes and 2.22), based on the known NH_2 termini and amino acid composition. The values of 205 minutes and 1.80 $R_{Ala}1$ establish Ile-Ala to be the second dipeptide released, but the values of 250 min and 2.30 $R_{Ala}1$ do not satisfy any of the above glutamic acid dipeptides. These values, however, do satisfy those known for Gln-Phe (250 minutes and 2.32). The sequence was therefore concluded to be Gln-Phe-Ile-Ala-Lys.

Sequencing Using Membrane Diffusion to Recover Products. Time-course digestions have been conducted in a small flow-through, diffusion cell that provides for the continuous removal of dipeptide products for subsequent identification. This approach has three advantages: the substrate is not consumed in the sampling procedure, the dipeptide products can be quantitated (by hydrolysis of the dialyzate) without the necessity of separating them from unreacted substrate by column chromatography, and the undigestible residue is retained by the membrane for further study. The methodology has been applied successfully to the sequencing of moderately sized peptides (>25 residues) that are only partially digested by DAP I, thereby leaving a sizable undigested fragment. Care should be taken when applying this procedure to peptides that are completely digested since it is not known at what residual size the substrate itself will diffuse through the membrane at a rate sufficient to obscure product identification.

a. Apparatus. The device utilizes an Amicon Diaflo membrane in a flow-through cell with a 1 ml internal volume. It has provision for shear-type mixing and for continuous pumping of the buffer-activator through the cell. The retention membrane used is an Amicon UM-2 Diaflo membrane chosen over the UM-05 for its faster flow rate.

b. The buffer-enzyme-substrate. The buffer requires little buffering capacity since it is continually flushed through the reaction cell. Further, since the reaction is performed in a closed system, the enzyme need only be activated with a sulfhydryl reagent just prior to initiating the reaction. The closed system obviates the need for having a sulfhydryl reagent in the buffer supply used for the flow-through cell. The buffer, which is maintained under nitrogen, is composed of 928 mg of NaCl and 0.2 ml of glacial acetic acid diluted to 3000 ml with distilled water and adjusted to pH 4.2 with approximately 0.32 ml of 2 *M* NaOH. It is not necessary to inhibit dipeptidase activity with EDTA since the dipeptide products are continuously removed. The buffer is pumped through the cell at 0.65 ml per minute (producing approximately 50 psi upon the membrane) and the substrate (100 nmoles or less dissolved in the buffer) is injected through the pump into the cell. Two units of enzyme (per 100 nmoles of substrate) are diluted to 1 ml with 3-fold concentrated buffer, activated with 0.04 ml of 0.1 *M* dithiothreitol, and added to the reaction cell through the pump. Four-minute fractions (about 2.5 ml) are collected.

c. Dipeptide identification. Half of each fraction is taken to dryness over NaOH pellets and extracted with 2 portions (0.01 ml) of 95% isopropanol–5% 12 *M* HCl (to exclude the NaCl) for spotting on thin layers of microcrystalline cellulose. The remaining half of selected fractions (5 to 15 are sufficient) are hydrolyzed and their amino acids are quantitated. The remainder of other selected fractions are dansylated[7,13] to determine their NH_2-terminal residue or pooled and applied to the 5- and 60-cm columns to determine the elution times of the product dipeptides.

d. Ordering of peptide sequence. Steps a, b, and c provide elution times, chromatographic R_f values, quantitative sequential information on the dipeptides, the amino acid composition of the dipeptides, and the NH_2-terminal residues of the dipeptides. This information is usually sufficient to establish the amino acid sequence for a sizable substrate. For example, when 100 nmoles of glucagon was used as a substrate in the reaction cell, the information gained made it possible to obtain the sequence of the first 16 amino acids which is the extent to which glucagon (a hormone with 29 residues) is degraded by DAP I.[3]

[24] Degradation of Peptides by the Edman Method with Direct Identification of the Phenylthiohydantoin-Amino Acid

By Walter A. Schroeder

The Edman degradation[1] for the sequential removal of amino acid residues is carried out in consecutive reactions, thus:

[1] P. Edman, *Acta Chem. Scand.* **10**, 761 (1956).

$$C_6H_5\text{—NCS} + H_2N\text{—CHR—CO—NH—CHR'—CO—NH}\cdots$$

$$\downarrow (OH^-)$$

$$C_6H_5\text{—NH—CS—NH—CHR—CO—NH—CHR'—CO—NH}\cdots$$

Phenylthiocarbamyl (PTC) derivative

$$\downarrow (H^+)$$

$$H_2N\text{—CHR'—CO—NH}\cdots + \text{5-Thiazolinone}[2] \longrightarrow \text{Phenylthiohydantoin (PTH-amino acid)}$$

5-Thiazolinone[2]

Phenylthiohydantoin PTH-amino acid

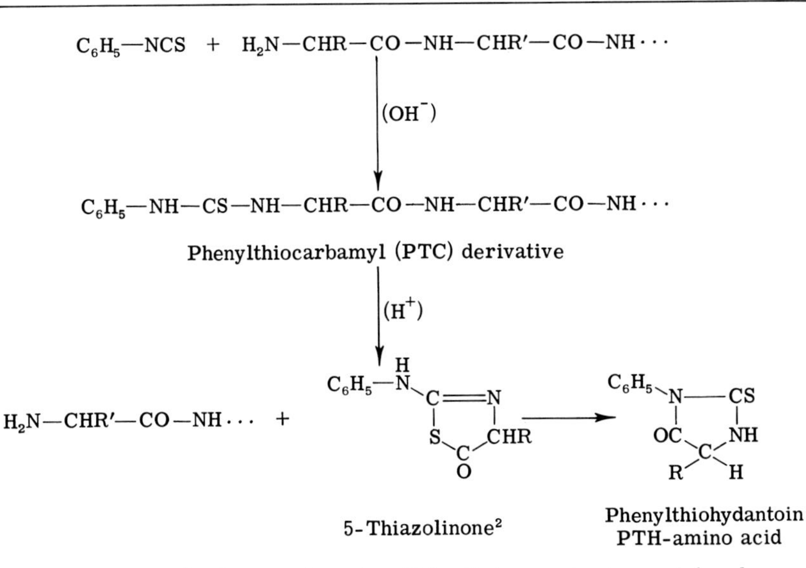

The original method proved to be difficult to apply to proteins because of the insolubility of the phenylthiocarbamyl (PTC) protein. In the Fraenkel-Conrat modification,[3] the protein is absorbed on paper strips, and all reactions are carried out on this support. When this modification was applied to peptides, the PTC-peptide was removed from the paper strip by certain extracting solvents. The procedure described below is an elaboration of the Fraenkel-Conrat modification and permits both peptides and proteins to be degraded sequentially.

Although basically qualitative, roughly quantitative data may be obtained. Indeed, with experience, it is readily apparent to the worker whether or not the results are normal and whether the peptide is degrading essentially quantitatively. By this method the amino acid residue that is removed at any step is identified directly rather than by the difference in amino acid composition between the original and the degraded peptide. Because of simplicity of operation, 8–12 peptides may easily be degraded simultaneously.

Methods

Materials

Whatman No. 1 filter paper in 1×7-cm strips is needed for the support of the peptide. Through each strip a small hole should be made with

[2] This intermediate usually is not isolated.

[3] H. Fraenkel-Conrat, *J. Amer. Chem. Soc.* **76**, 3606 (1954); and H. Fraenkel-Conrat, J. I. Harris, and A. L. Levy, *Methods Biochem. Anal.* **2**, 393 (1955).

a paper punch so that the strip may readily be suspended in reactive atmospheres. It is convenient to identify the strips by clipping off one or more of the corners. Thus, strips for one compound might be clipped at a corner nearer the punched hole, for a second at a corner further from the punched hole, for a third at two corners, etc. Whatman No. 1 filter paper is also used for the chromatography of the phenylthiohydantoin (PTH) amino acids; an 11 × 14-inch sheet is convenient for ascending chromatography and can carry 12 samples and standards.

Phenyl isothiocyanate, Eastman No. 1484, is used.

Dioxane must be distilled to remove peroxides. Distillation does not destroy the peroxides in the still, and care must be taken not to distill more than two thirds of the fluid in the still. This has been done for years without mishap. It is preferable to distill only enough dioxane to provide a 2-week supply, which is stored in 20-ml portions in the refrigerator.

The following chemicals are needed: benzene, xylene, pyridine, acetone, 90% formic acid, glacial acetic acid, propionic acid, concentrated hydrochloric acid, *n*-butanol, *n*-butyl acetate, formamide, soluble starch, potassium iodide, iodine (all reagent quality), heptane (technical), and sodium azide (practical).

PTH-Amino acids for reference standards may be prepared according to directions in the literature[4–6] or purchased from a chemical firm.[7]

Apparatus

A supply of 13 × 100-mm test tubes and 8-oz screw-cap jars, a desiccator (Corning No. 3120, 200 mm), and two ovens are needed.

For paper chromatography, it is necessary to have some type of equipment for applying the samples and for spraying with reagents. Cylindrical jars (6 × 14 inches) with glass covers are convenient for ascending chromatography with chromatographic papers of the size mentioned above. Chromatographic paper is easily marked for sample application with the device shown in Fig. 1. The side AD is placed along the shorter edge of an 18¼ × 22½-inch sheet of Whatman No. 1 filter paper. Penciled lines are made along AB, CD, EF, and GH; this marking is repeated on the other half of the paper. The paper is then cut along GH. After the paper has been starched and dried as described below, the edges along AB and DC with their extensions to GH are cut off. The notches along EF provide a marking for the placing of samples.

[4] P. Edman, *Acta Chem. Scand.* **4**, 277, 283 (1950).
[5] P. Edman and K. Lauber, *Acta Chem. Scand.* **10**, 466 (1956).
[6] A. A. Levy and D. Chung, *Biochim. Biophys. Acta* **17**, 454 (1955).
[7] For example, Nutritional Biochemicals Corporation, Cleveland, Ohio.

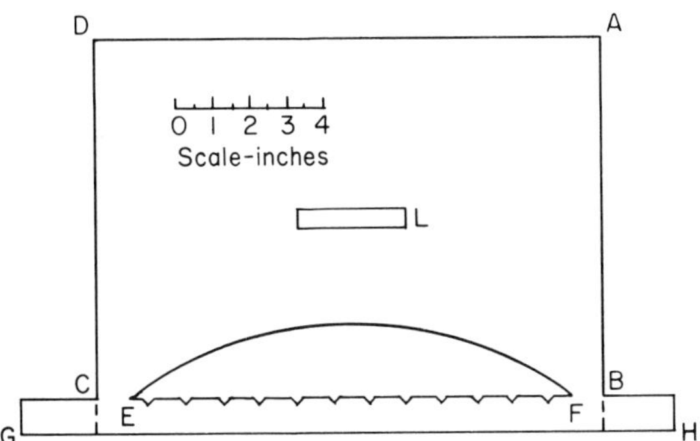

FIG. 1. Device for making chromatographic papers. The material is ¼-inch Lucite. L is a handle.

Glass racks with hooks are used to hold the paper strips in various stages of the procedure. Two convenient types are shown in Fig. 2.

The procedures described below require an atmosphere free of atmospheric pollution (smog) of the type that exists in the Los Angeles area. In order to provide satisfactory conditions, it may therefore be necessary to use a smog-free chamber of the type in Fig. 2 where the humidity also is controlled. Compressed air is passed through two filters (Commercial Filters Corporation, Melrose, Massachusetts, Model 25a, and Koby Corporation, Boston, Massachusetts, Model S) before dehumidification with a Lectrodryer (Model B-6-A, Pittsburgh Lectrodryer Division, McGraw-Edison Company, Pittsburgh, Pennsylvania). The apparatus is more than adequate for maintaining a relative humidity of about 25% in a volume of approximately 25 ft^3.

Procedure

Application of the Peptide to the Paper Strips. It is convenient to use 0.3–1.0 μmole of peptide, which is applied in aqueous solution over a sufficient number of strips so that each carries about 0.2 μmole or less. (In order to prevent a total loss of sample by accident, it is advisable to use four strips regardless of the amount.) The strips are hung on a glass rack and allowed to dry for half an hour in the smog-free box (which is convenient but not essential at this stage).

Reaction with Phenyl Isothiocyanate. Each strip is thoroughly saturated with a 20% solution of phenyl isothiocyanate in dioxane. The reagent is applied with a small dropper while the strip is held with tweezers. Although immersion of each strip in solution will accomplish this saturation, it may also wash off some of the peptide. Reagent is

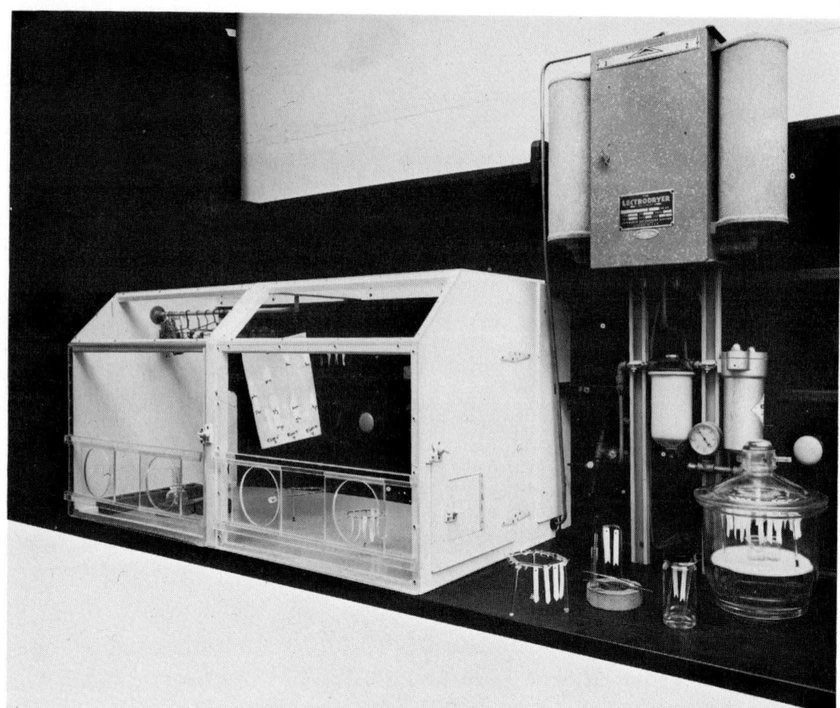

FIG. 2. Smog-free and humidity-controlled chamber with accessory apparatus. Glass racks and other items for the procedure are also shown.

added to the strip slowly, and only 1 or 2 drops are allowed to drip off. About 0.2 ml of solution per strip is required.

Two strips are suspended on a U-shaped rack (Fig. 2) and placed in one of the 8-oz screw-cap jars to which has been added 15 ml of a mixture of equal volumes of pyridine, dioxane, and water.[8] After the jar is covered with aluminum foil, the cover is tightly closed. The jars and contents are heated at 40° for 3 hours in an oven. For an efficient schedule, the heating should begin at 11 AM.

Extraction of Excess Reagents. In the original modification by Fraenkel-Conrat,[3] the paper strips were extracted first with benzene and then with alcohol–ether. Many PTC-peptides are soluble in alcohol–ether, and hence are extracted from the strips. However, extraction with

[8] This volume may seem excessive for two strips. There is some evidence, however, that the reaction is incomplete if more strips and more solvent are placed in a large container and heated. Perhaps this occurs because the larger container is brought to temperature more slowly in the oven.

benzene alone is adequate and does not remove the PTC-peptide. It is nevertheless fairly critical to observe precautions during the extraction.

After removal from the reaction atmosphere, the strips are somewhat translucent. They must be dried only to such a degree that they lose translucency and must then be dropped into 13 × 100-mm test tubes that contain sufficient benzene to cover them. If drying is omitted completely, some PTC-peptide will be lost. If the strips are dried too thoroughly, benzene does not completely extract diphenylthiourea which is a product of a side reaction. Unless diphenylthiourea is removed at this step, it will cause confusion in the chromatographic identification because it cannot be distinguished from PTH-methionine.

After extraction for 1.5 hours the benzene is poured off, discarded, and replaced with fresh benzene for an equal period of time. Finally, a third extraction is made overnight.[9]

Degradation of the PTC-Peptide. After extraction with benzene, the strips are aerated for approximately 1 hour in a smog-free atmosphere and then transferred to a desiccator into which have been placed 15 ml of glacial acetic acid and 15 ml of 6 N hydrochloric acid in individual beakers (Fig. 2). After the pressure in the desiccator has been reduced to 100 Torr, the degradation[10] is carried out for 7 hours at room temperature. This time of degradation is generally useful but may have to be altered in special cases (see further discussion below). This degradation is most efficiently scheduled to be completed at 4 PM.

Extraction of the PTH-Amino Acid. After removal from the degradation atmosphere, the strips are aerated in the smog-free chamber overnight. A smog-free atmosphere is essential at this point. As a substitute, the strips may be placed in a desiccator over Drierite. Two-hour-long extractions with acetone are used to remove the PTH-amino acid. It is important that the strips be very dry during the extraction. In a very humid atmosphere, they may absorb so much water that the peptide is partially extracted. The extractions are conveniently made in 13 × 100-mm test tubes.[11] The acetone extract is then evaporated under reduced pressure at 40° to leave the residue of PTH-amino acid. If the identification is not carried out immediately, the PTH-amino acid should be stored in a smog-free chamber or in a freezer.

[9] The timing of these extractions is not critical, but it is convenient for routine work. Furthermore, the tubes need not be shaken either continuously or periodically during the extraction.

[10] For a discussion of the mechanism of this degradation and for other references, see D. Ilse and P. Edman, *Aust. J. Chem.* **16,** 411 (1963).

[11] After the extractions, the next degradation may be started as soon as the acetone has evaporated from the strips.

Identification of the PTH-Amino Acid. The chromatographic procedures are essentially those of Sjöquist and Edman.[12]

STARCHING OF CHROMATOGRAPHIC PAPER. Starched paper has been used in the chromatographic procedures as part of the reagent for detecting the PTH-amino acids.[13] A 0.5% solution of soluble starch is made by pouring a thin slurry of 5 g of soluble starch in 100 ml of cold water into 900 ml of boiling water and allowing the solution to boil until the starch is dissolved. After cooling, any water lost by evaporation is replaced. This starch solution is then poured into a tray, and the 11 × 18-inch sheets of marked Whatman No. 1 paper are passed through the solution and allowed to dry in air. The liter of solution will starch about 40 sheets. The starched sheets need not be specially protected, nor need care be taken to avoid touching them because the reagents do not detect fingerprints as does ninhydrin. However, because they are wrinkled after starching and drying, the sheets should be pressed flat under weights when dry.

SAMPLES AND STANDARDS. Adequate standards of known PTH-amino acids are used side by side with the samples from the degradation. Solutions of standards at a concentration of 2 mg/ml in acetone are stable for at least 2 months at room temperature; an exception is PTH-histidine, which decomposes in about a week. Deterioration becomes evident through the appearance of extraneous spots on the chromatogram. It is most convenient to prepare solutions of the individual PTH-amino acids rather than a complete mixture. Because only 5 μl of each standard is needed, a mixture appropriate to the particular peptide under investigation may be made up by application of several standards consecutively in the same place. In some instances, for example with solvent D (see below), a standard mixture of the PTH derivatives of proline, leucine, valine, alanine, and glycine is very useful. This subject will be discussed more thoroughly below.

After the solvent has been evaporated from the acetone extract, the residue is dissolved in 0.5 ml of acetone, and a portion equivalent to 0.1 μmole of peptide is applied to each paper. The application of the sample may precede or follow certain treatments of the paper as described below.

DEVELOPERS. Five developers (of which A, C, D, and E were devised by Sjöquist and Edman[12]), are used for the paper chromatographic identification of the PTH-amino acids. The preparation and use of these

[12] J. Sjöquist, *Acta Chem. Scand.* **7**, 447 (1953); and P. Edman and J. Sjöquist, *Acta Chem. Scand.* **10**, 1507 (1956).

[13] PTH-amino acids may also be detected with ultraviolet light and a fluorescent screen [see footnote 12 as well as J. Sjöquist, *Biochim. Biophys. Acta* **41**, 20 (1960)].

developers will be given, and their application to specific identifications will be described.

Solvent A is a 7:3 (v/v) mixture of heptane and pyridine, which may be prepared in 1-liter quantities and used over a period of time. When cylindrical jars of the type described above are used for ascending chromatography, the samples are applied to the paper and a cylinder of paper is formed and stapled together. Approximately 50 ml of solvent A is poured into the jar, and the paper cylinder is placed in the liquid. Equilibration is unnecessary. With this developer, the solvent reaches the top of an 11-inch paper in approximately 3 hours.

Atmospheric humidity has much influence on the quality of the separations with solvent A. When the humidity is very low, almost all of the PTH-amino acids will move with the front of the developer. On the other hand, if the relative humidity is above 60%, solvent A forms multiple fronts with attendant poor separations. Both these difficulties may be overcome if the paper is equilibrated at 45–50% relative humidity[14] for 3 hours after the samples have been applied. Other methods of countering these difficulties may also be used. Thus, if the humidity is low, the starched paper may be dipped in a 10% solution of water in acetone. During the evaporation of the acetone, the samples are applied and then the chromatogram is started. On the other hand, if the humidity is high, the paper may be dried under a heat lamp after the sample has been applied. Obviously, both these methods require some experience in order to be able to judge when the paper is in the proper state for chromatography and is neither too limp nor too crisp. These methods are quicker than equilibration at 45–50% relative humidity and are equally satisfactory if time is taken to gain sufficient experience to judge the condition of the paper.

Solvent C is composed of heptane, *n*-butanol, and 90% formic acid in the ratio 2:2:1 (v/v/v). It is of less general usefulness, is less frequently used than the others, and is made up in 50-ml quantities as needed. The solvent may separate into phases if the temperature varies a few degrees. Phase separation (with subsequent erratic results if the solvent is used for chromatography) can be prevented by placing the lower part of the chromatographic jar in a water bath at 30° during the development (3.5–4 hours).

Solvent D is xylene, but prior to the application of the samples the sheet is dipped in a 20% solution of formamide in acetone. As the acetone evaporates, the samples are applied and the chromatogram is begun.

[14] A closed chamber that contains a saturated solution of potassium carbonate is a ready means for providing an atmosphere at 45–50% relative humidity.

Here again some experience is necessary to know how long the paper should be dried before the development is started. Development time is about 2 hours.

Solvent E is prepared by shaking 970 ml *n*-butyl acetate thoroughly with water in a separatory funnel and allowing the two phases to equilibrate overnight. Propionic acid is next added to the water-saturated butyl acetate to a concentration of 3%. Finally, this mixture is saturated with formamide (40 ml will provide an excess for the quantities given above). The starched paper is also dipped in 20% formamide in acetone prior to application of the samples. Development requires about 3.5 hours.

A detectable separation of PTH-leucine and PTH-isoleucine can be achieved with benzene–heptane in the ratio 3:2 (v/v) on paper that has been dipped in 20% formamide in acetone. Because this separation is marginal at best, several standards for each PTH-amino acid should be run on either side of the sample so that thorough comparison can be made. Under these conditions the R_f values of PTH-leucine and PTH-isoleucine are 0.71 and 0.66, respectively. It is best to use descending chromatography on a sheet at least 40 cm long. The solvent requires about 2 hours to reach the lower edge of the sheet, and development should be continued for another hour.

Detection of the PTH-amino acids on the chromatograms. After the chromatographic procedures have been completed, the sheets are aerated in a hood for at least 5–10 minutes. The chromatograms from those procedures that used formamide or formic acid are then heated in an oven at 100° for 30 minutes. Subsequently the sheets are sprayed heavily[15] (although not to the extent that liquid begins to run down the paper) with the iodine-azide reagent of Sjöquist[12] (about 15 ml per sheet). This reagent is a freshly prepared 1:1 mixture of two aqueous solutions: one contains 82.5 g of potassium iodide and 2.5 g of iodine per liter, the other 32.5 g of sodium azide per liter (the individual solutions are stable). It is imperative that this reagent be used in a well-ventilated area because inhalation of the vapors causes nasal bleeding. The PTH-amino acids appear as white spots on a dark purple background that rapidly changes to brown. Although the color gradually fades, the spots are still readily visible after a week at room temperature. If interleaved with white paper[16] and kept in a freezer, the chromatograms are stable indefinitely.[17]

[15] The advantage of heavy spraying will be discussed below.

[16] Chromatograms in contact with each other transfer images.

[17] For preservation and comparison of data, it is convenient to photograph the chromatograms by transmitted light and attach the photographs in the notebook with other pertinent data about the degradation.

Comments

Identification of the PTH-Amino Acids. In the above procedure, the PTH-amino acid from each degradation is identified solely by its chromatographic behavior, not by conversion to the amino acid or by difference from the original peptide. The chromatographic procedures must therefore be unequivocal. It is assumed that the investigator is in possession of certain facts about the peptide to be degraded: presumably he knows the type of peptide (tryptic, chymotryptic, or other), the amino acid composition, and the amount. Therefore he will know whether or not there are residues that require special care. Thus, histidine can produce special problems, and certain pairs of amino acids (for example, PTH-valine and PTH-phenylalanine) behave similarly on chromatography. However, if histidine and/or valine is absent, these problems are nonexistent. Likewise, if a tryptic peptide is under investigation, one expects lysine or arginine to be at the COOH-terminus.

One portion of the unknown PTH-amino acid is developed with solvent A for tentative identification of the PTH-amino acid, and on the basis of this information another solvent is chosen as the developer for a second portion. For example, in Fig. 3a good development of fifteen PTH-amino acids by solvent A is shown. If, in addition to these standards, an unknown sample had been chromatographed, it would be easy to limit the identity of the unknown to one of several amino acids despite the fact that many pairs move at similar or identical rates. Thus, a limited number of standards can be chosen for comparison with the unknown in the second development. Accordingly, solvent D would be chosen as the second developer for unknowns that move as rapidly as, or more rapidly than, PTH-glycine; and solvent E for those that move more slowly than PTH-glycine. Representative chromatograms for solvents D and E are depicted in Figs. 3b and 3c. Examination of these figures will show that pairs with identical movement in solvent A are frequently very well separated with solvent D or E (for example, PTH-proline, PTH-leucine, and PTH-valine or PTH-alanine and PTH-methionine). On the other hand, pairs distinctly different in movement in solvent A may be identical in solvent D or E (PTH-phenylalanine and PTH-valine or PTH-serine and PTH-aspartic acid).

In special instances, these three solvents are inadequate for identification. Thus, solvent C must be used to distinguish PTH-histidine from PTH-arginine and PTH-methionine sulfone from PTH-glutamic acid. Likewise, the special developer must be used for PTH-leucine and PTH-isoleucine.

In a few instances the iodine-azide spray will produce colors that are an added aid in identifying the PTH-amino acid. The standards are in

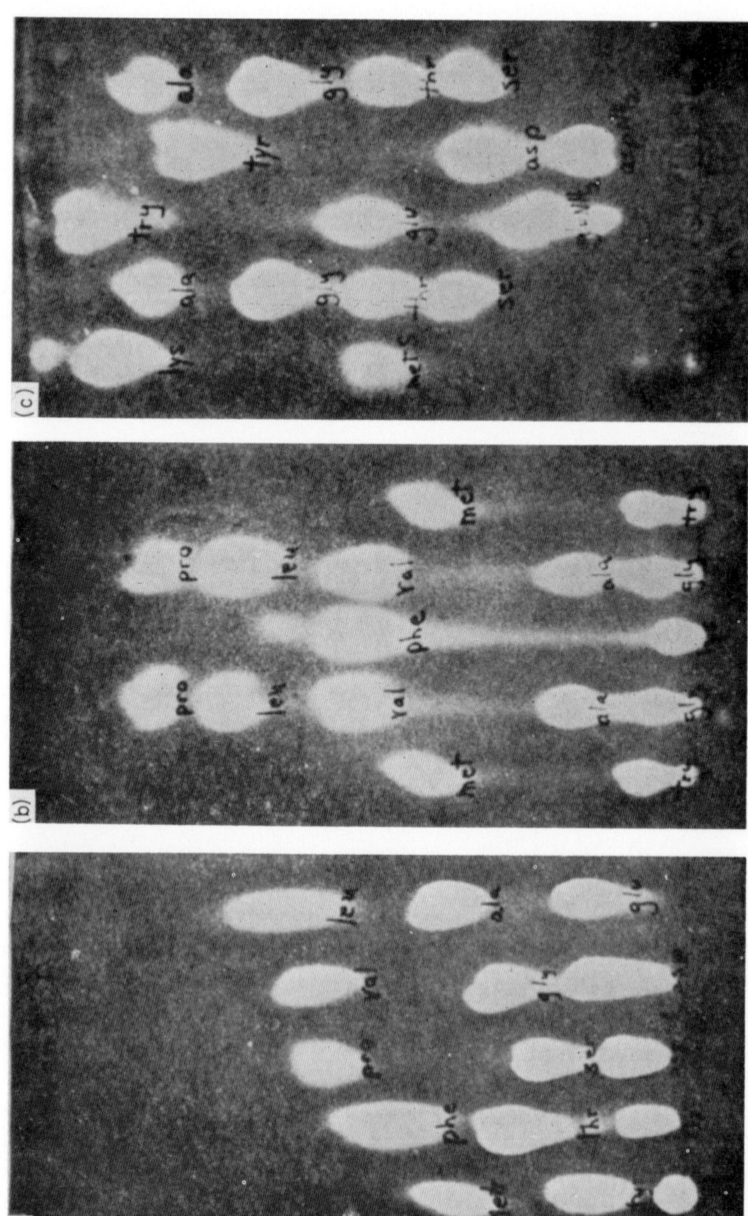

FIG. 3. Representative behavior of PTH-amino acids with different chromatographic solvents: (a) solvent A, (b) solvent D, and (c) solvent E.

sufficient amount that PTH-tyrosine, PTH-phenylalanine, and PTH-tryptophan give spots with yellow centers, and PTH-glycine, PTH-serine, and PTH-threonine give spots with pink centers. Unknowns may not produce colors unless the amount is large.

Heavy spraying of reagent is advised to provide a realistic estimate of relative quantity if more than one PTH-amino acid is present in an unknown. If a chromatogram is sprayed only lightly, even a small amount of PTH-amino acid will bleach the iodine-starch complex as completely as a large amount and thereby produce as prominent a spot: it might even be concluded that equal quantities are present. If heavy spraying is used, the PTH-amino acid in larger amount may again bleach the complex completely, but lesser bleaching by a smaller amount will produce a decided contrast.

Two-Dimensional Chromatography. If less than the recommended quantity of peptide is available for degradation, it may not be possible to divide the available PTH-amino acid so that two chromatograms can be made. However, a two-dimensional procedure, which in most instances gives reliable results, may be carried out as follows. An 11 × 14-inch paper is ruled as shown in Fig. 4. After the unknown sample has been placed at ⊙ and appropriate standards at ×, the whole sheet is developed with solvent D in the direction shown. Part E is then separated from the wet chromatogram by cutting along the dot-dash line, appropriate standards are applied at △, and after the xylene has evaporated part E is chromatographed with solvent E as shown. (The procedure must be carried out in this way: the whole sheet is dipped in formamide solution for solvent D, but part E cannot be dried and reimmersed in formamide before the use of solvent E.) Part D is cut off along the dotted line, and both parts D and E are dried and sprayed as usual.

By this method, PTH-alanine and those PTH-amino acids that move

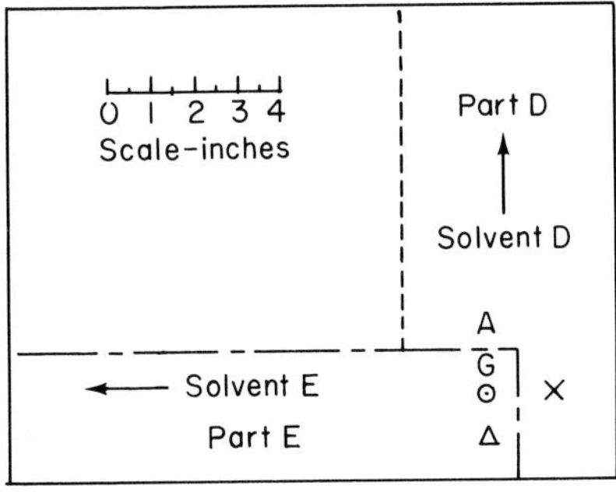

FIG. 4. Chromatographic paper marked for two-dimensional chromatography.

more rapidly than PTH-alanine will be identified in solvent D (with the exception of PTH-valine and PTH-phenylalanine). Thus, PTH-alanine will have moved approximately to position A, and PTH-glycine to position G with solvent D. Therefore PTH-glycine and more slowly moving PTH-amino acids will be identified with solvent E. Ambiguities in identification (PTH-threonine and PTH-glutamic acid, etc.) can frequently be eliminated on the basis of the amino acid composition of the peptide.

Examples of the application of this procedure are presented in Fig. 5.

Quantitation. Reasonably satisfactory quantitative estimations may be made by chromatographing serial dilutions of sample and standards side by side and matching the spots.

Special Cases. Peptides known to contain histidine or non-COOH-terminal arginine must be treated specially. These PTH-amino acids are not extracted by acetone after the degradation. Although they may be extracted with a 5% solution of water in acetone, further degradation is impossible because the remaining peptide is also extracted. This problem is best solved by clipping off a portion of each strip (so that the total is about 0.1 μmole) followed by extraction of these portions with this solvent. Sufficient material is thus obtained for identification, but the degradation may be continued on the remainder of the strips without problem. If, then, one or more of these amino acids is present (unless a tryptic peptide with presumably COOH-terminal arginine is being degraded), the schedule should be altered. Thus, the next degradation should not be started until the presence or absence of an acetone-soluble

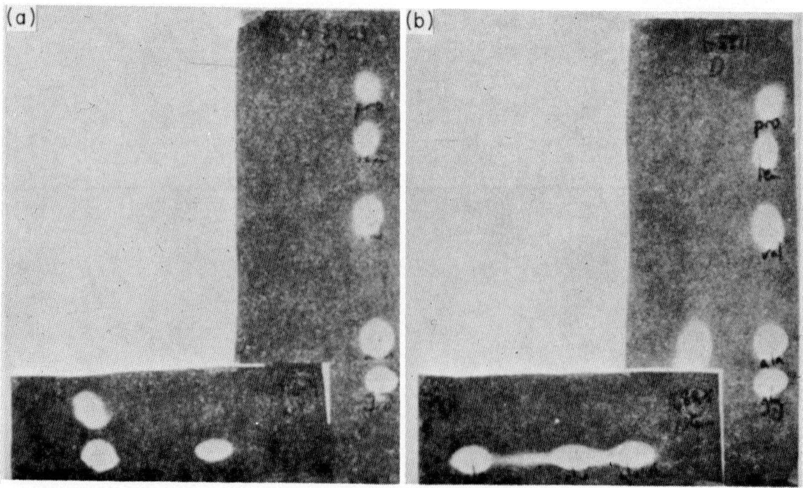

FIG. 5. Identification of (a) PTH-tyrosine, and (b) PTH-alanine, by the two-dimensional procedure.

PTH-amino acid has been established. If one is absent, a portion of the strips should be extracted as above and the identification should be made with solvent C. If the extract is divided into two parts before chromatography with solvent C, the completed chromatogram may be cut in two so that one part can be sprayed with the iodine–azide reagent and the other with a specific reagent for histidine or arginine.

Electrophoresis may also be used for identification of these PTH-amino acids. Starched Whatman 3 MM paper is dipped in pyridine-acetic acid buffer at pH 6.4 (190:10:0.4 water–pyridine–glacial acetic acid, v/v/v). Portions of strips from the procedure are then placed in good contact with the wet paper, and equilibration is carried out for 15 minutes. After electrophoresis at 15–20 V/cm for 2 hours and drying, the PTH-amino acids are detected with the iodine–azide spray.

Variations. The procedures described have been routinely used in several thousand degradations. On occasion, for special problems, variations in procedure have been made. Thus, the duration of heating with phenyl isothiocyanate has been increased; or double treatment with phenyl isothiocyanate has been carried out (that is, treatment, extraction of excess reagents, reapplication of reagent, reheating, and reextraction); or the time of extraction of reagents with benzene has been lengthened; or the time of cyclization has been changed. None of these changes has had major effect. Yields are not perceptibly better on double treatment with phenyl isothiocyanate. Increased extraction with benzene will not remove the diphenylthiourea of the strips are allowed to dry too long before being placed in benzene. Increased time of cyclization tends only to produce some scission at bonds especially sensitive to acid.

Applicability, Expectations, and Problems. The Edman degradation as described above may be applied without problem to most of the common amino acids that occur in proteins. In contrast to other published modifications of the procedure, no difficulties are encountered when the method is applied to threonine, serine, glutamine, and asparagine. These amino acids degrade normally and can be identified readily. Glutamine under certain conditions is an exception. Thus, if the peptide contains NH_2-terminal glutamine and this residue has been converted to pyrrolidonecarboxylic acid by acidic conditions prior to application of the PTH procedure, no reaction with phenyl isothiocyanate and no degradation of the peptide occur. Accordingly, if the peptide is found to contain glutamic acid by analysis and if degradation does not occur, it may be tentatively concluded that an NH_2-terminal glutamine residue has cyclized. The absence of a blocking group must, of course, be ascertained and the presence of NH_2-terminal pyrrolidonecarboxylic acid must be substantiated by other experiments. On the other hand, if glutamine is not originally NH_2-terminal in the peptide, its identification usually proceeds normally despite the fact that the removal of the preceding residue will bring it into NH_2-terminal position.

The behavior of lysine is somewhat erratic for reasons yet unknown. Usually PTH-ϵ-PTC-lysine will be present and detectable in the ether extract, but on occasion it is absent and a gap in sequence appears because no residue can be identified. If the degradation is continued, succeeding steps usually proceed without difficulty. The certainty that the "gap" indeed contains lysine must come from other data.

Histidyl residues are somewhat less consistent in behavior than other types of residue. Nevertheless, their placement in histidyl peptides that have been degraded during a 3-year period has been substantiated by other data. However, conflicting results then began to be obtained,[18] and the placement of histidyl residues became unreliable. From a detailed study with peptides of known sequence, the following conclusion was reached: when histidine is (or becomes) NH_2-terminal, the histidyl residue is removed during reaction with phenyl isothiocyanate and the succeeding residue then reacts also, so that, after cyclization, both residues are detected. Although variables such as humidity, temperature, pH, purity of chemicals have been carefully examined, the cause has not been identified. At times during these experiments, normal behavior was observed, but immediate duplication of the procedure has always ended in the undesired result. At the present time, the method cannot be relied on for histidyl peptides unless the user tests to show that histidyl peptides of known sequence can be satisfactorily degraded. The author has been told about similar problems with histidine when other modifications of the PTH method are used.[19]

Experience has led to the expectation that application of the procedure will elucidate the sequence of 6–8 residues, if the peptide is that long. This expectation may be exceeded, but the results on occasion may fall far short of the mark. There are, for example, several reasons to believe[18] that a certain tryptic peptide of catalase begins with the sequence Gly-Pro-Leu-Leu-. When the degradation is applied, PTH-glycine appears in excellent yield; repeated attempts even with the fragment (Gly,Pro,Leu_2) from the same peptide have not yielded a detectable PTH-derivative in the second step. The cause is unknown and is the more unexpected because many peptides with proline (and even with the sequence -Gly-Pro-) have been successfully degraded.

When residues of the same kind occur consecutively, the results must sometimes be regarded with skepticism, especially if several residues

[18] W. A. Schroeder, J. B. Shelton, and J. R. Shelton, unpublished data, 1964.

[19] R. F. Doolittle and W. R. Gray, personal communications.

have already been identified. Although the yield at each step is of the order of 80% or better, some evidence of the preceding residue is commonly seen at each step. Thus, when identical consecutive residues are present, an excellent yield must be obtained from both. In a peptide from catalase,[20] the sequence Thr-Thr-Gly-Gly-Gly- was regarded with some suspicion when it was first determined; other experiments, however, substantiated these data.

Scheduling. The method can be used most efficiently if six or more peptides are degraded simultaneously. The following schedule of operations has proved to be very satisfactory:

Day 1		
	10 AM– 11 AM	Apply peptides and dry strips
	11 AM– 2 PM	React with phenyl isothiocyanate
	2 PM–3:30 PM	First extraction with benzene
	3:30 PM– 5 PM	Second extraction with benzene
	5 PM	Third extraction with benzene
Day 2	– 8 AM	
	8 AM– 9 AM	Aeration to remove benzene
	9 AM– 4 PM	Cyclization
	4 PM	Aeration to remove acetic acid and hydrochloric acid
Day 3	– 8 AM	
	8 AM– 9 AM	First extraction with acetone
	9 AM– 10 AM	Second extraction with acetone
	10 AM– 11 AM	Evaporation of extracts and preparation of chromatograms
	11 AM– 2–3 PM	Chromatography
	3 PM– 4 PM	Drying, spraying, and recording

In practice, it is advantageous to divide the peptides under investigation into two equal groups. The first group would follow the above schedule on day 1 whereas the second group would begin the day 1 schedule on day 2. By day 3, the first group will be ready for the day 1 schedule of the second degradation, etc. Histidyl peptides should be placed on a 3-day schedule, as mentioned earlier.

Acknowledgments

The investigations that resulted in the development of the above procedure were supported by a grant (HE-02558) from the National Institutes of Health, U.S. Public Health Service. It is a pleasure to acknowledge the care, diligence, and thoughtfulness that J. Roger Shelton and Joan Balog Shelton have put into the development and continued use of the method.

Contribution No. 4140 from the Division of Chemistry and Chemical Engineering, California Institute of Technology.

[20] W. A. Schroeder, J. R. Shelton, J. B. Shelton, and B. M. Olson, *Biochim. Biophys. Acta* **89**, 47 (1964).

[25] Mass Spectroscopy of Methylthiohydantoin Amino Acids: Identification, Quantitation, and the Analysis of Mixtures

By FRANK F. RICHARDS and ROBERT E. LOVINS

The Edman reaction is a precise qualitative[1] and quantitative tool for the determination of successive N-terminal amino acid residues. Identification of the cyclized thiohydantoin derivatives may be accomplished by chromatography,[2] indirect methods,[3,4] or gas chromatography.[5,6] The advent of mechanized methods of protein sequential degradation[7] and the analysis of peptide mixtures without prior separation[8,9] has made accurate quantitative analysis of amino acid thiohydantoin derivatives necessary. This article will deal chiefly with mass spectroscopic identification and quantitation of methylthiohydantoin (MTH) amino acids and the interpretation of the mass spectra of MTH-amino acid mixtures.

The MTH derivatives rather than the phenylthiohydantoin (PTH) derivatives are used because they are rather more volatile than the PTH derivatives. This volatility enables excess reagent to be removed *in vacuo*. The wide range in volatilizing temperatures of these derivatives and the fact that most of the MTH derivatives give good molecular ions (M^+) and simple mass spectra make them particularly suitable for mass spectrometry.

In outline, the quantitative analysis of mixtures of MTH derivatives (derived from simultaneous analysis of peptide or protein mixtures) involves the following steps. Methyl isothiocyanate is used to degrade sequentially polypeptides (or mixtures of polypeptides) from the N-terminal end. Standard mixtures of ^{15}N-enriched amino acids or their MTH derivatives are added to the reaction. Increase in the $^{14}N:^{15}N$ isotope ratio is observed when MTH-amino acids derived from the N

[1] H. Fraenkel-Conrat, J. I. Harris, and A. L. Levy, *Methods Biochem. Anal.* **2**, 360 (1955).

[2] W. A. Schroeder, Vol. 11, p. 445.

[3] W. Konigsberg, Vol. 11, p. 461.

[4] W. R. Gray, Vol. 11, p. 469.

[5] J. J. Pisano and T. J. Bronzert, *J. Biol. Chem.* **244**, 5597 (1969).

[6] M. D. Waterfield and E. Haber, *Biochemistry* **9**, 832 (1970).

[7] P. Edman and G. Begg, *Eur. J. Biochem.* **1**, 80 (1967).

[8] F. F. Richards, W. T. Barnes, R. E. Lovins, R. Salomone, and M. D. Waterfield, *Nature (London)* **221**, 1241 (1969).

[9] F. F. Richards, T. Fairwell, and R. E. Lovins, *Proc. 2nd Amer. Peptide Symp.* (in press) (1971).

termini of polypeptide chains dilute the standard mixture. The extracted MTH derivatives are transferred to the sample probe of a single-focusing mass spectrometer; the temperature of the probe is gradually raised, the MTH-amino acids are distilled sequentially into the ion beam and identified, and their $^{14}N:^{15}N$ isotope ratio is established. Spectra are taken at 10 eV as well as 80 eV ionizing voltage, because at the lower energy, M^{+} ions are retained, while the fragmentation ion pattern becomes less complex. One or several polypeptides simultaneously may be sequenced by this method.

Apparatus

A single-focusing mass spectrometer with a resolution of 1:500 or better and a solid sample probe is required. For the analysis of MTH-amino acid mixtures (described below) a variable ionization voltage attachment and a probe capable of being heated slowly are desirable. Our experience is chiefly with the CEC 21-490. (Resolving power 1:1100.) Double-focusing mass spectrometers of high resolving power are at present not desirable instruments since the mechanics of sample analysis are too time-consuming. The CEC 21-490 has a variable ionizing voltage attachment, variable temperature solid inlet probe and electrical detection and is useful in the procedures described below. It is manufactured by the Analytical Instrument Division of the duPont Corporation.

Conditions for Edman Degradation Compatible with Mass Spectrometry

The conditions for carrying out Edman degradations have been reviewed in detail.[2] This section will deal merely with the modifications of the procedure that make it suitable for mass spectroscopic analysis of MTH-amino acids. Use of methyl isothiocyanate (MITC) instead of phenyl isothiocyanate (PITC) per se requires no alteration in reaction conditions. It has been shown[9,10] that there is relatively little difference in the efficiency with which MITC and PITC react to form TH derivatives.

Care must be taken not to subject the reaction mixture to conditions under which the more volatile MTH derivatives (alanine, valine, glycine) are volatilized and lost. Precautions must also be taken to avoid volatile contaminants which may show in the mass spectrum. The common ones are stopcock grease, pump oil, and solvent vapors contaminating the pump oil. Teflon stopcocks and a liquid nitrogen rather than a Dry Ice

[10] T. Fairwell, W. T. Barnes, F. F. Richards, and R. E. Lovins, *Biochemistry* **9**, 2260 (1970).

trap between the pump and the reaction mixture are both essential. Contact of the reagents with rubber or plastic should be avoided. All reagents are thoroughly deaerated by flushing with argon and are stored under argon (prepurified nitrogen is satisfactory). The reaction is carried out in a closed vessel under a positive argon pressure of 1–2 cm Hg to prevent oxygen diffusion through the glass joints. All reactions are carried out in the dark.

Reagents

Dioxane, redistilled from lithium aluminum hydride ($LiAlH_4$) and stored in aliquots of 20 ml at —20°C under argon

Methyl isothiocyanate (Eastman), twice redistilled (mp 35–36°) and stored under argon at —20°

Tetrahydrofuran (Eastman), redistilled from $LiAlH_4$ suspension (bp 66°). About 5.0 g of $LiAlH_4$ per liter of tetrahydrofuran is required.

Trifluoroacetic acid (Pierce, highest purity), twice redistilled. With some batches of the Pierce reagent no redistillation has been needed.

Pyridine (Spectro grade), twice redistilled.

The reagents are stored, sealed in glass, under argon.

Typical Degradation Conditions. A sample of protein or peptide (0.02–5.0 μmoles) is dissolved (or, if insoluble, suspended) in 50% aqueous dioxane containing 10% (v/v) pyridine. A higher percentage of pyridine may be used. If the MTH residues are to be quantitated, the requisite ^{15}N-enriched amino acids (whose exact $^{14}N:^{15}N$ ratio has been predetermined) are introduced at this stage in quantities roughly equimolecular to those of the sample.

The apparent pH of the solution is adjusted (if necessary) to 8.5 with pyridine. A 10- to 20-fold molar excess of MITC (based on estimated amino side-chain and amino terminal groups as determined by amino acid analysis) is added. For subsequent steps a 5-fold molar excess of MITC is used. The reaction mixture is flushed with argon, or prepurified nitrogen, and stirred at 60° in the dark for 2 hours under argon. If the polypeptide is insoluble, coupling time is increased to 3 hours.

The reaction mixture is cooled to 20° in the apparatus, and excess MITC, water, and pyridine are removed at 1×10^{-3} Torr for 30 minutes at 60°. This step is critical; all the pyridine must be removed, as trifluoroacetic acid is added later and the pyridinium salt of this acid is relatively nonvolatile. In addition, care must be taken to prevent bumping of the fluid in the flask. The dry methylthiocarbamylated polypeptide under argon is treated in the dark with 1.0 ml of trifluoroacetic acid, and the temperature is raised to 60° for 1 hour. During this process the mixture is stirred. The trifluoroacetic acid is then removed at 20° at 1×10^{-3} Torr. A few drops of dry, peroxide-free ethyl ether are added to break the protein-trifluoroacetic acid gel, and then vacuum is applied for another 30 minutes. The dried peptide is extracted three times with 1.5 ml of tetrahydrofuran. MTH-aspartic acid, -asparagine, and -arginine are relatively insoluble in tetrahydrofuran, but will dissolve well in a 1:1 v/v mixture of trifluoroacetic acid and tetrahydrofuran. Since this mixture may dissolve some peptides, a small aliquot of the reaction mixture is used for analysis; the rest is put unextracted through the next cycle. The next time one of these residues is present, cumulative extraction occurs. This presents no difficulty since [^{15}N]aspartic acid, deutero-asparagine, and arginine are added on the first occurrence. In subsequent occurrences no [^{15}N]MTH derivatives are added. The changing ^{14}N:^{15}N ratios clarify any possible ambiguities.

The extract is concentrated to 40 μl, and 10 μl is transferred to the bottom of a Kimax 0.9–1.1-mm diameter capillary sealed at one end. This is dried under vacuum and transferred to the solid-sample probe of the mass spectrometer. The probe is then slowly heated. If the sample cannot be analyzed at once, it should be stored in the sealed capillary *in vacuo* at —70°.

Identification of MTH-Amino Acids in the Mass Spectrometer

MTH derivatives are identified by a combination of three properties (Table I); (1) mass of the molecular ion (M^+); (2) the relative abundance of the M^+ ion to one or two principal fragment ions present at both 80 eV and 10 eV; (3) the temperature range at which the MTH derivative is volatile. The MTH-amino acids may be divided into three groups.

Group 1. This group includes MTH-valine, -leucine, -isoleucine, -glycine, -threonine, -serine, and -alanine. Group 1 MTH derivatives volatilize at 10^{-7} Torr at temperatures between 70° and 100°. Valine (70–80°) has an abundant M^+ (m/e 172), which is its criterion. Leucine and isoleucine both appear at 80–90°; both have an M^+ ion at m/e 186. Leucine, however, has a fragment ion peak at m/e 143, which is 30% of the M^+ peak at 80 eV and at 10 eV. No such peak appears in the isoleucine mass spectrum. The only other MTH derivatives with fragment ions of m/e 143 are methionine and lysine. Both of these appear only at higher temperatures. The M^+ ion of glycine has an m/e 130 peak. Most other MTH derivatives give a fragment peak of this mass, and these obscure the M^+ ion of glycine. However, the m/e 102 fragment ion (90–

TABLE I
MASS SPECTROSCOPIC IDENTIFICATION OF AMINO ACID METHYLTHIOHYDANTOIN DERIVATIVES[a]

Amino acid methylthiohydantoin derivative	m/e	Relative abundance (%)		Temp. range (°C)	Amino acid methylthiohydantoin derivative	m/e	Relative abundance (%)		Temp. range (°C)
		80 eV	10 eV				80 eV	10 eV	
Valine	172*	65	100	70–80	Glutamic acid	202*	90	100	140–160
	130	100	50			184	65	60	
Leucine	186*	70	100	80–90		156	20	8	
	143	30	30			142	100	30	
	130	100	72			130	5	2	
Isoleucine	186*	47	44	80–90	Tyrosine	236*	20	40	150–170
	130	100	100			152		3	
Glycine	130*	100	100	90–100		130	30	20	
	102	32	46			107	100	100	
Threonine	174*	4	2	90–100	Aspartic acid	188*	100	100	160–180
	156	76	20			170	3	3	
	130	100	100			142	80	28	
Serine	160*	0?	0?	90–100	Asparagine	187*	100	100	170–190
	142	100	100			170	48	30	

Alanine	144*	100	100	90–110
Proline	170*	100	100	100–110
Methionine	204*	100	100	110–125
	156	16	16	
	143	43	78	
	130	48	91	
S-Methylcysteine	190	50	50	120–130
	142	100	100	
Phenylalanine	220*	60	30	125–140
	129	6	6	
	91	100	100	
Glutamine	201*	75	100	125–150
	184	100	88	
	156	20	8	
	142	80	44	
	130	17	2	

	142	80	60	
Tryptophan	259*	20	6	170–190
	130	100	100	
Lysine	274*	3	4	190–210
	243	80	100	
	212	10	60	
	201	35	34	
	184	40	37	
	143	100	25	
	130	30	6	
Arginine	229*	10	9	190–210
	212	70	13	
	170	94	100	
	142	100	17	
	130	88	71	
Histidine	210*	100	100	220–230
	130	25	18	

[a] Ratios and relative abundances at 80 and 10 eV are given for the M^+ ions (*) and for those fragmentation ion peaks useful for identification. The temperature ranges given are those for the maximal ion current for each compound.

100°) has an abundance of 32% of the M^+ ion at 80 eV, of 46% at 10 eV, and this peak is unique to glycine. The M^+ ion of threonine is very small, and the ion is absent in the case of serine. The *m/e* 156 ion is used to identify threonine. Methionine, glutamic acid, and glutamine also produce *m/e* 156 ions, although at higher temperatures. A mixture of threonine and methionine is easily resolved since the ratio *m/e* 156:*m/e* 204 will be too high for methionine alone. Serine identification is made more complex by the fact that MTH-serine undergoes dehydration to the anhydro derivative. This derivative may undergo polymerization at elevated temperatures. Hence, the M^+ of MTH-dehydroserine at *m/e* 142 is not always intense. If there is only a small quantity of serine present, it may be difficult to identify owing to the low abundance of the mass ion. In case of uncertainty the problem ion would be resolved by quantitation with ^{15}N-serine. It is much easier to measure the change in $^{14}N:^{15}N$ ratio of the *m/e* 142 and 143 peaks than to decide whether a low peak at *m/e* 142 is significant. Alanine volatilizes at 90–110°; it has an abundant M^+ ion at *m/e* 144 and is easily identified.

Group 2. This group comprises the proline, methionine, *S*-methylcysteine, phenylalanine, glutamine, glutamic acid, and tyrosine MTH derivatives. There is little difficulty with this group. The members emerge into the ion beam at temperatures between 100° and 150°; without exception, their spectra exhibit abundant molecular ions which identify these MTH derivatives without ambiguity. In the proline spectrum the only important peak is the M^+ peak at *m/e* 170. Other amino acid residues giving a peak at *m/e* 170 are asparagine, aspartic acid, and arginine. These, however, are volatile at much higher temperatures and do not interfere with identification. Methionine gives an M^+ peak at *m/e* 204, and *S*-methylcysteine gives an M^+ peak at *m/e* 190. Phenylalanine gives an abundant M^+ peak at *m/e* 220 and also an intense fragment ion peak at *m/e* 91, which is due to the benzyl group and therefore characteristic of phenylalanine. The peaks at *m/e* 91 and *m/e* 220 form excellent markers near the upper and lower end of the mass range of interest. On prolonged heating in the probe, a new ion at *m/e* 218 (phenylalanine M-2) begins to appear, presumably as a result of the loss of H_2 to form a cinnamyl ion. Glutamine and glutamic acid give essentially the same spectrum except that glutamine has an abundant M^+ ion at *m/e* 201 and the M^+ ion of glutamic acid is at *m/e* 202. On ^{15}N quantitation, the two molecules have a common M^+ at *m/e* 203. The deutero derivative of either glutamic acid or glutamine avoids the difficulty which occurs when one residue is quantitated in the presence of the other. Lysine has a fragment ion at *m/e* 201, but this comes off at a much higher temperature than glutamine. Therefore, at 125–150° the presence of an ion of *m/e* 201 indicates the presence of glutamine, and at a slightly higher temperature (140–160°), an ion at *m/e* 202 identifies glutamic acid. Tyrosine volatilizes at 150–170° giving an abundant M^+ ion at *m/e* 236 and an intense fragment ion at *m/e* 107, which is the methylphenol fragment characteristic of tyrosine. On prolonged heating tyrosine also gives an M-2 ion.

Group 3. This group includes the MTH derivatives of aspartic acid, asparagine, tryptophan, lysine, arginine, and histidine. These derivatives volatilize into the ion beam between 160–230° and are the least volatile derivatives. Aspartic acid has an M^+ ion at *m/e* 188; the M^+ ion for asparagine is at *m/e* 187. Apart from this, these residues have similar spectra and a common ^{15}N-M^+ at *m/e* 189. The deutero derivative of either serves to quantitate one in the presence of the other (as described above in connection with glutamic acid). The tryptophan M^+ ion at *m/e* 259 identifies the molecule, its chief fragment peak at *m/e* 130 is shared by many amino acids. The lysine spectrum is complex and contains a weak M^+ ion at *m/e* 274 and an abundant and unique fragment ion at *m/e* 243. This ion is used for identification and quantitation. Arginine volatilizes at 190–210°, producing a weak, but easily recognizable, unique M^+ ion at *m/e* 229 plus strong fragment ion peaks at *m/e* 212 and *m/e* 170. (These fragmentation peaks are shared by lysine, aspartic acid, and asparagine at temperatures close to those of arginine.) Histidine has the lowest volatility (220–230°), but is easily identified by an abundant M^+ ion at *m/e* 210 and by its simple spectrum.

Quantitation

The preferred technique is to run the sequential degradation technique in duplicate. The first run determines the nature of the amino acid residues at each locus; in the second run those [^{15}N]amino acids are added at each residue which correspond to the ones found at the locus in the first run. It is possible to add mixtures of twenty [^{15}N]MTH-amino acid derivatives at each step. This speeds the process but is expensive. The ^{14}N:^{15}N ratio of each [^{15}N]amino acid must be determined; little reliance can be placed on the quoted atoms excess ^{15}N given by the manufacturers. [^{15}N]Amino acids are added at the beginning of each sequence cycle. It is possible to control the accuracy of the quantitation by the amount of [^{15}N]amino acids added. The highest accuracies are obtained when [^{15}N]amino acids are added in quantities equimolecular in quantity to the [^{14}N]amino acids obtained from the peptide N terminus. Under these conditions it is possible to measure peak height ratios to an accuracy of better than 2% on repetitive scanning. Usually it is not necessary to work at such accuracy.

An alternative technique is to add [^{15}N]MTH-amino acid derivatives immediately before the acid cyclization stage. This was the original procedure adopted and has the theoretical advantage that it is independent of rate differences in the coupling and cycling reaction between peptide-linked and free amino acids. However, the synthesis[10,11] of [^{15}N] MTH-amino acid derivatives is expensive and time-consuming. It has also been shown that conversion of the amino acids via the thiazolinone derivative is quantitative.[9,10] There is, therefore, little to be said in favor of adding the isotopic label in the form of the MTH derivative.

Since the quantity of [^{15}N]amino acid added is accurately known, the ratio of the height of the ^{14}N:^{15}N is a direct function of the amount of N-terminal residue. Allowance must be made for naturally occurring ^{15}N and the natural abundances of other heavy isotopes. These standard corrections may be obtained from published heavy isotope abundance tables.[12]

Procedures and Problems in the Analysis of Mixtures

A major advantage of accurate quantitative techniques in the measurement of Edman reaction products is that a number of peptides or proteins may be sequenced simultaneously without prior separation of the constituents of the mixture. Information concerning insoluble peptides obtained from protein degradation, mixtures of closely related proteins which are difficult to separate (e.g., the immunoglobulins), synthetic peptides containing "failure sequences," enzymes with identical or nonidentical subunits[13,14] may be obtained by any method which gives accurate quantitation.

The major problem in the interpretation of the mass spectrum of MTH-amino acid mixtures is the complexity of the patterns observed when many MTH derivatives are present at the same time. A reduction in ionizing voltage from 80 eV to 10 eV retains the M^+ ion, but greatly reduces the complexity of the fragment ion pattern. When in addition the mass spectrometer is used to distill differentially the MTH derivatives into the ion beam, the most complex mixtures of MTH derivatives can be resolved. The recent advances in chemical ionization techniques[15,16] should also be useful.

[11] V. M. Stepanov and V. F. Krivtsov, *J. Gen. Chem. USSR* **35**, 53, 556, and 988 (1965).

[12] K. Biemann, "Mass Spectrometry," p. 374. McGraw-Hill, New York, 1962.

[13] J. M. Brewer and G. Weber, *Proc. Nat. Acad. Sci. U.S.* **59**, 216 (1968).

[14] J. M. Brewer, T. Fairwell, J. Travis, and R. E. Lovins, *Biochemistry* **9**, 1011 (1970).

[15] H. M. Fales, G. W. A. Milne, and M. L. Vestal, *J. Amer. Chem. Soc.* **91**, 3682 (1969).

[16] G. W. A. Milne, T. Axenrod, and H. M. Fales, *J. Amer. Chem. Soc.* **92**, 5170 (1970).

Interpretation. When quantitative analysis of the N-terminal MTH-amino acids from a mixture of polypeptides is carried out, the raw data consist of the molar quantities of a number of amino acids found at each locus of the peptide.

Table II, section A shows the raw data from an "unknown" mixture of polypeptides. The MTH-amino acid derivatives at each locus are listed in decreasing order of quantity. The maximum number of different amino acids found at any one locus is four (residues 3, 4, 5, 8, and 10). Therefore, there are probably at least four peptide chains present. In residues 3, 4, 5, 8, and 10 the molar quantity of each amino acid must indicate the molar quantity of each peptide chain. The most abundant amino acids at these residues decrease in value from 4.7 μmoles at residue 3, to 3.4 μmoles at residue 10, due to sequential handling losses. We can thus extrapolate values for the most abundant amino acids at residues 1, 2, 6, 7, 9 where there are fewer than four amino acids and, therefore, more than one peptide has the same amino acid. At each of these loci there is only one amino acid residue in the necessary abundance. An unambiguous major sequence may therefore be constructed, and this is given as sequence 1. It represents in molar concentration 47% of the total polypeptide present.

When this sequence is subtracted from the amino acids initially present, the remainder is set out in section B of Table II, and the same analysis is repeated. A second sequence is obtained (25% of the total number of moles of proteins present) which is unambiguous except for residue 9. Here either alanine or serine might fit. In this instance alanine was chosen, despite the fact that the value is high, because the serine values then fit more closely to the extrapolated values for the remaining two sequences. There is, however, real ambiguity at this locus. *S*-Methylcysteine residues at loci 5 and 6 and the histidine at locus 10 were not quantitated. Nevertheless, it is possible to deduce accurately the polypeptide chains to which these residues must be assigned.

When the second sequence is subtracted, it is clear that the remainder, (section C of Table II), corresponds to two further sequences present in approximately equimolecular amounts, each representing about 14% of the polypeptides present. Since the same quantities of each sequence are present it is not possible to assign each residue to either one chain or the other.

This artificial mixture (section D, Table II) was made to demonstrate the fact that polypeptides may be identified, quantitated, and sequenced, provided that they are not present in equimolecular quantities. This fact does not usually present a difficulty. For instance, where a proteolytic agent splits a protein very efficiently, so that equimolecular quantities of each fragment peptide are obtained, two methods may be used to

TABLE II

SIMULTANEOUS ANALYSIS OF A MIXTURE CONTAINING FOUR REDUCED AND CARBOXYMETHYLATED POLYPEPTIDE CHAINS

	Amino acid residue										
	1	2	3	4	5	6	7	8	9	10	Conclusions
Section A[a] (μmoles)	Lys, 7.0 Gly, 1.3 Phe, 1.2	Glu, 4.7 Val, 3.2 Ile, 1.0	Thr, 4.7 Phe, 2.2 Val, 1.0 Asn, 1.0	Ala, 4.7 Gly, 2.2 Gln, 0.9 Glu, 0.8	Ala, 4.6 Arg, 2.1 His, 1.0 Gln, 0.8	Ala, 4.3 Leu, 1.0 CM-Cys, N/M	Lys, 3.8 Glu, 1.6 CM-Cys, N/M	Phe, 4.6 Leu, 1.7 Ala, 0.8 Gly, 0.8	Glu, 3.3 Ala, 2.3 Ser, 1.5	Arg, 3.4 Ala, 1.5 Val, 0.8 His, N/M	4 Polypeptide chains present
Sequence 1 (μmoles)	Lys, 4.7	Glu, 4.7	Thr, 4.7	Ala, 4.7	Ala, 4.6	Ala, 4.3	Lys, 3.8	Phe, 4.6	Glu, 3.3	Arg, 3.4	Major sequence (47%)
Section B[b] (μmoles)	Lys, 2.3 Gly, 1.3 Phe, 1.2	Val, 3.2 Ile, 1.0	Phe, 2.2 Val, 1.0 Asn, 1.0	Gly, 2.2 Gln, 0.9 Glu, 0.8	Arg, 2.1 His, 1.0 Gln, 0.8	CM-Cys, N/M Ile, 1.0	Glu, 1.6 CM-Cys, N/M	Leu, 1.7 Ala, 0.8 Gly, 0.8	Ala, 2.3 Ser, 1.5	Ala, 1.5 Val, 0.8 His, N/M	
Sequence 2 (μmoles)	Lys, 2.3	Val, 2.2	Phe, 2.2	Gly, 2.2	Arg, 2.1	CM-Cys, N/M	Glu, 1.6	Leu, 1.7	Ala, 2.3	Ala, 1.5	Second sequence (25%)
Section C[c] (μmoles)	Gly, 1.3 Phe, 1.2	Val, 1.1 Ile, 1.0	Val, 1.0 Asn, 1.0	Gln, 0.9 Glu, 0.8	His, 1.0 Gln, 0.8	Ile, 1.0 CM-Cys, N/M	CM-Cys, N/M	Ala, 0.8 Gly, 0.8	Ser, 1.5	Val, 0.8 His, N/M	
Unresolved sequences 3 and 4 (μmoles)	Gly, 1.3 Phe, 1.2	Val, 1.1 Ile, 1.0	Val, 1.0 Asn, 1.0	Gln, 0.9 Glu, 0.8	His, 1.0 Gln, 0.8	Ile, 1.0 Cys, N/M	CM-Cys, N/M CM-Cys, N/M	Ala, 0.8 Gly, 0.8	Ser, 0.8 Ser, 0.8	Val, 0.8 His, N/M	Unresolved sequences 14% each sequence
Section D[d]											
Ribonuclease (5 μmoles)	Lys	Glu	Thr	Ala	Ala	Ala	Lys	Phe	Glu	Arg	
Lysozyme (2.5 μmoles)	Lys	Val	Phe	Gly	Arg	Cys	Glu	Leu	Ala	Ala	
Insulin A (1.4 μmoles)	Gly	Ile	Val	Glu	Gln	Cys	Cys	Ala	Ser	Val	
Insulin B (1.4 μmoles)	Phe	Val	Asn	Gln	His	Leu	Cys	Gly	Ser	His	

[a] The number of micromoles of recovered amino acid methylthiohydantoin derivative at each locus are given. The inferred major sequence (47% of total moles of protein present) is given as sequence 1. N/M = residue identified but not quantitated.

[b] Micromoles of amino acid methylthiohydantoin at each locus after subtraction of sequence 1. The second (25%) sequence inferred is given. For ambiguity at residue 9, see text.

[c] The residues at each locus of the two minor sequences (14%) present in equimolecular proportions.

[d] Composition of test mixture. The composition of each protein added was checked by amino acid analysis.

change the proportions of the peptides. Unfractionated protein may be added back. This serves to increase the amount of N-terminal peptide relative to the other peptides and allows its identification (enrichment technique). An incomplete gel filtration separation may also serve to change the proportions of peptides. One or the other, or a combination of these two methods will enable simultaneous sequencing of equimolecular mixtures of peptides.

Common Artifacts. Since mass spectrometry is a sensitive technique, a number of peaks due to the reagents, peptide, or the machine oil will appear in the final spectrum. These do not interfere with identification and have a positive value: they are useful as mass markers.

Trifluoroacetic acid will give peaks at m/e 69, 95, 97, and 114. Methylthiourea gives peaks at m/e 101 and m/e 104. Where some peptide or protein has become soluble in the tetrahydrofuran used to extract the MTH-amino acid, "protein" peaks show up at m/e 128, 141, 170, and 172 at temperatures around 200°. (The M^+ ion for proline m/e 170 and the M^+ ion for valine m/e 172 come off at much lower temperatures.) A large peak at m/e 149 is due to diffusion pump oil, which is used to grease the sample-probe seal-holder O rings.

Component Complexity and Resolution. The number of peptides in the mixture and the resolution achieved are not independent variables. If a very large number of peptides are present, some must be present in very small amounts. Under such conditions it is possible to follow only the major peptides; the ones that exist in very low quantities will be lost in the "noise." A second limitation on the number of peptides which may be sequenced at once is illustrated in Table II. It can be seen that in five out of the first ten N-terminal loci different peptide chains have common amino acids. In order to calculate the sequential handling loss, it is necessary to find loci where all the amino acids are different. With four polypeptides there were five such loci in the first ten residues. If the number of chains to be sequenced simultaneously is increased, it may be necessary to sequence sufficient residues to be able to construct a "loss line" for each sequence. Simultaneous analysis of up to six polypeptides has proved to be no problem, and it is probable that perhaps eight polypeptides may be sequenced simultaneously. For larger numbers of determinations it is advantageous to transfer the sequence analysis to a machine capable of carrying out Edman degradation. We have adapted a BioCal ES 300 (BioCal G. M. B. H., Munich, Germany) for this purpose, and an interface to the mass spectrometer has been built by one of us (R.E.L.) and is in operation. A major advantage of the mass spectrometric method of analysis is that the output is capable of being fed online to a computer and that sophisticated data handling and analysis procedures for this output are available commercially.

Note

The mass spectra of some PTH-amino acid derivatives have been reported in the literature.[17,18] In a recent paper[19] the mass spectra of all the commonly occurring amino acids have now been reported. High intensity M^+ ions are found for most amino acids except threonine, cysteic acid, *S*-carboxymethylcysteine, arginine, lysine, tyrosine, and tryptophan. Probe temperatures, but no volatilization temperature ranges are given. The N-terminal fifteen amino acids from horse myoglobin were identified, but no quantitation was carried out. It seems likely that mass spectroscopy of PTH derivatives will also prove to be useful in the sequence determination of proteins.

[17] V. M. Stepanov, N. S. Vulffson, V. A. Puchov, and A. M. Zyakun, *J. Gen. Chem. USSR* **34,** 3822 (1964).

[18] B. W. Melvas, *Acta Chem. Scand.* **23,** 1679 (1969).

[19] H. Hagenmaier, W. Ebbighausen, G. Nicholson, and W. Vötsch, *Z. Naturforsch. B* **25,** 681 (1970).

[26] Subtractive Edman Degradation

By William Konigsberg

Amino acid sequences in proteins have been determined by first cleaving the protein into fragments by chemical or enzymatic means, separating the resulting peptides, and finally breaking them into pieces small enough to apply a stepwise degradative procedure such as the Edman degradation.[1] The sequences in the smaller peptides thus obtained can be used to reconstruct the complete amino acid sequence of the protein if overlap peptides (obtained by splitting the intact protein at sites other than those that gave rise to the original fragments) are available. The stepwise degradation of peptides proceeding from the amino- to the carboxyl-terminal residues has been the principal method used during the last 10 years for establishing amino acid sequences of the smaller peptides ranging in size from 2 to 30 residues. The degradation as originally proposed by Edman[1] has undergone many technical modifications since its inception, but the fundamental reactions remain the same. The first step involves the coupling of the peptide to a suitable alkyl or aryl isothiocyanate under alkaline conditions to form an N-substituted thiocarbamyl peptide. After complete removal of excess reagent and solvent, the N-substituted thiocarbamyl peptide is treated with an anhydrous

[1] P. Edman, *Acta Chem. Scand.* **4,** 277, 288 (1950).

acid, causing cyclization. This results in removal of the NH_2-terminal residue as the thiazolinone and the exposure of the α-amino group of the penultimate amino acid. The thiazolinone rearranges to the N-substituted thiohydantoin by heating or by treatment with aqueous acid. The N-substituted thiohydantoin can then be removed from the residual peptide by solvent extraction. Repetition of these reactions results in the stepwise degradation of the peptide. To obtain the amino acid sequence from successive application of these reactions, three principal methods can be used. (1) Direct: the N-substituted thiohydantoins can be identified directly by chromatography[2] or mass spectroscopy[3] or they can be reconverted to amino acids by reductive hydrolysis with HI.[4] (2) Dansyl-Edman: the α-amino group of the residual peptide which is exposed after cyclization can be dansylated, the dansylated peptide hydrolyzed, and the dansyl amino acid identified by thin-layer chromatography[5] or paper electrophoresis.[6] (3) Subtractive: a portion of the peptide remaining after cyclization can be taken for hydrolysis and amino acid analysis. Comparison of the amino acid compositions before and after each step should reveal a decrease in the level of the amino acid that occupied the NH_2-terminal position of the peptide being degraded.[7]

This article will be concerned only with the subtractive modification of the Edman degradation.

Procedure

Coupling

Phenyl isothiocyanate[1] and, more recently, methyl isothiocyanate[3] have been used for the Edman degradation. In the coupling reaction, there must be an appreciable concentration of the unprotonated alpha amino group of the peptide, since the first step involves nucleophilic attack of the amino group on the thiocarbamyl portion of the reagent. Volatile bases such as pyridine or *N*-ethylmorpholine have been useful since they are usually good solvents for both the peptide and the coupling reagent. A typical reaction can be carried out by dissolving the peptide in 0.5–1.0 ml of 70% pyridine (redistilled), adding 0.05–0.10 ml of phenyl

[2] J. Sjöquist, *Acta Chem. Scand.* **7**, 447 (1953); P. Edman and J. Sjöquist, *Acta Chem. Scand.* **10**, 1507 (1956).

[3] This volume [25].

[4] O. Smithies, D. Gibson, E. M. Fanning, R. M. Goodfleish, J. S. Gilman, and D. L. Ballantyne, *Biochemistry,* in press.

[5] K. R. Woods and K. Wang, *Biochim. Biophys. Acta* **133**, 369 (1967).

[6] W. R. Gray, Vol. 11, p. 139.

[7] W. Konigsberg, Vol. 11, p. 461.

isothiocyanate and allowing the mixture to stand at 37° under N_2 for 2 hours. This can be conveniently done in a 3-ml test tube with a ground-glass stopper. If the peptide is in solution, an appropriate aliquot should be introduced into the tube and lyophilized. The reagents are then added, and the tube is flushed with N_2, stoppered, sealed with Parafilm, and placed in a 37° water bath or heating block. After the reaction, the excess reagent and solvents are removed by evaporation under high vacuum. The dried phenylthiocarbamyl-peptide is then ready for cyclization. For successful coupling and subsequent degradation certain precautions should be observed: (1) The peptide should be free of salts of acids and bases since they will interfere with pH control in the coupling step. (2) The solvents and reagents should be free of even traces of aldehydes, as they will form Schiff bases with amino groups and block the reaction with phenylisothiocyanate. (3) The reaction should be kept under N_2 since replacement of some of the sulfur by oxygen in phenyl isothiocyanate or the phenylthiocarbamyl peptide can occur.[8] If oxygen replaces sulfur in the phenylthiocarbamyl peptide, cyclization will not take place at an appreciable rate.

Cyclization

Anhydrous trifluoroacetic acid (TFA) (0.5–1.0 ml) is added to the dry phenylthiocarbamyl peptide, and the tube is flushed with N_2, stoppered, sealed with Parafilm and allowed to stand at 37° for 1 hour. Under these conditions the phenylthiocarbamyl peptide cyclizes to form the 2-anilinothiazolinone derivative of the NH_2-terminal residue.[9] The cyclization results in cleavage of the NH_2-terminal residue, exposing the amino group of the penultimate amino acid. The TFA is removed by evaporation under a stream of N_2.

Extraction of Phenylthiohydantoins

It is necessary to convert the 2-anilinothiazolinones to phenylthiohydantoins to increase their solubility in organic solvents. This can be done by dissolving the residue remaining, after removal of the TFA, in 2 ml of 0.2 *M* acetic acid and heating the solution to 60° for 10 minutes. The phenylthiohydantoin derivatives of the neutral amino acids, with exception of serine and threonine, can then be extracted into benzene. The extraction is performed in the same stoppered tube, three 2-ml portions of benzene being used. In some cases it may be necessary to centrifuge the tube to break up emulsions.[5] The phenylthiohydantoins

[8] D. Ilse and P. Edman, *Aust. J. Chem.* **16**, 411 (1963).

[9] P. Edman, *Acta Chem. Scand.* **10**, 761 (1956).

of arginine, histidine, threonine, serine, aspartic, glutamic, and cysteic acids are only slightly soluble in benzene. They can be removed by extraction with ethyl acetate (except for PTH-arginine, PTH-histidine, and PTH-cysteic acid), but there is danger of extracting some of the peptide into this solvent; thus, it is recommended that extraction with solvents more polar than benzene be avoided. The PTH derivatives of arginine, histidine, aspartic acid and glutamic acid are reconverted to their constituent amino acids in low yield by hydrolysis with 6 *N* HCl at 110°, so their presence would not ordinarily cause difficulty in the interpretation of the results in the subtractive procedure. PTH-serine and PTH-threonine dehydrate easily under acidic conditions, and there is no possibility of regenerating either of these amino acids from their PTH derivatives.

Hydrolysis

A suitable portion of the peptide remaining after cyclization and benzene extraction is hydrolyzed with 6 *N* HCl at 110° for 22 hours. The remaining peptide is lyophilized prior to performing the next round of degradation. The amount used for hydrolysis at each step depends on the quantity of peptide available.

With a Beckman-Spinco amino acid analyzer equipped with 6.6-mm cuvettes and a scale amplifier, good results can be obtained using 10 nmoles of the residual peptide for hydrolysis at each step.

Interpretation of the Results

Under ideal conditions, successive application of the Edman degradation should cause an integral decrease of a single amino acid residue at each step. As mentioned in an earlier article on this subject,[7] side reactions occur which lead to the accumulation of products that will not react with phenyl isothiocyanate or undergo cyclization. As a result, the remaining peptides will show a non-integral decrease of the amino acids which are being removed. The nature of these side reactions is not understood. However, we have attempted to circumvent these problems by purifying the residual peptide at a stage when the results of degradation become ambiguous. Since the presumed impurities have no positively charged groups at pH 2 (exceptions are peptides containing arginine or histidine), they should not bind to the acid form of a cation-exchange resin such as AG50 (Bio-Rad Corporation, Richmond, California). A simple procedure for purification of the residual peptide follows: After TFA-cyclization and conversion of the thiazolinones to the phenylthiohydantoins, the solution of the peptide in 0.2 *M* acetic acid is placed on a small column (5 mm high, 2 mm diameter) of AG50-

X2 (50–100 mesh) in the hydrogen (acid) form. After the solution has run through, 2 ml of water are passed through the column, which removes the undesired side products. The peptide can then be eluted with 2 ml of 0.1 N NH_4OH. Since losses due to transfer and handling are inevitable, this procedure should be used only when necessary to resolve ambiguities in interpreting the results. It will fail entirely if the residual peptide is insoluble in 0.2 M acetic acid or if the peptide contains cysteic acid. If the peptide contains several aromatic residues it may be difficult to recover because of the high affinity of aromatic side chains for the polystyrene backbone of the resin.

Problems Encountered with the Subtractive Edman Degradation

While many of the obvious difficulties to be expected with Edman degradation were reviewed in the earlier article,[7] a few problems have arisen that were not commented on before: (1) The presence of traces of aldehydes or oxidizing agents at any stage of the reaction will lower the yield of potentially degradable peptide for reasons already mentioned. (2) The use of N-ethylmorpholine acetate as a buffer in the coupling step has been abandoned because decomposition of the base resulted in material which gave a ninhydrin-positive peak in the position of isoleucine after acid hydrolysis. (3) Even when highly purified pyridine, PITC and TFA are used, a gradual increase of ninhydrin-positive material eluting in the position of serine and glycine is often observed as the degradation proceeds. When low levels of peptides containing serine and glycine are used, the results can be difficult to interpret even though a comparable reagent blank has been run in parallel. (4) When an internal glutamine is reached during the degradation, we have often found that it cyclizes to form a pyrrolidonecarboxylic acid residue, stopping any further degradation. We have overcome this difficulty in many cases by repeating the entire degradation to the step just prior to the one that would expose glutamine as the NH_2-terminal residue. At the next step, which would give NH_2-terminal glutamine, only a 15-minute exposure to TFA at 0° is permitted. This is followed by immediate removal of TFA and continuation of the next coupling step without delay. (5) In the earlier article[7] we referred to the problems encountered due to the presence of a certain proportion of β-peptide linkages involving aspartic acid residues when HCl–acetic acid was used for cyclization. In subsequent sequence studies, we have isolated, by ion-exchange chromatography, peptides in which the aspartyl residue was presumably linked to the next residue via its β-carboxyl group.[10] When these peptides were

[10] K. Weber and W. Konigsberg, *J. Biol. Chem.* **242,** 3563 (1967).

subjected to the Edman procedure, the degradation proceeded normally up to the first β-aspartyl residue. At that point cyclization led to the phenylthiohydantoin of aspartic acid, but the β-carboxyl group remained attached to the next residue, preventing further degradation. In these cases a decrease in aspartic acid was still observed since acid hydrolysis of PTH-aspartic acid did not yield an appreciable amount of aspartic acid. We have also encountered aspartyl-containing peptides where degradation ceased at the Asx residue itself.[10] Most of these peptides contained an Asx-Gly sequence. We have rationalized the failure to observe a decrease in aspartic acid on the basis that the aspartyl residue was joined to the rest of the peptide via an imide bond. In this case the phenylthiocarbamyl peptide would not be expected to undergo rapid cyclization since the required transition state would involve the sterically unfavorable fusion of two five-membered rings. Upon subsequent acid hydrolysis the phenylthiocarbamyl group would be cleaved, regenerating aspartic acid. It might be noted that these peptides were isolated by ion-exchange chromatography on AG50, and it is probable that the rearrangement of Asp or Asn residues to the imides and β form was due to catalysis by the resin. (6) We have occasionally had trouble degrading peptides having an X-Gly sequence where X was not Asp or Asn. We have found that the penultimate glycine residue is removed together with the NH_2-terminal residue. Others have also had this problem.[6] As suggested by Gray this may be due to abnormal cyclization of the phenylthiocarbamyl peptide in which an 8 rather than a 5-membered ring is formed.[6]

Comparison of the Subtractive Method with Other Methods of Identifying Residues in Following the Edman Degradation

As indicated in the earlier article,[7] the main advantage of the subtractive method is that it provides quantitative information about the yield at each stage of the degradation. This can be very important when difficulties in the degradation arise which prevent the continuation of the Edman procedure. When a sensitive, but qualitative, method such as direct identification of the PTH or dansyl derivatives is used, there is a possibility that a spurious end group might be observed and assigned to the peptide, but which in reality might have been derived from the degradation of a peptide impurity in the sample. The subtractive method is also useful in confirming the assignment of particular residues that are difficult to extract and identify directly as either the PTH or dansyl derivatives. For instance, when the direct method is used, special precautions must be taken to obtain PTH-serine and PTH-threonine, and PTH-leucine and PTH-isoleucine are difficult to separate. The PTH's of arginine, histidine, and cysteic acid cannot be readily extracted into an organic solvent without also extracting some of the peptide. When the Edman-dansyl method is used, problems sometimes occur in resolving dansyl-Glu from dansyl-Asp and dansyl-Ser from dansyl-Thr. Dansyl-Glu and dansyl-Asp can be obscured by *O*-dansyl-Tyr if it is present in high concentration. Dansyl-Ala can be missed since it often runs in a position coincident with dansyl amide on polyamide thin-layer plates.[5]

There are disadvantages of the subtractive method: (1) it is destructive of the peptide undergoing degradation; (2) it requires more material than the dansyl-Edman procedure; (3) an amino acid analysis is required at each step; (4) its usefulness is limited in the case of large peptides since variations in the amino acid analyses may obscure the decrease of a single residue. This is particularly true, because of the reasons mentioned earlier, when serine or glycine are involved.

Conclusions

While the subtractive Edman procedure will continue to be useful in special situations, it is being replaced as a general method for degrading small to medium-sized peptides by the dansyl-Edman method which has the advantages of speed and sensitivity. For larger peptides and proteins, direct identification of the phenylthiohydantoins or the dansyl-Edman procedures are clearly preferable. The development of the automatic sequenator has made it possible to determine the sequence of the first 60 residues in myoglobin.[11] It has also been used routinely to remove up to 23 residues from the NH_2-terminal portion of Bence-Jones proteins.[12] At the moment it is applicable only to large peptides and proteins, but recent improvements will undoubtedly permit its extension to smaller peptides. There is still a need, however, for investigation into some of the anomalies that have been encountered in the Edman degradation for many steps so that it can be applied with uniform success to all proteins and peptides with free NH_2-terminal residues.

[11] P. Edman and G. Begg, *Eur. J. Biochem.* **7**, 463 (1969).

[12] H. D. Niall and P. Edman, *Nature (London)* **216**, 262 (1967).

[26a] Sequence Analysis with Dansyl Chloride

By WILLIAM R. GRAY

Sequential degradation plus dansylation has been one of the most productive methods of sequence analysis for small amounts of peptides.[1,2] Successful application of the direct Edman degradation[3] to small peptides has required immaculate laboratory technique and poses some specific problems. First, to avoid fouling the systems for separating PTH amino acids, by-products of the coupling reaction must be extracted from the PTC-peptide; severe losses occur for many PTC-peptides which have no specially hydrophilic group to anchor them in the aqueous phase. Second, chromatographic separations of PTH-amino acids have been inferior to those for other derivatives, such as DNS-amino acids or DNP-amino acids. This may be due partly to the fact that there is no carboxyl group to provide subtle discrimination between derivatives, such as those of leucine and isoleucine. Third, detection by ultraviolet absorption is not very sensitive. If the Edman chemistry is used merely as a degradative procedure and the process is followed by dansyl end-group methods,[1] these problems can be overcome. Extraction losses are largely eliminated if no extraction is carried out until after cleavage of the PTC-peptide to produce the free zwitterion; a lot of handling is also eliminated. Dansyl amino acids seem to be easier to resolve chromatographically or electrophoretically, and their yellow fluorescence provides a means of instant detection of tiny amounts. Also useful are the fairly wide tolerances of the procedure with regard to operator skill. For small peptides, the main disadvantage is the extra work involved in positioning amide groups. Identification of tryptophan is also a problem, but one which occurs less frequently, and for which one usually has forewarning.

Different methods are recommended for two main classes of problems: (1) Rapid determination of up to about five residues of sequence; this may be entire sequences of small peptides, or the first few residues of larger peptides or proteins. (2) More circumspect analysis of longer sequences in peptides and proteins. Both these methods require only simple equipment and handling procedures, so that many samples can be worked up in parallel.

[1] W. R. Gray and B. S. Hartley, *Biochem. J.* **89**, 379 (1963).
[2] W. R. Gray, Vol. 11, p. 469 (1967).
[3] P. Edman, *Acta Chem. Scand.* **10**, 761 (1956).

Rapid Degradation Procedure

In this approach,[4] the manipulations are arranged so that the sample remains in the same small tube throughout; each degradation cycle involves only the addition of reagents and their removal under vacuum; a single extraction step is performed on each. Because one is aiming only for four or five cycles, the repetitive efficiency does not have to be especially high, so no precautions need be taken to exclude oxygen. The key to the method is commitment of the sample at the outset, dividing it into as many tubes as needed for the expected number of residues; six is usually a practical upper limit because of the buildup of side products (see later). Suppose we have a tetrapeptide divided into four samples. The first will not be degraded, leaving the original end group exposed; the second will be degraded one step, leaving the second residue exposed; the third and fourth samples will be degraded two and three steps, revealing the third and fourth residues, respectively. Dansylation of the samples then gives the sequence of the original peptide:

Sample	*Cycles*	*Product*	*Extract, dansylate*	*Hydrolyze*
1 A.B.C.D	$\xrightarrow{0}$	A.B.C.D	⟶ DNS—A.B.C.D	→ DNS—A
2 A.B.C.D	$\xrightarrow{1}$	B.C.D + $DPTU_1$	⟶ DNS—B.C.D	→ DNS—B
3 A.B.C.D	$\xrightarrow{2}$	C.D + $DPTU_2$	⟶ DNS—C.D	→ DNS—C
4 A.B.C.D	$\xrightarrow{3}$	D + $DPTU_3$	⟶ DNS—D	→ DNS—D

By the sodium dodecyl sulfate/*N*-ethylmorpholine/dimethyl formamide system for dansylating proteins (see this volume [8]), this rapid degradation scheme has been applied successfully to a number of proteins.

Apparatus and Glassware

Bench-top centrifuge, with adapters for holding micro test tubes
10-inch vacuum desiccator fitted with heating mantle (Glas-Col Apparatus Co., Terre Haute, Indiana) set at 60°
Thermostatically controlled heating block (e.g., Technilab Model 212) set at 50°
Vibrating stirrer for vigorous mixing of contents of small tubes
Aspirator pump connected to trap
Polypropylene blocks 3 × 6 × ¾ inches, drilled to accept small test tubes

[4] W. R. Gray and J. F. Smith, *Anal. Biochem.* **33**, 36 (1970).

Pyrex glass test tubes, 4 mm i.d., 50 mm long; tubes are cleaned by heating overnight in a glassblower's oven at 500°, and then stored in a dust-free box.

Reagents

Phenylisothiocyanate (PITC solution), analytical grade, used without further purification. Make up as a 5% (v/v) solution in pyridine (analytical grade); store in a foil-wrapped bottle in a freezer (−10°). Make up fresh about once a month, or when there is evident discoloration.

Trifluoracetic acid (TFA), analytical grade, anhydrous. To keep the reagent dry, it is best to purchase it in batches of several small bottles and to keep the bottle in current use at room temperature.

Ethyl acetate/water. Equal volumes of ethyl acetate and water are shaken together and stored in a screw-cap bottle. Both phases are needed, and should be accessible to micropipettes.

Disposition of Samples. Approximately 2–10 nmoles of peptide or protein are placed in the appropriate number of tubes. These should be labeled near the *bottom* with a diamond pencil; identification must include a number to indicate which residue will be exposed for dansylation (e.g., AP1, AP2, AP3, AP4).

Coupling with PITC. In the first round, AP1 is set aside for dansylation. Fresh coupling mixture is made at each cycle by mixing together 600 μl of PITC solution (5% by volume in pyridine) and 400 μl of water. To each sample which is to be degraded is added 50 μl of the coupling mixture, and the contents of the tube are mixed thoroughly to ensure complete solution. This mixing is especially important as degradation proceeds because of the buildup of by-products; to achieve complete solution, it may be necessary to increase the reagent volume to 75 μl, or even 100 μl. Some proteins have not been fully soluble at this step, but coupling has usually been adequate. The tubes are covered with Parafilm and incubated for 45 minutes at 50°; they are then dried thoroughly in the desiccator at 60°. Drying normally takes 10–15 minutes on the first cycle, and should leave an off-white or lightly pinkish crust of crystals [phenylthiourea and diphenylthiourea (PTU and DPTU)] in the bottom of the tube. In succeeding cycles, the amount of this deposit increases, and the color darkens; drying usually takes longer. If the residue appears oily, it should be redissolved in 30 μl of ethanol, and the drying should be repeated.

Cleavage Reaction. To each tube is added 50 μl of anhydrous TFA, and the samples are left uncovered in the heated desiccator. After 10 minutes, the vacuum is applied, and the samples are dried for about 10 minutes. Again, this drying must be very thorough and may take longer in the later cycles. The deposit of DPTU is more cohesive and fine-grained after the cleavage step than after the coupling step.

Second Cycle. Sample AP2 is set aside for extraction and dansylation while AP3 and AP4 are put through one and two repetitions of the coupling and cleavage reactions. Total time for each cycle is 1.5–2 hours, depending on the number of samples to be handled and the time required for drying. All samples are now ready for the extraction procedure.

Removal of Nonvolatile By-products. In each tube is a peptide with free amino group, plus various amounts of PTU and DPTU and amino acid derivatives removed during degradation. There may also be some salts. The amino acid derivatives do not interfere with dansylation, but the other by-products do. These are removed by extraction with ethyl acetate. To each tube (except AP1), is added 100 μl of the organic phase (top) of the ethyl acetate/water mixture, followed by 40 μl of the aqueous phase (bottom). The tubes are then mixed very vigorously on the mixer; prolonged mixing may be needed for those samples containing much DPTU. The phases are then separated by centrifuging the tubes for 2 minutes at top speed in the bench-top centrifuge. The top phase is then removed with a fine-tipped micropipette attached to the aspirator pump. Extraction is repeated a further two times with fresh lots of 200 μl of the organic phase. The samples are then dried and ready for dansylation.

Identification of End Groups. This is carried out by the methods described in this volume [8]. All the reactions are done on the samples in the same tubes in which they were degraded. Proteins are dissolved in 50 μl of 1% SDS solution, and labeled with the *N*-ethylmorpholine and dimethyl formamide system. Peptides are labeled by the acetone/sodium bicarbonate method. The labeled samples are then hydrolyzed with 6 *N* HCl, and the dansyl amino acids are detected by electrophoresis or thin-layer chromatography, as described earlier.

Successive residues of a single peptide or protein are usually run side by side on the electrophoresis system so as to obtain a close comparison of the samples. In this way, one can assess more readily whether there is overlap between residues or a significant falloff in yield.

Extended Degradations

The rapid sequencing method was developed to handle large numbers of small peptides produced during sequence analysis of immunoglobulins. It is not readily applicable beyond about five or six residues because of the continuing buildup of by-products during successive steps of degradation, and because some of the speed and convenience is obtained at the expense of a high cleavage yield. Dansylation is readily used in conjunction with more efficient degradation methods, however, allowing one to establish sequences of up to about 20 residues. There is a point at which direct sequential degradation should be used in preference to the indirect method using dansyl chloride. This point will vary, depending on whether a laboratory is primarily set up for identifying PTH- or DNS-amino acids. In general, it will be about 12 to 15 residues. Beyond this the dansyl methods rapidly lose their prime advantage of sensitivity: to counteract dropping yields, and to allow for enough sample to be taken at the twentieth step, one would need to start with approximately 250 nmoles, well within the range of the direct PTH methods. The following method is recommended when samples are to be degraded through an intermediate number of steps (5–15). Several other degradation procedures are entirely suitable.

Apparatus and Glassware. Apparatus required is the same as for the rapid degradation method, plus:

- Heat-cleaned Pyrex glass test tubes (approximately 1×7 cm) and suitable racks, which do not have to withstand temperatures greater than 70°
- Nitrogen tank (prepurified grade) and fittings

Reagents. Reagents and solvents should be of the highest purity available, preferably purified especially for sequence analysis (e.g., Pierce Chemical Company's "Sequenal" grade, or Beckman Instrument Company's Sequencer grade).

- PITC solution, 10% in pyridine, made up fresh weekly. The solution is stored in a glass-stoppered bottle, wrapped in foil, at −10°. Newly made up solutions should be thoroughly flushed with N_2 in a fume hood; the bottle is flushed with N_2 every time before being restoppered.
- *N*-ethylmorpholine, "Sequenal" grade or equivalent
- Trifluoroacetic acid, "Sequenal" grade or equivalent. The reagent should be purchased in batches of several small bottles; bottles not in use should be stored at 0°. Before it is opened, a bottle should be allowed to warm to room temperature; once opened, it should be flushed with N_2 each time before it is resealed, and stored at room temperature for not more than a week or two.
- Ethyl acetate/water. Equal volumes of ethyl acetate and water are shaken in a stoppered bottle. Both phases are required, and should be accessible to pipettes.

Coupling with PITC. The peptide or protein is dried *in vacuo* in a 7 × 1-cm test tube. For peptides, an amount equal to 2–10 nmoles per step should be used, the larger amounts being taken for longer degradations. The peptide is then dissolved in 200 μl of aqueous pyridine (50% v/v, flushed with N_2). Proteins require larger volumes (e.g., 500 μl) and may be difficult to solubilize, especially after several cycles; addition of a tertiary base such as *N*-ethylmorpholine or dimethylallylamine (10% by volume) is often helpful.[5]

A sample of suitable size is taken out for dansylation, and the original volume is restored by the addition of 50% pyridine. This is normally the most convenient point at which to remove samples of proteins for dansylation in subsequent rounds also. An exception is made for those proteins that are much more soluble in TFA than in the coupling buffer —samples can be taken immediately after the cleavage step.

To the main sample is added a half-volume of PITC solution (100 μl for peptides, 250 μl for proteins), and the tubes are flushed with N_2, covered with Parafilm, stirred thoroughly, and left to react for 30 minutes at 50°. Some PITC comes out of solution at first, but usually it dissolves as the reaction progresses.

Removal of Excess Reagent and Pyridine. After the reaction period, the samples are dried thoroughly *in vacuo* at 60°. This removes pyridine, excess PITC, and volatile by-products of the reaction; it does not remove nonvolatile by-products such as diphenylthiourea (DPTU). It is best to leave removal of the DPTU until after the PTC-peptide has been cleaved, as this avoids extraction losses of the PTC-peptide. After drying down, the contents of the tubes should appear pinkish brown and crystalline; if they appear oily, they should be suspended in 50–100 μl of ethanol and dried again.

Cleavage of PTC-Peptide or Protein. The dried residues are dissolved or suspended in 50–100 μl of anhydrous TFA, and the tubes are flushed with N_2 and sealed with Parafilm. Reaction is allowed to proceed for 10 minutes at 45°, and then the TFA is removed *in vacuo*. Again, drying should be thorough. At this stage, the samples usually adhere as a film to the tubes.

Removal of Nonvolatile By-products. After the cleavage step, the peptide is much less susceptible to extractive losses than was the PTC-peptide. Nonvolatile by-products, such as DPTU, are removed by extraction with ethyl acetate. The dried residue is dissolved or suspended in 1 ml of the organic phase (top) of the ethyl acetate/water mixture, and then 250 μl of the aqueous phase (bottom) is added. After thorough

[5] M. E. Percy and B. Buchwald, *Anal. Biochem.* (in press).

mixing on a Vortex mixer, the two phases are separated by a few minutes' centrifugation at top speed in a clinical centrifuge. The organic phase is removed and discarded, and the aqueous phase is reextracted with two further 1-ml volumes of the organic phase. A sample of the peptide is removed for dansylation; the remainder is dried *in vacuo* and stored in readiness for the next cycle of degradation.

Proteins are not usually soluble at this stage, and a different procedure is used for removal of the DPTU. Percy and Buchwald[5] recommended the following procedure. After the cleavage step, the protein is dried thoroughly *in vacuo,* and then is washed with 95% ethanol to make it less adherent to the walls. The residue is then dried, broken up finely, and extracted three times with chloroform/benzene (1:1, v/v). After drying again, the protein is suspended in 0.4 ml of water and extracted three times with 2-ml portions of butyl acetate with vigorous stirring with a Vortex mixer. The phases are separated by centrifugation; care must be exercised in removing the top phase, for particles of protein often sit at the butyl acetate–water interface. After it has dried *in vacuo,* the protein is redissolved in the buffer used for the coupling reaction, and a sample is removed for dansylation. The remainder is put through the next cycle of degradation.

Dansylation of Samples. The size of sample to be taken should be the minimum necessary to obtain good end-group analyses. In the longer degradations, it is necessary to increase the nominal size of the sample to compensate for losses due to extraction, side reactions, etc. It is most useful to base the size of sample taken upon the results of the previous round of degradation, and not to increase it unless necessary. This makes the overall degradation slower, but provides a useful backup feature, in that the dansylation can be repeated if it was unsuccessful for any reason. Whenever a sample must be stored, it is best to do so in the dried state, under vacuum, as the free peptide, i.e., after the cleavage step, but before it is dissolved in coupling buffer. With proteins it is probably best to go directly into the next round of degradation on the main sample after removal of the dansylation sample.

Dansylation and identification of end groups are carried out as described in this volume [8]. It is especially important to remove pyridine and to check the pH of the peptide solution before the dansyl chloride is added (for details, see this volume [8]).

Investigation of Amide Groups

Since amino-terminal asparagine and glutamine are converted to DNS-Asp and DNS-Glu, respectively, amide groups must be investigated before the peptide has been hydrolyzed by acid. This is done by analyzing the electrophoretic mobility of the peptide or dansyl peptide at pH 6.5, a method that is not applicable to proteins. Although it uses extra material and takes time, it should be remembered that one does not have to examine every peptide in this way. One normally starts out with considerable information about a peptide and can use this to decide the best approach: (1) if the peptide contains no Asx or Glx, there is no problem; (2) if the peptide contains a single Asx or Glx, determination of its net charge indicates directly whether there is an amide group or not; (3) with more than one Asx and/or Glx, a net charge determination may show that all are free, or all are amidated; (4) when the net charge reveals that free carboxyls and amide groups are both present, the disposition of these is found by analyzing the *change* in electrophoretic mobility as amino acids are removed during degradation. In the last case, one still does not have to measure mobilities at every step of the degradation, but only when Asx or Glx has just been removed. In practice, therefore, mobilities are measured in only about 10% or less of cases.

At pH 6.5, the ionic charge of dansyl peptides is usually integral and usually negative. Only those containing unreacted histidine have fractional charges, and only those having one or more arginine residues can be neutral or basic, provided all amino groups have been allowed to react. At this pH, the dimethylamino function of the dansyl group is uncharged. By comparing the electrophoretic mobilities of the dansyl peptides at successive stages of the PITC degradation, it is usually possible to decide unambiguously which amino acids carried positive, negative, or zero charges.

Quantitatively, the mobilities of DNS-peptides can be described by Eq. (1)[6]

$$\log(\text{mobility}) = 1.44 + \log(-e) - 0.6 \log(\text{MW}) \tag{1}$$

where e is the net ionic charge at pH 6.5. Mobility is defined as (distance moved by DNS-X)/(distance moved by DNS-OH). Measurements are made from the true neutral point, which is approximately the position taken up by DNS-NH_2. When log (mobility) is plotted against log (MW), a series of parallel straight lines is obtained, with a slope of -0.6. The individual lines correspond to derivatives having $e = -1$, -2, -3, etc., and DNS-peptides of known composition can be assigned uniquely to one of these lines, thus defining the net ionic charge of the compound. Those peptides that contain cysteic acid have slightly higher

[6] R. E. Offord, *Nature* (*London*) **211**, 591 (1966).

mobilities than suggested by Eq. (1); this probably reflects a difference in hydration compared with those that contain only carboxyl groups.

For measurement of mobility, a sample of the DNS-peptide is submitted to electrophoresis at pH 6.5 (80 v/cm, 1–1.5 hours). The mobility reference markers, DNS-OH and DNS-NH_2, are present in all reaction mixtures and need not be added. If necessary, the DNS-peptide may be eluted from the paper with 6 N HCl and then subjected to hydrolysis, but it is more convenient to use a separate sample for hydrolysis and end-group determination.

The mobility measurements are unsatisfactory for peptides of more than about 8 residues—rather diffuse bands are obtained, and fluorescence yields are often very low.

Figure 1 and the accompanying table illustrate the method using the tetrapeptide (Thr, Asx, Glx_2). The mobility of the free peptide cor-

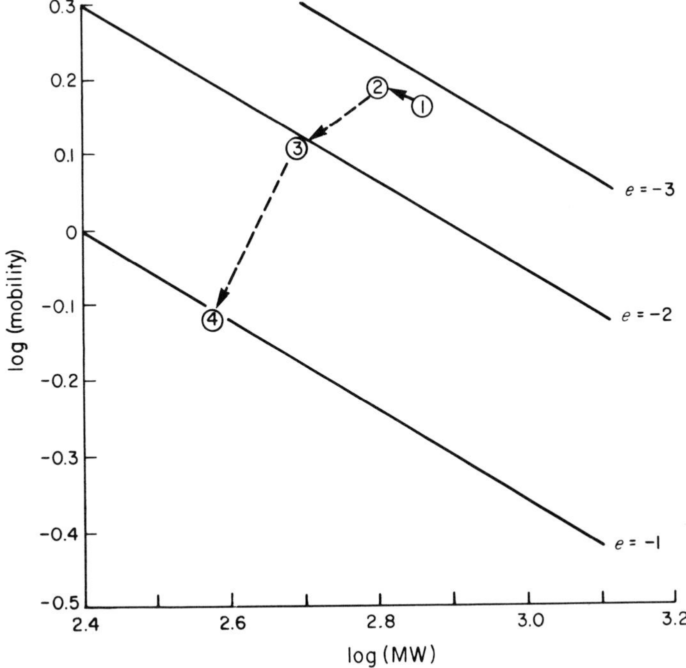

FIG. 1. Electrophoretic mobilities of dansylated peptides at successive stages of degradation of Thr.Asp.Glu.Gln. Mobilities were measured at pH 6.5 relative to that of DNS-OH. The lines on the graph represent solutions of Eq. (1) for peptides having net ionic charges of —1, —2, and —3. Points are numbered to correspond with the table.

SEQUENTIAL DEGRADATION OF TETRAPEPTIDE

Stage	Peptide	End group	m^a	e^a	Conclusions
1	Thr, Asp, Glu_2	DNS-Thr	1.43	−3	One amide present
2	Asp, Glu_2	DNS-Glu	1.50	−3	Glu had free —COOH
3	Asp, Glu	DNS-Asp	1.27	−2	Asp had free —COOH
4	Glu	DNS-Glu	0.77	−1	Same mobility as DNS-Glu (NH_2)
Markers	Glu (NH_2)		0.77	−1	
	Glu		1.37	−2	

[a] Mobility (m) and charge (e) refer to the DNS-peptide at pH 6.5 (see Fig. 1 and text).

responded to a net charge of —2, and thus indicated the presence of an amide group on one of the three side-chain carboxyls. The mobilities of the dansyl peptides at successive stages of degradation, plus the results of end-group analyses, established the complete sequence as Thr-Glu-Asp-Gln.

Comments on Special Problems

Many thousands of peptides have been sequenced using sequential degradation plus dansylation (often called the dansyl-Edman method). Most peptides can be expected to give little or no trouble, provided care is taken to follow the procedures carefully and to make changes only when all the implications are understood. A number of problems can arise, however, and deserve some comment.

Solubility of Samples. Solubility has rarely been troublesome with peptides, but it is one of the major problems with proteins. Occasionally, very hydrophobic peptides have been insoluble in the $NaHCO_3$/acetone system used for dansylation, even though they gave no trouble with the degradation as such. An example was the peptide Tyr.Pro.Lys.Asp.Ile.-Asn.Val.Lys.Trp, derived from chymotrypsin cleavage of a mouse Bence-Jones protein.[7] After one round of degradation, both lysine residues were converted to the much more hydrophobic ε-PTC-Lys, and the yields of end groups were very low. During extractions to remove DPTU, the peptide was insoluble in both the aqueous and organic phases, and remained at the interface. Fortunately, the remedy for this problem is simple—switch to a dansylation system similar to that used for proteins, but omitting SDS. The peptide will usually dissolve in 50% aqueous NEM (v/v), and can be labeled by the addition of DNS-Cl in DMF

[7] W. R. Gray, W. J. Dreyer, and L. Hood, *Science* **155**, 465 (1967).

(5 mg/ml). When the reaction is complete, the mixture should be dried *in vacuo,* and hydrolyzed in the usual way; do not attempt to precipitate the peptide with acetone.

Proteins pose much more severe solubility problems, during both the degradation and the dansylation steps. The systems described in the text have been the most widely successful that I have found so far. They generally work quite well with small to medium-sized proteins, especially those that are reasonably hydrophilic; larger hydrophobic proteins are much more variable.

Glycopeptides. These have not handled very well in the systems, and it has not usually been possible to analyze the sequence through the sugar-containing residue. There does not seem to be a single cause of the problem. Sometimes the reaction mixtures will not dry thoroughly and syrups are obtained; yields at all steps are then usually low. On other occasions, yields are good up to, but not including, the sugar-containing residue; sometimes no end group is obtained, and sometimes a repetitious Asp is obtained, similar to that found with free aspartic acid, or Asn.Gly sequences. The best hope of a solution to the problem is to attempt removal of the carbohydrate chain enzymatically, or with dilute acid.

Free Amino Acids. Most free amino acids are fully derivatized by the reaction with PITC, but Asp and Glu are exceptions. When present, they may give rise to spurious repetitive end groups. In such cases, it may be necessary to repurify the peptide. This is no problem with proteins, for the acetone precipitation of the DNS-protein leaves any free DNS-amino acids in solution.

Asn-Gly Sequences. These appear to be especially prone to undergo rearrangement to the β-peptide form.[8,9] This prevents further degradation. Partly, the terminal β-aspartyl group behaves as a free amino acid, so that one obtains Asp repetitiously with the dansyl end-group method. The link is susceptible to cleavage by hydroxylamine,[9] and this may provide a means of overcoming the problem.

Glutamic Acid and Glutamine. When exposed at the amino terminus of a peptide, either of these residues, but especially glutamine, may cyclize to pyrrolidone carboxylic acid. This has not usually been a serious problem. The effect of cyclization is to decrease yields at that step, and subsequent ones, but not to introduce ambiguity. Any decrease in yield can be compensated for by taking larger samples for dansylation.

Lysine, Histidine, and Tyrosine. The major products of these as

[8] W. Konigsberg, Vol. 11, p. 461 (1967).

[9] P. Bornstein and G. Balian, *J. Biol. Chem.* **245,** 4854 (1970).

amino terminal residues are α-DNS-Lys, α-DNS-His, and bis-DNS-Tyr, respectively. Even after two or three cycles of degradation, some ε-DNS-Lys is found. *O*-DNS-Tyr is obtained at all stages up to and including the one at which tyrosine is exposed as the terminal group; at subsequent stages none is obtained.

Tryptophan. DNS-Trp is destroyed by acid hydrolysis, so cannot be identified. It can also interfere with the degradation, especially when oxygen is present. Often the degradation is blocked when tryptophan is the second residue. The major cause of the problem is probably a combination of oxygen and TFA, leading to the production of 3-carboxy-β-carbolines.[10]

[10] R. A. Uphans, L. I. Grossweiner, and J. J. Katz, *Science* **129**, 641 (1959).

[27] Automatic Solid-Phase Edman Degradation

By Richard A. Laursen

A prerequisite to automation of the Edman degradation is a means of immobilizing the protein or peptide so that it can be separated from the reagents used and the anilinothiazolinone formed during the degradation. In Edman's automatic protein sequenator[1] proteins are immobilized as a thin film in a spinning cup, and separations are accomplished by a selective extraction procedure. The solid-phase[2,3] modification differs in that peptides are bound covalently to an insoluble resin which can be washed free of excess reagents.

The first step of the solid-phase method is to attach peptides by their carboxyl termini to the resin. Peptide amino groups are blocked using *t*-butyloxycarbonyl azide,[4] and the carboxyl groups are activated using carbonyldiimidazole.[5] The resulting peptide imidazolides are then coupled to an aminopolystyrene resin. Deblocking of the amino groups is accomplished by treating the resin with trifluoroacetic acid. Degradation of the peptide is carried out in the usual way (see previous articles) by treatment with phenyl isothiocyanate, followed by trifluoroacetic

[1] P. Edman and G. Begg, *Eur. J. Biochem.* **1**, 80 (1967).
[2] R. A. Laursen, *J. Amer. Chem. Soc.* **88**, 5344 (1966).
[3] R. A. Laursen, *Eur. J. Biochem.* **20**, 89 (1971).
[4] D. Levy and F. H. Carpenter, *Biochemistry* **6**, 3559 (1967).
[5] R. Paul and G. W. Anderson, *J. Amer. Chem. Soc.* **82**, 4596 (1960).

acid. The thiazolinones liberated in each stage are isomerized to phenylthiohydantoins and are identified using an isotope dilution procedure.[6]

The solid-phase peptide sequencer[3] was designed with the aim of providing a simple, rapid, and relatively inexpensive means for sequencing peptides of the size normally encountered in enzymatic digests of proteins. Although the sequencer is still in a period of development, and experience with it is limited, the initial results are encouraging. The scope and limitations of the method will be discussed.

Construction of the Sequencer

The general design of the sequencer is shown in Fig. 1. All reactions take place in the column (C). Reagents and solvents are removed from reservoirs (R) by pumps (P) and are forced via pneumatically actuated valves (V) into the reaction column (C). The column effluent is then directed by means of a 3-way valve (V) either to a waste bottle (W) or to the fraction collector (F). A thermostated bath (B) maintains a constant reaction column temperature. Solenoid valves (S) control the pneumatically actuated valves (V).

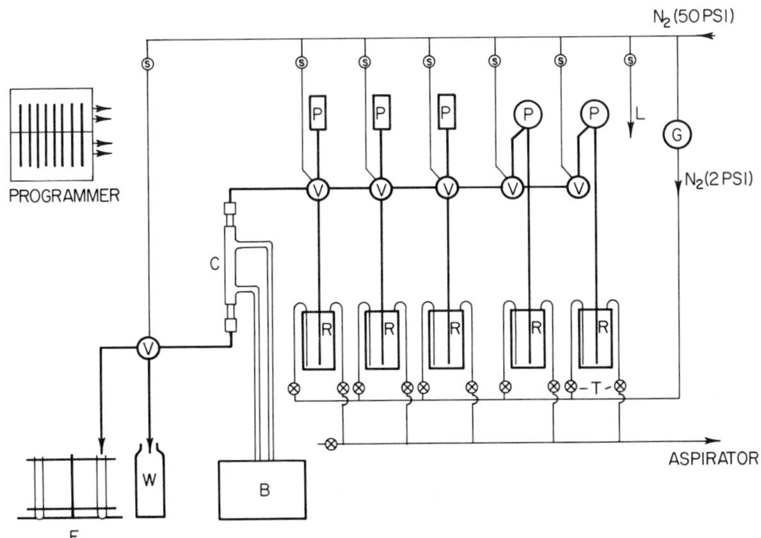

Fig. 1. Diagram of the sequencer: B, constant-temperature bath; C, reaction column; F, fraction collector; G, gas pressure regulator; L, auxiliary nitrogen line; P, pumps; R, reagent (solvent) reservoirs; S, solenoid valves; T, stopcocks; V, reagent (solvent) valves; W, waste bottle.

[6] R. A. Laursen, *Biochem. Biophys. Res. Commun.* **37**, 663 (1969).

All components that come in contact with reagents are constructed of Teflon, Kel-F, glass, or stainless steel.

A mechanical disk programmer controls the operation of the solenoid valves, pumps, and fraction collector.

Reagent Reservoirs. These are fabricated by attaching Ace Glass, Inc., threaded column fittings (No. 5820-28) to bottles of various sizes (200–500 ml). The Nylon plugs (Ace Glass No. 5345-15) used to seal the bottles are modified to accommodate four Chromatronix tube end fittings, thus providing nitrogen inlet and outlet tubes, a reagent outlet tube, and a reagent inlet or filling tube. Viton O rings are used to seal the reservoirs. A Teflon plug is fabricated for the trifluoroacetic acid (TFA) bottle. Reagents are kept under a positive pressure of 2 psi of prepurified nitrogen in order to keep oxygen out and to lift the reagents to the pumps located overhead. Unless the system is pressurized, bubbles tend to form in the syringe pumps. Check valves (Circle Seal, Model 259B-1PP) between the nitrogen manifold and each reservoir prevent cross-contamination of reagents. Nitrogen outlet tubes lead to a manifold connected to a water aspirator, which removes fumes during flushing operations. When radioactive phenyl isothiocyanate (PITC) is used, the PITC reservoir should be vented separately to prevent possible contamination of other reagents. All nitrogen inlet and outlet tubes can be closed by means of Teflon-and-glass stopcocks.

Reagent Valves. The arrangement of the reagent valves is indicated in Figs. 1 and 2. TFA, PITC, and buffer valves (Chromatronix type RV-4031, recycling) allow refilling of the syringe pumps in the "off" position. The slider for the PITC valve is modified by drilling an extra hole to permit simultaneous addition of buffer and PITC to the column. The methanol, dichloroethane (DCE), and waste-collect valves are of the 3-way type (Chromatronix CAV-3031). All valves are equipped with Chromatronix PA 875 pneumatic and SR-1 spring-return actuators. The body and actuators of the TFA and waste-collect valves and the bodies of the other valves are constructed of Kel-F. The pneumatic actuators

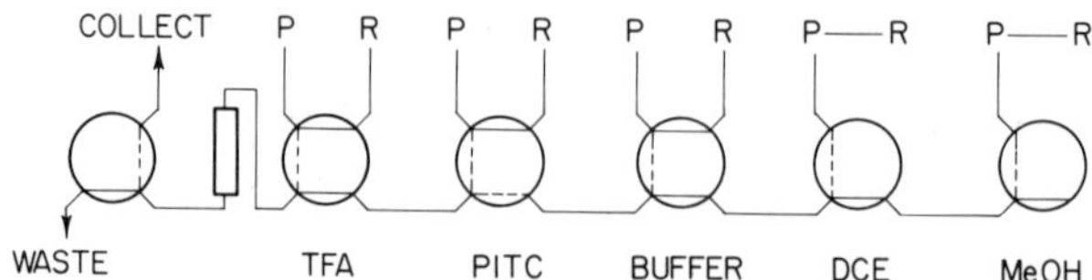

FIG. 2. Valve arrangement. Heavy lines indicate valve port configurations in the "off" position; broken lines, the "on" position. P, pumps; R, reagent reservoirs. See text for additional details. TFA, trifluoroacetic acid; PITC, phenyl isothiocyanate; DCE, dichloroethane.

are operated at a pressure of 50 psi of nitrogen and are controlled by solenoid valves (Skinner Electric Valve No. V53ADB1100).

Pumps. Milton-Roy "minipumps" (Model 196-31 recommended) are used for DCE and methanol. For the other reagents, Harvard Apparatus Co. portable infusion-withdrawal pumps (No. 1100) are used. These are modified by rewiring in the following way: in the "pump on" position, as signaled by the programmer, the pump operates in the infusion mode; in the "pump off" position, the pump is in the withdrawal or refill mode. When the syringe is full, a microswitch is tripped by the syringe carriage and the pump is turned off. The TFA and buffer pumps are operated at a motor speed of 0.5 rpm (about 0.057 ml per minute) and the PITC pump at 0.2 rpm (about 0.023 ml per minute). Methanol and DCE pumps are set at a pumping rate of 0.8 ml per minute. Hamilton gas-

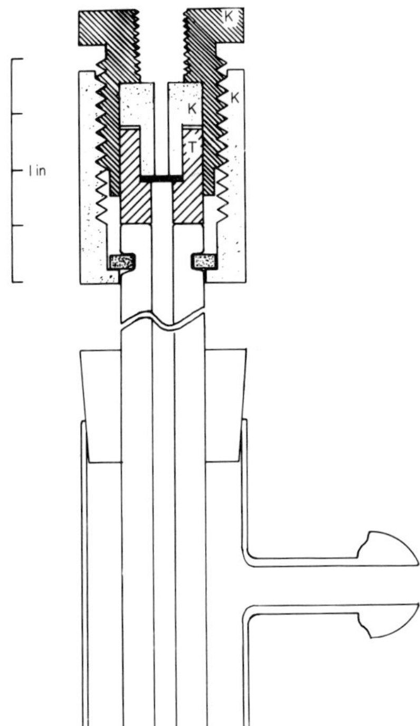

FIG. 3. Detail of reaction column and end fitting. Materials: K, Kel-F; T, Teflon. The glass reaction column has an inner diameter of 2.5 mm and an overall length of 240 mm. The water jacket is 140 mm long. A bed support of Teflon cloth is sandwiched between the two inner fittings, and a Kel-F split ring holds the column in place against the fitting. The uppermost part is threaded to accommodate a Chromatronix tube end fitting. The end of the glass column is fire-polished to assure a good seal.

tight syringes (5 ml) are modified by sealing Chromatronix glass-to-Cheminert fittings to the syringe. This is necessary because Hamilton Teflon Luer-lock fittings sometimes leak after moderate use.

Reaction Column. Details of the reaction column are shown in Fig. 3. The column is thermostated (usually at 45°) by passing water through the outer jacket. Any variable-temperature thermostated bath is suitable for this purpose. The column is readily demountable by disconnecting the ball joints on the jacket and the reagent inlet and outlet tubes. The

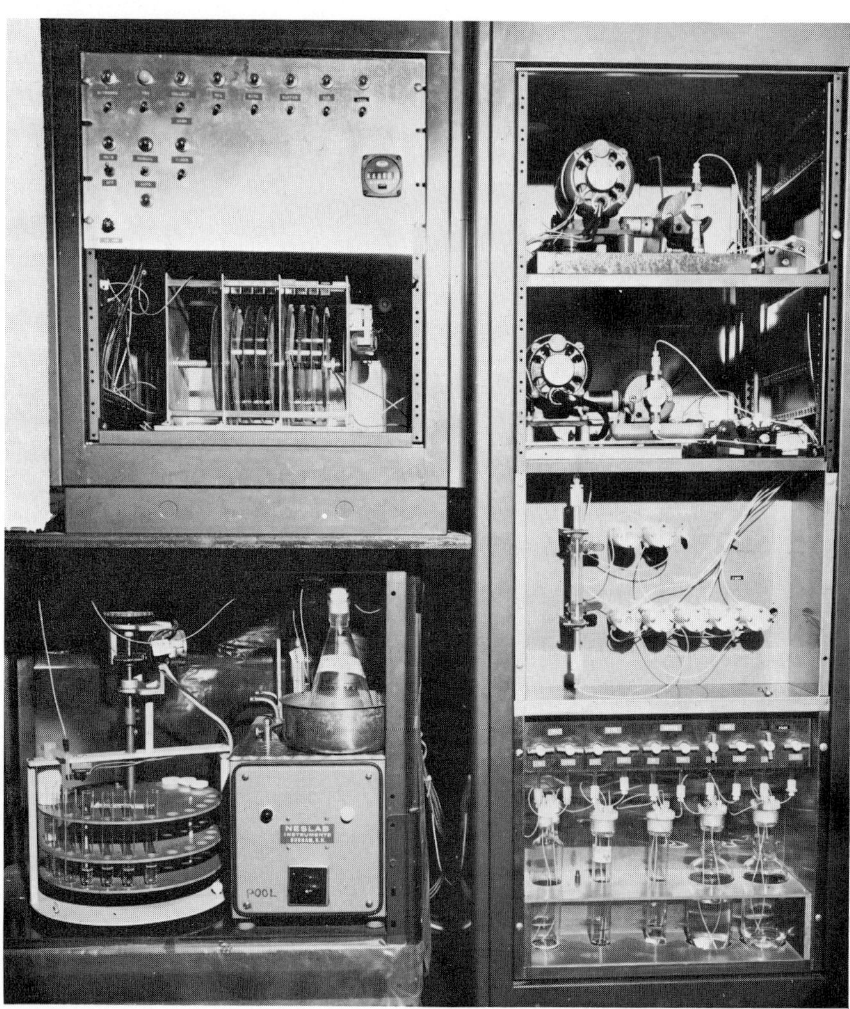

FIG. 4. Photograph of the sequencer.

unthermostated portions of the column are filled with glass beads to reduce the dead volume.

Fraction Collector. We have used a custom-made fraction collector capable of holding twenty-four 18 × 150-mm test tubes. The collector is advanced one tube during each cycle on a signal from the programmer. The instrument is shut down automatically at the end of a run by a plug, inserted in the turntable after the last tube, that activates a shutdown switch. Any fraction collector is suitable, provided it can withstand the corrosive vapors of trifluoroacetic acid. For this reason the motor should be well sealed or placed above the fraction collector.

Programmer. A Bliss Eagle Signal disk timer (Bulletin 1335), having a 2-hour cycle time and capable of being set to within 30 seconds, is used. Each of five disks controls the operation of one reagent pump and its respective valve. Two other disks control the waste-collect valve and the fraction collector.

Assembly. The sequence is assembled in electrical cabinet racks as shown in Fig. 4. Solvent reservoirs are placed at the bottom of the cabinet to minimize the danger of explosion or corrosion in the event of leakage. For the same reason, the pumps, whose motors may spark, are placed in the top of the cabinet. A 15-cfm flushing fan blows filtered air in the top of the cabinet over the pumps and then down through the cabinet and out the bottom, thus preventing the buildup of explosive or corrosive vapors. Ordinarily, the front of the pump compartment is covered with a plexiglass panel. The programmer is located in a separate cabinet.

Synthesis of the Peptide Resin

Chloromethylpolystyrene.[7] A stirred suspension of 27 g of dioxane-washed styrene–1% divinylbenzene copolymer (minus 400 mesh; Bio-Rad Laboratories BioBeads SX1) in 160 g of chloromethyl methyl ether is cooled to 0° in an ice–salt bath. A solution of 3.0 ml of anhydrous $SnCl_4$ in 40 g of chloromethyl methyl ether is added dropwise over a period of 15 minutes, the temperature being kept below 1°. Stirring is continued at 0° for 30 minutes more, and the resin is filtered off on a fritted-glass filter. The resin is washed with 400 ml of dioxane–water (3:1), and then alternately with dioxane and water until the washings give no precipitate with $AgNO_3$, and finally with methanol. The resin is dried under vacuum at 100°; yield, 25.5 g. Chlorine analysis: 1.23 meq/g.

Nitrochloromethylpolystyrene. Fifteen grams of chloromethylpoly-

[7] R. B. Merrifield, *Biochemistry* **3**, 1385 (1964).

styrene is nitrated according to Dowling and Stark[8] by adding it slowly with stirring, over a period of about 15 minutes, to 175 ml of fuming nitric acid (90%) that has been previously cooled to 0°, keeping the temperature below 2°. Stirring is continued for 1 hour at 0°. The mixture is poured onto about 1000 g of ice, and the resin is then filtered off and washed several times alternately with dioxane and water, and finally with methanol. The cream-colored product is dried under vacuum at 80°; yield, 20.5 g. Chlorine analysis: 0.78 meq/g.

Nitroethylenediamine Resin. A 5.0-g portion of the nitrochloromethyl–polystyrene is added to 15 ml of ethylenediamine. The mixture is stirred at room temperature for 15 minutes and is then heated on the steam bath for 15 minutes. The yellowish resin is filtered off and washed thoroughly with water and methanol, and is dried at 100° under vacuum.

Aminoethylenediamine Resin (Peptide Resin). A stirred suspension of 4.0 g of nitroethylenediamine resin in 75 ml of dimethylformamide (DMF) is heated to 75°, and a solution of 30 g of $SnCl_2 \cdot 2H_2O$ in 25 ml of warm DMF is added slowly (exothermic reaction).[8] The temperature is raised and maintained at 140–150° for 15 minutes; then 25 ml of concentrated HCl is added. Heating is continued at about 100° for 1 hour. The resin is separated on a fritted-glass filter and is washed thoroughly with water and HCl to remove stannic ions. Neutralization is effected by washing the resin with pyridine–water (1:1) followed by DMF–triethylamine (3:1) until chloride ions cannot be detected in the washings. After final washes with water and methanol, the yellow-gray product is dried at 100° under vacuum; yield, 3.4 g. The resin should swell to about 3.5 times its dry volume in DMF.

Reagents

Methanol (Baker and Adamson absolute, aldehyde-free)
1,2-Dichloroethane (Matheson, Coleman and Bell, Spectroquality)
Trifluoroacetic acid (TFA) (Eastman)
Methyl isothiocyanate (Eastman)
t-Butyloxcarbonyl (BOC) azide (Pierce)
Acetonitrile (Matheson, Coleman and Bell, Chromatoquality)
Pyridine (Fisher)
N-Methylmorpholine (Pierce, Sequenal grade)

(The above-listed reagents are used as purchased without further purification.)

Dimethylformamide (DMF) (Fisher), distilled from P_2O_5 at

[8] L. M. Dowling and G. R. Stark, *Biochemistry* **8**, 4728 (1969).

50°/20 mm Hg and stored over Linde type 3A molecular sieves in serum-capped bottles.

Buffer used in degradations: 3:2 mixture of pyridine and *N*-methylmorpholinium trifluoroacetate buffer (pH 8.1), prepared by diluting a mixture of 28 ml of *N*-methylmorpholine and 3.75 ml of trifluoroacetic acid to 200 ml with water. The resulting solution is 1 *M* in *N*-methylmorpholine and 0.25 *M* in the conjugate acid.

Radioactive phenyl isothiocyanate (PITC): ^{3}H-PITC (New England Nuclear) is diluted with unlabeled PITC (Eastman reagent) and the mixture is distilled at 90°/10 mm Hg. A sample converted to methylphenylthiourea, which was recrystallized and counted, showed a specific activity of 76,100 dpm/μmole. A 5% solution of PITC in acetonitrile is used for degradations.

Glass beads (200–325 mesh, Microbeads Division of Cataphote Corporation, Jackson, Mississippi) are washed with petroleum ether and are then heated in concentrated HCl for about 2 hours to dissolve metal particles. The beads are washed thoroughly with water, air-dried, and then silylated by treatment with 5% solution of trimethylsilyl chloride in benzene. The beads are filtered, washed with acetone, and dried in an oven. Silylation gives a more free-flowing product.

Carbonyldiimidazole, synthesized by the method of Paul and Anderson[5] or purchased (Pierce Chemical Co.). Stock solutions of carbonyldiimidazole in anhydrous DMF are kept in serum-capped vials. The titer of the solution is checked periodically by adding aliquots to excess aniline and weighing the diphenylurea formed. Solutions are stable for at least several weeks.

Procedure

Attachment of Peptide to Resin. NH_2-TERMINAL BLOCKING. A sample (up to about 2 μmoles) of salt-free peptide dissolved in triethylammonium bicarbonate buffer or dilute triethylamine is placed in a 1-dram snap-cap vial (Kimble No. 6075-L) and is lyophilized or evaporated in a vacuum desiccator. To the residue is added 1.0 ml of DMF, 0.1 ml of *N*-methylmorpholine, and 0.15 ml of BOC azide. The mixture is stirred by means of a small magnetic stirring bar until solution is complete. In cases where the peptide does not dissolve readily, 0.2 ml of water can be added. The vial is stoppered with a serum cap and is heated at 50° for at least 5 hours, after which the solvent and excess reagents are removed by evaporation in a vacuum desiccator using a good vacuum pump. Evacuation of the desiccator should be cautious at first to prevent bumping.

When the vial is dry, 0.5 ml more of DMF is added and is evaporated. Drying under vacuum is continued for 10 hours or more at room temperature.

ACTIVATION. The vial is removed from the desiccator and is immediately stoppered with a serum cap. A solution (see *Reagents*) of 8 mg of carbonyldiimidazole in 0.2 ml of anhydrous DMF is added by means of a syringe, and the mixture is stirred for 1–2 hours at room temperature. More anhydrous DMF may be added if necessary to dissolve the peptide. It is essential to maintain rigorously anhydrous conditions at this stage, as carbonyldiimidazole reacts almost instantaneously with traces of water.

ATTACHMENT. Peptide resin (200 mg/μmole; not more than about 300 mg) is allowed to swell in dry DMF (0.5 ml/100 mg) for 30 minutes, and one-third to one-half of the mixture is added[9] to the activated peptide solution to react with excess carbonyldiimidazole. The reaction mixture is stirred by means of a small magnetic stirring bar for 10 minutes, and the remainder of the resin is added. The mixture is stirred at room temperature for 2 hours, and for each 100 mg of resin there is added 0.35 ml of pyridine-*N*-methylmorpholinium trifluoroacetate buffer (see *Reagents*) and 0.1 ml of methyl isothiocyanate.[10] Stirring is continued for 2 hours at room temperature, and the resin is filtered and washed several times alternately with DMF and methanol in the apparatus shown in Fig. 5. The dried resin is stored in its original reaction vial in the refrigerator, as it darkens on contact with sunlight and air. The coupling yield may be obtained by amino acid analysis, following hydrolysis of a resin sample in 6 *M* HCl for 24 hours in a sealed tube containing an internal standard such as norleucine.

Packing the Reaction Column. A column end fitting (Fig. 3) is attached to one end of the reaction column, which is then placed in an upright position. Glass beads (200–325 mesh, see *Reagents*) are added by means of a small funnel until about 10 cm of unfilled column remain in the thermostated region. A sample of resin (30–50 mg) con-

[9] Swelling and transfer of the resin can be effected by means of a 1-ml syringe which has been cut off at the 0-ml graduation. The syringe plunger is withdrawn to the 1-ml mark, and the DMF and resin are added to the open end of the syringe barrel. When the resin is swollen, it is injected in portions into the reaction mixture.

[10] Treatment with methyl isothiocyanate serves to block excess amino groups on the resin which would otherwise react with the radioactive PITC used in the degradation. Consequently any breakdown of the methylthiocarbamyl groups on the resin will give rise to nonradioactive products, which will not be detected on thin-layer chromatography of the PTH's. For reasons not fully understood, PITC is not satisfactory for this purpose.

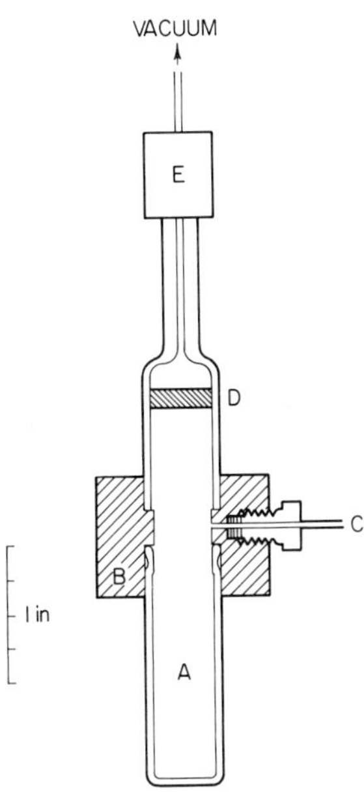

FIG. 5. Resin washing device. A, one-dram sample vial fitted into Teflon collar (B); C, solvent inlet tube; D, fritted-glass filter; E, Chromatronix glass-to-Cheminert fitting. Solvents are drawn into the apparatus by applying a vacuum to E and drawing solvents in through tube C.

taining up to about 0.35 μmole of peptide is mixed with 900 mg of glass beads, and the mixture is poured into the column with gentle tapping. Care should be taken that the packing is uniform and that resin bands do not appear, otherwise the column may become constricted during the degradation, because of swelling of the resin in TFA. Banding is avoided when the glass bead–resin mixture is slightly "sticky," rather than free-flowing. A trace of moisture usually imparts the proper stickiness. When all the resin has been added, the remainder of the column is filled with glass beads, the top end fitting is attached, and the column is mounted in the sequencer (Fig. 4). The reagent flow is downward.

Degradation of the Peptide. Degradation is carried out at 45°, as yields tend to decrease at higher and lower temperatures.[3] The operations controlled by the programmer are indicated in Fig. 6. In order to deblock the peptide amino groups, the programmer is set at mid-cycle (60 minutes) and the BOC groups are removed by pumping TFA through the column. Degradative cycles then occur automatically, and thiazolinones are collected in the fraction collector. After the last cycle, the instrument shuts down. The dead volume between the valves and

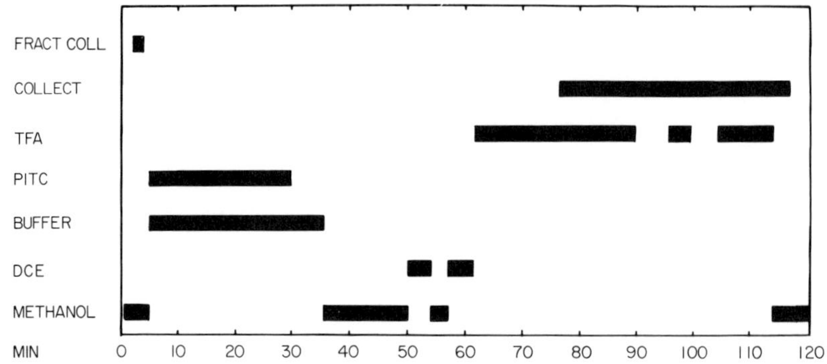

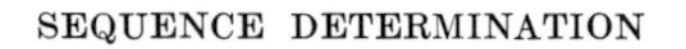

FIG. 6. Sequencer program. Darkened areas indicate times when a given function is turned on. With this program, 1.7 ml of buffer, 0.70 ml of 5% phenyl isothiocyanate (PITC) in acetonitrile, and 2.3 ml of trifluoroacetic acid (TFA) are pumped through the column during one cycle. DCE, dichloroethane.

resin is such that it requires about 15 minutes for the resin to be wetted completely by TFA or PITC-buffer. The difference in reaction time between the top and the bottom of the resin column is about 4 minutes. Invariably, bubbles form in the reaction column during the TFA cycle; this appears not to affect the results, however.

Analysis of Phenylthiohydantoins. PTH's are analyzed by the isotope dilution method.[6] Each tube (18 × 75 mm) in the fraction collector contains a standard mixture of unlabeled PTH's (about 0.1 μmole each) corresponding to the composition of the peptide. At the end of a cycle, each tube contains about 2.4 ml of TFA and 2.0 ml of methanol, which are partially converted to methyl trifluoroacetate and water on standing. The tubes are placed in a heated (50°) vacuum desiccator, which is evacuated first with a water aspirator to remove the more volatile components and then with a mechanical pump. A 0.2-ml aliquot of 20% TFA is added to each tube, the tubes are heated at 80° for 10 minutes to complete isomerization, and the contents are evaporated again. The residue is dissolved in 1 ml of methanol, and the solution is applied to a column (5 × 10 mm) of Dowex 50-X2 (previously washed with methanol) in a plugged disposable pipette. The original tube and resin column are rinsed with 4 × 1 ml of methanol, and the washings are collected in a small tube (18 × 75 mm; made from a 18 × 150-mm test tube) having a conical bottom. The tubes are evaporated on the rotary evaporator (using an adapter having a female standard taper joint on one end and a No. 1 rubber stopper on the other), and the residue is dissolved in 50–100 μl of acetonitrile or dichloroethane. Ten to 20% of the solution is applied to a precoated silica gel plate containing a fluorescent indicator (E. Merck and Co., No. F-254)[11] as a 1-cm streak. The plates are developed in an appropriate solvent (see this volume [6]), and spots are located under a UV lamp and are scraped off and transferred to scintillation vials.[12] Ethanol (0.5 ml), followed by 7 ml of toluene scintillator solution, is then added. It is advisable to wait for an hour or so before counting to allow the PTH's to be leached out of the silica gel completely. Quantitative estimation of PTH's can be made by measuring the ultraviolet absorption spectra and radioactivity of samples.

Sample Degradation

Figure 7 shows the results of degrading 0.24 μmole of oxidized A chain of insulin through 21 cycles at 45°. The PTH mixture from each cycle was dissolved in 100 μl of acetonitrile. Ten microliters (20 μl after the tenth cycle) of each solution was applied to a silica gel plate, and the plate was developed in chloroform–ethanol (98:2). The UV-absorbing spots were then located and counted (left-hand column of Fig. 7). In cases where the radioactive PTH remained at the origin, samples were also chromatographed in chloroform–ethanol (9:1) (right-hand column of Fig. 7). In the latter solvent, the PTH's of asparagine and glutamine are only partially resolved and some overlapping is observed. The PTH of glutamic acid methyl ester (gle), formed during the PTH work-up procedure by esterification of PTH glutamic acid or methanolysis of PTH glutamine, runs with PTH alanine in chloroform–ethanol (98:2).

PTH yields for each cycle are plotted in Fig. 8. Individual points are not corrected for mechanical loss, dilution errors, etc. These data show that degradation occurred with an average yield of about 94%.

Degradation of the A chain of insulin and analysis of the PTH's were completed within a period of about 70 hours. Included in this total were 43 hours for degrading the peptide on the sequencer, 16 man-hours for processing the PTH's, and 30 hours for counting 180 samples in the scintillation counter. Since the sequenator runs unat-

[11] Some precoated thin-layer plates contain a fluorescent indicator that phosphoresces during scintillation counting, resulting in an extremely high background.

[12] Transfer may be accomplished conveniently by means of a simple vacuum device consisting of a 10-mm diameter glass tube, ground square at one end attached to a vacuum line. The vacuum is turned on, and a 15-mm filter paper disk (e.g., Schleicher and Shuell No. 410 paper) is placed across the squared end of the tube. With the filter disk held in place by air pressure, the device can be used to pick up the loosened silica gel, which adheres to the filter paper. When the vacuum is released, both the disk and the silica gel fall into the scintillation vial.

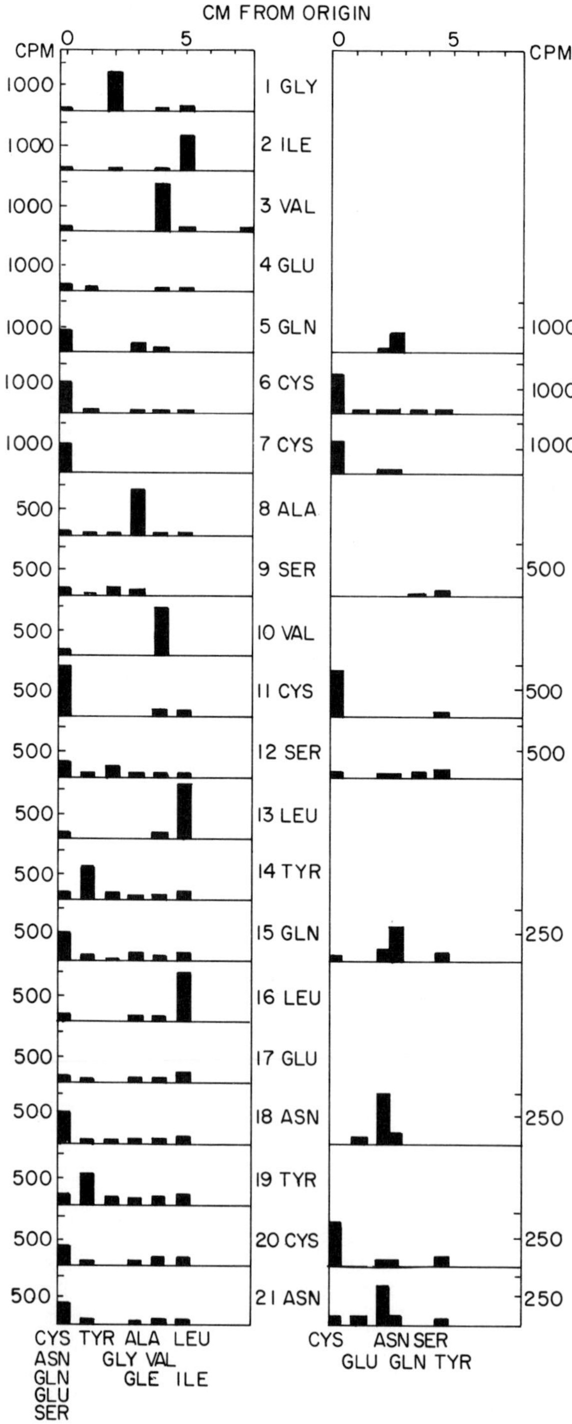

Fig. 7. Degradation of 0.24 μmole of oxidized insulin A chain (see text for details).

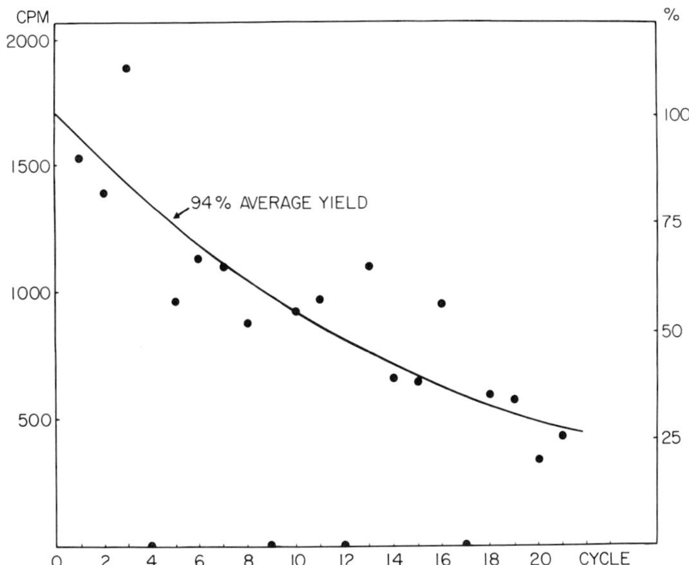

FIG. 8. Plot of PTH yields in the degradation of oxidized A chain of insulin. The solid line is that calculated assuming a degradative yield of 94% per cycle.

tended, processing of PTH's can be carried out simultaneously with degradation of the peptide.

Discussion

Scope of the Method. The data in Figs. 7 and 8 give some indication of the potential of the solid-phase method. It may be possible, even under present conditions, to perform as many as 30 degradative cycles on peptides. Of 20 common amino acids, only aspartic and glutamic acids present problems (see below) peculiar to the solid-phase method.

The sensitivity of the method has not been determined, but it has been possible to degrade a 100-nmole sample of the B chain of insulin through 18 cycles, with an average yield of 89%, and to identify the PTH's released unambiguously. The amount of degradable peptide remaining after the 18th cycle was about 12 nmoles. We have used the sequencer routinely to sequence 25- to 50-nmole samples of short (5–10 amino acids) peptides from ribosomal proteins.

Yields in attaching peptides to the resin have ranged from about 70% for the 30-amino acid B chain of insulin to nearly 100% for smaller peptides, using carbonyldiimidazole as the coupling agent. This includes blocking the N-terminal amino acid with BOC-azide. Loading ratios range from about 0.1 to 0.3 μmole of peptide per 50 mg of resin, the amount of resin ordinarily used in degradations. Doolittle and Mross[13] have achieved nearly quantitative attachment yields with a water-soluble carbodiimide for coupling.

Variables Affecting the Degradation. The optimum temperature for the reaction, under the present reaction conditions, is about 45°, as degradative yields decrease at higher and lower temperatures. At low temperatures, the reactions are incomplete, while at high temperatures a side reaction occurs (observable by amino acid analysis of undegraded peptide) which results in deactivation of the peptide to further degradation. Presumably this reaction is desulfurization[14] of the PTC peptide to the phenylcarbamyl peptide, which does not cyclize. It seems unlikely, however, that this is caused by oxidizing agents, since these should be preferentially scavenged by the several hundredfold excess of methylthiocarbamyl groups on the resin.

The protective action of the methylthiocarbamyl groups, as well, perhaps, as some unreacted resin amino groups, may explain why we have observed no effect of reagent and solvent purity on the degradation. Similar observations have been made with the solid-phase method of Dowling and Stark.[8]

The Sequencer. The solid-phase sequencer is simple in design and operation, and can be constructed, for the most part, from commonly available components at a relatively low cost. The absence of electronic components and moving parts (except for the pumps, pneumatic valves, and programmer motor) simplifies maintenance and modification of the apparatus. The reaction column is readily demountable, making it possible to start a new run within a few minutes after the first run has been finished. The sequencer also has the potential for being modified to sequence two or more peptides simultaneously. This can be done simply by adding more reaction columns and a valve that will distribute reagents to each column, and increasing the reagent pump rates.

Analysis of PTH's. The isotope dilution procedure[6] has been used for analysis of PTH's because it is sensitive, gives quantitative information, and requires no specialized instrumentation other than a scintillation counter. We have noted, however, under a rather wide variety of conditions, that the recovery of radioactive PTH's is only 60–70% that expected from amino acid analysis of the peptide on the resin. This may be due to acid-catalyzed exchange of tritium out of the phenyl ring of the intermediate anilinothiazolinone, although it remains

[13] R. Doolittle and G. Mross, personal communication (1970).

[14] D. Ilse and P. Edman, *Aust. J. Chem.* **16**, 411 (1963).

to be demonstrated. Air oxidation of the PTH's seems unlikely, as we have never observed any radioactive artifacts that could be attributed to PTH oxidation products.

Difficulties. When glutamic acid is present in peptides, its side-chain carboxyl becomes activated along with the COOH-terminal carboxyl and forms an amide bond with the resin in the peptide attachment step. As a result, the PTH of glutamic acid remains attached to the resin after Edman degradation and a gap appears in the PTH analysis (see Fig. 7, cycles 4 and 17). In simple cases, however, this does not present much of a problem, as often the position of glutamic acid can be deduced by difference. With aspartic acid the situation is worse, because degradation appears to stop completely at this amino acid. A likely explanation is that activation of the β-carboxyl by carbonyldiimidazole gives rise to an imide which cannot undergo further reaction (see discussion in this volume [26a]).

There are two possible solutions to this problem for tryptic peptides. Doolittle and Mross[13] have converted peptide carboxyl groups to amides using a water-soluble carbodiimide and glycinamide, and have cleaved the COOH-terminal amide selectively with trypsin. Preliminary results in our laboratory[15] indicate that tryptic peptides containing lysine can be attached by their lysine ϵ-amino groups to the peptide resin using *p*-phenyl diisothiocyanate, which also couples with the α-amino group of the *N*-terminal amino acid. Treatment of the resin with TFA releases the α-amino group of the second amino acid but leaves the peptide attached by the lysine ϵ-amino group. This procedure has the advantage that prior chemical modification of the peptide and anhydrous reaction conditions are not required.

Serine is degraded normally in the solid-phase degradation, but under present conditions, its PTH is completely destroyed by the TFA used in the cleavage step. For this reason, a gap also appears in the PTH analysis for serine. In some cases a distinction can be made between serine and glutamic acid, as the breakdown products of PTH serine give a characteristic pattern on thin-layer chromatography (compare, for example, cycles 4 and 17 with 9 and 12 in Fig. 7). A more satisfactory solution to this problem is needed, however.

PTH-threonine is also partially dehydrated in TFA to the dehydro derivative which usually runs with PTH-phenylalanine on thin-layer chromatography.

[15] M. Horn, unpublished results (1971).

[28] Subtractive Edman Degradation with an Insoluble Reagent

By GEORGE R. STARK

Principle[1]

An insoluble Edman reagent, analogous to phenylisothiocyanate, has been synthesized from polystyrene beads. The reagent contains covalently linked glucosaminol to increase its hydrophilic character. It may be used as a column to which peptides are attached during the addition reaction and from which shortened peptides are eluted during cleavage. Degradations are followed subtractively or with the dansyl end group procedure of Gray (this volume [8]). One complete cycle takes about 2 hours, and several peptides can be degraded in tandem easily. Yields are about 75% per cycle for peptides of about 12 residues or less. An advantage of the method is that the terminal amino acid is lost completely after each cycle, so that unambiguous results are obtained even after many degradative steps have been performed.

Preparation of the Reagent

The preparation is illustrated schematically in Fig. 1.

Nitropolystyrene. Polystyrene beads cross-linked with 0.25% divinylbenzene (No. L982) were obtained from Sondell Scientific Instruments, Palo Alto, California. We have used the fraction which passes through a 100-mesh screen when dry. Wash the beads several times with chloroform to remove residual linear polystyrene, then dry them in a hood after most of the solvent has been removed by suction filtration on a coarse sintered-glass funnel. Nitration and reduction are carried out as first described by Chen.[2] With rapid mechanical stirring, add 40 g of resin slowly, over a period of about 30 minutes, to about 1.5 liters of fuming nitric acid (specific gravity 1.5) precooled in an ice–water bath. Do not allow the temperature of the reaction mixture to rise above about 15°. As nitration proceeds, the resin will swell greatly. If stirring becomes too difficult, add small amounts of concentrated nitric acid. About 30 minutes after the last of the polystyrene has been added, remove the stirrer and allow the reaction mixture to stand overnight at about 4° to complete the nitration. Pour the reaction mixture slowly, with vigorous

[1] L. M. Dowling and G. R. Stark, *Biochemistry* 8, 4728 (1969).

[2] C. H. Chen, Ph.D. Dissertation, Polytechnic Institute of Brooklyn, Brooklyn, New York, 1955.

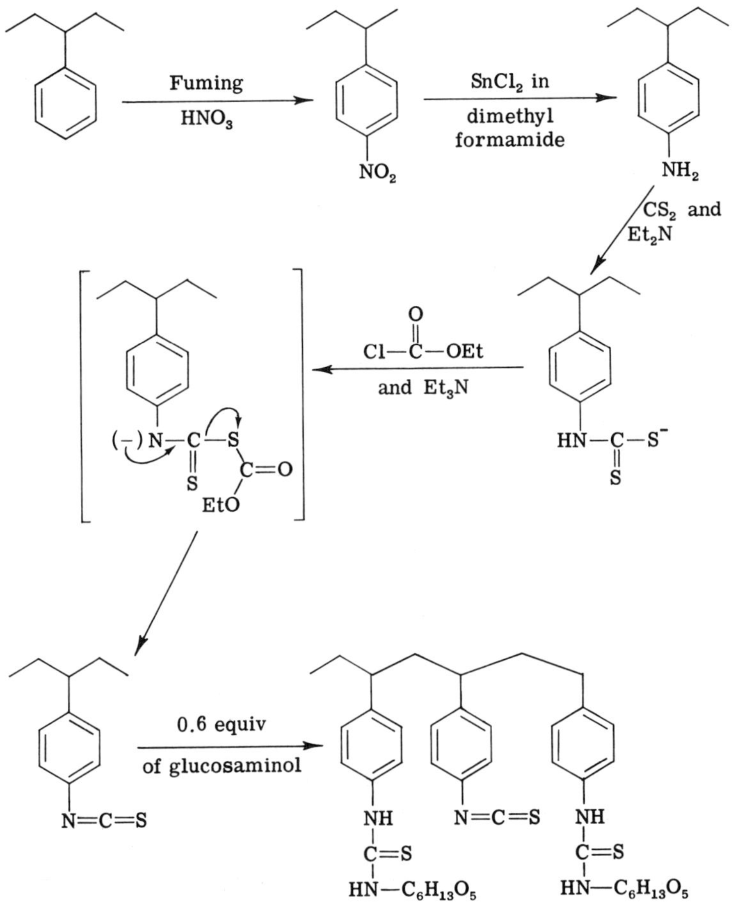

FIG. 1. Scheme for the synthesis of glucosaminol–NCS–polystyrene.

mechanical stirring, into an ice–water slurry, adding more ice as required (use a 4-liter beaker). Filter the resin on a coarse sintered-glass funnel and wash it five or six times with distilled water. Divide the damp nitropolystyrene into halves; store one half at 4° for a future preparation and use the other for the remainder of the reactions described below.

Aminopolystyrene. Wash the nitro resin on the sintered-glass funnel several times with reagent dimethyl formamide. The beads should collapse initially, then swell. Be sure that there are no clumps of unswollen resin remaining and that all the beads are transparent before proceeding with the reduction. Transfer the resin to a 2-liter Erlenmeyer flask in a total volume of about 1 liter of dimethyl formamide. Heat the suspension to 100° with stirring (a water bath atop a stir plate is convenient), and then add to it a solution of 200 g of $SnCl_2 \cdot 2H_2O$ in 400–500 ml of dimethyl formamide. Allow the mixture to stir at 100° for about 6 hours. Add 200 ml of 12 M HCl and continue to stir at 100° for about 6 hours more.

NCS–Polystyrene. The procedure for converting amino groups into NCS groups with CS_2, triethylamine, and ethyl chloroformate is adapted from Garmaise *et al.*[3] Filter the resin, using a sintered-glass funnel, and wash it alternately with dimethyl formamide and 6 M HCl, six times each. The beads should collapse in the HCl and swell again in dimethyl formamide. If they do not swell very much, add a little 6 M HCl. Finish the washes with dimethyl formamide and, this time, allow the beads to collapse partially. Add to the resin a mixture of dimethyl formamide and about 40 ml of triethylamine (redistilled from ninhydrin). Take about 10 minutes to break up the clumps of resin as well as possible with a large spatula, to ensure complete neutralization. Filter, then wash the resin with dimethyl formamide once or twice to remove excess triethylamine and to aid in dispersing the beads. Transfer the slurry of resin to a 2-liter three-necked flask in a total volume of about 1.2 liters of dimethyl formamide. Cool the flask in ice–water and, with rapid mechanical stirring, add dropwise and simultaneously 40 ml of CS_2 (stored in a freezer) and 90 ml of redistilled triethylamine. (CS_2 was purified according to Leonis and Levy[4] by shaking in turn with mercuric sulfate and concentrated nitric acid, then allowing it to stand over soda lime, followed by distillation. The purified CS_2 was stored in a freezer.) Remove the ice, fill the flask almost to capacity with dimethyl formamide, and allow the reaction mixture to warm up to room temperature over a period of about 4 hours, with continued rapid stirring. Cool the flask in ice–water again and add 85 ml of ethyl chloroformate (redistilled and stored in the freezer) drop-wise with stirring. Remove the ice–water and stir the reaction mixture slowly at room temperature overnight. Filter the resin and wash it with chloroform several times, until all visible triethylammonium chloride has been removed, then two or three times with pyridine (redistilled from ninhydrin and stored at 4°).

Glucosaminol–NCS–Polystyrene. After the final wash, allow the resin to remain on the filter with suction for about 5 minutes after the bulk of the solvent has passed through. Remove small portions of the damp resin for determination of capacity (see below) and, without delay, weigh the remainder into a 2-liter Erlenmeyer flask. Add about 1.2 liters of pyridine, and stir the resin with cooling for several hours while the

[3] D. L. Garmaise, R. Schwartz, and A. F. McKay, *J. Amer. Chem. Soc.* **80,** 3332 (1958).

[4] J. Leonis and A. L. Levy, *C. R. Trav. Lab. Carlsberg Sér. Chim.* **29,** 57 (1954).

determination of capacity is being completed, to disperse the beads as well as possible. Add all at once 75 ml of *N*-ethylmorpholine (redistilled from ninhydrin and stored in the freezer), then add a solution of glucosaminol hydrochloride in water dropwise with rapid stirring over a period of about 10 minutes, with enough cooling to prevent the temperature from rising above about 25°. The volume of this solution in milliliters should be about twice the weight of glucosaminol hydrochloride in grams. The number of millimoles of glucosaminol added should be 0.6 of the total number of milliequivalents of NCS, determined in the capacity experiment. After the glucosaminol has been added, continue to stir the solution for 1.5 hours at room temperature. Add dropwise, but rapidly, 60 ml of triethylamine (redistilled); after 15 minutes, filter the resin and wash it immediately three times with pyridine. Store the resin under pyridine at 4°. The capacity of the glucosaminol–NCS–polystyrene can be determined as described below, but remember to use about three times as much glucosaminol–NCS–resin as NCS–resin, since the capacity will be lower.

Determination of the Capacity of NCS–Polystyrene. Place about 1 g of damp resin into each of three tared 25-ml Erlenmeyer flasks and reweigh the flasks to determine the exact weight of the resin. Dry one of the samples in a vacuum desiccator over H_2SO_4 for several hours, using an oil pump, then reweigh the flask to determine the dry weight of the resin. To each of the other two flasks, add 10.0 ml of a mixture of nine volumes of pyridine and one volume of triethylamine, than 1.0 ml of 1.50 *M* glycine ethyl ester hydrochloride. Agitate the mixtures for several hours at room temperature along with a nonresin blank, dilute the supernatant solutions 1:500, and determine the amount of ninhydrin-positive material remaining in each. Calculate the capacity of the dry resin in milliequivalents per gram, including a correction for the volume of pyridine introduced along with the resin.

Glucosaminol Hydrochloride. Dissolve 50 g of commercially available glucosamine hydrochloride in 150 ml of water and, with cooling, adjust the pH to 10 with concentrated NaOH. Add 0.08 mole of solid $NaBH_4$ in small portions at room temperature with rapid stirring. One hour after the addition of $NaBH_4$ has been completed, adjust the pH to about 2 with concentrated HCl and evaporate the mixture under a vacuum until solid begins to separate. Add enough water to redissolve the solid, then precipitate the crude glucosaminol hydrochloride with nine volumes of alcohol. Recrystallize it from alcohol–water (final yield about 35 g).

Degradation of Peptides

Immediately before use, pipette the resin in pyridine suspension into a small jacketed tube, thermostated at 50°, about 0.6 cm i.d. A porous

Teflon disk, which is easily removed after use, is very convenient for closing the bottom of the tube. Wash the resin with addition solvent (75 volumes of pyridine, 5 volumes of *N*-ethylmorpholine, and 20 volumes of glass-distilled water) and adjust the height of the bed to about 5 cm. Dissolve the dry peptide in no more than 200 μl of addition solvent and pipette as much of the solution as possible onto the column of resin. (If the peptide contains lysine, it must be carried through one degradation with soluble reagent first, to block the ε-NH_2 group.) Allow the sample to flow into the resin by gravity or use gentle nitrogen pressure, then wash the remaining peptide onto the resin with two 100-μl portions of addition solvent. After 1 hour, wash the resin, first two or three times with addition solvent, then thoroughly with tetrahydrofuran, using nitrogen pressure to drive the solvents through rapidly. Stir the resin–tetrahydrofuran suspension with a glass rod to ensure that no regions of addition solvent remain through channeling. The same rod can be reinserted into the damp bed of resin and used to stir it again during cleavage. After the bulk of tetrahydrofuran has been driven off, place a clean heavy-walled glass tube under the column, then add about 1.5 ml of cleavage solvent (4 volumes of trifluoracetic acid and 1 volume of acetic acid, each redistilled), stir to suspend the resin, and blow the solvent into the tube with nitrogen. Resuspend the resin in about 1.5 ml more of cleavage solvent and allow the reaction to proceed for a total of 30 minutes at 50°. Blow the solvent into the tube, wash the resin three times with more solvent, and evaporate all to dryness, using an oil pump. Rotation during this procedure prevents bumping, and a warm-water jacket hastens evaporation. Redissolve the residue left in the tube in a known volume of addition solvent and remove a small aliquot for hydrolysis and amino acid analysis; place the remainder on a fresh column of resin for the next round of degradation. One complete cycle takes about 2 hours. It is convenient to carry out four cycles on a single peptide in a normal working day and it is easy to degrade several peptides in tandem. The resin is not used more than once: only about 1.5 ml of resin is required for a single degradation and the laboratory-scale preparation described above, which yields about 1 liter, suffices for about 600 experiments.

Discussion

Degradation of a Peptide. Even with recoveries of only about 75% per stage, it has been possible to degrade a dodecapeptide completely with glucosaminol–NCS–polystyrene (see the table). The peptide, derived from bovine catalase, was kindly provided by Dr. Walter Schroeder, who had previously determined its sequence independently by

DEGRADATION OF THR-THR-GLY-GLY-GLY-ASN-PRO-VAL-GLY-ASP-LYS-LEU

Amino acid	Theory	Initial peptide[b]	Molar ratios of amino acids[a]										
			Round										
			1[c]	2	3	4	5	6	7	8	9	10	11
Lys	1	0.9	1.0	—[d]	—[d]	1.0	1.2	—[d]	1.0	1.5	—[d]	—[d]	—[d,e]
Asp[f]	2	1.8	2.0	2.3	2.2	2.0	2.0	**1.2**	1.1	1.7[g]	1.0	**0.3**	0
Thr	2	1.4	**1.0**	**0.3**	0.2	0	—	—	—	—	—	—	—
Pro	1	0.9	0.9	1.2	1.3	0.9	0.9	0.8	**0**	—	—	—	—
Gly	4	3.7	3.5	3.8	**3.0**	**2.1**	**1.5**	1.6	1.2	1.7[g]	**0.5**	0.4	0
Val	1	1.0	0.9	0.9	0.9	1.0	1.1	0.8	1.1	**0.2**	0	—	—
Leu	1	*1.0*	*1.0*	*1.0*	*1.0*	*1.0*	*1.0*	*1.0*	*1.0*	*1.0*	*1.0*	*1.0*	*1.0*[e]
Yield (%)	—	—	100	18[h]	66	80	65	74	90	86	43	33	98
Amount hydrolyzed (%)	—	—	2.5	2.5	2.5	5.0	5.0	7.5	10	15	20	40	—[e]

[a] The number used as the basis of calculation of molar ratios is italicized. The residue that should be removed in each round is in boldface type.

[b] This analysis of the initial peptide was made shortly after isolation (November, 1965) by Dr. Walter Schroeder. Our own analysis differed by less than 0.1 residue for every amino acid *except* threonine, which had decreased to 1.1 residues.

[c] This round was carried out with phenyl isothiocyanate in solution; the yield at this stage is defined as 100%. Peptide (3.2 μmoles) was put onto the resin for round 2.

[d] Not analyzed; lysine values are somewhat high owing to an artifact from the resin.

[e] No hydrolysis.

[f] No attempt was made to distinguish between Asp and Asn.

[g] This particular sample was contaminated by appreciable amounts of serine and alanine, which are not present in the peptide. The high values for aspartic acid and glycine probably reflect the contamination.

[h] The low yield at this point probably reflects the low value for threonine in the initial peptide; see the text and footnote *b*.

other methods. The data reported in the table illustrate several points. (1) If a peptide contains lysine, the ϵ-amino group must be blocked before degradation with glucosaminol–NCS–polystyrene is begun.[5] This is easily accomplished by carrying out the first cycle of reactions with soluble reagents and, in the case illustrated in the table, the soluble round was followed by 10 rounds on the resin. Note that the carboxyl-terminal leucine was analyzed without hydrolysis after round 10. (2) In this particular case, the peptide was not pure, judging from the low recovery of threonine obtained by both Dr. Schroeder in 1965 and Dowling and Stark[1] in 1969. It appears that the amino terminus of a substantial fraction of the material has been altered. A blocked amino terminus is probably the cause of the exceptionally low recovery of peptide after the first round on the resin. Despite this difficulty, it was possible to achieve complete degradation because the minor fraction of peptide that did have a free amino terminus was separated from the major fraction of impurity during the first degradation with the resin. (3) The recoveries obtained require that an increasing fraction of peptide be used for analysis as a degradation progresses. The protocol used with the dodecapeptide is shown at the bottom of the table.

Apart from speed and convenience, the completeness of degradation evident in the data of the table is one of the principal advantages of the method. Side reactions or incomplete degradation can lead to a reduction in yield with the resin, but should not lead to contamination of the degraded peptide, because there is a selection for the desired material at two stages during each degradation. First, during the addition reaction, any compound that lacks a free amino group is not held covalently by the resin and will be washed away before the degraded peptide is released by trifluoroacetic acid. For example, if a peptide with a blocked amino group is present as a contaminant, it will be separated from peptides that do have free amino groups during the first step because it cannot add to the resin. Such a situation is sometimes encountered when a residue of glutamine is reached in degradations with soluble phenyl isothiocyanate, because glutamine cyclizes readily to pyrrolidonecarboxylic acid, which has no free amino group, under the alkaline conditions required for the addition reaction to occur. A second purification occurs during the exposure of the resin to trifluoroacetic acid, for no compound that has added to the resin will be released unless it is a peptide. For example, free amino acids would be removed from a peptide undergoing

[5] Digestion with carboxypeptidase B is an alternative procedure for preparing tryptic peptides for resin degradation, since such peptides are known to have their lysine residues at the carboxyl termini. No purification of the peptide should be necessary after the enzymatic digestion, since any free lysine produced, as well as any unreacted peptide, will be bound irreversibly to the resin through ϵ-amino groups.

degradation. Also, if the cleavage reaction is not complete, it is obvious that the uncleaved peptide will not be released and will not contaminate the product, as it would if the degradation were being carried out with soluble reagents.

The major limitation of the method is the failure to achieve complete addition of peptide to resin each time. The average recovery of peptide after each cycle varies from 69 to 78% in the degradations described by Dowling and Stark.[1] Although there is room for substantial improvement in this aspect, the procedure in its present state of development is useful for degradation of peptides of about ten residues or less.

Properties of the Resin. Polystyrene was chosen as the supporting matrix for NCS groups because of its resistance to chemical degradation and because it is readily available in stable bead form with a low degree of cross-linking, for maximum swelling. The degree of cross-linking chosen for glucosaminol–NCS–polystyrene, 0.25%, is the lowest degree compatible with physical stability: several experiments were attempted with 0.1% cross-linked polystyrene, but most of the beads disintegrated during synthesis and the rather mushy final resin was extremely difficult to work with in small columns. Some successful experiments were carried out with 0.25% cross-linked beads initially 25 μ in diameter, but these seemed to have no particular advantages over the larger beads recommended above (minus 100 mesh; i.e., smaller than 150 μ) and were much more difficult to use.

The capacity of a typical preparation of NCS–polystyrene was 5.4 meq/g (theory is 6.2). Since it is not unlikely that a small amount of triethylammonium chloride remains trapped in the resin after the chloroform washes, the actual capacity is probably somewhat higher. Note that residual salt does not affect the calculation of total capacity, necessary for an accurate estimation of the amount of glucosaminol to be added. Glycine ethyl ester was chosen for use in the determination of capacity because it is a small molecule and uncharged at alkaline pH. Use of an amino acid rather than an ester would have given a thiourea–resin with a high density of negative charge, which would have repelled additional molecules of amino acid, also negatively charged in the presence of triethylamine, resulting in a low estimate of capacity. Glucosaminol was chosen over glucosamine because of the observation[6] that aldehydes are deleterious since they can react with the terminal amino group of the peptide. However, no great difference was apparent when resins prepared with glucosamine were compared with resins prepared with glucosaminol.

The manner in which glucosaminol is added is the most crucial step

[6] P. Edman and G. Begg, *Eur. J. Biochem.* **1,** 80 (1967).

in the preparation of the final resin. If glucosaminol hydrochloride in water is added to a stirred suspension of resin in a mixture of pyridine and triethylamine, reaction of glucosaminol with resin is complete in less than 15 minutes at room temperature, but the product, although similar in appearance to resin prepared by slow addition of glucosaminol in the presence of *N*-ethylmorpholine, is very different in its properties. For example, glucosaminol (triethylamine) resin will not react with an appreciable amount of peptide in a solvent containing 20% water, whereas glucosaminol (*N*-ethylmorpholine) resin reacts with peptides readily under the same conditions. The glucosaminol (triethylamine) resin does react with some peptides when less water is present. The most probable explanation of the difference is that when glucosaminol is present as the free base (as in the presence of excess triethylamine), its rate of reaction with resin-bound NCS groups must be fast relative to its rate of diffusion into the resin beads. The result is a bead which has predominantly NHC(=S)–glucosaminol groups near the surface and predominantly NCS groups near the center. In polar solvents containing 20% water, the peptide cannot penetrate the hydrophobic interior of the bead, where the NCS groups are to be found, although it probably does penetrate the more hydrophilic exterior. When *N*-ethylmorpholine replaces triethylamine, the glucosaminol is present predominantly as the hydrochloride, slowing the reaction with NCS groups relative to diffusion and permitting a much more even distribution of NCS groups throughout the bead, so that a greater proportion of them are near the hydrophilic exterior. In the preparation described above, only about 60% of the glucosaminol has reacted with NCS groups after 1.5 hours at room temperature in the presence of *N*-ethylmorpholine. Triethylamine is added at this point to drive the addition to completion, because prolonged exposure of NCS groups to alkaline conditions in the presence of water results in hydrolysis and loss of capacity. It is probable that the properties of glucosaminol–NCS–polystyrene could be improved further by careful control of the mode of addition of glucosaminol.

In two preparations of glucosaminol–NCS–polystyrene, addition of 0.55 or 0.65 eq of glucosaminol per NCS group resulted in resins having a final capacity of 1.5 or 0.9 meq/g, respectively. These resins differ little in their physical properties (they both remained swollen during addition and during cleavage) or in the recoveries of peptides obtained. Hence, addition of 0.60 eq of glucosaminol, an intermediate amount, is recommended.

[29] Sequential Degradation of Peptides and Proteins from Their COOH Termini with Ammonium Thiocyanate and Acetic Anhydride

By GEORGE R. STARK

Principle[1,2]

Peptides and proteins react with ammonium thiocyanate and acetic anhydride under mild conditions to form peptidylthiohydantoins at their COOH termini. The peptidylthiohydantoin can be cleaved, also under mild conditions, to a thiohydantoin characteristic of the COOH terminus and an acetylated peptide or protein in which a new residue has become terminal (Fig. 1). Since internal peptide bonds are not cleaved in either step, the degradations can be performed sequentially. Degradations of peptides may be followed subtractively, by amino acid analysis of the peptide product after each stage. Alternatively, the thiohydantoins may be determined directly by thin-layer chromatography, sometimes in conjunction with hydrolysis and amino acid analysis. Direct identification of thiohydantoins is mandatory with proteins. The method described below is somewhat limited in that COOH-terminal aspartic acid and proline are not removed. (See the note added in proof.)

Procedure

Preparation of Reagents and Reference Thiohydantoins. Recrystallize reagent grade ammonium thiocyanate from absolute ethanol. Filter the hot solution, cool it quickly to complete the crystallization, and separate the product immediately by vacuum filtration. Solutions cooled too slowly develop a distinct pink color. Redistill acetic anhydride through a 2-foot Vigreux column; much better results are obtained with the distilled anhydride than with unpurified commercial reagent. The preparation should be stored at —20° and distilled again after several months. Reagent grade acetic acid may be used without purification for degradation of peptides. Purified acetic acid is required for thin-layer chromatography (see below). Reflux pyridine with excess phthalic anhydride, then redistill it through a Vigreux column. Distill hexafluoroacetone trihydrate, prepared from the sesquihydrate obtained from E. I. duPont

[1] G. R. Stark, *Biochemistry* **7**, 1796 (1968).

[2] L. D. Cromwell and G. R. Stark, *Biochemistry* **8**, 4735 (1969). See these papers for references to the previous literature.

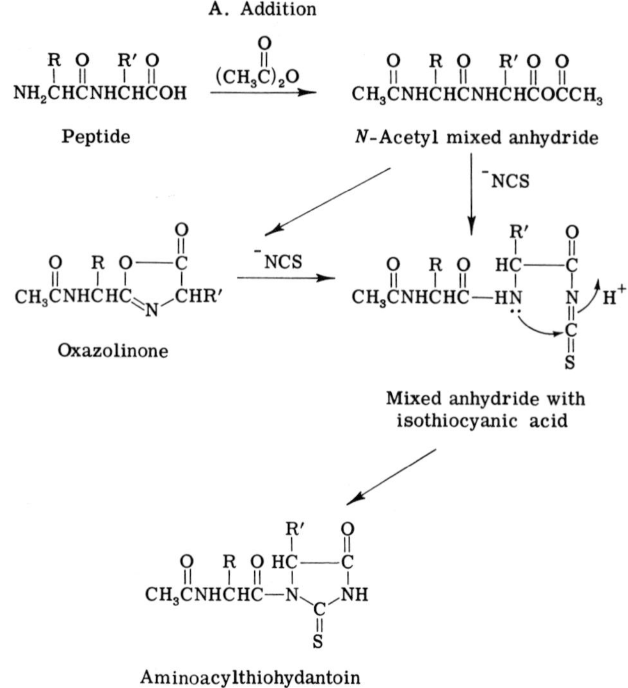

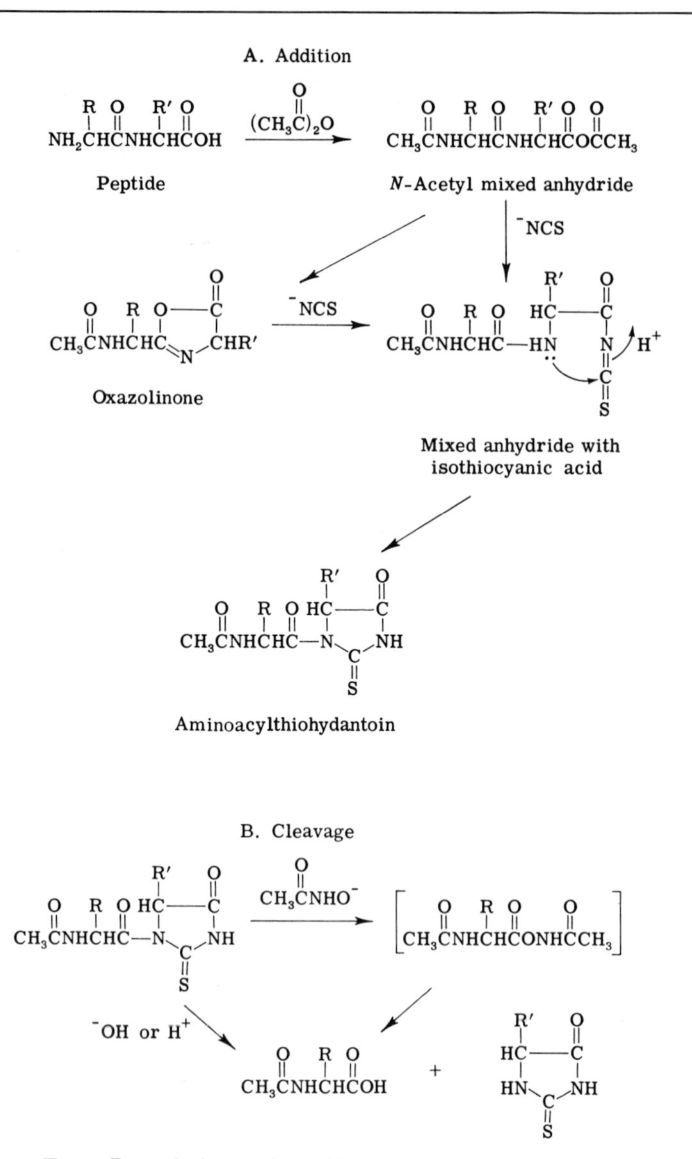

FIG. 1. Degradation of dipeptide from the COOH terminus.

de Nemours and Co., through a Vigreux column. Acetohydroxamic acid can be synthesized according to Blatt, and purified according to Wise and Brandt.[3]

[3] A. H. Blatt, "Organic Syntheses," Coll. Vol. 2, p. 67. Wiley, New York, 1944; W. M. Wise and W. W. Brandt, *J. Amer. Chem. Soc.* **77**, 1058 (1955).

Crystalline thiohydantoins derived from lysine, asparagine, threonine, glutamine, glycine, alanine, valine, methionine, methionine sulfone, isoleucine, leucine, tyrosine, phenylalanine, and *S*-carboxamidomethylcysteine may be synthesized by heating 10 mmoles of amino acid, 0.9 g of ammonium thiocyanate, 10 ml of acetic anhydride, and 1.3 ml of acetic acid to 100° for 30 minutes. After adding excess water, evaporate the solution to a solid or an oil. Redissolve the residue in 12 *M* HCl for 30 minutes at room temperature, evaporate to dryness again, and recrystallize from water. The thiohydantoin derived from histidine may be synthesized in the same way but is recrystallized from 95% alcohol. The thiohydantoin derived from tryptophan is synthesized similarly, but the intermediate acetylthiohydantoin is precipitated in water and purified by crystallization from 95% alcohol before cleavage in 12 *M* HCl; crystallize tryptophan thiohydantoin from ethyl acetate. Synthesize the thiohydantoins derived from arginine, glutamic acid, and *S*-aminoethylcysteine by dissolving the amino acid in a minimum volume of 50% acetic acid first, then add the other reagents specified above, plus enough additional acetic anhydride to react with the excess water. After cleavage in 12 *M* HCl, the thiohydantoin derived from arginine is purified by chromatography: the crude material is placed on a 2 × 10-cm column of Dowex 50-X8 in 0.2 *M* HCl and, after the column has been washed with several volumes of this solvent, the thiohydantoin is eluted with 100 ml of 3 *M* HCl. After removal of solvent in a vacuum, recrystallize the product from methanol. The derivative of glutamic acid is purified after cleavage in 12 *M* HCl as follows: Place the crude material on a 2 × 10-cm column of Dowex 1-X8 in 0.05 *M* HCl and elute the column with the same solvent. The thiohydantoin immediately follows a peak of yellow material which is not retarded by the resin; recrystallize it from ethyl acetate—petroleum ether (bp 30–60°). The acetylthiohydantoin derived from *S*-aminoethylcysteine is cleaved in 0.01 *M* NaOH at room temperature for 30 minutes, then recrystallized from water after neutralization. The properties of these reference thiohydantoins are summarized in Table I.

Degradation of Peptides

Method A

Formation of Peptidylthiohydantoin. Dissolve the peptide in 0.5 ml or less of 50% acetic acid. Prepare a mixture of 100 mg of recrystallized ammonium thiocyanate, 4.0 ml of redistilled acetic anhydride, and 1.0 ml of glacial acetic acid. When the salt has dissolved, add all the mixture slowly with swirling to the solution of peptide. Heat the mixture to 50°

TABLE I
PROPERTIES OF CRYSTALLINE THIOHYDANTOINS

Amino acid from which thiohydantoin is derived	Melting points (°C)		Yields of amino acid after alkaline hydrolysis–oxidation (%)[a]	
	Found	Literature	No chromatography	After thin-layer chromatography
Tryptophan	190–192	190[b]	Destroyed	Destroyed
Lysine[c]	189–191		88	71
Histidine[d]	220 dec		33	16
Arginine[d]	148–150		84[e]	47[e]
Asparagine	252 dec	246[f]	72[g]	65[g]
Threonine[h]	264[i]	264[b,f]	Destroyed	Destroyed
Glutamic acid	115–116		64	44
Glutamine	189–191	190–191[f]	62[g]	28[g]
Glycine	229–231	227[f]	102	112
Alanine	165–166	163–165[b]	67	45
Valine	137–140	139–140[b]	62	53
Methionine	147–149	148–149[b]	42[j]	20[j]
Methionine sulfone	233–236		22	26
Isoleucine	132–133	131–133[b]	70[k]	40[k]
Leucine	177–178	172–173[f]	48	35
Tyrosine	206–208	211[b]	108	59
Phenylalanine	178–180	184–184.5[b]	87	65
S-Carboxamidomethylcysteine[l]	167–168		Destroyed	Destroyed
S-Aminoethylcysteine[c]	180 dec		Destroyed	Destroyed

[a] Corrected for recovery of valine hydantoin (internal standard); average of duplicate experiments.

[b] M. Jackman, M. Klenk, B. Fishburn, B. F. Tullar, and S. Archer, (1948). *J. Amer. Chem. Soc.* **70,** 2884.

[c] The side-chain amino group is acetylated in the thiohydantoin.

[d] Hydrochloride salt of the thiohydantoin.

[e] Recovered as ornithine, which can be separated from lysine on the 55-cm column of the amino acid analyzer at pH 5.28.

[f] J. M. Swan, *Aust. J. Sci. Res.* **A5,** 711.

[g] Recovered as the acid.

[h] The product is 5-ethylidene-2-thiohydantoin. Anal. Calcd.: C, 42.24; H, 4.26. Found: C, 42.06; H, 4.33.

[i] Rapid heating.

[j] Recovered as methionine sulfone.

[k] Isoleucine plus alloisoleucine.

[l] Anal. Calcd.: C, 32.86; H, 4.14; N, 19.14. Found: C, 33.07; H, 4.17; N, 18.64.

for 6 hours. Add 100 mg of ammonium thiocyanate, swirl the liquid to dissolve the salt, then continue heating for 18 hours at 50°.

Isolation of Peptidylthiohydantoin. Add 3 ml of water to the reaction mixture, swirl it, then allow the solution to stand for a few minutes to allow hydrolysis of the excess acetic anhydride to be completed. Transfer the solution to a 2×50-cm column of Sephadex G-25 (fine beads in 50% acetic acid). Deaeration of the eluent and swollen Sephadex with an aspirator prevents accumulation of gas pockets in the column. Wash the column with 50% acetic acid at any convenient flow rate. Collect 5-ml fractions and pool those with ultraviolet absorbancy. An inexpensive ultraviolet flow cell and recorder is a great convenience in locating the peptidylthiohydantoin. Yellow, probably polymeric, products of the decomposition of thiocyanate emerge from the Sephadex column ahead of the usual position of salt; peptidylthiohydantoins containing fewer than five or six residues merge with the beginning of this peak. Peptidylthiohydantoins larger than about seven residues are resolved. Although the first 20–25 ml of the peak does contain material that absorbs in the ultraviolet region, it contains very little salt so that, when small peptides are degraded, these fractions can be pooled and used in the cleavage step. Other yellow reaction products adsorb strongly to the Sephadex and require about two column volumes of wash to be eluted. The column is ready for use again when no more ultraviolet-absorbing material emerges.

Cleavage of the Peptidylthiohydantoin. Pool the peak tubes quantitatively and remove the solvent by evaporation in a vacuum. Redissolve the residue in 0.5 ml of 0.1 *M* acetohydroxamic acid dissolved in 50% pyridine. (The reagent is stable at 4° for at least several months.) Heat the solution to 50° for 2 hours, then evaporate it to dryness. Redissolve the residue in 3 ml of 50% acetic acid, and desalt it again. Transfer a suitable portion of the solution for hydrolysis to a tube. The remaining peptide, after concentration, is used for the next round of degradation.

Method B

A somewhat shortened procedure may be used. In this, the sample for analysis is removed immediately after the first desalting. The small amount of amino acid liberated from the thiohydantoin during acid hydrolysis does not contribute significantly to the analysis of the degraded peptide. Following cleavage and evaporation of the cleavage solvent, the peptide is redissolved in 0.5 ml of 50% acetic acid and the next addition reaction is begun without desalting again.

Hydrolysis and Analysis. Evaporate the portion to be hydrolyzed to dryness in a tube, then add 1 ml of 6 *M* hydrochloric acid and a small crystal of phenol. Seal the tube under vacuum and heat it to 110° for about 16 hours. The amount of peptide chosen for hydrolysis at each stage will depend upon the sensitivity of amino acid analysis.

Degradation of Proteins

Formation and Isolation of Proteinylthiohydantoin. Dissolve about 1 μmole of protein in a mixture of 1.0 ml of hexafluoroacetone trihydrate and 0.35 ml of water. (Since 50 nmoles of thiohydantoin can be detected readily by the thin-layer chromatographic procedure described below, less protein can be used if material is scarce.) After solution is complete, add dropwise with swirling a homogeneous solution, prepared immediately before use, of 100 mg of ammonium thiocyanate in 1.0 ml of hexafluoroacetone trihydrate and 4.5 ml of acetic anhydride. Heat the mixture to 50° for 2 hours, add 100 mg more of ammonium thiocyanate, swirl the liquid to dissolve the salt, then continue to heat for about 18 hours. Add 3 ml of water, wait a few minutes for the hydrolysis of excess acetic anhydride to be completed, then desalt the proteinyl thiohydantoin on a 2 × 50-cm column of Sephadex G-25, fine beads, in 50% acetic acid.

Cleavage of the Proteinylthiohydantoin and Separation of the Products. Pool the peak tubes and remove the solvent by evaporation under vacuum. Redissolve the residues in 1 ml of 12 *M* HCl, allow the solution to stand at room temperature for 30 minutes, then remove the HCl rapidly with a good vacuum pump, not heating the sample above room temperature. Redissolve the residue in a small volume of 50% acetic acid and separate protein from thiohydantoin in the same solvent on a column of Sephadex G-25 identical with the one used for desalting. Remove the solvent from the thiohydantoin immediately, add a small volume of methanol, and proceed with thin layer chromatography without delay.

Thin Layer Chromatography. Plates precoated with silica gel and impregnated with fluorescent indicator (No. F-254, Brinkmann Instruments, Inc.) are used. Acetic acid should be refluxed with chromic acid and redistilled; all other solvents are Spectrograde.

System A. Activate the plate by heating it to 110° for 60 minutes just before use and develop it with a mixture of heptane (95 volumes), 1-butanol (65 volumes), and 99% formic acid (30 volumes).

System B. Do not activate the plate. Develop it with a mixture of chloroform (100 volumes), 95% ethanol (50 volumes), and glacial acetic acid (15 volumes). Spot the thiohydantoins onto the plates from methanol solution and run them along with appropriate standards. The spots are easily visible under ultraviolet light, since thiohydantoins quench fluorescence strongly; 50 nmoles of thiohydantoin can be detected readily.

Conversion of Thiohydantoins into Amino Acids. Remove solvent from the thin layer plate under vacuum without delay, scrape the spot from the plate into a small centrifuge tube, extract the thiohydantoin with methanol, pool the extracts (an internal standard may be added at this point if desired), and evaporate them to dryness in a heavy-walled hydrolysis tube. Flush the tube with a slow stream of nitrogen and, with nitrogen still blowing into the tube, add 0.2 ml of 0.2 *M* NaOH, previously degassed with nitrogen. Cover the tube with a small beaker as the nitrogen line is withdrawn and immediately place it in a boiling-water bath for 2 hours. Add 0.8 ml of water, then 10 μl of 30% H_2O_2 to the cooled tube. After 30 minutes at room temperature, evaporate the solution to dryness, add 0.4 ml of 1.5 *M* Na_2SO_3 and 0.4 ml of 2.0 *M* NaOH, seal the tube under vacuum, and heat it to 110° for 16 hours. Open the cooled tube, add 1.5 ml of 1 *M* HCl and evaporate the solution to dryness. Add 3.0 ml of water, stir, centrifuge the precipitated silicates, and analyze 1.0 ml of the supernatant solution.

Discussion

Some Representative Results. In Tables II and III are shown degradations of two peptides derived from insulin and ribonuclease. In the

TABLE II
DEGRADATION OF THE OXIDIZED PHENYLALANYL CHAIN, INSULIN, BY METHOD B
NH_2Phe . . . Thr-Pro-Lys-Ala-CO_2H

Amino acid	Theory	Molar ratios of amino acids[a]			
		Initial peptide	Round 1	Round 2	Round 3
Lys	1	0.9	0.8	**0.2**	0.2
His	2	1.7	1.7	1.8	2.0
Arg	1	0.9	1.0	1.0	1.0
$CySO_3H$	2	1.8	1.8	1.9	2.0
Asp	1	1.0	1.0	0.9	1.0
Thr	1	1.0	0.9	0.9	0.9
Ser	1	0.9	0.9	1.0	1.0
Glu	3	*3.0*	*3.0*	*3.0*	*3.0*
Pro	1	1.0	0.8	0.9	**1.0**
Gly	3	2.9	3.2	3.2	3.0
Ala	2	1.9	**1.2**	1.0	1.1
Val	3	2.9	2.9	2.8	2.7
Leu	4	3.9	4.1	4.2	4.1
Tyr	2	1.9	1.9	2.0	1.9
Phe	3	2.9	2.9	2.9	2.7

[a] The number used as the basis of the calculation of molar ratios is italicized. The residue that should be removed in each round is in boldface type.

TABLE III
DEGRADATION OF O-T-10, RIBONUCLEASE, BY METHOD B
NH_2Lys-Glu-Thr-Ala-Ala-Ala-LysCO_2H

Amino acid	Theory	Initial peptide	Molar ratios of amino acids[a] Round 1	2	3	4	5
Lys	2	1.7	**0.8**	0.9	0.8	0.9	0.8
Thr	1	0.9	0.9	0.9	0.9	0.7	**0.6**
Glu	1	*1.0*	*1.0*	*1.0*	*1.0*	*1.0*	*1.0*
Ala	3	2.8	2.7	**1.8**	**1.1**	**0.7**	0.7

[a] The number used as the basis of the calculation of molar ratios is italicized. The residue that should be removed in each round is in boldface type.

degradation represented in Table II, no change in amino acid composition is apparent after the third round. This result is expected since proline, the third residue in from the COOH terminus, is not removed with this method (see the discussion below and also the note added in proof). The degradation shown in Table III is interpretable for four rounds. It was found that 2 or 3 stages of degradation are usually possible with most peptides; the most obtained was 6.[1] Table IV shows the results of 1-stage degradations of several proteins, in which a qualitative identification of the COOH-terminal residue has been made.

Formation of Acylthiohydantoins. To ensure that the cyclization reaction is complete, the conditions used for peptides are more extreme than those required for complete reaction of small model compounds, such as acetyl amino acids. Under the conditions recommended, the COOH-terminal residue of the phenylalanyl chain of insulin forms a thiohydantoin to an extent greater than 95%.[1]

With proteins, complete solubility under anhydrous conditions is a prerequisite for complete reaction. The use of hexafluoroacetone is crucial (see the note added in proof). Every protein tested[2] dissolved readily in hexafluoroacetone–water and remained soluble in the mixture with excess acetic anhydride and ammonium thiocyanate. Apparently, the charged and polar groups of the protein are acetylated by acetic anhydride at the same time that excess water is removed. Therefore, it seems likely that solubility under these anhydrous conditions is a general property of proteins and that in most instances the conversion of a protein into a proteinylthiohydantoin can be carried out in a homogeneous solution. The extent of conversion of the COOH-terminal valine of ribo-

TABLE IV
RESULTS OF CARBOXYL-TERMINAL DEGRADATIONS WITH PEPTIDES AND PROTEINS OF KNOWN STRUCTURE. ANALYSIS BY DIRECT IDENTIFICATION OF THE THIOHYDANTOINS

Peptide or protein	Residues identified as carboxyl terminal
Oxidized phenylalanyl chain, bovine insulin[a]	Ala (I); Lys (II)
Tryptic peptide, residues 1–22 of the oxidized phenylalanyl chain[a]	Arg (I); Glu (II); Gly (III)
Bovine insulin[a]	Ala and Asn
Bovine pancreatic ribonuclease A[b]	Val
Sperm whale myoglobin[c]	Gly
Hens' egg white lysozyme[d]	Leu
Porcine glucagon[e]	Thr
E. coli aspartate transcarbamylase	
Catalytic subunit[f]	Leu
E. coli aspartate transcarbamylase	
Regulatory subunit[f]	Asn

[a] F. Sanger, *in* "Currents in Biochemical Research" (D. E. Green, ed.), pp. 434–459. Wiley (Interscience), New York.
[b] D. G. Smyth, W. H. Stein, and S. Moore, *J. Biol. Chem.* **238,** 227 (1963).
[c] A. B. Edmundson, *Nature* (*London*) **205,** 883 (1965).
[d] R. E. Canfield, *J. Biol. Chem.* **238,** 2698 (1963).
[e] W. W. Bromer, L. G. Sinn, and O. K. Behrens, *J. Amer. Chem. Soc.* **79,** 2807 (1957).
[f] K. Weber, *Nature* (*London*) **218,** 1116 (1968).

nuclease into a thiohydantoin was found to be essentially quantitative under the conditions tested.

Proline and aspartic acid are not degraded when they occur at the COOH terminus (see the note added in proof). In the case of proline, cyclization to form an acylthiohydantoin would require quaternization of the imino nitrogen (see Fig. 1). Aspartic acid forms a cyclic anhydride readily when heated with acetic anhydride, and this anhydride does not react with ammonium thiocyanate to form an acylthiohydantoin in good yield.

Usually, the initial round in the degradation of a peptide causes loss of nearly one residue of the terminal amino acid, but degradations eventually become uninterpretable because only partial removal of the terminal residue is obtained in subsequent rounds. This occurs even if cleavage is catalyzed by dilute alkali or concentrated HCl rather than acetohydroxamate (see below). Prolonging the cyclization, raising the temperature, or adding more thiocyanate does not improve the yield.

In some instances, premature deacylation of the peptidylthiohydantoin is observed during the cyclization reaction.[1] The most extreme example is given by α-aspartylalanine. When this peptide is subjected to the conditions of cyclization, only a product absorbing at 260 nm (a thiohydantoin rather than an acylthiohydantoin) can be obtained, strongly implying that the β-carboxyl group of the aspartic acid catalyzes precleavage. Appreciable precleavage of serine occurs in degradations of peptides O-T-7-9 and O-T-16 from ribonuclease.[1] In these cases, since the second residue can form an acylthiohydantoin, simultaneous removal of terminal and subterminal residues is observed by subtractive analysis. Amino acids that are neither terminal nor subterminal seem to be stable under the conditions of the cyclization and cleavage reactions. The hydroxyl groups of serine and threonine are undoubtedly acetylated during the first exposure to acetic anhydride, but apparently no β-elimination reaction ensues during cleavage. Histidine and carboxamidomethylcysteine may undergo very slow degradation, with loss of one-tenth residue or less per round; this has not obscured the results of any degradation in which these residues were present. No attempt has been made to degrade peptides containing cystine or cysteine. Methionine undergoes partial oxidation to methionine sulfoxide, which is not resolved from aspartic acid on the 55-cm column now most commonly used for analysis of neutral and acidic amino acids. This problem may be avoided by oxidizing methionine to methionine sulfone with performic acid before the degradation is begun.

Cleavage of Peptidylthiohydantoins. The most useful catalysts are 12 *M* HCl and oxygen-containing nucleophiles such as hydroxylamine and acetohydroxamate (see also the note added in proof).

Despite the fact that the reaction is rapid, hydroxylamine was not chosen for the cleavage of acylthiohydantoins. Jencks[4] has shown that the initial attack on a carbonyl group is by the oxygen of hydroxylamine, but that the *O*-acylhydroxylamine that is formed initially is rapidly converted into the thermodynamically more stable hydroxamic acid by attack of the nitrogen of a second molecule of hydroxylamine. The hydroxamic acid would not be degraded since, for example, alaninehydroxamic acid does not form a thiohydantoin. When hydroxylamine is used for the cleavage of a large peptidylthiohydantoin, some of the new carboxyl-terminal residue must be present as the hydroxamic acid, but not all, since partial removal of the second residue can be obtained. However, the yields in round 2 are much better when acetohydroxamate is used. In this case, nucleophilic attack by the nitrogen cannot occur, and the product must have the structure, peptide—NHCH(R)C(=O)-

[4] W. P. Jencks, *Biochim. Biophys. Acta* **27**, 417 (1958).

$ONHC(=O)CH_3$. It is not known whether this derivative is reactive enough to yield the acylisothiocyanate directly with thiocyanate anion or whether it first decomposes to generate a free carboxyl group. The second possibility seems likely, since it is well known that *O*-acylhydroxamic acids undergo the Lossen rearrangement, but under milder conditions than hydroxamic acids:

$$\text{Peptide-NH}\overset{\overset{\displaystyle R}{|}}{C}\text{H}\overset{\overset{\displaystyle O}{\|}}{C}\text{ONH}\overset{\overset{\displaystyle O}{\|}}{C}\text{CH}_3 \longrightarrow \text{Peptide-NH}\overset{\overset{\displaystyle R}{|}}{C}\text{HCO}_2\text{H} + [\text{CH}_2\text{NCO}]$$

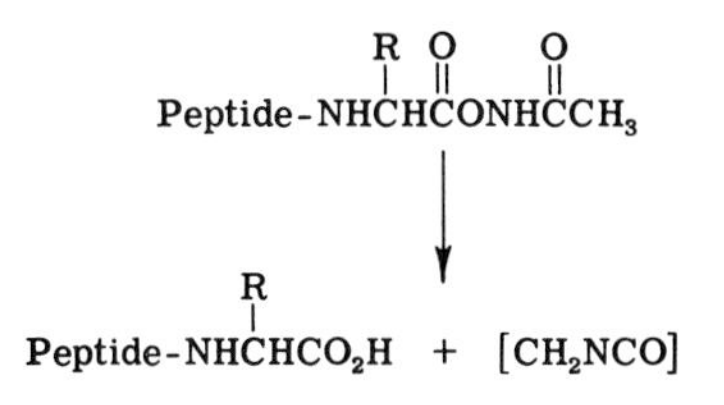

In any event, when cleavage is catalyzed by acetohydroxamate, the second round of degradation of large peptides proceeds with yields at least as good as those obtained when cleavage is catalyzed by HCl, which gives unambiguous formation of a free carboxyl group from the peptidylthiohydantoin.

The pK_a for dissociation of the imide proton from acetylalaninethiohydantoin, determined spectrophotometrically, is 6.9 at 25°. In the acetohydroxamate-catalyzed cleavage, reaction between the anionic nucleophile and uncharged acylthiohydantoin is favored. In this case, the optimum pH for cleavage would be the average of the pK_a's of the reagent and the thiohydantoin. Since the pK_a of acetohydroxamic acid is about 9.4, the average is about 8.2, and the rate is indeed fastest at about pH 8. The cleavage reaction with acetohydroxamate is much slower in 50% pyridine than in water, but the pyridine must be used since many peptides are not soluble in water near pH 7, especially after their polar functional groups have been acetylated. All the peptides tested dissolve readily in 50% pyridine.[1]

Cleavage of Proteinylthiohydantoins. HCl (12 *M*) was chosen as the reagent for this cleavage because (1) it is an excellent solvent for proteins; (2) it does not destroy thiohydantoins; (3) it is volatile; and (4) reaction is rapid. Apparently, very little unspecific hydrolysis of peptide bonds occurs during brief treatment of a protein with 12 *M* HCl at room temperature, since many thiohydantoins would be expected in the second cycle if new COOH termini were created during the first round cleavage, and this is not observed. Cleavage was also attempted with dilute NaOH in water and with acetohydroxamate in 50% pyridine.[2] Although 0.2 *M* NaOH was successful in a one-cycle degradation of insulin, several other proteinylthiohydantoins were insoluble in this reagent and could not be cleaved. Scoffone and Turco[5] have previously described the instability of thiohydantoins in dilute alkali. Acetohydroxamate is difficult to use with proteins since the thiohydantoins are not readily separated from the excess reagent and substantial amounts of acetohydroxamate interfere with thin-layer chromatography. Judging from the rate at which their spectra change, thiohydantoins are destroyed slowly by acetohydroxamate in 50% pyridine also, with a half-life of about 4 hours at 50°.

Separation of Thiohydantoins. Thin-layer chromatography of the crystalline thiohydantoins of Table I in two solvent systems is shown in Fig. 2. Most of the thiohydantoins can be identified unambiguously, and equivocal situations can be clarified by converting the thiohydantoin into an amino acid after extraction from the thin-layer plate (see below). Threonine, represented by 5-ethylidene-2-thiohydantoin in Fig. 2; and serine, from which no crystalline thiohydantoin was obtained, give rise to multiple spots. In Fig. 3 are shown the thin-layer chromatograms obtained from alanine, serine, and threonine after exposure to ammonium

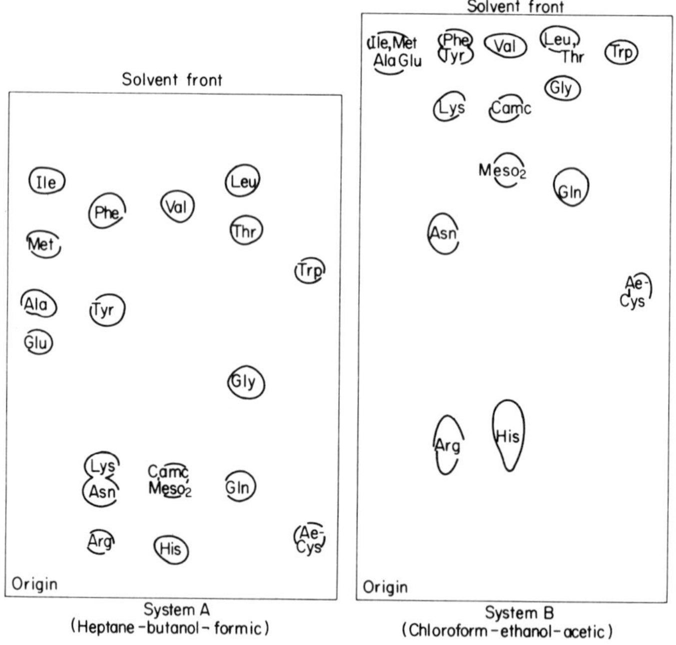

FIG. 2. Thin-layer chromatography of reference thiohydantoins.

[5] E. Scoffone and A. Turco, *Ric. Sci.* **26**, 865 (1956).

thiocyanate and acetic anhydride, then 12 M HCl. Alanine gives rise to the expected thiohydantoin plus an additional spot, marked "x" in Fig. 3, which probably results from not removing excess thiocyanate before cleavage and which is present in all three samples. Serine gives two spots. The one marked "Ser-1" has λ_{max} 267 nm and accounts for 22% of the maximum absorbancy expected; the one marked "Ser-2" has λ_{max} 265 nm and accounts for 6% of the maximum absorbancy expected. (The recovery of absorbancy was 88% for alanine in a parallel experiment.) One of these two spots is probably 5-acetoxymethyl-2-thiohydantoin (*O*-acetylserine thiohydantoin); the other is probably *not* 5-methylene-2-thiohydantoin, which would be obtained by β elimination of the *O*-acetyl compound, since an exocyclic double bond would be expected to shift the spectrum of the thiohydantoin about 60 nm toward the red, as it does in the case of 5-ethylidene-2-thiohydantoin (see below). Threonine gives rise to three spots. The one marked "Thr-1," obtained in 9% yield, corresponds in chromatographic behavior and spectrum to authentic 5-ethylidene-2-thiohydantoin. Spot "Thr-2," obtained in 12% yield, has λ_{max} 267 nm; spot "Thr-3," obtained in 3% yield, has λ_{max} 266 nm. As with serine, one of these two spots probably represents the *O*-acetylthiohydantoin.

Extraction and Quantitation. Extraction of thiohydantoins from the

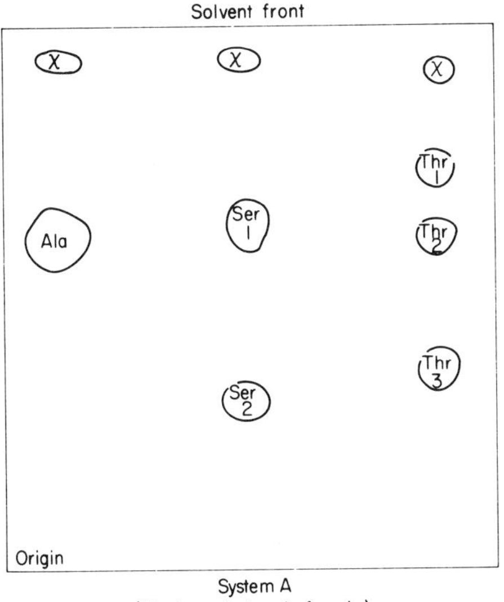

FIG. 3. Thin-layer chromatography of thiohydantoin products derived from serine and threonine.

thin-layer plates with methanol is virtually quantitative, judging from the recovery of absorbancy at about 262 nm in the supernatant solutions. Most of the thiohydantoins absorb light maximally at about 262 nm, with molar extinction coefficients close to 1.75×10^4. The exceptions are 5-ethylidene-2-thiohydantoin (from threonine) which, owing to the exocyclic double bond, has λ_{max} 319 nm and ϵ_M 2.5×10^4, and 5-(3-indolylmethyl)-2-thiohydantoin (from tryptophan) which, owing to the indole ring, has a shoulder at 288 nm, λ_{max} 266 nm, and ϵ_M 2.15×10^4.

Conversion of Thiohydantoins into Amino Acids. The best results in converting thiohydantoins into amino acids were obtained with a sequential procedure involving (1) hydrolysis to the thiohydantoic acid in dilute alkali,[5] (2) treatment with alkaline hydrogen peroxide; and (3) alkaline hydrolysis. Except for those amino acids known to be destroyed by alkaline hydrolysis or hydrogen peroxide, the conversion of thiohydantoins into amino acids is achieved in good yield (Table I). For reasons that are not clear, the yields are not as good when the procedure is applied to the methanol extract of a spot from a thin-layer chromatogram after removal of the methanol under vacuum (Table I).

Haurowitz and Lisie[6] had reported that thiourea could be quantitatively converted into urea by alkaline H_2O_2 and that the urea could be determined subsequently with urease. They obtained 0.2–0.5 mole of urea per mole of each of several thiohydantoins, implying that the α-carbon–nitrogen bond had been cleaved, since urease is highly specific for unsubstituted urea. In order to avoid this cleavage, we have used conditions milder than those described by Haurowitz and Lisie. Also, better yields are obtained if the thiohydantoins are first converted into thiohydantoic acids in alkali. It is well known that 5-alkyl-2-thiohydantoins can be oxidized readily by molecular oxygen in 0.5 *M* NaOH (see Edward and Nielsen,[7] for a detailed discussion and description of the products), and we have found that exclusion of oxygen until conversion into the thiohydantoic acid is complete improves the yields significantly.

Stability of Thiohydantoins. Oxidation by molecular oxygen probably accounts also for our observation that exposure of a peptidylthiohydantoin to 0.05 *M* $NaHCO_3$ at room temperature causes the acylthiohydantoin spectrum to disappear without concomitant appearance of a thiohydantoin spectrum, with a half-time of about 3 hours. It is probably wise to minimize the exposure of peptidylthiohydantoins and thiohydantoins to air, since desulfuration by oxygen, analogous to the well-known desulfuration of phenylthiocarbamylamino acids[8] may occur

[6] F. Haurowitz and S. G. Lisie, *Anal. Chim. Acta* **4,** 43 (1950).

[7] J. T. Edward and S. Nielsen, *J. Chem. Soc.* 2327 (1959).

[8] D. Ilse and P. Edman, *Aust. J. Chem.* **16,** 411 (1963).

with these compounds, reducing the yields. Exposure to light should also be minimized.

The recent finding[9] that dithioerythritol is extremely effective in protecting phenylthiohydantoins from oxidative desulfuration, suggests that inclusion of this compound in solvents for separation of thiohydantoins will improve recoveries greatly and perhaps even make possible the use of the method for quantitative estimation of protein COOH termini.

Prospects for Improving the Method. In addition to protection of the thiohydantoins, procedures for identification of thiohydantoins alternative to thin-layer chromatography would certainly improve the method for proteins. In most attempts to do a second cycle of reactions with a protein, a peak with a spectrum appropriate to a thiohydantoin was obtained from the Sephadex column after cleavage, but thin-layer chromatography gave only an intense spot at the origin. We have attempted to volatilize thiohydantoins for gas–liquid partition chromatography with silylating reagents or the methylating reagents diazomethane and dimethyl sulfate, but were frustrated by our finding that 2-thiohydantoins give rise to at least two major peaks, in contrast to 3-methyl-2-thiohydantoins, which give single derivatives.[2] We have not attempted to develop a system using liquid–liquid chromatography, but use of this new technique, which would circumvent the problem of converting thiohydantoins into volatile derivatives, seems especially promising. With such a system, cleavage with 12 *M* HCl may give results interpretable for several rounds, or perhaps other methods of cleavage, such as acetohydroxamate in pyridine, may become practical for proteins (see also the note added in proof). Another advantage of liquid–liquid chromatography is that quantitation is more reliable than with thin-layer chromatography.

Note Added in Proof

S. Yamashita, *Biochim. Biophys. Acta,* **229,** 301 (1971) has substantially improved the procedure for sequential degradation of proteins from their COOH termini and reports a 4-step sequential degradation of bovine pancreatic ribonuclease A. Proteinyl thiohydantoins are formed by first dissolving the protein in a mixture of acetic anhydride, trifluoroacetic anhydride and acetic acid, followed by addition of sodium thiocyanate. The isolated proteinyl thiohydantoin is cleaved by the H^+ form of Amberlite 1R-120. Yamashita's procedure (paraphrased) follows:

Dissolve about 1 μmole of protein in a mixture of 0.5 ml of acetic anhydride, 0.5 ml of trifluoroacetic anhydride and 0.5 ml of acetic acid. When solution is complete, add 100 mg of crystalline sodium thiocyanate with vigorous stirring. Keep the mixture at 60° for 6 hours, during which time the sodium thiocyanate dissolves and the

[9] M. A. Hermodson, L. H. Ericsson, and K. A. Walsh, *Fed. Proc. Fed. Amer. Soc. Exp. Biol.* **29,** 728 (1970).

colorless solution becomes yellow. Evaporate the solvent by lyophilization, add water to dissolve the residue, desalt the solution by dialysis, and lyophilize it again. Dissolve the residue in a minimal amount of water or in 0.1 *M* HCl and add 1 g of Amberlite 1R-120, H^+ form, 2 mequiv per gram. Shake the mixture vigorously on a test tube shaker for 2–6 hours at room temperature and collect the liquid phase by filtration. Wash the resin 3 times with water, then with 5 ml each of ethyl acetate and water, successively. Concentrate the combined washings to dryness under reduced pressure, dissolve the residue in 2 ml of 50% acetic acid and chromatograph the solution on a column of Sephadex G-25. Combine the peptide fractions, which emerge just after the void volume, lyophilize them and use them for the next cycle of reactions. Combine the thiohydantoin fractions (strong absorption at 260 nm), concentrate them under reduced pressure at 40°, and analyze them by thin layer chromatography.

Yamashita reports for the first time the synthesis of the 1-acetyl thiohydantoin derived from serine (m.p. 144–145) and the synthesis of the thiohydantoin derived from aspartic acid (m.p. 220), which had previously been made by F. Haurowitz, M. Zimmerman, R. L. Hardin, S. G. Lisie, J. Horowitz, A. Lietze, and F. Bursa, *J. Biol. Chem.* **224**, 827 (1957). His modified procedure apparently allows the formation of a thiohydantoin from COOH terminal aspartic acid, although only in 10% yield from position 4 of ribonuclease. Pure reference standard thiohydantoins derived from proline and hydroxyproline are also reported, although they have not been crystallized. Details are promised in a forthcoming publication. Yamashita reports a successful degradation of gly-pro, although he does not discuss in detail how this particular degradation was performed; proline and hydroxyproline must either form an acylthiohydantoin in which the acylated nitrogen atom is quaternary (which seems unlikely) or deacylation must be concomitant with cyclization.

Section VIII

Modification Reactions

[30] Reduction and Reoxidation at Disulfide Bonds

By FREDERICK H. WHITE, JR.

A protein disulfide bond may be cleaved by reagents containing the SH group as follows:

$$\underset{\text{S}\text{——}\text{S}}{\overline{\text{Protein}}} + \text{RSH} \rightleftharpoons \underset{\text{HS}\quad\text{SSR}}{\overline{\text{Protein}}} \overset{\text{RSH}}{\rightleftharpoons} \underset{\text{HS}\quad\text{SH}}{\overline{\text{Protein}}} + \text{RSSR} \qquad (1)$$

This reaction approaches completion with an excess of the reducing agent, which is specific for disulfide bonds.[1] For cleavage of all such bonds in a given protein, it may be necessary to employ a denaturing agent, such as urea, to render the protein into a conformational state in which the bonds are sufficiently exposed to the reducing agent. Concomitant with scission of the disulfide bonds is loss of the native secondary and tertiary structures as well as biological activity.

In general this reaction would be of use under circumstances necessitating the removal of native conformation with no deleterious effect upon the primary structure. Proteolytic cleavage, for example, may not proceed satisfactorily unless the coiling and folding of the native protein are first eliminated. This reaction therefore has found use as a preliminary step for enzymatic degradation associated with investigation of amino acid sequence. Reduction of chicken egg white lysozyme, for example, permitted Canfield[2] to obtain a form of this protein from which he was able to complete the determination of amino acid sequence. Reduction has further been of use in studies on the effects of ionizing radiation. For example, comparative studies of native and reduced proteins have permitted conclusions regarding the effects of conformation on the distribution of carbon free radicals, resulting from irradiation,

[1] F. H. White, Jr., *J. Biol. Chem.* **235**, 383 (1960).
[2] R. E. Canfield, *J. Biol. Chem.* **238**, 2698 (1963).

among the amino acid residues.[3] Haskill and Hunt[4] found that irradiated ribonuclease (RNase), after full reduction, could be reoxidized to yield an enzyme differing in its K_m value from the native enzyme, thus revealing a "hidden" damage due to irradiation that could not be detected in any other way.

Reoxidation of the reduced protein may be readily carried out by exposure in dilute, slightly alkaline solution, to atmospheric oxygen. It is especially significant that products are obtained in high yield from this reaction which appear to be identical with the native proteins. The reappearance of native conformation in the absence of any biological system demonstrates that the characteristic manner in which a protein chain coils and folds to produce its native conformation during biosynthesis is governed by its amino acid sequence. Beef pancreatic RNase was first observed to exhibit this behavior[5] and then a number of other proteins (see White[6] for summary). More recently the return of antibody specificity on oxidation of reduced antibody fragments[7,8] and of the whole protein[9] demonstrated that this property is controlled by the primary structure of the antibody chain. (For a more detailed discussion see Tanford.[10]) Gutte and Merrifield[11] and Hirschmann *et al.*[12] simultaneously accomplished for the first time the synthesis of an enzyme (RNase) by employing, in the final step, the principal that the primary structure governs formation of the secondary and tertiary structures. Thus, the fully synthesized protein chain was allowed to oxidize and fold to the native and enzymatically active conformation. Additional

[3] F. H. White, Jr., P. Riesz, and H. Kon, *Radiat. Res.* **32,** 744 (1967).

[4] J. S. Haskill and J. W. Hunt, *Biochim. Biophys. Acta* **105,** 333 (1965).

[5] F. H. White, Jr., *J. Biol. Chem.* **236,** 1353 (1961).

[6] F. H. White, Jr., *J. Biol. Chem.* **239,** 1032 (1964).

[7] E. Haber, *Proc. Nat. Acad. Sci. U.S.* **52,** 1099 (1964).

[8] P. L. Whitney and C. Tanford, *Proc. Nat. Acad. Sci. U.S.* **53,** 524 (1965).

[9] M. H. Freedman and M. Sela, *J. Biol. Chem.* **241,** 2383 (1966).

[10] C. Tanford, *Accounts Chem. Res.* **1,** 161 (1968).

[11] B. Gutte and R. B. Merrifield, *J. Amer. Chem. Soc.* **91,** 501 (1969).

[12] R. Hirschmann, R. F. Nutt, D. F. Veber, R. A. Vitali, S. L. Varga, T. A. Jacob, F. W. Holly, and R. G. Denkewalter, *J. Amer. Chem. Soc.* **91,** 507 (1969).

significance of this monumental achievement lies in the final elimination of any question that the refolding of a reduced protein might be initiated by a small amount of residual native structure.

Little is known of the mechanism by which refolding occurs. Evidence has been found that the process may depend upon attractions between residues[13] and upon hydrophobic interactions.[14] Friedman *et al.*[15] have implicated tyrosine at position 115 in the refolding of RNase. The process of refolding, however, must still be regarded as poorly understood.

Reduction and Reoxidation

Reagents. The reducing agent of choice is mercaptoethanol, which has the advantages of long shelf life, with little oxidation of SH groups, and no reacting groups by which side reactions might occur with the protein. Urea is used as the denaturing agent. Cyanate, which occurs in equilibrium with urea, will react with amino and SH groups in proteins to form the carbamyl derivatives[16] and must therefore be eliminated. Crystallization of urea from ethanol and water should be sufficient, but as an additional precaution, cyanate may be decomposed by acidification,[16] or removed by passage through a mixed-bed ion-exchange column.[7] A convenient colorimetric test for cyanate is given by Werner.[17]

Procedure for Reduction. Urea (2.4 g) and mercaptoethanol (0.1 ml) are dissolved in 2.5 ml of water. This solution is adjusted to pH 8.5 with 5% aqueous trimethylamine and used to dissolve 50 mg of RNase. After a final check on the pH, the solution is adjusted to 5 ml with water. Nitrogen is passed through the solution in a test tube for 5 minutes, and the tube is stoppered. Reduction of the four disulfide bonds of RNase is essentially complete after 4 hours at room temperature (22–25°).

To separate the reduced protein, the reaction mixture is first brought

[13] E. Haber and C. B. Anfinsen, *J. Biol. Chem.* **237,** 1839 (1962).

[14] C. J. Epstein and R. F. Goldberger, *J. Biol. Chem.* **239,** 1087 (1964).

[15] M. E. Friedman, H. A. Scheraga, and R. F. Goldberger, *Biochemistry* **5,** 3770 (1966).

[16] G. R. Stark, W. H. Stein, and S. Moore, *J. Biol. Chem.* **235,** 3177 (1960).

[17] E. A. Werner, *J. Chem. Soc.* **123,** 2577 (1923).

to pH 4 with acetic acid and then applied to a column (2.5 × 25 cm) of Sephadex (Pharmacia, Uppsala, Sweden; G-25, 100–270 mesh) at room temperature, previously equilibrated with 0.1 *M* acetic acid. For a column of this size, it is advisable not to exceed 5 ml of sample volume. The column is eluted with 0.1 *M* acetic acid, and 3-ml fractions are collected at a flow rate not greater than 250 ml per hour. The protein peak appears first, followed by the components of lower molecular weight in another peak. Both peaks may be detected at 280 nm, the concentration of reduced RNase being determined by an extinction coefficient of 8.55×10^3 $M^{-1}cm^{-1}$.[5] The volume of the resulting protein solution should be 15–20 ml, at approximately 2 mg/ml. This solution is stable at 5° for several days with no detectable loss of SH groups. Alternatively, the reduced protein may be precipitated by a mixture of acetone and HCl, as described earlier.[1]

Procedure for Reoxidation. An aliquot of the reduced RNase solution is added to 0.1 *M* tris(hydroxymethyl)aminomethane (Tris)–acetic acid buffer of pH 8, to produce a protein concentration of 20 μg/ml. Oxidation occurs at room temperature on standing in an open beaker. Typically more than half the specific activity of native RNase is regenerated within 1 hour, and the yield of active protein will approach 100% at 3 hours. The assay may be performed with ribonucleic acid or cyclic phosphates as substrates by methods reviewed earlier.[18]

To separate the reoxidized protein from the buffer, the volume may first be reduced by lyophilization and the buffer removed by dialysis, as was done previously.[5] More recently carboxymethyl (CM) Sephadex has been used in this laboratory for concentrating small amounts of reoxidized RNase from large volumes. The reoxidation mixture is diluted to a Tris concentration of 0.05 *M* and then adjusted to pH 5 by addition of acetic acid. The solution is passed through a column of CM-Sephadex (C-25, 100–270 mesh, 4.4 meq/g), previously equilibrated with 0.05 *M* ammonium acetate buffer of pH 5. Column dimensions of 2.5 × 10 cm are ample for adsorption of 1 g of RNase. The column is washed with water and then eluted with 0.2 *M* ammonium bicarbonate to remove the protein. The fractions containing protein, detected at 280 nm, may then be combined and lyophilized. The remaining bicarbonate may be removed by dialysis.

[18] C. B. Anfinsen and F. H. White, Jr., *in* "The Enzymes" (P. D. Boyer, H. Lardy, and K. Myrbäck, eds.), p. 95. Academic Press, New York, 1961.

Discussion

Partial Reduction. Reduction of specific disulfide bonds has been reported with other reducing agents. With phosphorothioate, the 4-5 and 3-8 bonds of RNase were selectively cleaved.[19] More recently Sperling *et al.*[20] reduced the 4-5 bond with either dithiothreitol or dithioerythritol, in the absence of a denaturing agent. Use of the former reagent is treated in detail elsewhere.[21] It may be of interest that, in the experience of this laboratory, mercaptoethanol, in the absence of urea but under conditions otherwise similar to those given herein for full reduction, reduces RNase to approximately the same extent. It remains a subject for further investigation to determine if the same or different disulfide bonds are attacked by these reagents.

Determination of the Extent of Reduction or Reoxidation. Most commonly these reactions have been followed by measuring the SH content. The spectrophotometric titration with *p*-chloromercuribenzoate developed by Swenson and Boyer,[22] especially as modified by Sela *et al.*,[23] has been satisfactory in this laboratory. Titration with 5,5′-dithiobis(2-nitrobenzoic acid)[24] also has been successful for this purpose.[25] An additional method involves reaction of SH groups with iodoacetate,[5] followed by analysis for *S*-carboxymethylcysteine; it has yielded results that agree well with those obtained by titration.[26] Other methods of SH determination are reviewed by Cecil and McPhee.[27]

Applicability to Other Proteins. It may be necessary to modify the above method for reduction of other proteins. Guanidine hydrochloride (a more effective denaturing agent than urea) may be used in place of urea. Variations in temperature and time of reaction may be necessary, depending upon the ease of solubilization and unfolding of the protein.

There are several critical factors to be considered for reoxidation. The concentration of the reduced protein should be as low as possible to minimize intermolecular disulfide bond formation. The yield of active, reoxidized RNase, however, diminishes below 10 μg/ml, since the reduced

[19] H. Neumann, I. Z. Steinberg, J. R. Brown, R. F. Goldberger, and M. Sela, *Eur. J. Biochem.* **3**, 171 (1967).

[20] R. Sperling, Y. Burstein, and I. Z. Steinberg, *Biochemistry* **8**, 3810 (1969).

[21] This volume [13].

[22] A. D. Swenson and P. D. Boyer, *J. Amer. Chem. Soc.* **79**, 2174 (1957).

[23] M. Sela, F. H. White, Jr., and C. B. Anfinsen, *Biochim. Biophys. Acta* **31**, 417 (1959).

[24] G. L. Ellman, *Arch. Biochem. Biophys.* **82**, 70 (1959).

[25] This volume [37].

protein appears to be adsorbed onto the surface of the glass reaction vessel.[28] Temperature must be controlled, since the correct refolding of RNase decreases markedly above room temperature, and at 37° the yield of soluble, active protein is only 35% of that at 24°.[28] The effects of temperature may be expected to differ with each protein. For chicken egg white lysozyme, for example, the efficiency of refolding is greater at 38° than at room temperature.[29]

The tendency of some proteins to aggregate in the reduced state may result in a low yield of correctly refolded protein, since intermolecular disulfide bond formation is enhanced by proximity of the reduced chains. The rate of refolding of lysozyme, as well as the yield of active enzyme, has been increased by the addition of mercaptoethanol, thus allowing the incorrectly formed disulfide bonds to break and then recombine correctly.[29] The aggregation of trypsin after reduction has been prevented by complexing the native protein with CM-cellulose.[30] Freedman and Sela[9] employed an extensive polyalanylation of the ϵ-amino groups of rabbit antibovine serum albumin. The heavy chain, which normally is insoluble after reduction, was thereby rendered soluble, and the 23 disulfide bonds of the native protein could then be reformed with regeneration of immunological activity.

A proteolytic enzyme must be safeguarded against autodigestion during reoxidation. The immobilization of trypsin by attachment to CM-cellulose has achieved this effect, in addition to prevention of aggregation, as mentioned above. Autolysis might also be prevented by including a specific inhibitor during oxidation, as has been tried by Epstein and Anfinsen[31] in the use of lima bean trypsin inhibitor for the oxidation of reduced trypsin.

[26] A. M. Crestfield, S. Moore, and W. H. Stein, *J. Biol. Chem.* **238**, 622 (1963).

[27] R. Cecil and J. R. McPhee, *Advan. Protein Chem.* **14**, 256 (1959).

[28] C. J. Epstein, R. F. Goldberger, D. M. Young, and C. B. Anfinsen, *Arch. Biochem. Biophys.*, Suppl. 1, **223** (1962).

[29] C. J. Epstein and R. F. Goldberger, *J. Biol. Chem.* **238**, 1380 (1963).

[30] C. J. Epstein and C. B. Anfinsen, *J. Biol. Chem.* **237**, 2175 (1962).

[31] C. J. Epstein and C. B. Anfinsen, *J. Biol. Chem.* **237**, 3464 (1962).

[31] Oxidation with Hydrogen Peroxide

By Norbert P. Neumann

Principle

Hydrogen peroxide is a relatively nonspecific oxidizing agent which reacts with a wide variety of organic compounds. This fact has not, however, hindered its use as a reagent for the chemical modification of numerous proteins. In recent studies the chemistry of the oxidation of proteins by this chemical has been extensively examined by a number of workers. It was found that under suitable, relatively mild conditions this reagent is highly specific for a small number of amino acid side chains. Possible sites of attack include the following functional groups: thioether, indole, sulfhydryl, disulfide, imidazole, and phenolic. Under acidic conditions the primary reaction is the conversion of methionine residues to the sulfoxide. Toennies and Callan[1] and more recently Caldwell and Tappel[2] have shown that methionine is much more reactive than either cystine or cysteine. Oxidation of the latter two substances is favored by increased pH and the presence of heavy metals in the case of cysteine. Tryptophan and tyrosine residues in proteins are generally more resistant to attack. Hachimori *et al.*[3] have shown for example that the oxidation of *N*-acetyl-L-tryptophan occurs optimally between pH 8 and 10 and that the rate falls off rapidly outside of this range. Although free histidine was attacked under the same conditions, it appeared to be resistant when present in polypeptide linkage.

Analysis of the Course of Reaction

Amino acid analysis of the reaction product is performed by standard quantitative chromatographic procedures after appropriate pretreatment of the material. The usual preliminary hydrolysis with hydrochloric acid prior to analysis does not suffice, since methionine sulfoxide is decomposed[4] largely to methionine. Numerous other products have been reported, including methionine methyl sulfonium salt, homocystine, and homocysteic acid.[5] The stoichiometry of the reaction is undoubtedly

[1] G. Toennies and T. P. Callan, *J. Biol. Chem.* **129**, 481 (1939).

[2] K. A. Caldwell and A. L. Tappel, *Biochemistry* **3**, 1643 (1964).

[3] Y. Hachimori, H. Horinishi, K. Kurihara, and K. Shibata, *Biochim. Biophys. Acta* **93**, 346 (1964).

[4] W. J. Ray, Jr. and D. E. Koshland, Jr., *Brookhaven Symp. Biol.* **13**, 135 (1960).

[5] W. F. Floyd, M. S. Cammaroti, and T. F. Lavine, *Arch. Biochem. Biophys.* **102**, 343 (1963).

dependent upon a number of factors, not the least of which is the chemical composition of the protein being studied. This limits the usefulness of these products as an index of the extent of chemical reaction.

Several alternatives are possible.[6] The first of these is an alkaline hydrolysis prior to analysis. The hydroxides of sodium, potassium or barium have been used as catalysts by various workers, usually at concentrations of 4–6 N at 110° for a period of 16–50 hours. Although a number of amino acids are destroyed by this treatment, methionine sulfoxide is generally stable. There is some evidence to suggest that potassium or barium hydroxide may be preferable, particularly for obtaining maximal recoveries of tryptophan. In addition, the use of sodium hydroxide has at times resulted in low recoveries of methionine sulfoxide from samples of oxidized ribonuclease.

The other alternative is to exhaustively alkylate the oxidized protein with iodoacetic acid or iodoacetamide. Suggested conditions are pH 2.0 at 40° for a period of 20 hours. The alkylating agent does not react with sulfoxide residues, but converts any unmodified methionine to the corresponding sulfonium salt. This alkylated product then is oxidized with performic acid,[7,8] which converts methionine sulfoxide residues to the sulfone, but has no effect[9] upon the residues present as the sulfonium salt. Analysis for methionine sulfone after acid hydrolysis gives a direct measure of the amount of methionine sulfoxide originally present in the protein.

A modification of this second procedure involves measurement of the incorporation of radioactivity into the protein after treatment with [^{14}C]-iodoacetic acid or iodoacetamide.[10,11] This gives a direct measure of unmodified methionine residues and eliminates the need for the extra step involving oxidation with performate. The reaction at acidic pH is highly specific for the thioether functional group, so that spurious uptake of label by other residues does not seem to be a problem. Although the reaction with iodoacetamide is slower than with iodoacetate, it has the advantage of introducing an additional positive charge for each residue reacted, which may be helpful in the isolation of derivatives or peptides containing the label.[12]

[6] N. P. Neumann, Vol. XI, 487.
[7] C. H. W. Hirs, *J. Biol. Chem.* **219,** 611 (1956).
[8] S. Moore, *J. Biol. Chem.* **238,** 235 (1963).
[9] H. G. Gundlach, S. Moore, and W. H. Stein, *J. Biol. Chem.* **234,** 1761 (1959).
[10] H. Schachter and G. H. Dixon, *Biochem. Biophys. Res. Commun.* **9,** 132 (1962).
[11] V. Holeyšovský and M. Lazdunski, *Biochem. Biophys. Acta* **154,** 457 (1968).
[12] H. Schachter and G. H. Dixon, *J. Biol. Chem.* **239,** 813 (1964).

Modification of Proteins

The following description of the method used for the oxidation of ribonuclease[13] will provide an example of the general techniques used.

Adjust a 1% solution of the protein under study to pH 2.5 with perchloric acid[14] and bring to constant temperature in a 30° water bath. Add 0.03 ml of 30% hydrogen peroxide (reagent Superoxol, Merck) per milliliter of enzyme solution to give a final concentration of approximately 0.3 *M* in peroxide. To follow the time course of inactivation of the enzyme, pipette 0.05-ml portions into 0.4 ml of 0.01% catalase solution (in 0.1 *M* phosphate buffer, pH 7.0) at room temperature. Then assay for enzymatic activity after excess peroxide has been destroyed (about 10 minutes). After a suitable time (usually 0.5–2 hours) as determined by the results of the assay and the extent of modification desired, adjust the reaction mixture to pH 6.5–7.0 with 6 *N* NaOH (pH meter) and add 0.02 volume of 0.1% catalase solution every 2 minutes until evolution of gas ceases. A total of five additions should be adequate, resulting in a contamination of the final product by catalase to the extent of only 1%. Then the preparation may be subjected to fractionation by chromatography or by a variety of other methods to separate it into components with varying degrees of modification. Alternatively, the modified enzyme may be dialyzed and lyophilized, or stored frozen prior to analysis for methionine sulfoxide content. If the presence of small amounts of catalase in the final product is undesirable, removal of the hydrogen peroxide can be accomplished either by dialysis or gel filtration on a small column of Sephadex G-25.[11]

Reversibility of the Modification

The ability to reverse a chemical protein modification by suitable procedures is a useful feature which is of value in verifying both the presumed specificity of the modifying reaction as well as assessing the contribution of a particular residue to various aspects of the physical structure of the protein. In a number of cases it has been possible to reverse the oxidation and regenerate methionine from the sulfoxide.

[13] N. P. Neumann, S. Moore, and W. H. Stein, *Biochemistry* **1**, 68 (1962).

[14] If the enzyme solution contains halide ions, these should first be removed by dialysis or some other means to avoid conversion to free halogen, which could react with the protein. For similar reasons it is best to avoid buffers containing organic acids, such as formic acid, because of the danger of forming peracids which could cause overoxidation of the methionine residues as well as attack tryptophan and cystine residues.

Thus treatment of oxidized forms of ACTH,[15] parathormone B,[16] and yeast invertase[17] with various sulfhydryl compounds has resulted in a regeneration of the activity lost by oxidation. Reduction of the sulfoxide to methionine was demonstrated in the first two examples and is probably true in the third. On the other hand, attempts to reactivate chymotrypsin[18] or glucose oxidase[19] by similar procedures have failed. The reason for this is not apparent, but it should be noted that in the case of yeast invertase, where mercaptoethanol was effective in restoring lost activity, both cysteine and dithiothreitol had no effect, a result indicating that the choice of a reducing agent for reactivation in a particular case can be of critical importance.

The reactivation of enzymes containing oxidized sulfhydryl groups is usually readily accomplished with appropriate reducing agents as long as oxidation has not proceeded to the sulfonic acid stage.[20]

Comments

The analytical methods described have been used successfully by a number of workers. Both procedures have general utility, and the method selected in a particular case appears to be related to individual preference as well as the nature of the protein under study. It should be emphasized that successful application of the indirect method involving secondary alkylation of the unmodified methionine residues is dependent upon the complete alkylation of all the unoxidized methionine. Experience with various proteins has shown that most methionine residues are unavailable to chemical reagents unless the protein structure is first disrupted (noteworthy exceptions include methionine-192 of chymotrypsin[21] and methionine-29 in ribonuclease[22]). The conditions recommended for alkylation, namely pH 2.0, 40° for 20 hours are drastic enough to accomplish this and studies with a number of proteins using similar, although not identical conditions verify this conclusion. Specific examples include trypsin and trypsinogen,[11] cytochrome *c*,[23] parathor-

[15] M. L. Dedman, T. H. Farmer, and C. J. O. R. Morris, *Biochem. J.* **66**, 166 (1957).
[16] H. Rasmussen, *J. Biol. Chem.* **234**, 547 (1959).
[17] S. Shall and A. Waheed, *Biochem. J.* **111**, 33 P (1969).
[18] J. R. Knowles, *Biochem. J.* **95**, 180 (1965).
[19] K. Kleppe, *Biochemistry* **5**, 139 (1966).
[20] C. Little and P. J. O'Brien, *Arch. Biochem. Biophys.* **122**, 406 (1967).
[21] D. E. Koshland, Jr., D. H. Strumeyer, and W. J. Ray, Jr., *Brookhaven Symp. Biol.* **15**, 101 (1962).
[22] T. P. Link and G. R. Stark, *J. Biol. Chem.* **243**, 1082 (1968).
[23] E. Stellwagen and S. Van Rooyan, *J. Biol. Chem.* **242**, 4801 (1967).

mone,[24] ribonuclease *S*-peptide,[25] and myokinase.[26] A possible exception is apparent in the case of a pancreatic trypsin inhibitor[27] where only approximately 10% of the methionine in the protein reacted with iodoacetate at pH 2.7, 40° in 3 hours. It is quite possible that at a lower pH and longer reaction time even this very resistant molecule might react completely. Other workers have performed the alkylation in the presence of denaturants, such as urea.[10,11,28] Under these conditions, reaction with methionine residues occurs rapidly and, as Stark and Stein[29] noted for ribonuclease in the presence of 8 *M* urea, the rate actually increases 2-fold with an increase in pH from 3.5 to 5.8. It is likely that similar conditions will suffice for most, if not all, proteins, and the simple expedient of running an experiment on the unoxidized protein to demonstrate complete reaction of all the methionine residues with the alkylating agent will serve as a useful control for the application of the method.

With regard to the selection of conditions for the reaction of hydrogen peroxide with proteins, a variety of procedures has been adopted which differ from those originally used for ribonuclease.[13] Most studies in which a specific reaction with methionine residues has been documented have been carried out at a pH below 4 at 25–30° with peroxide concentrations of 0.1–0.5 *M* and the protein being studied at a concentration of 0.1–1.0%. Two exceptions are the reports of Kleppe[19] on the inactivation of glucose oxidase at pH 5.8 and of Stauffer and Etson[30] on subtilisin at pH 8.8 where even at the higher pH only methionine appears to have been modified. A marked reduction in the concentration of hydrogen peroxide (down to 0.005 *M*) necessitated a much longer reaction time (22 hours) in the case of chymotrypsin in order to complete the oxidation of methionine-192. Several examples[11,31,32] have been reported where the addition of 6–8 *M* urea to the reaction medium has been found to increase, as might be expected, both the rate of reaction and the number of methionine residues affected. It is noteworthy that even under these relatively drastic conditions, tryptophan as well as other residues were resistant to attack. In contrast, the presence of substrate during oxidation of chymotrypsin[32] caused modification of trypto-

[24] A. H. Tashjian, Jr., D. A. Ontjes, and P. L. Munson, *Biochemistry* **3**, 1175 (1964).

[25] P. J. Vithayathil and F. M. Richards, *J. Biol. Chem.* **235**, 2343 (1960).

[26] L. F. Kress and L. H. Noda, *J. Biol. Chem.* **242**, 558 (1960).

[27] B. Kassell, *Biochemistry* **3**, 152 (1964).

[28] G. R. Stark, W. H. Stein, and S. Moore, *J. Biol. Chem.* **236**, 436 (1961).

[29] G. R. Stark and W. H. Stein, *J. Biol. Chem.* **239**, 3755 (1964).

[30] C. E. Stauffer and D. Etson, *J. Biol. Chem.* **244**, 5333 (1969).

[31] H. Weiner, C. W. Batt, and D. E. Koshland, Jr., *J. Biol. Chem.* **241**, 2687 (1966).

[32] H. Schachter, K. A. Halliday, and G. H. Dixon, *J. Biol. Chem.* **238**, PC 3134 (1963).

phan and cystine as well as methionine. Although the exact mechanism is uncertain it has been suggested that the formation of a peracid by the acylated enzyme effects reaction with tryptophan and cystine. This substrate promotion of the inactivation may be a useful tool for the selective modification of amino acid residues of enzymes belonging to this general class and possibly other types of enzymes as well.

At higher pH values, residues other than methionine become susceptible to reaction, particularly cysteine, cystine, and tryptophan. Thus Hachimori *et al.*[3] have developed a procedure which differentiates between various states of tryptophan residues in proteins. Their procedure is based upon reaction at pH 8.1–9.4 with relatively low concentrations of hydrogen peroxide in the presence of 10% dioxane, which markedly enhances the rate of reaction presumably due to the formation of an organic peroxide. Denaturation of several enzymes with alkali prior to treatment resulted in complete oxidation of the tryptophan, making it possible to measure quantitatively the total tryptophan content.

The reaction of sulfhydryl groups with hydrogen peroxide is a well known reaction.[33] Unlike low molecular weight compounds which readily form disulfides upon oxidation, the sulfhydryl groups in proteins, due to steric limitations, often produce instead acidic derivatives containing, initially, sulfenic acid which is susceptible to further oxidation ultimately to the sulfonic acid.[20] In recent studies it has been possible to demonstrate the reversible inactivation by hydrogen peroxide of L-histidine ammonia-lyase[34] and glyceraldehyde-3-phosphate dehydrogenase.[35] In the latter case, loss of enzymatic activity was paralleled by a loss in sulfhydryl titer. The inactivations were performed near pH 7 at relatively low peroxide concentration. Analyses for methionine sulfoxide were not reported. In another study with ferricytochrome *c*, O'Brien[36] noted destruction of methionine, cystine, tyrosine, and histidine residues by hydrogen peroxide at pH values from 2.2 to as high as 13. It is quite possible in this case that the heme group may have catalyzed the reaction, since an enhancement of the oxidation of glyceraldehyde-3-phosphate dehydrogenase by cytochrome has been noted.[35]

The previous examples illustrate the fact that optimal conditions for the oxidation of a particular protein by hydrogen peroxide will be dictated by a number of factors, including the presence or absence of sulfhydryl groups and particularly the relative accessibility of the various residues subject to attack. The usual parameters of temperature,

[33] R. Cecil and J. R. McPhee, *Advan. Protein Chem.* **14,** 255 (1959).

[34] A. Frankfater and I. Fridovich, *Biochim. Biophys. Acta* **206,** 457 (1970).

[35] C. Little and P. J. O'Brien, *Eur. J. Biochem.* **10,** 533 (1969).

[36] P. J. O'Brien, *Biochem. J.* **101,** 12 P (1966).

pH, and the presence of substrates, denaturants, or other activators or inhibitors can have marked effects upon both the extent and the specificity of the reaction.

Other Reagents

In addition to hydrogen peroxide, a number of other reagents have been used in recent years to modify methionine residues in proteins. Several of these are discussed in other sections of this monograph. Others include periodate, azide, β-propiolactone, and iodine as well as a host of others which are of limited interest either because of a lack of specificity or various difficulties in their application to protein modification.

Periodate, a reagent well known to carbohydrate chemists, has been shown[37] to attack a variety of amino acid residues in proteins, including cysteine, cystine, methionine, tryptophan, tyrosine, histidine, and N-terminal serine and threonine. In spite of this apparent lack of specificity, this reagent has been successfully applied to the specific oxidation of the methionine residues in chymotrypsin[18] and apomyoglobin.[38] The reaction conditions selected—pH 5.0 and 0°—appeared to be critical, since increases in pH and/or temperature caused destruction of tryptophan and tyrosine residues as well.

β-Propiolactone, a very reactive compound, which attacks many different functional groups[39] under neutral or alkaline conditions, has been utilized by Atassi[40] at pH 3 and 25° for a period of 5 hours to specifically modify the two methionine residues of apomyoglobin. The product of the reaction, presumed to be a sulfonium salt, is relatively unstable and reverts to methionine upon either alkaline or acidic hydrolysis. This necessitated the use of enzymatic cleavage and examination of peptide maps to establish the site of the modification. The instability of the sulfonium salt is a drawback to the general utility of this otherwise useful reagent.

Azide, which has long been known to associate reversibly with the heme group of hemoproteins, also reacts with methionine to produce methionine sulfoximine.[41] The postulated mechanism involves production of an unstable sulfidimine intermediate, which hydrolyzes to the sulfoxide. In turn this reacts with a second molecule of hydrazoic acid to produce the final methionine sulfoximine. Examination of this sequence indicates that the reaction would be facilitated by both a decrease in

[37] J. R. Clamp and L. Hough, *Biochem. J.* **94**, 17 (1965).
[38] M. Z. Atassi, *Biochem. J.* **102**, 478 (1967).
[39] M. A. Taubman and M. Z. Atassi, *Biochem. J.* **106**, 829 (1968).
[40] M. Z. Atassi, *Immunochemistry* **6**, 6 (1969).
[41] J. K. Whitehead and H. R. Bentley, *J. Chem. Soc.*, p. **1572** (1952).

pH (hydrazoic acid is the presumed reacting species) and the presence of peroxide or other oxidizing agents. Brill and Weinryb[42] have demonstrated a hydrogen peroxide-catalyzed inactivation of a hemoprotein, horseradish peroxidase, by azide which is favored by acidic conditions. Although the data are not unequivocal due to difficulties in analysis for sulfoximines, the authors have suggested that the modification involves a methionine residue and furthermore postulate the direct participation of a methionine residue as the sulfoxide in the active center of the enzyme. It is likely that the promotion of the azide inactivation by peroxide is a result of the catalytic effect of a heme-peroxide complex in this particular case. The possible application of azide with or without peroxide to other proteins remains to be explored. Again, as with β-propiolactone, the analytical difficulties remain a problem.

Although iodine has received most attention as a reagent for the modification of tyrosine at neutral or alkaline pH, Lavine[43] long ago demonstrated its ability to react with methionine. Koshland *et al.*[44] observed the appearance of methionine sulfoxide in an antibody preparation after treatment with iodine monochloride at pH 8.5, although at a greatly reduced rate compared to the modification of tyrosine and histidine residues. Tryptophan was attacked as well. Similar results have been noted by Filmer and Koshland[45] for chymotrypsin. On the other hand, treatment of yeast invertase with iodine under acidic conditions produces a partially inactive derivative[46] which does not appear to involve modification of tyrosine or histidine residues.[17] Recent data[47] implicate one or more methionine residues as the site of attack when the reaction is carried out at pH 5.0. Under these conditions the inactivation is reversible by treatment with mercaptoethanol. At pH 7.0 the inactivation becomes irreversible, and the implication is that probably tyrosine and histidine residues become involved. Although these findings are somewhat preliminary it is possible that further study would demonstrate the usefulness of iodine under acidic conditions as another specific reagent for methionine modification.

[42] A. S. Brill and I. Weinryb, *Biochemistry* **6**, 3528 (1967).
[43] T. F. Lavine, *J. Biol. Chem.* **151**, 281 (1943).
[44] M. E. Koshland, F. M. Englberger, and S. M. Gaddone, *J. Biol. Chem.* **238**, 1349 (1963).
[45] D. L. Filmer and D. E. Koshland, Jr., *Biochem. Biophys. Res. Commun.* **17**, 189 (1964).
[46] H. von Euler and S. Landergren, *Biochem. Z.* **131**, 386 (1922).
[47] A. Waheed and J. O. Lampen, personal communication, 1970.

[32] Dye-Sensitized Photooxidation

By E. W. Westhead

An earlier volume of this series contains an excellent discussion of this subject which should be consulted by anyone interested in using the technique.[1] The present contribution is intended to cover development in the technique since the previous article.[2] Dye-sensitized photooxidation of enzymes for identification of critical residues at the catalytic site was introduced 20 years ago[3] and appeared then to be a promising tool. The promise has been largely unfulfilled, partly because chemical reagents with equal or greater specificity have been developed which provide a convenient label for subsequent location of the critical residue in the peptide chain. It is also true, unfortunately, that the great majority of published experiments using this technique have been carried out under arbitrarily chosen conditions, have reported incomplete amino acid analyses, and have shown uncritical interpretations of the data. On the other hand, the best illustrations of the method indicate that for some enzymes and under some circumstances the method can indeed be a useful tool for the determination of mechanisms or functional groups.

The technique is basically no more complex than the irradiation of a solution containing protein, photosensitizing dye, and oxygen, long enough to measure a change in some property of the protein (usually catalytic activity). Ray has designed an apparatus to ensure a high degree of reproducibility and given a method for calibrating this system.[1] A far less sophisticated arrangement may be used for preliminary experiments, but the calibration procedure with histidine is a valuable adjunct to any set of experiments. The use of an oxygen electrode[4] instead of a Warburg apparatus for following oxygen uptake is also to be recommended for simplicity, speed, and economy. An oxygen electrode alone is not expensive and can be coupled to an ordinary laboratory recorder with a very simple circuit design given in reference 4. This design is reproduced in Figure 1.

Mechanism and End Products

Various authors have suggested that the active species in dye-sensitive photooxidation include the activated dye itself, a dye–oxygen

[1] W. J. Ray, Jr., Vol. XI, p. 490.

[2] References have been selected as illustrations or entries to the literature. A more complete bibliography to June 1970 is available from the author.

[3] L. Weil, W. G. Gordon, and A. R. Buchert, *Arch. Biochem. Biophys.* **33**, 90 (1951).

[4] E. W. Westhead, *Biochemistry* **4**, 2139 (1965).

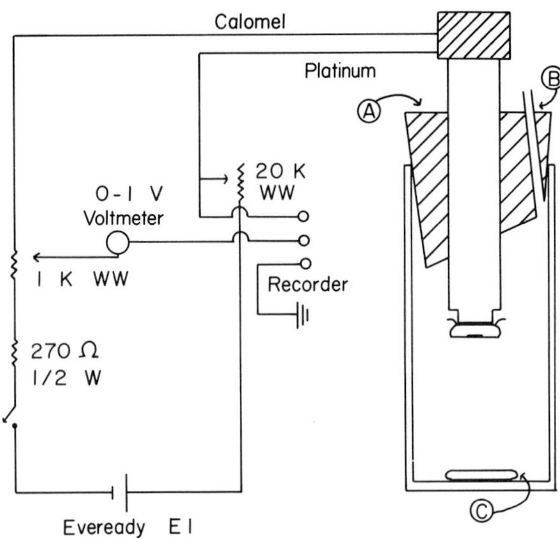

FIG. 1. Diagram of an oxygen electrode system. The electrode is inserted through stopper (A) of inert material such as polyethylene, cut on a bias to allow expulsion of an air bubble as the stopper is inserted. The air exits through a narrow (approximately 1 mm) hole or tubing (B); no significant diffusion of oxygen occurs through the column of liquid in this narrow opening over a period of 15 minutes or more. Samples of enzyme for analysis may be removed through this opening, while excluding oxygen, by pressing the stopper deeper into the solution while the sample is withdrawn. Rapid stirring of the solution past the electrode membrane by the magnetic stirbar is critical (C).

The Clark electrode [Yellow Springs Instrument Co., Yellow Springs, Ohio] is filled with KCl solutions in accord with the manufacturer's instructions. Very thin polyethylene film, of the type used by dry cleaners to cover clothes, makes an excellent, rapidly responding membrane. The voltmeter and wire-wound potentiometer are inexpensive types, used merely to adjust the voltage from the dry cell to 0.6 V, the reduction voltage for O_2. Any 10 mV laboratory recorder is suitable for measuring the current produced. To calibrate the recorder, air-saturated water at 25° is considered to be 0.25 mM in oxygen. On addition of a few milligrams of sodium hydrosulfite (dithionite) per milliliter, the oxygen content is taken to be zero. Between these two extremes, the recorded voltage is directly proportional to the oxygen content of the solution. To test the apparatus, one may photooxidize, for example, a solution of imidazole buffer. If there are no leaks in the system and no air bubbles have been trapped, the decrease in oxygen concentration should stop abruptly and remain constant when the light is turned off. With a rapidly responding recorder, a thin membrane, and rapid stirring of the solution, the lag in response to change in illumination should be less than 5 seconds.

complex and oxygen in the triplet or one of two singlet energy states.[5-9] Evidence has been shown for the participation of each of these species in a photooxidation of either proteins or small molecules, and there is ample evidence to indicate that several mechanisms may participate in a given photooxidation reaction. Three dyes at least, methylene blue,[10] eosin,[11] and riboflavin,[7] will participate in an anaerobic hydrogen transfer reaction from the substrate to the photoactivated dye.

A study of the oxidation pathways and products of histidine oxidation[12] makes it appear certain that the major product will be very dependent on the environment of histidine in a protein, and quite possibly on the presence of other species in solution. Under some conditions aspartic acid may be a major end product of histidine oxidation[12] and in one case a substantial amount of material absorbing at 320 nm was observed upon oxidation of histidine in a protein.[13] Oxidation products of tryptophan have never been quantitated, but careful purification of lysozyme derivatives after photooxidation yielded a group of monooxidized enzymes in which a single tryptophan was oxidized. Even this relatively homogeneous product could be separated into different species of enzyme with different degrees of residual activity and different ultraviolet spectra.[14] Only in the case of methionine can the end products be identified quantitatively. This amino acid is converted rapidly to the sulfoxide, in good yield, and then, much more slowly, to the sulfone.[1] The sulfoxide is readily reduced to the thioether again by sodium borohydride[15] or by sulfhydryl compounds like mercaptoethanol[16] or thioglycolic acid.[17] It cannot be stressed too strongly, however, that methionine sulfoxide is readily converted to methionine during hydro-

[5] L. Weil, *Arch. Biochem. Biophys.* **110,** 57 (1965).

[6] D. R. Kearns, R. A. Hollins, A. U. Khan, R. W. Chambers, and P. Radlick, *J. Amer. Chem. Soc.* **89,** 5455 (1967).

[7] D. R. Kearns, R. A. Hollins, A. U. Khan, and P. Radlick, *J. Amer. Chem. Soc.* **89,** 5456 (1967).

[8] J. S. Bellin and C. A. Yankus, *Arch. Biochem. Biophys.* **123,** 18 (1968).

[9] C. S. Foote, *Science* **162,** 963 (1968).

[10] D. Cavallini, R. Scandurra, and C. DeMarco, *Biochem. J.* **96,** 781 (1965).

[11] G. Reske, F. Nimmerfall, and J. Stauff, *Z. Naturforsch. B* **21,** 305 (1966).

[12] M. Tomita, M. Irie, and J. Ukita, *Biochemistry* **8,** 5149 (1969).

[13] M. Martinez-Carrion, R. Kuczenski, D. C. Tiemeier, and D. L. Peterson, *J. Biol. Chem.* **245,** 799 (1970).

[14] N. A. Kravchenko and V. Kh. Lapuk, *Biokhimiya* **34,** 832 (1969).

[15] Unpublished data of the author.

[16] G. Jori, G. Galiazzo, A. Marzotto, and E. Scoffone, *J. Biol. Chem.* **243,** 4272 (1968).

[17] U. Kenkare and F. M. Richards, *J. Biol. Chem.* **241,** 3197 (1966).

chloric acid hydrolysis of the protein.[1] Acid hydrolysis, therefore, is not a suitable way to measure the degree of methionine oxidation. Considering the change in polarity of methionine and tryptophan that result from their oxidation, it seems likely that the oxidation of one of these amino acids will produce some change in the conformation of a protein. Suitable methods for the measurement of these two amino acids have been described.[1]

Conditions Governing Specificity

The choice of dye and the pH of the reaction seem to be the most generally important determinants of specificity. Any conditions affecting protein conformation might have a critical effect on specificity of oxidation. A survey of conditions should certainly be made when work is begun on any enzyme. Since oxygen uptake is a useful index of the extent of oxidation, comparisons, under different conditions, of rate of activity loss vs. rate of oxygen uptake can provide a preliminary measure of specificity. For this purpose an oxygen electrode is highly recommended. More critical work depends on the use of an amino acid analyzer and other measurements of the destruction of specific amino acids. Spectral measurements after photooxidation must be interpreted with caution since the author and others have observed that dyes sometimes *cannot* be removed quantitatively from proteins after exposure of the mixture to light and air.

Choice of Dyes. A large number of dyes may be used for photosensitized oxidation,[18] but most authors have used the cationic dye methylene blue. Rose bengal, chosen for its anionic nature, was found to be much more specific for active site oxidation of histidine in yeast enolase[4] and has been used by a number of investigators since then. Photosensitizing dyes exhibit some specificity even toward free amino acids in solution. For example, below pH 5, rose bengal is able to oxidize tryptophan in buffered solutions while methylene blue is effective only in acetic acid solution.[19] Rose bengal at neutral pH shows higher specificity toward histidine, compared with other amino acids, than does methylene blue and some other dyes.[8]

With enzymes, however, the specificity of a dye depends on the individual enzyme and quite possibly on the mechanism through which oxidation takes place. In some cases[20,21] no difference is found between

[18] G. Oster, J. S. Bellin, R. W. Kimball, and M. E. Schrader, *J. Amer. Chem. Soc.* **81**, 5097 (1959).

[19] G. Jori, G. Galiazzo, A. Marzotto, and E. Scoffone, *Biochim. Biophys. Acta* **154**, 1 (1968).

[20] P. M. Burton and S. G. Waley, *Biochem. J.* **100**, 702 (1966).

rose bengal and methylene blue while the former more efficiently oxidizes enolase,[4] creatine kinase,[4] and aspartate transaminase.[22] While methylene blue will apparently oxidize carboxypeptidase only after removal of bound metal,[23] rose bengal, perhaps because of the charge difference, inactivates the zinc enzyme efficiently.[24] An increased specificity of rose bengal for histidine was also seen in the case of glyceraldehyde-3-phosphate dehydrogenase, where only histidine was oxidized by that dye[25]; when methylene blue was used, sulfhydryl groups were also oxidized.[26]

Other dyes have been used less frequently. Eosin oxidizes free tryptophan and histidine at exactly the same rate, whereas thionine under the same conditions oxidizes histidine four times as fast as tryptophan.[27] A study of the photodynamic action of porphyrins on amino acids showed that methionine could be oxidized below pH 7 fairly specifically by hematoporphyrin. With this dye the authors selectively oxidized a single methionine residue in hen egg white lysozyme.[28] FMN and FAD are photoreactive, and are discussed elsewhere in this series.[29] Model studies indicate that FMN is far more effective than FAD[30] and that free flavin may be able to bring about excitation of bound flavin when binding causes inactivation of the molecule.[31] A substrate analog, thiouridylic acid, has been used to photooxidize ribonuclease.[32] Inactivation, however, occurred through modification of a tyrosine residue which does not appear to be near the active site. The idea seems sound, and further work in this direction might prove to be valuable.

Other Conditions. The oxidation of histidine or histidine in proteins by methylene blue,[5] proflavin,[33] or rose bengal,[4] has been shown to be sharply pH dependent, the protonated form of histidine being resistant to oxidation. A pH profile of inactivation which follows the titration curve of histidine is thus reasonable first evidence for the destruction of

[21] G. C. Chatterjee and E. A. Noltmann, *Eur. J. Biochem.* **2,** 9 (1967).
[22] M. Martinez-Carrion, C. Turano, F. Riva, and P. Fasella, *J. Biol. Chem.* **242,** 1426 (1967).
[23] T. L. Coombs, T. Omote, and B. L. Vallee, *Biochemistry* **3,** 653 (1964).
[24] K. A. Freude, *Biochim. Biophys. Acta* **167,** 485 (1968).
[25] J. S. Bond, S. H. Francis, and J. H. Park, *J. Biol. Chem.* **245,** 1041 (1970).
[26] P. Freidrich, L. Polgar, and G. Szabolsci, *Nature (London)* **202,** 1214 (1964).
[27] T. Gomyo, Y. Yang, and M. Fujimaki, *Agr. Biol. Chem.* **32,** 1061 (1968).
[28] G. Jori, G. Galiazzo, and E. Scoffone, *Biochemistry* **8,** 2868 (1969).
[29] M. B. Taylor and G. K. Radda, Vol. 18b.
[30] G. R. Penzer, *Biochem. J.* **116,** 733 (1970); G. Weber, *Biochem. J.* **47,** 114 (1950).
[31] F. Y.-H. Wu and D. B. McCormick, *Biochim. Biophys. Acta* **236,** 479 (1971).
[32] F. Sawada, *J. Biochem.* **65,** 767 (1969).
[33] L. A. Æ. Sluyterman, *Biochim. Biophys. Acta* **60,** 557 (1962).

a critical histidine. If methylene blue is used, this interpretation has to be made cautiously since this dye and proflavin both show some pH effect in the oxidation of other amino acids.[5,33] Above pH 8, sulfhydryl groups become especially sensitive to oxidation, and tyrosine, which is ordinarily quite resistant to dye-sensitized photooxidation, also begins to become susceptible.

Ionic strength has been shown to affect photooxidation rates,[21,34] but it seems likely that these effects are only on the protein conformation and the accessibility of the oxidizable groups. It is worth noting that 8 *M* urea has no effect on the rate of oxidation of any of the amino acids by methylene blue.[35] With the same dye, Weil found that the number of moles of oxygen taken up per mole of amino acid increased with increasing temperature.[5] On the other hand, the rate of methionine destruction shows no temperature dependence, and the rate of tryptophan destruction shows only a moderate temperature dependence.[19] These results do not necessarily conflict, since the extra oxygen uptake at higher temperatures is probably due to further oxidation of the primary products of the reaction. In the oxidation of proteins, conformational effects are likely to be overwhelming; e.g., the photooxidation of bovine insulin at 10° caused the oxidation of only the two histidine residues, but above 10° tyrosines also were affected.[35]

In general, there seems to be very little point in using a higher concentration of dye than will adsorb, say, 90% of the incident light. In a 1-cm cell this obviously corresponds to an absorbancy of 1.0; correspondingly, for deeper solutions one may use a lower concentration of dye. Experiments with small molecules have shown that an excess of dye molecules may lead to secondary reactions.[9] On the other hand, when dyes are bound by large molecules their efficiency may be lowered, and this has been used to explain the inefficiency of low dye concentrations in the oxidation of ovalbumin.[36] Other things being equal, the rate of photooxidation will be directly dependent on the intensity of incident light. As Ray has pointed out,[1] shorter reaction times may circumvent artifacts caused by slow unfolding of partly reacted enzyme.

Whenever the rate of reaction has been examined as a function of oxygen concentration, it has been found that the systems are saturated with respect to oxygen in air-saturated water or buffers. At 25° the concentration of oxygen in water or dilute buffer saturated with air is about 0.25 m*M*. In systems so far examined,[4,18] the reaction rates are independent of oxygen down to at least one-tenth of this concentration.

[34] W. J. Ray. Jr. and D. E. Koshland, Jr., *J. Biol. Chem.* **237**, 2493 (1962).

[35] L. Weil, T. S. Seibles, and T. T. Herskovits, *Arch. Biochem. Biophys.* **111**, 308 (1965).

[36] T. Gomyo and Y. Sakurai, *Agr. Biol. Chem.* **31**, 1474 (1967).

Efficient oxygenation of the system therefore, is not critical in the early stages of a photooxidation reaction. Breaks in kinetics, however, must be examined carefully since the rate of oxidation will drop sharply if the oxygen in the solution approaches depletion. Again, the use of an oxygen electrode to monitor oxygen concentration during the reaction is a valuable aid in experiments of this type.

Newer Approaches to Specificity. Most authors have hoped to use photooxidation to demonstrate the presence of a critical histidine at an active center. Oxidation of cysteine or methionine residues, the most frequent side reactions, are then only troublesome. However, the oxidation of a sulfhydryl group in β-methylaspartase, together with the absence of other oxidative destruction, was useful in showing that a sulfhydryl group is in fact critical to the activity of this enzyme.[37] Other approaches have been used to circumvent the difficulties of oxidation of sulfhydryl groups or methionine residues. The presence of a critical histidine in myosin A has been shown by photooxidation after the highly reactive sulfhydryl groups of the enzyme were covered by disulfide exchange.[38] Photooxidation then led to the loss of a single histidine residue at the same rate as the loss of activity, with a much lower rate of loss of methionine. No tryptophan was lost. When the sulfhydryl groups were regenerated by reduction in a dark control, all activity was regained. An alternative approach was taken in the case of 6-phosphogluconate dehydrogenase, in which the sulfhydryl groups were reversibly masked with *p*-mercuribenzoate.[39] Histidine was then the only amino acid significantly affected by photooxidation. Again, removal of the protecting groups from the unoxidized enzyme led to essentially complete recovery of activity. Complications due to the oxidation of methionine have similarly been avoided during the photooxidation of lysozyme and the photooxidation of ribonuclease S and the S-peptide. In the former case, it was shown that the entire effect of photooxidation of lysozyme could be reversed by the reduction of oxidized methionine residues with mercaptoethanol.[16] In the case of ribonuclease,[17] the reduction of the methionine groups allowed the authors to attribute the loss of enzyme activity to a specific histidine group, histidine-119. A final example which deserves special comment is the rose bengal-sensitized oxidation of glyceraldehyde-3-phosphate dehydrogenase.[25] This work is notable because previous attempts to implicate histidine in the mechanism were unsuccessful: attempts to photooxidize the enzyme with methylene blue did not lead to a sufficiently specific oxidation.[26] Since the enzyme retained partial activity, the authors were able to show which step of the catalytic

[37] V. R. Williams and W. Y. Libano, *Biochim. Biophys. Acta* **118**, 144 (1966).
[38] A. Stracher, *J. Biol. Chem.* **240**, PC958 (1965).
[39] M. Rippa and S. Pontremoli, *Biochemistry* **7**, 1514 (1968).

reaction (acyl cleavage) was inhibited upon loss of the histidine residue. Furthermore, although the photooxidation technique does not yield a convenient marker for the derivatized peptide, the authors of this paper were able, by parallel separations of peptides from the oxidized and nonoxidized protein, to actually locate the peptide which contained the oxidized histidine.

Such results lead to the discussion of another observation with potential importance for certain enzyme systems. A number of dehydrogenase enzymes have been shown to bind certain dyes rather specifically and in some cases it has been shown also that these dyes bind in direct competition to nucleotide binding (see e.g., Brand *et al.*[40]). It seems very possible then that the cited results on glyceraldehyde-3-phosphate dehydrogenase derived their specificity from the fact that the dye is bound directly at the nucleotide binding site of the protein. Further support of this observation comes from the discovery that dye-sensitized photooxidation of phosphofructokinase leads to loss of the ATP control site with no change in the activity of the enzyme.[41] It will be interesting to see whether the binding of dyes to nucleotide sites is a general phenomenon that will open the way to study of the control sites of adenine nucleotide-sensitive enzymes and to the nucleotide binding sites of dehydrogenases.

Avoidance of Artifacts

Many papers have appeared showing only that loss of activity is "parallel" to the loss of, e.g., histidine. Without a complete and accurate analysis for all oxidizable amino acids, cysteine, methionine, tryptophan, histidine, and tyrosine, such data are quite uninformative. Interpretation of such results is particularly hopeless when it can be seen that, at 50% inactivation, half a dozen or more histidine residues have been oxidized. In addition to acceptable analyses, the quantitative correlation of rate of loss and rate of single residue destruction[34] can reasonably be expected.

A decreased inactivation rate caused by substrate or other catalytic affector may be interpreted as specific protection of a catalytic group, when in reality the affector is decreasing the efficiency of the dye or providing an alternate oxidation substrate. A control experiment can easily be designed to eliminate this artifact. If the protection is indeed enzyme specific, the effect will not be seen when the dye is used to oxidize some other substance, such as histidine.[1]

Artifacts are particularly likely when the photodynamic inactivation

[40] L. Brand, J. R. Gohlke, and D. S. Rao, *Biochemistry* **6**, 3510 (1967).

[41] C. E. Ahlfors and T. E. Mansour, *J. Biol. Chem.* **244**, 1247 (1969).

is slow. Molecules which are known to react at least slowly with photosensitized dyes include dipolar ionic buffers, such as tricine, TES, etc.,[42] Tris,[42,43] and EDTA.[44] The transfer of hydrogen from EDTA to methylene blue can inactivate the dye or, in the presence of reducible metal ions, lead to a cyclic reaction of the dye to give oxidized EDTA and reduced metal.[45]

[42] R. K. Yamazaki and N. E. Tolbert, *Biochim. Biophys. Acta* **197,** 90 (1970).
[43] D. B. McCormick, J. F. Koster, and C. Veeger, *Eur. J. Biochem.* **2,** 387 (1967).
[44] V. Massey and G. Palmer, *Biochemistry* **5,** 3181 (1966).
[45] G. K. Oster and G. Oster, *J. Amer. Chem. Soc.* **81,** 5543 (1959).

[33] Specific Modification of NH_2-Terminal Residues by Transamination

By HENRY B. F. DIXON and ROBERT FIELDS

Selective modification of the amino groups of proteins (recently reviewed[1]) usually presents little difficulty, since these groups are among the most powerful nucleophiles present. Some differentiation between terminal amino groups of the polypeptide chains (α-amino groups) and the lateral amino groups of lysine residues (ϵ-amino groups) is possible on the basis of their very different basic strengths, but a much more selective modification of α-amino groups can be achieved by using a reaction that depends upon direct participation of the adjacent peptide bond. Such a reaction is transamination.

Transamination specifically removes the α-amino group and so allows any function that this group may possess to be studied, but at the same time it also specifically introduces a highly reactive carbonyl group.

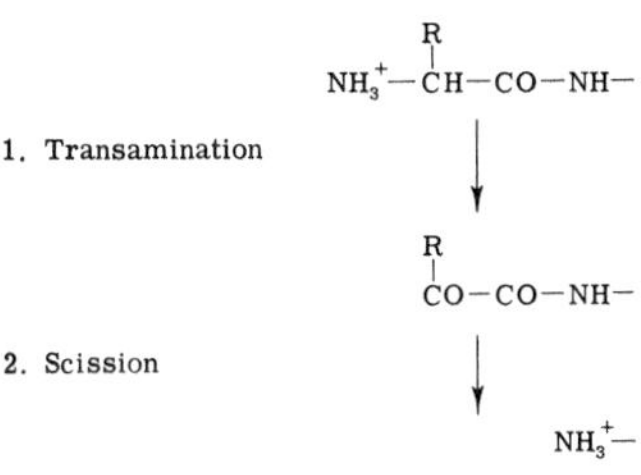

SCHEME 1

[1] G. R. Stark, *Advan. Protein Chem.* **24,** 261 (1970).

This allows further specific reactions to be carried out; the main one so far studied is the reaction under comparatively mild conditions with a bifunctional nucleophile which removes the ketoacyl residue and so leaves the polypeptide chain shorter by one residue (Scheme 1). Thus the function of the whole N-terminal residue may be studied, and the process of modification and scission of the terminal residue may be repeated.

Conditions for the transamination of proteins that were mild enough so that the protein would not be denatured were developed from the conditions found suitable for the transamination of amino acids.[2] These were applicable to peptides,[3] and the discovery[4] that pyridine greatly

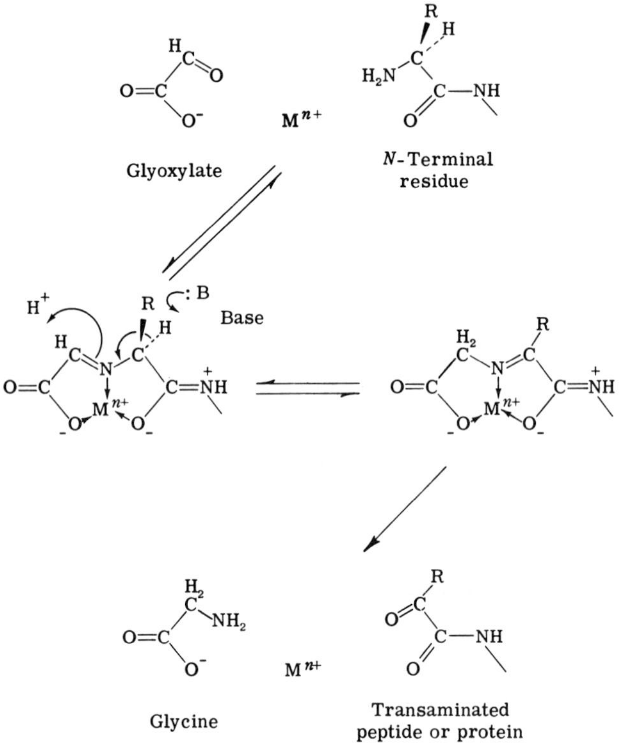

Glyoxylate M^{n+} *N*-Terminal residue

Base

Glycine M^{n+} Transaminated peptide or protein

SCHEME 2

[2] D. E. Metzler and E. E. Snell, *J. Amer. Chem. Soc.* **74,** 979 (1952).

[3] C. Cennamo, B. Carafoli, and E. P. Bonetti, *J. Amer. Chem. Soc.* **78,** 3523 (1956).

[4] H. Mix and F. W. Wilcke, *Hoppe-Seyler's Z. Physiol. Chem.* **318,** 148 (1960).

accelerated the transamination of amino acids led to the milder conditions of reaction that were necessary in the case of labile proteins.[5-8]

The need for each of the essential components of the system—(1) the acceptor of the amino group (usually glyoxylate), (2) the cation of a heavy metal (usually nickel or copper), and (3) a high concentration of base (usually acetate)—is explained in Scheme 2. This is proposed by analogy from the mechanisms for free amino acids.[9,10] The scheme also explains the need for the peptide carbonyl group and hence the specificity for α-amino groups, since the carbonyl oxygen plays a part in stabilizing the complex, and also because the electron attraction by the -CO-NH- system (enhanced by its chelation) helps in labilizing the α-proton. We do not know whether the uptake of a proton from the medium is concerted as drawn. More probably it follows the loss of the α-proton.

Transamination has also been reported without addition of metal ions or base with the much stronger amino acceptor phenylglyoxal.[11] Preliminary experiments[12] suggest that this reaction can be made specific for α-amino groups if a low pH is used.

Transamination

Conditions of Reaction

Typical reaction conditions for proteins are incubation for about 30 minutes at 20° in a solution of 2 *M* sodium acetate, 0.4 *M* acetic acid, 0.1–0.2 *M* sodium glyoxylate, and 2–10 m*M* $CuSO_4$ (pH 5.5). If cupric ions are likely to harm the protein, the copper sulfate may be replaced by nickel sulfate, but the pH should then be raised (to pH 7.0) by diminishing the concentration of added acetic acid to 10 m*M* and the time be increased to 1 hour.[8] Sodium glyoxylate is conveniently purified from commercial glyoxylic acid by the method of Radin and Metzler.[13] Protein concentration is not critical; up to about 1 mg/ml may be suitable. The protein may be separated from the reaction mixture by gel filtration. In the case of peptides, however, typical conditions are pyridine (1 *M*),

[5] H. B. F. Dixon, *Biochem. J.* **90,** 2C (1964).
[6] H. B. F. Dixon, *Biochem. J.* **92,** 661 (1964).
[7] H. B. F. Dixon and V. Moret, *Biochem. J.* **94,** 463 (1965).
[8] H. B. F. Dixon, *Biochem. J.* **103,** 38P (1967).
[9] A. E. Braunstein and M. M. Shemyakin, *Biokhimiya* **18,** 393 (1953).
[10] D. E. Metzler, M. Ikawa, and E. E. Snell, *J. Amer. Chem. Soc.* **76,** 648 (1954).
[11] K. Takahashi, *J. Biol. Chem.* **243,** 6171 (1968).
[12] R. Zito, personal communication.
[13] N. S. Radin and D. E. Metzler, *Biochem. Prep.* **4,** 60 (1955).

acetic acid (0.05 M), sodium glyoxylate (0.1–0.2 M) and nickel acetate (10 mM) at 20° for 1 hour. In this case much of the pyridine may be removed by rotary evaporation before the product is isolated by anion–exchange chromatography.

Many variants of these conditions can be used; some may be appropriate for proteins that have particular requirements for stability; e.g., the temperature can be diminished to 1°.[14] The amino acceptor may be changed to glyoxylamide or pyridoxal.[15] An early suggestion[6] that glyoxylamide was more satisfactory than glyoxylate for transaminating N-terminal glycine appears to have been mistaken, but it may in some cases be easier to remove from the product than glyoxylate. Since glyoxylamide is acid-labile, a simple way of preparing it is to pass a solution of tartramide through a column of a strongly basic resin (as the acetate), the top of which contains periodate in excess of the amount of tartramide. Hence a solution of a known amount of glyoxylamide without other components is obtained.

Of the metal ions tried, cupric, nickel, cobaltous, and zinc ions promote transamination.[8] Nickel, cobalt, and zinc all work best near neutrality and are inhibited by acidification. Zinc and cobalt, however, catalyze the reaction more slowly than nickel. Copper accelerates the reaction more than nickel, and faster still at a lower pH (5.5), but there is more danger of various side reactions. Several of the metal ions known to catalyze transamination between pyridoxal and free amino acids have not been tried with peptides for fear of oxidative side reactions. The catalysis by aluminum does not appear to be accelerated by bases.[16]

Pyridine accelerates the reaction more than acetate, but in at least one case it was more harmful to a protein.[17] Nevertheless we advise its use for peptides, especially since it is volatile.

Assessment of the Reaction

Since transamination is accompanied by a change of charge, a method of separation of the protein that is sensitive to its charge is desirable for the simultaneous *analysis* of the extent of transamination and *isolation* of the wanted product. Ion-exchange chromatography has been used in several cases,[6,7,15] and electrophoresis may also be suitable.

In the case of peptides, ion-exchange chromatography is likely to be

[14] S. van Heyningen, K. F. Tipton, and H. B. F. Dixon, *Biochem. J.* **108**, 508 (1968).

[15] H. B. F. Dixon, *in* "Pyridoxal Catalysis: Enzymes and Model Systems" (E. E. Snell, A. E. Braunstein, E. S. Severin, and Yu. M. Torchinsky, eds.), p. 99. Wiley (Interscience), New York, 1968.

[16] T. C. Bruice and R. M. Topping, *J. Amer. Chem. Soc.* **85**, 1480 (1963).

[17] S. van Heyningen and H. B. F. Dixon, *Biochem. J.* **104**, 63P (1967).

needed to separate the product not only from any unreacted peptide, but also from the other components of the reaction mixture. With proteins, however, isolation of a transaminated protein from unreacted protein may not be necessary since the reaction may be complete; its separation from the small molecules of the medium by gel filtration may be sufficient. In this case, however, some analytical method is required to demonstrate that transamination is complete and that the product is pure.

Methods based on the disappearance of amino groups will probably be insensitive, since only the one terminal group disappears in a polypeptide chain which may contain many lateral amino groups. Methods based on quantitative determination of the end group are laborious, and they are likely to be unreliable, since anything that interfered with the determination would increase the apparent degree of transamination. Hence demonstration of the appearance of a reactive keto group is usually the best way of characterizing the product. The following procedure may be used.

Determination of Reactive Carbonyl Groups in Peptides and Proteins[17a]

Reagents

2,4-Dinitrophenylhydrazine, 5 m*M*, in 2 *M* HCl; prepared by the method of Friedemann and Haugen.[18]

First Method. This method is rapid and useful as a semiquantitative test, but it is less accurate and sensitive than the methods below. The sample, 0.2 ml or less, is allowed to react with 0.2 ml of reagent for 5 minutes. The mixture is then diluted to 5 ml with 1 *M* HCl and the extinction at 370 nm is determined. A blank is also prepared and may be used in the spectrophotometer to give a difference reading directly. As the blank extinction is near unity, accurate dilution is required for determining small concentrations of carbonyl group.

Second Method. This method is indicated when many determinations are to be made, as in assaying fractions of effluent from a chromatographic column, or when the concentration of carbonyl group is high enough so that small volumes of sample may be taken. The reagent is diluted with HCl before reaction with the sample, thus eliminating errors that arise from individual dilutions. Samples (10–100 μl) are mixed with

[17a] R. Fields and H. B. F. Dixon, *Biochem. J.* **121,** 587 (1971).

[18] T. E. Friedemann and G. E. Haugen, *J. Biol. Chem.* **147,** 415 (1943).

2.5-ml portions of a solution of reagent diluted 25-fold with 1 *M* HCl. One hour later the extinction at 370 nm is determined.

Third Method. This is the most accurate and is recommended for use with proteins. The sample solution is adjusted with 10 *M* HCl to give an HCl concentration of 1 *M*. Of this solution, 2.4 ml is pipetted into a cuvette. Then 0.09 ml of reagent solution is mixed with 2.4 ml of 1 *M* HCl for use as a blank, and with it the spectrophotometer is adjusted to give a null reading at 370 nm. The reaction is initiated by adding 0.10 ml of the reagent to the sample and mixing the contents of the cuvette by inversion. A stopwatch is started, and several readings are taken during the first few minutes. These readings are extrapolated to give an accurate initial reading, which is subtracted from the final extinction reached after 1 hour.

Standardization. If possible, a standard curve should be prepared for the particular hydrazone that is being studied. If not (as in the case of a protein), a model compound is used. Such a compound is pyruvoylglycine, which was found to give a difference of extinction at 370 nm of 12,000 $M^{-1}cm^{-1}$.

Scope of the Reaction

General

So far the number of proteins whose transaminated derivatives have been characterized is small. They are *Pseudomonas* cytochrome *c*-551,[7] azurin,[19] carboxypeptidase,[14] insulin[20] and ribonuclease T_1.[17a] We may expect, however, that the technique should be applicable to many proteins, so the reasons for possible failure will be considered. Since the reaction is rapid at neutral pH and low temperature, the intrinsic stability of the protein under the reaction conditions should not usually be a problem. Nevertheless the metal ions may damage some proteins. Cases may occur in which the N terminus of the protein is masked in its native structure. The fact that the transamination reaction occurs in strong solutions of urea[21] may overcome this difficulty, at least for those proteins whose urea denaturation can be reversed. Another difficulty is that proteins may precipitate in the reaction medium; it was found that horse hemoglobin, with its reactive thiol groups masked as the disulfide with cysteamine,

[19] H. B. F. Dixon, *Abstr. 5th Meeting FEBS (Fed. Eur. Biochem. Soc.)*, p. 164. Prague, 1968.

[20] H. F. Bünzli and H.-R. Bosshard, *Hoppe-Seyler's Z. Physiol. Chem.* **352**, 1180 (1971).

[21] S. van Heyningen, Ph.D. Thesis, Cambridge University, Cambridge, England, 1968.

precipitated in strong acetate when both nickel sulfate and sodium glyoxylate were added, although not with either of them by itself.[22]

Apparently a wide range of N-terminal residues can be transaminated. Besides the alanine, asparagine, glutamic acid, glycine, and phenylalanine of the proteins listed above, and alanine in a number of model peptides,[23] isoleucine, the residue most likely to be resistant,[24] on both electronic[25] and steric grounds, was satisfactorily transaminated[26]; in a medium of pyridine (1 *M*), acetic acid (50 m*M*), sodium glyoxylate (0.2 *M*), and nickel acetate (10 m*M*), 92% of Ile-Tyr (40 m*M*) was transaminated in 1 hour at 20°. N-terminal glycine, which, by analogy with free glycine, might be expected to transaminate rapidly,[27] but with a less favorable equilibrium constant than other residues,[24] in fact reacts satisfactorily; Gly-Tyr appeared to be completely transaminated under these conditions in 10 minutes.

Clearly there are dangers of side reactions. Oxaloacetate, for example, is notorious for the ease of its reaction with aldehydes,[28] and this reactivity is likely to be enhanced, as is its decarboxylation,[29] in the presence of metal ions. In view of the ease of β-elimination reactions under transamination conditions,[30,31] such reactions may be expected in the case of cystine and conceivably some other residues. Hence reactions other than transamination may sometimes occur.[15]

Periodate Oxidation

Periodate oxidation provides an alternative to transamination for the conversion of an N-terminal residue into an α-oxoacyl group when the terminal residue is threonine or serine. This reaction has a long history in protein chemistry. Since periodate oxidizes the grouping $>C(-NH_2)-C(-OH)<$ about a thousand times faster than $>C(-OH)-C(-OH)<$, extremely mild conditions of periodate treatment may be used for this purpose. Thus, in the oxidation of the N-

[22] J. V. Kilmartin, unpublished observations.

[23] A. Levitzki, M. Anbar, and A. Berger, *Biochemistry* **6**, 3757 (1967).

[24] D. E. Metzler, J. Olivard, and E. E. Snell, *J. Amer. Chem. Soc.* **76**, 644 (1954).

[25] P. Hermann and I. Willhardt, *Hoppe-Seyler's Z. Physiol. Chem.* **349**, 395 (1968).

[26] H. B. F. Dixon, *Abstr. 7th IUPAC Symp. Chem. Natural Products,* p. 48. Riga, 1970.

[27] L. W. Fleming and G. W. Crosbie, *Biochim. Biophys. Acta* **43**, 139 (1960).

[28] A. Ruffo, E. Testa, A. Adinolfi, and G. Pelizza, *Biochem. J.* **85**, 588 (1962).

[29] H. A. Krebs, *Biochem. J.* **36**, 303 (1942).

[30] D. E. Metzler and E. E. Snell, *J. Biol. Chem.* **198**, 353 (1952).

[31] C. de Marco and A. Rinaldi, *Arch. Biochem. Biophys.* **140**, 19 (1970).

terminal serine of ribonuclease S-protein, the desired reaction proceeded over 1000 times faster than any other uptake of periodate.[32] Of course this specificity may be expected only for proteins without thiol groups or oxidizable prosthetic groups, such as heme. To obtain the maximum specificity, it seems wise to follow the uptake of periodate[32] and to destroy the excess as soon as the wanted reaction is complete. In the case of ribonuclease S-protein this was 5 minutes at room temperature in a 2-fold excess (50 μM) of periodate.

Reverse Transamination

It has not proved possible to reverse the transamination of proteins under mild conditions once the ketoacyl residue is formed. The marked exception to this is the glyoxyloyl group, which can easily be converted into a glycyl residue. Since the glyoxyloyl group may be produced by periodate oxidation of N-terminal serine or threonine, this gives a method of converting these residues into glycine. When corticotropin was subjected to the two steps of periodate oxidation and transamination,[33] it lost most of its activity on oxidation but recovered it when the amino group was restored in the second step. Much milder conditions for converting glyoxyloyl groups into glycine residues than those originally used are now available; these are incubation at 20° for 30 minutes in M pyridine, 0.2 M glutamic acid, and 10 mM $NiSO_4$.[34]

Uses of Transamination

The most obvious use of the reaction is the specific removal of a single amino group from the polypeptide chain. This enables the role of that group in any function of the protein to be assessed. But transamination also introduces the keto group into the protein, a group with electrophilic properties unlike any of the other groups of most proteins. Hence further reactions of the transaminated protein may be made to take place in a highly specific manner. Thus, a heavy atom could be introduced for crystallographic purposes into a specific position in a transaminated protein with a reagent such as *p*-iodobenzenesulfonylhydrazide. The molecules of this substance are somewhat large, and presumably smaller molecules could be found that possessed both a heavy atom and a nucleophilic group. Similarly the reaction may be used for the introduction, specifically at the N terminus, of groupings for various other purposes. But the main use so far made of the keto group is for

[32] R. Fields and H. B. F. Dixon, *Biochem. J.* **108**, 883 (1968).
[33] H. B. F. Dixon and L. R. Weitkamp, *Biochem. J.* **84**, 462 (1962).
[34] H. B. F. Dixon, *Biochem. J.* **107**, 124 (1968).

the scission of the terminal residue, and this is dealt with in the next section.

Scission of α-Ketoacyl Residues

The electrophilic carbon of the keto group of a transaminated protein can bind a reagent that will specifically attack and break the adjacent peptide bond. In this way the modified residue is selectively removed, so that the protein molecule is left shorter by one residue. This extends the transamination procedure to permit study of the role of the side chain of the first residue as well as the role of its α-amino group, and, of course, enables the whole two-step process to be repeated.

To be a scission reagent, a molecule must contain two nucleophilic groups with one or two atoms between them. One nucleophilic group attaches itself to the carbonyl carbon, and so places the other correctly for attack on the peptide carbon with formation of a 5- or 6-membered ring. In early experiments with model peptides we had some success with thio- and selenosemicarbazones,[5,6] but *o*-phenylenediamine proved far more satisfactory as a reagent.[7]

Conditions of Reaction

The transaminated protein is incubated in a solution of 2 *M* sodium acetate, 2 *M* acetic acid, and 40 m*M* *o*-phenylenediamine at 37° overnight. Under these conditions the half-time for the reaction of transaminated cytochrome *c*-551 is 1.6 hours.[17] The reaction can be stopped, and the protein can be removed from the reaction mixture by gel filtration.

At present the conditions of this reaction cannot be varied much with advantage. More acidic conditions may accelerate the scission, but they are likely to be harmful to most proteins. The sodium acetate is present to keep the pH reasonably high. The concentration of the diamine is not critical, at least within the limits of 10–50 m*M*. Exclusion of oxygen, although it diminishes the formation of colored oxidation products of the diamine, does not appear to improve the yield of the scission. These colored side products are adsorbed onto the Sephadex G-25 used for stopping the reaction, and may be washed off with 50% acetic acid. Other ortho diamines, such as 2,3-diaminobenzoic acid and 3,4-diaminobenzoic acid,[17] show no clear advantage. Although it has been reported[35] that *o*,*o*′-diaminodiphenylamine is superior to *o*-phenylenediamine, the conditions used were different from those we have found to be most

[35] Y. Degani and A. Patchornik, *Abstr. 7th Int. Congr. Biochem.*, p. 11. Tokyo, 1967.

satisfactory. With general acid catalysis, we could find little difference between the two.

Many side reactions are probable in the course of the scission (Scheme 3). As the yield of the desired product has exceeded 95%,[17] however,

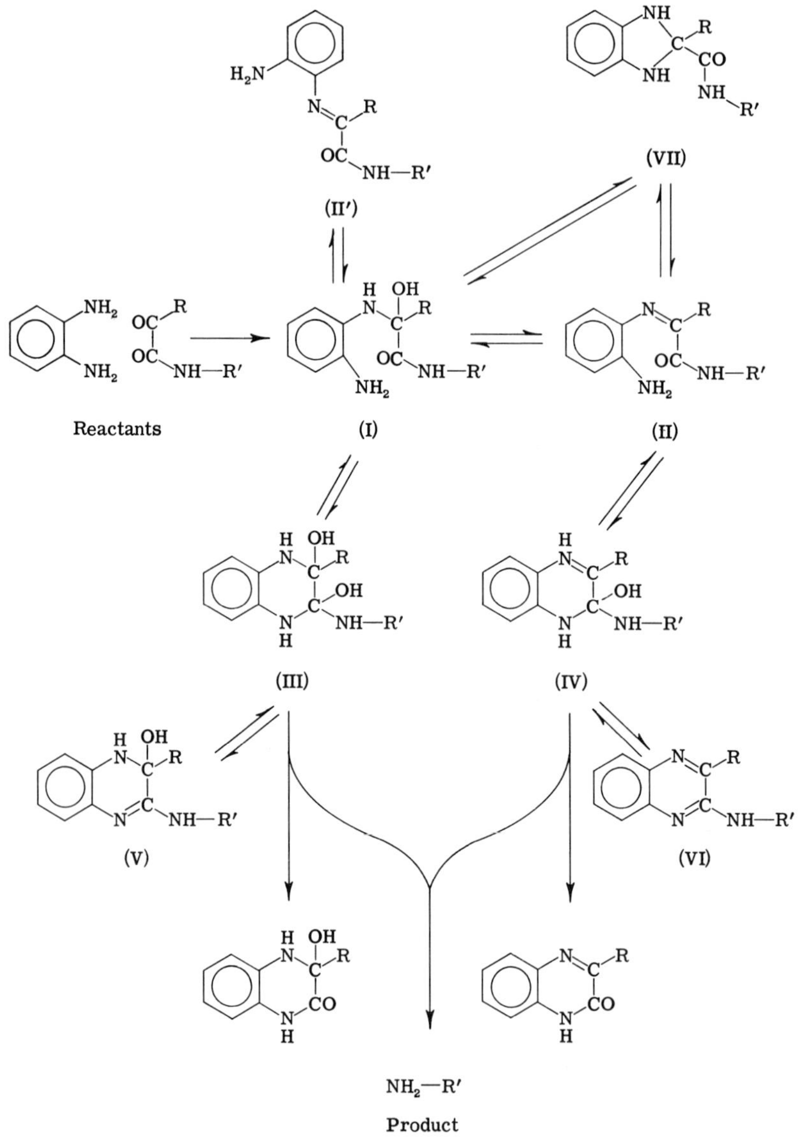

SCHEME 3

those that occur to a significant extent must be reversible, but they may greatly slow down the scission. Some would be excluded if the intermediate ketimine (II) could be hydrogenated, because then the 5-membered ring (VII) could not form by attack of both nucleophilic groups at the same carbonyl carbon, and the formation of the unproductive trans isomer (II′) of the ketimine would also be ruled out. The rapid course of the similar reaction of Holley and Holley[36] supports the view that such hydrogenation would accelerate the scission. We therefore attempted to reduce the intermediate (II) with cyanoborohydride[37] to form an *N*-(*o*-aminophenyl) peptide. We have not succeeded in finding conditions to make this quantitative, since considerable reduction of the ketone to the alcohol occurs. Nevertheless, some of the wanted compound did form, and it appeared to break down rapidly in the desired manner. It is therefore likely that even milder conditions of scission would be available if a satisfactory reductive amination of transaminated proteins with *o*-phenylenediamine could be developed.

Acknowledgment

We thank Dr. Simon van Heyningen for his advice and discussion as well as for his large contribution to the work.

[36] R. W. Holley and A. D. Holley, *J. Amer. Chem. Soc.* **74,** 5445 (1952).
[37] R. F. Borch and H. D. Durst, *J. Amer. Chem. Soc.* **91,** 3996 (1969).

[34] Modification of Tryptophan with BNPS-Skatole (2-(2-Nitrophenylsulfenyl)-3-methyl-3-bromoindolenine)

By ANGELO FONTANA

N-Bromosuccinimide (NBS)[1] is an extremely reactive reagent, which can cause modification, as well as peptide bond cleavage, not only of tryptophan, but also of tyrosine and histidine residues.[2-4] Thus, effects of modification of proteins with NBS must be interpreted with caution.[5,6]

[1] Abbreviations used: NBS, *N*-bromosuccinimide; NPS-skatole, 2-(2-nitrophenylsulfenyl)-3-methylindole; BNPS-skatole, 2-(2-nitrophenylsulfenyl)-3-methyl-3-bromoindolenine.
[2] B. Witkop, *Advan. Protein Chem.* **16,** 261 (1961).
[3] L. K. Ramachandran and B. Witkop, Vol. XI, p. 283.
[4] T. F. Spande and B. Witkop, Vol. XI, p. 506.
[5] M. J. Kronman, F. M. Robbins, and R. E. Andreotti, *Biochim. Biophys. Acta* **143,** 462 (1967).
[6] E. H. Eylar and G. A. Hashim, *Arch. Biochem. Biophys.* **131,** 215 (1969).

A much milder oxidizing agent than NBS has been recently introduced for the modification of the tryptophan residue in proteins.[7,8] On treatment of 2-(2-nitrophenylsulfenyl)-3-methylindole (NPS-skatole) with 1 eq of NBS in acetic acid, a crystalline bromine-containing compound is obtained, whose spectral properties (NMR, IR, UV) fit the structure below, 2-(2-nitrophenylsulfenyl)-3-methyl-3-bromoindole (BNPS-skatole).

NPS-Skatole BNPS-Skatole

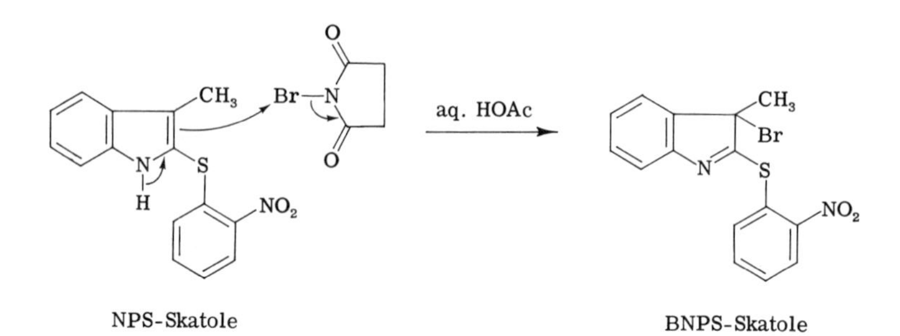

Selectivity of the Reagent. BNPS-skatole is a much more selective agent than NBS. After exposure of an amino acid mixture to 10 eq of BNPS-skatole in 50% acetic acid for 30 minutes at room temperature, tryptophan was completely absent, methionine was converted to methionine sulfoxide, and the other amino acids were recovered quantitatively. By contrast, in an amino acid mixture oxidized with NBS, not only tryptophan, but also methionine, tyrosine, histidine, and cystine were *completely* absent. Cysteine, as expected, is easily oxidized by BNPS-skatole to cystine and to some extent, by excess reagent, to cysteic acid. If present, sulfhydryl groups must be reversibly protected, as in selective modification with many other classes of reagents.[9]

Cleavage of the Tryptophanyl Peptide Bond. As anticipated from the action of NBS on tryptophan peptides,[2-4] BNPS-skatole is useful also for selective cleavage of tryptophanyl peptide bonds.[7,8] With a variety of tryptophan-containing peptides, 10-fold excess of reagent in 20 hours gave yields of cleavage (30–59%) comparable to those obtained with NBS. However, BNPS-skatole does cleave at sites other than tryptophanyl peptide bonds.

Since cleavage is time-dependent and since tryptophan reacts rapidly

[7] A. Fontana and T. F. Spande, 1969, unpublished results, cited by T. F. Spande, B. Witkop, Y. Degani, and A. Patchornik, *Advan. Protein Chem.* **24**, 97 (1970). See Section VII, B, 5.

[8] G. S. Omenn, A. Fontana, and C. B. Anfinsen, *J. Biol. Chem.* **245**, 1895 (1970).

[9] G. R. Stark, *Advan. Protein Chem.* **24**, 261 (1970).

with the reagent, modification without any significant cleavage is to be expected when the reaction is carried out with a low excess of reagent and for shorter times. This was clearly proved with staphylococcal nuclease[8] and with the A_1 protein from bovine and human myelin.[10]

Preparation

NPS-skatole.[11] 2-(2-Nitrophenylsulfenyl)-3-methylindole was prepared by reaction of equimolar quantities of skatole (3-methylindole; obtained from Eastman) and 2-nitrophenylsulfenyl chloride (obtained from Pierce Chemical Company, Rockford, Illinois) in glacial acetic acid for 1 hour at room temperature. The solution was then concentrated *in vacuo*, and the product was recrystallized from acetic acid–water, mp 125–127°.

BNPS-skatole.[8] To a solution of 1.4 g (5 mmoles) of NPS-skatole in 50 ml of acetic acid, 0.9 g (5 mmoles) of NBS (obtained from Aldrich, recrystallized from water) was added with stirring at room temperature. The color of the solution turned from deep to pale yellow. After 15 minutes, water was added (150 ml) and the oily precipitate was taken up in ethyl ether. The organic layer was washed with water, dried over Na_2SO_4, and evaporated *in vacuo*. The oily residue was redissolved in hot ligroin (bp 50–70°) and, after standing several hours at room temperature, the product crystallized in the form of yellow stars. The yield was 1.42 g (78%), mp 97–100°, with decomposition. It gave a single yellow spot upon thin-layer chromatography on silica and reacted strongly with starch–KI. The reagent is also now commercially available from Pierce Chemical Company, Rockford, Illinois.

Precaution. BNPS-skatole is stable for months at —20°, whereas at room temperature it decomposes after several days and becomes dark with release of bromine. Fresh solutions of the reagent must be used, and heating must be avoided in dissolving it.

Modification of Nuclease

Reaction with BNPS-skatole and subsequent reduction of the 4 partially oxidized methionine residues produced a derivative of nuclease with modification only of residue 140, the single tryptophan.

Procedure.[8] To a solution of 17 mg of nuclease (1 μmole) in 70% aqueous acetic acid, 3.4 mg (9 μmoles) of freshly prepared BNPS-skatole were added (final solvent conditions 2 ml of 80% acetic acid). This amount of reagent was calculated to give a 5-fold excess for reaction

[10] P. R. Burnett and E. H. Eylar, *J. Biol. Chem.* **246**, 3425 (1971).

[11] A. Fontana, F. Marchiori, R. Rocchi, and P. Pajetta, *Gazz. Chim. Ital.* **96**, 1301 (1966).

with the single tryptophan residue after reaction with the 4 methionine residues of nuclease. After 15 minutes of stirring, 20 μl of thioglycolic acid (1% v/v) was added to stop the reaction and to reduce methionine sulfoxide to methionine. After 3 hours, the mixture was applied to a column of Sephadex G-25 (0.9 $\times$ 40 cm) equilibrated with 0.05 M acetic acid. The peak containing the protein upon lyophilization gave a fluffy white product (14 mg). Since preliminary amino acid analysis showed incomplete reduction of methionine sulfoxide, 12 mg of the material was redissolved in 1 ml of distilled water with 20 μl of thioglycolic acid. The flask was flushed with a stream of nitrogen and held at 37° for 24 hours. The thioglycolic acid was removed by passage through Sephadex G-25.

BNPS-skatole is poorly soluble in water. The stability of nuclease in aqueous acetic acid permitted use of such acidic solvent conditions. If other proteins require a neutral pH range, the reagent may be dissolved in an organic solvent and then added to the protein solution. As with nuclease, reversibly denaturing conditions of pH or of added urea or guanidine may be required to expose the tryptophan residues to reagent.

Cleavage of Tryptophanyl Peptide Bond of Nuclease

When nuclease was allowed to react with 100 eq of BNPS-skatole in 50% acetic acid for 28 hours, cleavage at the tryptophanyl peptide bond occurred, in addition to modification. The C-terminal peptide 141–149 was detectable by electrophoresis after 7 hours reaction time and appeared maximal by about 21 hours. From the amino acid recovery on an analyzer of a hydrolyzate of the peptide isolated by gel filtration on Sephadex G-25, the yield of cleavage product was estimated to be approximately 15%. No additional peptides released by cleavage at any other site in the sequence were detected by electrophoresis.

Procedure.[8] To a solution of 11 mg of nuclease (0.6 μmole) in 66% acetic acid were added 22.5 mg (100 eq) of freshly prepared BNPS-skatole in glacial acetic acid (final volume 1.3 ml). The reaction was allowed to proceed at room temperature with stirring for 28 hours. Then the mixture was applied to a column of Sephadex G-25 equilibrated with 50% acetic acid. The protein and the peptide were eluted in two peaks prior to the reagent.

Recently,[10,12] the A_1 encephalitogenic protein was similarly allowed to react for 24 hours at 37° in 50% acetic acid with 10 eq of BNPS-skatole. Selective cleavage of the protein at the single tryptophanyl-bond occurred to an extent of 60%.

[12] H. Bergstrand, *Eur. J. Biochem.* **21,** 116 (1971).

Comments

The important feature of BNPS-skatole as a modifying reagent for tryptophan residues lies in its higher selectivity in comparison to NBS. BNPS-skatole rapidly (2–3 eq, 30 minutes in 50% acetic acid) achieves quantitative oxidation of the indole ring of tryptophan and the sulfur atom of methionine. Since methionine sulfoxide can be reduced to methionine, it is possible to obtain a protein derivative selectively modified at tryptophan.

The possibility of peptide bond cleavage must be also kept in mind, although modification can be carried out under conditions in which there is little risk of cleavage of peptide bonds. Much longer reaction time and excess reagent are needed for cleavage, as clearly indicated by the studies with nuclease[8] and with the A1 encephalitogenic protein.[10] The use of BNPS-skatole for the *selective* cleavage of the tryptophanyl peptide bond might prove useful in sequence studies. The yields so far obtained, up to 59% with peptides,[7,8] 15% with staphylococcal nuclease,[8] 60% with A_1-encephalitogenic protein,[10] suggest that the reagent will be useful for the nonenzymatic fragmentation of polypeptide chains, approaching in its utility the cyanogen bromide reaction.[13]

Note Added in Proof

Recently,[14] the single tryptophanyl peptide bond of horse heart cytochrome *c* was cleaved by BNPS-skatole treatment. The protein was allowed to react with 50 eq of BNPS-skatole in 50% acetic acid in presence of 50 eq of tyrosine. This amino acid was added to the reaction mixture as scavenger in order to prevent the modification of the tyrosine residues of the protein by the excess reagent. After 48 hours stirring in the dark at room temperature, the reaction mixture was diluted with water and several times extracted with ethyl ether in order to remove excess reagent and by-products. The lyophilized material was then fractionated by gel filtration. By recycling chromatography on Sephadex G-50 using 10% acetic acid as eluent, three peaks of polypeptide material were isolated, representing oxidized but uncleaved cytochrome, fragment 1-59 and fragment 60-104. The yield of cleavage, in three different preparations, was 55-65%, as judged from the dry weights of the lyophilized fragments. The heme group of cytochrome is oxidatively removed by the action of BNPS-skatole. The mechanism by which the thioether bridge linking the heme to the protein is cleaved is probably similar to the iodine cleavage recently described by Lederer and Tarni.[15]

[13] E. Gross, Vol. XI, p. 278.

[14] A. Fontana and C. Vita, unpublished results, 1971.

[15] F. Lederer and J. Tarni, *Eur. J. Biochem.* **20,** 482 (1971).

[34a] Carboxymethylation

By FRANK R. N. GURD

Principle

A variety of nucleophiles in proteins are carboxymethylated by monohaloacetates, usually bromoacetate or iodoacetate. Often, but not always, the corresponding acid amides are interchangeable for the acids. The haloacid reacts with the basic, unprotonated nucleophile with displacement of the halogen ion. Hence this modification reaction usually is most effective near and above the pH corresponding to the p*K* of the individual protein group. Unless the side chain is in a masked state, the ether sulfur in methionine is available for reaction over the entire normal pH range. In addition to control of pH, differences in rates of reaction with different nucleophiles can also be exploited to obtain selectivity, often by limiting the concentration of haloacid and the time of exposure. Although the reaction frequently can be restricted largely to a given class in the protein, such selectivity generally cannot be assumed.

Carboxymethylation of a protein almost always leads to the formation of stable derivatives that will remain intact during further study of the protein. Especially useful is the stability during the steps of sequence determination which makes possible the identification of the particular residues in a protein that have undergone carboxymethylation. The various reaction products are readily identified by the automatic amino acid analysis techniques following hydrolysis of the protein, with the exception of carboxymethylmethionine which undergoes various partial conversions during hydrolysis. Most frequently, the course of a carboxymethylation treatment is followed by amino acid analysis of the protein derivative so that the modification of the several types of protein groups can be observed concurrently.

Precautions

The usual precautions for modification reactions should be observed. These include allowing for instability of the protein under reaction conditions, and examination of the homogeneity of the reacted protein at the end of the reaction. Under most conditions carboxymethylation increases negative charge, so that the separation from unreacted protein is feasible chromatographically or electrophoretically. Complete removal of unreacted iodoacetate or bromoacetate and of iodide or bromide is necessary after the desired reaction is over. First, the reaction may continue slowly with the formation of unintended products. Second, excess unreacted alkylating agent that is present during hydrolysis with hot acid may react with methionyl residues or free methionine that had escaped reaction during the treatment procedure.[1] Third, traces of iodide or bromide ions present during the hydrolysis cause low recoveries of tyrosine. Furthermore, if suitable precautions such as exclusion of light are not observed during carboxymethylation with alkyl iodides, the released iodide ion may undergo oxidation to iodine which may cause other complications. Chromatography, precipitation, extraction, or dialysis may be used for removal of these substances.[1-4]

Reaction Products

The four equations (1–4) illustrated show the side-chain products of reaction with each group considered.

The first three sets of reaction products seem stable, within limits, in the protein and during hydrolysis conditions.[1,2,5] The fourth scheme for methionyl residues shows three alternative pathways of breakdown during hydrolysis of the protein or following release of the residue during the hydrolytic step.[6]

$$-CH_2S^- + ICH_2COO^- \longrightarrow -CH_2SCH_2COO^- + I^- \quad (1)$$

Carbosymethylcysteinyl residue[2]

$$-CH_2-C{=}CH \text{ (imidazole ring: HN, C(H), N)} + ICH_2COO^- \quad (2)$$

1-Monocarboxymethylhistidyl residue[7,8] ($^-OOCH_2C$ on N-1)

3-Monocarboxymethylhistidyl residue[7,8] (CH_2COO^- on N-3)

1,3-Dicarboxymethylhistidyl residue[7] ($^-OOCH_2C$ on N-1, CH_2COO^- on N^+-3)

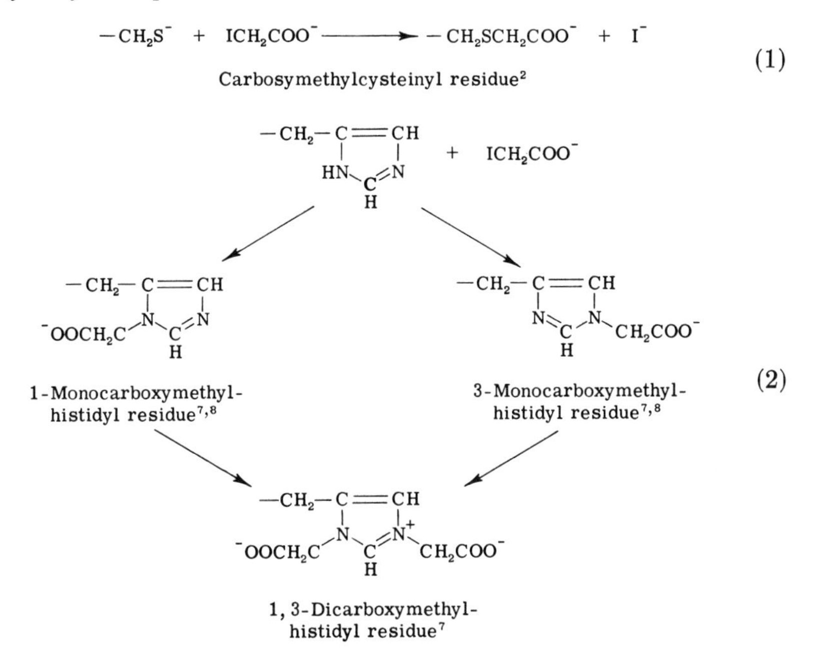

[1] H. G. Gundlach, W. H. Stein, and S. Moore, *J. Biol. Chem.* **234**, 1754 (1959).
[2] R. D. Cole, W. H. Stein, and S. Moore, *J. Biol. Chem.* **233**, 1359 (1958).
[3] P. J. Vithayathil and F. M. Richards, *J. Biol. Chem.* **235**, 2343 (1960).
[4] M. Sela, F. H. White, and C. B. Anfinsen, *Biochim. Biophys. Acta* **31**, 417 (1959).

$$-CH_2-CH_2-CH_2-CH_2-NH_2 + ICH_2COO^- \quad (3)$$

$$\swarrow \qquad \searrow$$

$-CH_2-CH_2-CH_2-CH_2-NH(CH_2COO^-)$

ε-Monocarboxymethyllysyl residue[1]

$-CH_2-CH_2-CH_2-CH_2-N(CH_2COO^-)_2$

ε-Dicarboxymethyllysyl residue[1]

$$-CH_2-CH_2-S-CH_3 + ICH_2COO^-$$

$$\downarrow$$

$$-CH_2-CH_2-S^+(CH_2COO^-)-CH_3$$

Carboxymethylsulfonium salt of methionyl residue[6]

$$\swarrow \quad \downarrow \quad \searrow \quad (4)$$

$CH_2OH-CH_2-CHNH_2-COOH$

Homoserine

$\rightleftharpoons$

$CH_2-CH_2-CHNH_2-C(=O)-O$ (ring)

Homoserine lactone

$CH_3-S-CH_2-CH_2-CHNH_2-COOH$

Methionine

$CH_2OOH-S-CH_2-CH_2-CHNH_2-COOH$

Carboxymethylhomocysteine

---

[5] L. J. Banaszak, P. A. Andrews, J. W. Burgner, E. H. Eylar, and F. R. N. Gurd, *J. Biol. Chem.* **238**, 3307 (1963).

[6] H. G. Gundlach, S. Moore, and W. H. Stein, *J. Biol. Chem.* **234**, 1761 (1959).

[7] A. M. Crestfield, W. H. Stein, and S. Moore, *J. Biol. Chem.* **238**, 2413 (1963).

[8] K. D. Hapner and R. W. Roeske, *Abstr. Amer. Chem. Soc. Meeting, 1964,* p. 38A (1964). (See also footnote 19.)

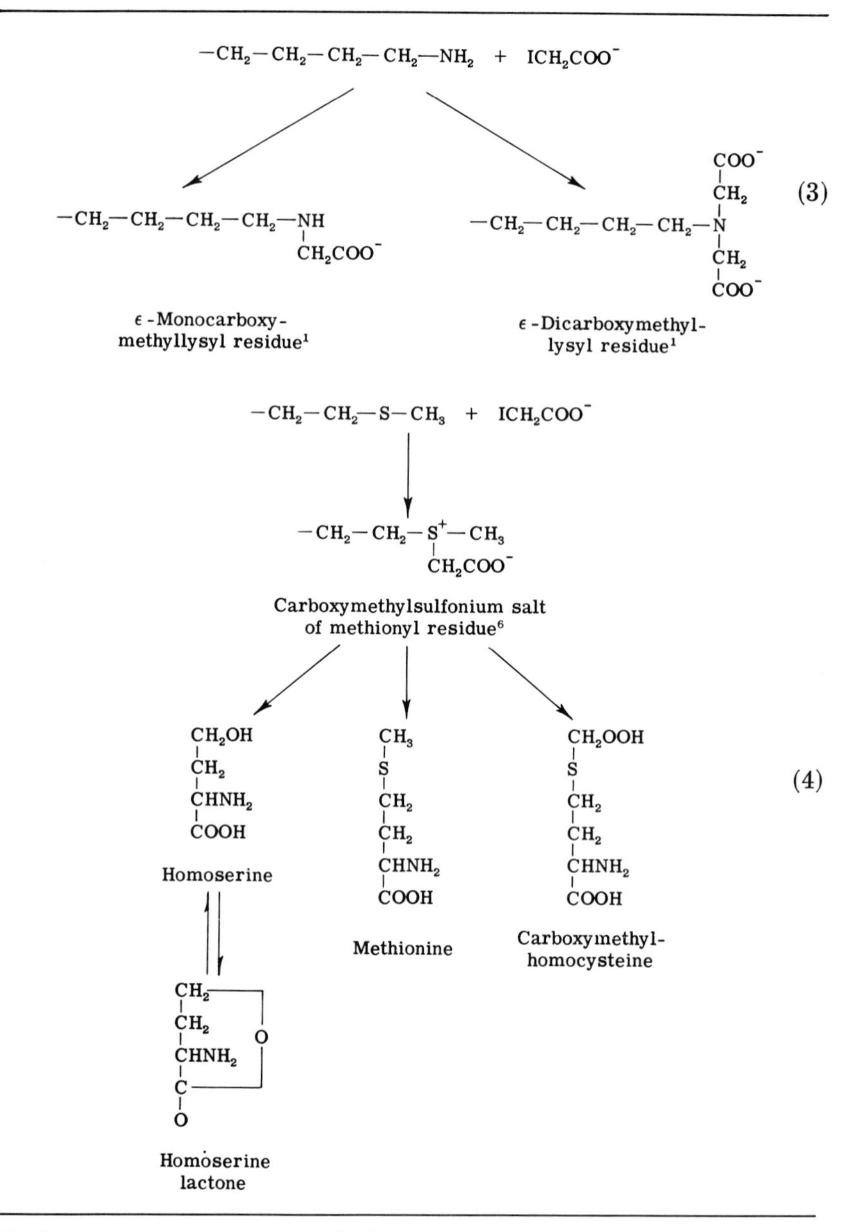

Analysis of the Course of Reaction

The general procedure is amino acid analysis after acid hydrolysis, to be discussed below. The reaction can be followed as it occurs by several methods that are only alluded to here. Two general methods are determination of the release of halide[9–11] from the α-haloacetate, or of uptake of alkali in a pH-stat, since all reactions except the first step in reaction (4) are accompanied by at least some displacement of protons.[1,5] In some cases spectrophotometric measurements are suitable.[12] The disappearance of sulfhydryl groups can be followed by the nitroprusside reaction,[9] and alteration of histidyl residues by the Pauly reaction[13] or by

TABLE I
CHROMATOGRAPHY OF CARBOXYMETHYL AMINO ACIDS AND THEIR DECOMPOSITION PRODUCTS

Carboxymethyl derivative or decomposition product	Emerges from chromatography column[a]: Volume (ml)	Near	Reference
S-Carboxymethylcysteine	96	Aspartic acid	*a*
1-Monocarboxymethylhistidine	180	Proline	*b*
3-Monocarboxymethylhistidine	280	Cystine	*b*
1,3-Dicarboxymethylhistidine	68	Aspartic acid	*b*
ε-Monocarboxymethyllysine	335	Methionine	*c*
ε-Dicarboxymethyllysine	90	Aspartic acid	*c*
Carboxymethylsulfonium salt of methionine	105–110	Aspartic acid	*d*
Methionine	330	—	*a*
S-Carboxymethylhomocysteine	195–205	Proline	*d*
Homoserine	155–165	Glutamic	*d*
Homoserine	78–80	Ammonia	*d*
Methionine sulfone	130	Aspartic acid	*d*

[a] D. H. Spackman, W. H. Stein, and S. Moore, *Anal. Chem.* **30,** 1190 (1958).
[b] A. M. Crestfield, W. H. Stein, and S. Moore, *J. Biol. Chem.* **238,** 2413 (1963). For an alternative system of separation, see L. J. Banaszak and F. R. N. Gurd, *J. Biol. Chem.* **239,** 1836 (1964).
[c] H. G. Gundlach, W. H. Stein, and S. Moore, *J. Biol. Chem.* **234,** 1754 (1959).
[d] H. G. Gundlach, S. Moore, and W. H. Stein, *J. Biol. Chem.* **234,** 1761 (1959). For an alternative system of separation, see H. J. Goren, D. M. Glick, and E. A. Barnard, *Arch. Biochem. Biophys.* **126,** 607 (1968).

[9] L. Rosner, *J. Biol. Chem.* **132,** 657 (1940).
[10] D. C. Watts, B. R. Rabin, and E. M. Crook, *Biochim. Biophys. Acta* **48,** 380 (1961).
[11] G. R. Stark and W. H. Stein, *J. Biol. Chem.* **239,** 3755 (1964).
[12] B. J. Finkle and E. L. Smith, *J. Biol. Chem.* **230,** 669 (1958).
[13] M. Nakatani, *J. Biochem. (Tokyo)* **48,** 469 (1960).

reaction with diazotized 1*H*-tetrazole.[14,15] Paper chromatographic measurements may be made after hydrolysis.

Table I lists the effluent volumes at which the carboxymethyl derivatives or decomposition products emerge from standard analytical columns.[16] The volumes are given (in milliliters), and the nearest regularly observed components are named. Homoserine lactone is the only component that is separated with the basic amino acids. Methionine sulfone is included because the identification of unreacted methionine in a carboxymethylated preparation can be made by treating the preparation with performic acid before hydrolysis and measuring the change in methionine sulfone content. Direct analysis for methionine cannot be used because of decomposition of the sulfonium salt during acid hydrolysis. The sulfonium salt is impervious, however, to performic acid.[3,17] It should be noted that the provisional identification of the 1- and 3-carboxymethylhistidines[18] has been confirmed by unambiguous synthesis.[8]

Electrochemical Properties of Derivatives

The introduction of the carboxymethyl group usually affects the net charge of the protein at a given pH. This in turn affects the choice of conditions for separation by electrophoresis or ion-exchange chromatography, either of the intact modified protein or of digestion products. Negative charges are in general borne by the dervatives at all but low pH values, as is implied by the emergent volumes in Table I. Special mention should be made of the methionine and histidine derivatives. The carboxymethylsulfonium salt of methionine bears an added positive charge as well as an added negative charge. The carboxyl group charge is not much suppressed even at pH 2.[3] The histidine derivatives all show side-chain carboxyl group pK' values of about 2 or below.[19] In the case of the dicarboxymethyl derivative, these two ionizations occur at sufficiently low pH values that the titration of the α-carboxyl group, pK' 2.61, is clearly discerned. With the monocarboxymethyl derivatives there is more overlap with the α-carboxyl dissociation. The imidazole ring in the dicarboxymethyl derivative is quaternized and bears a positive charge. The pK' values at ionic strength 0.16 and 25° for the imidazole groups of the 1- and 3-carboxymethyl derivatives are 6.35 and 5.70, respectively.[19]

[14] H. Horinishi, Y. Hachimori, K. Kurihara, and K. Shibata, *Biochim. Biophys. Acta* **86**, 477 (1964).

[15] B. L. Vallee and J. F. Riordan, *Annu. Rev. Biochem.* **38**, 733 (1969).

[16] D. H. Spackman, W. H. Stein, and S. Moore, *Anal. Chem.* **30**, 1190 (1958).

[17] G. R. Stark, W. H. Stein, and S. Moore, *J. Biol. Chem.* **236**, 436 (1961).

[18] H. Jaffe, *J. Biol. Chem.* **238**, 2419 (1963).

[19] K. D. Hapner, Ph.D. dissertation, Indiana Univ., 1965.

Examples of Experimental Conditions

In Table II are collected examples of the ranges of experimental conditions that have been used in various studies. Certain specific cases that have been most extensively studied are not included, but are dealt with separately. For simplicity, ranges of time and temperature have been omitted.

Sulfhydryl and Thioether Groups. These groups are the most easily modified. Accordingly, the experiments directed at these groups are often conceived in terms of simply attaining a small excess of reagent. The thioether group, if exposed in the protein structure, is reactive at all pH values over which the primary structure of the protein is stable. As shown in scheme (4), the carboxymethyl derivative undergoes partial destruction under the usual acid hydrolysis conditions so that the products include reformed methionine and the carboxymethyl thioether as well as carboxymethylmethionine. This difficulty is fairly well overcome by converting the unmodified methionine to the sulfone by performic acid oxidation prior to hydrolysis to determine the carboxymethyl derivative by difference.[3,20] Careful separation of radioactively labeled products has also been used for analysis.[21] The carboxymethylation of cysteine residues occurs so readily that it can be achieved with small excess of reagent even at pH 5 or less. In this case, even a very unfavorable proportion of protonated to unprotonated nucleophile is acceptable.

Imidazole Group. Modification of the imidazole group yields histidine residues alkylated at either or both nitrogen atoms. In general the reaction proceeds more readily at the 3-position than at the 1-position. All three products are stable during hydrolysis procedures and are readily measured by the standard analysis routines. The *N*-alkylation is sufficiently less rapid that much higher concentrations of reagents are required compared with the *S*-alkylation (Table II). The distinction is clear enough that modification can be fairly readily limited to the cysteine and methionine residues. In fact, the potentiality for imidazole alkylation was little appreciated before the work of Korman and Clarke.[22]

Amino Group. Modification of the amino group generally can be expected to overlap with imidazole modification, particularly with regard to the α-amino group whose pK is relatively close to neutrality. Normally, ϵ-amino groups undergo some small degree of modification under conditions causing extensive modification of α-amino and imidazole

[20] N. P. Neuman, S. Moore, and W. H. Stein, *Biochemistry* **1**, 68 (1962).

[21] H. J. Goren, D. M. Glick, and E. A. Barnard, *Arch. Biochem. Biophys.* **126**, 607 (1968).

[22] S. Korman and H. T. Clarke, *J. Biol. Chem.* **221**, 113, 133 (1956).

TABLE II
EXAMPLES OF APPLICATION OF CARBOXYMETHYLATION PROCEDURES*

Protein	Principal residues involved (numbers are per subunit where applicable)	Reagent concentration	pH	Remarks	References
Firefly luciferase	2 Sulfhydryl	IAM, 0.02 *M*	8	Other reagents more specific	*a*
Streptococcal proteinase	1 Sulfhydryl	ClAA, 10–20 m*M* max	6–8	Contrasts with ClAM†	*b, c*
		ClAM, 1–20 m*M*	>7	Contrasts with ClAA	*b, c*
Ficin	1 Sulfhydryl	ClAM, 1 m*M*	7–10	Accelerated by substrates, products†	*d, e*
		IAM, 1 m*M*	7–8		
		IAA, 1 m*M*	~6		*f*
		ClAA, 1 m*M*	6	See ClAM, IAM	*e*
Papain	1 Sulfhydryl	ClAM, 1 m*M*	7–9	Differences between reagents; many details†	*g*
		ClAA, 1 m*M*	6–9		*h*
Muscle phosphorylase	3 Sulfhydryl	IAM, 10 m*M*	7.5	Differential effects	*i*
Hemoglobin	Sulfhydryl	IAM, 10-fold	7.3	Ca. 2/mole in HbO_2 only	*j*
		IAA, 60 m*M*	7.1	Most rapid for β112 in COHbH	*k*
Lysozyme	1 Imidazole	IAA	7.2	1 Cm histidine isolated	*l*
		IAM	7.2		
3-Phosphoglyceraldehyde dehydrogenase	1 Sulfhydryl	IAA, 2m*M*	8.0	10 Min reaction time	*m*
Fumarase	1 Imidazole 3 Methionine	IAA, 20 m*M*	6.5	Reduced reaction with fumarase inhibitors	*n*

Isocitrate dehydrogenase	1 Methionine	IAA, 10 mM	5.5	Prevented by isocitrate	*o*
Malate dehydrogenase	1 Methionine	IAA, 0.16 M	7.5		*p*
Trypsinogen and trypsin	2 Methionine	IAM, 15 mM	2.2	45°; some with IAA	*q*
Enterotoxin B	8 Methionine	IAA, 40 mM	2.7	4 Residues react rapidly	*r*

* The halide symbols are followed by AA for the acetate and AM for the acetamides.

† Evidence for enhanced reactivity under at least certain conditions; see footnote *c*, also H. Lindley, *Biochem. J.* **74,** 577 (1960); **82,** 418 (1963).

[a] R. Lee and W. D. McElroy, *Biochemistry* **8,** 130 (1969).

[b] T.-Y. Liu, W. H. Stein, S. Moore, and S. D. Elliott, *J. Biol. Chem.* **24,** 1143 (1965).

[c] B. I. Gerwin, *J. Biol. Chem.* **242,** 451 (1967); see also T.-Y. Liu, *J. Biol. Chem.* **242,** 4029 (1967).

[d] M. R. Holloway, A. P. Mathias, and B. R. Rabin, *Biochim. Biophys. Acta* **92,** 111 (1964).

[e] J. R. Whitaker, *Biochemistry* **8,** 4591 (1969).

[f] S. Zuckerman-Stark, *Enzymologia* **32,** 380 (1967).

[g] I. M. Chaiken and E. L. Smith, *J. Biol. Chem.* **244,** 5087 (1969).

[h] I. M. Chaiken and E. L. Smith, *J. Biol. Chem.* **244,** 5095 (1969).

[i] M. L. Battell, C. G. Zarkadas, L. B. Smillie, and N. B. Madsen, *J. Biol. Chem.* **243,** 6202 (1968).

[j] R. E. Benesch and R. Benesch, *Biochemistry* **1,** 735 (1962).

[k] E. J. Neer, *J. Biol. Chem.* **245,** 564 (1970).

[l] S. M. Parsons, L. Jao, F. W. Dahlquist, C. L. Bordes, T. Groff, J. Racs, and M. A. Raftery, *Biochemistry* **8,** 700 (1969).

[m] J. S. Bond, S. H. Francis, and J. H. Park, *J. Biol. Chem.* **245,** 1041 (1970).

[n] R. A. Bradshaw, G. W. Robinson, G. M. Hass, and R. L. Hill, *J. Biol. Chem.* **244,** 1755 (1969).

[o] R. F. Colman, *J. Biol. Chem.* **243,** 2454 (1968).

[p] V. Leskovac and G. Pfleiderer, *Hoppe-Seyler's Z. Physiol. Chem.* **350,** 484 (1969).

[q] V. Holeyšovský and M. Lazdunski, *Biochim. Biophys. Acta* **154,** 457 (1968).

[r] F. S. Chu and M. S. Bergdoll, *Biochim. Biophys. Acta* **194,** 279 (1969).

groups. Amino groups can either undergo monoalkylation or dialkylation. Another potentially reactive group is the phenolic of tyrosine which appears to escape reaction in most studies. Bromide or iodide ions introduced during carboxymethylation must not be present during acid hydrolysis if loss of tyrosine and phenylalanine is to be avoided. Tryptophan residues and carboxyl groups are rarely affected in carboxymethylation procedures, although an interesting case of carboxyl group modification will be described below.

Apart from the examples in Table II, a large number of cases of inhibition observed by enzymologists could be cited. A very full review is given by Webb,[23] who also should be consulted for various comparisons of reactivity and stability of the different haloacetates.

Ribonuclease. Bovine pancreatic ribonuclease A has been the subject of the most intensive application of carboxymethylation procedures to any protein. With only minor differences in interpretation of results, the work has been pursued vigorously by both W. H. Stein and Moore and their collaborators[1,6,7,11,17,24] and by Barnard and his collaborators.[21,25-28] The results illustrate a number of important principles in the application of this protein modification procedure.

Most of the attention has been centered on the alkylation of two histidine residues implicated in the active site, numbers 119 and 12. Both iodoacetate and bromoacetate alkylated these two residues according to a quite remarkable pattern. (1) Reactions occur most rapidly near pH 5.5, where the rates are much higher than at pH 7, at which the greater reactivity would normally be anticipated. (2) The products are specific: nearly exclusively 1-carboxymethylhistidine-119 and 3-carboxymethylhistidine-12. (3) Reaction at either histidine residue appears to destroy the enhanced reactivity at the other site, so that the majority of the molecules are converted exclusively to one product or the other in the approximate ratios of 7 or 8 molecules modified at residue 119 to 1 molecule modified at residue 12. (4) The reaction is strongly inhibited by phosphate and sulfate ions, just those ions that have been present during X-ray crystallographic analysis. For this reason a direct correlation of the reactivity studies with the detailed crystallographic struc-

[23] J. L. Webb, "Enzyme and Metabolic Inhibitors," Vol. 3, Chap. 1. Academic Press, New York, 1966.

[24] A. M. Crestfield, W. H. Stein, and S. Moore, *J. Biol. Chem.* **238**, 2421 (1963).

[25] E. A. Barnard and W. D. Stein, *J. Mol. Biol.* **1**, 339 (1959).

[26] W. D. Stein and E. A. Barnard, *J. Mol. Biol.* **1**, 350 (1959).

[27] H. J. Goren and E. A. Barnard, *Biochemistry* **9**, 959 (1970).

[28] H. J. Goren and E. A. Barnard, *Biochemistry* **9**, 974 (1970).

ture[29,30] has been hindered, although Bello and Nowoswiat have shown that under appropriate conditions ribonuclease A alkylation follows a similar pattern in the crystalline state as in solution.[31]

The special pattern of alkylation of histidine residues 119 and 12 has been explained as involving initial binding of the haloacetate ion in a preferred orientation followed by the nucleophilic displacement reaction proper.[24] Good evidence for this stepwise type of process has since been obtained for the carbonic anhydrase system.[32] In the case of ribonuclease, the elegant suggestion is that the cationic form of one of the pair of imidazole groups is responsible for attracting the carboxyl group of the haloacetate in such a way as to facilitate the reaction at the other imidazole group.[33] Once modified, the imidazole group becomes a very poor site for the haloacetate binding, and the enhanced reactivity of the companion imidazole group is no longer evident. The nearby lysine-41 residue appears not to be an essential part of the effective anion binding site in this process.[34] The binding of phosphate and sulfate would readily explain their ability to inhibit the reaction strongly, an important practical point since these ions are frequently present during preparative procedures. The possibility of inhibition by other anions must be borne in mind.[24,27,28] Such inhibition may be the dominant factor in the observation that the reaction is slower as the concentration of sodium chloride is increased.

The concurrent requirement for the cationic, protonated form of one imidazole group to attract the carboxyl group of the haloacetate and for the basic, unprotonated form of the other imidazole to act as the nucleophile explains the observed pH dependence of the reaction.[35]

As would be expected according to the proposed mechanism, the corresponding amides, bromoacetamide and iodoacetamide, react substantially less rapidly with histidine-119 and histidine-12.[36] The effect is relatively greater for histidine-119 than for the less activated process at histidine-12. A number of larger alkylating agents of the same general

[29] H. W. Wyckoff, D. Tsernoglou, A. W. Hanson, J. R. Knox, B. Lee, and F. M. Richards, *J. Biol. Chem.* **245,** 305 (1971).

[30] F. M. Richards and W. H. Wyckoff, *in* "The Enzymes" (P. D. Boyer, ed.), 3rd ed., in press.

[31] J. Bello and E. F. Nowoswiat, *Biochemistry* **8,** 628 (1969).

[32] P. L. Whitney, *Eur. J. Biochem.* **16,** 126 (1970).

[33] R. L. Heinrikson, W. H. Stein, A. M. Crestfield, and S. Moore, *J. Biol. Chem.* **240,** 2921 (1965).

[34] R. L. Heinrikson, *J. Biol. Chem.* **241,** 1393 (1966).

[35] M. C. Lin, W. H. Stein, and S. Moore, *J. Biol. Chem.* **243,** 6167 (1968).

[36] R. G. Fruchter and A. M. Crestfield, *J. Biol. Chem.* **242,** 5807 (1967).

pattern have been compared, and differences in reactivity of optical antipodes have been found.[33,34] Rate constants for these compounds are included in Table III.

Native ribonuclease undergoes slow reaction at one or more methionine residues. Link and Stark[37] showed that the closely related alkylating agent methyl iodide formed a sulfonium salt primarily at methionine-29, which alone has a somewhat exposed ether sulfur atom. However, prolonged exposure to haloacids has usually produced some ill-defined, often aggregated products that may reflect reaction at other methionine residues. At extremes of pH the reaction of methionine residues is accelerated, presumably as these conformational changes become more marked, and in frankly denaturing conditions modification of methionine residues proceeds swiftly.[1,11,17,20] It is significant that under such conditions the specially high reactivity of histidine residues 119 and 12 is lost.

The patient development by Barnard and co-workers[21,27,28] of methods for sorting out minor products by labeling with [^{14}C]bromoacetate has led to the discovery that the apparently buried[29,30] methionine-30 residue undergoes slow modification over a wide pH range. The extent of

TABLE III
SECOND-ORDER RATE CONSTANTS FOR ALKYLATION OF RNASE A AND L-HISTIDINE AT pH 5.5 AND 25°[a,b]

Compound	At His 119	At His 12	Overall	L-Histidine
Iodoacetate	51.1	7.3	58.4	—
Bromoacetate	184.5	20.5	205.0	0.086
Chloroacetate	2.13	0.17	2.30	—
L-α-Bromopropionate	0.66	0.19	0.85	0.0027
DL-α-Bromopropionate	1.09	1.88	2.97	0.0027
D-α-Bromopropionate	1.84	4.16	6.00	0.0028
L-α-Bromo-*n*-butyrate	0.50	0.18	0.68	—
DL-α-Bromo-*n*-butyrate	0.79	1.61	2.40	—
D-α-Bromo-*n*-butyrate	1.11	3.60	4.71	—
DL-α-Bromovalerate	0.76	0.05	0.81	—
DL-α-Bromocaproate	0.89	—	0.89	0.0008
β-Bromopyruvate	911	—	911	—
β-Bromopropionate	6.33	—	6.33	0.0023

[a] R. L. Heinrikson, W. H. Stein, A. M. Crestfield, and S. Moore, *J. Biol. Chem.* **240,** 2921 (1965).

[b] Reactions were carried out in the dark in 0.10 *M* sodium acetate buffer at pH 5.50. With the enzyme, the ionic strength of the solutions varied from 0.11 to 0.13, depending upon the concentration of reagent used; with histidine, the ionic strength was 0.32.

[37] T. P. Link and G. R. Stark, *J. Biol. Chem.* **243,** 1082 (1968).

modification of this residue is not correlated with loss of enzymatic activity, and indeed an active protein fraction that showed this reaction product alone was separated chromatographically. Besides the increase in modification below pH 4, an unsuspected increase was also observed between pH 5.5 and about pH 6.0. Richards and Wyckoff[30] have summarized other evidence that the region of the molecule near methionine-30 may be easily deformable.

The carboxymethylation of lysine-41 is enhanced over that of other lysine residues,[34] in keeping with other evidence of particular reactivity associated with a relatively low pK for this amino group.[38] The carboxymethylation reaction has been followed at pH 8.5 by Heinrikson,[34] who showed that besides the preferential alkylation of lysine-41 there remained some slow reactivity of histidine residues 119 and 12 with the same specificity of product formation as was observed at pH 5.5. Goren and Barnard[27,28] have shown that the amino-terminal lysine residue in ribonuclease A is subject to modification at the α-amino group.

Myoglobins. Carboxymethylation of myoglobins has been studied in detail primarily to test the potentialities of this reaction as a measure of exposure of reactive groups. In concentrated urea solution all histidine residues are converted to the dicarboxymethyl form in a few days at pH 6.8 in 0.2 M bromoacetate at 23°.[5] The terminal α-amino group is also modified, along with the two methionine residues and a small fraction of the ϵ-amino groups.[39] This modification procedure has been useful in sequence studies[39-41] and renders the protein unable to refold in the native conformation.[5] The presence of urea has no striking effect on the reactivity at either α- or ϵ-amino groups, but the methionine and certain of the histidine residues are unreactive in the native state.[5,42,43] A similar pattern of reactivity is given by iodoacetamide.[44-47] The reaction of sperm whale myoglobin with bromoacetate has been studied with the protein in the crystalline state in order to test the correlation of the

[38] R. P. Carty and C. H. W. Hirs, *J. Biol. Chem.* **243,** 5254 (1968).
[39] R. A. Bradshaw, W. H. Garner, and F. R. N. Gurd, *J. Biol. Chem.* **244,** 2149 (1969).
[40] R. A. Bradshaw, R. H. Kretsinger, and F. R. N. Gurd, *J. Biol. Chem.* **244,** 2159 (1969).
[41] A. B. Edmundson, *Nature (London)* **205,** 883 (1965).
[42] T. E. Hugli and F. R. N. Gurd, *J. Biol. Chem.* **245,** 1930 (1970).
[43] T. E. Hugli and F. R. N. Gurd, *J. Biol. Chem.* **245,** 1939 (1970).
[44] C. R. Hartzell, K. D. Hardman, J. M. Gillespie, and F. R. N. Gurd, *J. Biol. Chem.* **242,** 47 (1967).
[45] J. F. Clark and F. R. N. Gurd, *J. Biol. Chem.* **242,** 3457 (1967).
[46] C. R. Hartzell, J. F. Clark, and F. R. N. Gurd, *J. Biol. Chem.* **243,** 697 (1968).
[47] R. H. L. Marks, E. H. Cordes, and F. R. N. Gurd, *J. Biol. Chem.* **245,** 466 (1971).

pattern of reactivity of the histidyl residues under the constraints of the crystalline structure.[42] When account is taken of the environment of each imidazole group, both in terms of the given protein molecule and neighboring protein molecules in the crystal array, the observed reactivity pattern appears to fit well. The test was made in terms of yields of each residue in the various carboxymethylated forms. These results indicate that, given a certain degree of inherent structural stability in the protein, the carboxymethylation reaction will give a realistic picture of the state of exposure of the various histidyl residues on the surface of the protein.

Carboxymethylation of sperm whale myoglobin in solution yielded generally similar results to those observed in the crystalline state.[43] Two points of difference were seen. First, examples of consecutive reactions[5] could be detected, such as the formation of small amounts of dicarboxymethyl derivative of histidine-119 in addition to the preponderant production of 1-derivative.[42,43] This observation implies that a local disruption can occur, possibly made easier by the initial monoalkylation of this residue. At the other end of the scale, it is known that histidine-82 escapes alkylation even when appreciable loss of native structure has resulted from prolonged and vigorous treatment with alkylating agent.[48] Second, the reaction pattern at histidine-36 is entirely different from that in the crystalline state. In solution no alkylation can be detected whereas in the crystalline state the 3-carboxymethyl derivative is obtained in high yield. These results imply a local rearrangement of the protein structure, which, however, does not appear to affect the properties of the heme group in the protein.[49] With due care to avoid excessive reaction conditions the carboxymethyl myoglobins retain native properties to a very satisfactory degree.[5,42–47,49,50]

Ribonuclease T_1: An Example of a Reactive Glutamic Acid Residue. Ribonuclease T_1 from *Aspergillus oryzae* shows a superficial resemblance to ribonuclease A in that it is inactivated by a 180- to 300-fold molar excess of iodoacetate at pH 5.5 and 37°.[51] Iodoacetamide is ineffective. However, the only residue that undergoes modification is glutamic acid-58 at the active center. It is possible that a positive charge at an appropriate distance attracts and orients the anionic reagent in the way suggested for the bovine enzyme. The native conformation of the en-

[48] L. J. Banaszak and F. R. N. Gurd, *J. Biol. Chem.* **239**, 1836 (1964).

[49] K. Wüthrich, R. G. Shulman, T. Yamane, B. J. Wyluda, T. E. Hugli, and F. R. N. Gurd, *J. Biol. Chem.* **245**, 1947 (1970).

[50] F. R. N. Gurd, A. Allerhand, D. M. Doddrell, V. G. Glushko, P. J. Lawson, A. M. Nigen, and P. Keim, *Fed. Proc., Fed. Amer. Soc. Exp. Biol.* **30**, 1046 (1971).

[51] K. Takahashi, W. H. Stein, and S. Moore, *J. Biol. Chem.* **242**, 4682 (1967).

zyme is required. The conversion —COO^- to —$COOCH_2COO^-$ is reversible in 1 *M* hydroxylamine at pH 9 in 8 *M* urea. Polyvalent anions, such as citrate, phosphate, or the substrate analogs 2′- or 3′-guanylic, are inhibitory, as are Cu(II) or Zn(II) ions.[51] The product of acid hydrolysis of the alkylation derivative in this case is glycolic acid. A possibly similar mechanism has been suggested for the reaction of iodoacetamide with the active center of γ-glutamyltranspeptidase.[52]

Cytochrome c. Studies of several cytochromes *c* have shown that the carboxymethylation of a particular methionine residue leads to drastic changes in the properties of the molecule. The interest in this modification reaction lies in the fact that the state of the iron in the heme moiety strongly affects the readiness of reaction of the key methionine residue. Schejter and George[53] showed that at pH 7 alkylation of the cyanide complex of horse heart ferricytochrome *c*, followed by removal of the cyanide, leads to marked changes in the visible absorption spectra. No change is obtained upon reaction in the absence of cyanide. Harbury and co-workers[54] found that these results reflected the effect of cyanide on the susceptibility to carboxymethylation of methionine-80, now known to be coordinated with the iron in the absence of cyanide.[55,56] Cytochrome *c* from tuna, cow, and moth give similar results.[54,57,58] In addition to the formation of the cyanide complex the process can be facilitated by carboxymethylation at low pH, where the methionine-iron linkage is weakened. Ferrocytochrome *c* was resistant so long as air was rigorously excluded.[59] Other residues of methionine (e.g., 65 in beef and man and also 12 in man) and histidine (e.g., 33 in beef) were modified at pH 5 to 6 regardless of the ligand state.[59] In all this work bromoacetate and iodoacetate or iodoacetamide could be used, at concentrations such as 0.16 *M*. A recent summary of the alterations in properties of the enzyme following modification of methionine-80 is provided by Schejter and Aviram.[60] *Pseudomonas* cytochrome *c* represents a particularly important case historically because it contains only one histidine

[52] A. Szewczuk and G. E. Connell, *Biochim. Biophys. Acta* **105**, 352 (1965).

[53] A. Schejter and P. George, *Nature* (*London*) **206**, 1150 (1965).

[54] H. A. Harbury, J. R. Cromin, M. W. Eauger, T. P. Hettinger, A. J. Murphy, Y. P. Myer, and S. N. Vinogradov, *Proc. Nat. Acad. Sci. U.S.* **54**, 1658 (1965).

[55] R. E. Dickerson, T. Takano, D. Eisenberg, O. B. Kallai, L. Samson, A. Cooper, and E. Margoliash, *J. Biol. Chem.* **246**, 1511 (1971).

[56] K. Wuthrich, *Proc. Nat. Acad. Sci. U.S.* **63**, 1071 (1969).

[57] K. Ando, H. Matsubara, and K. Okunuki, *Biochim. Biophys. Acta* **118**, 240 (1966).

[58] H. J. Tsai, H. Tsai, and G. R. Williams, *Can. J. Biochem.* **43**, 1995 (1965).

[59] K. Ando, H. Matsubara, and K. Okunuki, *Biochim. Biophys. Acta* **118**, 256 (1966).

[60] A. Schejter and I. Aviram, *J. Biol. Chem.* **245**, 1552 (1970).

residue (at position 16) but is susceptible to changes like those mentioned involving methionine residue 61.[61]

Carbonic Anhydrase. Although at the time of writing less information is available about the sequence and crystal structure of carbonic anhydrases than about ribonuclease, myoglobin, or cytochrome *c*, several studies on this enzyme are of great interest.[32,62-65] Human carbonic anhydrases A and B have been intensively studied with similar results. The following deals with the B enzyme. Carboxymethylation of a histidine residue by iodoacetate was extremely rapid, about 10^5 times faster than with histidine residues in the denatured enzyme.[65] Bromoacetate reacts extremely rapidly as well. Iodoacetate and bromoacetate are powerful effective competitive inhibitors of the enzyme.[32] It is clear that the reagents first bind within the active site region and then proceed to form the 3-carboxymethylhistidine derivative.[32,62] An interesting observation is that the carboxymethylated enzyme itself has activity despite the strong inhibitory effect of the merely bound form of the reagent. Bromoacetate and iodoacetate react at a rate controlled by an apparent pK of 5.6 or 5.8, whereas for iodoacetamide this apparent pK is close to 5.0. In each case the reactive form of the enzyme may itself be determined by the nature of the initial complex. A recent summary of the findings in this system has been published by Whitney.[32]

[61] M. W. Fanger, T. P. Hettinger, and H. A. Harbury, *Biochemistry* **6**, 713 (1967).

[62] S. I. Kandel, S.-C. C. Wong, M. Kandel, and A. G. Gornall, *J. Biol. Chem.* **243**, 2437 (1968).

[63] M. Kandel, A. G. Gornall, S.-C. C. Wong, and S. I. Kandel, *J. Biol. Chem.* **245**, 2444 (1970).

[64] P. L. Whitney, G. Fölsch, P. O. Nyman, and B. G. Malmström, *J. Biol. Chem.* **242**, 4206 (1967).

[65] S. L. Bradbury, *J. Biol. Chem.* **244**, 2002, 2010 (1969).

[35] Iodination—Isolation of Peptides from the Active Site

By OLIVER A. ROHOLT and DAVID PRESSMAN

Iodination of a protein can be carried out under relatively mild conditions and with useful and highly reproducible results. Ordinarily, the mono- and diiodination of tyrosyl residues are the principal modifications involved in the incorporation of iodine; to a lesser extent, iodohistidyl residues are formed. The oxidizing activity of iodine converts sulfhydryl groups to disulfides and may cause some modification of tryptophan. The presence of these or of other readily oxidizable components or impurities in the protein preparation can seriously decrease the extent of incorporation of iodine into tyrosyl and histidyl residues. Thus, the treatment of some proteins with an iodinating reagent may result in little or no iodine incorporation if there has been extensive oxidative modification or if the iodine has been reduced by impurities. It is therefore necessary to determine the level of iodination of the protein, i.e., the average number of iodine atoms covalently bound to the protein molecule after the iodination procedure.

For iodinating a protein, several reagents or combinations of reagents, including triiodide ion, iodine monochloride, iodine in alcohol, and iodide ion with an oxidizing agent such as nitrous acid, iodate ion, or chloramine T, have been used. In each case the reactive iodinating agent appears to be hypoiodous acid, HOI. For the procedures described here, ICl has been taken as the reagent of choice.

The effect of iodination of an enzyme on its activity depends on the level of iodination of the enzyme and on whether this level disturbs or disrupts the tertiary structure of the protein or causes iodination of a specific residue in the active site of the enzyme. If iodination does lead to a loss of activity and the activity can be preserved by carrying out the iodination in the presence of a competitive inhibitor of the enzyme, then the simplest conclusion is that there is an essential residue in the active site of the enzyme that was modified in the absence of the inhibitor, but was protected against this modification by the presence of the inhibitor. If the presence of the inhibitor during iodination does not protect the enzyme activity, then the inactivation may be due to tertiary structure alteration rather than attack at the site. However, care must be taken that an apparent lack of protection is not due to too extensive iodination and that protection can be observed at lower levels of iodination.

When an iodinatable residue is found to be in the active site of an enzyme, iodopeptides containing this residue can be obtained by proteolysis and the residue can be identified as described below.

Iodination causes the inactivation of bovine carboxypeptidase A; this inactivation is prevented by the presence of the competitive inhibitor, β-phenylpropionate.[1] The residue that is iodinated in the active site of the enzyme has been shown to be a tyrosine and the isolation of an iodopeptide, isoleucyl-(diiodo)tyrosyl-glutaminyl-alanine, indicates that the tyrosyl residue is at position 248[2,3] in the enzyme molecule.

Iodination may also be used to infer other aspects of the structure

[1] R. T. Simpson and B. L. Vallee, *Biochemistry* **5**, 1760 (1966).

[2] O. A. Roholt and D. Pressman, *Proc. Nat. Acad. Sci. U.S.* **58**, 280 (1967).

[3] R. A. Bradshaw, L. H. Ericsson, K. A. Walsh, and H. Neurath, *Proc. Nat. Acad. Sci. U.S.* **63**, 1389 (1969).

of enzymes and of proteins in general. This is based on the reasoning that a residue that reacts with a reagent must be exposed to the solvent, while a residue that is internally located will not react with the reagent. In this respect radiolabeled iodine is useful for studying tyrosine and histidine, the residues which incorporate iodine. For example, we found that the relative reactivities toward iodination of the four tyrosyl residues of chymotrypsin were in the order 146 > 94 > 171 > 229,[4] and there appears to be a parallel between the exposure and reactivity in the cases of tyrosyl residues 94, 146, and 229.[5] In the case of the six tyrosyls of bovine ribonuclease, two tyrosyls, 25 and 97, were found to react very slowly with iodine and four were found to be reactive toward iodination.[6] Examination of the crystal structure[7] shows that residues 25 and 97 indeed appear to be buried, while the other four are exposed to varying degrees. Similar studies have been carried out with D-glyceraldehyde-3-phosphate dehydrogenase.[8]

Iodination of an Enzyme

Principles

The iodination of the enzyme is carried out in a buffer of suitable pH using a solution of hypoiodous acid at ice-bath temperature. The use of radiolabeled hypoiodous acid of known specific activity permits accurate determination of the level of iodine incorporated into the enzyme molecule. If the level of iodine incorporation is the prime interest, the specific activity may be low; if the iodinated protein is to be used for the isolation of iodopeptides, then the specific activity should be high.

In order to know accurately the specific activity of the iodinating reagent, it is necessary that the radioactive iodine present be in a chemical form in equilibrium with the iodinating reagent and that no non-equilibrium chemical forms of the radioiodine be present. Therefore, the radioiodine preparation as purchased is treated as follows. An appropriate portion is exchanged with a very small amount of ICl in acidic solution. The radiolabeled ICl is allowed to react with KI to form radiolabeled I_2, which is extracted into CCl_4, leaving all other forms of radioiodine in the aqueous phase. The radiolabeled I_2 is then extracted

[4] S. K. Dube, O. A. Roholt, and D. Pressman, *J. Biol. Chem.* **241**, 4665 (1966).

[5] D. M. Blow, J. J. Birktoft, and B. S. Hartley, *Nature (London)* **221**, 337 (1969). See R. E. Dickerson and I. Geis, "The Structure and Action of Proteins." Harper, New York, 1969. (Stereo Supplement).

[6] C.-Y. Cha and H. A. Scheraga, *J. Biol. Chem.* **238**, 2958, 2965 (1963).

[7] G. Kartha, J. Bello, and D. Harker, *Nature (London)* **213**, 862 (1967).

[8] S. Libor and P. Elodi, *Eur. J. Biochem.* **12**, 336, 345 (1970).

into a buffer of the desired pH as radiolabeled HOI and iodide ion. The iodinating solution itself is prepared by adding a volume of this very dilute radiolabeled iodine solution, containing the desired amount of radioactivity, to an HOI solution of the desired concentration prepared by adding ICl solution to a volume of buffer. The calculated volume of iodinating solution is now added to the protein solution with good mixing.

Experimental

As a typical procedure, the iodination of porcine carboxypeptidase B at pH 9.0 is given below.

Reagents

Stock iodine monochloride solution: To 22.14 g (0.133 mole) of potassium iodide and 14.27 g (0.067 mole) of potassium iodate in a 100-ml volumetric flask, add 50 ml of concentrated (37%) HCl and water to about 90 ml. The contents are stirred magnetically until solution is complete; this requires several hours. The volume is then brought to 100 ml, and the contents mixed. This gives a 2 *M* solution of ICl_2^- in 2 *N* HCl. The reagent is stable for years.

Glycine buffer, 1 *M*: 7.5 g of glycine is adjusted to pH 9.0 with NaOH in a total volume of 100 ml.

Glycine buffer, 0.1 *M*: 0.75 g of glycine is adjusted to pH 9.0 with NaOH in a total volume of 100 ml.

Dilute iodine monochloride solution: Stock iodine monochloride solution is diluted to the desired concentration in 1 *M* NaCl or 2 *M* HCl as indicated.

Radioiodine, purchased as the iodide for chemical use, and free of carrier and reducing agent. Avoid the use of clinical radioiodide since this usually has an organic preservative, which may react with the ICl.

Iodination of Carboxypeptidase B. This procedure is carried out in a hood because of the volatility of I_2. To 1 ml of water in a 12-ml centrifuge tube, 0.01 ml of carrier-free ^{131}I solution (6 mCi/ml), 0.2 ml of concentrated HCl, and 0.2 ml of 4×10^{-4} *M* ICl (8×10^{-2} μmole) in 2 *M* HCl are added and mixed after each addition. Then 0.27 ml of KI solution (50 mg of KI per liter) (8×10^{-2} μmoles of KI) is added. Mixing is achieved by gently pumping the solution in and out of a disposable pipette several times, taking care not to bubble air through the solution. Then 2 or 3 ml of CCl_4 is added and the extraction of the iodine is carried out with the aid of a disposable pipette, as above. The tube is covered with aluminum foil and centrifuged (clinical centrifuge) briefly. With a new disposable pipette, the aqueous layer is removed and discarded. With another disposable pipette, the CCl_4 is transferred to a 12-ml centrifuge tube, taking care not to transfer any of the residual aqueous layer. The transfer of the CCl_4 need not be quantitative. To the tube containing the CCl_4, about 1.5 ml of pH 9.0 1 *M* glycine buffer is added, and, with the aid of a disposable pipette, the iodine is extracted into the glycine buffer as HOI and iodide ion. The tube is centrifuged and as much of the upper layer ("hot" glycine) as can be easily obtained is transferred to a clean test tube in an ice bath. The subsequent steps are all carried out in an ice bath.

To prepare iodinating solution for iodinating carboxypeptidase B to a level of about 6 iodine atoms per enzyme molecule, 0.60 ml of the "hot" glycine, containing approximately the desired cpm, e.g., 10^7 cpm, is added to a mixture of 0.10 ml of 2×10^{-2} *M* ICl in 2 *M* NaCl and 0.30 ml of pH 9.0, 1 *M* glycine buffer. To 3 mg of porcine carboxypeptidase B in 0.33 ml of solution, 0.73 ml of pH 9.0 0.1 *M* glycine buffer is added and mixed, and then 0.44 ml of the iodinating solution is added with thorough magnetic stirring. The solution is allowed to stand for 0.5 hour. Iodination of the enzyme in the presence of an inhibitor is carried out by iodination as above, but using 0.73 ml of an appropriate inhibitor solution at pH 9.0 in place of the buffer. A mock-iodinated control is prepared by adding pH 9.0 1 *M* glycine buffer to the protein solution instead of the iodinating solution.

A standard for determining the specific activity of the iodine is prepared by diluting a known volume of the iodinating solution with a known volume of potassium iodide (0.001 *M*), sodium thiosulfate (0.01 *M*) solution so that the radioactivity is about 4×10^4 cpm/ml. This solution is stored in a tightly sealed container to be used subsequently in the determination of the amount of iodine incorporated into the protein.

The protein solutions are dialyzed successively against several 1-liter portions of cold 0.005 *M* pH 9.0 glycine buffer. The first two dialysis solutions also contain 10^{-3} *M* KI to exchange with any adsorbed radioiodide. The protein solutions are then removed from the dialysis bags and the level of iodination of the protein is estimated. (In a particular experiment it was found that 6.5 iodine atoms were incorporated per enzyme molecule.)

Estimation of the Extent of Iodination. The extent of iodination of the protein is calculated from the protein concentration, the radioactivity incorporated, and the known specific activity of the iodine.

The most convenient method for determining the concentration of the iodinated protein is to measure the absorbancy in 0.1 *N* NaOH at 280 nm and, for the molar absorbancy of the protein, to use the value determined for the uniodinated protein under the same conditions. The contribution of the monoiodotyrosine and diiodotyrosine introduces an error which is small at low levels of iodination and increases with increased iodination, depending on the protein and the level of iodination. In the case of two preparations of a protein which have been iodinated to about the same level, the relative concentrations determined on this basis are accurate. For a more accurate value of the protein concentration, a nitrogen determination may be used. In this case the protein must be dialyzed against a nitrogen-free buffer.

The radioactivity in a known volume of the protein solution and in a portion of the standard prepared from the iodinating solution is now determined; from these values and the protein concentration, the number of iodine atoms incorporated per enzyme molecule is calculated.

Spectral methods have been developed for determining the amounts of tyrosine, monoiodotyrosine (MIT) and diiodotyrosine (DIT) in an iodinated protein,[9] but monoiodohistidine (MIH) and diiodohistidine (DIH) are not included, and thus an error in the total amount of incorporated iodine may be introduced.

Identification of the Iodinated Residues of the Iodinated Protein. This is based on a separation of the iodinated residues by high voltage paper electrophoresis following complete proteolysis of the iodinated protein. A portion of the iodinated protein containing about 10^5 cpm is digested for 5–6 hours in 0.1 ml of 0.5 *M* formic acid containing 0.1 mg of pepsin. The digest is lyophilized. To the dried residue, 0.1 ml of 0.1 *M* ammonium bicarbonate containing 0.1 mg of Pronase is added and the mixture incubated overnight at 37°. Then 0.1 ml of 0.1 *M* ammonium bicarbonate containing 1 mg of pancreatin is added and again incubated overnight at 37°. Of the digest, 25, 50, and 100 μl are applied 5 cm from one end of three strips (3×57 cm) of Whatman 3 MM paper, the samples are dried, and electrophoresis is carried out in 1 *M* formic acid for 2 hours at 40 V/cm. After drying, the strips may be scanned or a radioautograph prepared to locate the iodoamino acids. Under these conditions, the ratios of the distances moved from the origin for DIT: MIT:DIH:MIH are 1:1.3:1.9:2.7. Identification of the iodinated residue in an iodopeptide isolated from an iodinated protein is carried out similarly, starting with the Pronase digestion. Less radioactivity is required since identification involves mainly a single iodoamino acid.

[9] R. E. Perlman and H. Edelhoch, *J. Biol. Chem.* **242**, 2416 (1967).

Comments of Iodination

In determining the effect of iodination on the activity of an enzyme, it is important that the amount of iodine actually incorporated into the enzyme be known, not only the amount of iodine to which the enzyme was exposed. For a given protein, the amount of iodine incorporated under specific conditions is very reproducible, but it may differ from protein to protein. For example, with bovine carboxypeptidase A, bovine chymotrypsin, some immunoglobulins and immunoglobulin fragments, and with porcine carboxypeptidase B at pH 9, we have found that 65–80% of the added iodine is incorporated into the protein. In the case of porcine carboxypeptidase B at pH 7.5, 30–40% is incorporated.

As already pointed out, the "hot" glycine contains a small amount of iodide ion, which is completely exchangeable with the ICl added in preparing the iodinating solution. The amount of this iodide is not known and therefore is kept low in order to avoid an error in the specific activity of the iodination reagent. In the example given, a maximum of 0.08 μmole of this iodide ion would be present if all of the "hot" glycine were used in preparing the iodinating solution (a maximum of 0.08 μmole of I_2 was formed from the ICl and I^-, following the original exchange, and then passed through the CCl_4). However, the amount of this I_2 is less than 0.08 μmole, since it is apparently never formed quantitatively, probably because of the reduction of some of the ICl by traces of reducing substances in the radioiodine preparation as purchased. Furthermore, as a matter of convenience, more of the "hot" glycine is prepared than is needed for the iodinating solution, so that the amount of this iodide ion can be limited to a known maximum amount, e.g., 0.02 μmole. This is trivial compared to the total amount of HOI in the iodinating solution. The "hot" glycine solution is stable for at least several hours, since iodinating solution prepared with it during this period of time shows no loss in ability to incorporate radioactivity into protein. The iodinating solution is likewise stable.

Identification of an Iodinatable Residue in the Active Site of an Enzyme[2]

Principle

The identification of iodopeptides from the active site of an enzyme which contains an iodinatable residue in the site may be achieved by a procedure in which one portion of the enzyme is iodinated with ^{125}I-labeled iodine and a second portion is iodinated to the same iodine level with ^{131}I-labeled iodine, but in the presence of inhibitor. The two portions are mixed and then digested with a proteolytic enzyme. The peptides are separated on paper by high voltage electrophoresis and chromatography. A radioautograph is prepared, and areas on the paper corresponding to spots on the radioautograph are cut out. The ^{125}I and the ^{131}I of the iodopeptide on each piece of paper are counted, and the quotient obtained by division of the ratio of the cpm of ^{125}I to the cpm of ^{131}I in each iodopeptide by the ratio for the unfractionated digest is calculated. This quotient is called the relative ratio. A high relative ratio indicates that the iodinated residue in the iodopeptide represents a residue that is blocked during iodination by the presence of the specific inhibitor that was used. This entire procedure is referred to as the paired-label procedure.

If the amino acid sequence of the enzyme is known, then the position of the iodinated residue of the high-ratio iodopeptides in the protein sequence can be determined by sequencing the iodopeptide.

Paired-Iodination of the Enzyme

Experimental. The paired-iodination of porcine carboxypeptidase B is given as an example, and this is followed by some considerations relevant to its application to a study of other enzymes. The iodination was carried out as described in A, but using Tris buffer at pH 7.5.

Three milligrams of porcine carboxypeptidase B was iodinated at pH 7.5 in a total volume of 1.5 ml with ^{125}I-labeled hypoiodite (starting with 0.6 mCi of ^{125}I), and 3 mg was similarly iodinated with ^{131}I-labeled hypoiodite (starting with 0.6 mCi of ^{131}I), but in the presence of 0.1 M β-phenylpropionate. After exhaustive dialysis against 10^{-3} M formic acid, the two preparations were found to contain 4.5 ^{125}I-labeled iodine atoms per enzyme molecule (1.8×10^6 cpm/mg) and 4.0 ^{131}I-labeled iodine atoms per enzyme molecule (4.2×10^6 cpm/mg), respectively.

Considerations Concerning Iodination for the Paired-Label Procedure. In this procedure it is not essential that the enzyme be completely pure as long as there are no iodinatable impurities whose iodination is affected by the inhibitor. Then any alteration of the iodination pattern in the presence of the inhibitor is due to iodination of a residue in the enzyme site. The levels of iodination achieved in the presence and in the absence of the inhibitor should be within a few percent of each other. The inhibitor should not react with the iodine.

The inhibitor must bind strongly enough to block the iodination of the site; the constant for carboxypeptidase B and β-phenylpropionate is about 10^4, and inhibitor at 0.1 M concentration gives satisfactory protection. The pH for iodine incorporation should be 7.5 or above. Iodination proceeds much more slowly at lower values. The mock-iodination control will indicate the stability of the enzyme under these conditions. Iodination should be at a level that is known to reduce the enzyme activity appreciably and at which the inhibitor protects the activity effectively. For carboxypeptidase B we incorporated 4–6 atoms of iodine per protein molecule (20–21 tyrosines). The radioactivity incorporated should be sufficient to give a radioautograph of the peptide map on overnight exposure, i.e., 1 or 2×10^6 cpm/mg protein for ^{125}I and $4–8 \times 10^6$ cpm per milligram of protein for ^{131}I. The dialysis of the iodinated proteins must be against a volatile buffer, i.e., 10^{-3} *M* formic acid or ammonium hydroxide, so that salts that would interfere with the peptide mapping are removed.

Digestion of Protein and Separation of the Iodopeptides

Experimental. Volumes containing 2.5 mg of protein of each iodinated preparation are mixed and lyophilized; 0.50 ml of 0.5 *M* formic acid containing 0.16 mg of pepsin is added. The vessel is well capped. The dried material is wet by gentle shaking and then incubated at 37° overnight. During the first 1 or 2 hours, the vial is agitated from time to time to aid wetting and digestion. A clear digest should result.

A 1-mg portion of the digest (100 μl in four equal portions) is applied to a Whatman 3 MM paper (see Fig. 1, Vol. XI [36]), subjected to high voltage electrophoresis (1 *M* HCOOH, 40 V/cm, 2 hours), and then chromatographed in the second direction with *n*-butanol:acetic acid: water (4:1:5).

Comments. Pepsin is used for the digestion since most proteins are extensively digested by it in 0.5 *M* formic acid, and the resultant iodopeptides move from the origin and separate well in the peptide-mapping system used. In the paired-iodination procedure this is important in order that deviant-ratio iodopeptides are not lost in a composite spot. However, peptic digestion may release free iodotyrosine rather than an iodopeptide containing this residue. This becomes very important if the free iodotyrosine is derived from a tyrosyl residue in the active site. We have also used elastase in 0.1 *M* ammonium bicarbonate with satisfactory results.[2]

Tryptic digests of even completely reduced and alkylated iodinated proteins usually contain considerable iodinated material that does not move from the origin in the separation system used here. These iodopeptides will, however, frequently move well in a different system.[10]

[10] B.-K. Seon, O. A. Roholt, and D. Pressman, *Biochim. Biophys. Acta* **194**, 397 (1969).

Radioautography

A radioautograph of the peptide map is prepared using Kodak No-Screen, Medical X-ray film. A 17 × 21-inch sheet of film is made by butting 1½ sheets (17 × 14 inches) together and holding them by cellulose tape at the ends of the joint. The dried paper is trimmed at the top and ends to the film size, and about 6 spots (5 μl) of radioactive ink (India ink containing about 10^7 cpm of ^{125}I per milliliter) are placed at random around the edge of the paper as fiducial points for subsequent aligning of the paper and radioautograph. The paper is fastened to the film by a single staple near one edge. The film and paper are placed in a large envelope which is put between two layers of polyurethane foam. The complete sandwich is then clamped between two plywood sheets for 15–20 hours. When the developed film is viewed, a longer or shorter exposure time may be indicated.

Designation of Iodopeptides from the Site

Spots on the radioautograph, including the one at the origin, are encircled with a wax pencil and numbered sequentially. Each large spot should be subdivided into several areas; otherwise a high-ratio iodopeptide may be obscured by overlapping average-ratio iodopeptides.

The peptide map is aligned over the marked film according to the fiducial marks with the aid of a light-box. Areas on the paper are then outlined and numbered according to the marked radioautograph. These areas of paper are then cut out, transferred to tubes, and the counts per minute of ^{125}I and of ^{131}I in each tube are determined in a dual-channel γ-ray spectrometer. The counts per minute of ^{125}I and ^{131}I on bits of paper to which 5- or 10-μl portions of the unfractionated digest had been applied are also determined. The ratio, cpm ^{125}I:cpm ^{131}I, for the paper in each tube is divided by the ratio in the unfractionated digest, and this quotient is the relative ratio.

Discussion

In the pair-iodination procedure, the difference in the incorporation of iodine in a protected site compared to that in an unprotected site results in a high relative ratio and permits identification of the active-site iodopeptide. There may be minor variations, however, in the relative ratio among those iodopeptides not derived from the site, and the ratios may be either high or low. The reason for this variation is either a conformational change or the availability of additional iodine. When an enzyme with a protected residue in the site is iodinated with the same amount of iodinating reagent (not a saturating amount) as is the unprotected enzyme, an amount of iodine equivalent to that which would otherwise have reacted with the (protected) residue in the site is available to react with other tyrosyl (and histidyl) residues which are still available for iodination. Each one of these residues will thus be iodinated to a slightly higher extent (depending on its reactivity) than it is in the unprotected sample because of the altered distribution of iodine, so that the total iodine incorporation in the two preparations is approximately the same.

This increase in the iodination of each of these tyrosyl (or histidyl) residues in the protected preparation may lead to either an increase or a decrease in the yield of each corresponding iodopeptide, and consequently to a higher or lower cpm ^{125}I:cpm ^{131}I ratio, and, therefore, relative ratio, if the monoiodo derivative is involved. This is because during iodination the amount of the monoiodo compound that is formed rises to a maximum and then decreases as iodination increases and the monoiodo residue is converted to the diiodo residue. Thus, if the amount of the monoiodo derivative that has formed is still well short of the maximum amount, further iodination will increase its yield (until the maximum is reached) and in the protected preparation a low relative ratio will result for the monoiodo peptides derived from this residue. If, on the other hand, the amount of the monoiodo derivative is already near, or has decreased beyond, the maximum, further iodination will decrease the yield of this derivative in the protected preparation and a high relative ratio will result. For a diiodo derivative, further iodination in the protected preparation will cause the formation of more of the derivative, and the relative ratio will always decrease.

A consideration of these same factors indicates that, in a situation where the inhibitor only partially protects the site against iodination, a low-ratio iodopeptide from the site may appear and this peptide would necessarily be a monoiodo peptide.[11]

If a larger amount of the enzyme is to be pair-iodinated as a source of more of the iodopeptide from the site, e.g., for sequence studies, some modifications of the procedure are useful. Since most of the iodopeptide from the site is actually derived from the unprotected preparation, not from the protected enzyme, enzyme can be conserved by iodinating only a small amount of protected enzyme and several times this amount of the unprotected enzyme, but with comparable total counts per minute of ^{125}I and ^{131}I on the unprotected and protected preparations, respec-

[11] O. A. Roholt and D. Pressman, *Biochim. Biophys. Acta* **147,** 1 (1967).

tively. The number of iodine atoms incorporated per molecule of the two preparations should, of course, be the same.

The mixing of the two preparations, digestion, etc., are carried out in the same manner as described above, and the ratio for each iodopeptide is determined in the same way, i.e., from the $^{125}I:^{131}I$ ratio of each iodopeptide relative to the ratio for the unfractionated digest.

It is to be noted that when equal weights of a protected and an unprotected enzyme preparation, both iodinated to the same level, are used, the relative ratio for a particular iodopeptide—i.e., the quotient of the ratio of the cpm ^{125}I:cpm ^{131}I for the iodopeptide and the ratio for the unfractionated digest—is equal to the ratio of the *amount* of the ^{125}I-labeled iodopeptide to the *amount* of the ^{131}I-labeled iodopeptide.

However, when unequal amounts of the two preparations are used, the value of the relative ratio for an iodopeptide will be different than the ratio of the *amounts* of iodopeptide from the two preparations; the relative ratio must be multiplied by a factor correcting for the difference in the amounts of ^{125}I- and ^{131}I-labeled iodine in the unfractionated digest. This factor is simply the ratio of the weights of the protein used.

Any difference in the levels of iodination must also be taken into account as a second factor, i.e., the ratio of the number of iodine atoms per enzyme molecule in one of the preparations to the number in the other. Ideally, this factor should have a value of 1, but it may range between 0.9 and 1.1. This factor is present even in experiments in which equal weights of the unprotected and protected preparations are mixed.

Confidence in the reproducibility of the iodination, and hence in the paired label procedure, can be gained by the investigator by making a second unprotected iodination with ^{131}I and pairing this with the product from the unprotected iodination with ^{125}I. The iodopeptides will all have the same ratio; if not, some technical detail is at fault.

[36] Reactions with *N*-Ethylmaleimide and *p*-Mercuribenzoate

By J. F. Riordan and B. L. Vallee

The sulfhydryl groups of enzymes possess a high degree of reactivity as evidenced by their interaction with a wide variety of agents. This distinctive property has permitted their chemical modification through the design of several highly selective and specific reagents which react rapidly and stoichiometrically and may be employed under relatively mild conditions.[1]

The existent sulfhydryl reagents exhibit a range of reactivities which can be utilized either to increase the specificity of the reaction, or to limit it to one or just a few of the "accessible" groups. Conversely, the reactivity can be extended by inducing alterations in the three-dimensional protein structure. Many of these reactions can be reversed by addition of an excess of a sulfhydryl-containing compound. The number of sulfhydryl reagents is very large, and the chemistry of their interaction, their specificity and limitations have been the subject of several extensive and excellent reviews.[1]

The reactions of two reagents, *p*-mercuribenzoate (PMB)[2] and *N*-ethylmaleimide (NEM),[3,4] in particular, are readily quantitated because of an alteration in spectral characteristics and because the products can be measured directly as well. The former has long been thought to be a highly specific, but reversible sulfhydryl reagent and has been utilized successfully as a quantitative reagent for measuring sulfhydryl groups in proteins. NEM has received considerable attention, particularly because its reaction product is stable even to acid hydrolysis, and it, too, has been employed as a quantitative reagent. The methods by which these two compounds may be used to modify sulfhydryl groups of enzymes are essentially the same as those established for determining the number of such groups in nonenzymatic proteins, with the exception that prior denaturation is obviously omitted when the establishment of functional significance is the ultimate objective. Since quantitation of the reaction is one of the major prerequisites for defining the basis of altered activity consequent to chemical modification of an enzyme, the methods employed will be discussed from this aspect.

[1] There have been several recent reviews on the role of sulfhydryl groups in proteins, and methods of reaction and analysis have been described in various degrees of detail. The reader is referred to these reviews for a critical and exhaustive discussion of this topic.[1a–e]

[1a] R. Benesch, R. E. Benesch, P. D. Boyer, I. M. Klotz, W. R. Middlebrook, A. G. Szent-Györgyi, and D. R. Schwartz, eds., *in* "Sulfur in Proteins." Academic Press, New York, 1959.

[1b] P. D. Boyer, *in* "The Enzymes" (P. D. Boyer, H. Lardy, and K. Myrbäck, eds.), Vol. 1, p. 511. Academic Press, New York, 1959.

[1c] R. Cecil and J. R. McPhee, *Advan. Protein Chem.* **14,** 255 (1959).

[1d] R. Benesch and R. E. Benesch, *Methods Biochem. Anal.* **10,** 43 (1962).

[1e] R. Cecil, *in* "The Proteins" (H. Neurath, ed.), 2nd ed., Vol. 1, p. 379. Academic Press, New York, 1963.

[2] L. Hellerman, F. P. Chinard, and V. R. Deitz, *J. Biol. Chem.* **147,** 443 (1943).

[3] N. H. Alexander, *Anal. Chem.* **30,** 1292 (1958).

[4] E. Roberts and G. Rouser, *Anal. Chem.* **30,** 1291 (1958).

Reaction with *p*-Mercuribenzoate (PMB)

The sulfhydryl groups of proteins will react with a variety of metals to form mercaptides. Divalent mercury, for example, is known to react with the single sulfhydryl group of 1 molecule of mercaptalbumin or papain. The second coordination site of the metal is filled by a pair of electrons from the sulfhydryl group of a second protein molecule leading to dimer formation. When mercury is covalently bound to an organic residue only monomercaptides are formed with proteins. Several organic mercurials undergo an alteration in their spectral characteristics when they react with thiols. However, thus far, *p*-mercuribenzoate (PMB) is the only mercurial giving rise to an analytically adequate spectral change due to mercaptide formation.[5]

Initial studies with PMB employed nitroprusside titration to detect the number of sulfhydryl groups in proteins.[6] Other methods have determined the excess mercurial remaining after reaction with the protein either colorimetrically[7] or amperometrically.[1d] However, the spectrophotometric method introduced by Boyer[5] is the most generally useful technique, and this has greatly increased the precision of analysis. It is mild enough not to complicate studies of activity. PMB exhibits an absorption maximum at 233 nm with a molar absorptivity of 1.69×10^4. On formation of a mercaptide, the molar absorptivity increases to 2.2×10^4. However, this change is small in comparison with the maximal difference in absorbance which occurs between PMB and its mercaptide in the region from 250–255 nm and which directly reflects the amount of reagent reacted with the protein.

A standard procedure for determining sulfhydryl groups has been described in great detail by Benesch and Benesch.[1d] Because the change in molar absorptivity is a function of the protein under study, the analysis is carried out as a spectrophotometric titration. PMB is either titrated with protein or vice versa. A stock solution of approximately 10^{-3} *M* PMB is prepared by dissolving 9 mg of Na *p*-chloromercuribenzoate (Sigma Chemical Company) in a slight excess of alkali and diluting to 25 ml, followed by centrifugation. Working solutions are prepared by diluting 2 ml of the stock to 25 ml with 0.01 *M* phosphate buffer, pH 7.0, or with 0.33 *M* acetate buffer, pH 4.6. This solution can then be standardized on the basis of ϵ_{233}, or by titrating a solution of glutathione of known concentration. A 25-μl aliquot of a sufficiently concentrated solution of thiol-containing material is added to a known

[5] P. D. Boyer, *J. Amer. Chem. Soc.* **76,** 4331 (1954).

[6] F. P. Chinard ond L. Hellerman, *Methods Biochem. Anal.* **1,** 1 (1954).

[7] I. Fridovich and P. Handler, *Anal. Chem.* **29,** 1219 (1957).

amount of the buffered PMB solution (about $6 \times 10^{-5} M$) and to an equal volume of buffer in 1-cm quartz cuvettes. The cuvettes are covered with Parafilm, and the contents are mixed by inverting. The absorbance increase is measured at 255 nm if the titration is performed at pH 4.6, or at 250 nm if at pH 7.0. Additional 25-μl aliquots of thiol are added, and the procedure is repeated. The titration is continued in this manner until there is no further change in absorbance successive additions. The $\Delta\epsilon_{255}$ for monothiols at pH 4.6 is 6200, while at pH 7.0 $\Delta\epsilon_{250}$ is 7600. However, these values do not always pertain for proteins, hence the end point of the titration must be estimated graphically.

With proteins of low sulfhydryl content or of low solubility, the reverse procedure is employed. PMB, 10 μl at a time, is added to 4 ml of a 0.1–0.3% protein solution and to 4 ml of buffer blank. The cells are covered with Parafilm, mixed, and the increase of absorption at 255 nm is recorded. If the protein absorbs strongly in this region, a blank containing protein previously treated with, e.g., iodoacetate can be employed. In either procedure, the rate of reaction varies from protein to protein and readings must be made vs. time until constant before additional protein or PMB is added. Solutions to be used for enzyme assays should be dialyzed or separated from excess reagent by gel filtration.

Boyer pointed out in his initial description of the method that the choice of buffers and pH can be of critical significance not only when the reagent is employed for analytical purposes, but also in functional group studies.[5] This important detail should be kept in mind when the procedure is employed, and attention to its features will obviate otherwise puzzling problems. For most proteins, the rate of reaction is faster at pH 5 than at pH 7. High concentrations of phosphate, of chloride, or of nitrate inhibit or retard the reaction, but a number of other anions, such as pyrophosphate, sulfate, or perchlorate, may promote the reactivity of PMB. This latter effect may be due to an increased solubility of the reagent as occurs when pyrophosphate or sulfate replaces chloride, or anions may increase reactivity by being more readily displaced from the mercurial than is hydroxide. The nature of the anion may play an additional role in limiting the specificity of the reaction.

Several problems concerning this analytical method should be mentioned. Spectral measurements are made in a region where most proteins absorb strongly. The observed differences may be small and, hence, adequate corrections must be made but may prove difficult. In fairly high concentrations PMB may inhibit enzymes by reactions which do not involve sulfhydryl groups.[8] The spectral change observed can be a

[8] M. R. Sohler, M. A. Seibert, C. W. Kreke, and E. S. Cook, *J. Biol. Chem.* **198**, 281 (1952).

function of the protein under study and often differs from that found for simple monothiols. For instance, reaction of apocarboxypeptidase A with PMB gave spectral changes consistent with the presence of a single thiol group, while the enzyme containing 1 gram-atom of zinc per mole gave no reaction. This was taken to indicate that one of the metal-binding ligands of carboxypeptidase was a cysteinyl residue.[9] Subsequently, it was shown that the metal binding site did not contain a sulfhydryl group, but rather was composed of a glutamyl and two histidyl residues. From this it would appear that the reaction of PMB with such a site can mimic that with a thiol, although this has not been studied in detail.

The reaction of PMB with sulfhydryl groups of proteins can be reversed by adding an excess of a reagent such as cysteine or glutathione. This allows greater flexibility in studying the essentiality of sulfhydryl groups in enzymes. Failure of this treatment to restore activity to an enzyme inactivated with PMB may indicate either that irreversible denaturation or dissociation has occurred, or that PMB has reacted with groups other than sulfhydryl in the enzyme. The latter alternative can be examined by comparing the results of analysis determined from the change at 255 nm with those determined by measuring the amount of mercury incorporation.[10] Similarly, dialysis of the PMB-inactivated enzyme vs. cysteine, followed by mercury analysis of the protein should reveal any noncysteinyl-bound PMB.

Reaction with *N*-Ethylmaleimide (NEM)

In general, only the most reactive sulfhydryl groups of proteins combine with compounds containing an activated double bond to form stable thiol ethers. Maleic anhydride, maleic acid, and especially NEM have been shown to be particularly useful for this purpose.[3,11,12] The formation of a stable alkyl derivative that resists acid hydrolysis is an important feature of this reagent, particularly when verification of the reaction by amino acid analysis is contemplated.

NEM has received considerable attention because its absorption spectrum has a maximum at 305 nm which is abolished when the agent combines stoichiometrically with a compound containing sulfhydryl groups.[13] This has allowed a quantitative spectrophotometric determination of the number of groups reacting. The reaction can be specific

[9] B. L. Vallee and J. F. Riordan, *Brookhaven Symp. Biol.* **21**, 91 (1968).

[10] M. Suzuki, T. L. Coombs, and B. L. Vallee, *Anal. Biochem.* **32**, 106 (1969).

[11] E. Friedmann, D. H. Marrian, and I. Simon-Reuss, *Brit. J. Pharmacol.* **4**, 105 (1949).

[12] T.-C. Tsao and K. Bailey, *Biochim. Biophys. Acta* **11**, 102 (1953).

[13] J. D. Gregory, *J. Amer. Chem. Soc.* **77**, 3922 (1955).

for sulfhydryl groups, but under certain conditions other groups have been shown also to react.[14]

The analytical procedure is usually carried out at 25° with 10^{-3} *M* NEM in 0.1 *M* phosphate buffer, pH 7.0.[1d] Sufficient protein is added directly to this solution in a 1-ml cuvette to decrease the absorption at 305 nm, initially 0.620, by 0.1–0.5 absorption units. The blank consists of buffer containing the same amount of protein. The decrease in absorption, corrected for any dilution, is divided by 620, the molar absorptivity of NEM, to give the molar sulfhydryl concentration. Under these conditions, the reaction is generally quite rapid with small thiol-containing molecules; with glutathione, for example, the spectral change is completed within 2 minutes. However, with some proteins, up to 100 minutes, or even longer, has been required before a further decrease in absorbance is no longer observed. Excess reagent can be removed by reaction with β-mercaptoethanol, by gel filtration, or by dialysis.

As judged by these conditions, this reagent is well suited for modification reactions of proteins. Extremes of temperature and pH are avoided, and no changes in pH occur during the reaction. Not only is the formation of the thio-ether usually very fast but stoichiometric amounts of NEM are all that is required. Precise adjustment of pH is critical. Below pH 6 the rate of reaction falls off considerably. The reaction should not be attempted at pH values above neutrality. The reagent itself becomes unstable and undergoes hydrolysis to *N*-ethylmaleamic acid, leading to a loss in absorbance at 305 nm and yielding a product which reacts very slowly with sulfhydryl groups. Moreover, the tendency to react with other than sulfhydryl groups is increased under these conditions.[15]

In spite of its many desirable features, there are some disadvantages of the spectrophotometric assay procedure involving NEM relative to the PMB titration discussed above. Problems have been encountered with the purity of commerically available NEM. They can be obviated, however, by appropriate synthesis.[16] In addition, the sensitivity of the method is low; the change in molar absorptivity is only about one-twelfth that for PMB. Since the disappearance of an absorption peak is being observed, the measurement deals with small differences between relatively large numbers, and possible errors are thereby amplified. High concentrations of protein are required to avoid this problem. This presents additional difficulties if the enzyme under study is not extremely soluble, since relatively large volumes of protein solution must be added to the NEM. Moreover, since some proteins absorb significantly at

[14] D. G. Smyth, A. Nagamatsu, and J. S. Fruton, *J. Amer. Chem. Soc.* **82**, 4600 (1960).
[15] D. G. Smyth, O. O. Blumenfeld, and W. Konigsberg, *Biochem. J.* **91**, 589 (1964).
[16] C. F. Fox and E. P. Kennedy, *Proc. Nat. Acad. Sci. U.S.* **54**, 891 (1965).

305 nm, high blank corrections may result. In some instances this difficulty can be avoided by adding the protein to an NEM solution and then deproteinizing with, e.g., 2% perchloric acid.[17] The supernatant is compared to a blank treated similarly, but without NEM. Finally, the disappearance of the reagent rather than the formation of a product is the ultimate spectrophotometric criterion of the reaction. Hence, the evidence for reaction with a group on the protein is only inferential, and conclusions based solely on such data can be in doubt, particularly if extraneous cricumstances can result in analogous changes. Fortunately such difficulties can be eliminated by employing other than a spectrophotometric end point for studying the reaction of NEM with proteins. Radioactively labeled NEM is commercially available and uptake of ^{14}C can readily be measured. Importantly, the reaction product can be isolated and identified, since addition of NEM to a sulfhydryl group results in the formation of *S*-(ethylsuccinimido)cysteine. On acid hydrolysis this is converted to *S*-succinylcysteine and ethyl amine, both of which can be determined quantitatively by means of amino acid analysis.[18] This allows a direct measure of the amount of cysteine in the protein which has reacted with the reagent, and this can be compared with the indirect spectrophotometric analysis. Hydrolysis of the modified protein is carried out *in vacuo* in 6 *N* HCl at 110° for 72 hours. Yields of 88% have been reported, and apparently these can be increased to 94–95% if careful evacuation is performed.[19] *S*-Succinyl cysteine elutes at 91 ml (relative to 111 ml for aspartic acid) on the 150-cm column[18] using the analytical procedure of Spackman, Stein, and Moore.[20] Ethylamine elutes at 100 ml (relative to 52 ml for lysine) on the 15-cm column.[19] Since unreacted NEM liberates ethylamine, care must be taken to remove all traces of excess reagent prior to hydrolysis. It is important to keep in mind that ethylamine is liberated on hydrolysis not only from the reaction product with sulfhydryl groups, but also from that with amino groups. Thus by comparing the yields of *S*-succinylcysteine and ethylamine, the specificity of the reaction is immediately established.[15] The reaction of sulfhydryl groups with maleic anhydride also yields *S*-succinylcysteine on acid hydrolysis. Prolonged hydrolysis, which is required for conversion of the NEM adduct to succinylcysteine, is unnecessary,[21] but this reaction cannot be cross-checked with respect to ethylamine.

Mention should be made of a number of other substituted maleimides

[17] R. Benesch and R. E. Benesch, *J. Biol. Chem.* **236**, 405 (1961).

[18] G. Guidotti and W. Konigsberg, *J. Biol. Chem.* **239**, 1474 (1964).

[19] D. G. Smyth, F. C. Battaglia, and G. Meschia, *J. Gen. Physiol.* **44**, 889 (1961).

[20] D. H. Spackman, W. H. Stein, and S. Moore, *Anal. Chem.* **30**, 1190 (1958).

[21] G. R. Stark, *Advan. Protein Chem.* **24**, 261 (1970).

which have proved useful, since they introduce a chromophoric substituent into the protein. These include *N*-(dimethylamino-3,5-dinitrophenyl)maleimide,[22] *N*-(2,4-dinitroaniline)maleimide,[23] and *N*-(4-hydroxy-α-naphthyl)maleimide.[24] Such reagents allow labeling of reactive sulfhydryl groups in the protein and ready detection of the labeled peptide after proteolytic degradation and separation.[25,26]

Comments

The sulfhydryl groups of enzymes evidence a considerable variation in their reactivity, ranging from unreactive through several stages of sluggishness to free and immediately reactive. The reactivity of any given sulfhydryl group in a particular enzyme may be modified by adjustment of reaction conditions, by prior alteration of other protein groups, by suitable selection of the reagent employed, and by denaturation. The factors which determine sulfhydryl group reactivity include location in the three-dimensional structure of the protein, microscopic environment or neighboring group effects, and interaction with other functional groups.

These considerations are very pertinent to the interpretation of enzymatic activity changes consequent to reaction with sulfhydryl group reagents. Thus, it may be necessary to determine reactivity under assay conditions with a variety of reagents.

The precise assignment of a functional role also presents problems. More than a hundred enzymes have been assigned to the category of "sulfhydryl enzymes" based on their inactivation with one or another of the so-called sulfhydryl reagents.[1b] Boyer has enumerated the various possibilities which could result in loss of activity of an enzyme consequent to modification of its sulfhydryl groups.[1b] While caution is generally exercised in the interpretation of such data, it is not inappropriate to emphasize the need for continued conservatism. Any interpretations based solely on loss of enzymatic activity will always be in doubt. Consequently, having demonstrated an effect of either of the thiol reagents discussed above, the task still remains to establish the relationship of the chemical modification to the biological activity which is under study.

[22] A. Witter and H. Tuppy, *Biochim. Biophys. Acta* **45**, 429 (1960).

[23] G. D. Clark-Walker and H. G. Robinson, *J. Chem. Soc.* **547**, 2801 (1961).

[24] K. C. Tsou, R. J. Barrnett, and A. M. Seligman, *J. Amer. Chem. Soc.* **77**, 4613 (1955).

[25] K. A. Walsh, K. S. V. Sampath Kumar, J.-P. Bargetzi, and H. Neurath, *Proc. Nat. Acad. Sci. U.S.* **48**, 1443 (1962).

[26] A. H. Gold and H. L. Segal, *Biochemistry* **3**, 778 (1964).

[37] Reaction of Protein Sulfhydryl Groups with Ellman's Reagent

By A. F. S. A. HABEEB

The reagent 5,5′-dithiobis(2-nitrobenzoic acid) (DTNB) was developed by Ellman as a sulfhydryl reagent.[1] DTNB has been found to be a sensitive tool for the assay of thiol groups in tissues, body fluids, and proteins.

Ellman[1] described a method for its synthesis. However, the reagent is commercially available. DTNB is an aromatic disulfide, and, since it has a higher standard oxidation-reduction potential than aliphatic analogs, it will react with aliphatic thiols by an exchange reaction to form a mixed disulfide of the protein and 1 mole of 2-nitro-5-thiobenzoate per mole of protein sulfhydryl group (reaction 1).

$$P{-}SH + \begin{array}{l} S{-}C_6H_3(NO_2)(COO^-) \\ | \\ S{-}C_6H_3(NO_2)(COO^-) \end{array} \longrightarrow P{-}S{-}S{-}C_6H_3(NO_2)(COO^-) + {}^-S{-}C_6H_3(NO_2)(COO^-) \qquad (1)$$

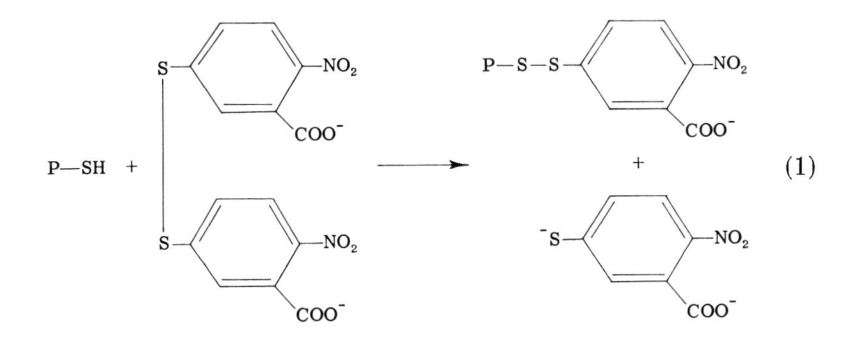

Sulfhydryl–disulfide exchange reactions between disulfide compounds with sulfur directly attached to aromatic groups and simple alkyl mercaptans should go to completion.[2] Protein SH groups may behave similarly to simple alkyl mercaptans unless steric factors interfere with the course of reaction. Therefore, reaction (1) could be followed by reaction (2). Intramolecular and intermolecular disulfide formation may be considered

$$P{-}S{-}S{-}C_6H_3(NO_2)(COO^-) + P{-}SH \longrightarrow {}^-S{-}C_6H_3(NO_2)(COO^-) + P{-}S{-}S{-}P \qquad (2)$$

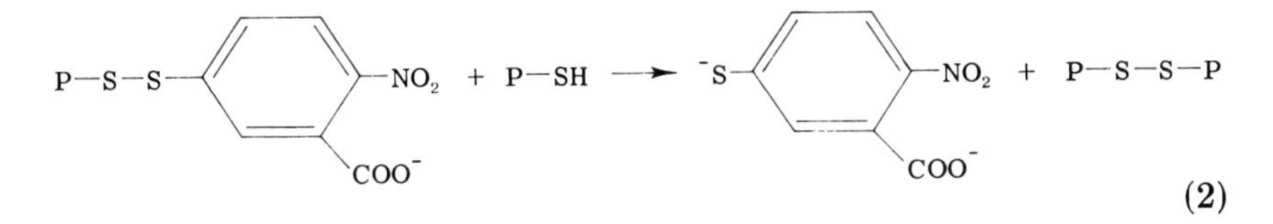

[1] G. L. Ellman, *Arch. Biochem. Biophys.* **82,** 70 (1959).

[2] A. J. Parker and N. Kharasch, *Chem. Rev.* **59,** 583 (1959).

a possibility. Indeed, Kleppe and Damjanovich[3] observed a heavy molecular weight component as a result of treatment of muscle phosphorylase *b* with DTNB, suggesting that inter- and intramolecular disulfide bonds formed during the reaction.

However, whether reactions (1) or (1) and (2) occur, the same stoichiometry applies, namely, 1 mole of 2-nitro-5-thiobenzoate anion is formed per mole of protein sulfhydryl. The nitromercaptobenzoate anion has an intense yellow color with a molar absorptivity of 13,600 $M^{-1}cm^{-1}$ at 412 nm. With DTNB, a solution of 0.01 μmole of sulfhydryl per milliliter gives an absorbance of 0.136 (1-cm light path) at 412 nm. Simple thiols, e.g., cysteine, give complete color development within 2 minutes and the color is stable for 2 hours.

Determination of Total Protein Sulfhydryl

The sulfhydryl groups in proteins exhibit variable reactivity toward DTNB owing to steric factors. Therefore, determination of total sulfhydryl content requires that the protein be denatured, preferably with sodium dodecyl sulfate.[4] About 0.01–0.04 μmole of protein is dissolved in 6 ml of solution containing 2% sodium dodecyl sulfate,[5] 0.08 *M* sodium phosphate buffer, pH 8, and 0.5 mg/ml EDTA. To 3 ml of the solution is added 0.1 ml DTNB solution (40 mg DTNB in 10 ml of 0.1 *M* sodium phosphate buffer, pH 8). The color is developed for 15 minutes and read at 410 nm against protein solution in SDS to give apparent absorbance. A reagent blank is subtracted from the apparent absorbance to give the net absorbance. For calculation of sulfhydryl content, the net absorbance is employed with a molar absorptivity value of 13,600 $M^{-1}cm^{-1}$. Protein concentration is derived from the absorbance at 280 nm and the known value of $E^{1\%}_{280\ \text{nm}}$.

Glaser *et al.*[6] used an alcohol–buffer solution of DTNB as a sensitive spray for the visualization of thiols on chromatograms. The spray consists of 0.1% solution of DTNB in a 1:1 mixture of ethanol and 0.45 *M* Tris buffer, pH 8.2.

Determination of Available Protein Sulfhydryl

The reaction is performed as described above, but in the absence of denaturing agents, and the absorbance at 412 nm is followed as a function of time. Thus it is possible to distinguish between the different classes of

[3] K. Kleppe and S. Damjanovich, *Biochim. Biophys. Acta* **185,** 88 (1969).

[4] M. J. Fernandez Diez, D. T. Osuga, and R. E. Feeney, *Arch. Biochem. Biophys.* **107,** 449 (1964).

[5] A. F. S. A. Habeeb, *Biochim. Biophys. Acta* **115,** 440 (1966).

[6] C. B. Glaser, H. M. Maeda, and J. Meienhofer, *J. Chromatogr.* **50,** 151 (1970).

sulfhydryl groups that exist in a protein. For example, tryptophan–transfer ribonucleic acid synthetase[7] contains 8 sulfhydryl groups, of which 3 react very rapidly and the remaining 5 do so more slowly. The 8 sulfhydryls in the enzyme are revealed in native as well as in 8 *M* urea-denatured protein, indicating the absence of any buried sulfhydryls. In the presence of tryptophan and ATP, the 3 fast-reacting sulfhydryls are still available, but 4 of the 5 slow-reacting sulfhydryls are no longer detected.

Phosphorylase *b*[3] contains 3 classes of sulfhydryl groups distinguished by their reactivity with DTNB. In the presence of a low concentration of DTNB (0.33 m*M*) the first class, consisting of 2 SH groups, reacts. As the concentration of DTNB is increased to 6.6 m*M*, the second class, consisting of 4 SH groups, reacts, followed by the third class, consisting of 10 SH groups, after further increase of the concentration of DTNB.

The rate of reaction of sulfhydryl groups has been used to detect conformational differences between proteins. Guidotti[8] attributed the difference in the velocity of the reaction of the sulfhydryl groups in oxyhemoglobin and carbon monoxide-hemoglobin with DTNB to differences in the conformation of the two proteins. Phillips *et al.*[9] used DTNB to determine the number of sulfhydryl groups in several β-lactoglobulins (cow, sheep, and goat) and to obtain evidence for conformational differences between them. To the protein solution in 9 ml of 0.01 *M* potassium phosphate, pH 7.6, containing 0.001 *M* EDTA, was added 0.2 or 0.5 ml of 0.01 *M* DTNB; after mixing, the absorbance at 412 nm was recorded at intervals. The results showed that native β-lactoglobulins from cows, goats, and sheep react with DTNB at the same rate and each contains 2 sulfhydryl groups per mole of protein. The similarity in availability of the sulfhydryls suggested that the gross conformations of these β-lactoglobulins are similar. When the COOH-terminal sequence His-Ile was removed, the modified β-lactoglobulins exhibited increased reactivity toward DTNB, indicative of conformational changes as a result of loss of His-Ile.

Determination of Sulfhydryl Groups in Animal Tissues

Autoxidation of thiols in tissue homogenates is frequently prevented by the presence of 0.2 *M* EDTA[10,11] or by the presence of EDTA and

[7] M. DeLuca and W. D. McElroy, *Arch. Biochem. Biophys.* **116,** 103 (1966).

[8] G. Guidotti, *J. Biol. Chem.* **240,** 3924 (1965).

[9] N. I. Phillips, R. Jenness, and E. B. Kalan, *Arch. Biochem. Biophys.* **120,** 192 (1967).

[10] H. Sakai and K. Dan, *Exp. Cell Res.* **16,** 24 (1959).

[11] G. S. Tarnowiski, R. K. Barclay, I. M. Mountain, M. Nakamura, H. G. Satterwhite, and E. M. Solney, *Arch. Biochem. Biophys.* **110,** 210 (1965).

α,α′-dipyridyl.[12] For determination of acid-soluble thiols,[11] 1.0-ml aliquots of tissue homogenates prepared in 0.2 *M* EDTA solution, pH 4.5, are mixed with 1 ml of 0.2 *M* EDTA, pH 4.5. Then 1 ml of 15% perchloric acid is added and the mixture is heated in a boiling water bath for 5 minutes and filtered through Whatman No. 54 paper. Precipitates are washed 3 times with 1-ml aliquots of 0.2 *M* EDTA, pH 4.5. The pH of the filtrates is adjusted to 7.5 and the samples are diluted to volume (10 ml) with 0.2 *M* EDTA, pH 7.5. Colorimetric assays are made on 2-ml aliquots by adding 1.8 ml of 0.1 *M* Tris buffer, pH 7.5, and 0.2 ml of 0.001 *M* DTNB solution in Tris buffer. To calculate the content of acid-soluble thiols in tissues, a molar absorptivity at 412 nm of 13,600 $M^{-1}cm^{-1}$ is used. With reduced glutathione as an internal standard the recovery is 83 to 89%.

Sedlak and Lindsay[13] devised a procedure for the determination of protein-bound, nonprotein, and total sulfhydryl groups (PB-SH, NP-SH, and T-SH, respectively) in various tissues, as follows. Tissues (200–400 mg) were homogenized in 8 ml of 0.02 *M* EDTA, pH 4.7, in an ice-bath. Under these conditions T-SH and NP-SH did not change during 2–4 hours. For determination of T-SH, a 0.5-ml aliquot of tissue homogenate was mixed with 1.5 ml of 0.2 *M* Tris buffer, pH 8.2, and 0.1 ml of 0.01 *M* DTNB in methanol. The mixture was brought to 10 ml with absolute methanol. A reagent blank (without sample) and a sample blank (without DTNB) were prepared. Color was developed for 30 minutes, after which the solution was filtered. The absorbance was read at 412 nm and a value for the molecular absorptivity of 13,600 $M^{-1}cm^{-1}$ was used for calculation. For NP-SH, a 5-ml aliquot of the homogenate was mixed with 4 ml of distilled water and 1 ml of 50% trichloroacetic acid. After shaking for 10–15 minutes, it was centrifuged. To 2 ml of supernatant was added 4 ml of 0.4 *M* Tris buffer, pH 8.9 and 0.1 ml DTNB, and the sample was shaken. The absorbance was read at 412 nm. The PB-SH was calculated by subtracting the NP-SH from T-SH.

This procedure requires that the pH of the final reaction mixture be above pH 8.0, since the intensity of color produced below pH 8 varies with pH. Color intensity was constant from pH 8 to 9, and its development from reduced glutathione and cysteine was linear and constant for identical concentrations, and was not altered when water, 0.5% sodium dodecyl sulfate, or 8 *M* urea were substituted for methanol. The color developed to maximum intensity within 5 minutes and remained stable for 1 hour in the three media. Color production in dodecyl sulfate was

[12] G. Calcutt and D. Doxey, *Brit. J. Cancer* **16**, 562 (1962).

[13] J. Sedlak and R. H. Lindsay, *Anal. Biochem.* **25**, 192 (1968).

unaffected by pH in the range of 7.5–9 and that in 8 *M* urea was unaffected from pH 7–9. Maximum absorbance from bovine serum albumin and reduced glutathione was identical with methanol, 0.5% dodecyl sulfate, and 8 *M* urea; however, the values obtained from rat liver were considerably lower with urea than with dodecyl sulfate or methanol. In addition, high tissue blanks with both dodecyl sulfate and urea in rat liver indicate that methanol is the most suitable diluent for T-SH determination in tissue preparations. The recovery of sulfhydryl added as albumin, reduced glutathione or cysteine to T-SH and NP-SH ranged from 96 to 102%. Whereas both bovine serum albumin and reduced glutathione were measured in T-SH, bovine serum albumin was not detected in NP-SH, while reduced glutathione was quantitatively recovered in the NP-SH fraction.

Determination of Disulfide Bonds with DTNB after Reduction

Disulfide bonds play an important role in stabilizing the three-dimensional structure of proteins, and therefore a knowledge of their number, reactivity, and availability to denaturing agents is important. Methods have been developed in which the protein is first reduced, then assayed for liberated sulfhydryl groups by DTNB.

Reduction with Sodium Borohydride. Cavallini *et al.*[14] used the following method: To a test tube containing 1.44 g urea was added 0.1 *M* sodium EDTA, 0.5–1 ml of protein sample, 1 ml of 2.5% $NaBH_4$, water to 3 ml, and a drop of octyl alcohol as an antifoaming agent. The tubes were mixed at 38°, and reduction was allowed to proceed for 30 minutes at 38°, after which 0.5 ml of 1 *M* KH_2PO_4 containing 0.2 *N* HCl was added. After 5 minutes, 2 ml of acetone was added, then nitrogen was bubbled through for 5 minutes. Then 0.5 ml of 0.01 *M* DTNB was added, and the volume was made to 6 ml with water. Nitrogen was bubbled for 2 minutes and the tube was stoppered, ensuring that the gas space was filled with nitrogen. After the preparation had stood for 15 minutes, the absorbance was determined at 412 nm. Blanks containing all reactants except the protein solution were included. A molar absorptivity of 12,000 $M^{-1}cm^{-1}$ was used for calculating the number of sulfhydryl groups formed after reduction. With the exception of chymotrypsinogen, which gave low values, trypsin, ribonuclease, lysozyme, insulin, and bovine serum albumin gave satisfactory results.

Maeda *et al.*[15] developed a method for visualization of cystine-containing peptides in peptide maps. The chromatogram was sprayed

[14] D. Cavallini, M. T. Graziani, and S. Dupre, *Nature* (*London*) **212**, 294 (1966).
[15] H. Maeda, C. B. Glaser, and J. Meienhofer, *Biochem. Biophys. Res. Commun.* **39**, 1211 (1970).

with a freshly prepared 0.4% $NaBH_4$ solution in 95% ethanol. After 20 minutes or more, the paper was dipped in an acid reagent (acetic acid, 6 *N* HCl, and acetone 8:2:90) to decompose excess $NaBH_4$. The paper was air-dried for 1.5 hours, followed by exposure to ammonia vapors to neutralize the acid reagent. Excess ammonia was removed by air drying for 15 minutes, then the paper was sprayed with DTNB solution (0.1% DTNB in a 1:1 mixture of ethanol and 0.45 *M* Tris buffer, pH 8.2). Yellow spots appeared immediately. These were eluted and analyzed to establish the amino acid composition of the various cystine-containing peptides in lysozyme.

Reduction with Dithioerythritol. Zahler and Cleland[16] developed a method for assay of disulfide groups based on reduction with dithioerythritol and determination of the resulting monothiols with DTNB in the presence of arsenite. The arsenite forms a tight complex with dithiols but not with monothiols. The disulfide in 0.2 ml of solution is mixed with 0.1 ml of 0.05 *M* Tris pH 9 and 0.1 ml of 0.003 *M* dithioerythritol. Reduction is allowed to proceed for 20 minutes or for a period of time sufficient for the disulfide to be reduced. After reduction, 0.2 ml of 1 *M* Tris, pH 8.1, 1.5 ml of 0.005 *M* sodium arsenite, and water to give 2.9 ml are added, and the solution is mixed and allowed to stand for 2 minutes. DTNB is then added (0.1 ml of 0.003 *M* solution in 0.05 *M* acetate, pH 5) and the absorbance at 412 nm is recorded for at least 3 minutes. The absorbance resulting from the monothiols is determined by extrapolation of the linear portion of the curve to the time of addition of DTNB and subtraction of a blank value for a sample containing no disulfide. Solutions of cystine, pantethine, oxidized glutathione, and hydroxyethyl disulfide were completely reduced in 20 minutes. On the other hand, oxidized coenzyme A required 90 minutes at 0.009 *M* dithioerythritol. Proteins may give doubtful results, as Zahler and Cleland[16] found for bovine serum albumin tested with and without urea in the solution. The resulting polythiol peptide chains may form arsenite complexes that are too stable to permit assay by this method. Walsh *et al.*[17] adapted the method of Zahler and Cleland[16] for continuous detection of cystinyl peptides in the effluent of ion-exchange chromatograms of peptide mixtures.

Reduction with β-Mercaptoethanol. Reduction can be done in the absence and presence of varying concentration of denaturing agent. The resulting sulfhydryl groups are determined by DTNB. This serves

[16] W. L. Zahler and W. W. Cleland, *J. Biol. Chem.* **243,** 716 (1968).

[17] K. A. Walsh, R. M. McDonald, and R. A. Bradshaw, *Anal. Biochem.* **35,** 193 (1970).

as a measure of disulfide groups that are available to reduction in the native molecule and also the ease with which the disulfides become exposed as a function of concentration of the denaturing agent. The procedure is as follows[18]: To about 6 mg of protein contained in 4 ml of Tris·glycine buffer (0.043 *M* Tris and 0.046 *M* glycine) adjusted to pH 7 and containing 0.5 mg of EDTA is added 1 ml of 0.25 *M* β-mercaptoethanol in the same Tris·glycine buffer. Several 6-mg samples of the protein are reduced in the presence of varying concentrations of a guanidine salt. At 1, 3, and 6-hour intervals, 1.5 ml is withdrawn and added to 10 ml of 5% trichloroacetic acid to precipitate the protein. The precipitate is washed twice with 10 ml of 5% trichloroacetic acid to remove β-mercaptoethanol. Then the precipitate is dispersed with a glass rod and dissolved in 1 ml of 8 *M* urea in Tris·glycine, pH 8, containing 0.5 mg of EDTA per milliliter. The solution is diluted with 5 ml of 2% sodium dodecyl sulfate in Tris·glycine buffer, pH 8, containing 0.5 mg/ml EDTA. To 3 ml of the solution is added 0.1 ml of DTNB (4 mg per milliliter of Tris·glycine, pH 8) and the solution is read at 410 nm against a blank with no DTNB. A reagent blank is subtracted to give net absorbance at 410 nm. The concentration of protein is obtained from the absorbance at 280 nm of the remainder of the protein solution (to which no DTNB was added) and the known $E^{1\%}_{280\ \mathrm{nm}}$. The sulfhydryl content is calculated from a molar absorptivity of 13,600 $M^{-1}\mathrm{cm}^{-1}$ at 410 nm for 2-nitro-5-thiobenzoate anion. The use of Tris·glycine buffer[18] instead of the phosphate buffer previously used[19] is found to be advantageous when guanidine salts are used as denaturants. There is a loss of sulfhydryl groups with time and with increased guanidine salt concentration in phosphate buffer. This is eliminated in Tris·glycine buffer. It is likely that guanidine hydrochloride can contain impurities which react with sulfhydryl groups, but which are eliminated in the presence of Tris·glycine buffer.

The difference in susceptibility of disulfide bonds to reduction in a native protein and its modified counterpart form the basis of a method to detect conformational changes that result from chemical modification.[5,20] Such conformational changes were found to increase in bovine serum albumin (BSA) after modification, as follows: succinyl BSA > acetyl BSA > nitroguanyl BSA > guanyl BSA > BSA.[5,20] An increase in reducible disulfides accompanied the nitration of tyrosine-20

[18] M. Z. Atassi, A. F. S. A Habeeb, and L. Rydstedt, *Biochim. Biophys. Acta* **200**, 184 (1970).

[19] A. F. S. A. Habeeb, *Arch. Biochem. Biophys.* **121**, 652 (1967).

[20] A. F. S. A. Habeeb, *Fed. Proc. Fed. Amer. Soc. Exp. Biol. Abstr.* **24**, 224 (1965).

and tyrosine-23,[21] and the modification of 6 tryptophan residues[22] in lysozyme, behavior indicative of conformational changes. The sequence homology between lysozyme and α-lactalbumin was found not to be reflected in a conformational homology,[18] as the two proteins differ in the reducibility of their disulfide bonds and in the ease with which they unfold in guanidine hydrochloride solutions.

[21] M. Z. Atassi and A. F. S. A. Habeeb, *Biochemistry* **8**, 1385 (1969).
[22] A. F. S. A. Habeeb and M. Z. Atassi, *Immunochemistry* **6**, 555 (1969).

[38] The Rapid Determination of Amino Groups with TNBS

By Robert Fields

TNBS[1] was originally proposed by Okuyama and Satake[2] for determining amino acids and peptides and for protein modification studies, as the conditions required for its reaction were mild compared with those with ninhydrin,[3,4] and the reaction was more specific than that of aryl halides: no reaction with tyrosine or with histidine side chains was detected.

Sulfite is displaced from TNBS by an attacking nucleophile.

$$O_2N\text{-}C_6H_2(NO_2)_2\text{-}SO_3^- + {:}NH_2\text{—} \longrightarrow O_2N\text{-}C_6H_2(NO_2)_2\text{-}NH\text{—} + SO_3^{2-} + H^+$$

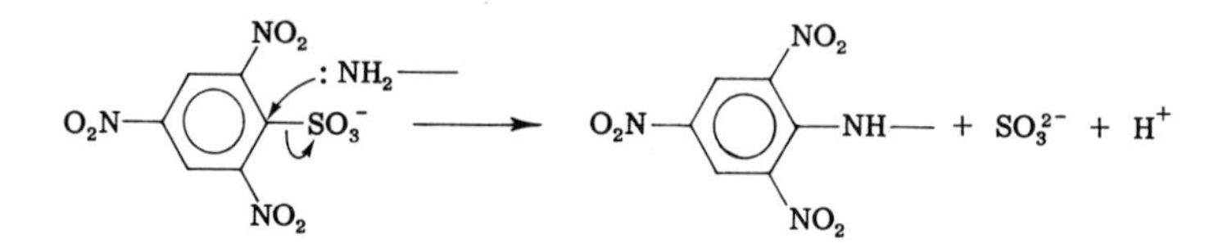

Its generation has presented a serious nuisance, since it associates reversibly with TNP-amino groups or TNP-thiol groups to form complexes whose absorption spectrum is altered, thus making quantitation of the reaction difficult. Reaction mixtures have been acidified to dissociate the sulfite complexes before reading,[5,6] empirical corrections have

[1] Abbreviations: TNBS, 2,4,6-trinitrobenzenesulfonic acid; TNP-, 2,4,6-trinitrophenyl-.
[2] T. Okuyama and K. Satake, *J. Biochem.* (*Tokyo*) **47**, 454 (1960).
[3] S. Moore and W. H. Stein, *J. Biol. Chem.* **211**, 907 (1954).
[4] H. Rosen, *Arch. Biochem. Biophys.* **67**, 10 (1957).
[5] K. Satake, T. Okuyama, M. Ohashi, and T. Shinoda, *J. Biochem.* (*Tokyo*) **47**, 654 (1960).
[6] A. F. S. A. Habeeb, *Anal. Biochem.* **14**, 328 (1966).

been applied to the absorption data,[7,8] or readings have been made at the isosbestic point for the TNP-amino group and its sulfite complex.[9] In the technique to be described,[10] however, measurements of the orange color of the sulfite complex are made at 420 nm. This wavelength is removed from the region of absorption of TNBS, thus permitting a high concentration of reagent to be used, to shorten the time required for reaction without increasing the blank readings. The reaction is carried out in borate buffer at pH 9.5; at this pH, however, there is a reaction with hydroxide ion to give a blank extinction; the latter reaction is stopped by lowering the pH to neutrality after the amino groups have been trinitrophenylated, or it is subtracted out continuously, using a split-beam recording spectrophotometer. A concentration of 0.5 to 2 mM sulfite has been found to be optimal for the full development of color associated with the TNP-amino group complexes.[10]

Materials

Recrystallization of TNBS. The quality of commercial samples of TNBS varies, but one recrystallization, carried out as follows, gave acceptable material in 50% yield. TNBS (1–25 g trihydrate or tetrahydrate) is dissolved in 1 part water (w/w) by heating, and HCl (sp gr 1.18) is then added to about 2 M. The fine crystalline solid that comes out on cooling is washed on the filter with cold 1 M HCl before drying in a desiccator. The appearance of the dry TNBS is flaky and white, resembling the trihydrate, and is unlike the large transparent yellow crystals of the tetrahydrate. When suitably pure, TNBS was found to have an ϵ_{340} not greater than 600 M^{-1}cm^{-1}.

Solutions. Note: As the oxidation of sulfite is catalyzed by traces of metal ions, all solutions should be made up in deionized, glass-distilled water.

A: 100 ml of 0.1 M Na_2SO_3 (made fresh weekly)
B: 1 liter of 0.1 M NaH_2PO_4
C: 1 liter of 0.1 M $Na_2B_4O_7$ in 0.1 M NaOH
D: 1.5 ml of solution A plus 98.5 ml of solution B (made fresh daily)
TNBS 1.1 M: 100–500 mg of TNBS is weighed accurately and

[7] A. R. Goldfarb, *Biochemistry* **5,** 2570 (1966); A. R. Goldfarb, *Biochemistry* **5,** 2574 (1966).
[8] R. B. Freedman and G. K. Radda, *Biochem. J.* **108,** 383 (1968).
[9] B. V. Plapp, S. Moore, and W. H. Stein, *J. Biol. Chem.* **246,** 939 (1971).
[10] R. Fields, *Biochem. J.* **124,** 581 (1971).

0.200 ml water added for each 100 mg TNBS. The solution is kept stoppered and is frozen when not in use.

Procedures

Procedure for the Determination of Amino Groups. The sample containing amino groups is added to 0.5 ml of borate buffer (solution C) and the volume is made up to 1.0 ml. Then, 0.02 ml of 1.1 M TNBS solution is added and the solution is rapidly mixed. After 5.0 minutes the reaction is stopped by adding 2.0 ml of 0.1 M NaH_2PO_4 which contains 1.5 mM sulfite (solution D), and the absorbance at 420 nm is determined. A blank is also prepared. When multiple determinations are made, it is convenient to add TNBS by microsyringe to successive samples at 15-second intervals, mixing the tubes in between. The reactions are stopped after 5 minutes in the same order at 15-second intervals, using a repeating 2.0-ml pipette.

Values for the molar absorptivity of TNP-α-amino groups are 22,000 $M^{-1}cm^{-1}$, for TNP-ϵ-amino groups, 19,200 $M^{-1}cm^{-1}$, and for TNP-thiol groups, 2250 $M^{-1}cm^{-1}$. The table shows the molar absorptivities of compounds determined by this procedure.

Procedure for Determining the Rate of Reaction of TNBS with Amino Groups. To 2.0 ml of a solution of substances that contain amino groups in 0.05 M $Na_2B_4O_7$ in 0.05 M NaOH (1.0 ml solution C, plus 1.0 ml sample) in a cuvette of 1-cm path length is added 20 μl of 0.1 M Na_2SO_3 (solution A); this addition is also made to a cuvette containing 2.0 ml of the diluted buffer. Then, 20 μl of 1.1 M TNBS solution is added

MOLAR ABSORBANCE COEFFICIENTS AT 420 NM AND RATES OF REACTION OF TNBS WITH VARIOUS COMPOUNDS

Compound		ϵ_{420} ($M^{-1}cm^{-1}$)[a]	Half-life (sec)[b]
N-Acetyl-L-cysteine		2,250	0.92
L-Alanine		22,000	36
Glycylglycine		22,400	29
N^α-Acetyl-L-lysine		19,150	31
Glycyl-L-lysine	α-amino group	22,000	23
	ϵ-amino group	19,200	105
L-Lysine	α-amino group	21,900	—
	ϵ-amino group	13,500	—

[a] Determined by the first method.

[b] Determined by the second method. It should be noted that the concentration of TNBS used in the second method is half that used in the first; hence the reaction times are twice as long.

to each cuvette. The solutions are rapidly mixed by inversion and placed in sample and reference compartments of a split-beam recording spectrophotometer, and the differences in extinction at 420 nm are recorded.

Values for the molar absorptivity of trinitrophenylated groups at pH 9.5 are about 5% higher than those determined by the preceding method. The table shows the rates of reaction with typical amino and thiol groups under these conditions.

Analysis of the Curves. Since the concentration of TNBS is much greater throughout the reaction than that of amino groups, it may be assumed that each class of amino group reacts according to pseudo first-order kinetics. A plot of log amino groups remaining vs. time should, therefore, give a straight line when only one amino group is involved, or a curved line if there is more than one group or class of groups that possess different kinetic constants. Graphical techniques have been described[7,10] for resolving component reactions, and these may be of use in analyzing the results of a reaction. In favorable cases, the pK of a reacting group may be determined by analyzing the experimental curves for constituent reactions, at several pH values.[10]

Comments

A check should be made that the reaction of TNBS with the amino groups that are being studied goes to completion in 5 minutes. Although the slowest reacting amino group shown in the table, the ϵ-amino group of glycyl-L-lysine, would be expected to undergo more than 98% reaction, groups that are partly buried in proteins may react more slowly. The length of incubation can be extended to 10 or 20 minutes without greatly affecting the accuracy of the assay, or a higher concentration of TNBS may be used. A solution of 1.8 M TNBS may be prepared by dissolving the solid in an equal part (w/v) of water with warming. This solution forms crystals on standing at room temperature, and should be heated to 40° before use to avoid the possibility of its crystallizing in a microsyringe or pipette.

Millimolar sulfite imparts a pink color to blank solutions used in the assay, by forming complexes with the TNBS. This color is a convenient indication that the sulfite concentration is adequate; however, the exact concentration of sulfite may be determined using Ellman's reagent.[11,12]

Precautions. Sulfite complexes of TNP-amino groups may interact with each other or with aromatic side chains in proteins to decrease the molar absorbances of these groups. The hypochromic interaction of the

[11] G. L. Ellman, *Arch. Biochem. Biophys.* **82,** 70 (1959).

[12] R. E. Humphrey, M. H. Ward, and W. Hinze, *Anal. Chem.* **42,** 698 (1970).

TNP-N$^\alpha$- and TNP-N$^\epsilon$-amino groups of L-lysine (see the table) provides a simple illustration of this. Such interactions make the interpretation of rate data difficult, since the molar absorbance value of an individual group changes during the course of the reaction. In proteins, any local or generalized unfolding of the protein structure that occurs during trinitrophenylation, e.g., as a result of the introduction of hydrophobic TNP-groups, may easily increase or decrease the rate of reaction of TNBS with partly buried residues. Accurate quantitative analysis of experimental curves in cases in which the reaction of TNBS with specific groups of a protein leads to an alteration in the rate of reaction with other groups is almost certain to be impracticable, owing to the distribution of modified species at any one time. The "kinetic sets" obtained by analysis of such curves cannot correspond truly to groups in the protein.

In view of the likelihood of one or more types of interference occurring during the trinitrophenylation of a protein of average size, the interpretation of absorption data should be cautious. Reactive groups that have been found by kinetic analysis ought to be shown to change their rate of reaction in a way that indicates that the group possesses a fixed pK value. If this is not possible, characterization of the groups, by peptide mapping[13,14] or by other chemical means seems desirable.

Sulfite is known to cleave disulfides,[15] and its presence in a reaction mixture with a protein at high pH means that the possibility of this reaction should not be ignored. A simple check that cleavage has not occurred would consist of a comparison of the results obtained in the presence of sulfite with those from an experiment in which no sulfite has been added; for example, the value obtained by the first procedure, in which sulfite is added only after the reaction with TNBS is complete, may be compared with that obtained by the second method.

[13] P. J. Anderson and R. N. Perham, *Biochem. J.* **117,** 291 (1970).
[14] B. S. Hartley, *Biochem. J.* **119,** 805 (1970).
[15] R. Cecil and J. R. McPhee, *Biochem. J.* **60,** 496 (1955).

[39] Modification of Proteins with Active Benzyl Halides

By H. R. Horton and D. E. Koshland, Jr.

In addition to such alkyl halides as iodoacetate and iodoacetamide, which are discussed elsewhere in this volume [34a], certain reactive benzyl halide derivatives have found application in recent years in protein modification studies.

To a considerable extent, earlier use of benzyl halides as protein alkylating reagents was complicated by their sluggish reactivities under mild conditions and the insolubility of reactants and products in aqueous media. For example, alkylation of the sulfhydryl group of glutathione with benzyl chloride was found to occur slowly and incompletely under conditions in which arylation with 2,4-dinitrochlorobenzene proceeded rapidly.[1]

The finding that the reactivity of benzyl halides could be enormously enhanced and their specificity sharpened by the introduction of an *ortho* hydroxyl substituent[2,3] has led to the synthesis of derivatives of considerable utility as protein reagents and reporter groups. The compound 2-hydroxy-5-nitrobenzyl bromide (HNB-Br)[4] was found to react very rapidly under mild conditions and to show a marked selectivity for tryptophan.[2,3] Under appropriate conditions of pH in the absence of sulfhydryl groups, the reagent appears to be completely specific for tryptophan. In strongly alkaline solutions, at pH values near or above the pK_a of the phenolic hydroxyl group, 2-hydroxy-5-nitrobenzyl bromide can also alkylate tryrosine,[3] and evidence has been presented for the partial modification of the α-amino group of carboxypeptidase A during its treatment with 170:1 molar ratio of HNB-Br at neutral pH.[5] Nevertheless, there is sufficient selectivity of HNB-Br for tryptophan under most conditions to provide a means for the quantitative determination of tryptophyl residues in proteins.[6]

The reactivity and selectivity of 2-hydroxy-5-nitrobenzyl bromide which have made it a useful protein reagent have led to investigations of related compounds. 2-Hydroxy-5-nitrobenzyl chloride[7] displays the same reactivity characteristics as the bromide. In contrast, 2-methoxy-5-nitrobenzyl bromide was found to have the more sluggish reactivity of a classical benzyl halide.[8] 2-Hydroxy-3,5-dinitrobenzyl chloride possesses considerably greater reactivity than the classical benzyl halides, but was found to be somewhat less reactive with tryptophan than either HNB-Br or HNB-Cl.[9] A class of reagents in which a neighboring hy-

[1] B. C. Saunders, *Biochem. J.* **28,** 1977 (1934).

[2] D. E. Koshland, Jr., Y. D. Karkhanis, and H. G. Latham, *J. Amer. Chem. Soc.* **86,** 1448 (1964).

[3] H. R. Horton and D. E. Koshland, Jr., *J. Amer. Chem. Soc.* **87,** 1126 (1965).

[4] The abbreviation used is: HNB-, 2-hydroxy-5-nitrobenzyl-.

[5] T. M. Radhakrishnan, R. A. Bradshaw, D. A. Deranleau, and H. Neurath, *FEBS (Fed. Eur. Biochem. Soc.) Lett.* **7,** 72 (1970).

[6] T. E. Barman and D. E. Koshland, Jr., *J. Biol. Chem.* **242,** 5771 (1967).

[7] C. A. Buehler, F. K. Kirchner, and G. F. Deebel, *Org. Syn.* **20,** 59 (1940).

[8] H. R. Horton, H. Kelly, and D. E. Koshland, Jr., *J. Biol. Chem.* **240,** 722 (1965).

[9] H. R. Horton and G. Young, *Biochim. Biophys. Acta* **194,** 272 (1969).

droxyl group is "masked" and then generated *in situ* is exemplified by 2-acetoxy-5-nitrobenzyl chloride. Acetoxynitrobenzyl chloride, like the methoxy analog, was found to react very sluggishly in acidic solutions. However, increased concentrations of OH^- catalyze a pH-dependent hydrolysis with the liberation of 2-hydroxy-5-nitrobenzyl alcohol. Another class of tryptophan-selective reagents analogous to the 2-hydroxy-5-nitrobenzyl halides is the water-soluble dimethyl-(2-hydroxy-5-nitrobenzyl)sulfonium salts.[10] The properties of each of these reagents provide a fairly good picture of their applicability and usefulness in protein modification studies.

2-Hydroxy-5-nitrobenzyl Bromide

Preparation. The synthesis of 2-hydroxy-5-nitrobenzyl bromide (or chloride) is based on the halomethylation procedure of Buehler *et al.*[7] A mixture of *p*-nitrophenol (0.18 mole), methylal or paraformaldehyde (0.27 mole), 2 ml of 36 *N* sulfuric acid, and 226 g of 48–50% hydrobromic acid is vigorously stirred at 66 ± 1° for 2 hours. (The temperature is critical: significant quantities of the bisbromomethyl derivative are formed when bromomethylation is allowed to occur at temperatures above 70°, and at lower temperatures the normal by-product of the reaction, 6-nitro-1,3-benzodioxan, is formed as the major product.[11,12] Considerable quantities of this contaminant have been detected in some commercially available preparations of HNB-Br.) The desired product begins to precipitate from the hot, clear solution after the first 10–15 minutes of reaction. After reaction is complete, the mixture is cooled and the HNB-Br is collected and extracted into methylene chloride (or diethyl ether). The extract is dried over anhydrous magnesium sulfate, after which the organic solvent is removed by flash evaporation. The isolated compound is recrystallized 2 or 3 times from hot benzene solution; it forms long, off-white needles, mp 146–147°.

Procedure for 2-Hydroxy-5-nitrobenzyl Bromide as a Protein Reagent. Since the reagent is relatively insoluble in water but highly reactive, it is convenient to dissolve it in an organic solvent that is inert to the reagent but miscible with water. The reagent solution can then be added with continuous mixing to an aqueous solution of the protein to be treated. In order to avoid partial solvolysis of the reagent, solvents such as methanol should not be used, although some studies have employed

[10] H. R. Horton and W. P. Tucker, *J. Biol. Chem.* **245**, 3397 (1970).
[11] F. D. Chattaway and R. M. Goepp, *J. Chem. Soc.*, 699 (1933).
[12] W. P. Tucker and H. R. Horton, unpublished results (1966).

this solvent.[2,13–15] The reagent has frequently been dissolved in acetone for use in protein labeling studies with apparent success,[3,5,6,15–19] but isolated hydroxynitrobenzyl adducts of substituted indoles undergo rearrangement when dissolved in acetone, even at low temperatures.[12] Such changes were not observed in dimethoxyethane or in dioxane, both of which have proved to be convenient solvents and can be added to numerous aqueous protein solutions without appreciable denaturation. The preferred concentration of the reagent in the solvent and particular solvent chosen can be ascertained by appropriate screening tests.

Since strong acid is liberated during the course of reaction, pH is maintained by adding sodium hydroxide, either automatically with a pH-stat, or manually in small increments. Final solvent concentrations of 3–15% and final reagent concentrations of up to 0.01 *M* have been successful in labeling a variety of proteins under "nondenaturing" conditions. In such studies, the organic solution of reagent (e.g., 0.2 *M*) is best prepared with dried solvent immediately before use and is protected from strong light. In other studies, solid reagent has been added to aqueous solution; reaction proceeds more slowly as reagent enters solution, and 2-hydroxy-5-nitrobenzyl alcohol, which is formed by hydrolysis, precipitates.[17,20] The reagent reacts extremely rapidly, either with protein or water, and is consumed in approximately 1 minute after dissolution.

The modified protein may be separated from the excess yellow compound by one or a combination of standard separation techniques. Of general applicability has been the gel filtration technique through appropriately equilibrated columns of dextran, e.g., Sephadex G-25. The dextran gel tends to adsorb 2-hydroxy-5-nitrobenzyl compounds as indicated by the delay in elution (see Fig. 1), and this property can be used to enhance the degree of separation. Extensive dialysis has frequently been employed, often in conjuction with gel filtration (see, for example, footnote 17 reference). Other methods of separation include differential precipitation of protein and excess alcohol, e.g., with ammonium sulfate, precipitation with organic solvent in which excess HNB-

[13] B. R. DasGupta, E. Rothstein, and D. A. Boroff, *Anal. Biochem.* **11**, 555 (1965).

[14] D. A. Boroff and B. R. DasGupta, *Biochim. Biophys. Acta* **117**, 289 (1966).

[15] D. Griffin, D. K. Tachibana, B. Nelson, and L. T. Rosenberg, *Immunochemistry* **4**, 23 (1967).

[16] R. F. Steiner, *Arch. Biochem. Biophys.* **115**, 257 (1966).

[17] K. Yamagami and K. Schmid, *J. Biol. Chem.* **242**, 4176 (1967).

[18] T. A. A. Dopheide and W. M. Jones, *J. Biol. Chem.* **243**, 3906 (1968).

[19] V. H. Paetkau, E. S. Younathan, and H. A. Lardy, *J. Mol. Biol.* **33**, 721 (1968).

[20] N. B. Oza and C. J. Martin, *Biochem. Biophys. Res. Commun.* **26**, 7 (1967).

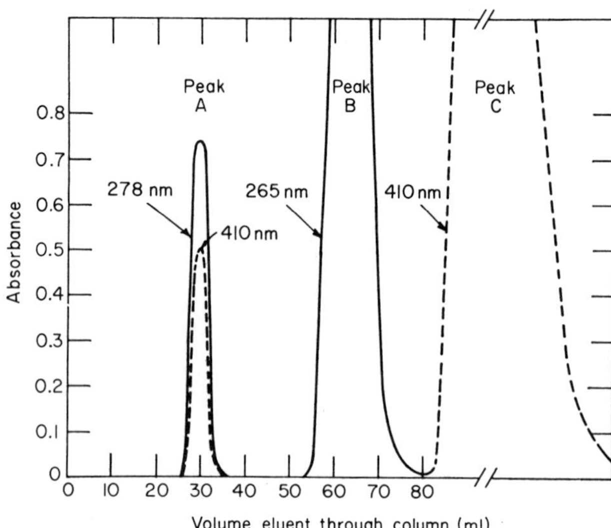

FIG. 1. Separation of 2-hydroxy-5-nitrobenzyl-substituted phosphoglucomutase from other components of reaction mixture on dextran gel. Conditions: 20 × 2-cm column of Sephadex G-25 (fine beads), equilibrated with 0.01 *M* Tris·chloride buffer (pH 7.5), and eluted with the same buffer. Peak A: labeled protein, absorbance maxima at 278 nm and 320:410 nm. Peak B: acetone from reaction mixture, absorbance maximum at 265 nm. Peak C: 2-hydroxy-5-nitrobenzyl alcohol, maxima at 320:410 nm.

alcohol remains dissolved, or reversible precipitation of the protein through pH adjustments.[19] Ion-exchange chromatography and precipitation with denaturants, such as trichloroacetic acid, may also be employed.

Determination of Residues Reacted. The number of *p*-nitrophenolic groupings introduced into the protein can sometimes be estimated by absorbance measurements alone, but in other cases requires further assays. The ionized form, the *p*-nitrophenoxide ion, absorbs strongly in the 410-nm region of the spectrum in which polypeptide structures are transparent (see Fig. 2). The molar absorptivity of the ionized form of the monosubstitution product of the reaction of tryptophan ethyl ester hydrochloride with HNB-Br has been determined to be 18,450 $M^{-1}cm^{-1}$ at 410 nm in 2 *N* NaOH.[21] The nonionized form of the *p*-nitrophenol group in 2-hydroxy-5-nitrobenzyl alcohol exhibits maximum absorbance at 320 nm with an absorptivity of approximately 9600 $M^{-1}cm^{-1}$. To avoid complications arising from the absorbance of the protein, the degree of labeling is best estimated by measuring the absorbance in the 410 nm region at a pH greater than 10.

[21] G. M. Loudon and D. E. Koshland, Jr., *J. Biol. Chem.* **245**, 2247 (1970).

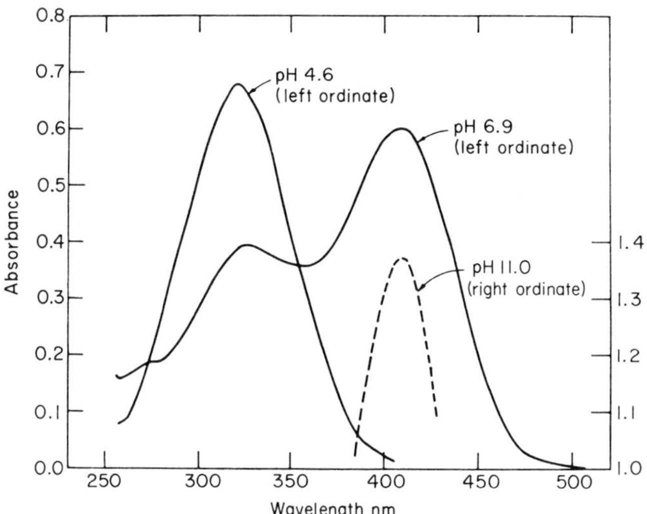

FIG. 2. Absorption spectra of 7.6×10^{-5} *M* 2-hydroxy-5-nitrobenzyl alcohol at various pH values.

The number of hydroxynitrobenzyl groups thus determined can be related to the protein concentration in several ways. A first approximation of total protein concentration can often be made by measuring its absorbance in the 278–280 nm region. This is satisfactory provided the number of nitrophenol groups introduced is low. Correction should be made for the contribution of the hydroxynitrobenzyl grouping to the 280 nm region absorbance, and is readily determined from appropriate absorbance measurements and reference to the HNB-OH spectrum (Fig. 2). A more reliable determination of protein concentration, particularly in instances in which the number of HNB groupings introduced is relatively large, can be made by amino acid analysis of acid hydrolyzates of suitable aliquots[22] or micro-Kjeldahl assays.[23]

To ascertain which amino acid residues are modified, the specificity properties of the reagent are helpful. In proteins lacking free SH groups, such as chymotrypsin, the number of tryptophan residues modified agrees with the number of nitrophenol groups incorporated as determined by spectrophotometry. 2-Hydroxy-5-nitrobenzyl bromide has also been shown to react with cysteine, but to a considerably smaller extent than it does with tryptophan.[3] To prevent such hydroxynitrobenzylation of sulfhydryl groups, SH-containing proteins have been carboxymethylated[6]

[22] D. H. Spackman, W. H. Stein, and S. Moore, *Anal. Chem.* **30**, 1190 (1958).
[23] D. J. Jenden and D. B. Taylor, *Anal. Chem.* **25**, 685 (1953).

and carbamidomethylated[19] prior to treatment with HNB-Br. However, it has been observed that the SH groups of a number of proteins do not react to a significant extent with 2-hydroxy-5-nitrobenzyl bromide at pH 2.7 in 10 *M* urea.[6] Accordingly, at least in acidic media, the prior blocking of SH groups in many proteins may be unnecessary.

Specific analyses can be performed to determine the extent to which cysteine residues in a given protein may have reacted with HNB-Br. The *S*-HNB-cysteine peak emerges from the short column of the amino acid analyzer[22] in the same region as histidine under standard conditions of analysis.[3] Determination of unreacted cysteine (and half-cystine) residues can be made by oxidizing the HNB-labeled protein with performic acid[24] and determining the amount of cysteic acid produced.[3] An alternate procedure is to treat the HNB-labeled protein with iodoacetate at pH 9 and to quantitate the *S*-carboxymethylcysteine formed.[10]

In alkaline solution it was found that tyrosine was modified as well as cysteine and trytophan.[3] In one instance, incomplete recovery of an N-terminal residue, aspartic acid, was obtained after treatment of carboxypeptidase A with HNB-Br.[5] Guanidine hydrochloride changes the specificity of the reagent in unknown ways and cannot be used when the reagent is being employed as a tryptophan modifier.[6] Accordingly, when proteins are treated with 2-hydroxy-5-nitrobenzyl bromide in neutral or alkaline solutions, complete amino acid analyses are required for unequivocal determination of the amino acid residues modified.

Procedure for Quantitative Determination of Tryptophan in Proteins. The specificity of 2-hydroxy-5-nitrobenzyl bromide for tryptophan in acidic media has led to its application in the quantitative estimation of tryptophan residues in a variety of proteins.[6,13,17,19,25] The following procedure has been shown to provide accurate results with a wide variety of proteins.[6]

The protein to be analyzed is allowed to "unfold" by incubation for 16–20 hours at 37° in 1 ml of 10 *M* urea (recrystallized from ethanol–water) adjusted to pH 2.7 with concentrated HCl. After incubation, the solution is cooled to room temperature and approximately 5 mg of HNB-Br in 0.1 ml of dried acetone is added, with continuous stirring, from a pipette immersed below the surface of the protein solution. During addition, the urea-containing solution usually remains clear; however, any 2-hydroxy-5-nitrobenzyl alcohol which precipitates can be readily removed by centrifuging.

The HNB-labeled protein is then separated from excess reagent by

[24] S. Moore, *J. Biol. Chem.* **238**, 235 (1963).

[25] T.-L. Chan and K. A. Schellenberg, *J. Biol. Chem.* **243**, 6284 (1968).

gel filtration through a column of Sephadex G-25 (23×1.1 cm) which has been previously equilibrated with 0.18 M acetic acid (pH 2.7) or 10 M urea (pH 2.7). (The urea is employed in those cases in which it is required to keep the modified protein in solution.) It is convenient to collect 1-ml fractions at flow rates of 100 ml per hour or greater. Fractions containing protein (usually 1–4 ml at elution volume) are pooled, and the protein is precipitated by addition of 50% trichloroacetic acid to a final concentration of 5%. (In the presence of 10 M urea, a 5-fold dilution with water is necessary before addition of the trichloroacetic acid.) Precipitation is usually complete within 30 minutes; however, the mixture can be stored overnight at 4° at this stage if desirable. The precipitate is collected by centrifuging and washed twice with 5 ml of ethanol–HCl (2 ml of concentrated HCl to 98 ml of 95% ethanol). The washed precipitate is dissolved in 1 ml of concentrated HCl. An aliquot (usually 0.1 ml) is adjusted to pH > 12 by addition of 2.5 M NaOH and diluted to 2.5 ml. The concentration of HNB groups in this solution is determined spectrophotometrically by measuring its absorbance at 410 nm (molar absorptivity 18,450 M^{-1}cm^{-1}). The remaining portion of the 12 M HCl solution can be analyzed for protein content by standard methods.[6,22,23] When acid hydrolysis is part of the procedure being used, it is convenient to transfer a measured volume of the concentrated HCl solution of labeled protein to the hydrolysis tube and then to add an equal volume of distilled water prior to evacuation and sealing of the tube.

Nature of Reaction with Tryptophan Derivatives. The utility of substituted benzyl halides as protein reagents ultimately depends upon an understanding of the nature of their reactivities with residues in proteins. While reaction of HNB-Br with cysteine appears to be a straightforward alkylation of the thiol group, reaction with tryptophan and its derivatives is complex, and has received study in several laboratories.[21,25-30]

Initial reaction of HNB-Br or HNB-Cl with indole or substituted indoles is at the 3 position.[21,25-30] With indole this can form a stable adduct (I) (Fig. 3), but with 3-substituted indoles, such as tryptophan

[26] T. F. Spande, M. Wilchek, and B. Witkop, *J. Amer. Chem. Soc.* **90**, 3256 (1968).

[27] M. Wakselman, G. Decodts, and M. Vilkas, *C. R. Acad. Sci. Ser. C* **266**, 1089 (1968).

[28] G. M. Loudon, D. Portsmouth, A. Lukton, and D. E. Koshland, Jr., *J. Amer. Chem. Soc.* **91**, 2792 (1969).

[29] B. G. McFarland, Y. Inoue, and K. Nakanishi, *Tetrahedron Lett.* **11**, 857 (1969).

[30] J. Wang, J. Zuniga, W. P. Tucker, and H. R. Horton, unpublished results; J. Zuniga, M. S. Thesis, North Carolina State University, Raleigh, 1968; H. R. Horton, *N.I.H. Symp. Recent Develop., 18th Bethesda, 1968,* p. 9.

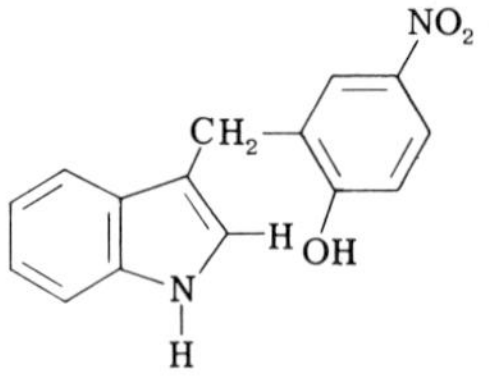

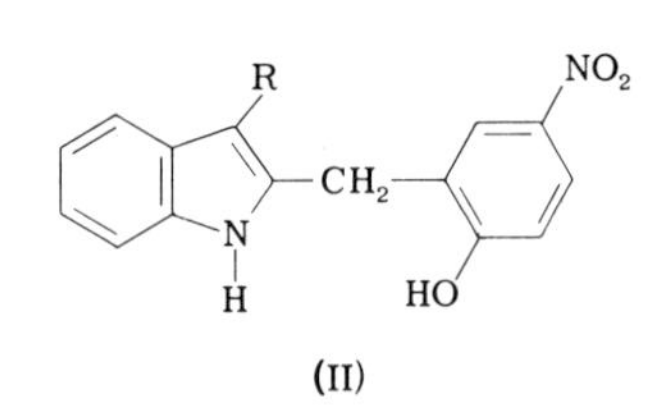

FIG. 3. Structures of substituted indole and tryptophan compounds.

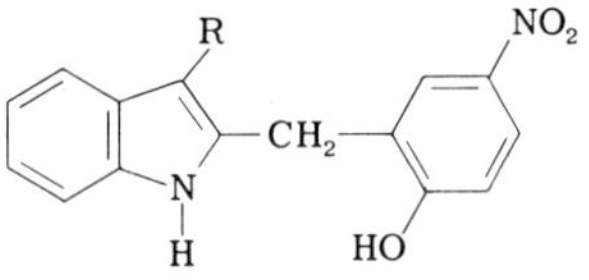

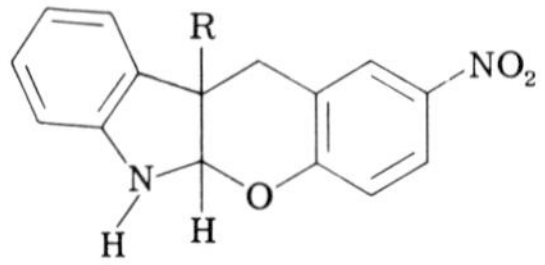

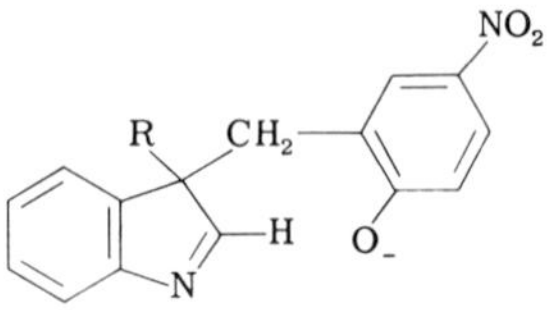

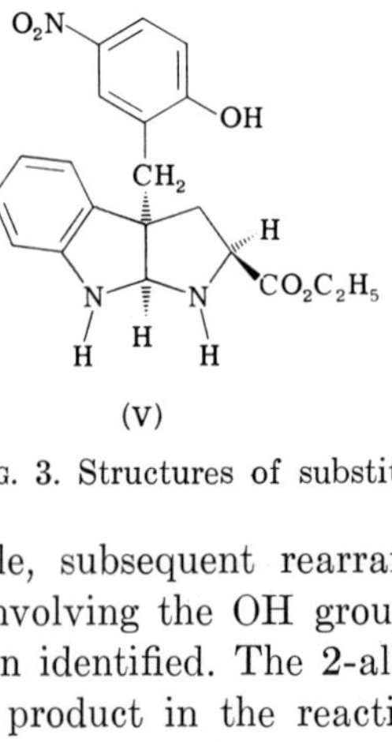

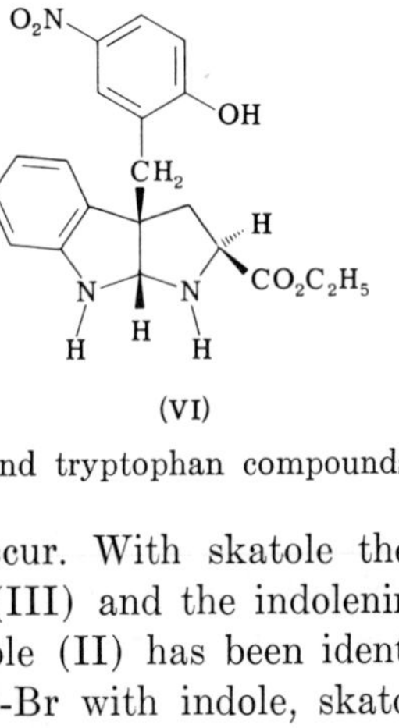

or skatole, subsequent rearrangements occur. With skatole the cyclic adduct involving the OH group of HNB (III) and the indolenine (IV) have been identified. The 2-alkylated indole (II) has been identified as a minor product in the reaction of HNB-Br with indole, skatole, and tryptophan. This derivative can be formed by thermal- or acid-catalyzed isomerization of compounds containing the benzyl group on the 3 position, e.g., compound (III). When tryptophan contains a free amino group, the cyclic diastereomeric products (V) and (VI) are obtained. Thus, the possibilities for reaction of 2-hydroxy-5-nitrobenzyl bromide with tryptophyl residues in proteins involve initially two diastereomeric modes of addition to position 3 of a given indole nucleus (one of which may be favored by a given residue's molecular environment), followed by rearrangements of the resulting indolenines. *O*-Cyclic derivatives (analogous to III) may be produced through attack by the 2-hydroxyl group, or *N*-cyclic structures (V and VI) through attack of the tryptophyl residue's α-amino nitrogen. Depending on the indolenine's environment, an acid-catalyzed migration to form a substituted indole (II) may also ensue.

Evidence for the formation of disubstitution products upon treatment of trytophan or its analogs with HNB-Br suggests that such additional substitution of tryptophyl residues in proteins may also occur when excesses of reagent are added.[6]

2-Methoxy-5-nitrobenzyl Bromide

Preparation. The preparative procedure for 2-methoxy-5-nitrobenzyl bromide parallels that of the 2-hydroxy derivative, except that *p*-nitroanisole is used as the starting material.[8] The product isolated is twice crystallized from heated benzene–petroleum ether solutions, resulting in white needles, mp 78–79°.

Procedure for Modifying Amino Acid Residues. In general, the procedures for modifying the amino acids in proteins are similar to those described for the hydroxy derivative, except that the time required for modification is far greater since the methoxyl reagent is less reactive. It is convenient to dissolve the reagent in a small amount of organic solvent, partly to facilitate its introduction into the aqueous protein solution and also, by adding organic solvent to the aqueous solution, to increase the solubility of the methoxynitrobenzyl bromide. With free amino acids at 10^{-3} *M* in a 5% acetone solution containing 0.01 *M* 2-methoxy-5-nitrobenzyl bromide, 57% of the tryptophan had reacted in 18 hours and 70% in 89 hours. In the same experiment 53% of the methionine had reacted in 18 hours and 63% in 89 hours. Cysteine also reacts with this reagent. Thus, methylation of the hydroxyl group changes the reactivity of the substituted benzyl halide back to that of a conventional benzyl bromide.[8]

Separation of unreacted reagent can be performed in a manner similar to the separations outlined for the reactive hydroxy compound. One notable difference exists, however. In the case of 2-hydroxy-5-nitrobenzyl halides, excess reagent reacts readily with water or hydroxide ions present in solution. Thus, once separation commences, there is little further chance for reaction of unreacted reagents. By contrast, 2-methoxy-5-nitrobenzyl bromide is so sluggishly reactive that, in the initial phases of separation of treated protein, excess unreacted reagent is usually present and precautions against further reaction during separation must be taken.

Since 2-methoxy-5-nitrobenzyl groups absorb in the 320-nm region there is considerable overlap with the absorption of the protein itself.[8] Consequently, the total number of residues introduced is best ascertained by amino acid analyses unless, of course, the reagent is made radioactive.

Usefulness of the Reagent. 2-Methoxy-5-nitrobenzyl bromide complements the 2-hydroxy derivative. It reacts in a fashion similar to a classical benzyl halide, forming a stable methionine product. It also reacts with tryptophan and cysteine like the 2-hydroxy derivative (though on a different time scale) and therefore can probably be used in conjunction with 2-hydroxy-5-nitrobenzyl halides for identification of residues playing a role in enzyme action. It has the further advantage that its absorption spectrum is not sensitive to pH, but is sensitive to changes in the polarity of the environment.[8] Significant shifts occur on changing the solvent from hexane to 50% aqueous dioxane. 2-Methoxy-5-nitrobenzyl bromide can thus be used as an environmentally sensitive probe in the study of conformational changes or other properties of proteins. The sluggish reactivity of this reagent makes it, in general, less valuable than the 2-hydroxy derivatives, but its spectral characteristics render it more valuable than the simple unsubstituted benzyl halides, which possess similar reactivities and specificity patterns.

2-Acetoxy-5-nitrobenzyl Chloride

Preparation. 2-Acetoxy-5-nitrobenzyl chloride is synthesized by treating 2-hydroxy-5-nitrobenzyl chloride with 20-fold molar excess of acetyl chloride under reflux.[9] After reaction is complete (thin-layer chromatography using silica gel/benzene), excess acetyl chloride is removed by evaporation, and the product is twice crystallized from warm benzene, resulting in white prisms, mp 76–78°.

Procedure for 2-Acetoxy-5-nitrobenzyl Chloride as an Enzyme-Modifying Reagent. The acetoxy reagent is designed to introduce the 2-hydroxy-5-nitrobenzyl chromophore selectively at tryptophyl residues in the vicinity of the active sites of enzymes such as chymotrypsin. The acetoxy derivative, like the methoxy derivative, reacts sluggishly, but acylation of the active site liberates HNB-Cl, which then has the full reactivity of the hydroxy compound. It will then react most readily with tryptophan residues in the immediate environment, i.e., near the active site. Conditions are selected such that the enzymatic hydrolysis of the nitrophenyl ester, to liberate the nitrophenol (2-hydroxy-5-nitrobenzyl chloride in this case) at the active site is maximum with respect to the nonspecific base-catalyzed hydrolysis.[31,32] In the case of chymotrypsin, such

[31] H. Gutfreund and J. M. Sturtevant, *Biochem. J.* **63**, 656 (1956).

[32] F. J. Kezdy and M. L. Bender, *Biochemistry* **1**, 1097 (1962).

selective labeling is best achieved at pH values between 7.2 and 8.0, with final concentrations of dimethoxyethane of 8–16%. At higher pH values, greater hydroxide ion-catalyzed hydrolysis occurs, leading to low yields of tryptophan modification. On the other hand, low pH results in low enzyme-catalyzed displacement of the acetate blocking group, permitting the slower and less selective modification of other amino acid residues by the acetoxy reagent. Modified protein is then separated from excess reagent (largely in the form of 2-hydroxy-5-nitrobenzyl alcohol) by means of gel filtration and dialysis procedures as outlined for the HNB-Br-treated proteins.

Usefulness of Acylated Benzylating Reagents. Like 2-hydroxy-5-nitrobenzyl halides, the 2-acetoxy derivative can be utilized to incorporate the environmentally sensitive *p*-nitrophenolic reporter group into protein structures. In general, the usefulness of a conformational probe depends upon its being selectively incorporated into a protein's structure at only one or a few sites. In the case of enzymes which possess nitrophenyl ester-hydrolyzing activities and contain rather large numbers of tryptophyl residues, acylated benzylating reagents offer a means of selective labeling of those residues in the vicinity of the active site at which the highly reactive 2-hydroxy-5-nitrobenzyl species is generated. The altered selectivity of such reagents, as compared to 2-hydroxy-5-nitrobenzyl halides, may also provide a further means of assessing the role of tryptophyl residues in substrate binding and enzyme catalysis.

Dimethyl-(2-hydroxy-5-nitrobenzyl)sulfonium Salts

Preparation. Dimethyl-(2-hydroxy-5-nitrobenzyl)sulfonium chloride is readily prepared by stirring 2.28 g of 2-hydroxy-5-nitrobenzyl chloride with 20 ml of dimethyl sulfide for 6 hours at room temperature.[10] Considerable product precipitates from solution during this period, and the rest is precipitated by adding 400 ml of diethyl ether. The solid is collected, washed well with ether, and dried, then dissolved in warm methanol and reprecipitated with ether. The resulting off-white material melts at 152–153°, decomposing with evolution of dimethyl sulfide. In a similar manner, dimethyl-(2-hydroxy-5-nitrobenzyl)sulfonium bromide is prepared by treating 2-hydroxy-5-nitrobenzyl bromide with dimethyl sulfide. The product melts at 172–173° with evolution of dimethyl sulfide. An alternate route to the preparation of such sulfonium salts involves synthesis of 2-hydroxy-5-nitrobenzyl methyl sulfide and its subsequent methylation.[10]

Procedure for Modifying Amino Acids in Proteins. The major advantage of the hydroxynitrobenzylsulfonium salts as protein modification reagents over the hydroxynitrobenzyl halides is the increased solubility of the reagent in the absence of organic solvents. Two methods of adding the sulfonium reagents have been found to give satisfactory results with chymotrypsin and carboxypeptidase A preparations.[10] The first involves weighing a suitable quantity of the reagent and adding it as a solid, with continuous stirring, to a solution of protein. During reaction the desired pH is maintained either by including appropriate buffers or by adding base from a pH-stat. (In the case of proteins such as carboxypeptidase A, which require high salt concentrations to avoid precipitation or denaturation, the solution of alkali to be added can be prepared with sufficient salt to provide constant ionic strength during the course of reaction.)

An alternative procedure involves dissolving the hydroxynitrobenzylsulfonium salt in water (or nonbuffered salt solution of appropriate ionic strength) and adding an appropriate volume to the solution of protein maintained at the desired pH. The success of this procedure results from the stability of dimethyl-(2-hydroxy-5-nitrobenzyl)sulfonium salts in aqueous solutions at pH < 3, which contrasts with the behavior of 2-hydroxy-5-nitrobenzyl halides.[3,10] However, it is to be noted that the sulfonium reagents are rapidly hydrolyzed in solutions of higher pH in the absence of tryptophan or cysteine. Accordingly, the reagent solution to be used for labeling proteins should not be buffered.

After reaction, the protein can be separated from the 2-hydroxy-5-nitrobenzyl alcohol which is formed by hydrolysis of excess reagent through the use of gel filtration, chromatography, fractionation, and dialysis procedures as outlined above. Dimethyl sulfide, which is formed as a by-product of reaction, is volatile and can be readily removed.

Determination of the amino acid residues which have reacted with the sulfonium reagent requires the same procedures as those given for proteins labeled with 2-hydroxy-5-nitrobenzyl bromide. The specificity of the sulfonium salts is similar to, although not identical with, that of HNB-halides. Less of the sulfonium reagents is required than of HNB-Br for a given degree of tryptophan modification, probably because of the lower hydrolysis rate of the sulfonium reagent. The reactivity of cysteine at neutral pH appears to be considerably greater, again because of the slower and less competitive hydrolysis of the sulfonium salts.

Reaction of dimethyl(2-hydroxy-5-nitrobenzyl)sulfonium halide with tryptophan ethyl ester in aqueous solutions at pH 4.7 produces the two diastereomeric monosubstitution products, (V) and (VI) (Fig. 3), derived from alkylation of position 3 of the indole nucleus followed by cyclization through the α-amino group. These products are formed in the same reproducible ratio, 1.4:1, as is characteristic of the reaction of 2-hydroxy-5-nitrobenzyl bromide. However, the monoalkylated products are formed in nearly quantitative yield when tryptophan ethyl ester is treated with an equimolar quantity of the sulfonium reagent, whereas the monosubstituted isomers account for about one-third of the reaction with hydroxynitrobenzyl bromide.[32a] Moreover, in contrast with the reported hydroxynitrobenzylation of the α-amino group in carboxypeptidase A treated with hydroxynitrobenzyl bromide in acetone,[5] treatment of this enzyme with dimethyl(2-hydroxy-5-nitrobenzyl)sulfonium chloride at pH 7.5 produced hydroxynitrobenzylated carboxypeptidase in good yield with no detectable N-terminal alkylation.[32b] Tryptophan appears to be the site of reporter group incorporation under these conditions. Based on these findings, it appears that dimethyl(2-hydroxy-5-nitrobenzyl)sulfonium salts may be employed to introduce the hydroxynitrobenzyl reporter group into proteins of biological interest without the complications which may arise from the use of acetone or other organic solvent and from excessive reagent hydrolysis experienced with 2-hydroxy-5-nitrobenzyl bromide.

Usefulness of the Reagents. In general, the sulfonium salts can be applied to the same purposes as the parent compound, 2-hydroxy-5-nitrobenzyl bromide, but their water solubility avoids certain complications, e.g., partial denaturation and unfolding effects, encountered when some proteins are exposed to organic solvents.

Benzyl Halides

Procedure for Reaction. In general, the conditions for reaction of benzyl bromide or benzyl chloride parallel those for 2-methoxy-5-nitrobenzyl bromide, i.e., the reagents react slowly and are largely insoluble. Benzyl bromide has been shown to react with the SH groups of wool proteins[33] and with the methionine residue of chymotrypsin.[34] Since these reactions correspond with those of 2-methoxy-5-nitrobenzyl bromide, it seems probable that unsubstituted benzyl halides similarly react with tryptophan also.

The total number of benzyl derivatives introduced cannot be determined by simple absorbance measurements and must be determined by amino acid analyses. Schramm and Lawson[35] have shown that performic acid treatment of methionine[24] can readily distinguish between

[32a] W. P. Tucker, J. Wang, and H. R. Horton, *Arch. Biochem. Biophys.* **144,** 730 (1971).

[32b] V. R. Naik and H. R. Horton, *Biochem. Biophys. Res. Commun.* **44,** 44 (1971).

[33] J. M. Gillespie, *Proc. Int. Wool Textile Res. Conf., Aust.* **1955B,** p. 35.

[34] H. J. Schramm and W. B. Lawson, *N.Y. State Dep. Health, Ann. Rep. Div. Lab. Res.* **1962,** p. 63.

[35] H. J. Schramm and W. B. Lawson, *Hoppe-Seyler's Z. Physiol. Chem.* **332,** 97 (1963).

modified and unmodified methionine residues. Modification of cysteine residues can similarly be determined using amino acid analysis.

In general, the long intervals of time for reaction require the precautions outlined for the case of the 2-methoxy-5-nitrobenzyl bromide.

Usefulness of the Reagent. Neither benzyl bromide nor benzyl chloride appears to offer any distinct advantages over the substituted benzyl halides for studies of protein reactivity. However, in some cases it may be desirable to block an amino acid side chain with the simplest aromatic group possible to avoid substituent interactions which provide a loss of activity not directly related to the amino acid residue alkylated. Furthermore, it may be desirable to block a reactive residue with a nonchromophoric group in order to subsequently cause a protein to react with a colored reagent, which can then be placed selectively. Finally, a sluggish reagent can provide information discriminating between various side chains of a protein which have different intrinsic reactivities.

[40] Sulfenyl Halides as Modifying Reagents for Polypeptides and Proteins

By ANGELO FONTANA and ERNESTO SCOFFONE

Sulfenyl halides have been found to be specific, mild reagents for modification of the tryptophan and cysteine residues of polypeptides and proteins in acidic media.[1-9] Tryptophan is converted by reaction with sulfenyl halides into a derivative with a thioether function in the 2 posi-

[1] A. Fontana, F. Marchiori, R. Rocchi, and P. Pajetta, *Gazz. Chim. Ital.* **96**, 1301 (1966).

[2] E. Scoffone, A. Fontana, F. Marchiori, and C. A. Benassi, *in* "Peptides." (*Proc. Eur. Peptide Symp. 8th, Noordwijk*), pp. 189–194. North-Holland Publ., Amsterdam, 1967.

[3] E. Scoffone, A. Fontana, and R. Rocchi, *Biochem. Biophys. Res. Commun.* **25**, 170 (1966).

[4] E. Scoffone, A. Fontana, and R. Rocchi, *Biochemistry* **7**, 971 (1968).

[5] A. Fontana, E. Scoffone, and C. A. Benassi, *Biochemistry* **7**, 980 (1968).

[6] A. Fontana, F. M. Veronese, and E. Scoffone, *Biochemistry* **7**, 3901 (1968).

[7] F. M. Veronese, A. Fontana, E. Boccù, and C. A. Benassi, *Z. Naturforsch. B* **23**, 1319 (1968).

[8] F. M. Veronese, E. Boccù, and A. Fontana, *Ann. Chim.* (*Rome*) **58**, 1309 (1968).

[9] A. Fontana and E. Scoffone, *in* "Mechanisms of Reactions of Sulfur Compounds" (N. Kharasch, ed.), Vol. IV, p. 15. Intra-Science Res. Found., Santa Monica, California, 1969.

tion of the indole nucleus (Eq. 1), and cysteine to an unsymmetrical disulfide (Eq. 2).

$$\text{Trp residue (-NHCHCO-, } CH_2\text{-indole, N-H)} + RSCl \longrightarrow \text{2-SR-Trp residue} + HCl \qquad (1)$$

$$\text{-NHCH(CH}_2\text{SH)CO-} + RSCl \longrightarrow \text{-NHCH(CH}_2\text{SSR)CO-} + HCl \qquad (2)$$

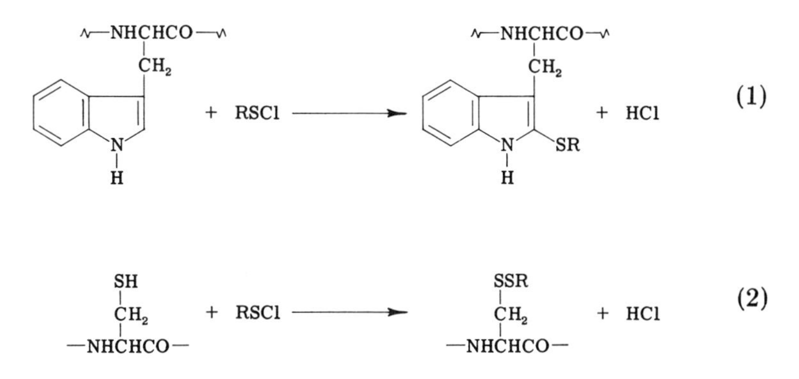

The high specificity of these reagents toward tryptophan and cysteine was established by treating an amino acid calibration mixture with 2-nitrophenylsulfenyl chloride (NPS-Cl)[10] in acetic acid.[4] Quantitative recovery of all tested amino acids was obtained, as determined by automatic amino acid analysis.

As a further check, ribonuclease A, a protein containing neither tryptophan nor cysteine, was allowed to react with 20 eq of NPS-Cl in 50% acetic acid. The protein was recovered unchanged and fully active toward RNA.[4]

By reaction with nitrophenylsulfenyl halides, a chromophore is generated in a protein which absorbs in the visible part of the spectrum.[4] This fact offers an easy quantitation of the reaction, allowing the determination of the tryptophan content of a protein. At the present time, the most used sulfenyl halide is the commercially available NPS-CL. The 2-(2-nitrophenylsulfenyl)tryptophan[11] derivative, obtained by reaction with tryptophan-containing proteins absorbs at 365 nm with a molar absorptivity of 4000.[4]

The sulfenylation of tryptophan can be accomplished in partly aqueous solvents such as 30–50% acetic acid, 30–50% dioxane, or 30–50% dimethylformamide, if hydrolysis-resistant sulfenyl halides, such as

[10] Abbreviations: NPS, 2-nitrophenylsulfenyl; pNPS, 4-nitrophenylsulfenyl; DNPS, 2,4-dinitrophenylsulfenyl; NCPS, 2-nitro-4-carboxyphenylsulfenyl; DNPDS, 2,4-dinitrophenyl-1,5-disulfenyl; ABS, azobenzene-2-sulfenyl; Trp(NPS), Trp(pNPS), Trp(DNPS), Trp(NCPS), reaction products of tryptophan with the corresponding sulfenyl halides. The amino acids, peptides, and peptide derivatives are of the L configuration. The abbreviations are those recommended by the IUPAC–IUB Commission on Biochemical Nomenclature, *Biochemistry* **5**, 1455 (1966).

[11] The most appropriate name is 2-thio-(2-nitrophenyl)tryptophan.

NPS-Cl, DNPS-Cl, and NCPS-Cl, are used.[12,13] In the case of the easily hydrolyzable sulfenyl halides, such as pNPS-Cl, an aqueous solvent can still be employed, provided that an excess of reagent dissolved in anhydrous solvent is used and added dropwise under stirring to an aqueous solution of the protein. In the table are reported the maximum of absorption and the molar extinction of several tryptophan-containing polypeptides treated with various sulfenyl chlorides.

It was shown[8] that nitrophenylsulfenyl halides react with the indole nucleus of tryptophan much faster when this residue is linked in the protein rather than in free tryptophan and in small peptides. This enhanced reactivity makes it possible to perform the sulfenylation reaction in aqueous solvents (50% acetic acid), in spite of the fact that a rapid (2–3 minutes) decomposition of the reagent occurs.[8] This observation may be explained by the selective adsorption of the reagent on the protein molecule prior to reaction with the tryptophan residues (see L. A. Cohen[14] for a discussion of similar proximity effects).

Quantitative modification of tryptophan in proteins is achieved readily by using low molar ratio of reagent to protein. The tryptophan residues of lysozyme, trypsin and α-chymotrypsin were fully modified by allowing the proteins to react with 20 eq of NPS-Cl in 50% acetic acid.[4] More recently,[15] a heptadecapeptide, α^{1-17}-ACTH, and human growth hormone also were successfully modified by performing the reaction in 0.2 *M* acetic acid, with the pH held at 4.0 with the use of pH-stat.

Cysteine residues, if present, react with sulfenyl halides to form unsymmetrical disulfides.[5] The thiol function can be restored easily by means of reducing agents (thiols, sodium borohydride), allowing labeling only of tryptophan. The mild conditions required to introduce and remove the sulfenyl groups at the level of SH-groups were tested on reduced ribonuclease, which was allowed to react with NPS-Cl and pNPS-Cl in 50% acetic acid. The modified protein was then reduced with β-mercaptoethanol in 8 *M* urea solution, and after air oxidation the recovered enzyme had the same properties as the native protein.[5]

2-Nitrophenylsulfenyl Chloride (NPS-Cl)

Preparation. NPS-Cl is synthesized by chlorinolysis of 2,2′-dinitrophenyl disulfide.[16] The product may be recrystallized from anhydrous

[12] N. Kharasch, W. King, and T. C. Bruice, *J. Amer. Chem. Soc.* **77**, 932 (1955).

[13] L. Di Nunno, G. Modena, and G. Scorrano, *Ric. Sci.* **36,** 825 (1966).

[14] L. A. Cohen, *Annu. Rev. Biochem.* **37,** 695 (1968).

[15] L. Brovetto-Cruz and C. H. Li, *Biochemistry* **8,** 4695 (1969).

[16] M. H. Hubacher, "Organic Syntheses," Coll. Vol. 2, p. 445. Wiley, New York, 1943.

WAVELENGTHS OF MAXIMUM ABSORPTION (λ_{max}) AND MOLAR ABSORPTIVITIES (ϵ) OF TRYPTOPHAN PEPTIDES TREATED WITH SULFENYL HALIDES

No.	Compound	Solvent	λ_{max}	$\epsilon \times 10^{-4}$	λ_{max}	$\epsilon \times 10^{-4}$
1	Trp(NPS)[a]	HAc, 80%	280	1.67	362	3.75
2	Trp(NCPS)[b]	HAc, 80%	283	2.25	353	4.65
		Buffer, pH 7	282	2.20	363	4.65
3	Z-Trp(NCPS)-Gly-OEt[b]	HAc, 80%	283	2.35	354	4.75
		MeOH	284	2.38	355	4.70
4	Z-Leu-Trp(NPS)OMe[c]	HAc, 80%	280	1.55	362	4.1
		MeOH	280	1.67	365	4.4
5	Z-Ala-Trp(NPS)-Gly-OEt[c]	HAc, 80%	280	1.61	363	4.05
6	Z-Leu-Trp(pNPS)-OMe[c]	HAc, 80%	290	1.50	328	11.7
7	Z-Leu-Trp(DNPS)-OMe[c]	HAc, 80%	278	1.80	340(sh)	8.3
8	Phe-Val-Gln-Trp(NPS)-Leu[d]	HAc, 80%	280	1.50	365	3.95
9	Phe-Val-Gln-Trp(NCPS)-Leu[b]	HAc, 80%	284	2.15	353	4.60
10	NPS-gramicidin A[c,e]	HAc, 80%	282	6.05	363	15.8
11	NPS-β^{1-24}-corticotropin[c,f]	HAc, 10%	280	1.69	363	4.1

[a] A. Fontana, F. Marchiori, R. Rocchi, and P. Pajetta, *Gazz. Chim. Ital.* **96,** 1301 (1966).

[b] F. M. Veronese, E. Boccù, and A. Fontana, *Ann. Chim.* (*Rome*) **58,** 1309 (1968).

[c] E. Scoffone, A. Fontana, and R. Rocchi, *Biochemistry* **7,** 971 (1968).

[d] E. Wünsch, A. Fontana, and F. Drees, *Z. Naturforsch. B* **22,** 607 (1967). This pentapeptide corresponds to the 22–26 sequence of glucagon.

[e] Gramicidin A contains four tryptophanyl residues per mole [R. Sarges and B. Witkop, *J. Amer. Chem. Soc.* **87,** 2011 (1965)]. For NPS-gramicidin A, a molecular weight of 2506 was assumed.

[f] β^{1-24}-Corticotropin contains two tyrosyl (ϵ = 1300 at 280 nm) and one tryptophanyl residue per mole [R. Schwyzer and H. Kappeler, *Helv. Chim. Acta* **46,** 1550 (1963)]. For NPS-β^{1-24}-corticotropin, a molecular weight of 3650 was assumed.

ethyl ether, mp 75–76°. The reagent is also marketed in analytical grade by several companies (Eastman Organic Chemicals; Pierce Chemicals Company; Fluka AG, Basle, Switzerland).

NPS-Cl as a Protein Reagent. The reagent has to be dissolved in an organic solvent and added with rapid mixing to an aqueous solution of the protein. Acetic acid and dioxane have both been found to be convenient solvents.[4] The choice of the reaction medium may depend to some extent on the nature of the protein under study, but must be mildly acidic, since amino groups can react with the halide, giving sulfenamides. Screening tests for denaturation effects of the solvent on the dissolved protein in the absence of reagent are recommended. The preferred concentration of the protein is about 1 m*M*. The modified protein may be separated from excess reagent by one or a combination of standard separation techniques. Gel filtration on Sephadex G-25 is the recommended procedure, although in some instances precipitation with organic solvents (acetone:1 *N* HCl, 39:1), in which the excess reagent and its decomposition products remain dissolved, may be sufficient to purify the labeled protein.[4] The completeness of removal of adsorbed reagent from labeled protein may be assessed by subjecting aliquots to repeated precipitation with trichloroacetic acid or an organic solvent, until a constant ratio of NPS groups to protein is revealed by spectrophotometry. Protein denaturants, such as trichloroacetic acid, cannot be used if modification of the protein is done to determine the effect of chemical modification on enzymatic activity and other biological properties of the enzyme.

Determination of Reacted Residues. The degree of sulfenylation may be conveniently assessed spectrophotometrically.[4] The Trp(NPS) residue, obtained by reaction of NPS-Cl with tryptophan-containing proteins, absorbs at 365 nm with a molar absorptivity of 4000. The labeled protein may be dissolved (about 0.1 μmole of protein) in aqueous acetic acid (50–80%) as well as in aqueous formic acid (3 ml), which in some instances behaves as a better solvent. The number of groups so determined can be related to the protein concentration in several ways, the most accurate being amino acid analysis of an acid hydrolyzate of an aliquot of the solution. Since the sulfenyl halide does not react with histidine or arginine, an analysis for basic amino acids with an analyzer is sufficient to establish the quantity of protein present. The micro-Kjeldahl analysis of exhaustively dialyzed aliquots or gravimetric analysis of salt-free lyophilized protein dried over P_2O_5 also may prove useful.

The following description of procedures will provide an illustration of some of the techniques that have been used for the labeling of tryptophan in peptides and proteins.

GRAMICIDIN A.[4] To a solution of 24 mg of gramicidin A in 1.5 ml of glacial acetic acid, 10 mg of NPS-Cl was added, and the mixture was allowed to stand at room temperature for 1 hour. The product was then precipitated with 20 ml of ethyl ether–petroleum ether (1:3), separated by centrifugation, and washed several times with the same solvent mixture. After drying under high vacuum over P_2O_5 the yield was 23 mg; single yellow, Ehrlich-negative spot (R_f 0.95) in TLC using chloroform–glacial acetic acid (1:2) as solvent. In the same solvent system gramicidin A had R_f 0.35.

The pentadecapeptide gramicidin A has four tryptophan residues per mole.[17] The molar extinctions in 80% acetic acid at 282 nm, 6.05×10^{-4}, and at 363 nm, 1.58×10^{-4}, account exactly for four residues ($\epsilon = 15{,}000$ at 280 nm and 4000 at 365 nm, mean values of the Trp(NPS) residue).

α^{1-17}-ACTH.[15] The single tryptophan of this 1–17 sequence of ACTH has been completely modified by NPS-Cl, using as a solvent mixture 0.2 *M* acetic acid, pH 4.0. The reagent in 20-fold molar excess was added as a solid to the reaction mixture. The pH was maintained at 4.0 in a pH-stat. The product was separated from the excess reagent by gel filtration on a Sephadex G-10 column (1.2 × 26 cm) preequilibrated with 0.2 *M* acetic acid. On the basis of a molar extinction of 4000 at 365 nm of the Trp(NPS) chromophore, 1.0 residues were modified per molecule.

HUMAN GROWTH HORMONE.[15] The single tryptophan residue in human growth hormone (HGH) has been allowed to react quantitatively and specifically with NPS-Cl. The NPS-derivative retained full growth-promoting activity and no lactogenic activity. Its physicochemical properties are similar to those of the native molecule.

Two different media were used for the reaction of HGH with NPS-Cl.

(i) Acetic acid, 50%. HGH (1 μmole) was dissolved in 2 ml of 25% acetic acid, and 20 μmoles of NPS-Cl dissolved in 1 ml of glacial acetic acid added with continuous stirring. After 1 hour, the protein was separated from the excess reagent by gel filtration on a column (2.2 × 20 cm) of Sephadex G-25, preeliquibrated with 0.2 *M* acetic acid. The protein peak was lyophilized or concentrated by ultrafiltration, and a final purification was achieved by gel filtration on a Sephadex G-100 column (3 × 60 cm), preeliquibrated, and eluted with 0.2 *M* acetic acid solution. The V_e/V_o, between 2.0 and 2.5, was concentrated by ultrafiltration or recovered by lyophilization.

(ii) Acetic acid, 0.2 *M*, pH 4.0. The protein was dissolved in 2.0 *M* acetic acid and a 20-molar excess of solid reagent was added. The pH

[17] R. Sarges and W. Witkop, *J. Amer. Chem. Soc.* **87**, 2011 (1965).

was maintained at 4.0 in a pH-stat. After 2 hours at room temperature, the insoluble excess reagent was removed by centrifugation and the supernatant was submitted to the purification procedure described under (i).

Determination of the Extent of the Reaction of NPS-Cl with HGH.[13,15] To estimate the extent of modification, the NPS-HGH derivative was dissolved in 80% acetic acid (0.5–1.0 mg/ml) and the amount of NPS-chromophore was determined spectrophotometrically ($\epsilon = 4000$ at 365 nm). The protein concentration was determined on the basis of the dry weight, or using the molar extinction coefficient of HGH at 277 nm corrected by the amount (molar extinction coefficient of Trp(NPS) in peptide linkage minus the molar extinction coefficient of Trp $= 1.51 \times 10^{-4} - 5.5 \times 10^{-3}$) to account for the change in extinction following derivatization of the Trp residue. When the reaction was performed in 50% acetic acid the tryptophan found (theory 1.0) was 1.1–1.2 residues (range of 5 different experiments) and 0.4–0.5 residue (3 experiments) when the reaction was performed in 0.2 *N* acetic acid, pH 4.0.

Other Nitrophenylsulfenyl Chlorides

In principle, all members of the wide class of sulfenyl halides could be used for the labeling of tryptophan. However, these reagents are highly unstable in aqueous solvents, with the exception of 2-nitrophenyl compounds, which are more resistant (1000 times) to hydrolysis in comparison to the other halides.[12,13]

2,4-Dinitrophenylsulfenyl Chloride (DNPS-Cl). DNPS-Cl may be prepared by bubbling chlorine through a suspension of 2,4-dinitrophenyl disulfide in ethylene dichloride, with a few drops of sulfuric acid as catalyst. A faster reaction and higher yields were obtained by performing the reaction at 80° instead at 25°, as suggested by Kharasch *et al.*[18] The compound is recrystallized from anhydrous chloroform–ethyl ether, mp 94–95°. The reagent is marketed by several companies (Fluka AG, Basle, Switzerland; Eastman Organic Chemicals; Aldrich Chemical Company).

4-Nitrophenylsulfenyl Chloride (pNPS-Cl). The reagent is prepared by chlorination in chloroform of 4,4′-dinitrophenyl disulfide.[19] For a successful preparation of pNPS-Cl, care needs to be taken to use a disulfide of good quality. A much more pure compound was obtained by chlorination of 4-nitrothiophenol in chloroform.[20] Since the reagent is

[18] N. Kharasch, G. I. Gleason, and C. M. Buess, *J. Amer. Chem. Soc.* **72**, 1796 (1950).

[19] T. Zincke and S. Lenhardt, *Justus Liebigs Ann. Chem.* **400**, 2 (1913).

[20] F. M. Veronese and E. Boccù, personal communication, 1970.

less stable than the corresponding 2-nitrophenyl compounds, only fresh preparations of the reagent should be employed.

2-Nitro-4-carboxyphenylsulfenyl Chloride (NCPS-Cl). The carboxy-substituted sulfenyl chloride was chosen for the modification of tryptophan, since the modified peptides and proteins would be expected to be more soluble.[8] NCPS-Cl was prepared from 2,2′-dinitro-4,4′-dicarboxyphenyl disulfide, as described in the literature.[21] This sulfenyl halide is commercially available from Nutritional Biochemical Corporation.

2,4-Dinitrophenyl-1,5-disulfenyl Chloride (DNPDS-Cl)

Taking advantage of the selective reaction of sulfenyl halides with tryptophan residues in acidic media, a new method was proposed[22] for the inter- and intramolecular cross-linking of tryptophan residues in peptides and proteins (for a review on bifunctional cross-linking reagents, see Wold[23]). The reaction employed involves treatment of the tryptophan-containing derivative with 2,4-dinitrophenyl-1,5-disulfenyl chloride (DNPDS-Cl) (I).

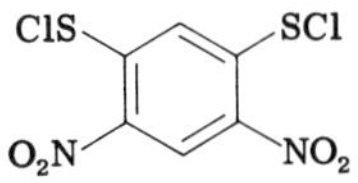

ClS SCl
O_2N NO_2

(I)

This bifunctional sulfenyl halide reacts selectively with two molecules of tryptophan at the 2-position of the indole nucleus, leading to a cross-linked compound (II).

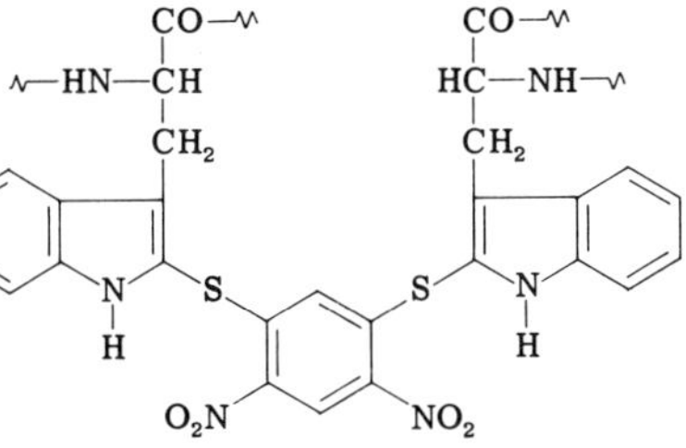

CO— CO—
—HN—CH HC—NH—
CH_2 CH_2
S S
N N
H H
O_2N NO_2

(II)

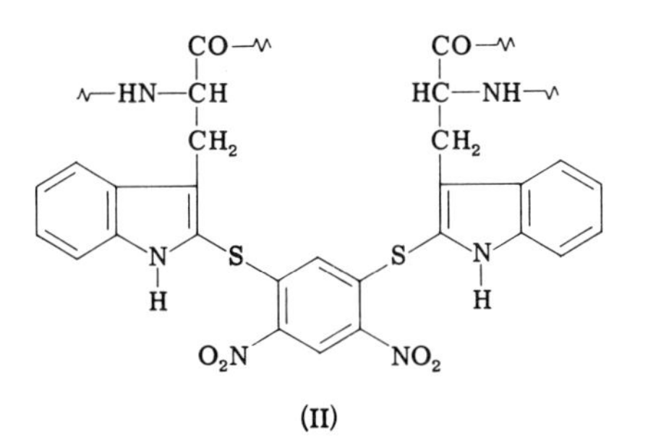

The only other amino acid affected is cysteine, which similarly can be cross-linked through formation of mixed disulfide bonds (III). How-

[21] A. J. Havlik and N. Kharasch, *J. Amer. Chem. Soc.* **77,** 1150 (1955).
[22] F. M. Veronese, E. Boccù, and A. Fontana, *Int. J. Protein Res.* **2,** 67 (1970).
[23] F. Wold, Vol. XI, p. 617.

ever, the reaction with the thiol function of cysteine can be reversed by reduction with mercaptans.

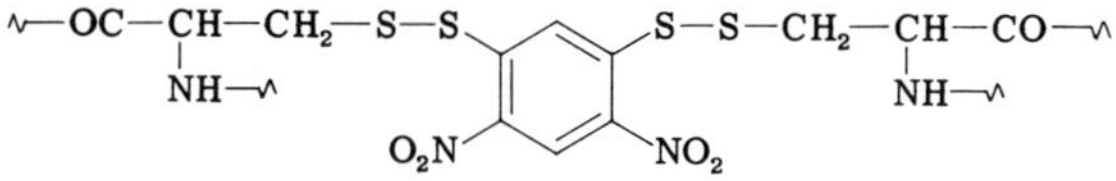

$\sim$OC—CH—CH$_2$—S—S—[$C_6H_2(NO_2)_2$]—S—S—CH$_2$—CH—CO$\sim$
NH$\sim$ (on each CH)
O_2N, NO_2 (ring substituents)

(III)

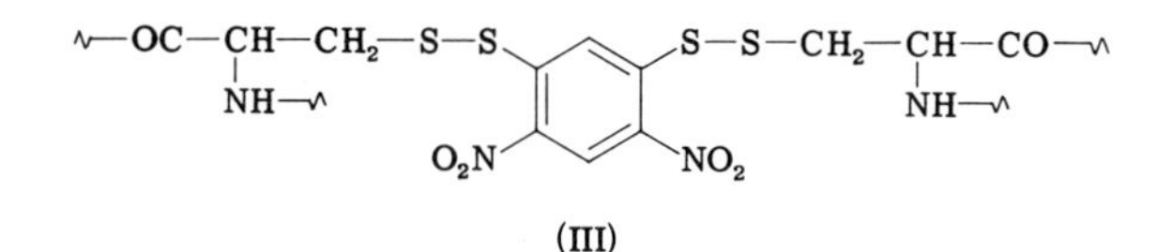

DNPDS-Cl was prepared by chlorination of 2,4-dinitro-1,5-dithiophenol in refluxing ethylene chloride, under catalysis of sulfuric acid. The thiol was in turn prepared from 2,4-dinitro-1,5-dichlorobenzene by sulfide displacement of the chlorine atoms. This particular reagent was specifically chosen, since the nitro group in the ortho position stabilizes the sulfenyl function against hydrolysis. In addition, the reagent offers various desired characteristics: easy crystallizability, good yields in the preparations, great stability in storage, and a high molar absorption of the derivatives with tryptophan.

The effectiveness of the method was tested with several tryptophan derivatives, which were cross-linked by reaction with DNPDS-Cl in acetic or formic acid, including short sequences of the pancreatic hormones glucagon and gastrin.

The reaction was also applied to glucagon, in which case the single tryptophan residue was allowed to react with the reagent and the dimer was isolated after gel filtration.

Cross-linking of Glucagon by Reaction with DNPDS-Cl.[22] To a solution of 7.2 mg (2 μmoles) of glucagon in 1 ml of glacial acetic acid, 0.36 mg (1.2 μmole) of DNPDS-Cl in 0.2 ml of the same solvent was added at room temperature under vigorous stirring. The stirring was continued for 2 hours and then 0.5 ml of water was added and the solution was stirred for an additional 10 minutes. The solution was lyophilized and the residue redissolved in 0.8 ml of 0.2 *N* acetic acid and poured into a Sephadex G-25 column (1.5 $\times$ 130 cm) equilibrated with 0.2 *N* acetic acid. The column was developed using the same eluent, and fractions of 3.75 ml were collected (flow rate 37.5 ml per hour). The fractions containing the glucagon dimer were combined and lyophilized (2.95 mg).

Azobenzene-2-sulfenyl Bromide (ABS-Br)

Azobenzene-2-sulfenyl bromide (ABS-Br) (IV) was found to react *selectively* with cysteinyl residues in polypeptides and proteins giving unsymmetrical disulfides.[6] Tryptophan and tryptophan peptides, as well as lysozyme, a protein containing tryptophan but not cysteine, were recovered unchanged after reaction with ABS-Br in buffer solution at pH 5 or in glacial acetic acid. The lack of reactivity of the indole nucleus toward this sulfenyl halide can be explained by the saltlike nature of this reagent, since its true structure has been shown to be that of 2-phenylbenzo-1-thio-2:3-diazolium bromide (V).[24,25] The reagent shows the property, unusual for a sulfenyl halide, of solubility and stability in water. The reagent is marketed by Nutritional Biochemical Corporation.

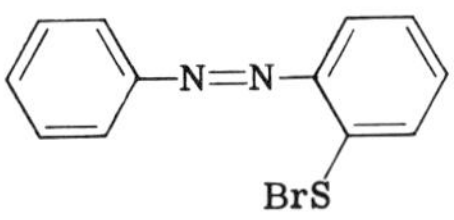

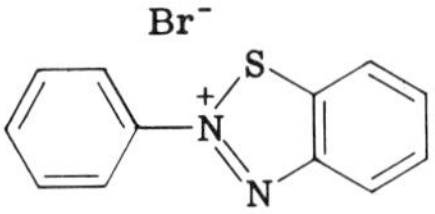

It was found[6] that the basicity of the azo function, determined spectrophotometrically, of the mixed disulfides obtained upon reaction of ABS-Br with SH-groups is quite different depending on whether the chromophore was covalently bound to simple SH-compounds like cysteine or to an SH-protein, e.g., reduced lysozyme. In view of the fact that the ionization of the ABS group is influenced by the protein structure, an interesting application might be its use as a reporter group for conformational studies. However, a limitation inherent in the method is the great instability of the disulfide linkage in alkaline media, thus precluding in some instances the possibility of studying the properties of the modified enzymes over the complete range of pH.

Analytical Aspects

For analytical purposes, the procedure for labeling the proteins is the same as described above, the solvent of choice being 50% acetic acid, with a protein concentration of about 1 mM and using approximately 20 eq of NPS-Cl.[26] Recently,[27] it was found that, when a protein solution of much lower concentration (0.018 mM) was used, a large excess of reagent (about 500 eq) was needed for complete labeling of tryptophan, as expected from the bimolecular nature of the reaction. Under these conditions, however, some nonspecific labeling was observed, the excess chromophores being removed by treatment with β-mercaptoethanol.

[24] A. Burawoy, F. Liversedge, and C. E. Vellins, *J. Chem. Soc.* 4481 (1954).

[25] A. Burawoy and C. E. Vellins, *J. Chem. Soc.* 90 (1954).

[26] E. Boccù, F. M. Veronese, A. Fontana, and C. A. Benassi, *Eur. J. Biochem.* **13**, 188 (1970).

[27] G. W. Robinson, *J. Biol. Chem.* **245,** 4832 (1970).

If the protein to be analyzed for tryptophan contains cysteine residues, reactive toward sulfenyl halides, the reaction can still be employed for the analytical determination of tryptophan.[26] The chromophoric groups bound to cysteine can be removed easily by dissolving the labeled protein in 0.1 *N* NaOH for 30 minutes at room temperature. The alkyl-aryl disulfides at the cysteine residues cleave quantitatively with formation of arylthiophenol, whereas the 2-thioaryltryptophan moiety is completely stable.[27a] The protein labeled only at the tryptophan residue is then separated by acid precipitation or gel filtration, and finally the tryptophan content is determined as usual, spectrophotometrically.

By reaction of pNPS-Cl with the SH-groups of a protein, the *p*-nitrothiophenol moiety can be covalently bound to the protein by a disulfide linkage. By dissolving the labeled protein in 0.1 *N* NaOH, the thiol ($\epsilon = 13{,}600$ at λ_{max} 412 nm) is quantitatively released, provided that deaerated solutions are employed.[26]

When a protein containing both tryptophan and cysteine is to be analyzed for SH content, the procedure involving the reaction of pNPS-Cl can still be employed, since Trp(pNPS) shows almost no absorption at the λ_{max} (412 nm) of *p*-nitrothiopenol. On the other hand a higher wavelength can be selected for absorption measurements, i.e., 450 nm, where the thiol still strongly absorbs, $\epsilon = 8500$. The procedure has been investigated with a wide variety of tryptophan and cysteine-containing proteins and shown to be accurate for both amino acids.[26]

The selective reaction of ABS-Br with cysteine residues in buffer solution, pH 5.0, did not always give quantitative results. Most probably the bulkiness of the reagent prohibits complete modification of SH-containing enzymes.[28]

Comments

Sulfenyl halides have several particularly valuable features for protein modification studies. In the first place, their high selectivity for trypto-

[27a] Most probably the alkaline cleavage of the *S*-nitroarylsulfenylcysteine occurs via β-elimination (known to occur in alkali-treated cystinyl compounds) producing a dehydroalanine residue and a persulfide ion, which further decomposes to a thiol and free sulfur [D. S. Tarbell and D. P. Harnisch, *Chem. Rev.* **49**, 1 (1951)]. An indication of this mechanism is given by the fact that the reaction product of reduced ribonuclease with 4-nitrophenylsulfenyl chloride (8 arylsulfenylcysteine residues per mole of protein), after standing at 22–24° in 0.1 *N* NaOH for 15 hours, produces lysinoalanine [*N*-(DL-2-amino-2-carboxyethyl)-L-lysine] (A. Fontana and E. Scoffone, unpublished). This amino acid is known to arise from a condensation between dehydroalanine and the ϵ-amino group of lysine in alkali-treated proteins [Z. Bohak, *J. Biol. Chem.* **239**, 2878 (1964)].

[28] E. Boccù and F. M. Veronese, unpublished results, 1970.

phan in proteins lacking SH-groups is remarkable. Even if cysteine residues are present, selectivity for tryptophan can be achieved, since the mixed disulfides formed from SH-groups are easily cleaved by reduction with β-mercaptoethanol. Other techniques for tryptophan modification, such as reaction with *N*-bromosuccinimide,[29] are less specific. The Koshland reagent, 2-hydroxy-5-nitrobenzyl bromide, upon reaction with tryptophan gives several derivatives[30] allowing formation of different species of the labeled protein, as shown with pepsin.[31]

The sulfenylation reaction increases the number of procedures available for the modification of tryptophan, providing greater experimental latitude in the study of the biological function of tryptophan-containing enzymes. One of the main advantages of this technique is the possibility of using sulfenyl halides carrying different groups, thus leading to a change in size, polarity, or other physicochemical properties of the labeled enzymes. The nitroaryl chromophoric groups, if optically active, might further reveal characteristic extrinsic Cotton effects as additional probes of protein conformation. Conformational studies on poly-L-tryptophan modified with NPS-Cl have been reported.[32]

Among the useful information on the involvement of tryptophan in enzymatic processes, the analytical utility of this reaction is another major application. The quantitative reaction with NPS-Cl of the tryptophan residues of several proteins appears to offer a new rapid and convenient spectrophotometric method for determining the tryptophan content of a protein.[26]

Several typical applications of the sulfenylation reaction have been reported. Lysozyme has been modified at the tryptophan residues with NPS-Cl and the sulfenylated derivative [6 Trp(NPS) residues per molecule] has been studied with respect to their biological, immunological, and physical properties.[33] The single tryptophan residue of human growth hormone has been modified with NPS-Cl; the modified protein retained full growth promoting activity.[15] Similarly, a polypeptide fragment of pepsin,[34] a bacterial proteinase,[27] the A_1-encephalitogenic protein[35] and neocarzinostatin[36] were treated with NPS-Cl for analytical purposes. More recently,[37] staphylococcal nuclease has been modified with several

[29] L. K. Ramachandran and B. Witkop, Vol. XI, 283.

[30] T. F. Spande, M. Wilchek, and B. Witkop, *J. Amer. Chem. Soc.* **90,** 3256 (1968).

[31] T. A. A. Dopheide and W. M. Jones, *J. Biol. Chem.* **243,** 3906 (1968).

[32] E. Peggion, A. Fontana, and A. Cosani, *Biopolymers* **7,** 517 (1969).

[33] A. F. S. A. Habeeb and M. Z. Atassi, *Immunochemistry* **6,** 555 (1969).

[34] V. Kostka, L. Moravek, and F. Šorm, *Eur. J. Biochem.* **13,** 447 (1970).

[35] E. H. Eylar and G. A. Hashim, *Arch. Biochem. Biophys.* **131,** 215 (1969).

[36] H. M. Maeda and J. Meienhofer, *Int. J. Protein Res.* **2,** 135 (1970).

[37] I. Parick and G. S. Omenn, *Biochemistry* **10,** 1173 (1971).

sulfenyl halides (NPS-Cl, DNPS-Cl, NCPS-Cl), and the effect of modification on the physical and biological properties of the molecule were examined. In this case 1 mole of the different nitrophenylsulfenyl chromophores was introduced into the protein molecule, as determined by spectrophotometry; and furthermore, no evidence of unreacted tryptophan could be detected in fluorescence emission spectra upon excitation at 295 nm. The modified nuclease retained full immunological reactivity and about 50% enzymatic activity against both DNA and RNA.

Finally, the yellow color of the tryptophan derivatives is a convenient marker for peptides in chromatography. In particular, the carboxy-substituted sulfenyl halide, NCPS-Cl has been used for identifying tryptophan peptides in the diagonal electrophoresis technique.[38]

Acknowledgment

This work was supported by the Istituto di Chimica delle Macromolecole of the Consiglio Nazionale delle Richerche.

Note Added in Proof

Recently several interesting applications of the sulfenylation reaction have appeared in the literature. The two tryptophan residues of dolphin myoglobin were labeled quantitatively with NPS-Cl [P. Nedkov and B. Meloun, *Collection Czechoslov. Chem. Commun.* **34,** 2021 (1969)]. From the tryptic digests of the modified protein peptides containing the labelled tryptophan residue were isolated and their structure determined. Reaction of the single tryptophan residue in the encephalitogenic protein with NPS-Cl leads to a modified protein which is no more encephalitogenic in guinea pigs [Li-Pen Chao and E. R. Einstein, *J. Biol. Chem.* **245,** 6397 (1970)]. The melanonophore stimulating activity of (5-glutamine)-α-melanotropin, a synthetic analog of α-melanotropin, was found to be undiminished by chemical modification with NPS-Cl [J. Ramachandran, *Biochem. Biophys. Res. Commun.* **41,** 353 (1970)]. The NPS-derivative of adrenocorticotropin in which the single tryptophan residue of the molecule is modified was found to be fully active in stimulating amphibian melanophores, whereas loss of its lypolytic activity occurred [J. Ramachandran and V. Lee, *Biochem. Biophys. Res. Commun.* **38,** 417 (1970)].

[38] M. Ohno, personal communication, 1970.

[41] Acetylation

By J. F. Riordan and B. L. Vallee

Acetylation of amino groups is one of the most common means employed for the chemical modification of enzymes. Several reasons account for this circumstance. First, amino groups tend to be located on the "surface" of the three-dimensional structure of proteins in contact with the ambient environment and, hence, readily accessible to chemical attack. Second, at least one acetylating agent is available which has the requisite properties of high specificity and rapid reactivity under mild conditions. Third, the extent of reaction can be assessed by relatively simple analytical means. Nonetheless, the analytical characterization of acetylation presents certain difficulties and, consequently, there are problems in interpretation of the relationship of acetylation to changes in biological activity.

Methods

Acetylation with Acetic Anhydride. This reagent remains the one of choice for introducing acetyl groups into proteins. The earlier procedure[1] is as follows: a 2–10% protein solution or suspension in half-saturated sodium acetate is cooled in an ice bath and stirred with a magnetic mixer. An equal weight of acetic anhydride is added in five equal portions over the course of 1 hour at 0° and the stirring is continued for an additional hour. The product is isolated by gel filtration or dialysis. Sodium acetate presumably functions to buffer the acetylation reaction mixture and as a catalyst[2]; but high concentrations of acetate may play an additional role by increasing the specificity of acetylation.

Acetylation under these conditions can often lead to protein denaturation.[3] The amount of anhydride employed may constitute one of the reasons. However, these concentrations are required because the reagent is rather unstable and undergoes spontaneous hydrolysis, thereby rapidly depleting the effective molar excess. In addition, the reaction is pH-dependent, since only unprotonated amino groups are acetylated, and soon after the addition of anhydride there is a fall in pH from about 8 to 5.5. An excess of anhydride is therefore required to achieve an appreciable rate of acetylation.

A preferred procedure is to carry out the reaction at constant pH using a pH-stat. In this instance a 30- to 60-fold molar excess of reagent is added to the protein solution (5–10 mg/ml) with or without an appropriate buffer at pH 7.5. The reaction mixture is well stirred and the temperature is maintained at 0–4° with an ice bath. The time course of acetylation is followed by recording the uptake of 1 *N* NaOH. The titration measures the release of protons due both to modification and to the spontaneous hydrolysis of the anhydride and, therefore cannot be used

[1] H. Fraenkel-Conrat, Vol. 4, p. 247.
[2] R. W. Green, K. P. Ang, and L. C. Lam, *Biochem. J.* **54**, 181 (1953).
[3] J. L. Bethune, D. D. Ulmer, and B. L. Vallee, *Biochemistry* **3**, 1764 (1964).

as a quantitative measure of the degree of acetylation. The reaction is generally completed within 20–30 minutes, and the product is isolated by gel filtration or dialysis. When large amounts of protein are to be modified it is best to add the anhydride in several portions.

Other Acetylating Agents. Several other acetylating agents have been employed for the modification of proteins. Acetyl chloride, while equally as effective in reacting with amino groups as acetic anhydride, is somewhat more vigorous and consequently its use might be expected to result in greater denaturation. For many years ketene was employed commonly as an acetylating agent, but a number of disadvantages in comparison with acetic anhydride have discouraged its use.[4] It is an unstable, toxic gas requiring special generating equipment. The reaction is carried out by bubbling the gas through the protein solution. Hence, quantitation of the amount employed depends on maintaining a known rate of gas flow and surface denaturation due to foaming becomes a major problem. Ketene is not recommended generally and should find its use largely for enzymes, such as pepsin,[5] which are particularly unstable at mildly alkaline pH.

Other acetylating reagents and procedures have been employed for several proteins. *N,S*-Diacetylthioethanolamine acetylates proteins at pH 9–10, but these conditions limit its use to a relatively small group of proteins, not being very favorable for the majority.[6] The technique employed is analogous to trifluoroacetylation described elsewhere (Vol. XI [34]).

N-Acetylimidazole has been used primarily for the acetylation of tyrosyl residues of proteins. However, in a number of proteins and polypeptides examined, acetylation of amino groups has been observed as well, although the degree of substitution was usually much less than with acetic anhydride.[7]

There is no evidence that *p*-nitrophenyl acetate acetylates amino groups of proteins in aqueous media, although *p*-nitrophenyl bromoacetate has been used to acylate ribonuclease in dimethylsulfoxide and about 90% of the ϵ-amino groups were modified.[8] The reagent is of interest since it introduces a cross-linking alkylating agent into the protein and presumably allows covalent bonding to antibody. This reaction exemplifies the use of organic solvents in allowing the exploration of

[4] F. W. Putnam, *in* "The Proteins" (H. Neurath and K. Bailey, eds.), Vol. 1B, p. 893. Academic Press, New York, 1953.

[5] R. M. Herriott, *Advan. Protein Chem.* **3**, 170 (1947).

[6] J. Baddiley, R. A. Keckwick, and E. M. Thain, *Nature* (*London*) **170**, 968 (1952).

[7] J. F. Riordan, W. E. C. Wacker, and B. L. Vallee, *Biochemistry* **4**, 1758 (1965).

[8] J. Guldalian, W. B. Lawson, and R. K. Brown, *J. Biol. Chem.* **240**, PC2758 (1965).

new means to chemically modify proteins, an approach which has been studied in detail in few instances[9] but which offers great potential.[10]

Several other anhydrides have been employed as acylating agents of enzymes by utilizing either of the two methods described above.[11,12] In general, the reactivity of these anhydrides decreases with increasing chain length, perhaps due to decreased water solubility. Dicarboxylic acid anhydrides offer unusual properties, not only because they have a much greater effect on protein charge, but because they allow a means of studying tyrosyl residues also. Maleic,[13] citraconic,[14] and tetrafluorosuccinic[15] anhydride are of particular interest because they act as reversible blocking agents for amino groups.

A number of other means of acetylation of proteins have been reported but, in general, these do not appear to be ideally suited for studies of structure–function relationships in enzymes. Thus, procedures employing acetic anhydride-ethylacetate–formic acid,[16] acetic acid–acetic anhydride[2] or hot 16% (v/v) acetic anhydride in acetic acid[17] for acetylation reactions are too much of an insult even for the most resistant of enzymes.

Analytical Characterization

The degree of acetylation of a protein can be determined either by measuring the number of acetyl groups introduced or by measuring the decrease of amino groups known to be present. By employing both methods, the specificity of the reaction can be assessed.

The use of ^{14}C-labeled reagents is an obvious means of measuring acetyl incorporation, though the incorporation itself does not identify the reactive group which is acetylated. Hence, additional chemical identification is always necessary.

Reaction with ninhydrin is a rapid chemical method for determining the decrease in amino groups as a consequence of acetylation.[18] Absorb-

[9] S. M. Vratsanos, *Arch. Biochem. Biophys.* **90,** 132 (1960).

[10] S. J. Singer, *Advan. Protein Chem.* **17,** 1 (1962).

[11] L. Terminiello, J. Sri Ram, M. Bier, and F. F. Nord, *Arch. Biochem. Biophys.* **57,** 252 (1955).

[12] J. F. Riordan and B. L. Vallee, *Biochemistry* **2,** 1460 (1963).

[13] P. J. G. Butler, J. I. Harris, B. S. Hartley, and R. Leberman, *Biochem. J.* **112,** 679 (1969).

[14] H. B. F. Dixon and R. N. Perham, *Biochem. J.* **109,** 312 (1968).

[15] G. Braunitzer, K. Beyreuther, H. Fujiki, and B. Schrank, *Z. Physiol. Chem.* **349,** 265 (1968).

[16] S. M. Bose and K. T. Joseph, *Arch. Biochem. Biophys.* **74,** 46 (1958).

[17] A. E. Brown and L. G. Beauregard, *in* "Sulfur in Proteins" (R. Benesch *et al.,* eds.), p. 59. Academic Press, New York, 1959.

[18] S. Moore and W. H. Stein, *J. Biol. Chem.* **176,** 367 (1948).

ance at 570 nm is an index of the free amino groups that can react with ninhydrin, and the degree of acetylation can therefore be calculated from the percentage decrease of this absorbance. Usually, the color yield of the protein is related to that for a standard amino acid, such as leucine, and is then expressed as leucine equivalents per gram or per micromole of protein. At best, the data are semiquantitative, but in most instances they serve as a convenient gauge of the extent of reaction. More accurate data are obtained by gasometric analysis of amino groups by the method of Van Slyke.[19] Formol titrations have been used in some instances. Free amino groups have also been determined by reaction of the protein with fluorodinitrobenzene followed by 18-hour hydrolysis of the DNP-protein *in vacuo* at 105° using 6 *N* HCl.[20] Chromatography of the aqueous phase after ether extraction allows determination of DNP-lysine corresponding to those residues that were not acetylated. Amino acid analysis[21] of the hydrolyzate identifies the number of lysine residues which were acetylated and hence protected against reaction with FDNB.

A number of procedures have been reported which are based on the titrimetric determination of the acetic acid released on hydrolysis of the acetylated protein with *p*-toluene sulfonic acid.[22] A sample containing 1–5 mg of acetyl groups is hydrolyzed for 2–4 hours under reflux with 20 ml of a 25% aqueous solution of *p*-toluene sulfonic acid. The hydrolyzate is quantitively transferred to a 100-ml micro-Kjeldahl flask, and the acetic acid is steam distilled. The distillate is titrated after removal of CO_2 by bubbling with nitrogen. A variation of this procedure has been reported which is suitable for micro amounts of acetylated protein.[23] Other variations involve the use of sodium methoxide[24,25] or sulfuric acid[2] as the hydrolyzing agent. The liberated acetate can also be assayed by gas–liquid chromatography.

If analysis demonstrates that more acetyl groups have been incorporated into the protein than can be accounted for by amino group modification, the possibility of *O*-acetylation must be examined (see this volume [42]). *O*-Acetyltyrosine can be deacetylated by hydroxylamine at neutral pH and can be quantitated either by spectral methods or by analy-

[19] D. D. Van Slyke, *J. Biol. Chem.* **83**, 425 (1929).

[20] H. Fraenkel-Conrat, J. I. Harris, and A. L. Levy, *Methods Biochem. Anal.* **2**, 359 (1955).

[21] D. H. Spackman, W. H. Stein, and S. Moore, *Anal. Chem.* **30**, 1190 (1958).

[22] E. A. Kabat and M. M. Mayer, "Experimental Immunochemistry," 2nd ed., p. 493. Thomas, Springfield, Illinois, 1961.

[23] P. Vithayathil and F. M. Richards, *J. Biol. Chem.* **235**, 1029 (1960).

[24] R. L. Whistler and A. Jeanes, *Ind. Eng. Chem. Anal. Ed.* **15**, 317 (1943).

[25] F. B. Cramer, T. S. Gardner, and C. B. Purves, *Ind. Eng. Chem. Anal. Ed.* **15**, 319 (1943).

sis of the resultant acethydroxamate. Esters of serine or threonine can be hydrolyzed with hydroxylamine above pH 10.5 and also quantitated from hydroxamate formation.[26] Thiol esters have a characteristic absorption maximum between 230 and 235 nm[27] and are also hydrolyzed by hydroxylamine. *N*-Acetylation of histidine is a possibility which cannot be dismissed even though acylimidazoles are quite unstable in aqueous solution and hydrolyze spontaneously. These can be detected by their absorbance at 245–250 nm.

Comments

Complete acetylation of amino groups in proteins occurs rarely; usually the reaction with acetic anhydride proceeds to between 60 and 90% substitution. It has been noted that complete acetylation of the S-peptide of ribonuclease required a 300-fold molar excess of anhydride.[23] When acetylation was performed with a 50-fold molar excess, mono-, di-, and triacetylpeptides were formed which could be separated by electrophoresis. Similarly, random acetylation of amino groups in proteins might be expected to yield a family of products wherein the number of acetyl groups as well as their distribution throughout the molecule could vary over a wide range.

Changes of enzymatic activity can be observed consequent to acetylation, which may be due to a variety of causes. Thus, loss of activity may be correlated with the modification of a single functional group of the enzyme.[28] However, since amino groups are so reactive, usually several of them become modified and it then becomes necessary to differentiate functionally active residues from others which are equally as reactive chemically. In such instances, protection experiments using substrates or inhibitors to prevent acetylation of active center residues have been most helpful in determining the number of critical groups.[12,29]

[26] A. K. Balls and H. N. Wood, *J. Biol. Chem.* **219**, 245 (1956).

[27] L. H. Noda, S. A. Kuby, and H. A. Lardy, *J. Amer. Chem. Soc.* **75**, 913 (1953).

[28] R. F. Colman and C. Frieden, *J. Biol. Chem.* **241**, 3661 (1966).

[29] C. H. W. Hirs, M. Halmann, and J. H. Kycia, *Arch. Biochem. Biophys.* **111**, 209 (1965).

[42] *O*-Acetyltyrosine

By J. F. RIORDAN and B. L. VALLEE

O-Acetyl groups are readily introduced into the tyrosyl residues of proteins, in some instances leading to functional consequences.[1,2,3] The reagents most commonly employed for this purpose are acetic anhydride and *N*-acetylimidazole. Acetic anhydride reacts with amino and sulfhydryl groups as well. *N*-Acetylimidazole also acetylates these residues, but it reacts with amino groups much less readily than does the anhydride.[2] In certain instances, as in carboxypeptidase A, where the protein does not contain any free sulfhydryl groups, *N*-acetylimidazole selectively acetylates tyrosyl residues.[3] Other mono- and dicarboxylic acid anhydrides also acylate tyrosyl residues. The latter provide a particularly interesting means of modification, since the products undergo spontaneous deacylation at rates characteristic for the anhydride employed.[4]

Experimental Procedure

Acylation

Acetic Anhydride. As described in a previous volume[5] acetylation of proteins with acetic anhydride is sometimes carried out in half to fully saturated sodium acetate. Under these conditions, amino groups are converted to acetamides and tyrosyl groups to acetic esters. However, such high concentrations of acetate ion may catalyze the hydrolysis of the phenolic esters, thereby increasing the specificity of the anhydride toward amino groups. Therefore, in order to introduce *O*-acetyl tyrosyl residues into proteins, the acetylation must be carried out under conditions that will be favorable for the stability of the resulting ester.

The protein is dissolved (5 mg/ml) in a suitable buffer (see below), such as sodium barbital at pH 7.5 or 8. Acetylation is carried out on a pH-stat at 0–4°. Redistilled anhydride is added to the protein solution with a syringe microburette (Micro-metric Instrument Company, Cleveland, Ohio), and the pH is kept constant by the automatic addition of 1 *N* NaOH. Since the anhydride is immiscible in water, efficient stirring with a magnetic mixer must be maintained. The optimal ratio of anhy-

[1] J. F. Riordan and B. L. Vallee, *Biochemistry* **2**, 1460 (1963).

[2] J. F. Riordan, W. E. C. Wacker, and B. L. Vallee, *Biochemistry* **4**, 1758 (1965).

[3] R. T. Simpson, J. F. Riordan, and B. L. Vallee, *Biochemistry* **2**, 616 (1963).

[4] J. F. Riordan and B. L. Vallee, *Biochemistry* **3**, 1768 (1964).

[5] H. Fraenkel-Conrat, Vol. 4, p. 247.

dride to protein should be determined empirically. In this laboratory we have found that a 48-fold molar excess of anhydride maximally acetylated the tyrosyl residues of carboxypeptidase A, as evidenced by activity, while minimally affecting the structure of the protein. An increase of the ratio of anhydride to carboxypeptidase much beyond this leads to gross changes in physical properties.[6] When more than 50 mg of protein is to be acetylated, the anhydride should be added in aliquots small enough to allow the pH-stat to maintain constant pH at all times.

The uptake of alkali is usually biphasic. The first phase corresponds to acetylation of the protein, and the second to the spontaneous hydrolysis of the anhydride. The reaction is judged to be complete either when alkali uptake ceases or when the second phase becomes linear. This usually occurs within 30 minutes. Excess reagent can be removed either by dialysis or by gel filtration.

Other Anhydrides. These are employed in the same manner as acetic anhydride. Succinic and other solid anhydrides are weighed out and transferred directly to the reaction vessel prior to addition of the enzyme solution. The electrode is inserted and titration is started as soon as possible after all components are mixed. Where the purity of the anhydride is in doubt, distillation or recrystallization from a suitable solvent, such as glacial acetic acid, is recommended. The presence of free acid can be checked by adding a small amount of the anhydride to cold, aqueous sodium bicarbonate.

N-Acetylimidazole. Acetylation is usually performed at room temperature by addition of a weighed amount of the reagent to the protein in buffer at pH 7.5. This pH is optimal for the stability of *N*-acetylimidazole[7]; however, the reaction can be performed at lower pH values with a proportionately greater amount of reagent. Higher pH levels should be avoided, since acetylimidazole is not only less stable, but *O*-acetyltyrosine is rapidly hydrolyzed under alkaline conditions. In most instances investigated so far, all reactive tyrosyl residues can be acetylated with a 60-fold molar excess of reagent although, in some instances even higher ratios have been found necessary. For optimal results, the ratio of acetylimidazole to protein should be determined in each instance. Where large excesses of reagent are used under conditions of low buffering capacity, pH should be maintained with a pH-stat.

N-Acetylimidazole is available commercially (Cyclo, Eastman, K & K, Pierce, Sigma) but can be synthesized readily.[8] The reagent is quite hy-

[6] J. L. Bethune, D. D. Ulmer, and B. L. Vallee, *Biochemistry* **3**, 1764 (1964).

[7] E. R. Stadtman, *in* "The Mechanism of Enzyme Action" (W. D. McElroy and B. Glass, eds.), p. 581. Johns Hopkins Press, Baltimore, Maryland, 1954.

[8] J. H. Boyer, *J. Amer. Chem. Soc.* **74,** 6274 (1952).

groscopic and should be stored *in vacuo* over a suitable desiccant. If it does become wet, it should be dissolved in dry benzene, the solution dried with anhydrous sodium sulfate, and the product recrystallized. Solutions of *N*-acetylimidazole in dry benzene have been kept for periods of at least 1 month without undergoing hydrolysis, and this procedure has proved to be convenient. An aliquot of this solution, containing the calculated excess of reagent, is transferred to a test tube, and benzene is removed by passing a stream of air over the solution while the tube is immersed in a beaker of warm water. The air should be dried by flowing it through a packed cylinder of Drierite. The protein solution is then added to the test tube to initiate the acetylation reaction. For spectral studies, excess reagent can be removed by gel filtration, dialysis, or by allowing it to undergo spontaneous decomposition.

Other Acetylating Agents. Ketene has been used to acetylate tyrosyl residues of proteins, but it is difficult to handle and often produces undesirable structural alterations. Studies with pepsin have shown that this reagent modifies tyrosyl residues, constituting one of the earliest indications of a functional role for such residues in enzymes.[9]

Transesterification using active esters such as *p*-nitrophenyl acetate can also be employed to acetylate tyrosyl residues. Experiments in this laboratory have shown that *p*-nitrophenyl acetate can acetylate the functional tyrosyl residues of carboxypeptidase. However, a 100-fold molar excess of reagent brought about only 40% of the modification achieved with *N*-acetylimidazole. The release of *p*-nitrophenol, which can be followed in the course of modification, offers an advantage when *p*-nitrophenyl acetate is employed. However, there is a large blank reaction, and protein groups other than tyrosine can also cause hydrolysis of the ester. Hence, the reagent is recommended only for special situations.

Acyl halides have also been employed as acylating agents, but these reactions are quite vigorous.[10] Not only is their reactivity less specific, but they are more likely to lead to denaturation. They should be used with caution.

Deacetylation

The *O*-acyl tyrosyl residue is readily hydrolyzed to regenerate the original functional group. This offers a distinct advantage in the use of acylating reagents for the modification of tyrosyl hydroxyl groups in proteins. Since this hydrolysis can be carried out under mild conditions, it is possible to assess the role of tyrosyl groups in protein function. The

[9] R. M. Herriott, *J. Gen. Physiol.* **19**, 283 (1935).
[10] G. H. Dixon and H. Neurath, *J. Biol. Chem.* **225**, 1049 (1957).

usual procedure is to incubate the acetylated protein with hydroxylamine at pH 7.5. The time required for complete deacylation is a function of the hydroxylamine concentration. In general, a 10-minute incubation is required with 1 *M* hydroxylamine, while at least 400 minutes would be necessary with a 0.01 *M* solution of this nucleophile. The choice of the concentration of hydroxylamine therefore allows control over the rate of the reaction, which can be studied in the desired detail. However, the lability of *O*-acyl tyrosine also presents some problems. Thus, extremes of pH lead to hydrolysis, which can be quite rapid even at pH 9; hence, the pH range within which the functional consequences of acetylation can be investigated are limited.

In addition, a large number of buffer anions are capable of promoting hydrolysis as a result of their nucleophile properties. In this connection the choice of buffers in which the acetylation is carried out can significantly affect the reaction. We have found that acetylation in 0.05 *M* Tris at pH 7.5 results in the formation of fewer *O*-acetyltyrosyl residues than in 0.02 *M* sodium barbital.[2] Tris is known to act as a nucleophile, causing hydrolysis of *O*-acetyltyrosine, and should therefore be avoided. Any buffer or condition employed should be tested for its effect on the stability of *N,O*-diacetyltyrosine prior to acetylation of the enzyme.

Estimation of O-Acetyltyrosyl Residues

Acethydroxamate Formation. Deacetylation of the acetylated protein is carried out at room temperature by mixing equal volumes of protein solution (5 mg/ml) and 2 *M* hydroxylamine, previously adjusted to pH 7.5. After the preparation has stood for 10 minutes, the liberated acethydroxamate is determined colorimetrically by the method of Balls and Wood.[11]

Spectral Analysis. O-Acetylation of tyrosine produces a characteristic decrease in absorption between 250 and 300 nm, and the molar absorptivity at 275 decreases.[1] The difference in molar absorptivity between the native and acetylated protein may be employed to calculate the number of *O*-acetyltyrosyl residues. The maximal change in absorption due to acetylation of proteins has been found to be at 278 nm, where the molar absorptivity of tyrosine decreases from 1230 to 70, a difference of 1160 per mole of *O*-acetyltyrosine formed. This method suffers in that the molar absorptivity of the *modified enzyme* must be determined carefully. This can be circumvented by measuring the increase in molar absorptivity at 278 nm on addition of hydroxylamine to a solution of the acetylated enzyme. The rate of deacetylation is proportional to the hy-

[11] A. K. Balls and H. N. Wood, *J. Biol. Chem.* **219,** 245 (1956).

droxylamine concentration. Full deacetylation occurs within 10 minutes with 1 M hydroxylamine, whereas it requires up to 400 minutes with 0.01 M hydroxylamine. When no further change in absorbance occurs the number of O-acetyltyrosyl residues, N, can be determined from the relationship

$$N = \frac{A_{278} \times M}{1160 \times C} \tag{1}$$

where A is the absorbance, M is the molecular weight of the protein, and C is its concentration in milligrams per milliliter. C can be determined from the known molar absorptivity of the native protein at 280 nm.

Applications

Site-Specific Modifications. While both acetic anhydride and N-acetylimidazole can be employed to acetylate tyrosyl residues of proteins, in most cases the latter will be the reagent of choice. It is not only more specific, but it is also less likely to induce unwanted structural modifications.[6] Two additional features of the acetylation reaction make it quite well suited to structure–function studies in enzymes. The course of acetylation can be followed readily by measuring the change in absorbance at 278 nm and, in addition, the reaction is reversible. With the procedures described here, it has been found that acetylation of carboxypeptidase A leads to virtual abolition of its peptidase activity while increasing esterase activity.[1] No changes in activity were observed when the acetylation was performed in the presence of a competitive inhibitor of the enzyme. Deacetylation of acetylcarboxypeptidase restored activities to the values characteristic of the native enzyme. Comparison of the molar absorptivities of enzymes acetylated in the presence and absence of inhibitor demonstrated that the activity changes were accompanied by acetylation of two tyrosyl residues presumed to be located at the active center.

Acylation with dicarboxylic acid anhydrides also provides a most convenient, albeit somewhat different, means for identification of functional tyrosyl residues in proteins. Succinylation of proteins results in the formation of O-succinyltyrosyl residues and brings about spectral changes analogous to those seen on acetylation.[4] However, O-succinyltyrosine is unstable and undergoes spontaneous deacylation due to intramolecular nucleophilic catalysis of ester hydrolysis by the free carboxyl group.[4,12] Consequently, the spontaneous restoration of activity altered by succinylation is a first indication of a functional tyrosyl residue. The

[12] T. C. Bruice and U. K. Pandit, *J. Amer. Chem. Soc.* **82**, 2860 (1960).

rate of deacylation of *O*-acyltyrosine depends on the nature of the acyl group. Thus, for a series of acyl carboxypeptidases, the rates of deacylation measured both by changes in absorptivity and in enzymatic activity were methylsuccinyl > succinyl > β,β-dimethylglutaryl > α,α-dimethylglutaryl > glutaryl, exactly the same as found in a series of *p*-bromophenyl esters.[4,12] This correspondence of the relative rates of deacylation of the acylcarboxypeptidases and the model compounds constituted yet additional evidence for the functional role of tyrosyl residues in this enzyme.

"Exposed" and "Buried" Tyrosoyl Residues. *N*-Acetylimidazole may be used to determine the number of "exposed" and "buried" tyrosyl residues of a protein.[2,3] Based on chemical reactivities, solvent perturbations, and pH titrations the designations, "exposed" and "buried"[13] were introduced to describe the state of amino acid side chains in proteins. These designations may have been taken literally at one time. However, they are employed here simply to denote the susceptibility of certain residues to chemical modification, but are not intended to define the location of residues on the interior or exterior of the three-dimensional protein structure, though this could obviously influence reactivity. Other environmental factors such as charge or polarity of neighboring residues are also recognized to contributions to chemical reactivity. By comparison with data on amino acid compositions, the number of "exposed" and "buried" tyrosyl residues can be determined with acetylimidazole in a manner that is technically less demanding but apparently quite as accurate as spectrophotometric titration.[14] In a recent study, all tyrosyls of synthetic random copolymers were acetylated readily.[2] However, at least two classes of tyrosyl residues could be detected in a number of proteins on the basis of their reactivity toward acetylimidazole. Those that were acetylated with a 60-fold molar excess of acetylimidazole under the conditions described above were considered to represent the "exposed" tyrosyls, and the numbers of such groups correlated quite closely to those determined by the procedure of Crammer and Neuberger[14] as having a p*K* less than 10.5. Those not acetylated under these conditions were considered "buried." Minor discrepancies between the values obtained by acetylation and spectrophotometric titration were thought to reflect tyrosyl residues which are partially buried or in "crevices."[2]

A number of other reagents have been employed to investigate "exposed" and "buried" tyrosyl residues in proteins.[15] Cyanuric fluoride, in particular, has been employed with a variety of proteins, and the results

[13] T. T. Herskovits and M. Laskowski, Jr., *J. Biol. Chem.* **237,** 2481 (1962).

[14] J. L. Crammer and A. Neuberger, *Biochem. J.* **37,** 302 (1943).

[15] B. L. Vallee and J. F. Riordan, *Annu. Rev. Biochem.* **38,** 733 (1969).

were compared with those obtained using acetylimidazole.[16] In most instances, these two reagents are quite comparable. However, there are marked differences in the experimental procedures, especially with regard to the pH at which modification is carried out, cyanuric fluoride being used at pH 10 or above, whereas acetylimidazole is used at pH 7.5. Moreover, since neither reagent is absolutely specific for tyrosine, modification of other residues might lead to alterations in protein conformation that could affect the "exposure" of tyrosyl residues, hence leading to different numbers of "exposed" residues by the two procedures.

Functional "Buried" Tyrosyl Residues. A number of enzymes are reversibly inactivated by denaturing agents such as urea or guanidine. With such proteins it is possible to study the role of "buried" residues by performing chemical modification reactions on the denatured form. Trypsin, which is reversibly denatured by urea, has been shown to contain 6 "exposed" and 4 "buried" tyrosyl residues. Acetylation of the active enzyme resulted in modification of 6 or 7 tyrosyl residues without alteration of activity. Acetylation in 8 *M* urea, however, led to modification of all the tyrosyl residues and to complete loss of activity.[17] The inability of the totally acetylated enzyme to regain activity on removal of urea was shown to be directly related to tyrosyl group modification. Addition of hydroxylamine hydrolyzed the *O*-acetyl linkages and simultaneously restored enzymatic activity. Similar results have been obtained on acetylation of ribonuclease. It would appear that in both these instances the "buried" tyrosyl residues participate in maintaining the tertiary structure of the enzyme.

[16] M. J. Gorbunoff, *Arch. Biochem. Biophys.* **138,** 684 (1970).

[17] J. F. Riordan, W. E. C. Wacker, and B. L. Vallee, *Nature (London)* **208,** 1209 (1965).

[43] Cyanuration[1]

By MARINA J. GORBUNOFF

Principle

Cyanuric fluoride (CyF) was introduced by Kurihara *et al.* in 1963 as a tyrosine modifying reagent.[2] Since CyF may be considered as a very

[1] This work was supported in part by the National Institutes of Health, Grant No. GM-14603.

[2] K. Kurihara, H. Horinishi, and K. Shibata, *Biochem. Biophys. Acta* **74,** 678 (1963).

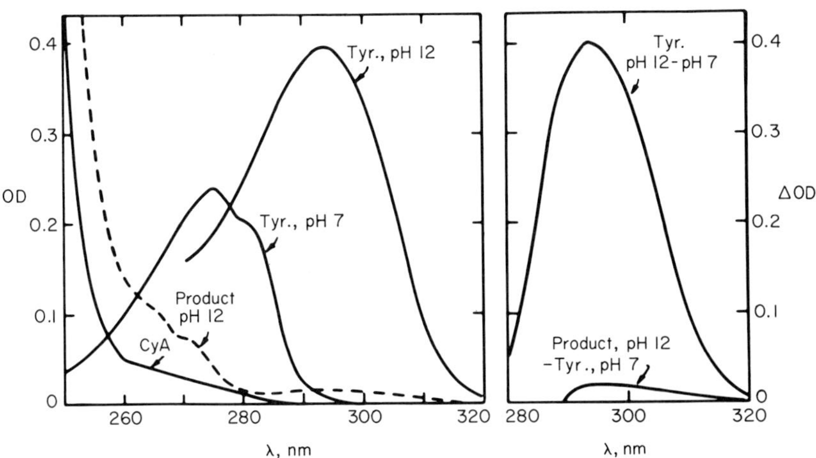

FIG. 1. Direct and difference ultraviolet spectra of tyrosine at pH 7 and 12, cyanuric acid (CyA), and the reaction product between tyrosine and CyF. Tyrosine concentration, $1.8 \times 10^{-4}\,M$; CyF concentration, $0.0232\,M$.

reactive acid halide, cyanuration of tyrosine should give the *O*-derivative, in complete analogy to the acylation of phenols with acid halides. This affects the π-π^* transition responsible for the 275-nm absorption band of tyrosine and causes a shift of this band to lower wavelengths (~260 nm).

The number of reacted residues may be estimated from changes in the tyrosine UV absorption upon cyanuration. For this, the 295-nm band of the phenolate ion is employed. It offers the advantages that none of the side products of the reaction absorb above 290 nm[2]; cyanuric acid, which is the hydrolysis product of CyF in aqueous medium, has absorption only below 280 nm. The principle is the same as that of the spectrophotometric titration of tyrosine residues.[3] Since the *O*-cyanurated derivative cannot ionize and, thus, does not contribute to the 295-nm band, the difference between the intensity of this band in the intact and reacted proteins at pH values of complete tyrosine ionization is a direct measure of the extent of the reaction. All the pertinent spectroscopic changes are shown in Fig. 1.

Procedure

Materials

CyF is highly hygroscopic. Therefore, no commercially available material should be used without prior purification. For best results,

[3] D. B. Wetlaufer, *Advan. Protein Chem.* **17,** 303 (1962).

fractionation on a Vigreux column is recommended. Only material boiling in the range of 70–80° (bp of CyF is 74°) should be used. Purified CyF should be stored in desiccators over calcium hydride or phosphorus pentoxide. The best grade of dioxane (Fisher Scientific Co.) is distilled twice from potassium hydroxide pellets and stored in the frozen state. The 1 *M* $KHCO_3$–KOH buffers used in the reaction should not be stored for a prolonged time, since, on standing, they tend to develop UV-absorbing backgrounds.

All CyF solutions are made up in volumetric flasks in the purified dioxane, which is redistilled from calcium hydride immediately before the preparation of solutions. As a stock solution, 2 ml (3.12 g) of CyF is dissolved in dioxane and brought to a total volume of 10 ml in a volumetric flask. Working solutions of CyF of needed concentrations are made up from this stock by appropriate dilutions. All solutions are kept in a desiccator over calcium hydride and are discarded after 1 week.

Methods

Prior to cyanuration, each protein is titrated by the difference spectral technique[3] in 1 *M* $KHCO_3$ buffer containing 10% dioxane in the pH range of 8.5–13.5 (or the pH of maximal ionization, if lower than 13.5) to determine whether the addition of 10% dioxane affects its ionization behavior. The results of this titration are compared with data obtained in the same buffers without dioxane. Cyanuration is carried out only under conditions in which the titration curves in the presence and absence of dioxane are identical and in which there are no observable time effects in the tyrosine titration.

Reaction with CyF. The protein solution (1 ml) at the desired concentration is mixed with 6.5 ml of 1 *M* $KHCO_3$ buffer at a given pH. To this, 1 ml of the CyF solution in dioxane of a given concentration is added, and the reaction flask is shaken well. After the mixture has stood at the desired temperature for a given length of time (usually 45 minutes),[4] the pH of the reaction mixture is adjusted to the pH found necessary for the maximal ionization of tyrosine residues in the titration experiments, and the solution is made up to 10 ml with a 1 *M* $KHCO_3$ buffer of the same pH.

[4] The length of the reaction time is not related to the time necessary for the completion of the reaction between CyF and the reactive ("available") tyrosine residues. This reaction is very rapid; it must take place before the occurrence of any considerable hydrolysis of CyF. Since the latter process is quite rapid (CyF is hygroscopic), the reaction of CyF with the tyrosine residues must be terminated immediately after the mixing of the two solutions. The reaction time is dictated by the necessity of the complete hydrolysis of excess CyF to the level of cyanuric acid, so as to preclude spectral interference in the region of 290–300 nm.

Determination of the Extent of Cyanuration. The number of cyanurated tyrosine residues is determined from the change in intensity of the phenolate ion absorption band in the 295-nm region. To do this, the UV spectrum of the reaction product at the pH of maximal tyrosine ionization is recorded vs. an unreacted protein standard of identical concentration in 0.1 *M*, pH 7, phosphate buffer containing 10% dioxane.[5] As a control it is necessary to determine whether the UV absorptions between 290 and 340 nm of the CyF-treated protein adjusted to pH 7 and of the untreated protein at pH 7 are identical. If this is not so, the difference in the UV absorptions of the two protein samples must be determined for all the conditions (combinations of pH and temperature) under which the cyanuration experiments had been carried out. This difference at 295 nm (or at the wavelength of the peak maximum as determined in titration experiments) must be introduced to correct for the absorption difference between the two reference solutions. In some cases it becomes impossible to obtain reliable UV spectra of the CyF-treated proteins at pH 7 because of strong time-dependent aggregation and resulting light scattering. This prohibits the use of the above described technique, making it necessary to use as reference standards solutions of untreated proteins at the pH of their maximal ionization.

The time selected for recording the UV spectra is dictated by the results of the preliminary titration of the untreated protein. It is the time necessary for the complete ionization of all buried tyrosine residues, i.e., the time necessary for the protein to unfold sufficiently to permit all buried tyrosines, if any, to come into contact with the medium and to become ionized. It has been found, however, that the presence of cyanuric acid at high pH causes aggregation of some proteins, with the result that the time-dependent increase in optical density is principally due to light scattering resulting from aggregation.[6] In such cases, the UV spectra are recorded for the minimal time after adjustment to high pH (2 minutes). Since this time frequently is not sufficient to produce complete ionization, a correction factor must be introduced. This factor is the inverse of the fraction of tyrosines ionizable in the given protein after 2 minutes at the high pH. This fraction is determined in triplicate for each set of reactions.

In a CyF-treated protein only those tyrosine residues which had not reacted with CyF under given conditions of pH and temperature will ionize. Thus, the number of tyrosine residues which reacted under a given

[5] In certain cases, in which standing at pH 7 affects the protein (e.g., autolysis of chymotrypsin), the reference solution is made up at conditions of the maximal stability of the protein, provided that under these conditions the UV spectrum of the protein does not differ from that at pH 7.

[6] M. J. Gorbunoff, *Arch. Biochem. Biophys.* **138**, 684 (1970).

set of conditions is given by the difference between the numbers of tyrosine residues which can be caused to ionize after standing at the pH of maximal ionization in the untreated and CyF-treated proteins. This is calculated in the same manner as in spectrophotometric titrations: the optical density at the wavelength of maximal absorption is divided by the product of the molar concentration of the protein with the molar extinction coefficient of phenoxide ion in the given protein. The result gives the number of moles of reacted tyrosine residues for experiments where a high pH reference was used, while for those recorded vs. a pH 7 reference, the result is the number of unreacted residues.

Method of Data Analysis. The tyrosine residues in a protein can be classified on the basis of changes in their reactivity with CyF caused by changes in pH, temperature, or both. This is done as follows. In preliminary experiments, the pH profile for a given protein is obtained by determining the average number of tyrosine residues in a protein which had reacted at the desired temperatures as a function of pH at a single constant concentration of protein and CyF, chosen at a level of reagent which is known to fall normally within the plateau range (0.023 M). Figure 2 shows the pH profiles for the α-, γ-, and δ-chymotrypsins.[7] On the basis of these profiles, concentration curves (experiments

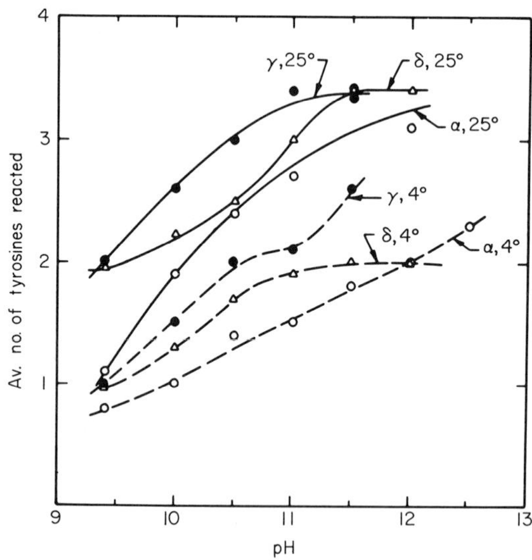

FIG. 2. Extent of average reactivity of the tyrosines of α-, γ-, and δ-chymotrypsins toward CyF at 4 and 25° as a function of pH. CyF concentration, 0.0232 M; protein concentration, 5.0 g/l.

[7] M. J. Gorbunoff, *Biochemistry* **10,** 250 (1971).

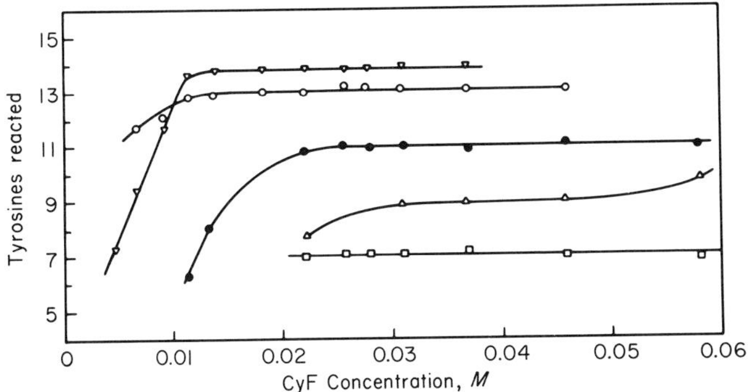

FIG. 3. Dependence of tyrosine reactivity in pepsinogen on CyF concentrations. □, 3°, pH 9.3; △, 3°, pH 9.7; ●, 25°, pH 9.7; ○, 25°, pH 10.8; ▽, 25°, pH 12.1.

with increasing concentration of CyF) are obtained. Combinations of pH and temperature under which whole numbers of tyrosines have reacted with 0.023 *M* CyF are then tried with 0.064 *M* CyF. Those giving the same whole numbers of reacted tyrosines with 0.064 *M* CyF as with 0.023 *M* CyF are selected for obtaining concentration curves. Figure 3 shows the concentration curves obtained with pepsinogen for several combinations of pH and temperature.[8] A plateau in a concentration curve indicates that a discrete level of tyrosine reactivity has been reached. These plateaus may be separated from each other by one or more

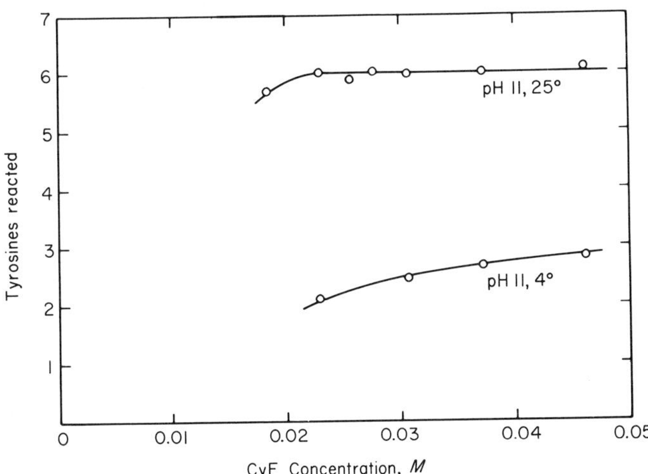

FIG. 4. Dependence of average tyrosine reactivity of bovine carbonic anhydrase B on CyF concentration at pH 11.0 and 4° and 25°. Protein concentration: 4.3 g/l.

[8] M. J. Gorbunoff, *Biochemistry* **7**, 2547 (1968).

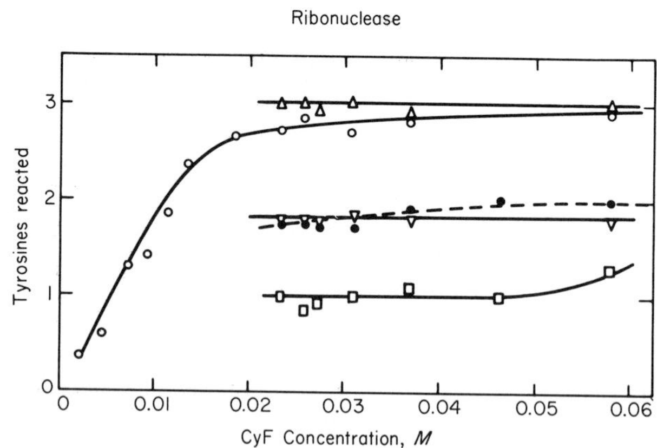

FIG. 5. Dependence of tyrosine reactivity in ribonuclease on CyF concentration. □, 3°, pH 9.5; ▽, 3°, pH 10.9; ●, 25°, pH 9.5; ○, 25°, pH 10.9; △, 25°, pH 11.2.

residues. The number of plateaus gives the number of states in which the tyrosine residues exist within the protein.[6] Obtaining an integral number of reacted tyrosines at a single concentration of CyF does not represent per se a discrete level of tyrosine reactivity, since in the absence of a plateau, the extent of reaction is a function of the amount of reagent used. This is shown in Fig. 4 for bovine carbonic anhydrase B.[6] There appears to be two reactive tyrosine residues at pH 11.0 at 4° and 6 residues at 25° when the enzyme is reacted with 0.023 *M* of CyF. Only pH 11.0 and 25° represent plateau conditions, however. The experiments at pH 11.0 and 4° illustrate conditions under which no plateau can be attained. Near a plateau, the concentration curves assume an asymptotic character. This is clearly shown by the results of the cyanuration of ribonuclease at 25° at pH 9.5 and 10.9 (Fig. 5).[9]

Comments

In a large excess of reagent, the cyanuration of tyrosine in 1 *M* $KHCO_3$ goes to completion in the pH range 9.6–12.6, both at 25°, and 4° as shown in Fig. 6 (curves C, D, E and F). It is independent of temperature and of pH in this range, which is essentially above the p*K* of the phenolic hydroxyl ionization. The decrease in reactivity below this pH range must reflect the decrease in the concentration of tyrosinate ion. In fact, when the reaction is carried out at such a concentration of CyF (0.0046 *M*) that modification is incomplete at all pH values (see curves A and B of Fig. 6), the shape of the pH-dependence of the extent of

[9] M. J. Gorbunoff, *Biochemistry* **6**, 1606 (1967).

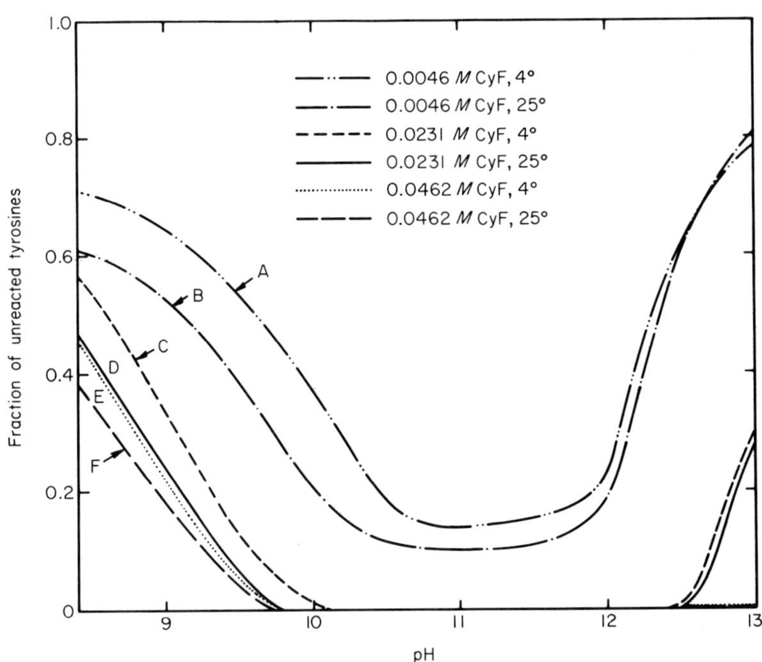

FIG. 6. Reactivity of L-tyrosine toward CyF as a function of pH, temperature, and reagent concentration. Tyrosine concentration: 3.13×10^{-4} M.

reaction at the low pH end is similar to that of the titration of an ionizable group. At pH 13 the rate of hydrolysis of CyF becomes much faster than that of the cyanuration reaction; this is suggested by the fact that a large excess of reagent is required to bring the reaction to completion (see curves E and F of Fig. 6).

The cyanuration of tyrosine is essentially temperature and pH independent in the pH region between 9.6 and 12.6, where cyanuration (substitution of F by tyrosinate ion) can compete with the hyrolysis of CyF (substitution of F by a hydroxyl ion).[10] At low CyF concentration, the hydrolysis of CyF is the predominant reaction and cyanuration cannot go to completion (see curves A and B of Fig. 6). The cyanuration of tyrosine residues in proteins, however, shows both pH and, with a few exceptions,[7,11] temperature dependence. This difference in behavior must reflect the constraints imposed on the tyrosine residues by their immediate environment, which reflects the secondary and tertiary structures of the proteins. These constraints may indicate either steric inter-

[10] M. J. Gorbunoff, manuscript in preparation.

[11] M. J. Gorbunoff, *Biochemistry* **8**, 2591 (1969).

ference (extent of exposure), or the nature of the polarity and charge distribution of the structural domain in the immediate vicinity of the tyrosine residues.

The pH profiles obtained at two temperatures give an indication of the temperature dependence of the conformational stability of the protein in question in the region vicinal to the reactive tyrosines. For example, a difference of one tyrosine residue between the reactivities at 25° and 4° is shown by γ-chymotrypsin (Fig. 2), suggesting a strong temperature dependence of the conformational stability of this protein in the alkaline region. It should be emphasized again, however, that only plateaus indicate discrete levels of reactivity.

Critique. CyF offers several advantages as a tyrosine modifier. It probably meets the criterion of nonrandom action. This is strongly supported by the appearance of discrete plateaus independent of the amount of reagent used at various levels of tyrosine residue reactivity.[2,6–9,11–15] Peptide studies in plateau regions of insulin[16] and glucagon[17] have resulted in the identification of the particular residues cyanurated, confirming that the reaction does not occur in random fashion.

The results of modification with CyF cannot be compared, without reservations, with modification results obtained with other reagents. Since the accessibility of a residue to a reagent is a function of the method of probing, comparison of modification studies can be meaningful only if the reagents are specific for the same position on the modified residue.[6] This is subject, however, to differences in the spacial requirements of the transition state complexes of the reactions, as well as in the actual reaction mechanisms. Since nitration with tetranitromethane and iodination attack the benzene ring of the tyrosine, while CyF attacks its oxygen atom, accessibility to the reagent does not have the same meaning for these two types of modifiers. This thesis is supported by the calculations of Lee and Richards,[18] who have shown that, in the case of ribonuclease S, the tyrosine hydroxyls and the ring carbons are exposed to nonidentical extents.

[12] Y. Hachimori, K. Kurihara, H. Horinishi, A. Matsushima, and K. Shibata, *Biochim. Biophys. Acta* **105,** 167 (1965).

[13] Y. Hachimori, A. Matsushima, M. Suzuki, Y. Inada, and K. Shibata, *Biochim. Biophys. Acta* **124,** 395 (1966).

[14] A. Tachibana and T. Murachi, *Biochemistry* **5,** 2756 (1966).

[15] O. Takenaka, H. Horinishi, and K. Shibata, *J. Biochem.* (*Tokyo*) **62,** 501 (1967).

[16] M. Aoyama, K. Kurihara, and K. Shibata, *Biochim. Biophys. Acta* **107,** 257 (1965).

[17] A. Matsushima, Y. Inada, and K. Shibata, *Biochim. Biophys. Acta* **121,** 338 (1966).

[18] B. Lee and F. M. Richards, *J. Mol. Biol.* **55,** 379 (1971).

[44] Nitration with Tetranitromethane

By JAMES F. RIORDAN and BERT L. VALLEE

Tetranitromethane, TNM, was first employed for the modification of proteins by Wormall[1] and subsequently by Ehrenberg *et al.*,[2,3] and Astrup.[4] However, details of the chemistry of the reaction were not investigated at that time, and biological activity served as the sole criterion for reaction. Subsequently, TNM was examined for its suitability to modify amino acid residues in peptides, polymers, and proteins.[5,6] It was shown that the reagent can be highly specific for the nitration of tyrosyl residues, and since reaction conditions are very mild, selective modification is achieved readily. The reaction can be quantitated easily by spectrophotometry, since nitrotyrosine absorbs in the visible region of the spectrum, and by amino acid analysis. Further, other products of the reaction, nitroformate and protons, can also serve as a means of quantitation.[5,6]

Nitration of proteins is generally carried out in 0.05 *M* Tris buffer at pH 8, at room temperature (20°). TNM (Aldrich Chemical Co. or Fluka, A.G.) is diluted 1:10 with 95% ethanol to give a solution which is 0.84 *M*. TNM itself is quite insoluble in water, and the reagent can be washed with water to remove any impurities arising from decomposition. Aliquots of the ethanolic solution of TNM are added to the protein, and the reaction mixture is stirred for about 1 hour. It is best to work with low concentrations of protein (10^{-4} *M*) and low molar excesses of TNM (1 to 10×) due to the limited solubility of the reagent. If TNM forms a separate phase, spurious reactions might occur either at the solvent interface or with protein which has been extracted into TNM. The reaction can be terminated by passing the mixture through a Bio-Gel P-4 column in 0.05 *M* Tris buffer at pH 8.

The degree of nitration is determined based on the absorption of the nitrotyrosyl residue at 428 nm, $\epsilon = 4200$ at pH 9.0. This absorption maximum shifts to 360 nm in acid with an isosbestic point at 381 nm and $\epsilon = 2200$, which can be used for quantitation independent of the pH of the solution being measured.

[1] A. Wormall, *J. Exp. Med.* **51,** 295 (1930).

[2] L. Ehrenberg, I. Fischer, and N. Lofgren, *Sv. Kem. Tidskr.* **57,** 303 (1945).

[3] L. Ehrenberg, I. Fischer, and N. Lofgren, *Nature* (*London*) **157,** 730 (1946).

[4] T. Astrup, *Acta Chem. Scand.* **1,** 744 (1948).

[5] J. F. Riordan, M. Sokolovsky, and B. L. Vallee, *J. Amer. Chem. Soc.* **88,** 4104 (1966).

[6] M. Sokolovsky, J. F. Riordan, and B. L. Vallee, *Biochemistry* **5,** 3582 (1966).

Amino acid analysis can be employed as another method to quantitate the nitration of tyrosyl residues. Nitrotyrosine is stable to the conditions normally employed for acid hydrolysis of proteins and it elutes from the amino acid analyzer as a discrete peak just after phenylalanine. Under the usual reaction conditions TNM also oxidizes sufhydryl groups. Nitration of tyrosine is dependent upon pH with little or no modification at pH 6 and very rapid modification at pH 9. Nitration and thiol oxidation can often be distinguished by performing the reaction at pH 6, where only oxidation takes place.

Nitration of phenol model compounds by TNM has been investigated in detail by Bruice *et al.*,[7] who have suggested the following mechanism:

$$XPhO^- + TNM \rightleftharpoons \text{Charge-transfer complex}$$

$$\text{Charge-transfer complex} \rightleftharpoons \text{(rate-determining)} \; [XPhO\cdot + NO_2\cdot] + C(NO_2)_3^-$$

$$[XPhO\cdot + NO_2\cdot] \longrightarrow NO_2XPhO^-$$

$$[XPhO\cdot + NO_2\cdot] \longrightarrow XPhO\cdot + NO_2\cdot$$

$$NO_2\cdot + XPhO^- \rightleftharpoons XPhO\cdot + NO_2^-$$

$$n\ XPhO\cdot \longrightarrow \text{Coupling products}$$

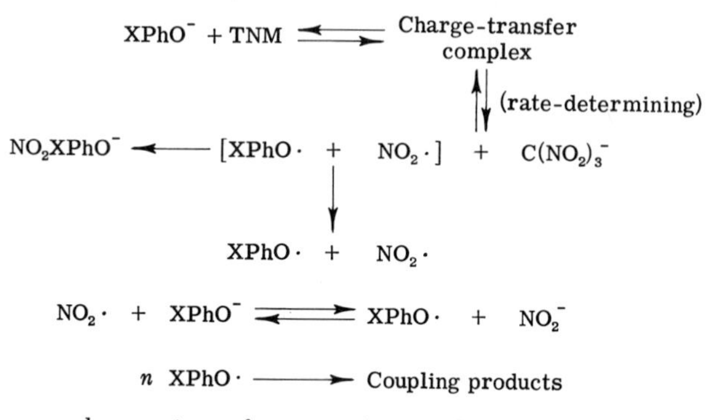

TMN forms a charge–transfer complex with phenoxide ion, in aqueous solution, thus accounting for the pH-dependence of the reaction. This is followed by a rate-determining electron transfer to generate the phenoxide and nitrite radicals plus the nitroformate anion. Nitration results from the coupling of $NO_2\cdot$ and $XPhO\cdot$, but additional reactions are made possible by the coupling of phenolate radicals with each other. Further, nitrite ions are formed by electron transfer between $NO_2\cdot$ and additional phenolate ions.

The reaction of TNM with sulfhydryl groups is thought to proceed by the following mechanism:[8]

$$R{-}SH + TNM \longrightarrow RSNO_2 + C(NO_2)_3^- + H^+$$

$$RSNO_2 \xrightarrow{RSH} RSSR + NO_2^- + H^+$$

$$RSNO_2 \xrightarrow{H_2O} RSOH + NO_2^- + H^+$$

$$RSOH \xrightarrow{[O]} RSO_2H$$

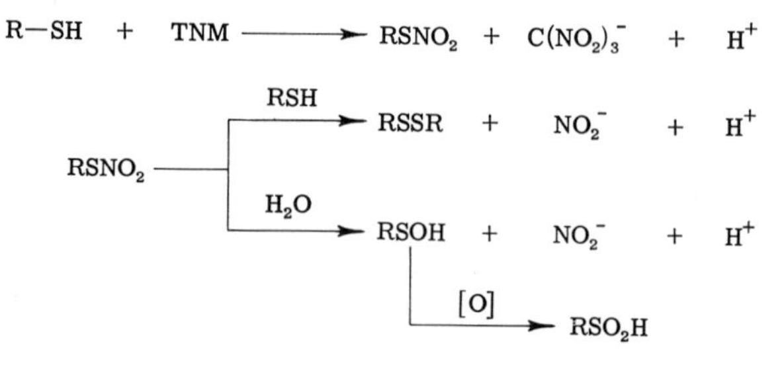

[7] T. C. Bruice, J. J. Gregory, and S. L. Walters, *J. Amer. Chem. Soc.* **90,** 1612 (1968).

[8] M. Sokolovsky, D. Harell, and J. F. Riordan, *Biochemistry* **8,** 4740 (1969).

An intermediate sulfenyl nitrate is formed which, in the presence of excess thiol, i.e., low molar excesses of TNM, gives the disulfide and nitrite ion. Alternatively, it can hydrolyze to give the sulfenic acid, which can air-oxidize to the sulfinic acid. The relative amounts of disulfide and sulfinic acid depend upon reaction conditions, the ratio of TNM to RSH, and the nature of the R group.[8]

As pointed out above, TNM can be highly specific for tyrosyl modification when it is employed under the mild conditions described. However, if it is used under more vigorous conditions, i.e., with high molar excesses at pH 8, or if the reaction is carried out at higher pH values, other residues such as histidine, tryptophan, and methionine will react.[8,9] The product of these reactions have not been identified thus far, but in the case of tryptophan a number of nitrated derivatives have been demonstrated.[9,10]

According to the above schemes, it should be possible to follow the modification of proteins with TNM by measuring the release of nitroformate anion. This species absorbs at 350 nm with an extinction coefficient of 14,000.[11] However, with a number of proteins the amount of nitroformate produced is greater than that which can be accounted for by tyrosyl modification, sulfhydryl oxidation, or loss of other residues, suggesting the possibility of a catalytic breakdown of TNM. Thus, the appearance of 350 nm absorption can serve only as a qualitative gauge for the reaction of TNM with proteins.

Nitration with TNM has been used successfully to study tyrosyl residues in many enzymes and proteins. Quite frequently the reaction is free from complications and the results can be interpreted without ambiguity. However, one should be aware of the fact that a number of circumstances can lead to a potentially significant side reaction, protein–protein interaction and polymerization.[12–14] In addition to the free radical mechanism for nitration,[7] the formation of nitrite ion, as a by-product, can also cause unwanted side reactions.[15] If, as has been done in some instances, modification is terminated by acidification, the resultant nitrous acid can (a) modify previously unreacted tyrosyl residues, (b) deaminate lysyl residues, (c) react with thiol and other side-chain groups, and (d) induce polymerization. Thus, it is critical that prior to lowering the pH,

[9] M. Sokolovsky, M. Fuchs, and J. F. Riordan, *FEBS (Fed. Eur. Biochem. Soc.) Lett.* **7**, 167 (1970).

[10] P. Cuatrecasas, S. Fuchs, and C. B. Anfinsen, *J. Biol. Chem.* **243**, 4787 (1968).

[11] D. J. Glover and S. G. Landsman, *Anal. Chem.* **36**, 1690 (1964).

[12] R. J. Doyle, J. Bello, and O. A. Roholt, *Biochim. Biophys. Acta* **160**, 274 (1968).

[13] J. P. Vincent, M. Lazdunski, and M. Dellaage, *Eur. J. Biochem.* **12**, 250 (1970).

[14] R. W. Boesel and F. H. Carpenter, *Biochem. Biophys. Res. Commun.* **38**, 678 (1970).

[15] M. Sokolovsky and J. F. Riordan, *FEBS (Fed. Eur. Biochem. Soc.)* **9**, 239 (1970).

excess reagent and by-products be removed from the reaction mixture by gel filtration or dialysis.

Protein polymerization may occur through another mechanism, i.e., the formation of nitrate esters.[8] Such active esters could react with, for example, amino groups, leading to inter- and intramolecular peptide bonds.

Nitration affords multiple opportunities for studying functional tyrosyl residues in proteins. In many cases modification is quite specific. Thus, the incorporation of about one nitro group per molecule of carboxypeptidase with a 4-fold molar excess of TNM enhances esterase and greatly reduces peptidase activity.[16] The presence of the inhibitor β-phenylpropionate during the nitration reaction prevents the changes in activity. Similarly, a low molar excess of tetranitromethane selectively nitrates the tyrosyl residue at position 85 in the extracellular nuclease of *Staphylococcus aureus*, resulting in loss of catalytic activity toward both DNA and RNA.[9] Reaction in the presence of Ca^{2+} and deoxythymidine 3′,5′-diphosphate, pdTp, selectively nitrates tyrosine 115, while tyrosine 85 is protected. This product is fully active toward DNA but only half active toward RNA.

Nitration of aspartate aminotransferase in the presence of the substrate pair, glutamate and α-ketoglutarate, abolishes activity within 1 hour concomitant with the modification of one tyrosyl residue. In the absence of substrate, there is virtually no inactivation, and in the presence of competitive inhibitors or either substrate alone only slight inactivation occurs. The tyrosyl residue only becomes reactive subsequent to the formation of the enzyme–substrate complex during the actual catalytic process. This *syncatalytic*, i.e., synchronous with catalysis, activation suggests that the tyrosyl residue plays a role in the mechanism of transamination.[17] By now at least two dozen other proteins and enzymes have been nitrated with TNM, with consequent alterations in biological activity. Many of these have been discussed in several recent reviews.[18,19]

TNM has been employed to examine intermediates in the enzyme-catalyzed reactions of yeast and muscle aldolases.[20] The high molar absorptivity of the product, nitroformate ($\epsilon_{350} = 14{,}400$), provides a convenient means to detect carbanionic enzyme–substrate complexes; in

[16] J. F. Riordan, M. Sokolovsky, and B. L. Vallee, *Biochemistry* **6**, 3609 (1967).
[17] P. Christen and J. F. Riordan, *Biochemistry* **9**, 3025 (1970).
[18] B. L. Vallee and J. F. Riordan, *Annu. Rev. Biochem.* **38**, 733 (1969).
[19] J. F. Riordan and M. Sokolovsky, *Accounts Chem. Res.* **4**, 353 (1971).
[20] P. Christen and J. F. Riordan, *Biochemistry* **7**, 1531 (1968).

fact, TNM has been suggested as a reagent to titrate carbanions, in general.[21]

The generation in a protein of a chromophore which absorbs in the visible region of the spectrum, i.e., nitrophenol, provides a number of convenient experimental approaches.

1. Since the nitrotyrosyl residue can ionize, it may serve to probe the microenvironment of active center residues by means of pertubation spectra and similar methods. Spectral titrations indicate an apparent p*K* of 6.3 for the nitrotyrosyl residue of mononitrocarboxypeptidase compared with the p*K* of 7.0 for *N*-acetyl-3-nitrotyrosine. Hence, the specific nitration of a single tyrosyl residue in carboxypeptidase would appear to be related to an abnormally low p*K* for that residue, perhaps owing to features of its chemical environment that induce ionization.

2. If the group modified is functional, substrates, substrate analogs, and inhibitors may affect its spectral properties. Addition of β-phenylpropionate to nitrocarboxypeptidase shifts the absorption spectrum. Spectral titrations in the presence of this inhibitor indicate a p*K* of 7.0 instead of 6.3 for the nitrotyrosyl residue, suggesting that the reagent alters the immediate chemical environment of the active center nitrotyrosyl residue. Further, studies on the nitration of tyrosyl copolymers have shown that the rate of nitration is a function of the net charge on the tyrosyl copolymer. Thus, if the tyrosyl residues are surrounded by positively charged lysyl groups, the rate of nitration is much faster and the p*K* of the resultant nitrotyrosyl group much lower than when they are surrounded by negatively charged aspartate or glutamate residues.

The specific nitration of tyrosine-115 in staphylococcal nuclease appears to be analogous to the situation in carboxypeptidase A. Thus, in the absence of nucleotide, the nitrotyrosyl residue has an abnormally low p*K* of 6.4. Addition of the inhibitor, deoxythymidine diphosphate, shifts the p*K* to 7.2, similar to that seen in model compounds. The p*K* of nitrotyrosyl residue 85 is also low, 6.6, but is altered only slightly by added nucleotide.

3. The nitrotyrosyl group, if optically active, might exhibit characteristic extrinsic Cotton effects as additional probes of protein conformation. Nitration of pancreatic trypsin inhibitor with TNM results in both a mono- and a disubstituted derivative in positions 10 and 10 and 21, respectively, both retaining full trypsin inhibitor activities. Optical rotatory dispersion measurements indicate that nitration does not affect the protein secondary structure but gives rise to a conformation-

[21] P. Christen and J. F. Riordan, *Anal. Chim. Acta* **51,** 47 (1970).

dependent, side-chain Cotton effect in the region of nitroaromatic absorption. A nitrotyrosine-containing pentadecapetide isolated from the tryptic digest of the nitrated and oxidized inhibitor does not exhibit such a Cotton effect.[22] Importantly, the spectral characteristics represent a guide to labeling and subsequently isolating nitrotyrosyl peptides.

4. The nitrotyrosyl residue is a potential site for derivatization since it can be reduced to an amino group using sodium hydrosulfite.[23] When a 5- to 6-fold molar excess of $Na_2S_2O_4$ is added to nitrotyrosyl derivatives at pH values between 6 and 9, nitrotyrosine disappears within a few minutes. The aminotyrosine derivative is a major product. The introduction of aminophenols into the primary sequence of proteins provides the opportunity for further specific chemical permutations since the aromatic amino group can be modified preferentially at slightly acid pH values.[23]

As determined by amino acid analysis, aminotyrosine has been generated virtually quantitatively in most proteins studied thus far, although in some model compounds, e.g., nitrotyrosine, glycylnitrotyrosine, and *N*-acetyl-3-nitrotyrosine, some other by-products have been observed, the major by-product presumably being a sulfamic acid derivative.[24] The mechanism of reduction by $Na_2S_2O_4$ is not yet clear, although participation of the SO_2 radical in the reduction could explain the formation of these products. Alternatively, or in addition, cyclization followed by hydrolysis might lead to such a product. In fact, yield of the by-product is increased if sodium sulfite is added prior to reduction of the nitrotyrosine.

While the pK' of the phenolic hydroxy group is lowered from near 10.0 to about 7.0 as a consequence of nitration, it is raised back to about 10.0 when the nitro group is reduced to an amino group. One of the mechanisms by which nitration can affect the biological activity of a protein is by altering the pK' of an essential tyrosyl residue. In instances where this mechanism prevails, reduction, because of its reverse effect on the phenolic hydroxyl pK', might be expected to restore, either fully or partially, activity changed as a result of nitration. Thus, in the case of *E. coli* alkaline phosphatase, nitration reduces hydrolase activity but leaves transferase activity unchanged. On reduction of the nitro to the aminotyrosyl enzyme, hydrolase activity is restored to that

[22] B. Meloun, I. Fric, and F. Šorm, *Eur. J. Biochem.* **4**, 112 (1968).

[23] M. Sokolovsky, J. F. Riordan, and B. L. Vallee, *Biochem. Biophys. Res. Commun.* **27**, 20 (1967).

[24] J. F. Riordan and M. Sokolovsky, *Biochim. Biophys. Acta* **236**, 161 (1971).

characteristic of the native enzyme and transferase activity increases 3-fold.[25] It is possible that alterations in pK' could account for these activity changes.

[25] P. Christen, B. L. Vallee, and R. T. Simpson, *Biochemistry* **10**, 1377 (1971).

[45] Diazonium Salts as Specific Reagents and Probes of Protein Conformation

By James F. Riordan and Bert L. Vallee

Diazonium salts have long been employed to modify proteins.[1] They are known to react readily with various amino acid side chains to form covalently bound chromophoric derivatives which absorb light between 300 and 600 nm. Such properties should make this class of chemical reagents particularly attractive since the absorption spectra of the azo-derivatives of tyrosine, histidine, and lysine appear sufficiently distinct to allow direct qualitative and quantitative identification of the corresponding residues in modified proteins. However, in practice, azoproteins have been found to exhibit multiple, broad, overlapping spectral bands whose lack of resolution has limited their value for investigations of structure–function relationships. This, together with a lack of specificity, formerly restricted interest in the use of diazonium salts as site-specific reagents, although they found widespread application, for instance, in the well-known Pauly test for imidazoles.[1]

Recent studies have again directed attention to the particular advantages offered by these reagents. Several years ago, diazonium-1*H*-tetrazole was introduced as a new coupling reagent to examine the degree of reactivity of histidyl residues in proteins and to differentiate those which are "buried" from those which are "free."[2] While not as specific for this purpose as implied initially, this reagent has been found to be useful for determining both the total tyrosyl and histidyl content of proteins and the preferential reactivities of some of these residues.[3,4] Furthermore, since azochromophores, covalently attached to proteins,

[1] H. Z. Pauly, *Hoppe-Seyler's Z. Physiol. Chem.* **42**, 508 (1904).

[2] H. Horinishi, Y. Hachimori, K. Kurihara, and K. Shibata, *Biochim. Biophys. Acta* **86**, 477 (1964).

[3] M. Sokolovsky and B. L. Vallee, *Biochemistry* **5**, 3574 (1966).

[4] A. Takenaka, T. Suzuki, O. Takenaka, H. Horinishi, and K. Shibata, *Biochim. Biophys. Acta* **194**, 293 (1969).

have been found to be both optically active and sensitive to changes in their environment, modification with diazonium salts can offer yet additional opportunities for examining changes in protein conformation related to biological function.[5,6,6a]

This article will cover procedures for the chemical modification of proteins with the common aromatic diazo reagents, for the determination of histidyl and tyrosyl residues in proteins with diazo-1*H*-tetrazole, and for employing azochromophores as probes of local protein conformation.

Procedures

Coupling with Diazonium Salt. The desired aromatic amine, 0.5 mmole, is weighed and dissolved in 20 ml of 0.15 *N* HCl. After this solution has cooled in an ice bath, 50 mg of sodium nitrite is added slowly with continuous stirring, and reaction is allowed to proceed for 15 minutes at 0°. These conditions are generally sufficient for complete diazotization as determined by coupling the reagent with excess of a phenol (e.g., *N*-chloroacetyltyrosine) for 30 minutes at pH 8.8 and room temperature. The yield of diazonium salt is estimated from the absorbance at 490 nm in 0.1 *N* NaOH, using a molar absorptivity determined with an authentic sample of the monoazotyrosyl derivative. Since the diazo reagent is unstable in acid media, the pH of the solution is increased to between 5.0 and 5.5 by the gradual addition of 3 *N* NaOH from a long-tipped syringe while the solution is stirred continually at 0°. The volume of the solution is then adjusted to 25 ml with water to give a reagent concentration of almost 2×10^{-2} *M*. Suitable aliquots of the reagent mixture are used immediately for protein modification, although the reagent is probably stable for at least 1 hour at 0°. In some instances it may be necessary to prepare a high concentration of the diazonium salt. Solutions 0.2 *M* in salt have been prepared without difficulty.[3]

Coupling is carried out with protein dissolved in 0.1–0.67 *M* sodium bicarbonate buffer, pH 8.8, 0°, generally at a concentration of 2×10^{-4} *M*. The concentration of buffer has not been found to be critical thus far. The reaction is initiated by the addition of a suitable aliquot of the diazonium salt, prepared as above, to the protein solution. Coupling is allowed to proceed at 0° for the desired length of time, usually 30–60 minutes, and is then terminated by the addition to the reaction mixture of sufficient 0.1 *M* aqueous phenol to give a final concentration, equal to that of the initial diazonium salt. The modified protein is transferred

[5] H. M. Kagan and B. L. Vallee, *Biochemistry* **8**, 4223 (1969).

[6] G. F. Fairclough, Jr. and B. L Vallee, *Biochemistry* **9**, 4087 (1970).

[6a] J. T. Johansen and B. L. Vallee, *Proc. Nat. Acad. Sci. U.S.* **68**, 2532 (1971).

into a suitable buffer solution by dialysis or gel filtration in order to remove any noncovalently bound azochromophores.

Diazonium salts couple readily with lysyl, tyrosyl, and histidyl residues of proteins to yield colored derivatives. The product of the reaction with lysyl residues has been thought to be a bisazo, i.e., pentazine, derivative.[7] However, recent evidence suggests that it may be a triazine.[5] Mono- and bisazo derivatives of both tyrosine and histidine can form, dependent upon the molar ratios of reactants. When the diazo reagent is used at 10 to 50 times the protein concentration, the monoazo species usually predominate. Studies with model compounds have shown that biscoupling normally requires as much as a 300-fold molar excess of reagent.

Quantitation of the modification can be carried out by spectral analysis[8,9] if data are available on the absorption characteristics of the proper model compounds. Since the spectra are dependent on the degree of ionization of the phenolic hydroxyl group of tyrosine, measurements are generally made with the protein dissolved in 0.1 N NaOH. Some typical spectral parameters for the azo derivatives of arsanilic, sulfanilic, and p-aminobenzoic acid are listed in the table. Because of the overlap of the absorption spectra for azotyrosine and azohistidine, it is necessary to set up the following simultaneous equation in order to determine each of these derivatives in an azoprotein[9]:

$$16.50X + 9.60Y = A_{460}$$
$$2.65X + 10.50Y = A_{500}$$

where X = moles of monoazohistidine per milliliter and Y = moles of monoazotyrosine per milliliter. The numerical coefficients are the molar absorptivities for the respective arsanilazoderivatives at 460 and 500 nm. These wavelengths are selected for measurement, since bisazolysine, if it is formed, interferes minimally in this region. The values for other azo derivatives will differ; hence, for accuracy, they should be determined using authentic model compounds. Solutions of azoproteins generally obey Beer's law.

The mono- and bisazo derivatives of histidine, tyrosine, and lysine are destroyed under the usual conditions for acid hydrolysis of proteins and the original amino acids are not regenerated. Hence, the number of such residues modified in a protein or peptide can also be determined by amino acid analysis. It is possible to measure directly the incorporation of certain diazo reagents by independent means, for example, determining the arsenic content of proteins labeled with diazotized arsanilic

[7] A. N. Howard and F. Wild, *Biochem. J.* **65**, 651 (1957).

[8] M. Tabachnick and H. Sobotka, *J. Biol. Chem.* **234**, 1726 (1959).

[9] M. Tabachnick and H. Sobotka, *J. Biol. Chem.* **235**, 1051 (1960).

SPECTRAL CHARACTERISTICS OF AZO DERIVATIVES OF AMINO ACIDS[a,b]

	Monoazotyrosine		Monoazohistidine		Bisazotyrosine		Bisazohistidine		Bisazolysine	
Diazo reagent	λ_{max}	*A*	λ_{max}	*A*	λ_{max}	*A*	λ_{max}	*A*	λ_{max}	*A*
Arsanilic acid	330	14,200	420	22,300	325	26,000			378	30,800
	490	11,000			545	17,500				
Sulfanilic acid	330	14,000	423	23,400			490	22,800	363	30,000[c]
	490	11,500								
p-Aminobenzoic acid	330	13,900	420	21,700						
	488	10,700								
5-Amino-1*H*-tetrazole[c,d]	480	8,400			570	12,200	480	25,700		

[a] From M. Tabachnik and H. Sobotka, *J. Biol. Chem.* **234,** 1726 (1959).

[b] In 0.1 *N* NaOH unless otherwise specified.

[c] At pH 10.0.

[d] A. Takenaka, T. Susuki, O. Takenaka, H. Horinishi, and K. Shibata, *Biochim. Biophys. Acta* **194,** 293 (1969).

acid.[5,10] There is some indication that other amino acid residues can also react with diazo reagents.[7] The NH_2-terminal α-amino group, and the NH group of NH_2-terminal proline or hydroxyproline can possibly be substituted. Certainly, free sulfhydryl groups can react, although the initial product, likely an *S*-azo derivative, probably decomposes to form a thio ether.[11] The guanido group of arginine may react with diazonium compounds, the product containing at least one, and perhaps up to three, diazo groups. There is even some evidence that aliphatic hydroxyl groups may undergo oxidation to aldehydes in the presence of excess diazo reagent.[12]

While this list includes many of the amino acid residues in proteins which have functional groups, it should be emphasized that in studies with proteins usually only tyrosine, histidine, and lysine (and cysteine, if present) will be modified. This, of course, assumes that low molar excesses of reagent will be employed under conditions that are not too alkaline and for a limited period of exposure.

It is quite possible that the specificity of the reaction can be increased further by the proper selection of pH, molar excess, and type of diazo reagent. Thus, for example, while only a 10% yield of 2-arylazoimidazole was obtained using *o*-methoxyphenyldiazonium chloride, an 85% yield was obtained with the *p*-bromophenyl reagent.[13] Moreover, some diazonium salts couple readily with phenols and more slowly with imidazoles or amines, while for others the reverse is true.

Coupling with Diazonium-1H-tetrazole (DHT). This reagent is prepared by slowly adding 0.7 g of sodium nitrite dissolved in 10 ml of water with constant stirring to 1 g of 5-amino-1*H*-tetrazole (Aldrich Chemical Corp., Milwaukee, Wisconsin, or Tokyo Kasei Co., Japan) dissolved in 23 ml 1.6 *N* HCl and kept on an ice bath. The derivative forms within 6–8 minutes with a yield of about 80% using a test reaction (see above) with a suitable tyrosine analog. The molar absorptivity of the monoazoderivative at pH 10 or above is 8400 at 480 nm.[4] Since DHT decomposes in acid solution, the mixture is adjusted to pH 5 after 6–8 minutes and is then diluted to the desired volume with cold water. The product is stable for at least 1 hour at 0°.

CAUTION: DHT is violently explosive both in the dry state and in concentrated solutions. It must therefore be handled with extreme care and solutions above 0.2 M should be avoided. Care should be taken to

[10] E. B. Sandell, "Colorimetric Determinations of Traces of Metals," pp. 141–147. Wiley (Interscience), New York, 1944.

[11] O. Stadler, *Chem. Ber.* **17**, 2075 (1884).

[12] A. Hantzsch, *Chem. Ber.* **31**, 345 (1898).

[13] F. L. Pyman and L. B. Timmis, *J. Soc. Dyers Colour.* **38**, 269 (1922).

immediately rinse all pipettes, electrodes, beakers, etc., which come in contact with DHT solutions. If the DHT solution is spilled on the floor, clothing, or skin, quickly rinse with water before it has a chance to dry. Discard any unused solution by flushing into a drain with large volumes of water.

DHT can be employed to examine the role of tyrosyl and histidyl residues in the catalytic properties of enzymes.[14] It appears to have been accepted by some as a reagent specific for histidyl modification, and, indeed, its proclivity toward this residue in certain instances is higher relative to that of other diazonium salts. However, it displays the same broad reactivity as all such reagents and there is no evidence of absolute specificity for histidyl residues.

It was believed initially that the absorption bands of the various mono- and bisazo tetrazolyl derivatives of tyrosine and histidine were resolved, apparently obviating the spectral shortcomings of similar reagents employed in the past.[2] Since histidine was converted rapidly to the bisazoderivative, DHT was suggested as a reagent for discrimination of the various states of histidyl residues in proteins. Subsequently, it was shown that tyrosyl modification interferes with quantitation of azohistidine, and a revised method of analysis was proposed.[3] It is now possible to determine the concentrations of biazohistidine and mono- and bisazotyrosine in a protein from spectral data using a set of simultaneous equations.[4] Hence, this reagent can be employed to determine the total tyrosyl and histidyl content of proteins and also to examine the so-called "free" and "buried" tyrosyl and histidyl residues in proteins. Two procedures have been proposed for the first of these objectives.[3]

Method A. An aliquot of protein (0.1–0.2 ml) is diluted to 1 ml with 0.5 N NaOH and allowed to remain at room temperature for 30 minutes. The sample is then diluted with 7 ml of 1 M $KHCO_3$, pH 8.8, and 2 ml of 0.16–0.2 M DHT solution. The final protein concentration should be from 6 to 10×10^{-5} M. After 10 minutes at room temperature, another 2-ml aliquot of DHT solution is added and the reaction is allowed to proceed for 90 minutes. If necessary the pH is adjusted to 8.8 after each addition. The absorbance at 480 and 550 nm is measured against a buffer blank. The tyrosine content is calculated from the absorbance at 550 nm with a molar absorptivity at pH 8.8 of 1.38×10^4. The histidine content is determined from the absorbance at 480 nm after subtracting the contribution of bisazotyrosine at that wavelength (50% of the A_{550}) and using a molar absorptivity of 2.05×10^4.

Method B. Protein (0.3–0.5 μmoles) is kept in 1 ml of 0.5 N NaOH

[14] M. Sokolovsky and B. L. Vallee, *Biochemistry* **6**, 700 (1967).

for 30 minutes and then diluted to 3 ml with 0.001 M Tris, pH 7.5. The solution is cooled to 0°, and 35 μl of acetic anhydride is added. The pH is maintained at 7.5 by means of a pH-stat. After 20 minutes another 25 μl of anhydride is added. After another 15 minutes the pH is raised to 13. The reaction mixture is kept at room temperature for 20–30 minutes to allow deacetylation of O-acetyltyrosine. The protein solution is then adjusted to a final volume of 5 ml. A 1-ml aliquot is diluted with 7 ml of 1 M $KHCO_3$, pH 8.8, and 2 ml of 0.16–0.2 M DHT solution. The final pH should be 8.8–9.0. After 20 minutes at room temperature, another 2 ml of the DHT solution is added and the reaction is allowed to stand at room temperature for 90 minutes. The absorbancies at 480 and 550 nm are used to calculate the histidyl and tyrosyl content as above.

In both these methods alkali denaturation is employed to ensure complete availability of all the tyrosyl and histidyl residues of the protein. Urea and guanidine cannot be used for this purpose since these agents themselves react with DHT. Method B includes an acetylation step. In at least one instance this was found necessary to expose these residues fully. The 4000-fold molar excess of DHT should be sufficient to convert all tyrosyl and histidyl residues to the bisazoderivatives, an assumption basic to both methods. It is possible that the bis derivatives of some proteins will not be formed even under the conditions outlined and, hence, this means of calculation may not give accurate quantitative data in such instances.

Yet another method has been proposed which can be applied to native proteins and where some of the histidyl and tyrosyl residues form only monoazo derivatives.[4] Owing to overlapping spectral properties it is necessary to solve a set of three simultaneous equations which take into account the contributions of mono- and bisazotyrosine and bisazohistidine to the absorbance of 480 and 570 nm, measured at pH 8.0 and pH 10.0.[15] These equations are as follows:

$$\begin{aligned} A_{480}\ (\text{pH } 8) &= 20{,}200X + 1500Y + 5900Z \\ A_{570}\ (\text{pH } 8) &= 400X + 10{,}100Z \\ A_{480}\ (\text{pH } 10) &= 25{,}700X + 8400Y + 6400Z \end{aligned}$$

where X, Y, and Z are the molar concentrations of bisazohistidine, monoazotyrosine, and bisazotyrosine, respectively, and the numerical coefficients are the molar absorptivities at the wavelengths and pH indicated for each species. Since this method of analysis can be applied to

[15] Monoazohistidine has only very low absorption in this region and is, therefore, neglected in the calculations. It can be determined as the difference between the total histidine lost as shown by amino acid analysis and the amount of biazo histidine formed as shown by spectral analysis.

proteins modified in their native state, it provides a means to determine the number of tyrosyl and histidyl residues exposed at the surface of the protein.

It is possible to compare the numbers of residues modified with the total amount of DHT incorporated into the protein using ^{14}C-labeled reagent (Schwarz BioResearch, New York). In this way, one might detect modification of residues such as tryptophan, which is normally destroyed on acid hydrolysis, or arginine, which might be regenerated by such treatment.

Azochromophores as Optically Active Probes of Protein Conformation

Native proteins exhibit optical activity in the spectral region between 200 and 250 nm, and the resulting *intrinsic* Cotton effects are characteristic of their macromolecular conformation. The interaction of proteins with chromophoric molecules, such as coenzymes, substrates, inhibitors, or metal atoms at specific sites, can result in *extrinsic* Cotton effects, often in the visible region of the spectrum and, hence, not superimposed on the intrinsic *spectrum* of the protein. Such Cotton effects reflect structural features of the chromophore combining sites and their environments.[16] *Extrinsic* Cotton effects in proteins, generated as a result of covalent modification, have been observed in a few sporadic instances.[17–21] However, except for the reaction of diazonium salts with proteins, which also generate extrinsic Cotton effects,[5,6] none of these has been examined for their general potential as conformational probes. Modification of about a dozen different proteins with a given diazonium salt results in a series of circular dichroic spectra, each different from the others.[6] The details of the Cotton effects, their wavelength maxima, and the signs and magnitudes of the composite bands are seemingly quite specific for each protein even though the corresponding absorption spectra are all very similar.

Modification of a single protein with a series of different diazonium salts, however, gives a closely related family of similar dichroic spectra. This suggests that protein structure must play a major role in determining the details of these spectra. The abolition of the dichroic spectrum on denaturation is consistent with this idea.

[16] D. D. Ulmer and B. L. Vallee, *Advan. Enzymol.* **27,** 37 (1965).
[17] R. M. Dowben and S. H. Orkin, *Proc. Nat. Acad. Sci. U.S.* **58,** 2051 (1967).
[18] G. H. Beaven and W. B. Gratzer, *Biochim. Biophys. Acta* **168,** 456 (1968).
[19] G. F. Johnson, G. Philip, and D. J. Graves, *Biochemistry* **7,** 2101 (1968).
[20] B. Meloun, I. Fric, and F. Šorm, *Eur. J. Biochem.* **4,** 112 (1968).
[21] D. S. Sigman, D. A. Torchia, and E. R. Blout, *Biochemistry* **8,** 4560 (1969).

Most of the optically active absorption bands of the azoproteins cluster between 320 and 340 nm, 375 and 395 nm, and 420 and 455 nm. These wavelength ranges correspond to those of the known absorption spectra of the azotyrosyl derivatives. However, the overlap of these spectra with those of azohistidyl residues complicates a definitive assignment of any one band to a particular residue.

The circular dichroic spectra of azoproteins offer a novel means to examine the effects of environmental influences on modified residues. They provide signals which are more specific than the intrinsic Cotton effect since they often arise from a single chromophore, whereas the latter is a composite of many, chromophores.[22] Further, azochromophores can be introduced into proteins in different strategic positions, where they can variously signal overall and local conformational changes that relate to activity. They can be located at the active centers of enzymes, at regulatory sites, or on highly reactive surface residues where they may be used to monitor conformational changes accompanying such processes as zymogen activation. Thus, the circular dichroic spectrum of arsanilazochymotrypsinogen, containing about one azotyrosyl residue per molecule, exhibits an extremum at 428 nm with a molar ellipticity of *minus* 8000. On activation, this is replaced by a *positive* band at 448 nm with a molar ellipticity of 7000. The changes in sign, position, and magnitude of the circular dichroic band correlate closely with the appearance of enzymatic activity.[23]

The activation of arsanilazoprocarboxypeptidase is also accompanied by striking changes in its circular dichroic spectrum.[24,25] Over the time course of activation the positive Cotton effect at about 395 nm progressively shifts to 438 nm and increases from a molecular ellipticity of approximately 5000 to about 10,000 degrees, a 2-fold change in magnitude. Again, the appearance of esterase and peptidase activity coincides with these circular dichroic changes indicating that the optical probe reflects structural changes which accompany zymogen activation.

The initial stage of proteolytic modification of carboxypeptidase A by a number of bacterial proteases is analogous, in many ways, to zymogen activation except that in this case activity is altered rather than induced. Under the proper conditions digestion of carboxypeptidase with subtilisin Carlsberg forms a derivative, carboxypeptidase S, which

[22] B. L. Vallee, *Abstr. Symp. Papers, 8th Int. Congr. Biochem., Interlaken, Switzerland, 1970*, **58.**

[23] G. F. Fairclough and B. L. Vallee, *Biochemistry* **10,** 2470 (1971).

[24] W. D. Behnke and B. L. Vallee, *Abstr. 158th Meeting Amer. Chem. Soc., Sept., 1969,* No. 249.

[25] W. D. Behnke and B. L. Vallee, *Biochem. Biophys. Res. Commun.* **43,** 760 (1971).

has two new N-terminal residues, Ser and Gly. The esterase activity of this derivative is almost five times higher and peptidase activity about two times lower than those of the native enzyme. In analogy to the studies on zymogen activation, the circular dichroic spectrum of arsanilazocarboxypeptidase, specifically labeled at tyrosine-248, undergoes marked changes when this enzyme is subjected to limited subtilisin cleavage. The changes in molar ellipticity at 320 nm correlate directly with changes in activity indicating that the structural events underlying subtilisin modification are localized to the vicinity of the active center of the enzyme.[26]

Removal of the zinc atom from azoprocarboxypeptidase eliminates the long wavelength dichroic band at 500 nm. It also abolishes the corresponding band in the absorption spectrum, indicating the presence of a metal-azotyrosyl interaction which accounts for these spectral signals. The zinc atom is essential to the catalytic mechanism of carboxypeptidase; hence, these data would suggest a close relationship between the arsanilazochromophore in the zymogen and some of its components destined to become part of the active site region of the enzyme.

Coupling of carboxypeptidase crystals with diazotized arsanilic acid has also been employed to examine differences between the enzyme in the crystalline state and in solution. In the crystalline state the zinc arsanilazoenzyme, labeled at tyrosine-248, is yellow, but it turns red when it is dissolved. The 510-nm absorption band, responsible for the red color is abolished on removal of zinc from the dissolved enzyme but it is restored on readdition of zinc. These color changes are indicative of the reversible formation of a zinc-azophenol complex.[6a] The yellow color of zinc arsanilazocarboxypeptidase crystals is characteristic of the arsanilazotyrosyl group, not of the zinc complex. This is consistent with X-ray data on native crystals of carboxypeptidase which demonstrate that tyrosine-248 and the zinc atom are too far apart to form a complex. However, the red color of the enzyme in solution denotes formation of a complex between zinc and arsanilazotyrosine-248. The most likely interpretation of these data is that the orientations of arsanilazotyrosine-248 in solution and in the crystal are different.[6a]

Chemical modification, employed extensively in the past to assign functional roles to specific amino acid residues, can also produce "chemo-optical" probes,[27] when the products exhibit environmentally sensitive absorption, nuclear magnetic resonance, or fluorescence spectra. The optical activity of covalently bound azochromophores constitutes an

[26] J. F. Riordan and D. M. Livingston, *Biochem. Biophys. Res. Commun.* **44,** 695 (1971).

[27] G. M. Edelman and W. O. McClure, *Accounts Chem. Res.* **1,** 65 (1968).

important addition to this class of reagents. An optically active probe displaying these characteristics should be of great value in exploring the interactions of a variety of substrates, inhibitors and modifiers with the active sites of enzymes and, further, might be useful in broader explorations of molecular topology of proteins.

Optical activity of macromolecules arises from the interaction and relative orientation of groups, and, hence, relates to molecular geometry. In simple systems, changes in sign and rotatory power of Cotton effects can be accounted for on the basis of symmetry rules and one electron, dipole coupling and electric-magnetic coupling mechanisms, although even in such instances specific assignment is difficult and often conjectural.[28] Assignments are clearly even more difficult in such complex systems as here described. Yet it has been feasible to assign circular dichroic bands to transitions of myoglobin on the basis of information available from studies of the crystal structure and absorption and circular dichroic spectra in solution. Since substitutions with diazonium salts can be highly selective, modifying one or a very limited number of groups, assignments ultimately adding to our understanding of the relationship between protein structure and function can be anticipated.

[28] J. A. Schellman, *Accounts Chem. Res.* **1,** 144 (1968).

[46] Acylation with Dicarboxylic Acid Anhydrides

By Michael H. Klapper and Irving M. Klotz

Succinic and maleic anhydride, or analogs of these two compounds, have been utilized in a variety of protein modification studies. These reagents are nonvolatile, reasonably stable solids, and therefore, easily handled. The anhydrides react with protein amino groups as follows:

$$\mathrm{P{-}NH_3^+} + \mathrm{R(CO)_2O} \longrightarrow \mathrm{PNH\overset{\displaystyle O}{\overset{\|}{C}}{-}R{-}COO^-} + 2\,\mathrm{H^+}$$

Substitution at amino groups is strongly preferred[1-3] but reaction with

[1] A. F. S. A. Habeeb, H. G. Cassidy, and S. J. Singer, *Biochim. Biophys. Acta* **29,** 587 (1958).

[2] A. W. Kenchington, *Biochem. J.* **68,** 458 (1958).

[3] P. J. G. Butler, J. I. Harris, B. S. Hartley, and R. Leberman, *Biochem. J.* **112,** 679 (1969).

hydroxyl groups occurs under suitable conditions,[4-6] and sulfhydryl groups may also be susceptible to attack.[7] Since the reaction can be carried out between 0 and 25°C and pH 7 and 9, there is little danger of nonspecific protein denaturation.

A typical detailed procedure is illustrated by the following example. Bovine serum albumin (1 g) and 50 ml of distilled water are placed in a beaker. (For proteins such as β-lactoglobulin or hemerythrin, which are insoluble in distilled water, 0.1 M NaCl may be used as the solvent.) A magnetic stirrer is placed in the beaker, which is then set on a magnetic stirring apparatus. Gentle stirring dissolves the protein and is continued throughout the preparation procedure. Glass electrodes are placed in the solution, and the pH is adjusted to 7 with 0.2 M NaOH or Na_2CO_3 added from a syringe microburette. (If the reaction does not proceed conveniently at pH 7, the pH may be set at a higher value, anywhere in the range 7–10.) Solid succinic anhydride is now added in small increments to the protein solution over a period of 15 minutes to 1 hour. The amount of anhydride added depends on the extent of succinylation desired and the reactivity of the protein. Typical quantities for 1 g of protein may vary from 5 mg to 5 g. With each addition of anhydride the pH will drop; it is returned to pH 7 by addition of NaOH or Na_2CO_3. If large quantities of anhydride are used, the concentration of base should be increased so that excessive volumes are not required. In general the volume of added base is kept well below 10 ml. After all the anhydride has been added, the solution is allowed to stand for about 30 minutes. The reaction may be carried out equally well at 0° or 25°. The reaction of maleic anhydride with chymotrypsinogen[3] has been shown to proceed optimally at pH 9 with a 3–20 molar excess of anhydride over protein amino groups. Variation of the temperature from 2° to 20° showed little apparent effect.

During the course of the reaction most of the anhydride is hydrolyzed. The dicarboxylic acid side product may be separated from the modified protein by dialysis, by passage through a column of anion exchange resin (e.g., Amberlite IRA-400 in the chloride form), or by gel exclusion chromatography (e.g., with Sephadex G-25).

The stability of the half amide adduct depends on the anhydride used in the reaction. The attached succinyl residue is stable to most normal protein treatments, but may be cleaved off with 6 N HCl at 100°. Suc-

[4] I. M. Klotz and V. H. Stryker, *Biochem. Biophys. Res. Commun.* **1**, 119 (1959).
[5] J. F. Riordan and B. L. Vallee, *Biochemistry* **3**, 1768 (1964).
[6] A. D. Gounaris and G. E. Perlmann, *J. Biol. Chem.* **242**, 2739 (1967).
[7] E. A. Meighen and H. K. Schachman, *Biochemistry* **9**, 1163 (1970).

cinylation is, therefore, the method of choice when a stable linkage is required.

The maleyl half amide is stable under alkaline conditions, but is slowly hydrolyzed below pH 5. This acid lability may be utilized to regenerate free amino groups under relatively mild conditions. As an example, the maleyl groups were removed from maleylchymotrypsinogen by the following procedure.[3] The modified protein was adjusted to pH 3.5, with formic acid and aqueous ammonia, and to a final concentration of 0.3 mg/ml. The solution was kept at 37° for varying times, and the hydrolysis was stopped by addition of two volumes of 0.2 *M* NaOH. Since the absorption of the maleyl adduct is greater than that of maleic acid in the ultraviolet region, the extent of hydrolysis was monitored at 260 nm. Approximately 90% of the maleyl residues were removed from chymotrypsinogen in 30 hours.

Other half amides are also easily hydrolyzed. The citraconic anhydride adduct is acid labile,[8] whereas the half amide derived from reaction with tetrafluorosuccinic anhydride is stable at low pH, but may be hydrolyzed at pH 9.5.[9]

Five purposes for which the dicarboxylic anhydrides have been utilized are enumerated:

1. *Protein Dissociation.* Extensive reaction with a dicarboxylic anhydride places substantial negative charge on the modified protein, generally an increment of —2 per acylated amino group, since the $—COO^-$ of the half amide product replaces a positively charged $—NH_3^+$. This large alteration of charge should introduce an increase in electrostatic repulsion. With a single polypeptide chain protein, an expansion of the macromolecule may occur. With proteins composed of subunits the increased electrostatic repulsion may cause the aggregate to dissociate into its component chains. Dissociation has been found to occur, for example, in the succinylation of hemerythrin[10] and glyceraldehyde-3-phosphate dehydrogenase,[11] and in the maleylation of aldolase, transaldolase, and fructose diphosphatase.[12] Subunit dissociation may be due not only to electrostatic effects, however, for the anhydrides may react with sulfhydryl groups as well, and these are often intimately involved in the maintenance of macromolecules in the aggregated state.

When succinylation or maleylation is used for dissociation, sedi-

[8] H. B. F. Dixon and R. N. Perham, *Biochem. J.* **109,** 312 (1968).

[9] G. Braunitzer, K. Beyreuther, H. Fujiki, and B. Schrank, *Hoppe-Seyler's Z. Physiol. Chem.* **349,** 265 (1968).

[10] I. M. Klotz and S. Keresztes-Nagy, *Nature (London)* **195,** 900 (1962).

[11] R. Jaenicke, D. Schmid, and S. Knof, *Biochemistry* **7,** 919 (1968).

[12] C. L. Sia and B. L. Horecker, *Biochem. Biophys. Res. Commun.* **31,** 731 (1968).

mentation measurements are frequently made to test for dissociated subunits. Almost invariably the introduction of a large negative charge reduces $s_{20,w}$ markedly (e.g., succinylation of bovine serum albumin results in a decrease from 4.0 S to 1.0 S). Such a drop in $s_{20,w}$ is not necessarily a demonstration of disaggregation. It is necessary, therefore, to carry out sedimentation velocity experiments in a series of solutions of increasing ionic strength (up to about 1 *M*) to swamp out possible electrostatic artifacts.

Disaggregation induced by maleylation may be reversed. Sia and Horecker[12] have reported that inactive, monomeric, maleylated rabbit muscle aldolase can be reaggregated to the tetrameric protein with recovery of 46% activity by incubation of the modified protein at pH 4.5.

2. *Protein Hybridization.* A protein modified by reaction with a dicarboxylic anhydride will have an electrophoretic mobility different from that of the unmodified protein due to the increase in negative charge. This alteration of electrophoretic mobility has been utilized to study the hybridization of oligomeric proteins. A detailed discussion of the technique is presented elsewhere in this volume [47].

3. *Mapping of Lysine Peptides.* Butler and co-workers[3] have suggested that maleylation might be useful in mapping lysine-containing peptides. The β-melanocyte-stimulating hormone (MSH) was acylated with maleic anhydride, after which the maleyl derivative was digested with chymotrypsin. The chymotryptic peptides were separated by paper electrophoresis, and the strip obtained from the electrophorogram was placed in a desiccator over a solution of acetic acid (5% v/v) and pyridine (1% v/v). The desiccator was heated at 60° for 6 hours so as to deblock the chymotryptic peptides absorbed to the paper strip. After drying, the strip was sewn to a fresh sheet of paper, and the unblocked peptides electrophoresed again under the identical conditions, but at right angles to the direction of the first separation. Those peptides not originally maleylated should migrate identically in both directions, and therefore, will lie on a 45-degree diagonal line. The maleylated peptides will be found off the diagonal due to the change of charge which occurs with deblocking. In this manner, the $—NH_3^+$ containing peptides specifically may be mapped. None of four chymotryptic peptides of the small MSH were found on the diagonal line as expected from the known sequence.

4. *Peptide Sequencing.* Blocking the ε-amino group of lysine with either a succinyl or maleyl residue renders the lysine residue immune to trypsin-catalyzed hydrolysis. Thus, succinylation or maleylation may be utilized in establishing the sequence of tryptic peptides. While other

reagents such as *S*-ethyl trifluorothioacetate have been used for this purpose,[13] peptides with the attached half amides generally remain soluble, which is often not the case in other modification procedures. Maleic anhydride is particularly impressive since the maleyl residue is so easily removed. A typical procedure has been described by DeLange and coworkers.[14] Maleylated calf histone was hydrolyzed with TPCK-treated trypsin.[15] The hydrolysis was halted by addition of glacial acetic acid to reach a final pH of 3.0. The peptide mixture was heated at 40° for 40 hours to remove the maleyl residues; after the solution had been dried, the unblocked lysine-containing peptides were separated and purified. These authors reported that no peptides were formed by hydrolysis occurring adjacent to a lysine group.

5. *Lysine Side-Chain Reactivity and Function.* The relatively high specificity of succinic and maleic anhydride for primary amino groups suggests that these reagents may be useful for studying the reactivities of lysine side chains and for elucidating the roles these residues may play in the biological activities of proteins. For example, reaction of spinach leaf aldolase with maleic anhydride resulted in the incorporation of 8–9 moles of the maleyl residue per mole of protein, i.e., approximately 2 acylated amino groups per protein subunit.[16] When the protein was denatured by urea prior to reaction, approximately 72 maleyl residues were incorporated, corresponding to reaction of all the lysine groups.

Freedman and co-workers[17] have found that an antibody directed against *p*-azobenenearsonate was completely inactivated by treatment with maleic anhydride, while a second, different antibody directed against the same antigen was not. Hydrolysis of the maleyl group off the inhibited antibody restored activity completely. These authors concluded that a primary amine is probably involved in the antigen–antibody interaction.

6. *Introduction of New Functional Groups.* The succinic anhydride nucleus may provide a vehicle for the introduction of various functional groups into proteins under very mild conditions. Sulfur in various

[13] R. F. Goldberger, Vol. XI [34].

[14] R. J. DeLange, D. M. Fambrough, E. L. Smith, and J. Bonner, *J. Biol. Chem.* **244**, 319 (1969).

[15] F. H. Carpenter, Vol. XI [26].

[16] G. Rapoport, L. Davis, and B. L. Horecker, *Arch. Biochem. Biophys.* **132**, 286 (1969).

[17] M. H. Freedman, A. L. Grossberg, and D. Pressman, *J. Biol. Chem.* **243**, 6186 (1968).

forms[18,19] (protected mercaptan, free mercaptan, disulfide, or thioether), additional carboxyl groups, or dimethyl amino groups have been attached to different proteins.[18,19] Dixon[20] has synthesized *N*-glyoxylarginine by reaction of the primary amino group with maleic anhydride, with a subsequent cleavage across the double bond by a mixture of sodium metaperiodate and osmium tetroxide. It should be possible, therefore, to introduce new reactive groups into proteins and to examine the effects of these groups on the conformation and interactions of the modified macromolecules.

[18] I. M. Klotz and R. E. Heiney, *Arch. Biochem. Biophys.* **96,** 605 (1962).

[19] I. M. Klotz, Y. C. Martin, and B. L. McConaughy, *Biochim. Biophys. Acta* **100,** 104 (1965).

[20] H. B. F. Dixon, *Biochem. J.* **107,** 124 (1968).

[47] Hybridization of Chemically Modified Proteins

By MICHAEL H. KLAPPER and IRVING M. KLOTZ

Principle

Mixing of variants, or isozymes, of the same protein may lead to the formation of one or more new species. This phenomenon, called *hybridization*, occurs with proteins composed of two or more noncovalently linked polypeptide chains. It may be detected only if the various hybrid species differ sufficiently from one another in some property so that they may be separated. Electrophoresis on a solid support is the most convenient separation technique because of the ease and rapidity of operation, and of the small amount of protein required. Thus, differences in net charge between variant proteins is the property most often exploited in hybridization experiments.

For successful hybridization experiments, the variant proteins must have sufficiently different electrophoretic mobilities. The variants commonly used are isolated from natural sources. There are proteins, however, for which such variant isolation is not feasible. In these cases, a charge alteration may be artificially introduced by chemical means. Protein modification with succinic anhydride (see this volume [46]) can provide the necessary chemically modified species. Hybridization of succinylated protein with unmodified protein has been used to detect association–dissociation equilibrium in an oligomeric protein[1] and to determine the number of subunits in an oligomer.[2,3]

Association–Dissociation Equilibrium

When testing the association–dissociation equilibrium mobility of an oligomeric protein, there is no requirement for a homogeneous population of chemically modified species. Succinylated hemerythrin, prepared without any attempt to obtain a homogeneous protein, was used to determine whether the octamer is in a mobile equilibrium with its subunits near neutrality.[1] A 1:1 mixture of succinylated and native aggregates was stored at pH 8 for 24 hours and 4°, and then placed on starch gel and subjected to electrophoresis. Control experiments were carried out with the native protein alone and with the succinylated protein alone. Each of the three substances was exposed to the same electric field. The mixture showed a diffuse band with a range of mobilities extending from that of native protein to that of the succinylated octamer.

In general the formation of any new species, or any alteration in electrophoretic mobilities following the mixing, indicates that the rate of equilibrium is comparable to, or faster than, the experimental separation time. Absence of changes in electrophoretic mobilities is not, however, conclusive evidence for a slow equilibrium.

Determination of Number of Subunits

Statistics of Hybridization

If each of the variants contains identical subunits, then it is easy to see that the number of possible hybrids, N, is one more than the number of subunits, S, in the oligomer. For example, for a dimeric protein ($S = 2$), AA, mixed with the variant, A′A′, it is obvious that after hybridization the following three different species will be present: AA, A′A, A′A′. (Note that A′A and AA′ would have the same physical properties and would be indistinguishable.) Similarly for any S-mer, since in regard to physical properties it does not matter in which position an A′ takes the place of an A, the hybrids will consist of the species A_s, $A_{s-1}A'$, $A_{s-2}A'_2$, . . . , A'_s; that is, there will be $S + 1$ of them. Hence we may write[4]

$$N = S + 1 \tag{1}$$

If the subunits are not identical even though they are structurally

[1] S. Keresztes-Nagy, L. Lazer, M. H. Klapper, and I. M. Klotz, *Science* **150,** 357 (1965).

[2] E. A. Meighen and H. K. Schachman, *Biochemistry* **9,** 1163 (1970).

[3] E. A. Meighen and H. K. Schachman, *Biochemistry* **9,** 1177 (1970).

[4] C. R. Shaw, *Brookhaven Symp. Biol.* **17,** 117 (1964).

and functionally equivalent, then the number of hybrids obtainable on mixing two forms is larger than specified by Eq. (1), and is given by Eq. (2).[4,5]

$$N = \frac{(M + S - 1)!}{S!(M - 1)!} \tag{2}$$

where M is the total number of different types of subunit.

An even more complicated situation arises when the protein system consists of two (or more) ensembles of subunits, each ensemble being structurally and functionally distinct from the others. An enzyme such as aspartate transcarbamylase[6] containing catalytic and regulatory subunit ensembles, is an example of such a situation. In essence, for each ensemble Eq. (2) is applicable. Hence, in the general case, for the entire system of j ensembles

$$N = \prod_j \frac{(M_j + S_j - 1)!}{S_j!(M_j - 1)!} \tag{3}$$

Limitations

Certain precautions must be exercised in the interpretation of hybridization patterns. An erroneous number of protein bands, N, may be obtained if the resolution of the protein species is inadequate or if the concentration of one or more species is very low. This problem may be somewhat alleviated by using different initial ratios of the protein variants since the amount of each randomly formed hybrid depends on the initial concentrations. Under certain conditions Eqs. (1)–(3) will not apply, and erroneous interpretations may result. Hybridization of hemoglobin variants does not yield the expected number of new species.[7] This may be due to the nonrandom substitution of the chains in each subunit class. It is also essential that the association–dissociation equilibrium of the aggregate be slow relative to the time in which separation of hybrids occurs. When the rates of aggregate association–dissociation are sufficiently rapid, the number of bands formed on hybridization will not be described by Eqs. (1)–(3), but will instead be dictated by the equilibrium dynamics and the transport properties of the system.[8]

In order to describe the quaternary structure of a protein preparation, the variables M, S, and J of Eq. 3 must be specified. Hybridization experiments yield N in favorable cases. However, there is no unique set

[5] C. L. Markert and G. S. Whitt, *Experientia* **24**, 977 (1968).

[6] J. C. Gerhart and H. K. Schachman, *Biochemistry* **4**, 1054 (1965).

[7] H. A. Itano and S. J. Singer, *Proc. Nat. Acad. Sci. U.S.* **44**, 522 (1958).

[8] G. A. Gilbert and R. C. Jenkins, *Proc. Roy. Soc. Ser. A* **253**, 420 (1959).

of values M, S, and J for low values of N. For example, hybridization of the dimer AA with A′A′ yields three species, but three would also be expected from the hybridization of AABB with A′A′BB. Therefore, a unique determination of the quaternary structure requires experiments in addition to hybridization.

Since use of Eqs. (1)–(3) depends on both a very slow protein association–dissociation reaction, and a complete equilibration between chain variants, the mixture of the variants generally must be reversibly denatured. Among the techniques which have been used are lowering the pH,[7] freezing and thawing in concentrated salt,[9] and addition of denaturants, such as urea.[2] Thus, successful hybridization depends also on finding conditions of reversible denaturation.

Chemical Modification

The criteria for a satisfactory chemical modification have been enumerated by Meighen and Schachman.[2] The modification should yield a homogeneous population, and should, therefore, be fairly specific. The modified protein should have an electrophoretic mobility substantially different from that of the native protein. The quaternary structures of modified and native protein must be identical; and the altered enzyme must be reconstitutable after dissociation.

These criteria are met by protein modification with succinic anhydride (see this volume [46]). Substitution at amino groups occurs preferentially, and a charge difference of close to —2 is introduced per succinyl residue. While extensive succinylation may lead to protein dissociation, conditions for partial succinylation with no disaggregation have been found for four proteins, hemerythrin,[1] aldolase,[2] glyceraldehyde-3-phosphate dehydrogenase,[3] and aspartyl transcarbamylase.[10]

Partial succinylations may yield nonhomogeneous preparations. If a protein contains i different classes of reactive amino groups, each of which reacts with a characteristic rate, and if the reaction of groups within each class is random, then the distribution of protein molecules with varying amounts of attached succinyl residues is characterized by the standard deviation

$$\sigma = \sum_i (n_i f_i (1 - f_i))^{1/2} \tag{4}$$

where n_i is the total number of amino groups in each class, and f_i the fraction of reacted groups in each class.[2] Inspection of Eq. (4) reveals

[9] C. L. Markert, *Science* **140,** 1329 (1963).

[10] E. A. Meighen, V. Pigiet, and H. K. Schachman, *Proc. Nat. Acad. Sci. U.S.* **65,** 234 (1970).

that the greatest homogeneity will be found at high and low levels of f, i.e., when very few or almost all the residues are modified. The greatest inhomogeneity will occur at intermediate values of f. If extensive succinylation results in dissociation, preparations with high values of f are of no use. On the other hand, succinylation of a small fraction of the residues introduces only a small charge alteration leading to less resolution.

To obtain discrete bands for the hybrids it is essential to find conditions for the production of a homogeneous preparation. This has been achieved with aldolase.[2] The tetrameric protein was mixed with varying amounts of succinic anhydride, and the sedimentation behavior of each sample was determined. The most succinylated protein sample which exhibited the least disaggregation was then purified by ion-exchange chromatography. Those fractions displaying a single band on cellulose acetate electrophoresis were pooled and used for hybridization.

Application

In a typical experiment,[2] aldolase and succinylaldolase were mixed (at varying molar ratios) in buffer at pH 6.50 containing 4.0 *M* urea and 0.1 *M* dithiothreitol, at 4°. After 30 minutes of standing, the solution was dialyzed against buffer to remove the urea. Zone electrophoresis was then carried out on cellulose acetate strips, and the proteins were stained to make their positions visible. With aldolase, five bands were observed. This corresponds, see Eq. (1), to a tetrameric oligomer for aldolase.[11]

The use of succinylated protein for detecting subunits in equilibrium with aggregated protein or for determining the quaternary structure of a protein should be quite general in applicability. Almost all proteins contain lysine groups which should react with succinic anhydride, producing macromolecules with altered electrophoretic mobilities. Hybridization experiments are, therefore, feasible with any subunit-containing protein.

[11] E. Penhoet, M. Kochman, R. Valentine, and W. J. Rutter, *Biochemistry* **6**, 2940 (1967).

[48] Thiolation

By FREDERICK H. WHITE, JR.

Thiolation, or introduction of the sulfhydryl (SH) group, may occur by aminolysis[1] of the thiolactone bond of *N*-acetylhomocysteine thiolactone (AHTL). A new residue, *N*-acetylhomocysteine, thereby becomes attached through a peptide bond to either the ϵ-nitrogen of a lysine residue or the α-nitrogen of the NH_2-terminal residue [Reaction (1)].

$$\text{Protein-NH}_2 + \begin{array}{l} H_2C-S \\ \;|\qquad\; \backslash \\ \;|\qquad\;\; CO \\ \;|\qquad\; / \\ H_2C-CH \\ \qquad\quad \backslash \\ \qquad\quad NHCOCH_3 \end{array} \longrightarrow \text{Protein-NHCO}\underset{\displaystyle \text{NHCOCH}_3}{\underset{|}{\text{C}}}\text{HCH}_2\text{CH}_2\text{SH} \qquad (1)$$

The SH group may be of use as a point of attachment for heavy metals. Proteins so labeled have been subjected to electron microscopy[2] and X-ray diffraction studies.[3-5] Thiolation may also be relevant to considerations of structural and functional relationships, since SH groups, as well as the amino groups that are blocked by thiolation, have been implicated in the active sites of some enzymes. Further alterations of molecular structure and biological function could result from interchange of the newly introduced SH group with disulfide bonds within the protein.

Thiolation Procedures

Of the two methods given below, method A is the simplest and has yielded interesting results, bearing on enzyme structure and function relationships.[6] Method B is more difficult to carry out but may result in a more homogeneous product. It has been useful in the highly selective labeling essential for X-ray diffraction studies.[3-5] For comparative purposes the same protein, beef pancreatic ribonuclease (RNase), was chosen for each method.

[1] The SH group may also be introduced by succinylation (see this volume [46]). The present discussion is limited to thiolation by aminolysis.

[2] P. A. Kendall, *Biochim. Biophys. Acta* **97**, 174 (1965).

[3] H. P. Avey, M. O. Boles, C. H. Carlisle, S. A. Evans, S. J. Morris, R. A. Palmer, B. A. Woolhouse, and S. Shall, *Nature* (*London*) **213**, 557 (1967).

[4] S. Shall and E. A. Barnard, *Nature* (*London*) **213**, 562 (1967).

[5] H. P. Avey and S. Shall, *J. Mol. Biol.* **43**, 341 (1969).

[6] F. H. White, Jr. and A. Sandoval, *Biochemistry* **1**, 938 (1962).

Method A

Procedure. Nitrogen is passed through a solution of 0.1 *M* NH_4HCO_3 for 15 minutes to exclude oxygen. Five milliliters of this solution is added to 50 mg each of RNase and AHTL. The final pH of this solution is 8.4. The solution is kept under a nitrogen atmosphere at room temperature (23–25°) throughout the reaction period.

After reaction for the desired length of time, the protein may be separated by application of the reaction mixture (after adjustment to pH 3–4 with acetic acid), to a column (2.5×30 cm) of Sephadex G-25, previously equilibrated with 0.1 *M* acetic acid. The fractions containing the protein, detected at 280 nm, may then be combined and lyophilized. Alternatively, the protein has been separated by precipitation with ice-cold acetone–1 *M* HCl (49:1),[6] or, in the case of insulin,[7] with acetone.

Results. After 100 minutes of reaction, 1.2 moles of SH per mole of protein as introduced, and after 180 minutes, 2.1 moles of new SH groups were found. A limiting value of approximately 3 is approached at 1200 minutes.

The cyclic phosphatase activity (determined as described earlier[6]), decreased with approximate linearity to 20% of the specific activity of native RNase in 8690 minutes. However, the activity toward ribonucleic acid increased to a maximum of 125% of the native activity after 600 minutes; it continued downward thereafter at a rate nearly parallel to that of the cyclic phosphatase activity. A summary of methods of assay for RNase activity was given earlier.[8]

Heterogeneity with regard to SH content has been indicated by chromatography of the carboxymethylated product on CM-cellulose.[6] After 7.5 hours of reaction, five components appeared with *S*-carboxymethylhomocysteine (SCHC) content ranging from 1 to 5.

Disulfide interchange occurs concomitantly with thiolation by this procedure. However, after 180 minutes only 0.095 mole of SH per mole of protein had undergone interchange. After 600 minutes, 0.7 mole had so reacted. This reaction continues slowly thereafter at a rate of approximately 0.5 SH within 8000 minutes. Disulfide interchange, therefore, is minimized at low reaction times.

Method B

When total exclusion of disulfide interchange is essential, the following procedure, essentially that of Shall and Barnard,[9] would be the method

[7] T. K. Virupaksha and H. Tarver, *Biochemistry* **3**, 1507 (1964).

[8] C. B. Anfinsen and F. H. White, Jr., *in* "The Enzymes" (P. D. Boyer, H. Lardy, and K. Myrbäck, eds.), Vol. V, p. 95. Academic Press, New York, 1961.

[9] S. Shall and E. A. Barnard, *J. Mol. Biol.* **41**, 237 (1969).

of choice. With this procedure, however, precautions must be taken to exclude traces of halide and phosphate ions, which would react with silver, present as a catalyst, and could adversely affect the reproducibility of this reaction.

Procedure. Any such trace contaminants present in commercial preparations of RNase are removed by chromatography on Amberlite IRC-50 according to the procedure of Hirs *et al.*[10] It is then deionized by dialysis, followed by passage through a small column of Amberlite MB3 mixed bed resin. The effluent protein solution is adjusted to pH 5–6 with 1 *M* HNO_3, giving RNase nitrate. Ion-free water is used throughout the above preparation as well as in the thiolation procedure described below.

The reaction is carried out in a pH-stat at room temperature with dilute NaOH and HNO_3 for adjustment and maintenance of the reaction mixture at pH 7.5. A calomel electrode cannot be used, since it may leak traces of chloride and therefore is replaced by an agar bridge in saturated KNO_3. The reaction mixture is stirred magnetically and is continuously flushed with nitrogen throughout the reaction period.

Fifty milligrams of RNase nitrate (prepared as above) and 2 mg of AHTL are dissolved in 5 ml of deaerated water. The solution is adjusted to pH 7.5 in the pH-stat, and 3 mg of $AgNO_3$ in 5 ml of deaerated water are added in 10 equally divided portions over a period of 15 minutes. The ensuing reaction is complete in 5–10 additional minutes.

The silver ion is removed by addition of 5 ml of 0.1 *M* KCN, followed by dialysis against deaerated water in the cold. A sensitive test for completeness of removal of silver is given by Benesch and Benesch.[11] The resulting solution of thiolated RNase in the SH form may then be subjected to carboxymethylation to cover the newly introduced SH groups.[12]

Results. This procedure yields essentially 1 mole of SH per mole of RNase. Chromatography of the carboxymethylated product on IRC-50 has produced two thiolated components, one of which was totally inactive. The other was enzymatically active, but its activity in the free SH form was completely blocked by the presence of silver.[9] Shall and Barnard[9] have shown that the inactive component is thiolated on the ε-amino group of the lysine residue at position 41.

In a comparison of these results with those of method A, it is obvious that the reaction in the presence of silver ions, in addition to

[10] C. H. W. Hirs, S. Moore, and W. H. Stein, *J. Biol. Chem.* **200,** 493 (1953).

[11] R. Benesch and R. E. Benesch, *Proc. Nat. Acad. Sci. U.S.* **44,** 848 (1958).

[12] F. H. White, Jr., *J. Biol. Chem.* **236,** 1353 (1961).

catalysis and prevention of disulfide interchange, is, at least for RNase, associated with greater selectivity.

Comments

Catalysis of Thiolation

Benesch and Benesch[11] first observed catalysis of thiolation of several proteins by silver ions. According to Schwyzer and Hürlimann[13] this effect is characteristic of any heavy metal ion having affinity for sulfur. Silver, however, is the most effective. Its catalysis, according to Benesch and Benesch,[11] is dependent upon the initial formation of a silver complex with AHTL. They reported that the extent of thiolation of gelatin increased with the ratio of moles of this complex to amino groups and that, at a ratio of 10:1, approximately 30 of the 36 amino groups initially present per protein molecule had reacted. However, in the special case of RNase, Shall and Barnard[9] have suggested that the high selectivity of thiolation may be the result of complex formation of silver with the protein, either between two imidazole groups or between one imidazole and the ϵ-amino group of a lysine residue. It would be an interesting topic for further investigation to determine whether there are other proteins that exhibit a similar selectivity with AHTL under these conditions.

In addition to catalysis, silver would be expected to prevent disulfide interchange during this reaction, since it would complex with the newly introduced sulfur. Furthermore, catalysis permits thiolation at a much faster rate and at pH values below those at which interchange would easily occur. However, it may be of interest that Virupaksha and Tarver,[6] who thiolated insulin in the absence of silver ions, reported no disulfide interchange.

More recently Elfbaum[14] has reported the catalysis of thiolation by a number of nucleophilic reagents. Since it is known that acylation may proceed through a high-energy imidazole intermediate,[15] it was postulated that imidazole should act as a catalyst in this manner after breaking the thiolactone ring. It was demonstrated that several amines, in addition to imidazole, have this effect. Also catechol, *m*-nitrophenol, and dibasic potassium phosphate were found to catalyze protein thiolation.

Kendall[2] used imidazole as a buffer during the thiolation of γ-globulin in the presence of silver ions, to react approximately 40% of the amino

[13] R. Schwyzer and C. Hürlimann, *Helv. Chim. Acta* **37**, 155 (1954).

[14] S. G. Elfbaum, Doctoral Dissertation, Northwestern University, Evanston, Illinois, 1966.

[15] E. R. Stadtman and F. H. White, Jr., *J. Amer. Chem. Soc.* **75**, 2022 (1953).

groups. The catalytic effect of imidazole, however, was apparently not suspected at that time.

Determination of the Extent of Reaction

The increase in SH content may be followed by any of a number of quite satisfactory procedures (see Cecil and McPhee[16] for summary). In particular the spectrophotometric titration with *p*-chloromercuribenzoate of Swenson and Boyer,[17] especially as modified by Sela *et al.*,[18] (see also Riordan and Vallee, this volume [36]), and reaction with Ellman's reagent[19] (see Habeeb, this volume [37]), are worthy of mention. The SH content may also be quantitated by carboxymethylation[12] to attach the carboxymethyl group to the homocysteine residue. The *S*-carboxymethylhomocysteine content may then be determined, after acid hydrolysis of the protein, by amino acid analysis.[20]

The loss of amino groups in the thiolated protein should correspond to the increase in SH content. Benesch and Benesch[11] confirmed this by formol titration[21] of several thiolated proteins. Abadi and Wilcox[22] distinguished between the α- and ϵ-amino groups of chymotrypsinogen by reaction with carbon disulfide which, at pH 6.9, is specific for the α-amino group. Kendall[2] caused the remaining amino groups of thiolated γ-globulin to react with ninhydrin by the method of Moore and Stein.[23]

Disulfide interchange may be followed by the extent of formation of cysteine. The SH group belonging to this residue is subjected to carboxymethylation, and the resulting content of *S*-carboxymethylcysteine is determined by amino acid analysis after hydrolysis of the protein sample.

Other Thiolating Agents

As a thiolating agent, AHTL has proved to be the most popular, and all the foregoing results were obtained with it. There are, however, certain other reagents containing the thiolactone or thiolester bond that have been employed in protein thiolation.

Dithioglycolide, first introduced by Schöberl,[24] may be of historical

[16] R. Cecil and J. R. McPhee, *Advan. Protein Chem.* **14,** 255 (1959).
[17] A. D. Swenson and P. D. Boyer, *J. Amer. Chem. Soc.* **79,** 2174 (1957).
[18] M. Sela, F. H. White, Jr., and C. B. Anfinsen, *Biochim. Biophys. Acta* **31,** 417 (1959).
[19] G. L. Ellman, *Arch. Biochem. Biophys.* **82,** 70 (1959).
[20] See Spackman, Vol. XI [1].
[21] D. French and J. T. Edsall, *Advan. Protein Chem.* **2,** 277 (1945).
[22] D. M. Abadi and P. E. Wilcox, *J. Biol. Chem.* **235,** 396 (1960).
[23] S. Moore and W. H. Stein, *J. Biol. Chem.* **211,** 907 (1954).
[24] A. Schöberl, *Angew. Chem.* **60,** 7 (1948).

interest. This reagent, however, has the disadvantage that it easily polymerizes to form polythioglycolides with a widely variable chain length and a corresponding variability in the length of the substituent groups. Thiolation of RNase with this polymer has been investigated in comparison to AHTL.[25] Benesch and Benesch[26] originally employed benzoyl homocysteine thiolactone for thiolation. However, they eventually settled upon the acetyl derivative as the reagent of choice because of its greater solubility in water. Klotz and Elfbaum[27] have used sodium thioparaconate in thiolation. Its reaction proceeds analogously to Reaction (1), and the results are similar to those obtained with AHTL.

[25] F. H. White, Jr., *J. Biol. Chem.* **235,** 383 (1960).
[26] R. Benesch and R. E. Benesch, *J. Amer. Chem. Soc.* **78,** 1597 (1956).
[27] I. M. Klotz and S. G. Elfbaum, *Biochim. Biophys. Acta* **86,** 100 (1964).

[49] Reaction of Proteins with Citraconic Anhydride

By M. Z. Atassi and A. F. S. A. Habeeb

Reversible masking of amino groups is an extremely valuable procedure for protecting these groups from side reactions which might, in certain cases, take place during modification of some other functional groups in proteins and peptides. Also, reversible masking of amino groups is useful for rendering hydrolysis with trypsin specific for cleavage at arginine residues. Several such reversible masking reagents have been reported. Amino groups have been reversibly modified by trifluoracetylation,[1] amidination,[2] maleylation,[3] acetoacetylation,[4,5] tetrafluorosuccinylation,[6] and, more recently citraconylation.[7] A careful investigation of the applicability of these reactions and comparison of their specificity, ease of reversal, homogeneity of the masked and unmasked derivatives, and recovery of conformation and biological properties has recently been

[1] R. F. Goldberger and C. B. Anfinsen, *Biochemistry* **1,** 401 (1962).
[2] M. J. Hunter and M. L. Ludwig, *J. Amer. Chem. Soc.* **84,** 3491 (1962).
[3] P. J. G. Butler, J. I. Harris, B. S. Hartley, and R. Leberman, *Biochem. J.* **103,** 78 P (1967).
[4] A. Marzotto, P. Pajetta, and E. Scoffone, *Biochem. Biophys. Res. Commun.* **26,** 517 (1967).
[5] A. Marzotto, P. Pajetta, L. Galzigna, and E. Scoffone, *Biochim. Biophys. Acta* **154,** 450 (1968).
[6] G. Braunitzer, K. Beyreuther, H. Fujiki, and B. Schrank, *Hoppe-Seyler's Z. Physiol. Chem.* **349,** 265 (1968).
[7] H. B. F. Dixon and R. N. Perham, *Biochem. J.* **109,** 312 (1968).

carried out by Singhal and Atassi,[8] using myoglobin as the protein and allowing it to react with diketene, maleic anhydride, tetrafluosuccinic anhydride, or citraconic anhydride. A similar series of derivatives was prepared by Habeeb and Atassi[9] using lysozyme as the protein model and studying all the foregoing criteria of reversibility. The results both with myoglobin and lysozyme showed that of the reagents studied citraconic anhydride was the most satisfactory, yielding upon deblocking (for both myoglobin and lysozyme) homogeneous preparations identical with the respective native proteins in biological properties and in conformational and hydrodynamic parameters.

This article deals with the reaction of proteins with citraconic anhydride and the properties of blocked and deblocked citraconyl (CT-) proteins. Reference to other reversible masking reagents will be made where it is useful to compare the results with such reagents with those obtained with citraconic anhydride on the *same* protein.

Citraconic anhydride (2-methylmaleic anhydride) reacts with amino groups to give two reaction products.

H_3C—C(=CH)—C(=O)—O—C(=O) (citraconic anhydride) + H_2N-Protein

↓

H_3C—C(—C(=O)—NH-Protein)=CH—C(=O)—OH + H_3C—C(—C(=O)—OH)=CH—C(=O)—NH-Protein

(A) (B)

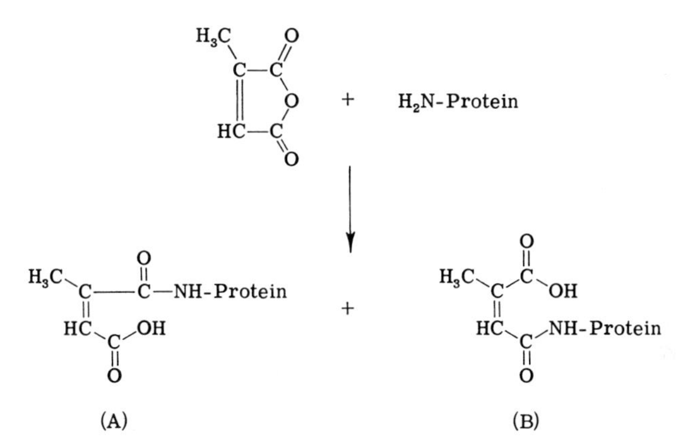

This apparent complication is not serious, since it was found[7] that the two reaction products do not differ greatly in their stabilities. As with all amino group acylation reactions, citraconylation can be carried out at pH 8–9. The following reaction with lysozyme gives an example of the procedure.

Aspects of the Reaction

Reaction of Lysozyme with Citraconic Anhydride.[9] The reaction is carried out under conditions that will lead to complete modification of

[8] R. P. Singhal and M. Z. Atassi, *Biochemistry* **10,** 1756 (1971).

[9] A. F. S. A. Habeeb and M. Z. Atassi, *Biochemistry* **9,** 4939 (1970).

all the amino groups. Lysozyme (1.96 g) is dissolved in water (50 ml) and the pH is adjusted to 8.20. Aliquots (100 μl) of citraconic anhydride are added to the magnetically stirred solution, at 30-minute intervals. A total of 800 μl of citraconic anhydride is added. The reaction proceeds at room temperature and a pH of 8.2 is maintained by the addition of 5 *N* NaOH with a pH-stat. When the addition of citraconic anhydride is completed, the reaction mixture is allowed to stir at room temperature for 2 more hours at pH 8.20. The solution is then dialyzed at 0° against several changes of water, preadjusted to pH 8.5–8.8 with NH_4OH, and finally freeze-dried.

Removal of Masking Groups from Citraconyl Proteins. Removal of citraconyl (CT) blocking groups is carried out at acid pH. In a representative experiment CT-myoglobin[8] (5 mg/ml) undergoes complete deblocking within 4 hours when kept at pH 3.5 and 30°. Deblocking of CT-lysozyme[9] may be carried out in 0.05 *M* acetate buffer at pH 4.2 and 40°. Complete deblocking is achieved within 3 hours. For deblocking CT-aldolase, Gibbons and Perham[10] dialyzed the protein (4 mg/ml) against 10 m*M* HCl, pH 2.0, at room temperature (20°). Dialysis for 6 hours regenerated more than 95% of the amino groups.

Methods of Assay. It may be helpful to give reference here to procedures that may be used to test for the specificity and complete blocking or deblocking of a protein with citraconic anhydride. Free amino groups may be determined by reaction with trinitrobenzene sulfonic acid.[11] Also, free amino groups may be made to react with fluorodinitrobenzene and an acid hydrolyzate of the DNP-protein analyzed with an amino acid analyzer. Acylated phenolic hydroxyl groups may be determined spectrophotometrically by using the procedure of Riordon and Vallee[12] which depends on the lability of ester linkages to hydroxylamine. Esterified hydroxyamino acids may be determined by the hydroxylamine–ferric chloride procedure.[13] Free sulfhydryl groups may be determined by reaction with 5,5′-dithiobis-(2-nitrobenzoic acid)[14] (see this volume [37]).

Specificity of Reaction with Citraconic Anhydride and Homogeneity of the Product. The specificity of the reaction was studied in detail with lysozyme by Habeeb and Atassi.[9] No *O*-acylation of tyrosine residues was detectable in CT-lysozyme. On the other hand, 2–3 residues of the hydroxyamino acids were esterified in CT-lysozyme. Gibbons and

[10] I. Gibbons and R. N. Perham, *Biochem. J.* **116**, 843 (1970).
[11] A. F. S. A. Habeeb, *Anal. Biochem.* **14**, 328 (1966).
[12] J. F. Riordan and B. L. Vallee, *Biochemistry* **3**, 1768 (1964).
[13] A. F. S. A. Habeeb and M. Z. Atassi, *Immunochemistry* **6**, 555 (1969).
[14] G. L. Ellman, *Arch. Biochem. Biophys.* **82**, 70 (1959).

Perham[10] reported a side reaction with sulfhydryl groups upon reaction of aldolase with citraconic anhydride. The nature of the reaction product with sulfhydryl groups has not been determined.

CT-myoglobin migrated as a single, strongly negatively charged band on starch-gel electrophoresis.[8] The homogeneity of CT-lysozyme was examined in detail by Habeeb and Atassi.[9] They showed that CT-lysozyme, completely blocked at all amino groups, was homogeneous in starch gel electrophoresis, but exhibited appreciable heterogeneity on disc electrophoresis. The corresponding maleyl (ML-) and succinyl derivatives of lysozyme showed similar electrophoretic heterogeneity, strongly suggesting that heterogeneity of CT-lysozyme was not caused by the type of reaction product (A and B mentioned earlier) of citraconic anhyride with amino groups. Heterogeneity was attributed[9] to a non-uniform acylation of hydroxyl groups with citraconic anhydride (or with succinic or maleic anhydrides).

Ease of Removal of Citraconyl Groups from Citraconylated Proteins. It is of value to comment on the lability of the citraconyl blocking group under various conditions of manipulation. Habeeb and Atassi[9] studied the rate of removal of the citraconyl blocking groups in CT-lysozyme at pH 4.2, 5.2, 7.2, or 8.0 and 40°. Results are shown in Fig. 1. At pH 4.2, complete (100%) deblocking of amino groups was achieved within 3 hours. Deblocking at pH 5.2 leveled off in 6 hours at a maximum of 6.7 amino groups deblocked (96%) per mole of lysozyme. At pH 6.4, deblocking was slower, and in 6 hours an average number of 4 amino groups were unmasked. After 2 days at pH 6.4, an average

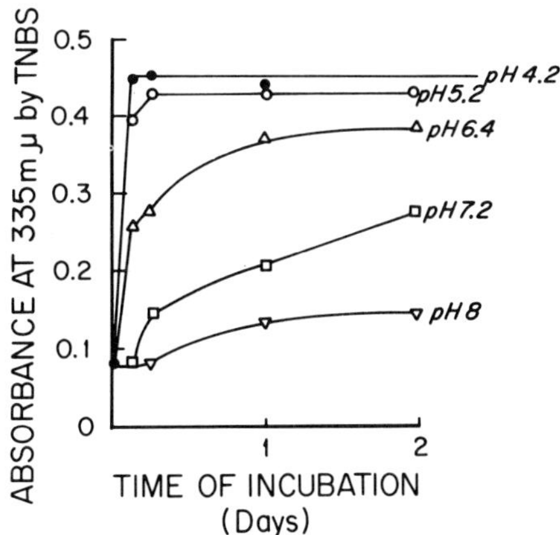

FIG. 1. Removal of citraconyl groups from completely citraconylated lysozyme at different pH values. Results are from A. F. S. A. Habeeb and M. Z. Atassi, *Biochemistry* **9**, 4939 (1970).

number of 5.9 amino groups were unmasked. It is interesting that citraconyl blocking groups are removed at neutral pH at an appreciable rate. In Fig. 1, it can be seen that even at pH 8.0, removal of citraconyl groups was still detectable at 40°. Habeeb and Atassi[9] reported that the best procedure for storage of citraconyl proteins is as freeze-dried products at 0° or lower. No unmasking of amino groups was observed in CT-myoglobin[8] or CT-lysozyme[9] on long (several months) storage in the freeze-dried state at 0°.

Properties of Deblocked Citraconyl Proteins

In this section some of the physicochemical and biological properties of deblocked CT-proteins are considered. In certain cases, the corresponding properties of preparations of the *same* protein after removal of other blocking groups are given, when available, to facilitate direct comparison and to emphasize the merits of citraconic anhydride.

Homogeneity. It has already been mentioned that with CT-lysozyme, complete deblocking of the amino groups was achieved within 3 hours at pH 4.2 and 40°. Habeeb and Atassi[9] reported that the resulting preparation showed minimal heterogeneity by disc electrophoresis after 24 hours' exposure to pH 4.2. Further exposure of this deblocked preparation to 1 *M* hydroxylamine at pH 10 for 1 hour abolished electrophoretic heterogeneity almost completely. However the single band obtained was slightly less electronegative (with mobility of 0.98 relative to native lysozyme = 1.00). Singhal and Atassi[8] examined the homogeneity of deblocked CT-myoglobin by starch-gel electrophoresis. Deblocking of CT-myoglobin was carried out at pH 3.5 and 30°. Complete deblocking of all amino groups was achieved within 4 hours. Starch-gel electrophoresis of deblocked material gave a single band migrating like native myoglobin. In heavily loaded gels a trace (1%) of a more negatively-charged band with mobility of 2.2 (relative to myoglobin) was obtained.

Conformation of Deblocked Preparations. The conformation of deblocked CT-myoglobin was examined by Singhal and Atassi,[8] using optical rotatory dispersion. They also investigated the conformations of the corresponding deblocked preparations from maleyl (ML-) and tetrafluorosuccinyl (TFSu-) myoglobins. The results show that native conformation is recovered completely *only* on deblocking of CT-myoglobin. Habeeb and Atassi[9] examined the conformation of blocked and deblocked lysozyme preparations by measurement of the reducibility of the disulfide bonds and of the susceptibility to tryptic attack. They reported that the native conformation is recovered upon deblocking, with CT-lysozyme, but not with ML-lysozyme. Lysozyme undergoes no change in sedimentation properties upon complete blocking with citraconic anhydride.[9] With proteins composed of subunits, dissociation into the subunits takes place, as expected, upon citraconylation. Gibbons and Perham[10] showed that aldolase blocked to the extent of 80% or more will undergo complete dissociation into subunits. Deblocking of CT-aldolase gave preparations with little or no material approaching the sedimentation coefficient of the native enzyme.

Influence of Blocking and Deblocking on Enzymatic Activity. The enzymatic activity of blocked and deblocked CT-lysozyme has been studied by Habeeb and Atassi.[9] Comparisons were made with the corresponding maleylated derivatives. Complete recovery of activity was obtained upon deblocking of CT-lysozyme. On the other hand, deblocking of ML-lysozyme for 5 days resulted in the recovery of only 83% of the enzymatic activity. With aldolase that had been blocked at 50% of its amino groups, the activity is recovered to the extent of 75% upon deblocking.[10] Aldolase in which 80% or more of the amino groups were blocked, showed little or no enzymatic activity upon deblocking, and this was attributed to an irreversible modification of some sulfhydryl groups.[10]

Immunochemistry of Deblocked Preparations. The completeness of removal of the citraconyl blocking group has also been followed immunochemically. Singhal and Atassi[8] studied the reactivity of completely blocked and of deblocked CT-myoglobin with antisera to the native protein. They also compared this with the behavior of the corresponding maleylated or tetrafluorosuccinylated derivatives. Their results show that the antigenic reactivity of the protein is fully recovered *only* upon deblocking of CT-myoglobin, but not with maleylated or tetrafluorosuccinylated derivatives. Similar studies have been carried out by Habeeb and Atassi[9] on lysozyme. In contrast to myoglobin, where antigenic reactivity is entirely eliminated upon complete blocking, CT-lysozyme retains appreciable (40–50%) reactivity with antisera to native lysozyme. Complete reactivity is recovered upon deblocking of CT-lysozyme, but not upon deblocking of ML-lysozyme.

Comparison with Other Reversible Blocking Reagents

Until recently, no systematic study of reversible blocking reagents for amino groups has been carried out on the *same* protein. The behavior of diketene and tetrafluorosuccinic, maleic, and citraconic anhydrides has now been studied in detail both with myoglobin[8] and with lysozyme.[9] The results can be mentioned briefly.

With myoglobin,[8] derivatives completely blocked by reaction with diketene or with maleic or tetrafluorosuccinic anhydrides showed single strongly negatively charged bands on starch-gel electrophoresis and had no antigenic reactivity with antisera to myoglobin. Deblocking of acetoacetyl-myoglobin (from reaction with diketene) for 16 hours resulted in appreciable precipitation. Starch-gel electrophoresis of the soluble fraction showed that little or no reversion to native myoglobin had occurred; 10% of the amino groups were unmasked, and only 10% of the antigenic reactivity was recovered. Deblocking of TFSu-myoglobin for 4 days resulted in a heterogeneous preparation with 65% of its amino groups unmasked. The native conformation was not recovered, and its antigenic reactivity was 61%. Deblocking of ML-myoglobin[8] for 92 hours resulted in a heterogeneous preparation with 91% of its amino groups free, 93% antigenic reactivity, and incomplete recovery of native conformation.

Acetoacetylated lysozyme[9] was electrophoretically heterogeneous by disc electrophoresis and was also modified at 5–6 hydroxyl groups. It had no enzymatic activity and reacted 39–44% with antisera to lysozyme. Deblocking for 3 days gave a heterogeneous preparation with 5.6–5.9 unmasked amino group, 82–89% enzymatic activity, and 78–81% antigenic reactivity. TFSu-lysozyme[9] was electrophoretically heterogeneous and had 2–3 esterified hydroxyl groups, no enzymatic activity, and 42–48% antigenic reactivity. Deblocking of TFSu-lysozyme for 4 days gave a heterogeneous preparation with 4.6 unmasked amino groups, 64% enzymatic activity, and 73–78% antigenic reactivity. Completely maleylated lysozyme was electrophoretically heterogeneous.[9] Demaleylation, even for 5 days at pH 3.5, gave an electrophoretically heterogeneous preparation with 90% deblocking of amino groups, 83% enzymatic activity, and incomplete recovery of antigenic reactivity and native conformation.[9]

Conclusions

Results obtained from studies with myoglobin and with lysozyme show that in derivatives prepared by reaction with tetrafluorosuccinic or maleic anhydrides or with diketene, complete unmasking of amino groups is not achieved. The deblocked proteins are highly heterogeneous, with partial recovery of enzymatic activity (for lysozyme), immunochemical properties and native conformation. In contrast, the corresponding citraconyl derivatives give, on deblocking, homogeneous preparations with 100% recovery of free amino groups, enzymatic activity, immunochemical properties, and native conformation.

From the foregoing results, it is clear that citraconic anhydride should be ideal for effecting specific cleavage with trypsin at arginine peptide bonds. Citraconylation of myoglobin has been employed by Singhal and Atassi[8] for tryptic cleavage specifically at arginine peptide bonds. The resultant peptides were isolated, deblocked, and their immunochemistry studied. Atassi and Singhal[15] also studied the conformation of these peptides by optical rotatory dispersion and circular dichroism measurements.

The usefulness of a reversible reagent depends on the ease of removal of the blocking groups under conditions that will not lead to denaturation, on the homogeneity of the preparation on deblocking, and on the extent of recovery of the biological properties and reversion to the native conformation. From the foregoing, it is clear that these criteria are satisfied by citraconylated derivatives.

[15] M. Z. Atassi and R. P. Singhal, *J. Biol. Chem.* **245,** 5122 (1970).

[50] Reaction with N-Carboxy-α-Amino Acid Anhydrides

By MICHAEL SELA and RUTH ARNON

N-Carboxy-α-amino acid anhydrides readily undergo polymerization, with carbon dioxide evolution, to yield the corresponding poly-α-amino acid.[1] Amino groups are among the best initiators for this polymerization. Proteins, containing numerous free amino groups, may serve as multifunctional initiators, thus yielding polypeptidyl proteins.[2] The polymerization on proteins proceeds under mild conditions (aqueous media, low temperature, and neutral pH range), which do not, as a rule, cause denaturation of most proteins. The only N-carboxy anhydride which does not react with the amino groups of proteins in aqueous solution is that of proline. In this particular case the solvent of choice is dimethylsulfoxide.[3] It is thus possible by this approach to prepare chemically modified proteins which maintain the principal structural features of the native macromolecule.

A considerable number of polypeptidyl enzymes have been prepared. These often retain enzymatic activity, although differing markedly from the native protein in their physicochemical properties, such as solubility and electrophoretic mobility. The peptidylation of enzymes may change their substrate specificity as well as pH and ionic strength dependence.[4-6]

[1] E. Katchalski, M. Sela, H. I. Silman, and A. Berger, *in* "The Proteins" (H. Neurath, ed.), Vol. II, p. 405. Academic Press, New York, 1964.

[2] R. R. Becker and M. A. Stahmann, *J. Biol. Chem.* **204,** 745 (1953).

[3] J.-C. Jaton and M. Sela, *J. Biol. Chem.* **243,** 5616 (1968).

[4] D. Wellner, H. I. Silman, and M. Sela, *J. Biol. Chem.* **238,** 1324 (1963).

[5] T. Isemura, T. Fukushi, and A. Imanishi, *J. Biochem.* (*Tokyo*) **56,** 408 (1964).

[6] T. Yoshimura, A. Imanishi, and T. Isemura, *J. Biochem.* (*Tokyo*) **63,** 730 (1968).

The investigation of a series of polypeptidyl derivatives of a given enzyme may shed light on the effect of charge, steric hindrance, and hydrophilic and hydrophobic groups on the catalytic activity of the enzyme.

In studies of the conformation of an enzyme molecule and its relationship to enzymatic activity, reactions are often used which cause insolubilization of the products obtained. Peptides of DL-alanine, when attached to the molecule, have acted as a "solubilizing agent" and, therefore, recourse may be made in such studies to modification of the enzymes with *N*-carboxy-α-amino-DL-alanine anhydride. A case in point is the alanylation of trypsin, which enabled a study of the reduction and reoxidation of this enzyme. In contrast to native trypsin, poly-DL-alanyl trypsin yielded a soluble product upon complete reduction of its disulfide bridges.[7] Another example, which is even more dramatic, is the alanylation of myosin.[8] In this case the native protein is completely insoluble in solutions of low ionic strength but becomes soluble after polyalanylation and can then be subjected to chemical and enzymatic studies.

In connection with the preparation of water-insoluble enzymes, it is often necessary to obtain a derivative of the enzyme which is enriched with a particular amino acid to facilitate binding to the insoluble carrier. The preparation of insoluble trypsin can serve as an example. The enzyme is coupled to the diazotized copolymer of *p*-amino-DL-phenylalanine and L-leucine. Since coupling is achieved through phenolic groups, the enzyme is reacted with *N*-carboxy-L-tyrosine anhydride, to yield polytyrosyl trypsin, prior to coupling to the polymer.[9]

The study of a polypeptidyl enzyme may also help to elucidate the role played by ε-amino groups in the unmodified enzyme. In this context the effects of the modification may be studied with regard to the catalytic activity, the physical and chemical properties, as well as the immunological reactivity of the enzyme.[5,6,10–12]

The Reaction of Enzymes with *N*-Carboxy-α-amino Acid Anhydrides

In this reaction both the α-amino and the ε-amino groups of the enzyme serve as initiators for the polymerization. The reaction proceeds according to the scheme given in Fig. 1.

When the *N*-carboxy-α-amino acid anhydrides contain blocking groups,

[7] C. J. Epstein and C. B. Anfinsen, *J. Biol. Chem.* **237**, 3464 (1962).

[8] I. S. Edelman, E. Hoffer, S. Bauminger, and M. Sela, *Arch. Biochem. Biophys.* **123**, 211 (1968).

[9] A. Bar-Eli and E. Katchalski, *Nature* (*London*) **188**, 856 (1960).

[10] C. B. Anfinsen, M. Sela, and J. P. Cooke, *J. Biol. Chem.* **237**, 1825 (1962).

[11] J. P. Cooke, C. B. Anfinsen, and M. Sela, *J. Biol. Chem.* **238**, 2034 (1963).

[12] R. Arnon and H. Neurath, *Immunochemistry* **7**, 241 (1970).

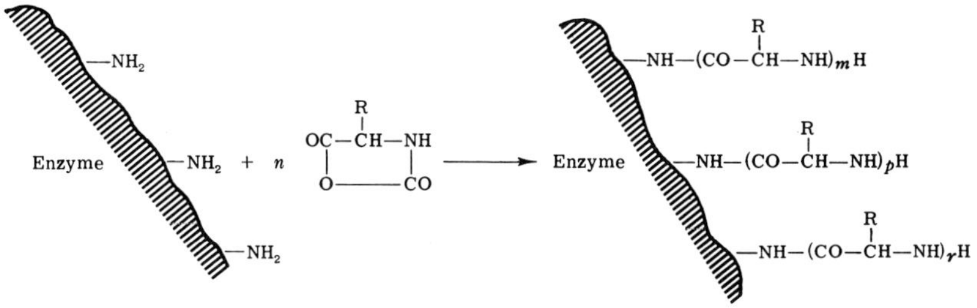

FIG. 1. Reaction of an enzyme with an N-carboxy-α-amino acid anhydride.

these groups should be removed, after the polymerization under mild conditions, to prevent the inactivation of the enzyme. Such conditions are available for preparing polylysyl enzymes (via the poly-N^{ϵ}-trifluoroacetyllysyl derivatives[13]). Although the peptide chains attached may differ somewhat in size, from a statistical analysis of the molecular weight distribution of multichain polyamino acids, it may be predicted that the homogeneity of the protein will not be greatly affected by the peptidylation reaction.[1]

The detailed preparation of four different polypeptidyl enzyme derivatives is given below.

Preparation of Polytyrosyl Trypsin[14]

A solution of 1 g of trypsin–50% $MgSO_4$ (i.e., 500 mg of trypsin) in 36 ml of 0.0025 N HCl is introduced into a 250-ml flask, and 36 ml of 0.1 M phosphate buffer, pH 7.6, is added. The final pH of the mixture is 7.2. The mixture is chilled in an ice bath to 2°. A solution of 0.8 g of N-carboxy-L-tyrosine anhydride[15] in 16 ml of anhydrous dioxane is added dropwise with vigorous stirring to the cold protein solution. The milky reaction mixture formed is stirred magnetically for 16 hours at 4° and then dialyzed for 7 days against daily changes of 6 liters of

[13] M. Sela, R. Arnon, and I. Jacobson, *Biopolymers* **1**, 517 (1963).

[14] A. N. Glazer, A. Bar-Eli, and E. Katchalski, *J. Biol. Chem.* **237**, 1832 (1962).

[15] N-carboxy-L-tyrosine anhydride,[16] N-carboxy-DL-alanine anhydride[17] and N^{ϵ}-trifluoroacetyl-N^{α}-carboxy-L-lysine anhydride[13] are prepared by reacting L-tyrosine, DL-alanine, or N^{ϵ}-trifluoroacetyl-L-lysine, respectively, with phosgene, according to Katchalski and Berger.[18] N^{ϵ}-Trifluoroacetyl-L-lysine is prepared by coupling L-lysine with ethyl thiotrifluoroacetate.[13]

[16] A. Berger, J. Kurtz, T. Sadeh, A. Yaron, R. Arnon, and Y. Lapidoth, *Bull. Res. Counc. Is. Sect. A* **7**, 98 (1958).

[17] M. Sela and S. Fuchs, *in* "Methods in Immunology and Immunochemistry" (M. W. Chase and C. A. Williams, eds.), Vol. 1, p. 185. Academic Press, New York, 1967.

[18] E. Katchalski and A. Berger, Vol. 3, p. 546.

0.0025 N hydrochloric acid. The resultant clear solution (any precipitate formed is centrifuged off)[19] is lyophilized and stored at 4°.

The enrichment in tyrosine is determined by ultraviolet absorption measurements; the number of moles of tyrosine attached per mole of trypsin is calculated from the spectra of trypsin and polytyrosyl trypsin in acid and alkali. Terminal amino groups are quantitated by dinitrophenylation; the number of moles of dinitrophenyltyrosine which are obtained, after hydrolysis, per mole of polytyrosyl trypsin indicates the number of moles of peptide chains attached, thus permitting calculation of the average length of the polypeptide side chain.

Poly-DL-alanyl Ribonuclease

Several poly-DL-alanyl derivatives of ribonuclease, differing in the extent of the peptidylation, have been prepared and characterized.[3,10,11] The enrichment in alanine peptides is a function of the ratio of the N-carboxy-DL-alanine anhydride to ribonuclease used for the reaction.

For a typical preparation, a solution of 1 g of bovine pancreatic ribonuclease in 150 ml of 0.05 M phosphate buffer, pH 6.8, is introduced into a 500-ml Erlenmeyer flask, and cooled in an ice bath to 2°. A solution of 3 g of N-carboxy-DL-alanine-anhydride[15] in 100 ml of anhydrous dioxane is added dropwise while the mixture is stirred vigorously. The reaction proceeds at 4° with stirring for 24 hours. The entire mixture is then dialyzed in heat-treated Cellophane tubing against several changes of distilled water to remove salts, alanine, small oligopeptides of alanine, and dioxane. The resultant solution is lyophilized (1.4 g). Any native ribonuclease or polypeptides of alanine present in the reaction product can be removed by purification on a column of phosphorylated cellulose with a combined gradient of a salt and pH.[10] No native enzyme was ever found in the reaction product. The enrichment in alanine is determined by amino acid analysis, and the number of peptide chains attached is quantitated by dinitrophenylation or by deamination. The average chain-length of the peptides attached may thus be calculated. Under the conditions cited above, 8 out of the 11 amino groups of the protein are alanylated, and the length of the side chains, found to be relatively uniform, is approximately 5. Two additional amino groups may be partially attacked by N-carboxyalanine anhydride upon more extensive alanylation, yielding a material that still retains enzymatic activity. The last lysine residue, identified as residue 41 in the polypeptide chain of the enzyme, is vulnerable to the N-carboxyanhydride only when the phosphate buffer is

[19] A more extensive tyrosylation, achieved by using larger amounts of the N-carboxyanhydride, will lead to the formation of insoluble material.

replaced by bicarbonate. Its alanylation brings about total loss of enzymatic activity.[11]

Poly-L-lysyl Ribonuclease[20]

The trifluoroacetyl function was found to be a useful reversible blocking agent for amino groups, being stable to the conditions of *N*-carboxyanhydride formation and its polymerization, and easily removed under mild conditions.[21] Consequently, it was used in the synthesis of poly-L-lysyl ribonuclease. In a typical preparation, a solution of 0.5 g of bovine pancreatic ribonuclease in 70 ml of 0.05 *M* phosphate buffer, pH 7, is introduced into a 250-ml Erlenmeyer flask and cooled with ice to 2°. A solution of 0.5 g of N^{α}-carboxy-N^{ϵ}-trifluoroacetyl-L-lysine anhydride[15] in 20 ml of anhydrous dioxane is added, with stirring. The reaction is allowed to proceed for 24 hours in the cold room (4°) and the entire mixture is dialyzed in heat-treated Cellophane tubing for 3 days against several changes of 6 liters of distilled water at 4°. During the reaction or the dialysis, a precipitate often appears which contains a fraction of the protein which has been more extensively peptidylated. The precipitate is centrifuged off, and the supernatant fluid is lyophilized. The yield is approximately 0.45 g. The extent of the reaction may be followed by fluorine determination.[13]

To remove the trifluoroacetyl groups, 100 mg of the poly-N^{ϵ}-trifluoroacetyllysyl ribonuclease is introduced into a 25-ml flask and suspended in 7 ml of 1 *M* aqueous piperidine. The suspension is stirred at 4° for 30 hours.[22] The clear solution obtained is neutralized with cold 0.5 *N* acetic acid, dialyzed in heat-treated Cellophane tubing at 4° for 3 days against daily changes of 1 liter of distilled water, and lyophilized. Determination of the enrichment in lysine and of the average length of the peptide side chains is carried out by amino acid analysis of native and polylysylribonuclease before and after deamination. The number of moles of lysine residues obtained after hydrolysis of the deaminated polypeptidylated material indicates the actual number of moles of peptide chains attached.

It should be noted that the availability of a modified enzyme preparation enriched with a basic amino acid, such as polylysyl ribonuclease, was of primary importance in the study of the effect of net electrical

[20] A. Frensdorff and M. Sela, *Europ. J. Biochem.* **1**, 267 (1967).

[21] F. Weygand and E. Csendes, *Angew. Chem.* **64**, 136 (1952).

[22] The material dissolves in the piperidine solution already after 1 hour, but at this stage the protein derivative still contains about half of the original trifluoroacetyl groups. The fluorine content drops to less than 1% only after 30 hours in piperidine.

charge of antigens on the type of the antibodies they elicit.[23] These studies demonstrated that the antigenic control of the antibody type occurs at the level of the complete antigenic molecule, and is thus dictated by the overall net charge of the molecule, rather than by the charge within the areas around the haptenic determinants.

Poly-DL-alanyl Myosin[8]

Myosin A, prepared from rabbit psoas muscle and freed from actomyosin, is dissolved in 0.3 *M* KCl. Ten milliliters of a 2–3-mg/ml solution are mixed with 10 ml of 0.1 *M* phosphate buffer, pH 7.0; a solution of *N*-carboxy-DL-alanine anhydride (100 mg) in 1 ml of anhydrous dioxane is added. The contents of the flask are stirred magnetically for 2 hours at 4° and then dialyzed against multiple changes of 0.3 *M* KCl for at least 48 hours. To remove any residual poly-DL-alanine chains, the dialyzed solution is concentrated by ultrafiltration and then applied to a Sephadex G-50 column (2.5×35 cm) and developed with 0.3 *M* KCl. The degree of alanine enrichment is determined by amino acid analysis, and the number of peptide chains attached is quantitated by deamination. Under the conditions specified, 40% of the ϵ-amino groups of myosin react with the *N*-carboxy-DL-alanine anhydride, and 600–700 alanine residues are added per molecule of myosin, an average chain length of 3.7 alanine residues. The modified myosin is completely soluble in 0.05 *M* KCl, whereas the native protein is completely insoluble at this ionic strength. The polyalanyl myosin is also 80% soluble in distilled water and in 0.005 *M* Tris buffer, pH 7.5. Despite the drastic change in solubility, it retains all the major features of myosin Ca-ATPase activity.

[23] E. Rüde, E. Mozes, and M. Sela, *Biochemistry* **7**, 2971 (1968).

[51] Guanidination of Proteins

By A. F. S. A. Habeeb

Guanidination of proteins by reaction with *O*-methylisourea has been reviewed by Kimmel.[1] The reagent is used at 0.2–1 *M* concentration, pH 10.2–11, at temperatures ranging from 2° to 25° and for 2–5 days. The rather high optimum pH necessary for extensive guanidination may be a limitation for some proteins. 1-Guanyl-3,5-dimethylpyrazole nitrate

[1] J. R. Kimmel, Vol. XI, p. 584.

(GDMP) was developed by Habeeb[2,3] to guanidinate proteins and has been found to be the reagent of choice. It permits reaction under more gentle conditions than does *O*-methylisourea. The optimum pH for guanidination of bovine serum albumin, β-lactoglobulin, and ovalbumin is 9.5.[2-4] This article presents data on the use of GDMP for guanidinating various proteins and, wherever possible, compares the results with those obtained with *O*-methylisourea (MIU).

Preparation and Stability of 1-Guanyl-3,5-dimethylpyrazole Nitrate

The reagent is commercially available and is conveniently prepared from aminoguanidine nitrate and 2,4-pentanedione as described by Bannard *et al.*[5] GDMP exists as colorless prisms, mp 166–168°. Scott and Reilly[6] found GDMP to be susceptible to hydrolytic deguanylation with the formation of 3,5-dimethylpyrazole and urea. Slow destruction of the reagent followed a first-order reaction mechanism and was temperature-dependent. Insignificant destruction[3] of GDMP occurred after 24 hours at room temperature, and 10% was decomposed after 42 days at 0°. On the other hand, MIU showed 41% destruction after 24 hours at room temperature.[7]

Reaction of GDMP with Proteins

As in the guanidination of proteins with MIU, the reaction with GDMP depends on several factors, e.g., pH, temperature, and concentration of reagent. The behavior of proteins toward GDMP reflects their unique structural features and may result in the development of physicochemical differences between native and guanidinated proteins.

$$\text{Protein-NH}_2 + \underset{\substack{|\\ \text{C=NH}\\ |\\ \text{NH}_2\cdot\text{HNO}_3}}{\text{H}_3\text{C-C}_5\text{(H)C=C(CH}_3\text{)N=N-}} \xrightarrow{\text{pH 9.5}} \text{Protein-NH-}\underset{\text{NH}}{\overset{}{\underset{\|}{\text{C}}}}\text{-NH}_2 + \text{H}_3\text{C-C=C-C(CH}_3\text{)=N-NH (3,5-dimethylpyrazole)}$$

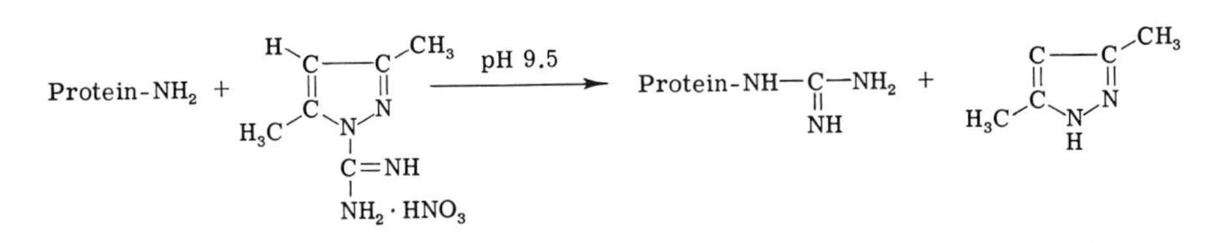

The rate of reaction of GDMP with the free amino groups of proteins increases rapidly with increase in pH,[2-4] indicating that the reaction in-

[2] A. F. S. A. Habeeb, *Biochim. Biophys. Acta* **34,** 294 (1959).
[3] A. F. S. A. Habeeb, *Can. J. Biochem. Physiol.* **38,** 493 (1960).
[4] A. F. S. A. Habeeb, *Can. J. Biochem. Physiol.* **39,** 729 (1961).
[5] R. A. B. Bannard, A. A. Casselman, W. F. Cockburn, and G. M. Brown, *Can. J. Chem.* **36,** 1541 (1958).
[6] F. L. Scott and J. Reilly, *J. Amer. Chem. Soc.* **74,** 4562 (1952).
[7] W. A. Klee and F. M. Richards, *J. Biol. Chem.* **229,** 489 (1957).

volves the uncharged form of one or both of the reactants. Owing to differences in behavior of proteins on guanidination with GDMP, representative conditions for three proteins are presented.

Reaction of GDMP with Bovine Serum Albumin[2,3]

By varying the concentration of GDMP (0.1, 0.2, and 0.5 *M*) and the pH (8.5, 9, 9.5, 10, and 10.5) the reaction was found to be maximal when protein at 5% concentration was reacted in 0.5 *M* GDMP at pH 9.5 and 0° for 7 days. GDMP is dissolved in few milliliters of water (in an ice bath), the pH is adjusted to pH 9.5 with 1 *N* NaOH (magnetic stirring), and the solution is diluted with water to give 0.5 *M* GDMP. Bovine serum albumin (BSA) is dissolved in the aqueous solution of GDMP to 5% concentration and allowed to react for 7 days at 0° with pH adjustment. After reaction, the solution is brought to pH 7.5 and dialyzed exhaustively against phosphate buffer, pH 7.5, ionic strength 0.1. An aliquot is used to determine protein concentration and estimate the free amino groups. Under these conditions 92% of the free amino groups are modified as determined by ninhydrin colorimetric analysis.

Reaction of GDMP with β-Lactoglobulin

The reaction of β-lactoglobulin in 0.5 *M* GDMP is complicated by gelling[2,3] under conditions that allow BSA to react. Similar results were found with MIU.[3] Gelling is avoided at 3% protein solution if the reaction is performed in 0.2 *M* GDMP at pH 9.5 and 0° for 7 days. Guanidinated β-lactoglobulin thus prepared shows polydispersity upon ultracentrifugation. Gelling may be attributed to formation of intermolecular disulfide bonds and is prevented by sulfhydryl reagents, e.g., *N*-ethylmaleimide and iodoacetic acid. To prevent aggregation, β-lactoglobulin is allowed to react with *N*-ethylmaleimide or iodoacetic acid for 30 hours under conditions that allow complete reaction of sulfhydryls.[8] The material after exhaustive dialysis and lyophilization is allowed to react in 0.2 *M* GDMP at pH 9.5 and 0° for 7 days. Guanidinated β-lactoglobulin thus prepared shows a single boundary on ultracentrifugation.

Reaction of GDMP with Ovalbumin

Preliminary results showed that although gelling occurred in 0.5 *M* GDMP at pH 9.5 and 0°, no gelling took place in 0.2 *M* or 0.3 *M* GDMP. To obtain guanidinated ovalbumin suitable for physicochemical studies, guanidination is performed in 0.25 *M* GDMP pH 9.5 and 0° for 7 days.[4]

[8] A. F. S. A. Habeeb, *Can. J. Biochem. Physiol.* **38**, 269 (1960).

Determination of Extent of Modification of Free Amino Groups

The Van Slyke method[9] for free amino groups and the Sakaguchi method for arginine and homoarginine were used by Hughes *et al.*[10] and by Chervenka and Wilcox.[11] Ninhydrin analysis by the method of Harding and MacLean[12] was used by Habeeb[2-4] to quantitate the decrease in free amino groups on guanidination of bovine serum albumin, β-lactoglobulin, and ovalbumin. These methods have been superseded by the use of 2,4,6-trinitrobenzenesulfonic acid[13,14] to determine free amino groups. The method is sensitive (0.5–1 mg protein), and the relationship between concentration and absorbance is linear.

Direct Determination of Unreacted Lysine as ε-DNP-Lysine

Guanidinated protein is allowed to react with fluorodinitrobenzene,[15] the modified protein is separated and hydrolyzed, and ε-DNP-lysine is either determined by two-dimensional paper chromatography[2,3] or by automatic amino acid analysis.[16]

Direct Determination of Homoarginine

Guanidination of proteins converts internal lysine residues into homoarginine residues which can be estimated conveniently in a protein hydrolyzate by chromatography on the short column of an automatic amino acid analyzer.[17] Various workers[3,4,11,18,19] have used this method for determination of homoarginine.

Determination of Extent of Reaction of GDMP with NH_2-Terminal Amino Acids of Proteins

In contrast to *O*-methylisourea, which does not react with NH_2-terminal amino acids of various proteins,[11,18] partial reaction with GDMP occurs at the NH_2-terminal amino groups of bovine serum albumin and β-lactoglobulin.[2,3] The extent of reaction is determined by reacting the

[9] D. D. Van Slyke, *J. Biol. Chem.* **83,** 425 (1929).

[10] W. L. Hughes, Jr., H. A. Saroff, and A. L. Carney, *J. Amer. Chem. Soc.* **71,** 2476 (1949).

[11] C. H. Chervenka and P. E. Wilcox, *J. Biol. Chem.* **222,** 635 (1956).

[12] V. J. Harding and R. M. MacLean, *J. Biol. Chem.* **24,** 503 (1916).

[13] A. F. S. A. Habeeb, *Anal. Biochem.* **14,** 328 (1966).

[14] A. F. S. A. Habeeb, *Arch. Biochem. Biophys.* **119,** 264 (1967).

[15] H. Fraenkel-Conrat, J. I. Harris, and A. L. Levy, *Methods Biochem. Anal.* **2,** 359 (1955).

[16] L. Wofsy and S. J. Singer, *Biochemistry* **2,** 104 (1963).

[17] D. H. Spackman, W. H. Stein, and S. Moore, *Anal. Chem.* **30,** 1190 (1958).

[18] G. S. Shields, R. L. Hill, and E. L. Smith, *J. Biol. Chem.* **234,** 1747 (1959).

[19] A. Nureddin and T. Inagami, *Biochem. Biophys. Res. Commun.* **36,** 999 (1969).

TABLE I

CONDITIONS OF GUANIDINATION AND ANALYSIS OF AMINO GROUPS, ε-DNP-LYSINE AND NH_2-TERMINAL AMINO ACID IN NATIVE AND GUANIDINATED PROTEINS

	Protein conc. (%)	Reagent conc. (*M*)	Temperature (°C)	Days of reaction	% Free amino groups by ninhydrin	Moles amino acid per mole protein			
						Lysine	Homoarginine	ε-DNP-lysine	NH_2-terminal residues
Bovine serum albumin[a]	—	—	—	—	100	56	0	56	Aspartic acid, 0.96
Guanidinated bovine serum albumin[a]	5	0.5	0	7	10	6	52	5.2	Aspartic acid, 0.37
	5	0.5	25	1	10				
Native β-lactoglobulin[a]	—	—	—	—	100	28.8	0	21–22[d]	—
Denatured β-lactoglobulin[a]	—	—	—	—	100	28.8	0	28	Leucine, 2.76
Guanidinated β-lactoglobulin (NEMI)[e]	3	0.2	0	7	10	1.54	17.4[g]	0.7	Leucine, 1.98
Guanidinated β-lactoglobulin (IAc)[f]	3	0.2	0	7	10	1.52	21.0[g]	0.7	Leucine, 1.98
Ovalbumin[b]	—	—	—	—	100	20.7	0	—	—
Guanidinated ovalbumin[b]	3	0.25	0	7	15	1.8	18.3	—	—
Guanidinated trypsin[c]	0.1	0.3	4	5	—	4	10	—	—

[a] A. F. S. A. Habeeb, *Can. J. Biochem. Physiol.* **38,** 493 (1960).

[b] A. F. S. A. Habeeb, *Can. J. Biochem. Physiol.* **39,** 729 (1961).

[c] A. Nureddin and T. Inagami, *Biochem. Biophys. Res. Commun.* **36,** 999 (1969).

[d] ε-Amino groups of lysine in β-lactoglobulin were not completely reactive with fluorodinitrobenzene except after denaturation.

[e] β-Lactoglobulin was treated with *N*-ethylmaleimide before guanidination (see text).

[f] β-Lactoglobulin was treated with iodoacetic acid before guanidination.

[g] The incomplete recovery of modified lysine as homoarginine was associated with the appearance of unidentified ninhydrin-positive peaks [see A. F. S. A. Habeeb, *Can. J. Biochem. Physiol.* **39,** 729 (1961).

guanidinated protein with fluorodinitrobenzene, followed by separation and acid hydrolysis. The DNP amino acids are extracted and subjected to two-dimensional chromatography. Table I gives a summary of guanidination conditions for various proteins using GDMP.

Specificity of Reaction of GDMP with Proteins

The reaction of GDMP with the ε-amino groups of lysine in proteins is never complete, possibly because of steric hindrance. The number of resistant lysines is 5–6 residues per mole of bovine serum albumin,[3] 1.5 in β-lactoglobulin,[3] 2 in ovalbumin,[4] and 3–4 in trypsin.[19] Partial reaction occurs with NH_2-terminal amino groups in bovine serum albumin (60%) and β-lactoglobulin (30%). The amino acid analysis of guanidinated ovalbumin compares favorably with that of ovalbumin,[4] suggesting specific reaction at lysine residues without affecting other amino acid residues in the protein.

Physicochemical Studies on Guanidinated Proteins

Data on sedimentation coefficient, diffusion coefficient, and electrophoretic mobility of some proteins are presented in Table II. Since reaction with GDMP replaces the positively charged NH_3^+ group of lysine residues with a positively charged $—NH—C(=NH_2^+)—NH_2$ group, no change of the net charge of the protein occurs around pH 8 as a result of guanidination. However, other factors may intervene. Table II indicates that the electrophoretic mobility of guanidinated bovine serum albumin is identical with that of the native protein. On the other hand, guanidinated ovalbumin is slightly more, and guanidinated β-lactoglobulin is appreciably more, electronegative than are the native proteins. The greater negative electrophoretic mobility of guanidinated β-lactoglobulin is

TABLE II
PHYSICOCHEMICAL DATA ON GUANIDINATED PROTEINS

Protein	$s_{20,w}$ (S)	$D^{\circ}_{20,w}$ ($cm^2 \cdot sec \times 10^7$)	Electrophoretic mobility ($cm^2/V/sec \times 10^5$)
Bovine serum albumin[a]	4.18	—	−6.6
Guanidinated bovine albumin[a]	4.18	—	−6.6
β-Lactoglobulin[a]	2.41	—	−5.4
Guanidinated β-lactoglobulin[a]	1.78	—	−7.0
Ovalbumin[b]	3.45	7.52	−6.22
Guanidinated ovalbumin[b]	3.35	7.21	−6.76

[a] A. F. S. A. Habeeb, *Can. J. Biochem. Physiol.* **38,** 493 (1960).
[b] A. F. S. A. Habeeb, *Can. J. Biochem. Physiol.* **39,** 729 (1961).

analogous to the behavior of guanidinated chymotrypsinogen,[11] which has been attributed to the greater binding of anions by guanidino groups.[11] Native proteins have an ordered conformation under physiological conditions. The conformation is determined by the stabilizing effect of noncovalent interactions in the direction that minimizes the total free energy. It is to be expected that chemical modification by covalent introduction of new groups will in effect generate a new protein. As a result, the noncovalent interactions may be expected to undergo reorganization. The magnitude of this reorganization will depend on the nature of the groups introduced and on the nature and location of the groups substituted. With guanidination, where the positively charged NH_3^+ is replaced by a positively charged guanidino group, little or no disruption of the noncovalent interactions may be expected, with minimal conformational changes.

The sedimentation coefficients,[3] intrinsic viscosity,[10,20] and Stokes radii[21] of guanidinated and native bovine serum albumin are identical. However, a subtle conformational difference in the proteins is revealed by a change in the reactivity of the disulfide bonds. Guanidinated bovine serum albumin contains two disulfide bonds reducible with β-mercaptoethanol[20] as compared to 0.5 reducible disulfide bond in bovine serum albumin. Other guanidinated proteins, e.g., ovalbumin[4] and ribonuclease,[7] have sedimentation coefficients identical with those of the native proteins. However, the sedimentation coefficient is a less sensitive index with which small conformational changes can be revealed than are the Stokes radius[22,23] or reducibility of disulfide bonds.[20,22] Considerable conformational differences between guanidinated chymotrypsinogen,[11] mercuripapain,[18] and insulin[24] and their native counterparts are apparent from the sedimentation coefficients.

Immunochemistry of Guanidinated Proteins

The immunochemical behavior of globular proteins is influenced highly by changes in the native conformation.[25,26] Any change in conformation due to chemical modification results in a decrease in the ability of the modified protein to react with antibodies to native protein (provided that the modified group does not form a part of an antigenic de-

[20] A. F. S. A. Habeeb, *Biochim. Biophys. Acta* **115,** 440 (1966).
[21] A. F. S. A. Habeeb, *Biochim. Biophys. Acta* **121,** 21 (1966).
[22] A. F. S. A. Habeeb, *Arch. Biochem. Biophys.* **121,** 652 (1967).
[23] M. Z. Atassi and D. B. Caruso, *Biochemistry* **7,** 699 (1968).
[24] R. L. Evans and H. A. Saroff, *J. Biol. Chem.* **228,** 295 (1957).
[25] M. Z. Atassi, *Biochem. J.* **103,** 29 (1967).
[26] A. F. S. A. Habeeb, *J. Immunol.* **99,** 1264 (1967).

terminant). Guanidinated bovine serum albumin (54 amino groups modified) has 91% of the precipitability with antibovine serum albumin possessed by the homologous system,[26] indicating very limited conformational changes on guanidination as compared to other modifications, e.g., succinylation. Guanidinated bovine serum albumin is itself antigenic in rabbits.[27] Only 42% of the antibodies to guanidinated bovine serum albumin are precipitated and absorbed with bovine serum albumin in early bleedings whereas with late bleedings 67% of the antibody is precipitable. These data indicate that the introduction of guanido groups imparts a new specificity to the protein. New antigenic determinants are created by guanidination, and at the same time antigenic determinants which are characteristic of bovine serum albumin are still maintained. In agar double-diffusion, the precipitin lines formed between bovine serum albumin, guanidinated bovine serum albumin, and antisera to guanidinated bovine serum albumin show a reaction of partial identity.[27]

Effect of Modification by Guanidination on Biological Activity of Proteins

Chemical modification of biologically active proteins often is employed to characterize the functional groups involved at the active site. However, interpretation of such studies may be difficult because conformational reorganization, which often accompanies the modification, may influence the observed biological activity. Also, the nature of a chemical modification at a given site determines its effect on biological activity.[28] The role of lysine residue has been investigated in many biologically active proteins by use of guanidination or reaction with acid anhydrides. On acetylation of lysozyme, growth hormone and lactogenic hormone, complete loss of activity, is observed.[29] In contrast, guanidination of these proteins does not affect their activity.[29] This indicates that the loss of activity on acetylation is nonspecific, and due to conformational changes that result from a loss of positively charged ϵ-amino groups. The conservation of biological activity on guanidination is evidence that the ϵ-amino groups are not important for the functional activity of lysozyme. Ribonuclease is not inactivated by guanidination of 9 lysine residues,[7,30] but becomes inactivated on guanidination of the tenth lysine residue, namely, lysine-41.[30] The importance of lysine-41 for the activity of ribonuclease is thus demonstrated. Guanidination

[27] A. F. S. A. Habeeb, *J. Immunol.* **101,** 505 (1968).
[28] M. Z. Atassi and A. F. S. A. Habeeb, *Biochemistry* **8,** 1385 (1969).
[29] I. I. Geschwind and C. H. Li, *Biochim. Biophys. Acta* **25,** 171 (1957).
[30] D. M. Glick and E. A. Barnard, *Biochim. Biophys. Acta* **214,** 326 (1970).

of chymotrypsinogen,[11] mercuripapain,[18] horse heart cytochrome c[31,32] and tuna heart cytochrome c[33] is without effect on their activity. The positive charges of the lysine residues in horse heart cytochrome c are important in maintenance of the native conformation. Partial acetylation or succinylation leads to a loss of activity.[34] Lysine residues in the ε-toxin of *Clostridium perfringens* type D are considered necessary for the toxic effect.[35] It was found that acetylation or guanidination are equally effective in inactivating the toxin. The toxicity drops to 7% when 50% of the free amino groups are either acetylated or guanidinated.[35]

Comments

Guanidination of proteins with GDMP occurs under conditions milder than those necessary with *O*-methylisourea. The reaction is specific for ε-amino groups of lysine. Partial reaction occurs with NH_2-terminal amino groups of some proteins. Differences exist with regard to the magnitude of conformational changes in proteins as a result of guanidination. Whereas some guanidinated proteins show little conformational change compared to their native counterparts, others exhibit considerable conformational reorganization because of unique structural features. Where correlation of biological activity to the presence of lysine residues is needed, guanidination may be the method of choice, since conformational and charge changes are minimal.

[31] K. Takahashi, K. Titani, K. Furuno, H. Ishikura, and S. Minakami, *J. Biochem. (Tokyo)* **45,** 375 (1958).
[32] T. P. Hettinger and H. A. Harbury, *Proc. Nat. Acad. Sci. U.S.* **52,** 1469 (1964).
[33] T. P. Hettinger and H. A. Harbury, *Biochemistry* **4,** 2585 (1965).
[34] S. Takemori, K. Wada, K. Ando, M. Hosokawa, I. Sekuzu, and K. Okunuki, *J. Biochem. (Tokyo)* **52,** 28 (1962).
[35] A. F. S. A. Habeeb, *Biochim. Biophys. Acta* **74,** 113 (1963).

[52] Modification of Arginine by Diketones

By J. A. YANKEELOV, JR.

The guanidinium group of arginine ($pK_a = 12.5$) is a planar, resonance-stabilized structure which exists in its ionized form over the usual pH range of protein stability. The marked resistance of this nitrogen function to acylation and other common amine reactions except under disruptive conditions prevented, until recently, its chemical investigation in proteins. Known reactions of alkyl guanidines[1,2] have been extended to the modification of arginine in proteins. Modification with malonaldehyde requires strong mineral acid and is accompanied by peptide bond cleavage.[3] However, the reaction of 1,2-cyclohexanedione with arginine,[4] which requires relatively strong alkali, has been useful in studies of antigenicity of myoglobin.[5]

The observation that arginine could be modified near neutral pH with 2,3-butanedione[6] opened the way for the application of various forms of the reagent in the study of arginine in antibodies,[7] carboxypeptidase-A,[8] bovine plasma albumin (BPA),[9] bovine pancreatic ribonuclease A (RNase A),[9,10] α_1-acid glycoprotein,[11] and low density lipoprotein.[12] Glyoxal[13] and phenylglyoxal[14] have also been used as mild arginine reagents. A summary of these reagents with corresponding structures, reaction conditions and references is given in Table I. The preparation of certain arginine reagents and conditions for their use are discussed in detail below.

Preparation of Crystalline Oligomers of 2,3-Butanedione[10]

Trimeric 2,3-Butanedione (Table I) or 2,5-Diacetyl-3a,5,6,6a-tetrahydro-6a-hydroxy-2,3a,5-trimethylfuro-[2,3-d]-1,3-dioxole (TB). Powdered glass (available from E. H. Sargent and Co., SC-12300) is cleaned by boiling briefly in 3:1 (v/v) HCl-HNO_3,[15] followed by thorough washing with water. All transfers of dry powdered glass are made in a fume hood to avoid inhalation of the particles. The glass is alkalinized by suspending the cleaned material in excess 0.1 *N* NaOH. The glass is col-

[1] M. Lempert-Stréter, V. Solt, and K. Lempert, *Chem. Ber.* **96,** 168 (1963).

[2] D. J. Brown, "The Pyrimidines," p. 32. Wiley (Interscience), New York, 1962.

[3] T. P. King, *Biochemistry* **5,** 3454 (1966).

[4] K. Toi, E. Bynum, E. Norris, and H. A. Itano, *J. Biol. Chem.* **242,** 1036 (1967).

[5] M. Z. Atassi and A. V. Thomas, *Biochemistry* **8,** 3385 (1969).

[6] J. A. Yankeelov, Jr., M. Kochert, J. Page, and A. Westphal, *Fed. Proc. Fed. Amer. Soc. Exp. Biol.* **25,** 590 (1966).

[7] A. L. Grossberg and D. Pressman, *Biochemistry* **7,** 272 (1968).

[8] B. L. Vallee and J. F. Riordan, *Brookhaven Symp. Biol.* **21,** 91 (1968).

[9] J. A. Yankeelov, Jr., C. D. Mitchell, and T. H. Crawford, *J. Amer. Chem. Soc.* **90,** 1664 (1968).

[10] J. A. Yankeelov, Jr., *Biochemistry* **9,** 2433 (1970).

[11] M. Ganguly and U. Westphal, *Biochim. Biophys. Acta* **170,** 309 (1968).

[12] R. S. Levy and C. E. Day, *in* "Proceedings of the Second International Symposium of Atherosclerosis," p. 186. (R. J. Jones, ed.). Springer-Verlag, New York, 1970 (in press).

[13] K. Nakaya, H. Horinishi, and K. Shibata, *J. Biochem.* (*Tokyo*) **61,** 345 (1967).

[14] K. Takahashi, *J. Biol. Chem.* **243,** 6171 (1968).

[15] L. A. Carlson, *Clin. Chim. Acta* **5,** 528 (1960).

TABLE I
DIKETONES USED AS ARGININE REAGENTS

Reagent and references	Structure	Conditions (25°)	Side reactions
2,3-Butanedione[a]	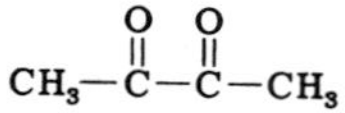 $CH_3-C(=O)-C(=O)-CH_3$	Borate buffer, 1.0 *M* NaCl–0.05 *M*, pH 7.5	—
Dimeric 2,3-butanedione[b,h] (DB)	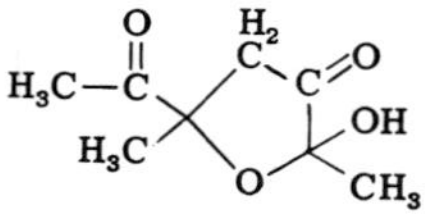 $H_3C-C(=O)$; H_3C; H_2C; $C=O$; OH; O; CH_3 (ring)	Phosphate buffer, 0.5 *M* pH 7.0	Amino groups
Trimeric 2,3-butanedione[b] (TB)	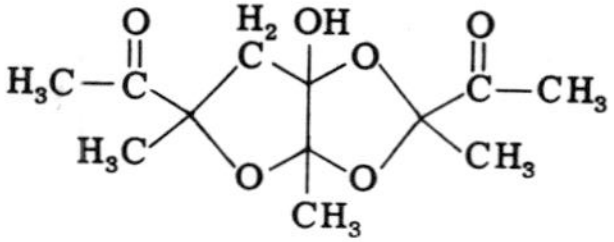 $H_3C-C(=O)$; H_3C; H_2C; OH; O; O; O; O; CH_3; $C(=O)-CH_3$; CH_3 (fused rings)	Phosphate buffer, 0.5 *M* pH 7.0, 8.0	Amino and SH groups
Glyoxal[c]	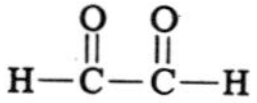 $H-C(=O)-C(=O)-H$	Bicarbonate buffer, pH 9.2	Amino groups

Phenylglyoxal[d]	C_6H_5—CO—CO—H	*N*-Ethylmorpholine–acetate buffer, 0.2 *M* pH 8.0, 7.0	Amino and SH groups
1,2-Cyclohexanedione[e]		NaOH, 0.2–0.05 *N*	—
Malonaldehyde[f,g]	H—CO—CH_2—CO—H	HCl, 10 *N*	Peptide cleavages

[a] B. L. Vallee and J. F. Riordan, *Brookhaven Symp. Biol.* **21,** 91 (1968).

[b] J. A. Yankeelov, Jr., *Biochemistry* **9,** 2433 (1970).

[c] K. Nakaya, H. Horinishi, and K. Shibata, *J. Biochem.* (*Tokyo*) **61,** 345 (1967).

[d] K. Takahashi, *J. Biol. Chem.* **243,** 6171 (1968).

[e] K. Toi, E. Bynum, E. Norris, and H. A. Itano, *J. Biol. Chem.* **242,** 1036 (1967).

[f] T. P. King, *Biochemistry* **5,** 3454 (1966).

[g] Used in the form of its ethyl acetal: 1,1,3,3-tetraethoxypropane.

[h] See also W. Y. Huang and J. Tang, *Fed. Proc. Fed. Amer. Soc. Exp. Biol.* **30,** 1183Abs (1971).

lected on a Büchner funnel, rinsed thoroughly with water until the washings are no longer alkaline and oven-dried overnight at 110°. 2,3-Butanedione (196 g, 2.28 mole) containing 0.1% (v/v) water is mixed with 415 g of the prepared glass in a closed vessel to prevent loss by evaporation. The mixture which initially separates into two layers is mixed with a spatula twice daily. After 5–7 days at room temperature (23–25°) the mixture usually hardens. The product is extracted with ether, and the extract is dried over $MgSO_4$.

After removal of ether by rotary evaporation at 25° and overnight storage of the product in the cold, 76 g (39%) of TB is collected. Recrystallization from ether gives colorless needles of mp 112.5–114°. Analytical data: λ_{max}^{KCl} 2.86, 5.82 μ; nuclear magnetic resonance (NMR) ($CDCl_3$) τ 5.52 (1H), 7.66 (3H), 7.75 (3H), 8.54 (3H), 8.60 (3H), 8.66 (3H), 6.74 (d, J = 13.5 cps, 1H), 8.09 (d, J = 13.5 cps, 1H); $\lambda_{max}^{1\,mM\,HCl}$ 286 nm (ϵ 84), MW 258.

This trimer has been prepared routinely in the above manner in our laboratory; however, several comments may be useful to ensure success. Since alkaline glass surfaces appear to promote depolymerization of the trimer, surfaces coming in contact with the ether solution should be prerinsed with 1 *N* HCl followed by water and dried before use. The melting point of TB will be depressed and not reproducible unless glass in contact with the compound during melting is also dealkalinized. This melting point behavior is known for other ketonic solids.[16] The reagent also forms crystals of a higher melting point (121–122°). Both forms are stable at room temperature if kept dry. Glass beads (0.2 mm, Schwarz BioResearch, Inc.) may be substituted for powdered glass in the procedure, but the reaction time is longer (15–20 days). Seed crystals may be useful in certain instances.

Dimeric 2,3-Butanedione (Table I) or 5-Acetyltetrahydro-2-hydroxy-2,5-dimethyl-3-oxofuran (DB). A solution of 1.0 *N* KOH (350 ml) is added dropwise to a stirred solution of 100 ml of 2,3-butanedione (1.14 mole) in 400 ml of water chilled to 0°. The addition is performed over a 55-minute period while maintaining the temperature at 0 ± 2° by ice–salt bath cooling. Fifteen minutes after the KOH addition has been completed, the solution is brought to pH 2.3 by addition of concentrated H_2SO_4 while maintaining temperature control. The resulting solution is saturated with NaCl and extracted with ether. The extract is dried over $MgSO_4$. Removal of solvent produces a yellow oil which is distilled through a 10-cm Vigreux column at 0.2–0.3 mm Hg. The fractions boiling

[16] R. L. Shriner, R. C. Fuson, and D. Y. Curtin, "The Systematic Identification of Organic Compounds," 5th ed., p. 30. Wiley, New York, 1964.

at 95–102° are collected to give 67.5 g of purified dimer. In the absence of seeds, crystallization occurred after 19 days storage at —20° with periodic thawing. Vacuum filtration followed by pressing with filter paper gave 26 g (27%) of crystalline dimer. When distillates are seeded immediately the yield is raised to 45–48%. Recrystallization from ligroin (60–90°) produces colorless needles, mp 59–61°. λ_{max}^{KCl} 2.97, 5.68, 5.82 μ; NMR ($CDCl_3$) τ 5.60 (1H), 7.68 (3H), 8.53 (6H), 6.66 (d, J = 19 cps, 1H), 7.80 (d, J = 19 cps, 1H), MW 172. This compound is stored with desiccant at —20°.

Modification of Proteins with Oligomers of 2,3-Butanedione[10]

Arginine can be modified by dimeric or trimeric 2,3-butanedione to give three ninhydrin-positive final products. The reaction corresponds to an overall addition of 3 moles of 2,3-butanedione per mole of arginine according to Eq. (1).

$$\text{Arginine} + \begin{matrix} (C_4H_6O_2)_3 \text{ (1 mole)} \\ \text{or} \\ (C_4H_6O_2)_2 \text{ (1.5 mole)} \end{matrix} \xrightarrow[25^\circ]{\text{pH 7.0 or 8.0}} \underbrace{\text{arginine} \cdot (C_4H_6O_2)_3}_{(Y,\ Z,\ \text{and } Z_0)} \qquad (1)$$

Procedures for Modifying Proteins with Trimeric 2,3-Butanedione. Since aqueous buffered solutions of up to 0.4 *M* TB can be prepared at 25°, reactions with proteins can be initiated by simply dissolving a weighed amount of protein in a freshly prepared solution of reagent in sodium phosphate buffer. When employing reagent concentrations of 0.2 *M* or less, a solution of approximately 12 mg of protein per milliliter in buffer is diluted with an equal volume of buffered 0.4 *M* reagent. The reaction mixture is capped, protected from light, and kept in a 25° bath for the desired period. Reaction times of 6–48 hours will generally result in modification of 50–100% of the arginine present depending on the nature of the protein and the conditions used. Table II includes examples of typical reaction conditions.

The reagent may be removed by dialysis (aged size 8 dialysis tubing,

TABLE II
EXAMPLES OF TYPICAL REACTION CONDITIONS

	Extensive, irreversible modification	Limited, reversible modification
Sodium phosphate buffer	0.5 *M*, pH 7.0	0.2 *M*, pH 6.0
Temperature (°C)	25	25
Time (hours)	24–48	6
Protein concentration (mg/ml)	6	6
TB concentration (*M*)	0.4	0.2

Union Carbide Corporation, Food Products Division) against distilled water or buffer, or by gel filtration. In a single experiment, trichloroacetic acid precipitation of BPA followed by washing the precipitate with absolute ethanol was successfully employed to remove reagent. It is desirable to exclude light during removal of reagent as well as during the reaction to avoid possible photochemical effects.

If reactions at several pH intervals are contemplated, it is convenient to prepare 0.5 *M* stock solutions of Na_2HPO_4 and H_3PO_4. One solution is titrated with the other to obtain the desired pH. More dilute buffer may be prepared by diluting the stock solutions separately prior to titration so that pH changes due to lowering the ionic strength are avoided. In solution the reagent is somewhat yellow, the intensity increasing with pH.

Determination of Groups Modified in Amino Acids and Proteins. Study of the reaction of 0.4 *M* TB with mixtures of 18 standard amino acids (1 m*M*) at pH 7.0 by amino acid analysis showed that 97% of the arginine reacts after 1 hour.[9] The next most readily altered amino acid was serine (18% after 1 hour). Other susceptible amino acids showing more limited reactivity were threonine, glycine, cystine, methionine, lysine, histidine, and tryptophan. The data on free amino acids do not fully resolve the question of side-chain reactivities, but studies on reduced glutathione revealed a definite side reaction with SH groups. At high concentrations of reagent, guanidinium group modification of arginine is accompanied by a side reaction of the amino group. With stoichiometric quantities of TB, reaction occurs more slowly with little or no modification of the amino group to give the ninhydrin-positive products Y, Z, and Z_0. These derivatives will be discussed in more detail below.

Modified proteins which have been freed of reagents are hydrolyzed in 6 *N* HCl for 24 hours and subjected to amino acid analysis.[17] Irreversibly modified arginine is converted by acid hydrolysis largely to materials not seen during automatic chromatography. However, in hydrolyzates of extensively modified proteins, additional components have been detected in the region of the basic amino acids of the 22-hour separation[10] of the Technicon system. Arginine itself is released in 12% yield from the purified derivatives (Y, Z, and Z_0). Performic acid-oxidized RNase which had been exhaustively treated with TB showed 12% residual arginine after hydrolysis. Accordingly, the percentage of arginine loss from a TB modified protein may be divided by 0.88 to correct for regeneration of the parent amino acid. Losses in lysine can

[17] S. Moore and W. H. Stein, Vol. 6, p. 819.

also be traced by amino acid analysis, but the extent of regeneration of the parent amino acid during acid hydrolysis has not been determined accurately. Regeneration of lysine, however, appears to be approximately 50%. Study of the detailed nature of the reversible products from arginine obtained with TB at pH 6.0 is yet to be performed.

Although amino acid analysis is the preferred method for measuring arginine loss, loss may also be estimated in protein hydrolyzates by the Sakaguchi reaction as described by Izumi.[18] These estimates have been found to be in satisfactory agreement with those determined by automatic chromatography, provided that excess reagent is thoroughly removed from the modified protein by dialysis prior to hydrolysis. Hydrolyzates of control samples of unmodified proteins as well as standards should be included to permit difference analysis. Water is used to dissolve amino acids of the hydrolyzate rather than pH 2.2, citrate buffer.[17]

Amino group recovery may be estimated on modified proteins by means of 2,4,6-trinitrobenzenesulfonic acid (TNBS) (available from Eastman Kodak, Organic Chemical Division, Rochester, New York) according to the method described by Habeeb[19] using borate buffer.[20] Any color present in the modified protein must be compensated for by incorporating a protein blank in which water is substituted for TNBS reagent. Protein concentration may be estimated from total nitrogen.[21]

A study of the recovery of arginine and amino groups from bovine plasma albumin as the molar ratio of TB to protein arginine is increased is shown in Fig. 1. The data suggest that reagent concentration may be lowered to approximately 0.2 *M* without loss of efficacy for arginine. High concentrations of phosphate buffer as well as higher pH favors irreversible blocking of arginine, but increased pH also enhances reactivity of lysine. After application of the more restrictive conditions (6 hours, pH 6.0), 75–80% of the modified arginines can be released as the parent side chain by dialysis against 1 *M* NaCl–0.05 *M*, pH 7.0 sodium phosphate buffer for 48 hours at 25°.

Tyrosine recovery from hydrolyzates of modified proteins may be as low as 65–75% of the control sample. Since the free amino acid is resistant to the reagent at pH 7.0 and quantitative recovery of tyrosine can be selectively restored by hydrolyzing the protein in the presence of phenol, low tyrosine recovery may be considered normal for the procedure in the absence of other implicating evidence.

[18] Y. Izumi, *Anal. Biochem.* **10**, 218 (1965).
[19] A. F. S. A. Habeeb, *Anal. Biochem.* **14**, 328 (1966).
[20] A. F. S. A. Habeeb, *Arch. Biochem. Biophys.* **119**, 264 (1967).
[21] C. A. Lang, *Anal. Chem.* **30**, 1692 (1958).

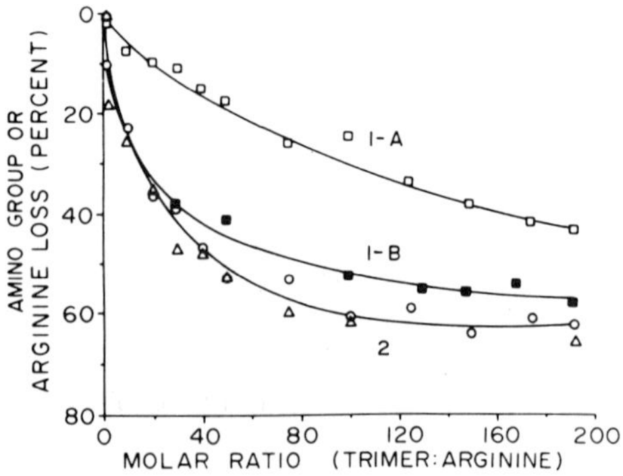

FIG. 1. Dependence of amino group and arginine loss from bovine plasma albumin (BPA) on molar ratio of trimeric 2,3-butanedione (TB) to protein arginine. Ratios were calculated from an assumed molecular weight of 66,000 and an arginine content of 23 residues for BPA. Weighed quantities of TB were added to BPA (6 mg/ml) in pH 7.0, phosphate buffer and the solutions were maintained at 25° for 24 hours. Curves 1-A (□) and 1-B (■) are data for amino groups (TNBS) and arginine (amino acid analysis), respectively. Reaction in 0.2 *M* buffer was followed by dialysis against 1 *M* sodium chloride–0.05 *M*, pH 7.0 phosphate buffer with a terminal dialysis against 0.05 *M* phosphate buffer. Curve 2 shows arginine loss when reaction occurred in 0.1 *M* buffer and dialysis was restricted to water: ○, Sakaguchi assay; △, amino acid analysis. [Reprinted from *Biochemistry* 9, 2433 (1970). Copyright (1970) by The American Chemical Society. Reproduced by permission of the copyright owner.]

Specific, Irreversible Modification of Guanidinium Groups by Successive Reaction of Proteins with Citraconic Anhydride and TB.[22] In many ways the serial reaction of citraconic anhydride and TB at pH 8.0 followed by deacylation at acid pH appears to be the most effective method of attaining specific, irreversible modification of arginine. An aqueous solution of protein (15 mg/ml) is treated with citraconic anhydride (100 mg/ml) by adding small enough increments so the pH may be maintained at 8 by the addition of 5 *N* NaOH according to the method of Dixon and Perham.[23] Aliquots containing approximately 1 mg of protein are removed before and after acylation and assayed with TNBS. After the pH has stabilized, the protein solution is diluted with an equal volume of freshly prepared 0.4 *M* TB in 1 *M*, pH 8.0 sodium phosphate buffer and kept for 24 hours at 25° in the dark. The protein is then dialyzed against 0.1 *M*, pH 3.5 ammonium formate buffer for an additional 24 hours. The protein is then exhaustively dialyzed against

[22] J. A. Yankeelov, Jr. and D. Acree, *Biochem. Biophys. Res. Commun.* **42**, 886 (1971).

[23] H. B. F. Dixon and R. N. Perham, *Biochem. J.* **109**, 312 (1968).

0.1 M, pH 7.0 sodium phosphate buffer. Aliquots are removed for nitrogen and TNBS assays. In preparation for standard acid hydrolysis the proteins may be subsequently dialyzed against water. After hydrolysis the HCl is removed by rotary evaporation and the amino acids dissolved in water to 1 mg/ml based on the amount of protein originally introduced. Aliquots (0.5 ml) are taken for Sakaguchi analysis. If the amount of protein taken for hydrolysis was not accurately known it may be estimated in the hydrolyzate by ninhydrin or total nitrogen analysis. When the procedure was applied to BPA and RNase A the acylation and deacylation steps were found to be quantitative. The corrected percentages of arginine modified in BPA and RNase A as determined by Sakaguchi assay were 92 and 75%, respectively. The corresponding values determined by amino acid analysis were 91 and 69%.

Specific Modification of Arginine in a Mixture of Amino Acids.[22] A 5.0 ml aliquot of 18 standard amino acids (Beckman Type 1 Amino Acid Calibration Mixture) was treated with 75 μl of citraconic anhydride[23] (Aldrich Chemical Company, Inc.) in 25-μl portions. The pH was maintained at 8.0 by addition of 5 N NaOH. A 1.0-ml sample of the citraconylated amino acids was treated with an equal volume of 0.4 M TB in 1 M, pH 8.0 phosphate buffer. The reaction mixture was kept in the dark for 24 hours at 25°. At the end of this period 1 ml of the reaction mixture was diluted with an equal volume of 0.5 M HCl followed by 3.0 ml of pH 2.2, citrate sample dilution buffer for the amino acid analyzer.[17] The sample (pH 2.8) was stored an additional 24 hours at 25°, after which the pH was carefully adjusted to 2.1 with 6 N HCl. An aliquot (0.8 ml) was applied to the 130-cm column of a Technicon Amino Acid Analyzer. A second aliquot of the citraconylated mixture of amino acids was carried through the entire procedure omitting only TB and was used as the control sample. Only 3% arginine remained in the mixture while recoveries of other amino acids were from 96 to 104%.[24] The only new peaks present in the chromatogram were in the expected positions of Y, Z, and Z_0. The combined yield of these derivatives was equivalent to the loss of arginine.

Use of Dimeric 2,3-Butanedione. As shown in Eq. (1) dimeric 2,3-butanedione reacts with arginine to give the same final products as the corresponding reaction of TB. Although the dimer is less stable and more laborious to prepare, it has the advantage that it reacts with the guanidinium group of free arginine without blocking the α-amino group. Consequently, the reaction can be followed directly by amino acid analysis.

[24] Integration of histidine was complicated by elution of product Z on the leading edge of this amino acid. The recovery of histidine based on double the half-width derived from the descending limb of this peak was 96%.

As shown in Fig. 2, the reaction occurring in phosphate buffer produces an intermediate designated as X, which is funneled to two major products and one minor product Y, Z, and Z_0, respectively. On the B column of an amino acid analyzer intermediate X is eluted prior to lysine, Y in the

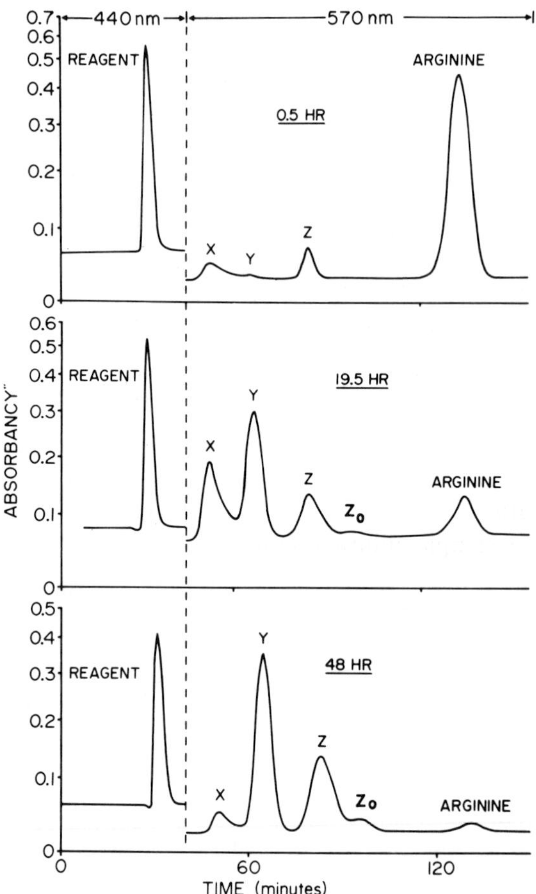

FIG. 2. Chromatographic profiles from reaction mixtures of dimeric 2,3-butanedione (DB) with free arginine. Reaction was allowed to occur in 0.5 *M*, pH 7.0 phosphate buffer (25°). Arginine and DB concentrations were 10 m*M* and 80 m*M*, respectively. At the times indicated, samples were removed and treated with 3 volumes of 0.25 *N* HCl. The acid-quenched mixtures (0.08 ml) were analyzed on a 12-cm column of an amino acid analyzer. Excess reagent appears at column volume, producing greater color at 440 nm than 570 nm. Z contains ammonia present in buffers. [Reprinted from *Biochemistry* **9**, 2433 (1970). Copyright (1970) by The American Chemical Society. Reproduced by permission of the copyright owner.]

position of lysine (or ornithine), and Z in the position of ammonia. Phosphate buffer has an important effect in directing the reaction toward irreversible products. In the absence of buffer, but with maintenance of pH 7.5 by the addition of NaOH, arginine (10 m*M*) is converted by 5% aqueous solution of DB to X in about 70% yield after 30 minutes of reaction as followed by B column analysis. Thin-layer chromatography of the same reaction mixture with ammonia–propanol (3:7, v/v) as developing solvent showed no detectable modification of arginine. Evidently the basic solvent removes the dimer moiety from the guanidinium group. Similarly, 87% of the arginine was lost from a mixture of standard amino acids (1 m*M*) treated with 0.1 *M* DB in *N*-ethylmorpholine–acetate buffer (pH 8.0, 0.2 *M* in acetate) for 2 hours. With the exception of threonine and serine which were 9–10% modified, recoveries of other amino acids were 97% or better. Under the same conditions 80% of the guanidinium groups were modified in the protamine salmine as determined by Sakaguchi assay.

The problem of following the formation of X in proteins containing ordinary amounts of arginine is complicated by the ease of reversibility of this reaction step. This behavior, however, may be potentially valuable in directing tryptic digestion in sequence studies. In contrast the family of products Y, Z, and Z_0 discussed earlier are not subject to reversible reaction. Thus after 50 hours' incubation at 37° in 1 *M* NaCl–0.05 *M*, pH 7.0 phosphate buffer, arginine (measured by Sakaguchi color) was released in only 4–5% yield from the purified free amino acid derivatives.[10]

The action of DB on proteins in phosphate buffer is similar to that of TB. For example, after 24 hours' reaction of DB (0.4 *M*) with RNase A (pH 7.0, 0.5 *M* phosphate buffer, dialysis against H_2O, 4°) arginine and lysine losses were 70 and 22%, respectively. Under identical conditions TB modified 85 and 25% of arginine and lysine, respectively. These data were corrected for a 12% release of arginine during acid hydrolysis of the protein. No correction was applied to lysine recoveries.

Use of Monomeric 2,3-Butanedione

General. Since 2,3-butanedione undergoes rapid self-condensation in mildly alkaline or even neutral phosphate buffer,[25] modification of arginine employing the monomer in this buffer is extremely complex. However, direct reaction of 2,3-butanedione with the guanidinium group in the absence of buffer or in other buffer systems gives a highly re-

[25] Unpublished observations with D. N. Robinson (1968).

versible modification. The reaction of monomer with carboxypeptidase A has been reported by Vallee and Riordan.[8]

Modification of Carboxypeptidase A with 2,3-Butanedione.[8] Carboxypeptidase A can be modified with a 150-fold molar excess of 2,3-butanedione in pH 7.5, 0.05 *M* borate–1 *M* NaCl buffer at 20°. Over a period of 1–2 hours there is a 300% increase in esterase activity and an approximately 90% loss in peptidase activity. Dialysis of the reaction mixture against 0.05 *M* Tris–1 *M* NaCl, pH 7.5 buffer (4°) causes rapid return of the esterase and peptidase activities to their original values. Under the above conditions the reaction is said to be specific for arginyl residues.

Evaluation of Procedures

Trimeric 2,3-butanedione is an easily prepared crystalline reagent of excellent shelf life. The reagent permits convenient and extensive modification of arginine in proteins. It can be used in phosphate buffer at neutral, mildly alkaline, or mildly acid pH. At neutral or alkaline pH irreversible modification of arginine is accompanied by amino group side reactions. These side reactions may be minimized by performing the reaction at pH 6.0 or eliminated by a stepwise procedure involving (1) protection of amino groups by citraconylation; (2) modification of arginines with TB at pH 8.0; and (3) liberation of the protected amino groups at acid pH. Since the introduction of carboxylate groups by the protecting group tends to swell the protein, arginine side chains are more likely to be accessible. The protecting group also reduces the likelihood of intermolecular cross-linking. Proteins modified in this serial fashion also suffer minimal discoloration during irreversible modification. Reversible precipitation of the protein was encountered during the deacylation step with RNase A and BPA at pH 3.5, which may be a disadvantage in some instances.

For reasons stated earlier dimeric 2,3-butanedione compliments the properties of the trimeric form. The dimer, however, is more laborious to prepare, less stable and somewhat less effective than the trimeric form, particularly at pH 6.0. The fact that two major products (Y, Z) rather than a single product are formed may be a disadvantage in some studies. The products, however, have similar properties. For example in thin-layer chromatography on silica gel in ammonia–propanol (3:7 v/v), the derivatives Y, Z, and Z_0 migrate as a single band with an R_f of 0.28–0.39.

The mechanisms of these reactions are yet to be defined, but it is clear from the identity of the products formed that reactions of DB and TB with arginine follow similar paths. Since TB is reported to behave like a mixture of DB and 2,3-butanedione in aqueous solution above pH 8.0,[26] some reaction of TB with arginine may occur through the dimeric form even at pH 7.0. The paths of both reactions are complex and are distinct from the reaction of monomeric 2,3-butanedione with arginine.[27]

The methods described here are reproducible provided the conditions for modification and removal of reagent are maintained constant. Discrepancies have resulted only when conditions are altered. For example, when one sample of BPA was modified with TB at pH 7.0 for 24 hours and exhaustively dialyzed against distilled water at 4°, its arginine content was 15–20% less than in an identically modified sample dialyzed against 1 *M* NaCl–0.05 *M*, pH 7.0 sodium phosphate buffer for 2 days at 25°. The greater extent of modification of the water-dialyzed sample was in part due to the presence of some reversibly modified arginines not released during the milder treatment.

Note Added in Proof

Nitromalondialdehyde as a reagent for arginine under alkaline conditions was described too recently to be included in the manuscript.[28] This reagent should be considered in addition to the others summarized in Table I.

[26] R. M. Cresswell, W. R. D. Smith, and H. C. S. Wood, *J. Chem. Soc.*, p. 4882 (1961).
[27] J. Yankeelov, Jr., unpublished observations (1968).
[28] A. Signor, G. M. Bonora, L. Biondi, D. Nisato, A. Marzotto, and E. Scoffone, *Biochemistry* **10,** 2748 (1971).

[53] Modification of Proteins with Cyanate

By George R. Stark

Principle

Cyanate is capable of reaction with amino,[1] sulfhydryl,[2] carboxyl,[1] phenolic hydroxyl,[3] imidazole,[4] and phosphate[5] groups in proteins to yield carbamyl derivatives according to the general scheme:

[1] G. R. Stark, *Biochemistry* **4,** 1030 (1965).
[2] G. R. Stark, *J. Biol. Chem.* **239,** 1411 (1964).
[3] D. G. Smyth, *J. Biol. Chem.* **242,** 1579 (1967).
[4] G. R. Stark, *Biochemistry* **4,** 588 (1965).
[5] For example, see C. M. Allen, Jr. and M. E. Jones [*Biochemistry* **3,** 1238 (1964)] for references on the synthesis of carbamylphosphate from cyanate and phosphate and for a discussion of the decomposition of carbamylphosphate.

$$\mathrm{RXH + HNCO \rightarrow RX\overset{\overset{\displaystyle O}{\|}}{C}NH_2}$$

The reagent has the advantage of being easily soluble in water, and in most cases reactions with cyanate occur with maximum velocity near pH 7 and are quite insensitive to moderate changes in pH. Selective modification of amino groups can be achieved for most proteins, since all the other reactions can be reversed under mild conditions or occur to an appreciable extent only at low pH. At neutral pH, α-amino groups react much more rapidly than ϵ-amino groups, so that α-monocarbamyl proteins may be prepared in good yield.[1]

All carbamyl derivatives are inert carbamylating agents, but the carbamylcarboxylates are reactive acylating agents. Unwanted reactions of carboxylate groups can be avoided by carrying out modifications with cyanate at neutral or mildly alkaline pH and by avoiding carboxylate buffers, particularly below pH 7.

Materials and Methods

Reagent grade potassium cyanate can be freed of the small amount of insoluble material usually present by recrystallization from water–ethanol at a temperature not above 50°.

Since cyanate reacts with itself to form cyanuric acid and cyamelide, it is usually best to perform reactions at total cyanate concentrations of about 0.2 *M*. In any case, concentrations above 1 *M* should be avoided. Hydrolysis and polymerization of cyanate result in the uptake of protons, so that it is usually necessary to control the pH in some way during reactions with proteins. Addition of mineral acids or small amounts of carboxylic acids at pH 7 or above by means of a pH-stat is convenient, although buffers with tertiary amines, such as *N*-ethylmorpholine, may be used instead. Accurate control of pH is not usually necessary since the rate at which a particular group is modified is essentially independent of pH in the range from 1 unit below its own pK_a to pH about 5. (The pK_a of cyanic acid[6] is 3.73 at 27°.)

Reactions with cyanate should *not* be terminated by addition of acid, for, although the reagent is destroyed rapidly, acidic conditions are ideal for modification of carboxyl groups[1] and for further carbamylation of ureas to biurets.[3] Cyanate can be removed by dialysis or gel filtration at neutral pH, or quenched in the reaction mixture by addition of excess glycylglycine, which reacts rapidly to give a stable derivative.

Rate constants for the reaction of some functional groups with cyanate are summarized in the table.

[6] M. W. Lister, *Can. J. Chem.* **33**, 426 (1955).

REACTION RATES WITH KNCO

Group modified	pK_a	$k_i{}^a$ ($M^{-1} \cdot \text{min}^{-1}$)	Temperature (°C)
α-NH_2 (glycine)	9.60	2.1×10^{-2}	30
α-NH_2 (glycylglycine)	8.17	1.4×10^{-1}	30
ϵ-NH_2 (ϵ-NH_2 caproic acid)	10.75	2.0×10^{-3}	30
SH (cysteine)	8.3	4.0	25
Imidazole	7.15	1.8×10^{-1}	25
Phenol	10.0	2.9×10^{-1}	30

[a] Constants are given for a bimolecular reaction according to the equations: rate = $k_i[NCO^-][RNH_3^+]$ (for the amines), rate = $k_i[NCO^-][RSH]$ (for cysteine), or rate = $k_i[NCO^-][ROH]$ (for tyrosine). Above pH about 5.5, the ionization of cyanic acid is essentially complete and the total concentration of cyanate can be substituted for $[NCO^-]$. The reactions could have been described equally well by the kinetically equivalent equations: rate = $k_m[HNCO][RNH_2]$ and rate = $k_m[HNCO][RS^-]$ or $k_m[HNCO][RO^-]$. Although these latter expressions reflect the actual mechanism [G. R. Stark, *Biochemistry* **4,** 1030 (1965). A. A. Frost and R. G. Pearson "Kinetics and Mechanism," 2nd ed., pp. 307–316. Wiley, New York, 1961], the rate constants are more convenient to use in the form given in the table. The two sets of constants are related by $k_m = k_i\, K_{NCO}/K_a$, where K_a is the acid dissociation constant of the protonated amine, mercaptan, or phenol, and K_{NCO} is the dissociation constant of cyanic acid.

Comments

Reaction with Amino Groups. The velocity with which an unhindered, primary amino group can be carbamylated can be estimated if its pK_a is known, since logarithms of the rate constants for a series of such amines are related linearly to their pK_a values.[1] The empirical equation which fits the data best for eight amino acids and peptides at 30° is

$$\log k_i\ (M^{-1} \cdot \text{min}^{-1}) = 7.94 - 0.71\ pK_a$$

At pH 7 or below, the α-amino groups of peptides and proteins can be expected to react with cyanate about 100 times faster than ϵ-amino groups and can be modified selectively. In attempting to carry out such a selective modification, it may be advantageous to ensure that the α-amino groups are accessible to the reagent by using a denaturing solvent.

Cyanate, in contrast with many other reagents, reacts rapidly with amino groups even at acidic pH values. This property is of considerable advantage in the procedure for specific cleavage of peptide bonds at serine and threonine following N → O acyl migration.[7] The amino groups of peptides in which the migration has occurred can be blocked quantita-

[7] D. G. Smyth and G. R. Stark, *Anal. Biochem.* **14,** 152 (1966).

tively with cyanate at pH 5, thereby preventing any reversal of the migration at alkaline pH during subsequent hydrolysis of the newly formed esters.

Carbamylamino groups are stable even in dilute NaOH at room temperature. The amino group may be regenerated quantitatively by hydrolysis overnight in 0.2 M NaOH at 110°, although of course such treatment causes hydrolysis of peptide bonds and results in extensive decomposition of derivatives of such amino acids as serine, threonine, and cysteine. Homocitrulline reverts to lysine slowly in 6 M HCl at 110°, whereas α-carbamylamino compounds cyclize to hydantoins under these conditions.[8]

Anaylsis for Homocitrulline. Homocitrulline may be converted to lysine quantitatively by hydrolysis in dilute alkali,[9] but the ϵ-carbamyl group is relatively stable to acid hydrolysis: only 24% is cleaved upon treatment with 6 N HCl at 110° for 22 hours.[10] However, the recovery of lysine from several proteins exhaustively carbamylated was 17–30% under the same conditions.[9] Homocitrulline immediately precedes valine and is well resolved from it under the usual conditions for amino acid analysis of protein hydrolyzates.

Reaction with Sulfhydryl Groups. Sulfhydryl groups are carbamylated much more rapidly than any other functional group (see the table). Selective modification is possible and has been demonstrated with reduced ribonuclease.[2] In contrast to carbamylamino groups, carbamylmercaptans decompose readily to free mercaptan and cyanate according to the equation:

$$\mathrm{HNCO} + \mathrm{RS}^- + \mathrm{H_2O} \rightleftharpoons \mathrm{RS\overset{\displaystyle O}{\overset{\|}{C}}NH_2} + \mathrm{OH}^-$$

The velocity of decomposition can be expressed by

$$k_D[\mathrm{RS\overset{\displaystyle O}{\overset{\|}{C}}NH_2}][\mathrm{OH}^-]$$

where $k_D = 7.2 \times 10^4\ M^{-1} \cdot \mathrm{min}^{-1}$ at 25°. Consequently, cyanate can be used as a reversible blocking reagent for SH groups: the carbamylmercaptans are relatively stable at pH 5 but decompose with $t/2 \cong 11$ minutes at pH 8 and 25°.

Reaction with Carboxyl Groups. Interaction of a carboxyl group with

[8] This volume [7].

[9] G. R. Stark, and D. G. Smyth, *J. Biol. Chem.* **238,** 214 (1963).

[10] G. R. Stark, W. H. Stein, and S. Moore, *J. Biol. Chem.* **235,** 3177 (1960).

$$R\cdot\overset{\overset{O}{\|}}{C}\cdot NH\cdot\overset{\overset{R'}{|}}{C}H\cdot CO_2^- \underset{OH^-}{\overset{HNCO\,+\,H_2O}{\rightleftharpoons}} R\cdot\overset{\overset{O}{\|}}{C}\cdot NH\cdot\overset{\overset{R'}{|}}{C}H\cdot\overset{\overset{O}{\|}}{C}\cdot O\cdot\overset{\overset{O}{\|}}{C}\cdot NH_2$$

$$\downarrow NCO^-$$

$$R\cdot\overset{\overset{O}{\|}}{C}\cdot NH\cdot\overset{\overset{R'}{|}}{C}H\cdot\overset{\overset{O}{\|}}{C}\cdot NCO \longrightarrow R\cdot\overset{\overset{O}{\|}}{C}-N\langle\text{hydantoin ring: } N-CH(R')-C(=O)-NH-C(=O)-N\rangle$$

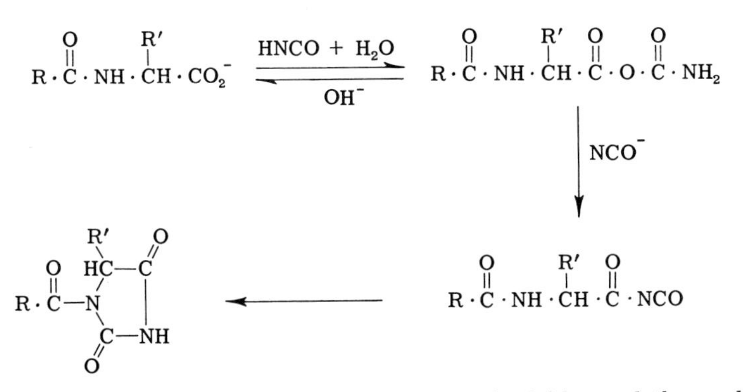

cyanate results in the formation of a mixed anhydride, and the carboxyl is thereby activated.[1] The anhydride can react with many nucleophiles, and, for example, formation of amides has been demonstrated. The carbamylcarboxylate can react further with cyanate to yield a mixed anhydride of the carboxylic acid and isocyanic acid. If the carboxyl is α to an acyl or a carbamylamino group, cyclization to an acyl- or a carbamylhydantoin can occur.[11]

Because of the pH dependence of both the forward and reverse reactions, an increase of 1 pH unit decreases the equilibrium concentration of carbamylcarboxylate 100-fold. Therefore reactions with carboxylic acids in the presence of cyanate that proceed briskly at pH 5 can be avoided entirely at pH 7 or 8. Reaction with ureas to form biurets is also insignificant at pH 7, but may be appreciable at low pH, where the concentration of HNCO is high.[3]

Reaction with Hydroxyl Groups. The aliphatic hydroxyl groups of several compounds are completely resistant to carbamylation even when high concentrations of cyanate are employed at low pH.[11] However, the reactive hydroxyls of chymotrypsin and other proteases react with cyanate to give urethans,[12] and the urethans, once formed, are relatively stable.[11] Modification of phenolate anion occurs much more readily, in a reversible reaction that is quite analogous to the one that occurs with SH groups. The half-time for the reverse reaction is about 30 minutes at pH 7.4 and 37°.[3]

Reaction with Other Functional Groups. Imidazole reacts with cyanate reversibly, but carbamylimidazole is quite unstable in aqueous solution at pH 7 and is inert as a carbamylating agent.[4] The reaction of cyanate with inorganic phosphate is well known, and it seems reasonable to assume that protein-bound phosphate esters would react similarly,

[11] G. R. Stark, *Biochemistry* **4**, 2363 (1965).

[12] D. C. Shaw, W. H. Stein, and S. Moore, *J. Biol. Chem.* **239**, PC 671 (1964).

although this has not been tested experimentally. On the other hand, it is most unlikely that amide, guanido, or indole groups in proteins would react with cyanate to a significant extent even under extreme conditions.

Some Representative Results with Proteins. Selective carbamylation of the α-NH_2 groups of hemoglobin has been achieved by Kilmartin and Rossi-Bernardi.[13] Modified hemoglobin tetramers were isolated in which all four of the NH_2 termini, or only the 2 β-chain termini, or only the 2 α-chain termini were carbamylated. The derivatives were used to show that all 4 α-NH_2 groups are sites at which CO_2 binds and that the α-NH_2 groups of the α chains are responsible for about one-fourth of the alkaline Bohr effect. Avramovic and Madsen[14] have carbamylated phosphorylase *b* both in the presence and in the absence of allosteric ligands and substrates. The results are interpreted in terms of an allosteric model in which each of 2 conformational states of the protein is inactivated at a different rate. Several proteases are affected in interesting and different ways by cyanate. Papain is inactivated rapidly and reversibly by reaction at the essential sulfhydryl group; the rate is about 3000 times faster than the reaction of cyanate with cysteine.[15] The activity of pepsin is affected reversibly by carbamylation of the 6 tyrosines, whereas carbamylation of the α-NH_2 group is without effect and the single lysine residue does not react.[16] Subtilisin (type Novo) can be carbamylated completely on all the amino groups without loss of activity.[17] However, the enzyme is inactivated by cyanate in a reaction which depends on a group with pK 7.7 and which can be reversed by hydroxylamine. The situation is very reminiscent of the reaction of serine-195 of α-chymotrypsin with cyanate.[12] All the amino groups of chymotrypsinogen can be carbamylated without effect on the potential activity upon activation with trypsin,[18] a result analogous to the effect of acetylation.

Carbamylation of proteins by the cyanate that is slowly formed from urea[10] has been observed during gel electrophoresis and chromatography in the presence of urea.[19] Carbamylation with $[^{14}C]NCO^-$ of the amino groups of peptides from L-arabinose isomerase has been useful in locating the peptides in maps by autoradiography.[20]

[13] J. V. Kilmartin and L. Rossi-Bernardi, *Nature (London)* **222**, 1243 (1969).
[14] O. Avramovic and N. B. Madsen, *J. Biol. Chem.* **243**, 1656 (1968).
[15] L. A. Æ. Sluyterman, *Biochim. Biophys. Acta* **139**, 439 (1967).
[16] S. Rimon and G. E. Perlmann, *J. Biol. Chem.* **243**, 3566 (1968).
[17] I. Svendsen, *C. R. Trav. Lab. Carlsberg* **36**, 235 (1967).
[18] B. H. J. Hofstee, *J. Biol. Chem.* **243**, 6306 (1968).
[19] Y. K. Kim, M. Yaguchi, and D. Rose, *J. Dairy Sci.* **52**, 316 (1969); J. Čejka, Z. Vodrážka, and J. Salák, *Biochim. Biophys. Acta* **154**, 589 (1968).
[20] J. W. Patrick and N. Lee, *J. Biol. Chem.* **244**, 4277 (1969).

[54] Amidination

By M. J. Hunter and M. L. Ludwig

The reaction of imidoesters with amino groups to form amidines is formally analogous to the reaction of *O*-methylisourea with amino groups to form guanidines.[1] The amidination reaction can be written as follows:

$$\mathrm{R{-}\overset{\overset{\displaystyle NH_2^+}{\|}}{C}{-}OR' + NH_2R'' \rightarrow R{-}\overset{\overset{\displaystyle NH_2^+}{\|}}{C}{-}NHR'' + R'OH} \tag{1}$$

A variety of R groups can be introduced into protein amino groups by this reaction, which proceeds more rapidly and at a lower pH than the corresponding reaction with *O*-methylisourea or other guanidinating reagents. The resulting amidines are stronger bases than the parent amines. Studies with both model compounds[2] and proteins[2-5] have provided no evidence for reaction of imidoesters with protein R groups other than the amino group when the reaction is performed in aqueous solution between pH 7 and 10 at room temperature.

In this article the class of compounds, RC(=NH)OX, where X is the ester R group, will be referred to as imidoesters; specific compounds will be called imidates, e.g., methyl acetimidate. The amidination of proteins has been reviewed several times in the last few years.[6-8] In this discussion we shall make only brief mention of certain subject matter which was covered in some detail in Volume 11[6] and the reader is referred to this earlier article for further information on these topics.

Reaction of Imidoesters with Amino Groups

The rate of reaction (1) is dependent on the pH. From their studies on benzimidates, Hand and Jencks[9] have concluded that the sharp pH optima can be quantitatively described if the rate-determining step

[1] W. L. Hughes, Jr., H. A. Saroff, and A. L. Carney, *J. Amer. Chem. Soc.* **71,** 2476 (1949).

[2] M. J. Hunter and M. L. Ludwig, *J. Amer. Chem. Soc.* **84,** 3491 (1962).

[3] L. Wofsy and S. J. Singer, *Biochemistry* **2,** 104 (1963).

[4] F. C. Hartman and F. Wold, *Biochemistry* **6,** 2439 (1967).

[5] A. Nilsson and S. Lindskog, *Eur. J. Biochem.* **2,** 309 (1967).

[6] M. L. Ludwig and M. J. Hunter, Vol. 11, p. 595.

[7] L. A. Cohen, *Annu. Rev. Biochem.* **37,** 695 (1968).

[8] A. N. Glazer, *Annu. Rev. Biochem.* **39,** 101 (1970).

[9] E. S. Hand and W. P. Jencks, *J. Amer. Chem. Soc.* **84,** 3505 (1962).

changes with pH. At lower pH values the breakdown of a tetrahedral intermediate is rate determining, whereas at higher pH values the rate is determined by the rate of reaction of amine free base with the cationic form of the imidoester. The position of the pH optimum is thus a complicated function of the p*K*s and structures of both imidoester and amine.

To provide the investigator with approximate guidelines in the selection of appropriate reaction conditions, some kinetic data have been compiled in the table:

Optimal Apparent Rate Constants and pH Values for Certain Imidoester–Amine Reactions

	Methyl benzimidate[a] and		Methyl acetimidate[b] and	
	ϵ-NH_2-Caproic acid	Glycylglycine	ϵ-NH_2-Caproic acid	Glycylglycine
$k_{(app)}$ (max)[c]	0.34	0.73	10	25
pH (max)	9.5–10.0	7.6	10–10.5	9.0

[a] 39°.
[b] 25°.
[c] M^{-1} min^{-1}.

As shown in the table, k_{app}(max) and pH (max) are dependent on the natures of both the imidoester and amine R groups. From these and other data[2,9] the following two generalizations are probably true in most instances: (1) For a given amine: as the pK_a of the imidoester increases, pH (max) and k_{app}(max) increase. (2) For a given imidoester: as the pK_a of the amine increases, pH (max) increases and k_{app}(max) decreases. It should be mentioned, however, that steric or other factors involved in the formation of the tetrahedral intermediate may on occasion result in deviations from the above relationships.

The pH profiles of the rate of reaction of α- and ϵ-amino groups with an imidoester are sufficiently different that preferential reaction of the α- or the ϵ-amino group can be achieved. For example, at pH 9.7 and 39° the rate of reaction of the ϵ-amino group of ϵ-aminocaproic acid with methyl benzimidate was 7 times greater than that of the α-amino group of glycylglycine. At pH 7.5, the α-amino group reacted about 30 times faster than the ϵ-amino group. The temperature dependence of k_{app} has been measured for the methyl benzimidate–ϵ-aminocaproic acid system and is presumably similar for other systems. An increase of 20° in temperature increases k_{app} by a factor of 4–5.

In calculating the constants in the table, no correction has been made for concomitant hydrolysis of the imidoester. Under the conditions of the experiments, the hydrolysis rate was over a 100-fold less than the rate of amidine formation. In reactions involving the complete amidination of a protein, however, the necessary presence of large excesses of imidoester will result in a large fraction of the imidoester being consumed by reaction with water. Since the hydrolysis may proceed by several, pH dependent, simultaneous pathways it is not possible to predict the resultant pH changes.[10] In modification of proteins above pH 8 with methyl acetimidate and methyl benzimidate, acid must be added to maintain the pH.

Preparation of Imidoesters

The most widely used method for the preparation of imidoesters is that described by Pinner[11] at the end of the last century. In this method, an imidate salt is formed by the reaction of a nitrile with an alcohol in the presence of a halide acid, generally HCl.

$$\mathrm{RCN} + \mathrm{R'OH} \xrightarrow{\mathrm{HCl}} \mathrm{R{-}\overset{\overset{\displaystyle NH{\cdot}HCl}{\|}}{C}{-}OR'} \qquad (2)$$

Strict precautions must be taken to ensure that the reagents are anhydrous and are protected from atmospheric moisture during the reaction. The reaction should always be carried out below 5°; at higher temperatures other reaction products are formed. Most aliphatic and aromatic nitriles can be converted to imidoesters by this procedure. In general, the ease and rapidity of imidoester formation decreases with increasing size[12] and increasing electronegativity of the nitrile R group. In the preparation of imidoesters as intermediates in amidine formation, it is recommended that the alcohol R′ group be methyl or ethyl. The yield of product can often be increased by variation of the mole ratios of the various reactants, by alteration of the time of reaction, or by the addition of inert solvent (ether, dioxane, chloroform, benzene, or ethyl acetate) either prior to the addition of HCl or later after considerable imidoester formation has occurred. For the reaction to proceed well the reactants should be in true solution, and this requirement is generally the deciding factor in the choice of solvent. Most diimidoesters have been prepared by modifications of the Pinner procedure.[13-15] The reader is referred to the work of McElvain and

[10] R. Roger and D. G. Neilson, *Chem. Rev.* **61,** 179 (1961).
[11] A. Pinner, "Die Imidoäther und Ihre Derivate." Oppenheim, Berlin, 1892.
[12] S. M. McElvain and J. W. Nelson, *J. Amer. Chem. Soc.* **64,** 1825 (1942).
[13] S. M. McElvain and J. P. Schroeder, *J. Amer. Chem. Soc.* **71,** 40 (1949).
[14] Von E. Hieke, *Pharmazie* **18,** 653 (1963).
[15] H. J. Schramm, *Hoppe-Seyler's Z. Physiol. Chem.* **348,** 289 (1967).

Schroeder[13] and to the article by Wold in this volume (cf. [57]) for further information on the synthesis of these reagents.

Imidoesters may also be prepared by the base-catalyzed reaction of nitriles with alcohols.

$$\mathrm{RCN + R'OH \underset{}{\overset{R'O^-}{\rightleftharpoons}} R{-}\overset{\overset{\displaystyle NH}{\|}}{C}{-}OR'} \tag{3}$$

This reaction was first studied about the turn of the century,[16] but virtually no further use was made of it until 1960, when Schaefer and Peters[17] undertook a systematic study of the reaction with a variety of nitriles. Their results show that although simple aliphatic and aromatic nitriles give no or very low yields of imidoesters by this procedure, many electronegatively substituted nitriles may be converted to imidoesters extremely easily in useful yield. Since these electronegatively substituted nitriles generally give poor yields of imidoester in the Pinner procedure, the two methods thus complement one another well. Many imidoesters with substituents of intermediate electronegativity can be prepared equally well by either method.

Since nitriles are sometimes difficult to obtain or prepare, a general method for the preparation of imidoesters from amides or thioamides should be mentioned. The *O*- and *S*-alkylation of amides and thioamides has been achieved with ethyl chloroformate, dimethyl sulfate, or triethyloxonium fluoroborate.[18–21]

$$\mathrm{R\overset{\overset{\displaystyle NHR'''}{|}}{C}{=}O \xrightarrow{EtOClOCl,\ Me_2SO_4\ or\ Et_3O^+BF_4^-} R{-}\overset{\overset{\displaystyle NHR'''}{\|}}{C}{-}OEt\ (or\ Me)} \tag{4}$$

This method has proved to be especially valuable for the synthesis of many *o*-substituted benzimidates which are not formed or are obtained in very poor yield by the Pinner procedure. Unsubstituted and methyl and ethyl *N*-substituted imidoesters can be prepared by this procedure. If other *N*-substituted or *N,N*-disubstituted reagents are required, however, they should be prepared by suitable reaction of the unsubstituted imidoester[10] or by the use of yet another method of imidoester preparation.[10]

[16] J. U. Nef, *Justus Liebigs Ann. Chem.* **287**, 265 (1895); B. C. Hess, *Amer. Chem. J.* **18**, 723 (1896); J. C. Hessler, *Amer. Chem. J.* **22**, 169 (1899); J. Steiglitz and H. I. Schlesinger, *Amer. Chem. J.* **39**, 738 (1908); and E. K. Marshall, Jr. and S. F. Acree, *Amer. Chem. J.* **49**, 127 (1913).

[17] F. C. Schaefer and G. A. Peters, *J. Org. Chem.* **26**, 412 (1961).

[18] A. Bühner, *Justus Liebigs Ann. Chem.* **333**, 289 (1904).

[19] W. Reid and E. Schmidt, *Justus Liebigs Ann. Chem.* **695**, 217 (1966).

[20] L. Weintraub, S. R. Oles, and N. Kalish, *J. Org. Chem.* **33**, 1679 (1968).

[21] F. H. Suydam, W. E. Greth, and N. R. Langerman, *J. Org. Chem.* **34**, 292 (1969).

The synthesis of imidoesters from aldehydes, ketones, or unsaturated systems has been known for some time,[10] and a new general method for the conversion of α,β-unsaturated aldehydes into saturated imidoesters via α-cyanoamines has recently been reported.[22] This method may prove to be the method of choice in the preparation of certain imidoesters.

For further information on the preparation of imidoesters, the reader is referred to the excellent review by Roger and Neilson.[10]

Two methods for the preparation of imidoesters are given in the following paragraphs. The first procedure is a modification of the Pinner method which is applicable to the synthesis of imidoesters from nitriles of high reactivity. The second procedure describes the synthesis of an imidoester from a nitrile by base catalysis. The particular nitrile used in this illustration could not be converted to its imidoester by various modifications of the Pinner procedure.[23] Insoluble nitrile hydrochloride was always the end product of the reaction in the presence of HCl.

Preparation of Methyl Acetimidate·HCl[2]

Dry methanol (40 ml, 1.0 mole) was cooled in a Dry Ice–Cellosolve bath during the addition of 40–50 g of dry HCl gas (1.1–1.4 moles). Precautions were taken to maintain anhydrous conditions. Dry acetonitrile (40 ml, 0.75 mole) was quickly added to the methanolic HCl after it had cooled to the temperature of the Dry Ice–Cellosolve bath. After the acetonitrile had frozen, the flask was allowed to warm to 0° in an ice bath. The reaction mixture was kept in the ice bath in a cold room overnight. Crystals separated out during this time. After 48 hours, 1–2 volumes of dry ether were added with mixing. About 1 hour later, the solid was filtered from the reaction mixture, washed with dry methanol–ether (1:2) and then with ether, and stored in an evacuated desiccator over desiccant. The filtering operations were carried out as rapidly as possible, preferably in a cold room. The crystals had a melting point of 93–95° with decomposition and gas evolution. Approximately 90% yields of acetimidate were obtained.

Preparation of Methyl Picolinimidate[17]

A solution of 26.0 g (0.25 mole) of 2-cyanopyridine and 1.35 g (0.025 mole) of sodium methoxide in 225 ml of methanol was allowed to stand overnight at room temperature. Acetic acid (1.5 ml, 0.025 mole) was

[22] J. S. Walia, P. S. Walia, L. Heindl, and H. Lader, *Chem. Commun.* 1290 (1967).
[23] W. Sclar, M. L. Ludwig, and M. J. Hunter, unpublished observations.

then added, and the solution was distilled. The material boiling at 118–122° at 28 mm weighed 31.4 g and was 97% methyl picolinimidate by titration. The yield was therefore 90% based on the amount of nitrile used.

Imidoester hydrochlorides are generally stored in evacuated desiccators over some desiccant, as they are readily decomposed by atmospheric moisture. Even under these conditions the salts decompose quite rapidly and should be used within a few weeks of their preparation. The imidoester free bases are considerably more stable than their salts. Stored over desiccant at 0° they can often be kept for several months without serious decomposition.

Amidination of Proteins

General Procedures

Proteins react readily with imidoesters in aqueous solution. By appropriate choice of pH the ratio of the reaction rates of the α- and ϵ-amino groups can be varied (see the table), but, because of the preponderance of ϵ-amino groups in proteins, it is difficult to modify the α-amino groups selectively. If complete reaction of all amino groups is desired, large excesses of imidoesters (concentrations as great as 1 M) can be used with impunity, provided the reactants can be kept in solution. To obtain high concentrations of imidoesters with large R groups it is often necessary to add organic solvents. Since most proteins are less stable in organic solvent–water systems, it is recommended that reactions in these systems be carried out at 0°. Methanol, ethanol, and acetone are good solvents for the imidoester free bases, and their addition to aqueous protein solutions at low temperatures does not, in general, result in denaturation of the protein. Removal of the products of reagent hydrolysis by dialysis and repetition of the reaction is an effective procedure for exhaustive amidination.[3]

The details of two methods for determining the extent of amidination, namely, the reaction of unmodified amino groups with DNFB followed by amino acid analysis[2,3] and formol titration,[2] were included in the previous discussion of amidination.[6] Three other methods are given below:

1. ^{14}C-labeled imidoesters can be allowed to react with proteins, and the extent of modification can subsequently be determined from radioactivity measurements on the dialyzed, amidinated proteins.[4,24] Noncovalent binding of side products, such as primary amidines, which has

[24] A. Nureddin and T. Inagami, *Biochem. Biophys. Res. Commun.* **36**, 999 (1969).

been observed after modification of albumin,[25] will introduce errors into these measurements.

2. The extent of amidination cannot be determined directly by analysis of the amidino-amino acid content of protein hydrolyzates, since the α-amidino group is completely hydrolyzed and the ϵ-amidino group partially hydrolyzed in 6 N HCl at 110°. However, by determining the lysine content of hydrolyzates as a function of time, Reynolds[26] was able to estimate the initial lysine content of acetamidino-ribonuclease by extrapolation to zero time. He found that the decomposition of ϵ-acetamidinolysine during ribonuclease hydrolysis was a first-order reaction with a half-time of 46 hours.

3. Reaction with trinitrobenzenesulfonate (TNBS) according to the method of Habeeb[27] can be used to measure the free amino groups remaining after amidination. To 1 ml of protein solution is added 1 ml of 4% $NaHCO_3$, pH 8.5, and 1 ml of 0.1% TNBS in water. The solution is held at 40° for 2 hours. Then 1 ml of 10% SDS and 0.5 ml of 1 N HCl are added, and the absorbance is read at 345 nm.[28] The extinction coefficient for the ϵ-TNP group in proteins at acid pH has been reported to be 1.0×10^4 to 1.3×10^4 [26–29] at wavelengths near 340 nm, and that for ϵ-TNP-lysine, 1.45×10^4.[30]

Properties of Amidinated Proteins

One significant advantage of the amidination reaction is that it results in minimal nonspecific perturbations of protein structure. Many proteins have been subjected to extensive amidination (greater than 80%) by reaction with acetimidate with few if any accompanying changes in physical or biological properties.[31,32] For example, carbonic anhydrase,[5,33] carboxypeptidase A,[34] trypsin,[24] and lysozyme[35,36] have been converted to their acetamidino derivatives without any observed

[25] J. R. VanAtta, Ph.D. Thesis, University of Michigan, Ann Arbor, Michigan, 1968.

[26] J. H. Reynolds, *Biochemistry* **7**, 3131 (1968).

[27] A. F. S. A. Habeeb, *Anal. Biochem.* **14**, 328 (1966).

[28] R. Haynes, D. T. Osuga, and R. E. Feeney, *Biochemistry* **6**, 541 (1967).

[29] K. Satake, T. Okuyama, M. Ohashi, and T. Shinoda, *J. Biochem.* (*Tokyo*) **47**, 654 (1960).

[30] A. R. Goldfarb, *Biochemistry* **5**, 2570 (1966).

[31] F. M. Robbins, M. J. Kronman, and R. E. Andreotti, *Biochim. Biophys. Acta* **109**, 223 (1965).

[32] B. Kassell and R. B. Chow, *Biochemistry* **5**, 3449 (1966).

[33] P. L. Whitney, P. O. Nyman, and B. G. Malmström, *J. Biol. Chem.* **242**, 4212 (1967).

[34] J. F. Riordan and B. L. Vallee, *Biochemistry* **2**, 1460 (1963).

[35] A. McCoubrey and M. H. Smith, *Biochem. Pharmacol.* **15**, 1623 (1966).

[36] R. C. Davies and A. Neuberger, *Biochim. Biophys. Acta* **178**, 306 (1969).

alterations in enzymatic activity. Several antigens and antibodies retain their native antigenic behavior,[3,37,38] optical rotation,[3] ultraviolet absorbance, and $s_{20,w}$[3] after extensive reaction with acetimidate. Titration curves of acetamidino- and unmodified bovine carbonic anhydrase B are identical up to the pH region in which lysyl residues titrate.[5] The absence of any change in the electrostatic parameter, w, strongly suggests that amidination has not resulted in any significant structural changes.

Various studies (X-ray crystal structures, titration data, the availability of amino groups to reagents) have indicated that the lysyl residues of globular proteins are generally located on the surface of the molecule and often protrude into the solvent rather than interact with other residues.[39] It is therefore not surprising that acetamidination of such residues, which adds the relatively small group, $CH_3C(=NH)$—, and preserves the charge of the residue, does not perturb the physical or biological properties of the protein. The minimal effects following modification of these "uninvolved" lysyl residues provide some justification for the tenet that significant changes observed after acetaminidation of a particular protein may be attributed to the involvement of one or more residues at an active site or in the maintenance of a particular protein conformation. For example, the effect of amidination of Ile_{16} on the pH-activity profile of chymotrypsin has led to inferences about the role of this residue in maintaining the active conformation of chymotrypsin.[40] A few further illustrations of alterations in physical or biological properties following amidination of lysyl residues are discussed in the following paragraphs.

The Use of Amidination to Assess the Function of Lysyl Residues

Amidination can provide information about the role of lysine in stabilizing tertiary or quaternary structures. For example, essentially complete modification of bovine serum albumin with acetimidate[3,25,41,42] and butyrimidate[25] has been achieved. Such modification affects the stability of the pH 4.5–8.5, or N form, of the molecule, which is characterized by the presence of many masked carboxylate ions and tyrosyl

[37] A. Dutton, M. Adams, and S. J. Singer, *Biochem. Biophys. Res. Commun.* **23,** 730 (1966).

[38] A. F. S. A. Habeeb, *J. Immunol.* **99,** 1264 (1967).

[39] L. Stryer, *Annu. Rev. Biochem.* **37,** 25 (1968).

[40] B. Labouesse, H. L. Oppenheimer, and G. P. Hess, *Biochem. Biophys. Res. Commun.* **14,** 318 (1964).

[41] J. Avruch, J. A. Reynolds, and J. H. Reynolds, *Biochemistry* **8,** 1855 (1969).

[42] A. F. S. A. Habeeb, *Biochim. Biophys. Acta* **115,** 440 (1966).

residues in some crevice(s) of the molecule. In the pH region between 4.5 and 3.5, the N form is reversibly converted to an expanded or F form, with an unmasking of the buried carboxyl and phenol groups. Various physical measurements on the native protein and the amidinated derivatives have shown the derivatives to undergo this transition more readily than the native molecule,[25,41] the derivative with the larger R group being the more effective destabilizer of the N form.[25] It would therefore appear that —$COO^- \cdots {}^+H_3N$— interactions contribute to the stability of the N form and that the presence of the amidino R groups in the crevice interferes with the charge–charge interactions.

Amidination has also been employed in studies of the aggregation of protein monomers.[31] Comparison of amidination with other lysyl modification reactions suggests that charge plays a crucial role in the aggregation of TMV protein monomers into long rods.[43] Reaction of TMV A protein, which contains 2 lysyl residues, 53 and 68, and an acetylated α-amino group, with acetimidate and picolinimidate does not interfere with the aggregation reaction, whereas modification of the amino groups with reagents which destroy the positive charge at the lysyl residues (*S*-ethyl trifluorothioacetate or maleic anhydride) inhibits assembly of the virus particle. Furthermore, since the reaction of intact virus with picolinimidate is restricted to residue 68, it appears that lysine-53 is the residue involved in the specific charge interaction in the aggregated rods.

Amidination has been useful in investigations of enzymes containing lysyl residues at the active site. One such enzyme is cytochrome b_5 reductase.[44] The presence of the cosubstrate, NADH, protects approximately one lysyl residue from modification during extensive (95%) reaction with acetimidate. The charge on this lysyl residue appears critical for NADH binding. Thus, further modification of acetamidinocytochrome b_5 reductase in the absence of NADH by reaction with acetic or succinic anhydrides resulted in an enzyme with a K_m for NADH approximately 300-fold that of the native protein, whereas acetamidination of the active center lysine increased K_m by a factor of only 3. In the case of ribonuclease,[26] amidination leads to complete inactivation. The physical properties (circular dichroism, s_{20}, and the anomalous pKs of three tyrosyl residues) of the 97% acetamidinated, inactive ribonuclease remain essentially the same as those of the native protein. Moreover, partial amidination of lysyl residues 1, 7, 31, and 37 by reaction with dimethyl adipimidate[4] in the presence of an inhibi-

[43] R. N. Perham and F. M. Richards, *J. Mol. Biol.* **33**, 795 (1968).
[44] A. Loverde and P. Strittmatter, *J. Biol. Chem.* **243**, 5779 (1968).

tor, phosphate, which protects lysyl residue 41 from reaction, does not cause proportionate decreases in activity. The results obtained with fully amidinated ribonuclease are therefore in good accord with earlier chemical modifications which had established an essential role for lysyl residue 41.[45] The steric requirements at lysyl residue 41 are apparently too rigorous to permit the inclusion of even the relatively small $CH_3C(=NH)$— group.

Cross-linking

Diimidoesters have proved to be useful reagents for inter- and intramolecular cross-linking of lysyl residues.[4,37,46,47] Because of the competing hydrolysis reaction, not every diimidoester in a position to do so actually forms a cross-link; monofunctional substitution also occurs at sites where no intramolecular cross-links can form. However, no change in the charge of the modified protein should ensue following hydrolysis of the unreacted functional group of the reagent. A few applications are presented here; for a more complete evaluation of the various reagents for cross-linking, the reader is referred to Article [57] and to Volume XI.[48] Two illustrations of the use of diimidoesters for the formation of interchain cross-links are summarized below.

In an effort to prepare ferritin-labeled antibodies with undiminished antibody activity, γ-globulins have been covalently linked to ferritin by reaction with diethyl malonimidate.[37] Immunoelectrophoresis demonstrated that 15–20% of the total protein had been converted to ferritin-γ-globulin cross-linked product in reaction mixtures initially consisting of 6% ferritin and 2% γ-globulin. In the second investigation the diimidoester, dimethyl suberimidate, has been employed to study the subunit structure of multichain, or oligomeric, proteins.[47] Subunits within an oligomer can be covalently cross-linked under conditions in which little or no interparticle cross-linking will occur. Disc electrophoresis of the products in SDS-containing gels yields a series of bands with molecular weights which are various multiples of the subunit molecular weights.

Intrachain cross-links have been used to establish the proximity of lysyl residues in the three-dimensional structure of ribonuclease. Reaction of ribonuclease A with [^{14}C]adipimidate yielded 47% intramolecularly linked monomer, 30% dimer, and 24% higher aggregates, based

[45] C. H. W. Hirs, M. Halmann, and J. H. Kycia, *Arch. Biochem. Biophys.* **111**, 209 (1965).
[46] F. C. Hartman and F. Wold, *J. Amer. Chem. Soc.* **88**, 3890 (1966).
[47] G. E. Davies and G. R. Stark, *Proc. Nat. Acad. Sci. U.S.* **66**, 651 (1970).
[48] F. Wold, Vol. XI, p. 617.

on recoveries from Sephadex G-75 columns.[4,46] Under the conditions employed, the monomer fraction had reacted with approximately 2 moles of reagent with the disappearance of about three amino groups. Separation of the radioactive tryptic peptides showed that the cross-link was not at a single unique position, but that some of the molecules contained a bridge between lysines 7 and 37, and others, between 31 and 37. Monosubstitution occurred at lysine 1.

Introduction of Chelating Groups

The ease with which a variety of R groups (Eq. 1) can be incorporated into imidoesters makes possible the introduction of specialized functional groups. One potential application of the amidination reaction is in the preparation of heavy atom isomorphs for the crystal structure analysis of proteins. If the product amidines are sufficiently powerful chelating agents, heavy atoms can be bound selectively at the sites of amidination. Benisek and Richards[49] have proposed the use of picolinimidate for this purpose. Picolinamidines bind Ni^{2+} and Pd^{2+} with association constants greater than 10^7, and compounds which are structural analogs of picolinamidine are known to complex other metal ions.[50] Moreover, the extent of amidination can readily be determined from spectral measurements due to the absorption of the pyridine moiety. Although crystals of fully amidinated lysozyme are not isomorphous with crystals of the unmodified protein,[49] it is possible that picolinamidino derivatives of other proteins may prove to be suitably isomorphous.

Reversal of the Amidination Reaction

The acetamidino group can be removed from intact proteins by reaction with ammonia–ammonium acetate (concentrated ammonia:glacial acetic acid, 15:1, v/v). The peptide bond adjacent to an ϵ-amidinolysyl residue is resistant to hydrolysis by trypsin, and it was therefore suggested that in sequence analysis amidination might be useful for the preparation of large tryptic peptides, which could be redigested after removal of the blocking group.[6,51] Some side reactions have been observed during the reversal step.[6,26] Ribonuclease activity can be fully regenerated by removing the acetamidino group, but on column chromatography the product does not behave like untreated ribonuclease. The altered properties are not attributable to deamidation of asparagine or glutamine, since the number of carboxylate groups is the same as in

[49] W. F. Benisek and F. M. Richards, *J. Biol. Chem.* **243**, 4267 (1968).
[50] T. R. Harkins and H. Freiser, *J. Amer. Chem. Soc.* **78,** 1143 (1956).
[51] M. L. Ludwig and R. Byrne, *J. Amer. Chem. Soc.* **84,** 4160 (1962).

untreated ribonuclease.[26] Such changes in the physical properties of the regenerated protein need not invalidate the use of amidination in sequence and cross-linking determinations. For example, removal of the amidino group has facilitated identification of the lysine residues cross-linked by reaction of ribonuclease with adipimidate.[4] In certain situations, however, it appears that reaction with maleic anhydride[52] may have advantages over reaction with imidates.

Procedures for the Amidination of Proteins

1. A method for exhaustive modification by reaction with methyl (or ethyl) acetimidate[3] has been given in Volume 11.[6]

2. The procedure for reaction of cytochrome b_5 reductase with acetimidate[44] is useful for other small-scale preparations: Enzyme (0.03 μmole, 22 lysine/FAD) is dissolved in 0.3 ml of pyrophosphate buffer, pH 8.5. Ethyl acetimidate (300 μmoles in 0.30 ml of 0.45 M borate buffer, pH 9.2) is added, and the mixture is left at 0° for 20 hours. The derivative can be separated from contaminating small molecules by passage through a Sephadex G-25 column.

3. Cross-linking using dimethyl suberimidate[47]: The pH is maintained at 8.5 (with 0.2 M triethanolamine·HCl) to favor amidination over reagent hydrolysis. Dimethyl suberimidate, adjusted to pH 8.5 in the above buffer, is mixed with a solution of the protein to be cross-linked to give a final protein concentration of 0.5–5 mg/ml and a dimethyl suberimidate concentration of 1–12 mg/ml, and the reaction is allowed to proceed for 3 hours.

[52] P. J. G. Butler, J. I. Harris, B. S. Hartley, and R. Leberman, *Biochem. J.* **112**, 679 (1969).

[55] Esterification

By PHILIP E. WILCOX*

In the short time that has elapsed since the publication of Volume 11, methods for the esterification of proteins have been extended substantially. Interest in modification of carboxyl groups has increased with each new discovery of the functional or structural role of a glutamic or aspartic acid side chain in an enzyme. The growing list of enzymes includes lysozyme, carboxypeptidase A, pepsin, trypsin and other serine proteases, ribonuclease T1, and triosephosphate isomerase.

In Volume 11 [74], it was pointed out that only two methods of

* Deceased.

esterification were known then to be group specific, namely, reaction with methanol–HCl or ethanol–HCl and reaction with derivatives of diazoacetic acid near pH 5. It now appears that reaction with triethyloxonium fluoroborate (Meerwein reagent) at pH values between 4 and 5 may be added to the repertoire of group-specific reagents. Furthermore, the use of aliphatic diazo compounds has been extended to include diazo ketones containing the grouping —CO—CH=N_2.

A remarkable advance in methodology has come from the discovery that copper ions promote the reaction between the carboxyl group of a specific aspartic acid residue in pepsin and a variety of aliphatic diazo compounds.[1,2] The mechanism is not presently clear, but it is likely that the method will be applicable to other enzymes. Indeed, the reaction has been shown to occur with a pepsinlike enzyme from the mold *Penicillium janthinellum*.[3]

Experience has furthermore shown that specific esterification of particular enzymes can be obtained with reagents that are not ordinarily specific for carboxyl groups alone. In most of these cases the reagent has been an active halogen compound, such as iodoacetate or a bromomethyl ketone. Specificity may result either from an enhanced reactivity of the carboxyl group due to the local environment in the native protein, or from specific binding of a particular reagent to a site in the enzyme, or perhaps from a combination of both effects.

This article will be focused mainly on new methods that have been developed for reagents that are relatively specific for carboxyl groups. The other category of reagent, the reactive halogen compounds, will be given a more cursory treatment because the particular structure of the binding site in the enzyme must be of primary importance in each case. Esterification may be regarded as a special case of alkylation. It is therefore likely that particular compounds, which would ordinarily be classified among typical alkylating reagents, will be found to esterify uniquely situated carboxyl groups in enzymes and other proteins as the result of specific interactions. The reader is referred to other articles on alkylation (this volume [34a] and [39]).

Aliphatic Diazo Compounds

The reactions of aliphatic diazo compounds are manifold and complex. A useful reference is the treatise by H. Zollinger.[4] In spite of their

[1] T. G. Rajagopalan, W. H. Stein, and S. Moore, *J. Biol. Chem.* **241,** 4295 (1966).
[2] G. R. Delpierre and J. S. Fruton, *Proc. Nat. Acad. Sci. U.S.* **56,** 1817 (1966).
[3] J. Šodek and T. Hofmann, *J. Biol. Chem.* **243,** 450 (1968).
[4] H. Zollinger, "Azo and Diazo Chemistry: Aliphatic and Aromatic Compounds," Wiley (Interscience), New York, **1961.**

reactivity, aliphatic diazo compounds show a high degree of specificity for carboxyl groups in an aqueous medium when they are partially stabilized by the presence of a carbonyl group attached to the diazotized carbon atom, as in the diazoacetyl or the diazomethyl ketone moieties. The only exception is the ability of these reagents to alkylate free sulfhydryl groups (see Vol. 11 [74]). Diazo compounds not so stabilized, for example, diazomethane, tend to be nonspecific.

The enzymatic properties of pepsin suggested to a number of investigators that the active site of this enzyme contained at least one carboxyl group as an important component. Therefore, a variety of diazo compounds were synthesized as site-specific reagents. Reliance was

TABLE I
ALKYL DIAZO COMPOUNDS FOUND TO INHIBIT PEPSIN

	Compound	^{14}C-labeled	Reference
I	Diphenyldiazomethane	Yes	*a*
II	*N*-Diazoacetyl-DL-norleucine	—	*b, c*
III	*N*-Tosyl-L-phenylalanyldiazomethane	Yes	*d*
IV	*N*-Tosyl-D-phenylalanyldiazomethane	—	*d*
V	*N*-Diazoacetyl-L-phenylalanine methyl ester	Yes	*d, e, j*
VI	*N*-Diazoacetyl-D-phenylalanine methyl ester	—	*d*
VII	*N*-Diazoacetylglycine methyl (ethyl) ester	Yes	*c–e*
VIII	*N*-Benzyloxycarbonyl-L-phenylalanyldiazomethane	Yes	*f*
IX	α-Diazo-*p*-bromoacetophenone	—	*g*
X	1-Diazo-4-phenylbutanone-2	Yes	*h*
XI	1-Diazo-3-dinitrophenylaminopropanone-2	—	*e*
XII	Ethyl 2-diazo-3-(*p*-hydroxyphenyl) propionate	—	*e*
XIII	Phenylbenzoyldiazomethane	—	*e*
XIV	*N*-Diazoacetyl-*N*′-2,4-dinitrophenylethylenediamine	—	*i*
XV	Methyl (or ethyl) diazoacetate	Yes	*c*

[a] G. R. Delpierre and J. S. Fruton, *Proc. Nat. Acad. Sci. U.S.* **54,** 1161 (1965).
[b] T. G. Rajagopalan, W. H. Stein, and S. Moore, *J. Biol. Chem.* **241,** 4295 (1966).
[c] R. L. Lundblad and W. H. Stein, *J. Biol. Chem.* **244,** 154 (1969).
[d] G. R. Delpierre and J. S. Fruton, *Proc. Nat. Acad. Sci. U.S.* **56,** 1817 (1966).
[e] L. V. Kozlov, L. M. Ginodman, and V. N. Orekhovich, *Biokhimiya* **32,** 1011 (1967).
[f] E. B. Ong and G. E. Perlmann, *Nature (London)* **215,** 1492 (1967).
[g] B. F. Erlanger, S. M. Vratsanos, N. Wassermann, and A. G. Cooper, *Biochem. Biophys. Res. Commun.* **28,** 203 (1967).
[h] G. A. Hamilton, J. Spona, and L. D. Crowell, *Biochem. Biophys. Res. Commun.* **26,** 193 (1967); K. T. Fry, O.-K. Kim, J. Spona, and G. A. Hamilton, *Biochemistry* **9,** 4624 (1970).
[i] V. M. Stepanov, L. S. Lobareva, and N. I. Mal'tsev, *Biochim. Biophys. Acta* **151,** 719 (1968).
[j] R. S. Bayliss and J. R. Knowles, *Chem. Commun.* (1968) p. 196; R. S. Bayliss, J. R. Knowles, and G. B. Wybrandt, *Biochem. J.* **113,** 377 (1969).

placed on the fact that the specificity of pepsin is directed toward aromatic amino acid residues in peptide substrates. Although specific binding of a hydrophobic moiety may be one factor contributing to the reaction of many of these compounds with a specific aspartic acid residue in the enzyme, recent evidence suggests that the formation of a copper complex with the reagent is a much greater determinant of specificity.

The procedures to be described have been selected from the work of a number of laboratories in order to provide a background that should be useful in planning a study of the modification of carboxyl groups in other enzymes. The basic procedure for the synthesis of derivatives of diazoacetic acid may be found in Volume 11 [74]. The application of this procedure to the synthesis of diazoacetyl-L-phenylalanine methyl ester is given below, and references to similar compounds may be found in Table I.

The synthesis of one diazo ketone, tosyl-L-phenylalanyldiazomethane, is given below as an exemplary case. A variety of analogous reagents may be synthesized by the same procedure.

Synthesis of Reagents

N-Diazoacetyl-L-phenylalanine Methyl Ester[5]

Glycyl-L-phenylalanine methyl ester hydrobromide is prepared by removing the blocking group from *N*-benzyloxycarbonyl-glycyl-L-phenylalanine methyl ester with HBr in glacial acetic acid. The blocked peptide ester may be synthesized from *N*-benzyloxycarbonylglycine and phenylalanine methyl ester by one of the mixed anhydride methods.[6] The starting materials are readily obtained commercially.

Dissolve 8.2 g (24 mmoles) glycyl-L-phenylalanine methyl ester hydrobromide in 40 ml of 2 *M* sodium acetate and add 2.0 ml of glacial acetic acid. Cool the mixture in an ice bath and add 3.0 g (43 mmoles) of $NaNO_2$ in small portions over 30 minutes. The mixture is allowed to stand for 3 hours in the cold, and the product is extracted with three 50-ml portions of ice-cold chloroform. The extracts are combined and dried over $MgSO_4$. Light petroleum ether (bp 40–60°) is added until the solution becomes slightly turbid, and the mixture is allowed to stand overnight. The resulting yellow needles are collected by filtration and dried. The yield is about 70% (4.5 g), mp 126–128°. The recrystallized material has an $[\alpha]_D^{20} = +186°$ ($CHCl_3$).

[5] R. S. Bayliss, J. R. Knowles, and G. B. Wybrandt, *Biochem. J.* **113**, 377 (1969).
[6] N. F. Albertson, *Org. React.* **12**, 157 (1962).

This procedure is readily adapted to the preparation of *N*-diazo-[1-^{14}C]-acetyl-L-phenylalanine methyl ester by starting with 1-^{14}C-labeled glycine in the preparation of benzyloxycarbonylglycine.

Tosyl-L-phenylalanyldiazomethane (*L-1-Diazo-4-phenyl-3-tosylamidobutanone-2*)

The necessary intermediate, tosyl-L-phenylalanyl chloride, may be readily prepared from a commercial sample of tosyl-L-phenylalanine. Suspend 3.2 g (10 mmoles) of the tosyl amino acid in 50 ml of anhydrous ethyl ether at 0°, and add 2.3 g (11 mmoles) of PCl_5. Shake the mixture at 0° for 10 minutes and at room temperature for 10 minutes. Allow the product to crystallize during 1 hour at 0°. Collect the crystalline material on a filter, wash it quickly with a small amount of cold ether and then with ice water, and dry the product in a desiccator under high vacuum. Yield, 80–90% (approximately 3.0 g), mp 128–129° (decomp.).

The diazo ketone is prepared by adding 2.0 g (6.6 mmoles) of tosyl-L-phenylalanyl chloride to a solution of diazomethane (20 mmoles) in 50 ml of ethyl ether at 0°. Keep the mixture in the dark at room temperature for 16 hours, evaporate the solvent under vacuum, and recrystallize the residue from a mixture of 1-butanol and cyclohexane. Yield, 0.8 g (35%), mp 109–111° (decomp.). A second recrystallization gives a product melting at 112–114° (decomp.), and $[\alpha]_D^{25} = -122°$ (approximately 0.4, ethanol).

Procedures

Reaction of Alkyl Diazo Compounds with Pepsin (Porcine)

The conditions chosen by several groups of investigators have been similar. The range of conditions may be summarized as follows:

Protein concentration	0.03–0.3 m*M* (1–10 mg/ml)
Inhibitor concentration	0.8–4.5 m*M*
Cu^{2+} concentration	1.0–4.5 m*M*
pH	5.0–5.5 (acetate buffer)
Temperature	15–38°
Time of reaction	10–90 minutes

Within this range of conditions, the inhibition of pepsin was found to be greater than 90%. In many cases for which estimates were made, the reaction was specific for esterification of one carboxyl group with a stoichiometry close to 1.0. Exceptions are noted below. The following procedure is taken from one of the successful experiments.[2]

Pepsin is prepared by activation of pepsinogen (see Vol. XIX [20]).

The esterification reaction is carried out at 15°. Prepare a solution of the protein at a concentration of 1.25 mg/ml in a 6.25 mM acetate buffer, pH 5.4. To 4.0 ml of this solution, add 0.5 ml of 0.01 M $CuCl_2$ followed by 0.5 ml of a freshly prepared solution of the inhibitor in ethanol, in this example, a 1.43 mM solution of tosyl-L-phenylalanyldiazomethane. The molar ratios of protein:reagent:Cu are 1:5:35. The reaction is essentially complete in 45 minutes. At that time, the activity of the enzyme should be less than 10%, and the extent of incorporation of the inhibitor should be 1.0 ± 0.1 mole per mole of protein.

Assay methods for pepsin may be found in Vol. 19 [20]. Methods for determination of the extent of esterification will be described below.

Comments

Subsequent to investigations of the reaction of diazo compounds with pepsin in a number of laboratories, a study by Lundblad and Stein[7] shed additional light on this reaction. Most significantly, they showed that preincubation of the diazo compound with cupric ion will eliminate a time lag in enzyme inhibition which is found when the order of addition is the one prescribed above, namely, addition of cupric ion followed by the inhibitor. The change in the order of addition results in a shift in the maximum rate of reaction from pH 5.7 to 5.0, and the reaction is complete within 10 minutes at 15°. It is concluded that the formation of a copper complex with the diazo compound is a rate-limiting step. Very likely this is a carbene complex which reacts very rapidly and specifically with the enzyme. A plot of the rate of this reaction as a function of pH is a bell-shaped curve with the maximum at pH 5.0, suggesting that the reagent reacts preferentially with a nonionized carboxyl group that in this case has an abnormal pK, at least as high as 5.6. Such a behavior, that is, a much higher rate of reaction with a nonionized carboxyl group compared with the ionized form, is consistent with what is known about the reactivity of aliphatic diazo compounds (see Vol. 11 [74]).

Denatured pepsin is not esterified by diazo compounds under the conditions of the procedure given above. There is evidence that in the native enzyme a second carboxyl group is situated close to the aspartic acid residue that becomes esterified. It is postulated that this second carboxyl is ionized at pH 5 and that the negative charge may be involved in binding the copper complex, thus enhancing the rate of the specific reaction. This effect appears to predominate over stereochemical interactions which depend upon the structure of the inhibitor. Many

[7] R. L. Lundblad and W. H. Stein, *J. Biol. Chem.* **244**, 154 (1969).

different diazo compounds have been found to react with pepsin in the presence of cupric salts, and to inhibit activity. These various inhibitors are listed in Table I. Some degree of stereospecificity has been observed, however, in the case of tosylphenylalanyldiazomethane, in that the L-isomer reacts more rapidly with the enzyme than does the D-isomer. It will be noted that almost all the compounds in Table I incorporate at least one hydrophobic moiety, frequently a phenyl group. In general, they have been designed as affinity labels for pepsin, taking into account the known specificity of the enzyme toward aromatic side chains in peptide substrates.

Incorporation of more than 1 mole of reagent per mole of pepsin is found in the case of diazoacetic acid methyl ester (compound XV). Evidently this compound, described in Vol. 11 [74] as a group specific reagent for carboxyl groups, possesses a reactivity that is great enough to esterify not only the specific aspartic acid residue, but also other carboxyl groups as well. Diphenyldiazomethane (compound I) also shows a lack of specificity for the active site carboxyl group. Incorporation of greater than 1 mole of reagent has also been found with some of the most specific reagents if the conditions are much more vigorous, for example, if *N*-diazoacetyl-L-phenylalanine methyl ester is initially 80 m*M*, the protein concentration is 10 mg/ml, and the time of reaction is 90 minutes.[5] Finally, the homogeneity of the enzyme is a factor in the specificity. Some commercial samples of pepsin incorporate as many as 4 moles of reagent per mole of enzyme.

Reaction with Triethyloxonium Fluoroborate

Triethyloxonium fluoroborate (Meerwein reagent) is a strong alkylating agent. In nonaqueous media, the reagent will ethylate such compounds as ethers, sulfides, nitriles, ketones, esters, and amides on oxygen, nitrogen, or sulfur, to give onium fluoroborates. In an aqueous medium, hydrolysis of the reagent occurs rapidly, the half-life being about 10 minutes at room temperature and pH values around 6. For this reason, only a small fraction of the reagent can be utilized for alkylation or esterification in aqueous medium, and it is difficult to cause any such reaction to go to completion. In this respect, the behavior of the Meerwein reagent resembles that of aliphatic diazo compounds (cf. Vol. 11 [74]).

Although the Meerwein reagent is very reactive, recent investigations have shown that it will specifically esterify carboxyl groups in lysozyme[8]

[8] S. M. Parsons, L. Jao, F. W. Dahlquist, C. L. Borders, Jr., T. Groff, J. Racs, and M. A. Raftery, *Biochemistry* **8**, 700 (1969).

and in trypsin[9] when the reaction is carried out at pH 4.5. Specificity was indicated by the observation that the amino acid analysis of the acid hydrolyzate of each protein gave results that were not appreciably different from the analysis of the corresponding native protein. Furthermore, estimates of the incorporation of ethyl groups into lysozyme were in agreement with measurements of the conversion of carboxyl groups into ester groups.

Specificity may not be restricted, however, to carboxyl groups for all proteins or under different sets of conditions. Yonemitsu *et al.*[10] have studied the esterification of a number of model peptides dissolved in bicarbonate buffer. With a molar ratio of Meerwein reagent to peptide of 20:1, esterification of carboxyl groups was obtained in yields greater than 80%. The side chains of serine, threonine, tyrosine, lysine, and arginine were completely nonreactive under these conditions. However, the side chains of methionine and histidine were ethylated to give sulfonium and quaternary imidazolium salts, respectively. It would also be expected that a free sulfhydryl group of a cysteine residue would be ethylated. Although these possibilities for alkylation exist, one may expect that reaction at methionine in native enzymes will be only infrequently observed because this amino acid residue is usually buried, and reaction at exposed histidine residues can probably be inhibited by working at pH values between 4 and 5.

Esterification by triethyloxonium fluoroborate proceeds by a nucleophilic attack of the negatively charged carboxylate group on the positively charged oxonium ion.

$$R{-}C(=O){-}O^{\ominus} + (CH_2{-}CH_3)_2\overset{\oplus}{O}{-}CH_2{-}CH_3 \longrightarrow R{-}C(=O){-}OCH_2CH_3 + O(CH_2{-}CH_3)(CH_2{-}CH_3)$$

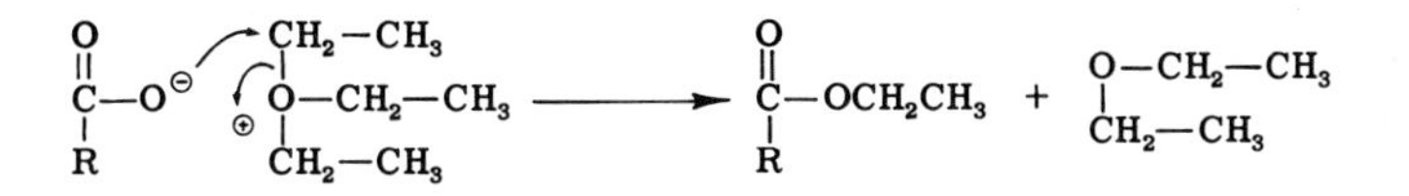

Therefore, one might expect that the reagent at a given pH would preferentially esterify carboxyl groups that are more highly ionized in comparison with those that are less ionized. Also, one might expect that an ionized carboxyl group in a hydrophobic pocket might be esterified preferentially because of a strong electrostatic interaction. The experimental results with lysozyme and trypsin are consistent with these postulated properties of the reagent.

It is noteworthy that the apparent preference of the Meerwein reagent for ionized carboxyl groups over nonionized groups is just the

[9] H. Nakayama, K. Tanizawa, and Y. Kanaoka, *Biochem. Biophys. Res. Commun.* **40,** 537 (1970).

[10] O. Yonemitsu, T. Hamada, and Y. Kanaoka, *Tetrahedron Lett.* **23,** 1819 (1969).

reverse of the behavior of alkyl diazo compounds, which react preferentially with the nonionized carboxyl group. These contrasting modes of reaction have a counterpart in the hydrolytic reactions. The reaction of diazo compounds with water is catalyzed by hydrogen ions, and there is not net consumption or production, whereas the reaction of water with the oxonium ion produces hydrogen ions. It appears that these two different types of reagents may be potentially useful for distinguishing and characterizing enzyme carboxyl groups with abnormal properties, the diazo compounds for such groups with abnormally high pK values, and the Meerwein reagent for those with abnormally low pK values.

Synthesis of Reagents

Triethyloxonium Fluoroborate

A detailed procedure for the synthesis of the oxonium salt is given by Meerwein.[11] The starting materials are available from commercial sources. In brief, epichlorohydrin (1.51 moles) is added dropwise to a solution of freshly distilled boron fluoride etherate (2.00 moles) in 500 ml of anhydrous ethyl ether. After the exothermic reaction has subsided, the mixture is refluxed for 1 hour and is allowed to stand overnight at room temperature. The solvent is removed from the crystalline product by filter stick, and the product is washed by the addition of fresh portions of dry ether. The yield of colorless crystals [mp 91–92° (decomp.)] is about 90%.

The salt is very hygroscopic. It may be dried with a stream of nitrogen in a dry box and stored in a tightly closed bottle, but the material in this form should be used within a few days after bottling. Alternatively, the oxonium salt may be stored for longer periods under ether at a low temperature. If there is reason to use the methyl analog for esterification, a procedure is available for converting the triethyloxonium salt.[12]

^{14}C-Labeled Triethyloxonium Fluoroborate[8]

Reagent labeled with ^{14}C may be prepared by exchange of ethyl groups between the oxonium salt and ^{14}C-labeled diethyl ether obtained from a commercial source. About 0.8 g of the oxonium salt is dissolved in 7 ml of methylene chloride containing 1.0 mmole of ^{14}C-labeled ethyl ether (1.0 mCi/mmole). The solution is placed in a heavy-walled glass

[11] H. Meerwein, *Org. Syn.* **46,** 113 (1966).
[12] H. Meerwein, *Org. Syn.* **46,** 120 (1966).

ampoule, all manipulations being carried out under a dry nitrogen atmosphere and at a temperature of about −80° (solid CO_2–methylene chloride). The ampoule is sealed by torch and then warmed for about 20 hours in refluxing methylene chloride. After the ampoule has been cooled and opened, 1 ml of the solution is removed and added to 100 mg of 3,5-dinitrobenzoic acid dissolved in 15 ml of methylene chloride. The mixture is refluxed for 4 hours, the solvent is evaporated, and the residue is washed with 5% $NaHCO_3$ solution. The ethyl 3,5-dinitrobenzoate (mp 83–84°) is recrystallized from ethanol–water to provide material for the determination of specific activity of the ethyl group (approximately 5×10^6 dpm/mole).

The main bulk of the ^{14}C-labeled oxonium salt is precipitated from methylene chloride solution as an oil by the addition of 20 ml of hexane. The supernatant solvent is drawn off and the oil is taken up in a small amount of dry acetonitrile. This solution, containing about 0.5 g of ^{14}C-triethyl-oxonium fluoroborate, is the form of the labeled reagent that is added to an aqueous solution of protein for the esterification of carboxyl groups.

Procedures[8]

The conditions used for the esterification of lysozyme and trypsin are as follows:

Protein concentration	10–12.5 mg/ml
pH	4.5 or 4.0
Temperature	20–25°
Initial reagent concentration	0.1 or 0.2 *M*

The protein is dissolved in water, and the pH is adjusted to the desired value by the addition of dilute perchloric acid. The pH during the reaction is maintained by a pH-stat using a solution of 4 *N* NaOH in the syringe. A portion of triethyloxonium fluoroborate is dried under a stream of nitrogen in a dry box and is weighted. An amount of acetonitrile equal to about one-fourth of the weight of the salt is added, and the resulting solution is weighed. All the solution is taken up into a dry, calibrated syringe, and the volume of solution that contains the amount of oxonium salt (MW 189.8) needed to bring the concentration of the reagent in the protein solution to 0.2 *M* (or 0.1 *M*) is calculated. Inject this amount of reagent into the protein solution while it is vigorously stirred in the pH-stat. The reaction and the consumption of base proceed for about 20 minutes. The modified protein is recovered by dialysis against distilled water (or 10^{-3} *M* HCl) in order to remove small molecular reaction products, followed by lyophilization. In order to obtain more extensive esterification, the dialyzed solution may be repeatedly treated with the oxonium salt, following the same procedure that was used for the first treatment. It may be advisable after each dialysis to remove any denatured protein that has precipitated.

Comments

As in the case of esterification with diazoacetyl compounds (in the absence of cupric ions), an estimate of the extent of esterification as a function of the consumption of the reagent, expressed as millimoles per milliliter of reaction mixture, gives a useful comparative measure of the efficiency of modification (see Vol. 11 [74], p. 616). For example, the consumption of oxonium salt in the amount of 1.2 mmoles/ml at pH 4.5 resulted in the esterification of 22% (2.6 residues) of the carboxyl groups in lysozyme, and the consumption of 0.2 mmoles/ml resulted in the esterification of 12% (1.7 residues) of the carboxyl groups in trypsin. By comparison, the consumption of 0.6 mmoles of one or another of the diazoacetyl reagents resulted in the esterification of about 20% of the carboxyl groups in each of the proteins, chymotrypsinogen, ribonuclease, γ-globulin, and serum albumin. It appears that the efficiency is about the same for the two types of reagents. However, this comparison does not take into account the likelihood noted above, that the specificities of the two types of reagent are directed toward two different classes of carboxyl groups (high pK and low pK), and furthermore, that one or two carboxyl groups in a particular protein may possess an enhanced reactivity toward one type of reagent or the other (cf. Vol. 11 [74] pp. 605–606).

The investigation of the esterification of lysozyme by triethyloxonium fluoroborate showed, indeed, that there are two carboxyl groups that have an enhanced reactivity toward this reagent.[8] By limiting the extent of modification to a level between 0.5 and 1.5 ester groups per mole, it was possible to obtain high yields of monoesterified derivatives. One of these, obtained in high yield by esterification at pH 4.0, proved to be a labile ester that completely reverted to native lysozyme during incubation at pH 7.2 and room temperature for 48 hours. A second specific carboxyl group became esterified in high yield by treatment with the oxonium salt at pH 4.5. After the labile ester group had been removed at pH 7.2, the stable monoesterified derivatives could be isolated and purified by chromotography on a cation-exchange resin.

The monoesterified derivative which was labile at pH 7.2 showed a reduced enzymatic activity which resulted from a lowered binding affinity for substrates compared with the affinity of native enzyme.

The second, stable monoesterified derivative appeared to be devoid of enzymatic activity. It was possible to demonstrate that the modified residue was β-ethylaspartic acid by means of total enzymatic hydrolysis and amino acid analysis. An examination of the peptide fragments produced by chymotryptic cleavage of the derivative showed that this residue was Asp-52 in the native enzyme. X-ray crystallographic studies had previously suggested that Asp-52 and Glu-35 may be involved in the catalytic mechanism of the enzyme.

The location of the principal site of esterification in trypsin has not been determined at the time of this review. However, the extent of reaction at pH 4.5 is reduced from 1.7 to 1.0 ester groups per mole by the presence of 50 m*M* β-naphthamidine. It is proposed that this inhibitor blocks esterification of Asp-177 which is situated in the binding pocket of this serine protease.[9]

Reaction with Active Halogen Compounds

A growing number of enzymes are now known that are inhibited in each case by esterification with an active halogen compound. Although reactivity of halogen compounds is ordinarily directed toward amino acid side chains containing nitrogen or sulfur atoms as nucleophiles, in each example of inhibition presented in Table II the carboxyl group of an aspartic or glutamic acid residue acts as the nucleophile.

The reactions listed in Table II all have a high degree of specificity for the structure of the inhibitor. In some instances, the structure has been designed as a substrate analog, and it seems clear that the reagent is bound specifically to the active site of the enzyme. In other instances, such as ribonuclease T1, a relationship to the substrate binding site is not presently obvious, although a similarity to the inhibition of pancreatic ribonuclease with iodoacetate is suggestive. The inhibitor of elastase has no obvious structural similarity to good substrates of the enzyme, and the site of reaction is far removed from the substrate binding site, yet the stoichiometric incorporation of one equivalent of inhibitor results in complete inactivation of the enzyme. Furthermore, the rate of the inhibition reaction is reduced by the presence of substrates. It would appear that this is an example of a conformational interaction between the substrate binding site and another site in the enzyme that specifically binds the inhibitor.

The reaction of pepsin with *p*-bromophenacyl bromide is another interesting example. The aspartic acid residue that is specifically esterified has been shown to be distinct from the aspartic acid that reacts with a variety of diazo compounds (see Table I). The site of reaction of the latter compounds has been located in the sequence, Ile-Val-Asp-Thr-Gly-

TABLE II
INHIBITION OF ENZYMES BY ESTERIFICATION OF ASPARTIC OR GLUTAMIC ACID RESIDUES WITH ACTIVE HALOGEN COMPOUNDS

Enzyme	Inhibitor	Conditions for 95% inhibition[a]	Residue esterified	Reference
Pepsin (porcine gastric mucosa)	*p*-Bromophenacyl bromide	pH 2.8, 37°, 3 hours	Asp	*b*
Ribonuclease T_1 (*Aspergillus oryzae*)	Iodoacetate, bromoacetate	[I] = 0.09 m*M*, [I]/[E] = 1.1 pH 5.5, 37°, 5 hours	Glu-58	*c*
Triose phosphate isomerase (rabbit muscle)	1-Hydroxy-3-chloro-2-propanone phosphate	[I] = 80 m*M*, [I]/[E] = 300 pH 8.0, room temp, 5 min	Glu	*d*
Elastase (porcine pancreas)	1-Bromo-4-(2,4-dinitrophenyl)-2-butanone	[I] = 2.0 m*M*, [I]/[E] = 4 pH 7.0, 37°, 10 hours	Glu-6	*e*
Carboxypeptidase A (bovine pancreas)	*N*-Bromoacetyl-*N*-methyl-L-phenylalanine	[I] = 1.0 m*M*, [I]/[E] = 200 pH 7.5, 25°, 25 min [I] = 10 m*M*, [I]/[E] = 170	Glu-270	*f*

[a] In the case of pepsin inhibited by *p*-bromophenacyl bromide, greater than 95% inhibition of activity is observed when benzyloxycarbonyl-L-glutamyl-L-phenylalanine is used as substrate. However, inhibition of activity toward denatured hemoglobin has not been observed to exceed 80%.

[b] B. F. Erlanger, S. M. Vratsanos, N. Wassermann, and A. G. Cooper, *J. Biol. Chem.* **240,** PC 3447 (1965); *Biochem. Biophys. Res. Commun.* **23,** 243 (1966).

[c] K. Takahashi, W. H. Stein, and S. Moore, *J. Biol. Chem.* **242,** 4682 (1967).

[d] F. C. Hartman, *J. Amer. Chem. Soc.* **92,** 2170 (1970).

[e] L. Visser, D. S. Sigman, and E. R. Blout, *Biochemistry* **10,** 735 (1971).

[f] G. M. Haas and H. Neurath, *Biochemistry* **10,** 3535 (1971); *ibid.* 3541 (1971).

Thr-Ser-Leu,[5-13] whereas a small peptide esterified at aspartic acid with *p*-bromophenacyl bromide was isolated and shown to have the composition, (Gly_2,Asp_1,Ser_1,Glu_1).[14] Compounds known to be competitive inhibitors of pepsin protect the enzyme from esterification by either diazo compounds or halogen compounds, and therefore both aspartic acid residues are presumed to be close to the substrate binding site. However, it is possible to prepare a diesterified derivative by using, successively, *p*-bromophenacyl bromide and α-diazo-*p*-bromoacetophenone.[15]

An examination of the optimal conditions for obtaining nearly complete inhibition and stoichiometric esterification with active halogen compounds indicates that at the present time no useful generalizations can be made concerning such specific reactions. It appears that each different enzyme must be considered as a unique case in which the enzyme structure and the special properties of certain carboxyl groups have a predominant influence.

Methods of Analysis

Although esterification of carboxyl groups in enzymes has been observed with very different types of reagents, the derivatives all have the ester linkage in common. In this section, methods that have been found useful for characterization of the various derivatives will be reviewed.

Estimation of the Extent of Incorporation of Reagent

The most common method for estimation of incorporation of reagent has utilized a ^{14}C-labeled form of the reagent and application of scintillation spectrometry. The diazo compounds which were used in the study of pepsin and which have been synthesized with ^{14}C-label are so indicated in Table I. The synthesis of [^{14}C]triethyloxonium fluoroborate is described above. Among the reagents listed in Table II, the inhibitors for ribonuclease T1, elastase, and carboxypeptidase A were labeled with ^{14}C. A method used in the study of triose phosphate isomerase is of special interest. In this case, the label was introduced after the inhibitor had been reacted with the enzyme by reducing the derivative with sodium borohydride containing tritium label. The carbonyl group of the inhibitor moiety was thereby reduced to a secondary alcohol group. The result was the simultaneous incorporation of a tritium atom and the stabilization of the ester linkage. Since a ketonic function is

[13] K. T. Fry, O.-K. Kim, J. Spona, and G. A. Hamilton, *Biochemistry* **9**, 4625 (1970).

[14] B. F. Erlanger, S. M. Vratsanos, N. Wassermann, and A. G. Cooper, *Biochem. Biophys. Res. Commun.* **23**, 243 (1966).

[15] B. F. Erlanger, S. M. Vratsanos, N. Wassermann, and A. G. Cooper, *Biochem. Biophys. Res. Commun.* **28**, 203 (1967).

present in many of the compounds that have been used for the esterification of proteins, this method may have wide applicability.

Another method for measuring the extent of incorporation of reagent has depended upon the synthesis of a reagent that contains an unnatural amino acid residue, namely norleucine.[1] Conventional amino acid analysis will give an accurate determination of the number of reagent molecules added to the enzyme.

In some cases, advantage has been taken of some other type of label in the reagent. For example, bromine has been determined by elemental analysis[15] or a chromophoric group has been determined by spectrometry.[16]

Characteristics and Estimation of Ester Groups

Analysis for Ethanol or Glycolic Acid after Hydrolysis

Lysozyme esterified with triethyloxonium fluoroborate has been analyzed for ethoxyl content by mild basic hydrolysis in sealed ampoules followed by gas–liquid partition chromatography using 10% Carbowax 20M on a 6-foot, 60–80-mesh column at 90°.[8]

Esterification by haloacetate, as well as by diazoacetic acid derivatives, may be estimated after base hydrolysis by the method described in Volume 11 [74], p. 615.[17]

Conditions for Basic Hydrolysis

Typical esters are characteristically hydrolyzed by dilute sodium hydroxide, and lability to dilute base is indicative of an ester bond. Saponification of esterified lysozyme in a pH-stat at pH 10 has even been used as a quantitative method.[8] However, the stability of ester groups in protein derivatives has been shown to vary widely. At one extreme, it has been observed that the labile ethoxy group that is introduced into lysozyme at pH 4.0 by triethyloxonium fluoroborate is removed by incubation at pH 7.2 and room temperature for 48 hours.[8] A study of the rate of hydrolysis at pH 8.0 of the ester bond in pepsin inhibited with 1-diazo-4-phenyl-2-butanone gave a half-life of 5.5 hours at 30°, comparable to half-life of hydrolysis of the model compound, 1-acetoxy-4-phenyl-2-butanone, under the same conditions.[13] At the other extreme of behavior, it was observed that incubation at pH 10 did not regenerate the activity of lysozyme esterified at Asp-52, whereas incubation at pH 2 resulted in reactivation of this derivative to the extent of 60%.[8]

[16] V. M. Stepanov, L. S. Labareva, N. I. Mol'tsev, *Biochim. Biophys. Acta* **151,** 719 (1968).

[17] K. Takahaski, W. H. Stein, and S. Moore, *J. Biol. Chem.* **242,** 4682 (1967).

Reaction with Hydroxylamine and Estimation of Hydroxamate

Ester groups in enzyme derivatives have been determined by conversion to hydroxamic acid groups using 1 *M* hydroxylamine at pH values between 7.0 and 9.0, followed by estimation of the hydroxamate by the well-known colorimetric method that makes use of the ferric complex (e.g., see Vol. 3, 323). A positive color test for hydroxamate using *N*-1-naphthylethylenediamine in the Segal test[18] is also indicative. However, care must be taken because it is not possible to devise a control that will make certain that reaction of hydroxylamine with particular ester groups in a protein derivative or peptide fragment is quantitative.

The most sensitive method of estimating hydroxamic acid groups is the one devised by Yasphe *et al.*[19] and effectively applied to the determination of ester groups in esterified lysozyme.[8]

Reagents

- Sodium acetate, 5%
- Sulfanilic acid reagent (dissolve 10 g of sulfanilic acid in 1 liter of 30% acetic acid)
- Iodine reagent (dissolve 1.3 g I_2 in 100 ml of glacial acetic acid)
- Sodium thiosulfate, 2.5%
- α-Naphthylamine reagent (dissolve 3 g of α-naphthylamine in 1 liter of 30% acetic acid)
- Hydrochloric acid, 3 *N*

Procedure. To 1.0 ml of sample containing from 0.05 to 0.1 μeq of hydroxamate, add 5 ml of 5% sodium acetate, 1 ml of sulfanilic acid reagent, and 0.5 ml of iodine reagent. Shake the mixture and allow it to stand for 5 minutes. Add 0.5 ml of 2.5% thiosulfate in order to remove excess I_2, add 0.5 ml of 3 *N* HCl, and finally, add 1 ml of naphthylamine reagent. Make up the volume to 10 ml and determine the optical density at 530 nm after 90 minutes. Construct a calibration curve using a standard solution of acetylhydroxamic acid. The optical density for 0.1 μmole is about 0.35.

Estimation of Esterified Residues by Difference, Using the Lossen Rearrangement

The conversion of ester groups in protein derivatives to hydroxamic acid groups, as indicated above, allows the identification of the residues

[18] F. Bergmann and R. Segal, *Biochem. J.* **62**, 542 (1956).

[19] J. Yasphe, Y. S. Halpern, and N. Grossowicz, *Anal. Chem.* 32, 518 (1960).

as aspartic or glutamic esters by means of the Lossen rearrangement.[20-23] The hydroxamic acid groups are reacted with 2,4-dinitrofluorobenzene, the rearrangement is brought about by treatment with 0.1 *N* NaOH, the modified protein is hydrolyzed with 6 *N* HCl, and a sample is analyzed on the automatic amino acid analyzer. Those residues of aspartic acid which had been esterified in the protein derivative appear in the analysis as 2,3-diaminopropionic acid, and those residues of esterified glutamic acid appear as 2,4-diaminobutyric acid. Details of the procedure may be found in Volume 11 [74], pp. 610–611.

Estimation of Esterified Residues by Difference, Using the CDI Method for Free Carboxyl Groups

A method for determination of the total content of free carboxyl groups in a protein has been developed by Hoare and Koshland[24] (see this volume [56]). The carboxyl groups are activated in the presence of an excess of glycine methyl ester by the use of a water-soluble carbodiimide in a denaturing solvent, such as 8 *M* urea or 6 *M* guanidinium chloride. After recovery of the modified protein, the incorporation of glycine may be determined by amino acid analysis. When this method is applied to a sample of esterified protein, the decrease in glycine incorporation compared to the amount obtained with native protein will give an estimate of the total number of esterified residues.[8,9]

Total Enzymatic Hydrolysis and Estimation of Aspartic or Glutamic Esters

Methods are now known by which esterified aspartic or glutamic acid residues may be liberated from certain modified proteins and identified by conventional chromatographic analysis.[17,25] It appears that several proteases attack the bonds on either side of the esterified residues at least as readily as the bonds to aspartic and glutamic acid residues themselves. For purposes of comparison, samples of ethyl or methyl esters of the acidic amino acids may be obtained commercially. However, two useful laboratory methods are given below, along with the procedure for the synthesis of the γ-carboxymethyl ester of glutamic acid.

A sample of ribonuclease T1 that had been inhibited by esterification with [^{14}C]iodoacetate was fragmented by treatment with subtilisin BPN′

[20] O. O. Blumenfeld and P. M. Gallop, *Biochemistry* **1**, 947 (1962).
[21] E. Gross, and J. L. Morell, *J. Biol. Chem.* **241**, 3638 (1966).
[22] L. Visser, D. S. Sigman, and E. R. Blout, *Biochemistry* **10**, 735 (1971).
[23] F. C. Hartman, *J. Amer. Chem. Soc.* **92**, 2170 (1970).
[24] D. G. Hoare and D. E. Koshland, Jr., *J. Biol. Chem.* **242**, 2447 (1967).
[25] S. M. Parsons and M. A. Raftery, *Biochemistry* **8**, 4199 (1969).

(Nagarse) for 3 hours at pH 8 and 37°. A radioactive peptide was isolated by paper chromatography. The peptide consisted of the sequence of residues 57-62, Try-Glu-Trp-Pro-Ile-Leu. A sample of the peptide was hydrolyzed with aminopeptidase M for 24 hours at pH 7 and 37°. Analysis of the hydrolyzate on the automatic amino acid analyzer (0.9 × 60-cm column) showed the presence of tyrosine, isoleucine, leucine, small amounts of tryptophan and proline, and a new ninhydrin-positive peak which emerged at 43 ± 3 ml, just ahead of the position for *S*-carboxymethylcysteine. The new component was identical in its properties with a synthetic sample of the γ-carboxymethyl ester of glutamic acid.

The stable, monoethyl ester of lyozyme inhibited by Meerwein's reagent has been hydrolyzed with subtilisin (Carlsberg) for 4 hours at pH 7 and 25°, followed by aminopeptidase M for 20 hours at pH 7.0 and 25°. An aliquot of the hydrolyzate was subjected to amino acid analysis using 0.3 *M* lithium citrate buffer, pH 2.80 (Spinco Technical Bulletin A-TB-044), at 25° in a Beckman-Spinco Model 120B analyzer. Compared with the analysis of native lysozyme, the derivative gave one less aspartic acid residue and a new peak appeared in the chromatogram between the positions of glutamic acid (plus glutamine) and proline. The position of this peak corresponded to that found for an authentic sample of β-ethyl aspartate. This compound is completely resolved from peaks of the common amino acids by using the lithium buffer system.

A chymotryptic peptide was isolated from a sample of the monoethyl ester of lysozyme. This peptide corresponded to fragment C-15 obtained by the action of chymotrypsin on native enzyme and was composed of residues 45–53, Arg-Asn-Thr-Asp-Gly-Ser-Thr-Asp-Tyr. Treatment of the esterified peptide with carboxypeptidase A resulted in the rapid liberation of the COOH-terminal tyrosine residue, followed by a somewhat slower liberation of β-ethyl aspartate from the penultimate position. This result is in contrast to the slow liberation of the penultimate aspartic acid residue from the peptide obtained from native protein. The rate of formation of free amino acids was determined by amino acid analysis with the automatic analyzer.

Synthesis of the γ-Carboxymethyl Ester of Glutamic Acid.[17] Suspend 1.0 g of L-glutamic acid and 20 g of sodium glycolate in 60 ml of dioxane in a round-bottom flask. Chill the mixture in an ice bath and add 10 ml of thionyl chloride. Stopper the flask and allow it to stand for 5 hours at 25°. An additional 5 ml of thionyl chloride are then added, followed 6 hours later by 25 ml of thionyl chloride and 5 g of sodium glycolate. After the mixture has stood overnight, the dioxane is removed in a rotary evaporator and about 100 ml of ethyl ether is added. The precipitate is collected and washed with ether. About one-sixth of the product is purified on a column (4 × 19 cm) of Dowex 50-X2 (200–400 mesh) using 0.1 *M* pyridinium acetate buffer, pH 3.2, as eluent at a flow rate of 20 ml per hour. The elution of the product is followed by analyses for amino acid with ninhydrin and for glycolic acid with chromotropic acid reagent (see Vol. 11 [74] p. 615). The peak fractions which give positive tests for both amino acid and glycolic acid are pooled and evaporated to dryness in a rotary evaporator at 40°. The residue is transferred to a test tube with 1 ml of water, and about 5 volumes of acetone is added. The precipitate is collected, and the material is crystallized from a small amount of water and acetone at 0°. The product is collected by centrifugation, washed with a little acetone and ether, and stored in a vacuum desiccator over sulfuric acid.

Synthesis of β-Ethyl Aspartate Hydrochloride.[25] Reflux a mixture of 5.0 g of DL-aspartic acid hydrochloride and 1.1 g of anhydrous HCl in 15 ml of absolute ethanol for 15 minutes. Add dry ether to the warm solution until a slight turbidity develops, and allow the product to crystallize at room temperature. The crystals are collected on a filter, washed with dry ether, and stored in a vacuum desiccator. Yield, 2.8 g, mp 174–177°.

Synthesis of γ-Ethyl L-Glutamate.[25] This compound may be readily synthesized by esterification of poly-L-glutamic acid with triethyloxonium fluoroborate according to the procedure described above for the esterification of lysozyme. The polymer is then hydrolyzed with subtilisin and aminopeptidase M to yield a solution containing essentially only L-glutamic acid and γ-ethyl L-glutamate in a ratio of 1:3. An aliquot of this solution may be used as a convenient reference for the amino acid analysis of enzymatic hydrolyzates of modified protein.

Stability of Ester Linkages during Fragmentation and Isolation of Peptides

Because the stability of the ester linkage in different enzyme derivatives has been shown to vary considerably depending upon the nature of the enzyme and the type of esterifying reagent, no general procedures for the fragmentation of these derivatives can be given. Nevertheless, a summary of successful methods may be useful to other efforts directed toward the location of the site of esterification in an enzyme derivative.

The ester linkage in pepsin inhibited with diazo compounds appears to be labile to incubation at pH values above 7.[5,13] The linkage becomes even more labile upon denaturation of the protein. Therefore, the derivative has been denatured either with acetone at low pH or by treatment at pH values around 7 for only 2 hours at 25°. The denatured protein can then be fragmented by digestion with native pepsin at pH values near 3 with no cleavage of ester bonds. Esterified peptides have been isolated by gel filtration, paper electrophoresis, and paper chromatography, using acidic buffers, several containing pyridine and acetic acid. The lability of the ester bond was put to some use by adapting the cleavage to the method of "diagonal electrophoresis." The mobility of the esterified peptide was changed between two electrophoretic runs by treating the paper with an aqueous solution saturated with triethylamine.[5]

The ester linkage in several other enzyme derivatives is more normal in stability. The linkage in lysozyme ethylated at Asp-52 is stable to the conditions used for oxidation of cystine by performic acid at —10°. The oxidized derivative is stable to incubation for 9 hours at pH 7.2 and 25°. Therefore, enzymatic digestion with trypsin, chymotrypsin, and carboxypeptidase A could be usefully applied.[25]

The ester bond at Glu-6 in elastase reacted with a bromoketone is stable to cyanogen bromide cleavage in 70% formic acid, followed by chromatography on SE-Sephadex using a pH gradient from 4.5 to about 7.0.[22]

In the case of ribonuclease T1, the ester bond in γ-carboxymethyl-Glu-58 was found to be cleaved to the extent of 30% during reduction and carboxymethylation of the protein at pH 8.5, using mercaptoethanol and iodoacetate in 8 *M* urea over a period of 20.5 hours. The ester bond was stable during digestion with substilisin for 3 hours at pH 8.0.[17]

Addendum[26]

In a study of the esterification of bovine insulin by treatment with anhydrous methanol–HCl (0.1 *M*), a specific N-to-O acyl shift was found to occur at the bond between Tyr-B26 and Thr-B27. The equilibrium mixture at 25° contained about 60% of the protein with an acyl ester linkage to the oxygen of the threonine residue. This derivative could be isolated by chromatography in 7 *M* urea at pH 4.75. Deamination with nitrous acid and treatment with 0.1 *M* sodium carbonate resulted in a specific cleavage of the ester bond between the tyrosine and threonine residues. On the other hand, treatment of the derivative in aqueous solution at pH values above 2.2 resulted in an irreversible O-to-N acyl shift, giving a derivative of the native structure with all six carboxyl groups esterified.

[26] D. Levy and F. H. Carpenter, *Biochemistry* **9**, 3215 (1970).

[56] Carbodiimide Modification of Proteins

By K. L. Carraway and D. E. Koshland, Jr.

Water-soluble carbodiimide reagents were first used for the modification of carboxyl groups in proteins in a mild procedure by Sheehan and Hlavka for the synthesis of small peptides[1] and the cross-linking of gelatin.[2] The method was extended to the investigation of an enzyme by Riehm and Scheraga,[3] who studied the effect of carboxyl modification on ribonuclease under acidic conditions to minimize cross-linking of carboxyls with lysine residues. The structures resulting from the modified carboxyls were not specified, but two possibilities were suggested—an imide and an *N*-acylurea. A modification reaction resulting in a stable defined product was achieved by Hoare and Koshland using a nucleophilic reagent to displace the carbodiimide group from the initially formed carbodiimide-carboxyl adduct.[4]

The rationale behind the carbodiimide-nucleophile approach can be seen from Scheme 1. The reaction sequence is initiated by addition of the

$$RCO_2H + R_1N{=}C{=}NR_2 \xrightarrow{H^+} R{-}C({=}O){-}O{-}C({=}\overset{+}{N}HR_1){-}NHR_2$$

$$\xrightarrow{HX} RC({=}O){-}X + R_1{-}NH{-}C({=}O){-}NHR_2$$

$$\longrightarrow RC({=}O){-}N(R_1){-}C({=}O){-}NHR_2$$

Scheme 1

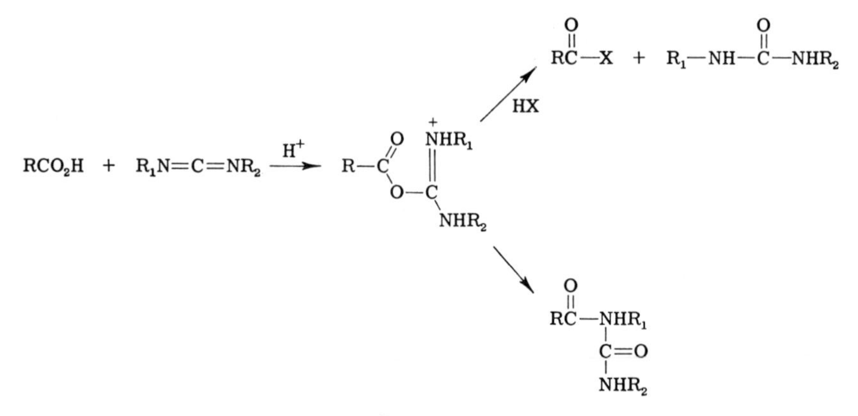

carboxyl across one of the double bonds of the diimide system to give an *O*-acylisourea.[5] The activated carboxyl group of this adduct can then react by one or two routes. First an attack by a nucleophile HX will

[1] J. C. Sheehan and J. J. Hlavka, *J. Org. Chem.* **21**, 439 (1956).
[2] J. C. Sheehan and J. J. Hlavka, *J. Amer. Chem. Soc.* **79**, 4528 (1957).
[3] J. P. Riehm and H. A. Scheraga, *Biochemistry* **5**, 99 (1966).
[4] D. G. Hoare and D. E. Koshland, Jr., *J. Amer. Chem. Soc.* **88**, 2057 (1966).
[5] H. G. Khorana, *Chem. Rev.* **53**, 145 (1953).

yield an acyl-nucleophile product plus the urea derived from the carbodiimide. Second, the *O*-acylisourea can rearrange to an *N*-acylurea via an intramolecular acyl transfer. In the special case where the nucleophile is water, the carboxyl will be regenerated with the conversion of 1 molecule of carbodiimide to its corresponding urea. Kinetic studies on model carbodiimide–carboxyl–nucleophile systems have shown that the rearrangement can be made slow compared to nucleophilic attack if the concentration of nucleophile is sufficiently high.[6] Therefore the coupling reaction of carboxyl and nucleophile can be driven essentially to completion in the presence of excess carbodiimide and nucleophilic reagent. An example of this is shown in the table, in which the carbodiimide is benzyldimethylaminopropyl carbodiimide (BDC) and the nucleophile is glycine methyl ester.[6] The reaction follows first-order kinetics under the stated conditions.

The advantage of the two-stage reaction sequence for modification of protein carboxyls lies in the versatility of the reaction. By varying

RATE OF MODIFICATION OF *m*-NITROBENZOIC ACID WITH BENZYLDIMETHYLAMINOPROPYL CARBODIIMIDE (BDC) AND GLYCINE METHYL ESTER[a]

	Yield of *m*-nitrohippurate	
Reaction time (min)	Observed (%)	Calculated[b] (%)
1	8.3	10.3
2	18.4	18.8
3	27.7	26.8
4	35.3	34.0
5	41.9	40.3
10	64.9	64.3
15	77.9	78.5
20	84.9	87.2
25	91.0	92.3
30	95.5	95.4

[a] Initial concentrations were 10^{-3} *M* *m*-nitrobenzoic acid, 0.10 *M* BDC, and 1.0 *M* glycine methyl ester, at 25°, pH 4.75; pH was maintained with 0.5 *M* HCl. Aliquots of 100 μl of 1.0 *M* acetate buffer, pH 4.75; 200 μl of 4.0 *M* HCl were added. The solution was extracted with 5 ml of chloroform, and the chloroform was back-extracted with 4 ml of 1.0 *M* carbonate buffer, pH 9.5. Yield was estimated from absorbance at 255 nm of the chloroform layer with an extinction coefficient of 7750, measured for net *m*-nitrohippuric acid in chloroform.

[b] Values calculated for the first-order reaction, in which k_1 is assumed to be 1.71×10^{-3} sec^{-1}.

[6] D. G. Hoare and D. E. Koshland, Jr., *J. Biol. Chem.* **242**, 2447 (1967).

the reaction conditions and the two reagents, one can potentially use the basic technique in many diverse ways, as described below.

Procedures

Determination of Number of Carboxyls in a Protein. A typical procedure for the quantitative modification of carboxyl groups in protein is as follows: A solution of the protein (1–10 mg/ml) and glycinamide hydrochloride or glycine methyl ester hydrochloride (1.0 *M*) is prepared in 7.5 *M* urea or 5.0 *M* guanidine hydrochloride. The pH is adjusted to 4.75 with a pH-stat, and sufficient ethyldimethylaminopropyl carbodiimide (EDC) is added as concentrated solution or solid to bring its concentration to 0.1 *M*. The pH of the reaction is maintained by automatic titration with 1.0 *M* hydrochloric acid. After 1 hour, the reaction is stopped by addition of an excess of 1.0 *M* acetate buffer, pH 4.75. The reagents are removed by dialysis or gel filtration, and the protein is prepared for amino acid analysis by standard procedures.[7] Extreme care must be exercised to assure that all free glycine is removed from the sample. The number of free carboxyls in the protein can be determined by the difference in the number of glycines observed in reacted and unreacted protein samples. When sufficient protein is available, aliquots may be taken at timed intervals during the reaction to assure that incorporation of glycine is complete. For proteins that have high glycine contents, the error involved in the determination of free carboxyls by this technique will be large. In these cases the incorporation of [^{14}C]-glycine methyl ester, or [^{14}C]-glycinamide, or an unnatural amino acid, such as norleucine, can be utilized for the analysis.

Determination of "Buried" Carboxyl Groups. The procedure is essentially identical to that above except the denaturant is not used initially. Because the reaction rate of some carboxyls is slower in the absence of denaturant, a longer reaction time is used. Solid carbodiimide is added at hourly intervals to maintain its concentration near 0.1 *M*. Aliquots should be taken at timed intervals to determine the extent of incorporation by amino acid analysis. When a saturation level of nucleophile, e.g., glycine methyl ester, is incorporated, a denaturant such as 8 *M* urea or 4 *M* guanidine hydrochloride, [^{14}C]-glycine derivative and carbodiimide are added to modify the previously buried residues. In this manner two buried carboxyls in α-chymotrypsin and chymotrypsinogen were identified.[8,9]

[7] D. H. Spackman, W. H. Stein, and S. Moore, *Anal. Chem.* **30,** 1190 (1958).

[8] K. L. Carraway, P. Spoerl, and D. E. Koshland, Jr., *J. Bol. Biol.* **42,** 133 (1969).

[9] J. P. Abeta, S. Maroux, M. Delaage, and M. Lazdunski, *FEBS* (*Fed. Eur. Biochem. Soc.*) *Lett.* **4,** 203 (1969).

Accessibility of protein carboxyl groups can be altered by changing the structure of either of the two reagents or by the denaturing properties of the reaction medium. An example is shown in Fig. 1, in which it can be seen that the number of lysozyme carboxyl groups modified increases as one goes from water to 7 *M* urea to 5.0 *M* guanidine hydrochloride.[6]

Side Reactions. Carbodiimides react with a number of organic functional groups. In aqueous solutions at acidic pH values the predominant protein groups reacting with carbodiimides are carboxyls, sulfhydryls,[10] and tyrosines.[11] The rates of reaction of model sulfhydryl and carboxyl compounds with ethyldimethylaminopropyl carbodiimide (EDC) are approximately equal, while tyrosine reacts more slowly. The contribution of tyrosine modification to activity losses can be estimated by treating the enzyme with hydroxylamine to regenerate tyrosine.[11] All of the initially reacted tyrosines in chymotrypsinogen could be freed by a 5-hour treatment with 0.5 *M* hydroxylamine at pH 7.0 and 25°. The number of modified tyrosines can be approximated by amino acid analysis, since the product is relatively stable to acid hydrolysis conditions.[11]

Attempts to regenerate sulfhydryl groups by nucleophilic displace-

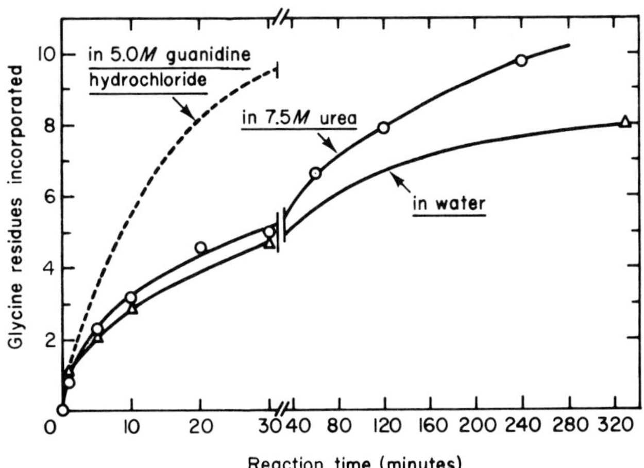

FIG. 1. Incorporation of glycine residues into lysozyme during its reaction with benzyldimethylaminopropyl carbodiimide (BDC) and glycine methyl ester. The reaction mixture contained 10 mg of lysozyme per milliliter, 0.10 *M* BDC, and 1.0 *M* glycine methyl ester hydrochloride at 25°; pH was maintained at 4.75 with 4.0 *M* HCl.

[10] K. L. Carraway and R. B. Triplett, *Biochim. Biophys. Acta* **200,** 564 (1970).
[11] K. L. Carraway and D. E. Koshland, Jr., *Biochim. Biophys. Acta* **160,** 272 (1968).

ment have not been successful.[10] Protection of sulfhydryl groups by prior reaction with SH reagents is one way of avoiding such modification. Protection with R-S-S-R type compounds (e.g., R = $—CH_2CH_2$-COOH or R = $—CH_2CH_2NH_3^+$) allows regeneration of the SH group after carboxyl modification. The report of the reaction of carbodiimide with the active site serine of chymotrypsin[12] suggests that other enzyme nucleophiles may also react. One must therefore be careful in interpreting activity changes resulting from carbodiimide treatments until possible side reactions can be ruled out. This is often possible by observing the effects of carbodiimide in the absence of nucleophile, since most of the side reactions are dependent only on carbodiimide.

Activity Studies. Correlation of carboxyl group modification with loss of enzyme activity allows the assessment of the importance of carboxyls as binding or catalytic groups. Care must be taken in these cases to eliminate possible artifacts which may lead to erroneous conclusions. In the case of activity studies it is usually inadvisable to quench the reactions with acetate buffer, since this procedure results in the formation of the acetylisourea which may acylate protein side chains. In the case of chymotrypsin, for example, treatment of the enzyme with EDC at pH 4.75 followed by quenching with acetate results in formation of acetylchymotrypsin. For this reason enzyme assays should be performed directly on the reaction mixture or appropriate controls should be run to eliminate the effects due to the potential acetylation reaction.

Other Applications

Protection by substrate or inhibitor can be utilized in a procedure analogous to that of detecting buried residues. The reaction procedure in aqueous solution is performed in the presence of saturating levels of substrate or competitive inhibitors. Subsequent treatment of this modified protein with radioactive nucleophile in the absence of substrate or inhibitor will permit labeling of carboxyls which have been "blocked" by substrate or inhibitor binding. In the absence of complicating conformational changes induced by substrate or inhibitor binding these labeled carboxyls can be assumed to be present at the active site. If conformational changes occur, the groups involved in such changes can be identified.

This method has been used successfully to investigate the nature of carboxyl groups in lysozyme[13] and trypsin.[14] The basic carbodiimide-

[12] T. E. Banks, B. K. Blossey, and J. A. Shafer, *J. Biol. Chem.* **244,** 6323 (1969).

[13] T. Y. Lin and D. E. Koshland, Jr., *J. Biol. Chem.* **244,** 505 (1969).

[14] A. Eyl and T. Inagami, *Biochem. Biophys. Res. Commun.* **38,** 149 (1970).

nucleophile reaction system has also been used to investigate the roles of carboxyls in the polymerization of β-lactoglobulin,[15] the activation of trypsinogen,[16] the structure of ribonuclease,[17] the mechanism of α-lactalbumin involvement in the lactose synthetase reaction,[18] and the mechanism of the sucrose phosphorylase reaction.[19]

Reaction with carbodiimide alone is most useful in cross-linking protein molecules or in coupling proteins to cell surfaces or insoluble supports. Since protein amino groups are not reactive at the lower pH values used for the carbodiimide-nucleophile system, cross-linking and coupling reactions are performed near pH 8. These reactions have been of value in immunochemical studies for which haptenic polypeptides were coupled to proteins[20] or for which antigens were coupled to red blood cells for passive hemagglutination procedures[21] and studies of complement fixation.[22] Coupling of antigens or antibodies to insoluble supports containing free carboxyls or amino groups can be used for the isolation and purification of the complementary binding protein.[23]

Another carbodiimide reaction with important implications for biochemical studies is the Lossen rearrangement of hydroxamates to yield the corresponding amines.[24] The reaction proceeds via a hydroxamate-carbodiimide adduct and an isocyanate intermediate, as shown in Scheme 2. Although quantitative conversion of simple hydroxamates to amines can be accomplished in this manner, it has not been possible

$$R{-}\overset{O}{\overset{\|}{C}}{-}NHOH \; + \; R_1N{=}C{=}NR_2 \longrightarrow$$

$$R{-}\overset{O}{\overset{\|}{C}}{-}NHO{-}\underset{NHR_2}{\underset{|}{\overset{\overset{+}{NHR_1}}{\overset{\|}{C}}}} \longrightarrow R_1NH\overset{O}{\overset{\|}{C}}NR_2 \; + \; RNCO$$

$$RNCO \longrightarrow CO_2 \; + \; RNH_3$$

SCHEME 2

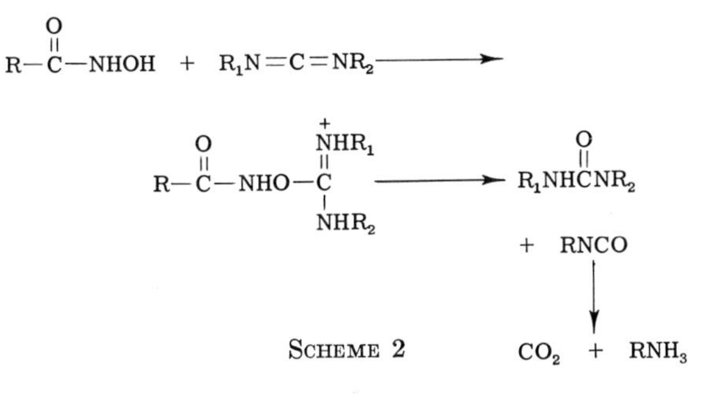

[15] J. M. Armstrong and H. A. McKenzie, *Biochim. Biophys. Acta* **147,** 93 (1967).
[16] T. M. Radhakrishnan, K. A. Walsh, and H. Neurath, *Biochemistry* **8,** 4020 (1969).
[17] M. Wilchek, A. Frensdorff, and M. Sela, *Biochemistry* **6,** 247 (1967).
[18] T. Y. Lin, *Biochemistry* **9,** 984 (1970).
[19] F. DeToma and R. H. Abeles, *Fed. Proc. Fed Amer. Soc. Exp. Biol.* **29,** 461 (1970).
[20] T. L. Goodfriend, L. Levine, and G. D. Fasman, *Science* **144,** 1344 (1964).
[21] H. M. Johnson, K. Frenner, and H. E. Hall, *J. Immunol.* **97,** 791 (1966).
[22] W. D. Linscott, W. P. Faulk, and P. J. Perucca, *J. Immunol.* **103,** 474 (1969).
[23] N. Wiliky and H. H. Weetall, *Immunochemistry* **2,** 293 (1965).
[24] D. G. Hoare, A. Olsen, and D. E. Koshland, Jr., *J. Amer. Chem. Soc.* **90,** 1638 (1968).

to convert carboxyl directly to amines in good yields by treatment with carbodiimide and hydroxylamine because of complicating side reactions. However, the reaction is potentially useful in proteins which have hydroxylamine-sensitive bonds that can be converted to hydroxamates. Such bonds are found in structural proteins,[25] in certain protein intermediates,[26] and in enzymes that have been modified by selective esterification.[27-29] Treatment of the protein hydroxamates with carbodiimide yields 2,4-diaminobutyric and 2,3-diaminopropionic acids in the protein from glutamic and aspartic acids, respectively. The diamino acids can be determined using a lengthened short column on an amino acid analyzer at pH 4.25.

In modification of specific carboxylic acid groups, for example, those at active sites or involved in crucial features of tertiary structure, certain features of the carbodiimide reagent make it particularly desirable. Since a variety of carbodiimides have the kinetic features requisite for quantitative yields, these activating agents may be varied. For example, the benzyl carbodiimide has an aromatic side chain which might attract it preferentially to the active site of enzymes operating on hydrophobic substrates, such as chymotrypsin. In that case the initial carboxyl groups modified might be those in the immediate vicinity of the active site, as in the case of histidine modification in the experiments of Schoellman and Shaw.[30] However, since the carbodiimide is ultimately replaced by the nucleophile, the bulk or special properties of the site-selecting carbodiimide need not be present in the finally modified carboxyl group. By displacing the activating carbodiimide by a much smaller group, the advantages of site selection can be obtained without the ambiguities of steric hindrance in the interpretation of the role of the modified residue.

A convenient device for modifying carboxyl groups but regenerating a negative charge is that of Wilchek, Frensdorff, and Sela,[17] who have used glycine-*N*-phthalimidomethyl ester as nucleophile and then removed the phthalimido group by conventional means. Incorporation of taurine also maintains a negative charge.

Variation in the size, charge, and nature of the activating diimide can be used to probe the three-dimensional structure of the protein. There may be positions in the protein, for example, in which an aryl

[25] H. B. Bensusan, *Biochemistry* **8**, 4723 (1969).

[26] L. E. Hokin, *J. Gen. Physiol.* **54**, 327S (1969).

[27] K. Takahaski, W. H. Stein, and S. Moore, *J. Biol. Chem.* **242**, 4682 (1967).

[28] E. Gross and J. L. Morell, *J. Biol. Chem.* **241**, 3638 (1966).

[29] A. A. Aboderin and J. S. Fruton, *Proc. Nat. Acad. Sci. U.S.* **56**, 1252 (1966).

[30] G. Schoellmann and E. Shaw, *Biochemistry* **2**, 252 (1963).

or substituted aryl carbodiimide could not react, whereas an ethyl derivative could. Selectivity in positions of attack might, therefore, be achieved by discriminating use of modified carbodiimides.

Summary

A carbodiimide-nucleophile procedure for modifying carboxyl groups in proteins can be utilized under mild conditions for the quantitative determination of all carboxyl groups or for the modification of selected residues in activity studies. The method allows numerous variations both in the activating carbodiimide and in the nature of the nucleophile. Side reactions do occur but can be corrected for.

[57] Bifunctional Reagents

By Finn Wold

Scope

The only direct chemical method for study of the folding of the polypeptide chains in proteins is based on the elucidation of the location of intra- or interchain chemical bonds. The disulfide bonds are the major naturally occurring links of this type, and the determination of the locations of disulfide bonds in several proteins (including insulin, ribonuclease, lysozyme, and chymotrypsin) has been an important step toward the ultimate goal of building exact three-dimensional models of these proteins. Other naturally occurring cross-links (ester linkages in collagen, or carbohydrate bridges) should potentially be equally useful as markers in the determination of tertiary and quaternary structure of proteins. It follows therefore that the introduction of stable covalent bridges between amino acid residues in the "native" structure of a protein should provide an important extension of this type of information, and a large volume of work directed toward the development of methods to achieve this goal has been carried out during recent years. The most versatile method is to react the protein with bifunctional reagents, reagents with two reactive groups capable of reacting with, and forming bridges between, the side chains of the amino acids in the protein. It is the purpose of this article to indicate in a general manner the type of information that can be obtained from such studies, and to survey the various approaches and reagents employed in this type of study.

The use of bifunctional reagents is obviously not restricted to proteins, and several of the early cross-linking reagents, such as mustards and epoxides,[1] have recently been used to cross-link DNA.[2] Since it has been demonstrated that at least some of the cross-links join opposite strands of the reacted DNA, very interesting structural and functional properties of double-stranded nucleic acids may be explored by the use of bifunctional reagents. This area, however, is beyond the scope of an article concerned with protein reactions.

In the reaction of proteins with bifunctional reagents, three general types of product can be considered, each product providing different ways of gathering information about the structure and function of proteins. The types of product and their potential uses are as follows.

1. Intramolecularly cross-linked proteins. The determination of the location of each cross-link will give specific information about inter-residue distances in the protein, with direct bearing on its three-dimensional structure. Studies of the physical properties of protein derivatives with different extents of cross-linking should also provide a measure of the effect of covalent bonds in stabilizing the tertiary structure. For the proper interpretation of this type of information, it is essential that control experiments be conducted with analogous monofunctional reagents (a derivative formed by reaction with a moles of X-R-R-X should be compared to a derivative formed by reaction with $2a$ moles of R-X) in order to separate the effects of chemical modification (by introducing R) and the effects of introducing cross-links (-R-R-).

2. Intermolecularly cross-linked proteins (homopolymers). These derivatives should provide excellent models for study of protein–protein interactions (e.g., cross-linking of subunits), and could also have practical use in providing stable (e.g., undialyzable) active proteins of high molecular weight.

3. Intermolecularly cross-linked protein complexes, composed of different proteins (heteropolymers) or of proteins and other reactive molecules. Again, these derivatives may provide models for study of protein–protein interactions (e.g., antibody–antigen interactions, multienzyme complexes, or cell-membrane lipoprotein structures), or highly practical reagents for the binding of biologically active proteins to matrices to form insoluble enzymes or enzymes of high molecular weight.

In all these considerations, it is assumed that the various cross-linked derivatives retain their biological activity. In view of all the

[1] P. Alexander, M. Fox, K. A. Stacey, and L. F. Smith, *Biochem. J.* **52**, 177 (1952).

[2] See for example P. D. Lawley and P. Brookes, *J. Mol. Biol.* **25**, 143 (1967); P. D. Lawley, J. H. Lethbridge, P. A. Edwards, and K. V. Shooter, *J. Mol. Biol.* **39**, 181 (1969).

information available to date, this appears to be a safe assumption, as it is probably always possible to select the proper reagent and reaction conditions to avoid loss of activity.

Selecting Reagent and Reaction Conditions

The selection of the proper reagent must be determined primarily by the specific product desired, along the lines discussed above. In addition, the choice is obviously limited by the stability range of the protein and the reagent in question. It should be emphasized that, although this discussion focuses on homobifunctional reagents where the two reactive groups are identical, it is generally feasible to make heterobifunctional reagents where the two groups are sufficiently different to permit well-controlled sequential reactions of each group in turn. Several heterobifunctional reagents have been tabulated by Zahn,[3] and some will be considered in this article. A very promising new group of heterobifunctional reagents has been explored recently: reagents in which one of the reactive functions is a relatively unreactive group which can be photoactivated to give a highly reactive photolysis product. With these reagents a very well defined sequence of reaction steps can be achieved by carrying out the first step in the dark and the second one in the presence of activating light. Diazoalkyl derivatives[3a] and aryl azides[3b,c] are examples of photoactivatable groups which have been explored as protein reagents in recent years. A quite general problem in all work with bifunctional reagents is the low solubility of many of the available reagents in the aqueous buffer solutions used for protein reactions. Some reactions, especially those involving structural proteins, such as wool, silk, and gelatin, or those in which exhaustive reaction with all protein sites is desired, are in fact conducted in mixed solvent systems (dimethyl formamide–water, acetone–water, dioxane–water). If conditions are restricted to aqueous solutions, the favored method is to add to the protein solution a concentrated solution of the reagent in a suitable organic solvent (acetone, alcohol, dioxane, etc.) in such a way that the concentration of organic solvent in the reaction mixture is kept low (less than 5%). It is interesting that optimal rates of reaction are obtained when the reagent is added slowly enough (continuously or in small increments) to just maintain saturation. Although this may not always be critical, a much lower rate of reaction is often obtained

[3] H. Zahn, *Abstr. 6th Int. Congr. Biochem.* 1964.

[3a] C. S. Hexter and F. H. Westheimer, *J. Biol. Chem.* **246,** 3928, 3934 (1971).

[3b] G. W. J. Fleet, J. R. Knowles, and R. R. Proter, *Nature* **224,** 511 (1969).

[3c] H. Kiefer, J. Lindstrom, E. S. Lennox, and S. J. Singer, *Proc. Nat. Acad. Sci. U.S.* **67,** 1688 (1970).

when the reagent is permitted to precipitate in the reaction mixture, or is added as a solid.

The main variables to consider in choosing the general reaction conditions are protein concentration, protein-to-reagent concentration ratio, pH, and ionic strength. The idealized scheme shown (Scheme 1)

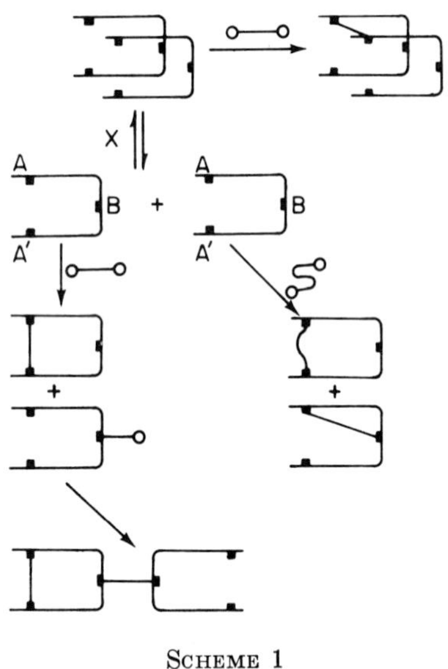

SCHEME 1

illustrates to a first approximation the kinds of consideration that may be useful in selecting reagent and reaction conditions favoring different reaction products. Although the equilibrium (X) has been represented as a formal monomer–dimer equilibrium, it should be considered as the most general case, with a high probability of two protein molecules being found associated at the top and a low probability of association at the bottom. If intramolecular cross-links of a simple nonassociated protein is desired, reaction conditions which favor the monomeric forms are chosen: low protein concentration, high net charge on the protein (pH different from the isoelectric point); and in addition a high ratio of protein sites to reagent concentration should also minimize the probability of intermolecular cross-link formation. If, on the other hand, intermolecular cross-links between the monomers in an associated complex is desired, the reaction conditions are selected to stabilize the associated system. Thus, in attempts to link like molecules (P–P) the reaction should be carried out at high protein concentration and at the pH of minimum net charge; in attempts to link different proteins (P–Q) the pH and ionic strength of the reaction mixture should be adjusted to give maximum opposite charge on P and Q to favor P^{+}–Q^{-} interactions, and at the same time reduce the probability of P^{+}–P^{+} and Q^{-}–Q^{-} interactions.

Other more indeterminate variables will obviously affect each individual reaction, depending on the proteins and reagents used [relative rate of reaction and reagent hydrolysis, the fit of reagent width with relative distances in the proteins (A-B compared to A-A′ in Scheme 1)]. Most of these can probably be controlled only empirically. It appears, however, that on the basis of these rather simple rules intra- and intermolecular reactions can be controlled quite well.

Characterization of the Product

In general the characterization of the product requires the same procedures as the characterization of proteins modified with monofunctional reagents. Excess reagent is eliminated by dialysis, gel filtration, or chromatography, the reaction mixture is fractionated into its component derivatives on the basis of size (gel filtration or sedimentation) or on the basis of ionic or solubility properties (ion-exchange, electrophoresis, or partitioning systems), and each derivative is then characterized in terms of as many biochemical, physical, and chemical parameters as desirable.

There are, however, two questions unique to the characterization of proteins reacted with bifunctional reagents: (1) Did both functional groups of each reagent molecule react with the protein (monofunctional or bifunctional reaction)? (2) Were the cross-links formed intra- or intermolecularly? The ideal approach to the first question is to degrade the protein and isolate and quantitate the individual cross-linked residues. If this is not possible, because of limited stability of the derivatives and difficulties in the attainment of quantitative recoveries of the cross-linked amino acid pairs, very accurate analyses must be made, both of the amount of reagent incorporated into the protein derivative and of the number of amino acid residues reacted per mole of reagent incorporated. By far the most accurate method of determining the number of moles of reagent incorporated utilizes a radioactive reagent and determination of total radioactivity in the purified protein derivative. Methods based on ultraviolet or visible absorption or fluorescence of the reagent have at least two distinct limitations, namely, if the reagent reacts with more than one kind of residue in the protein, each derivative may have different spectral characteristics, and because of microenvironmental effects in the protein, identical derivative forms may have different molar absorptivities or emission quantum yields in different proteins. The extent of reaction can also be determined by directly monitoring chemical changes associated with the reaction (e.g., acid is produced when alkyl halides or acyl halides react with proteins). In most cases where this approach is applicable the reagent will also react with solvent, and solvent blanks have to be included in parallel experiments.

With an accurate evaluation at hand of the number of moles of reagent incorporated per mole of protein, the next step is to determine the number of moles of amino acid residues reacted per mole of protein. If the amino acid derivative is stable to acid hydrolysis, quantitative amino acid analysis after complete hydrolysis can be carried out with ±2% accuracy. This is sufficient if the reaction involves approximately 20% or more of the total of any given amino acid (a reaction involving 2 of the 10 lysine residues in ribonuclease may be determined satisfactorily while a reaction involving 2 of the 60 lysine residues in serum albumin may not be established unequivocally by amino acid analysis). Where the derivatives are readily cleaved, other methods must be used, such as quantitative preparation of stable derivatives of the unreacted residues and evaluation of the number of residues reacted with the bifunctional reagent from the number released upon hydrolysis.

If the ratio of residues reacted to moles of reagent incorporated is 2, it may be concluded that the reagent reacted bifunctionally, and that each reagent molecule has formed a covalent bridge. With a ratio of 1, the reagent has behaved exclusively in a monofunctional way, and the second functional group of each reagent molecule has either not reacted at all, or reacted with solvent to give a stable monofunctionally substituted product. Ratios between 1 and 2 obviously signify a mixture of bridged and nonbridged derivatives. It is interesting to note that in most cases reported the observed ratios are close to 2, and that bridge formation thus appears to be favored.

As to the second question, if dimers or higher polymers have been formed by intermolecular cross-linking, these are readily detected by any one of the conventional methods, such as hydrodynamic measurements, light scattering, gel filtration, etc., and the characterization of each derivative should certainly include this type of measurement. The main difficulty encountered in distinguishing between intra- and intermolecular bridges is caused by false positive observations. In many cases even subtle chemical modifications of a protein will cause aggregation of the product, and therefore it is important to distinguish between noncovalent aggregation and covalent cross-links in the case of proteins reacted with bifunctional reagents. The most reasonable approach to the solution of this problem involves the use of solvent or agents known to disrupt noncovalent forces, such as guanidinium salts, urea, detergents, dioxane, extremes of pH. With these precautions, it should be possible to establish unequivocally whether or not intermolecular bridges have been formed in the reaction.

Degradation and Identification of Cross-Linked Peptides

Determination of the location of the covalent cross-links introduced into the protein is based on the general techniques for the location of disulfide bonds. Any method of peptide bond cleavage that does not disrupt the bridge can be used to accomplish a partial digestion of the protein derivative. Digestion with the selective enzymes trypsin, chymotrypsin, or pepsin is probably the most generally useful technique. After isolation of the cross-linked peptide or peptides, which is facilitated by any characteristic marker in the reagent (radioactivity, color, etc.), the problem of identifying the two covalently linked peptides remains. At this stage the work is greatly simplified if it is possible to disrupt or remove the synthetic bridge and work with each of the individual peptides separately. In selecting the proper reagent for a reaction in which the location of the bridge is to be determined, this aspect should be given substantial emphasis.

Individual Reagents

In the following survey, examples of the types of bifunctional reagent used to date are given brief consideration as to availability, properties, specificity, and reported uses.

Bifunctional Maleimide Derivatives

A number of *N*-substituted maleimide derivatives are well known as specific reagents for sulfhydryl groups (see [36], this volume). They are probably among the most specific protein reagents, reacting under mild conditions and with a minimum of side reactions. Furthermore, the characteristic changes in ultraviolet absorbance associated with the reaction of maleimide derivatives with sulfhydryl groups provide a convenient method for monitoring the reaction. *N*-substituted bismaleimide derivatives thus provide highly specific and mild bifunctional reagents. Five such derivatives have been synthesized and studied as protein reagents. The reagents are all insoluble in water and have generally been added to the aqueous protein solution (pH 7–8) as the solid. Since no significant side reactions occur, stoichiometric amounts can be added, and generally very slow reaction can be allowed to proceed without forcing the conditions. Several *N*-aryl- and *N*-alkyl maleimides have been prepared and are readily available as convenient and monofunctional analogs. The reaction products of compounds (I), (II), (IV), and (V) cannot be cleaved readily into the component half-molecules as the imide bond is susceptible only to treatments that will also cleave peptide bonds; however, compound (III) and its reaction products are

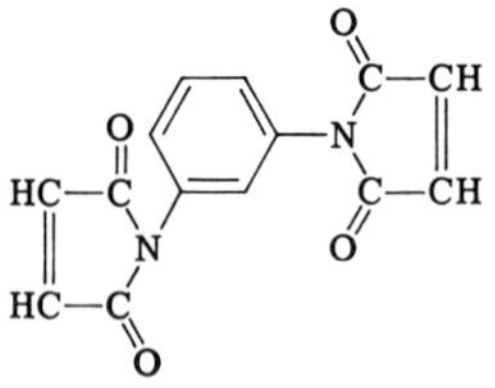

N, N′-(1, 3-Phenylene)bismaleimide[4]

(I)

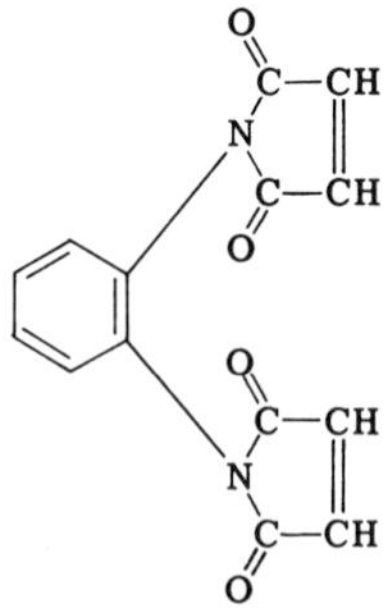

N, N′-(1, 2-Phenylene)bismaleimide[4]

(II)

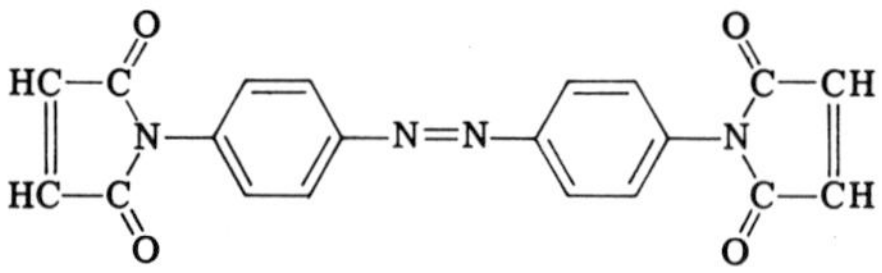

Azophenyldimaleimide[5]

(III)

readily cleaved by reduction of the azo groups with dithionite. When either (I), (II), or (IV) were allowed to react with BSA, an intermolecular cross-link was formed between free SH groups (BSA contains about 0.65 SH per mole), and about 40% of the protein was recovered as a stable dimer. Mercuric ion gave about the same amount of dimer, but reaction with *N*-phenylmaleimide did not result in dimer formation.[4,7] Similarly, when reduced wool was exhaustively reacted with (I) and (II), the properties of the product (in comparison with the product from *N*-phenylmaleimide) indicated that extensive cross-linking had taken place. Acid hydrolysis of wool modified with either (I), (II), or *N*-phenylmaleimide, gave 2-(2-amino-2-carboxyethylmer-

[4] J. E. Moore and W. H. Ward, *J. Amer. Chem. Soc.* **78,** 2414 (1956).
[5] H. Fasold, U. Groschel-Stewart, and F. Turba, *Biochem. Z.* **337,** 425 (1963).

```
        O          O
       //          \\
H—C—C              C—C—H
  ||  \            /  ||
  ||   N—(CH2)6—N    ||
  ||  /            \  ||
H—C—C              C—C—H
      \\          //
        O          O
```

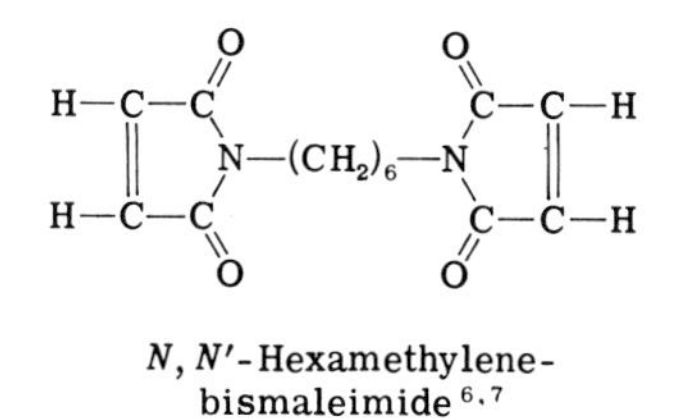

N, *N'*-Hexamethylene-bismaleimide[6,7]

(IV)

```
        O                      O
       //                      \\
H—C—C                          C—C—H
  ||  \                        /  ||
  ||   N—CH2—O—CH2—N    ||
  ||  /                        \  ||
H—C—C                          C—C—H
      \\                      //
        O                      O
```

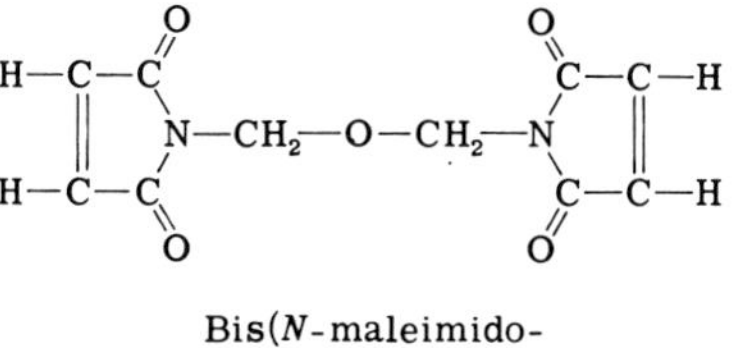

Bis(*N*-maleimido-methyl)ether[8,9,9a]

(V)

capto)succinic acid as the only reagent derivative (the imides are opened), thus demonstrating the specificity of these reagents.[4]

Compound (III) has not been made to react with proteins, but both mono- and bifunctional reaction products of (III) with cysteine, benzoylcysteine, and glutathione have been studied. The reagent was added as a dioxane solution to the aqueous solution of the SH compound at pH 7, and paper chromatography, in dioxane:0.8 N ammonia (10:3 v/v), gave good separation of the products.[5]

Compound (V) has been used to probe the subunit interaction in hemoglobin,[9] and in tryptophan synthetase.[9a] The cross-linking of oxyhemoglobin with (V) appears to cause a conformational restraint on the protein which is reflected in a deoxyhemoglobin-like behavior, accompanied by a decrease in the normal linked functions (Bohr effect and cooperative oxygen binding). The cross-linked derivative does, however, exhibit a very high affinity for oxygen. Although the observations are not yet complete enough for a clear interpretation, the experiment very nicely illustrates the broad spectrum of problems that can be explored with bifunctional reagents.

Reagent II and the corresponding 1,4-derivative are commercially

[6] P. Kovacic and R. W. Hein, *J. Amer. Chem. Soc.* **81,** 1187 (1959).

[7] H. Zahn and L. Lumper, *Hoppe-Seyler's Z. Physiol. Chem.* **349,** 485 (1968).

[8] P. O. Tawney, R. H. Snyder, R. P. Conger, K. A. Lubbrand, C. H. Stiteler, and A. R. Williams, *J. Org. Chem.* **26,** 15 (1961).

[9] S. R. Simon and W. H. Konigsberg, *Proc. Nat. Acad. Sci. U.S.* **56,** 749 (1966).

[9a] W. B. Freedberg and J. K. Hardman, *J. Biol. Chem.* **246,** 1439 (1971).

available (Aldrich Chemical Company, Inc., Milwaukee, Wisconsin; and K and K Laboratories, Inc., Plainview, New York).

Bifunctional Alkyl Halides

The alkylating reagents illustrated, bifunctional alkyl halides (I-VI), react primarily with sulfhydryl groups, sulfides, imidazole, and amino groups (see [34a], this volume). At neutral to slightly alkaline pH, the reaction with SH is favored to the extent that the reagents can be considered specific for these groups in the protein. At higher pH, reac-

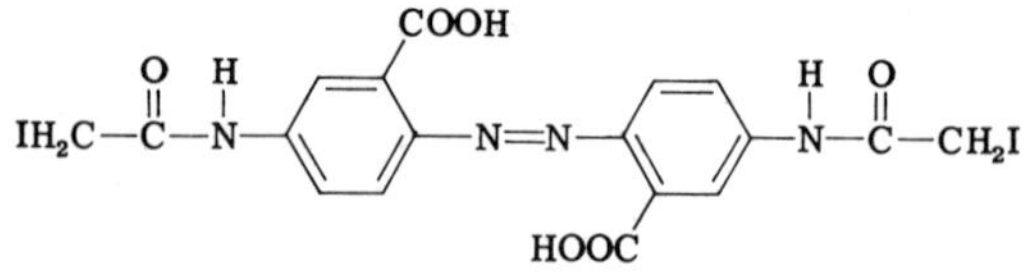

2, 2′-Dicarboxy-4, 4′-diiodoacetamidoazobenzene[10]

(I)

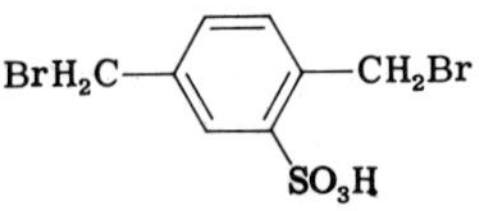

α, α′-Dibromo (or diiodo) *p*-xylenesulfonic acid[11]

(II)

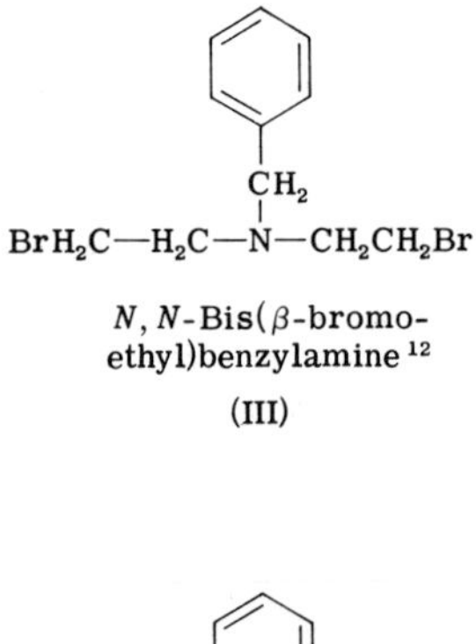

N, *N*-Bis(β-bromoethyl)benzylamine[12]

(III)

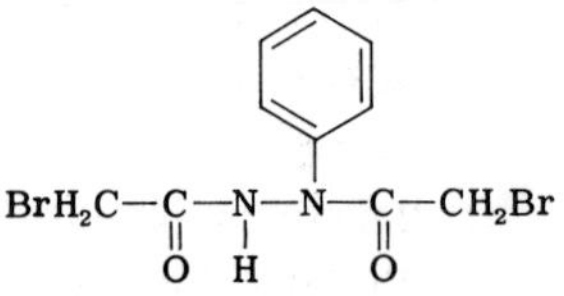

N, *N′*-Dibromoacetyl)-phenylthydrazine[12]

(IV)

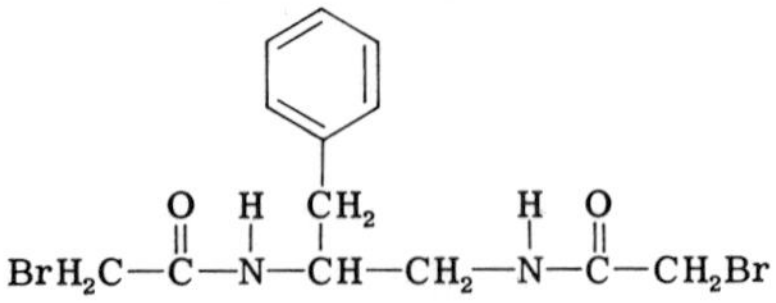

1, 2-Di(bromoacetyl)amino-3-phenylpropane[12]

(V)

[10] H. Fasold, U. Groschel-Stewart, and F. Turba, *Biochem. Z.* **339**, 487 (1964).

[11] C. B. Hiremath and R. A. Day, *J. Amer. Chem. Soc.* **86**, 5027 (1964).

[12] H. G. Gundlach, *Habilitationsschrift,* University of Wurzburg, p. 44, 1965.

tion with amino groups is favored. Quantitative symmetrical cleavage of the reaction products from reagent I can be achieved by careful reduction of the azo group with dithionite, while the secondary amines formed in the reaction of reagent II with amino groups can be cleaved by catalytic hydrogenolysis with Pd in glacial acetic acid, to liberate the original component peptides.

The reaction with proteins can readily be followed by KOH uptake in a pH-stat. In addition, reagent I has an absorption maximum at 370 nm (the reaction product between reagent I and glutathione has an absorbance of 22.6 for a 1 mg/ml solution), and reagent II has been synthesized with ^{35}S, which allows rapid and accurate characterization of reaction products.

When 3 ml of H-meromyosin solution (13 mg/ml in 0.1 M acetate buffer, pH 7.6) was treated with stirring at 4° with 3.3 μmole of (I) for 3 hours, 0.0246 μmole of SH groups (44% of the titratable SH groups) disappeared and spectrophotometric measurements showed that 0.0105 μmole of reagent had been taken up, clearly demonstrating the

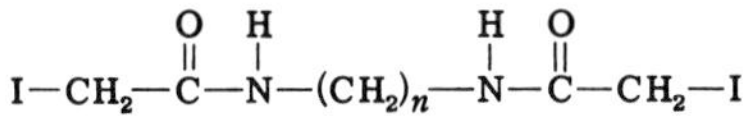

$$I-CH_2-\overset{\overset{O}{\|}}{C}-\overset{\overset{H}{|}}{N}-(CH_2)_n-\overset{\overset{H}{|}}{N}-\overset{\overset{O}{\|}}{C}-CH_2-I$$

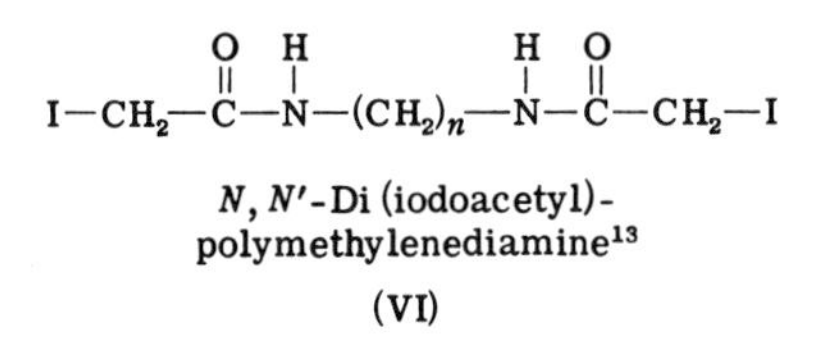

***N*, *N'*-Di (iodoacetyl)-polymethylenediamine**[13]

(VI)

bifunctional nature of the reaction. At pH 8, compound (I) was also found to react rapidly and quantitatively with the SH of reduced glutathione.

The reaction of (II) with lysozyme and the determination of the location of the cross-links[11] will be considered in some detail. A dilute solution (0.25 mg/ml) of lysozyme in 0.1 M borate buffer (pH 9.1) was allowed to react with ^{35}S-(II) (0.5 mg/ml) for 48 hours at 37°, and the product was freed from excess reagent on Sephadex G-25 (0.1 M acetic acid as eluent). The product showed essentially full activity, was homogeneous in the ultracentrifuge, and had a molecular weight very similar to that of the untreated enzyme. Electrophoresis showed that the product contained at least two derivatives, 70% of a presumed mono-(II) derivative and 30% of a presumed di-(II) derivative.

In order to identify the bridges introduced, the modified lysozyme was degraded according to Scheme 2. From Scheme 2 it may be concluded that the major cross-links in lysozyme were formed between

[13] H. Ozawa, *J. Biochem.* (*Tokyo*) **62**, 531 (1967).

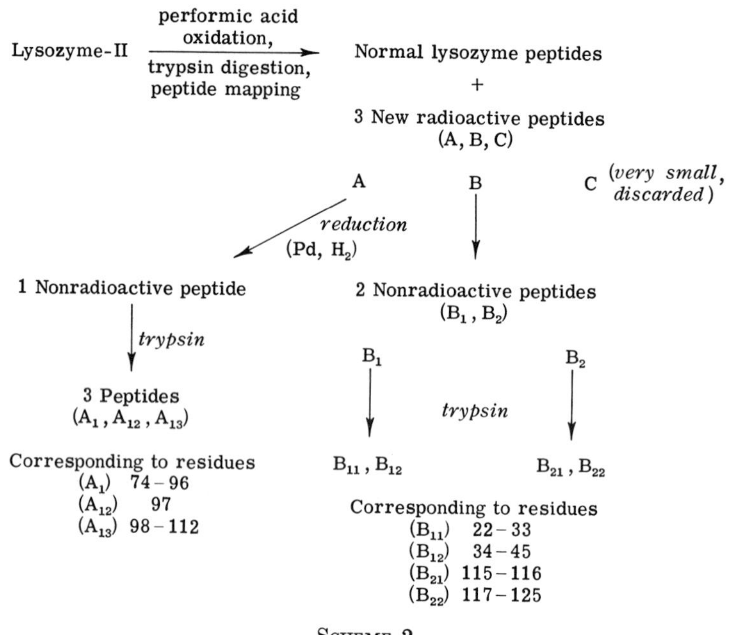

SCHEME 2

Lys_{96} and Lys_{97} and between Lys_{33} and Lys_{116}. The latter assignment has also been proposed for the cross-link formed by phenol-2,4-disulfonyl chloride with lysozyme.[14] It is of interest to note that one of the disulfide bonds in lysozyme is Cys_{30}-Cys_{115}, and the 33-116 cross-link, therefore, seems reasonable. This cross-link also appears acceptable from X-ray crystallographic data.[15]

Compounds (III), (IV), and (V) (bifunctional alkyl halides) represent a very interesting group of bifunctional reagents in that they were prepared specifically as reagents for the active site of chymotrypsin. The idea of introducing substrate affinity into the reagent has many exciting possibilities. In the case of (III), (IV), and (V), the aromatic ring constitutes the affinity group and the two alkylating functions should thus ideally bridge peptide sequences closely associated with the substrate binding site. The distance between the reactive functions in the three compounds differs—maximum span 6.5 Å in (III), 7.3 Å in (IV), and 10 Å in (V)—and comparative studies with the three reagents therefore provides a promising approach to the topographical descrip-

[14] D. J. Herzig, A. W. Rees, and R. A. Day, *Biopolymers* **2**, 349 (1964).

[15] C. C. F. Blake, D. F. Koenig, G. A. Mair, A. C. T. North, D. C. Phillips, and V. R. Sarma, *Nature* (*London*) **206**, 757 (1965).

tion of the active site of chymotrypsin. Preliminary studies show rapid inactivation of chymotrypsin with all three reagents.[12]

Compound (VI), specifically the hexamethylene ($n = 6$) derivative, was allowed to react with rabbit muscle aldolase, and formed cross-links primarily between sulfhydryl groups in a rather slow reaction. The extent of reaction was evaluated by direct analysis of *S*-carboxymethylcysteine after acid hydrolysis. An important point illustrated in this work is the use of substrate or substrate analogs to protect the active site and prevent inactivation during the cross-link formation. In this case reaction in carbonate buffer led to an inactive derivative, while in parallel experiments in phosphate buffer (phosphate has been shown to protect aldolase against sulfhydryl reagents[16]) the cross-linked derivative retained full activity.[13]

Bifunctional Aryl Halides

Reagents such as the ones illustrated are insoluble in water; they react preferentially with amino groups and tyrosine phenolic groups, but also with sulfhydryl and imidazole groups. Relatively high pH values are required for rapid reaction.

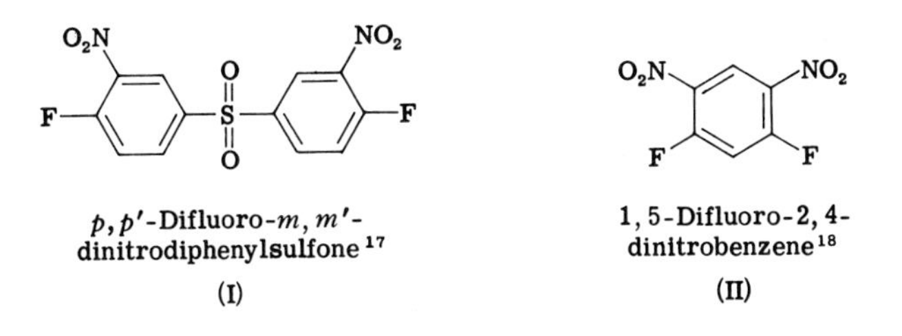

_p,p'_-Difluoro-_m,m'_-dinitrodiphenylsulfone[17]

(I)

1,5-Difluoro-2,4-dinitrobenzene[18]

(II)

The reagent is generally added to an aqueous solution of protein or amino acids as a concentrated acetone solution, and the reaction is followed by the release of HF. The products formed with both (I)[19] and (II)[20] also have characteristic visible and ultraviolet spectra. Alkaline hydrolysis of the derivatives will liberate the reacted amino acids[21,22] and in theory it should be possible to cleave (I) and its derivatives by catalytic reduction (Ni) of the sulfone. The lysine and tyrosine deriva-

[16] J. Kowal, T. Cremona, and B. L. Horecker, *J. Biol. Chem.* **240**, 2485 (1965).

[17] H. Zahn and H. Zuber, *Chem. Ber.* **86**, 172 (1953).

[18] H. Zahn and H. Stuerle, *Biochem. Z.* **331**, 29 (1958).

[19] F. Wold, *J. Biol. Chem.* **236**, 106 (1961).

[20] H. Zahn and J. Meienhofer, *Makromol. Chem.* **26**, 126 (1958).

[21] H. Zahn, H. Zuber, W. Ditscher, D. Wegerle, and J. Meienhofer, *Chem. Ber.* **89**, 407 (1956).

[22] G. L. Mills, *Nature (London)* **165**, 403 (1950).

tives are stable to normal acid hydrolysis. Fluorodinitrobenzene is a good monofunctional analog for both reagents, and the analog of (I) (*p*-fluoro-*m,m'*-dinitrodiphenylsulfone) has also been synthesized.[23] Compounds (I) and (II) are both commercially available (Pierce Chemical Company, Rockford, Illinois; Aldrich Chemical Company, Inc., Milwaukee, Wisconsin; and K and K Laboratories, Inc., Plainview, New York).

Compound (I) can be synthesized in good yield,[17] and ^{14}C-labeled (I) has been prepared.[24] The reagent is insoluble in water, moderately soluble in alcohols, and soluble in acetone and dioxane. Contrary to earlier proposals the two halves of the reagent do not react independently. In reaction with either amine and hydroxyl functions, the first half reacts faster than the second.[25] Both are less reactive than FDNB. A number of amino acid derivatives have been prepared and characterized.[20,21,23] *sec*-Butanol:formic acid:water, 750:135:115,[24] is a good solvent system for both paper chromatography and thin-layer chromatography on Silica Gel G. The di-*O*-tyrosine derivative is quite insoluble in a number of solvent systems.

The di-ϵ-lysine derivative of (I) (N^{ϵ},N'^{ϵ}-(*m,m'*-dinitrodiphenylsulfone-*p,p'*)bislysine) can be quantitatively determined in a normal amino acid analyzer short column run, by elution with a 0.1 *M* sodium borate–0.35 *M* sodium acetate buffer (pH 10.6) containing 10% *n*-propanol, after the normal analysis of the basic amino acids is completed.[25]

Reaction with different types of collagen[23,26] was carried out with excess reagent in a mixture of dimethyl formamide and water. An excess of sodium bicarbonate was used to neutralize the acid produced, and the reaction mixture was left at 35° for up to 200 hours. Comparison of the products from (I) with those from the monofunctional *p*-fluoro-*m,m'*-dinitrodiphenylsulfone analog clearly demonstrated the introduction of cross-links in the collagen fiber. The product after 120 hours' reaction time (with ^{14}C-labeled (I)) showed that reaction had occurred primarily at lysine and hydroxylysine.

A number of soluble proteins have also been made to react with (I).[19,27,28] In these cases an acetone solution of (I) was added to the aqueous protein solution maintained at pH 10.5. Under these conditions

[23] H. Zahn and D. Wegerle, *Kolloid-Z.* **172,** 29 (1960).
[24] H. Zahn and E. Nischwitz, *Kolloid-Z.* **172,** 116 (1960).
[25] L. C. Jen, Ph.D. Thesis, University of Illinois, Urbana, Illinois, 1967.
[26] R. L. Sykes, *Makromol. Chem.* **27,** 157 (1958).
[27] L. C. Jen and F. Wold, *Abstr. 7th Int. Congr. Biochem. Tokyo, 1967,* p. 596.
[28] S. S. Tawde, J. Sri Ram, and M. R. Iyengar, *Arch. Biochem. Biophys.* **100,** 270 (1963).

bovine serum albumin (BSA) reacted very rapidly with up to 15 moles of reagent per mole of protein. The reaction was found to be quantitative, with the reaction at lysine and tyrosine residues accounting for the theoretical total amount of reagent added. With 1% protein solution very little polymeric (intermolecularly cross-linked) BSA resulted from the reaction with 10 moles of reagent. With a 20% protein solution and with much longer reaction times, however, up to 55% of the protein was converted to a rapidly sedimenting polymeric derivative.

The monomeric cross-linked BSA derivatives were denatured under a number of conditions and the behavior compared with that of the corresponding DNP derivatives and native BSA. From the results it is clear that the covalent bridges introduced contribute stability to the derivative, and aid in maintaining the folded structure even after the natural intramolecular S-S bonds are broken.[19,29] Neither the cross-linked derivatives nor the homologous DNP derivatives show significant loss of normal antigenic properties in the reaction with anti-BSA,[29] but the antigenic properties of both derivatives are destroyed on denaturation by various procedures, thus showing that the cross-links do not provide protection of the structural determinants of the antigenic sites.

When yeast enolase (9 mg/ml) was reacted with 5 moles of (I) per mole of protein, 90–100% recovery of activity[19] was attained. Similarly, after reaction with 1 or 2 moles of (I) per mole of protein, ribonuclease A gave derivatives with a specific activity as great or greater than the starting material. On the basis of amino acid analyses of the ribonuclease derivatives, two lysine residues reacted with the first mole of reagent, while two lysines or one lysine and one tyrosine reacted with the second mole of reagent. Work on the location of the cross-link in a ribonuclease derivative with full enzymatic activity toward cyclic cytidylic acid has demonstrated a Lys_{31}-Lys_{98} cross-link as the major bridge in that derivative.[25,27] The same cross-link is formed in ribonuclease S′ (a derivative in which the peptide bond between residues 20 and 21 has been cleaved), while in parallel experiments reaction of S-protein alone (S-protein, residues 21–124, is ribonuclease S′ with the S-peptide, residues 1–20, removed) did not yield any of the 31–98 cross-link.[25] This experiment illustrates yet another use of bifunctional reagents in probing the conformational status of structurally analogous proteins. In assessing the contribution of the cross-link to the stability of ribonuclease A, it was found that while the rate of denaturation was significantly slower after cross-linking, the denatured states of the cross-linked derivative was indistinguishable from those of unreacted ribonuclease A.[25]

[29] F. Wold, *Biochim. Biophys. Acta* **54**, 604 (1961).

Chymotrypsinogen can be reacted with 3 moles of (I) to give a cross-linked derivative which generates 90–100% activity upon subsequent activation with trypsin. (Chymotrypsin suffers a higher loss of activity when reacted under the same conditions.) Lysine was found to be the primary site of reaction, but tyrosine also reacted to some extent. The incorporation of the reagent did not appear to interfere with normal activation, as demonstrated by N-terminal analysis of the activated product.[30] Molecular weight measurements with the cross-linked chymotrypsinogen indicated that highly polymerized material was formed in the reaction. However, on closer examination it was found that polymers are also formed when the analogous monofunctional reagent (FDNB) is substituted for (I). It was subsequently found that in the proper solvent (detergent, high pH, urea) both the DNP and the cross-linked derivatives behaved as monomers on gel filtration, while hydrodynamic measurements indicated the presence of some molecules of higher molecular weight. It was concluded that the aggregates of high molecular weight found in neutral buffers must be due primarily to protein–protein interactions promoted by the incorporated reagents. Along these lines, it must also be emphasized that one of the primary difficulties encountered in fractionating peptides cross-linked with (I) on a number of different resins and adsorbents has been the irreversible binding of the peptides to the column matrix, leading to extremely poor recoveries. This binding must be due to strong interaction between the derivative and the column material or to formation of aggregates within the column matrix. Peptides containing (I) have been found to fractionate well on Sephadex (G-10 and G-25) and on acrylamide gels (Biogel P-2 and P-4) with 1 *M* NH_4OH as solvent. Electrophoresis on a polyvinyl chloride–polyvinyl acetate resin (Geon, B. F. Goodrich and Co.) in a variety of solvent systems, paper electrophoresis, and disk electrophoresis have also afforded good resolution and recovery of purified fractions. Conventional resins are not suitable for these derivatives because of the strong binding, however.

Compound (I) has also been used to form covalently linked conjugates of several pairs of proteins, the formation of conjugates being favored by high protein concentration, high reagent concentration, and high pH.[31,32,32a] The protein conjugates were identified and distinguished from the individual component proteins by electrophoresis. In the case of (anti-BSA) γ-globulin–ferritin conjugate (produced in up to 50%

[30] R. T. Havran, Ph.D. Thesis, University of Illinois, Urbana, Illinois, 1965.

[31] J. Sri Ram, S. S. Tawde, G. B. Pierce, Jr., and A. R. Midgley, *J. Cell Biol.* **17**, 673 (1963).

[32] J. Sri Ram, *Biochim. Biophys. Acta* **78**, 228 (1963).

[32a] R. R. Modesto and A. J. Pesce, *Biochim. Biophys. Acta* **229**, 384 (1971).

yield) the reaction proceeded with a 40% loss of antibody activity, but the presence of ferritin associated with the BSA-precipitating activity could be demonstrated.

Difluorodinitrobenzene (II) has been studied quite extensively as a protein reagent. The two fluorine groups are not equal in reactivity, the difluoro derivative being considerable more reactive than the mono-fluoro–monosubstituted derivative obtained after the first step in the reaction. The maximum bridge span of (II) is 5–6 Å. A large number of mono- and bisamino acid derivatives have been prepared and characterized.[18,33]

Compound (II) has been shown to react primarily with lysine and tyrosine residues in wool, and all the possible derivatives (FDNP-N^ϵ-Lys, FDNP-*O*-Tyr, DNPene-bis-N^ϵ-Lys, DNPene-bis-*O*-Tyr, and DNPene-N^ϵ-Lys-*O*-Tyr) were identified by paper chromatography after acid hydrolysis (solvent systems: *sec*-butanol:90% formic acid: water, 750:135:115; and *sec*-butanol:10% ammonia, 85:15).[34] The reaction of II with insulin[35] illustrates very well the type of information that can be obtained by the use of bifunctional reagents. Insulin monomer (one glycine and one phenylalanine chain) is known to form stable aggregates in aqueous solution, and it was of interest to learn how the monomer units are arranged in the aggregates. Amorphous ox insulin (Zn content 0.03%) in 1% bicarbonate was treated with an acetone solution of (II) and stirred at room temperature for different lengths of time. The resulting insulin derivatives were subjected to hydrolysis and the hydrolyzates were separated into water- and ether-soluble fractions, which were examined by paper chromatography with reference to synthetic standards. In addition to the expected DNPene-Gly-Phe formed by cross-linking the two N-terminal groups of the monomer, high yields of DNPene-bis-Gly, DNPene-bis-Phe, and DNPene-bis-N^ϵ-Lys were also found. Since there is only one lysine residue in insulin, these three derivatives could be formed only in the aggregate of insulin, and it could thus be concluded that the aggregate is formed by parallel stacking of monomer units in such a way that the N-terminal groups are close enough to be bridged by the reagent.

Compound (II) has also been used successfully in the study of the three-dimensional structure of ribonuclease.[36,37] A 0.016% solution of RNase A in 1% sodium bicarbonate solution was treated with 1.8 moles

[33] H. Zahn and J. Meienhofer, *Makromol. Chem.* **26**, 126 (1958).

[34] H. Zahn and J. Meienhofer, *Melliand Textilber.* **37**, 432 (1956).

[35] H. Zahn and J. Meienhofer, *Makromol. Chem.* **26**, 153 (1958).

[36] P. S. Marfey, H. Nowak, M. Uziel, and D. A. Yphantis, *J. Biol. Chem.* **240**, 3264 (1965).

[37] P. S. Marfey, M. Uziel, and J. Little, *J. Biol. Chem.* **240**, 3270 (1965).

of (II) per mole of protein. The reagent was added slowly as a methanolic solution, and the reaction mixture was held in the dark for 17 hours. No intermolecular bridges were formed under these conditions, but the presence of DNPene-bis-N^{ϵ}-Lys showed that intramolecular cross-links had been introduced. Ion-exchange chromatography allowed the separation of three cross-linked derivatives that showed 15–49% of native RNase activity toward cytidine-2′,3′-cyclic phosphate and each of which contained one DNPene cross-link (by spectral analysis). Trypsin digestion of one of these derivatives (15% active) after reduction and carboxymethylation, followed by isolation and analysis of the DNPene-peptides showed unequivocally that the cross-link was located between lysine-7 and lysine-41. Reaction of RNase A crystals with (II), by allowing the reagent to diffuse into the crystals, led to alteration of the crystal lattice parameters. The derivatives formed under these conditions were essentially devoid of all enzymatic activity.[38]

Compound (II) has been used likewise as a cross-linking reagent for synthetic polypeptides (poly-Glu-Lys-Tyr).[39] A very interesting application of bifunctional reagents to the study of complex structures is heralded by a recent report of the reaction of (II) with erythrocyte membranes.[40]

A very elegant study involving both (I) and (II) will be briefly considered to illustrate one manner in which the cross-linking reaction with these two homobifunctional reagents was made to go in a well-controlled stepwise fashion.[41] Staphylococcal nuclease[42] was first nitrated with tetranitromethane in the absence and presence of deoxythymidine 3′,5′-diphosphate to yield mononitrotyrosine in position 85 or 115, respectively. These two derivatives were then reduced with sodium dithionite to give the corresponding monoamino tyrosine derivatives. Because of the low pK (4.7) of the aromatic amino group, the preceding reactions have in effect created a group in the enzyme with uniquely reactive properties toward I and II. Thus at pH 5 both reagents are stable and do not react with any of the normal functional groups in a protein, but a derivative with the aminotyrosine group formed readily. After removal of the excess reagent, the pH was raised to 9.4 and the normal reaction of the second reagent function could take place to form a cross-link. In this way the following specific cross-links were formed in good

[38] P. S. Marfey and M. V. King, *Biochim. Biophys. Acta* **105**, 178 (1965).
[39] P. S. Marfey, T. J. Gill, III, and H. W. Kunz, *Biopolymers* **3**, 27 (1965).
[40] H. C. Berg, J. M. Diamond, and P. S. Marfey, *Science* **150**, 64 (1965).
[41] P. Cuatrecasas, S. Fuchs, and C. B. Anfinsen, *J. Biol. Chem.* **244**, 406 (1969).
[42] For a review of structure and function of this enzyme, see P. Cuatrecasas, H. Taniuchi, and C. B. Anfinsen, *Brookhaven Symp. Biol.* **21**, 172 (1968).

yield: aminotyrosine-85-(I)-tyrosine-115, aminotyrosine-85-(II)-lysine-116, aminotyrosine-115-(I)-lysine-53 and aminotyrosine-115-(II)-lysine-136.

Bifunctional Isocyanates

Several diisocyanates are available, and the six compounds illustrated have been studied in reactions with amino acids and proteins.[43–45] Isocyanates in general react with amines to form substituted ureas, with alcohols to form urethans, and in aqueous solutions hydrolysis to the amine and CO_2 must also be considered. In reactions with proteins, especially at neutral to alkaline pH, the major reaction is with amino groups. The resulting urea derivatives are stable enough to allow mild degradation of the modified proteins, but are cleaved to different extents by acid hydrolysis (6 *N* HCl, 105°, 20 hours). Monofunctional analogs are readily available for most of these reagents.

Of the six isocyanate compounds illustrated, only (V) is water-soluble. It also has the advantage that the bridge it forms in a bifunctional reaction can be split readily by reductive cleavage of the azo group. The aliphatic isocyanate groups in (I) and (VI) are somewhat less reactive than the aromatic ones in the other compounds listed. In compounds (III) and (IV) there is a differential reactivity of the two functional groups, the isocyanate being more reactive than the isothiocyanate in (III), and the 4′-isocyanate more reactive than the hindered 4-isocyanate in (IV).

It is obvious that, as a group of reagents, the diisocyanates provide a wide variety of bridge lengths, reactivities, and solubilities, as illustrated by the compounds listed above.

Reagents I–IV have been used to form covalent bonds between two different protein molecules, namely, between bovine serum albumin and bovine γ-globulin on the one hand, and between ferritin and rabbit γ-globulin on the other.[43] In order to improve the yield of conjugate and to avoid loss of antibody activity, these reactions were carried out in two steps. In the first step a 1.5% solution of BSA (or ferritin) in sodium borate buffer, pH 9.5 for the aliphatic reagent (I), or in phosphate or acetate buffer, pH 7.5–5.5 for the more reactive aromatic isocyanates, was treated with a large excess of reagent at 0°. After vigorous stirring for 15–45 minutes, the excess reagent was removed and the modified protein was allowed to react with γ-globulin (1.5% solution) at pH 9.5. The formation of the desired conjugates was demonstrated by

[43] A. F. Schick and S. J. Singer, *J. Biol. Chem.* **236,** 2477 (1961).

[44] H. Fasold, *Biochem. Z.* **339,** 482 (1964).

[45] H. Ozawa, *J. Biochem. (Tokyo)* **62,** 419 (1967).

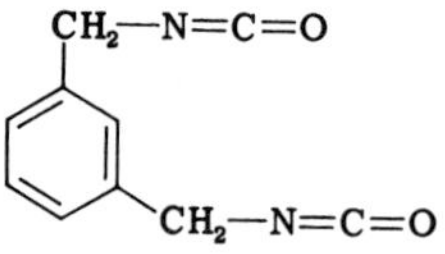

Xylylene-diisocyanate[46]
(I)

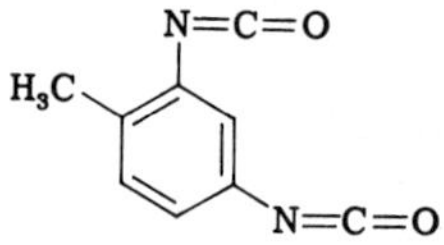

Toluene-2, 4-diisocyanate[46,47]
(II)

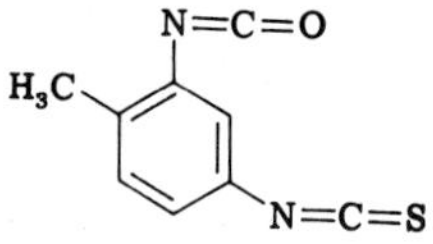

Toluene-2-isocyanate-4-isothiocyanate[47]
(III)

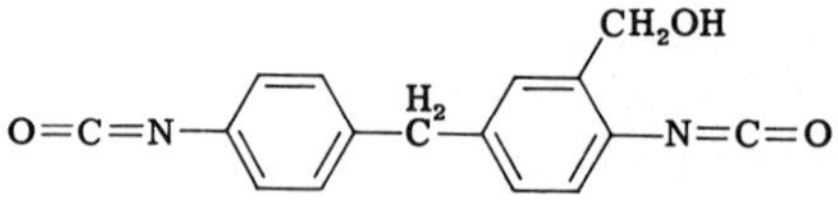

3-Methoxydiphenylmethane-4, 4′-diisocyanate[47]
(IV)

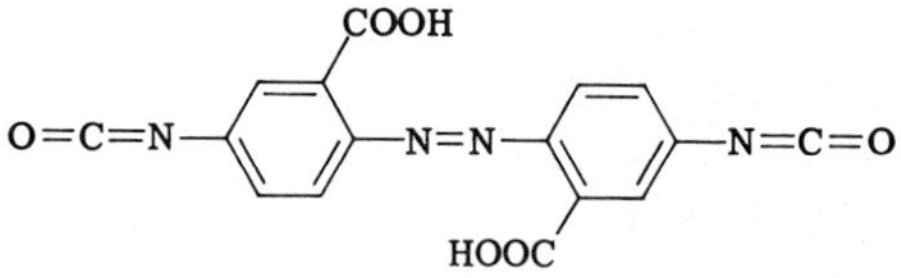

2, 2′-Dicarboxy-4, 4′-azophenyldiisocyanate
(V)

$O{=}C{=}N{-}(CH_2)_6{-}N{=}C{=}O$

Hexamethylenediisocyanate[48]
(VI)

electrophoresis, and evidence for formation of a covalent bond was obtained by the absence of conjugates in samples treated with ethylenediamine or NH_4^+ ion prior to addition of γ-globulin. Up to 41% of conjugate was formed.

Several observations from this work should be noted as typical of the problems likely to be encountered in work with bifunctional reagents. Under the reaction conditions employed in the first step (large excess of reagent), γ-globulin was rapidly inactivated. It is likely that extensive intramolecular cross-linking occurred under these conditions, and that this caused activity loss. Some self-coupling of BSA and ferritin was also observed in this reaction. The fact that relatively small amounts

[46] Available from K and K Laboratories, Inc., Plainview, New York.
[47] Available from Carwin Chemical Co., North Haven, Connecticut.
[48] Available from several commercial sources.

of cross-linked homopolymers were found may be explained on the basis of the high charge on the two acidic proteins under the conditions of the reaction, leading to electrostatic repulsion between like molecules. By contrast, the more basic γ-globulin molecule has a lower charge, and interaction between BSA (or ferritin) and γ-globulin would therefore be favored.

In the reaction of (I) with BSA, 600 moles of (I) were bound per mole of BSA after 45 minutes of reaction, indicating that side reactions take place that lead to extensive polymerization of the reagent on the protein surface. The most likely side reaction is the hydrolysis of the second isocyanate function in the reagent to an amine, which in turn can react with another diisocyanate molecule. With this large accumulation of aromatic residues on the surface of BSA, conjugates were found to form with γ-globulin even after all the free isocyanate groups in BSA had been blocked. These conjugates were dissociated at pH 9.5, however, and were therefore probably noncovalent complexes formed by hydrophobic interactions facilitated by the extensive incorporation of aromatic residues in BSA. This type of interaction is probably completely analogous to the aggregation observed with arylated proteins (above).

When (V) was allowed to react with isoleucine (phosphate buffer, pH 7.5), the product consisted of approximately equal amounts of the disubstituted isoleucine derivative (2,2-dicarboxyl-4,4′-diureidoisoleucineazobenzene) and the monosubstituted derivative in which the second isocyanate group had hydrolyzed to the amine. The two were separated by paper chromatography (dioxane:1 *M* ammonia, 10:4) and by paper electrophoresis (0.1 *M* sodium acetate buffer, pH 6), and were easily distinguishable by a difference in color; the disubstituted compound is light yellow and turns dark red in acid, the monosubstituted compound is dark yellow. Acid hydrolysis (22 hours at 105° with 6 *N* HCl) liberated isoleucine from the disubstituted derivative in good yield. The disubstituted isoleucine derivative was dissolved in water by addition of 1 *M* ammonia and treated with sodium dithionite. After 10 seconds in a water bath at 50°, the reductive cleavage of the azo group, to give 2 moles of *p*-amino-*o*-carboxybenzeneureidoisoleucine, was complete.[44]

Compound (V) has also been used successfully in the study of myoglobin.[49,50] A 7.5 mg/ml solution of whale metmyoglobin in 0.62 *M* phosphate buffer (pH 7.05) was treated with a large excess of (V) in dimethyl formamide (concentration about 250 mg/ml) in an ice bath. The product was found to be monomeric, and was indistinguishable from the unreacted starting material by several criteria (spectral

[49] H. Fasold, *Biochem. Z.* **342**, 288 (1965).
[50] H. Fasold, *Biochem. Z.* **342**, 295 (1965).

changes during denaturation and reactivity of histidine with bromoacetate). Separation of the chymotryptic peptides from the modified protein could be achieved by gel filtration and adsorption chromatography on talc. Four cross-linked peptide derivatives were eluted from peptide maps and, after cleavage of the bridge with dithionite, the component peptides of each derivative were identified. The cross-links thus established were Lys_{145}-Lys_{147}, Lys_{14}-Lys_{34}, Lys_{56}-$Lys_{62(63)}$, and Lys_{34}-Lys_{47}, all consistent with the interresidue distances deduced from X-ray crystallographic data.[51]

Compound (VI) has been used in reactions with chymotrypsin and ribonuclease and the dilysine derivative (N^{ϵ},$N^{\epsilon'}$-hexamethylenebiscarbamoyllysine) has been synthesized and characterized.[45] Some of the monofunctional aliphatic isocyanates (octyl, butyl, and propyl isocyanate) have also been studied as protein reagents and found to be specific inhibitors of chymotrypsin and elastase.[52] It may be of interest to note that this latter study showed that the half-life of the aliphatic isocyanates in aqueous solutions at pH 7.6 is less than 2 minutes.

Bifunctional Acylating Reagents

This group of reagents includes a very wide variety of compounds of different dimensions and reactivities, with monofunctional analogs readily available for control experiments. Potentially, any of the aliphatic or aromatic dicarboxylic acids or disulfonic acids can be activated to provide bifunctional acylating reagents capable of reacting under mild conditions to form covalent bonds with the protein. Two groups of compounds have been tested and will be considered briefly.

Nitrophenyl Esters of Dicarboxylic Acids. Synthetic methods and

$$O_2N-C_6H_4-O-\overset{O}{\overset{\|}{C}}-R-\overset{O}{\overset{\|}{C}}-O-C_6H_4-NO_2$$

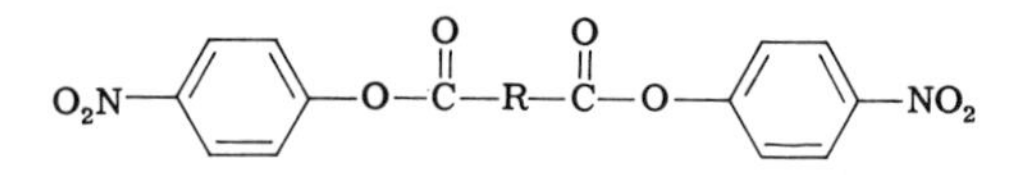

the properties of a large number of nitrophenyl esters of mono- and dicarboxylic acids have been summarized by Zahn and Schade,[53] and the acylation of several proteins with these reagents has been described.[54-56] Although the reagent specificity is not very high, α- and

[51] J. C. Kendrew, H. C. Watson, B. E. Strandberg, R. E. Dickerson, D. C. Phillips, and V. C. Shore, *Nature (London)* **190**, 666 (1961).

[52] W. E. Brown and F. Wold, *Science,* in press.

[53] H. Zahn and F. Schade, *Chem. Ber.* **96**, 1747 (1963).

[54] H. Zahn, F. Schade, and E. Siepmann, *Leder* **14**, 299.

[55] H. Zahn and F. Schade, *Angew. Chem.* **75**, 377 (1963).

[56] J. Schnell, F. Schade, and H. Zahn, *Abstr. Int. Congr. Biochem., 6th New York, 1964,* Vol. II, p. 179 (1964).

ε-amino groups appear to react most rapidly. In the studies reported, large quantities of reagent were introduced into the proteins and the reaction conditions were fairly drastic: protein (insulin, silk fibroin, wool, and collagen) in 4:1 dimethyl formamide–water was treated with an excess of nitrophenyl esters without addition of base for as long as 120 hours at room temperature to achieve optimal incorporation of the acyl group. Measurements of the shrinkage temperature showed that the collagen had been cross-linked to a significant extent.

Aromatic Sulfonyl Chlorides. These represent the second group of bifunctional acylating reagents used in protein studies. Examples of this type of reagent are illustrated.[14] Both (I) and (II) are quite insoluble

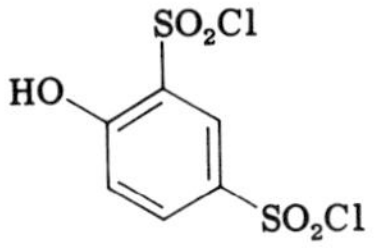

SO_2Cl HO SO_2Cl

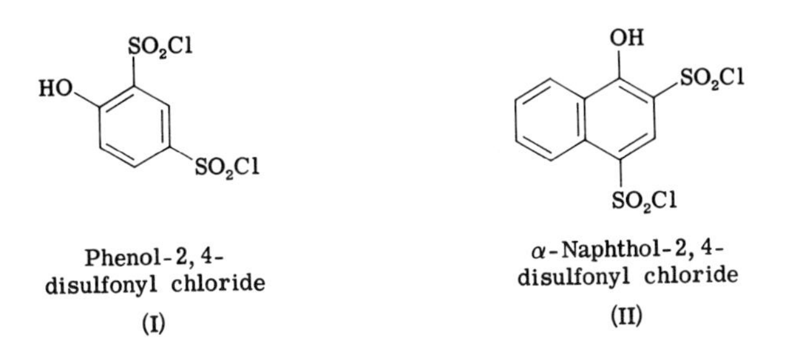

Phenol-2, 4-
disulfonyl chloride
(I)

OH SO_2Cl SO_2Cl

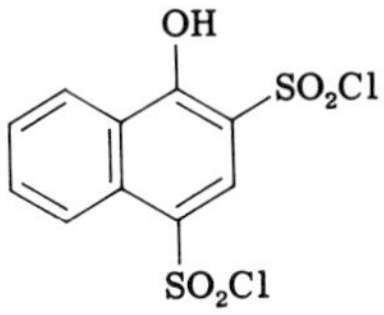

α-Naphthol-2, 4-
disulfonyl chloride
(II)

in water, and the sulfonyl chlorides hydrolyze rapidly. Reaction with protein amino groups is the most important, and the resulting stable sulfonamide linkage can subsequently be cleaved with HBr in glacial acetic acid under conditions that will not break peptide bonds. A large number of potentially useful bifunctional reagents of this type, as well as the analogous monofunctional ones, should be readily available from commercial starting materials.

Lysozyme was allowed to react[14] with (I) and (II) at a protein concentration of 1 mg/ml in 0.025–0.1 *M* borate buffer (pH 9.0). The required amount of reagent was added by pipetting an ether or methylene chloride solution of the reagent into an empty reaction flask, and the solvent was removed under a stream of nitrogen. The protein solution was then added, and the reaction mixture was stirred for 1 hour. The product contained intramolecular cross-links, and showed some loss in activity and an increased stability against denaturation when compared to native enzyme. More important, upon degradation of the derivative obtained from the reaction with 1 equivalent of phenoldisulfonyl chloride, evidence was obtained for the location of the cross-link. After tryptic digestion and peptide mapping, the unique peptides were eluted, and treated with HBr to cleave the sulfonamides. After a series of steps very similar to those described above (under Alkylating Reagents),[11] evidence was obtained that lysine residues 13, 33, and 116 were involved in the reaction. This is in agreement with evidence obtained by reaction with a water-soluble bifunctional alkylation reagent of similar dimensions.[11]

A very interesting heterofunctional acylating reagent has been reported recently.[56a] The compound, *p*-nitrophenyl chloroformate, reacts very fast with amino groups to form *p*-nitrophenyl carbamyl derivatives which are stable enough to be isolated. Subsequent exposure at slightly alkaline pH leads to cross-linking through reaction with a second amine with the release of the nitrophenyl group. The reagent has been used to convert methionyl-tRNA to the corresponding *p*-nitrophenyl carbamyl derivative, which in the subsequent reaction formed a specific cross-link to a single lysine residue in methionyl-tRNA synthetase.

Bifunctional Imidoesters

$$R'O-\overset{\overset{\oplus}{NH_2}}{\overset{\|}{C}}-R-\overset{\overset{\oplus}{NH_2}}{\overset{\|}{C}}-OR'$$

The synthesis of a number of bifunctional imidoester has been described,[57] and preliminary results on their reaction with proteins are available.[58-60] Imidoesters are soluble in water and react under mild conditions with a high degree of specificity with amino groups in the protein (see [54], this volume). The resulting amidine is quite stable to acid hydrolysis, but can be cleaved with ammonia (treatment with ammonia:acetic acid, 30:2, for 8 hours at 25°). Monofunctional imidoesters have been studied as protein reagents[61-63] and a typical degradation procedure for monofunctional protein derivatives has been described,[63] which should in principle apply to bifunctional derivatives as well. Since both monofunctional and bifunctional imidoesters are easily synthesized from the analogous nitriles, a large variety of reagents can be obtained from readily available starting materials. The nitriles can also be prepared with ease from the corresponding halides

[56a] B. S. Hartley, *Abstr. Int. Congr. Biochem., 8th Switzerland 1970,* p. 61.

[57] S. M. McElvain and J. P. Schroeder, *J. Amer. Chem. Soc.* **71,** 40 (1949).

[58] A. Dutton, M. Adams, and S. J. Singer, *Biochem. Biophys. Res. Commun.* **23,** 730 (1966).

[59] F. C. Hartman and F. Wold, *Biochemistry* **6,** 2439 (1967).

[60] G. E. Davies and G. R. Stark, *Proc. Nat. Acad. Sci. U.S.* **66,** 651 (1970).

[61] M. J. Hunter and M. L. Ludwig, *J. Amer. Chem. Soc.* **84,** 3491 (1962); cf. this volume [54].

[62] L. Wofsy and S. J. Singer, *Biochemistry* **2,** 104 (1963).

[63] M. L. Ludwig and R. Byrne, *J. Amer. Chem. Soc.* **84,** 4160 (1962).

by treatment with cyanide,[64] which permits synthesis of ^{14}C-labeled reagents. The di-ϵ-lysine derivative of adipimidate has been prepared; it is quantitatively eluted in an amino acid analyzer (short column) at pH 9.5, and can be determined in the usual manner.[59]

Since the amidines formed in the reaction between imidoesters and primary amines carry a formal positive charge (the pK_a of the amidines is considerably higher than that of the ϵ-amino groups of lysine), extensive reaction with lysine residues can be carried out without any change in the net charge of the protein.

Diethyl malonimidate has been studied in the reaction with BSA and with γ-globulin. In both cases extensive reaction with lysine residues could be accomplished (85% of the total free lysines) without destroying any of the antigenic determinants of the two proteins. Amidination of 84% of the lysine residues in anti-DNP antibody resulted in no decrease in its specific binding capacity for the hapten. Cross-linking of γ-globulin to BSA or to ferritin was also accomplished with an overall yield of 10–20% of the cross-linked pairs, and both components of the pairs retained their capacity to react with their respective antibodies. Sheep red cells modified by reaction with malonimidate showed greatly increased resistance to lysis in comparison with unreacted cells or cells modified with the analogous monofunctional reagent, ethyl acetimidate.[58]

Dimethyl adipimidate has been used to determine interresidue distances in ribonuclease.[59] A monomeric amidinated ribonuclease derivative, with a specific activity of 160 compared to 100 for the native enzyme, was degraded and two major cross-linked peptides were obtained. Quantitative amino acid analysis of the original cross-linked peptides and of the peptide components obtained upon cleavage of the amidine groups with ammonia and subsequent trypsin hydrolysis demonstrated that the two cross-links had formed between Lys_{31}-Lys_{37} and Lys_7-Lys_{37}.[59]

The same reagent has also been used to explore the architectural arrangement of proteins in erythrocyte membranes.[65] A preliminary analysis of the reacted membranes clearly demonstrated that cross-links had been formed between different membrane proteins, thus suggesting a possible approach to the elucidation of the manner in which the different proteins are arranged in the lipoprotein membrane structure.

An interesting use of bifunctional reagents in the study of the sub-

[64] R. A. Smiley and C. Arnold, *J. Org. Chem.* **25**, 257 (1960).

[65] W. G. Niehaus, Jr. and F. Wold, *Biochim. Biophys. Acta* **196**, 170 (1970).

unit structure of oligomeric proteins has recently been proposed.[60] Dimethyl suberimidate was reacted with a number of oligomeric enzymes (aldolase, glyceraldehyde-3-phosphate dehydrogenase, tryptophan synthetase B protein, L-arabinose isomerase, and the catalytic subunit of aspartate transcarbamylase). Reaction within each oligomer was favored, leading to different extent of interprotomer cross-linking. When the cross-linked derivatives were subsequently subjected to polyacrylamide gel electrophoresis in sodium dodecyl sulfate (see Part C, this volume), a set of bands corresponding to integral multiples of the protomer molecular weight was obtained. For the cases where the oligomer is made up of identical protomers, the number of bands was identical to the number of protomers in the oligomer. (A cross-linked tetramer would give 4 possible molecular weight species, monomer, dimer, trimer, and tetramer.)

Aliphatic Dialdehydes

Several simple dialdehydes, such as glyoxal, malondialdehyde, and glutaraldehyde, are readily available commercially and have been investigated as possible cross-linking reagents. The shorter ones have been allowed to react with DNA as model reactions in studies of the mechanism of radiation damage (glyoxal and malonaldehyde are produced in the radiolysis of carbohydrates) and cross-link formation has been demonstrated.[66] The most useful and extensively studied protein reagent is glutaraldehyde, however. This reagent is commercially available as a 25% aqueous solution. It can be purified by recrystallization as the bisulfite addition compound.

$$\overset{\displaystyle O}{\underset{\displaystyle HC}{\|}}\!-\!(CH_2)_3\!-\!\overset{\displaystyle O}{\underset{\displaystyle CH}{\|}}$$

The specificity for any particular type of functional group in a protein is probably low, although sulfhydryl and amino groups should be the primary points of attack. In analogy with formaldehyde (which was one of the first cross-linking reagents used with proteins), the alkylol group formed in the initial reaction (with SH or NH_2) will interact with a number of other functional groups in the protein. Derivatives of this particular reagent have not been compared or characterized, and little is known about the details of its reaction with proteins. A very interesting mechanism for the reaction of glutaraldehyde with proteins has been proposed.[67] This mechanism explains the many rather

[66] B. R. Brooks and O. L. Klamerth, *Eur. J. Biochem.* **5**, 178 (1968).
[67] F. M. Richards and J. R. Knowles, *J. Mol. Biol.* **37**, 231 (1968).

puzzling observations made with this reagent in the early work, namely the high stability of the product (too stable for simple Schiff bases), the rapid and very efficient reaction, the acid-base properties of the product, etc. The clue to the proposed mechanism was based on NMR studies which demonstrated that commercial aqueous glutaraldehyde contains virtually no free glutaraldehyde, but rather consists of a very complex mixture of polymeric material rich in α,β-unsaturated aldehydes. This explains the ready reaction with amino groups to give stable derivatives (only one possible polymer is used to illustrate the reaction):

$$\mathrm{H{-}\overset{\displaystyle O}{\overset{\|}{C}}{-}CH_2{-}CH_2{-}CH_2{-}CH{=}\overset{\displaystyle HC{=}O}{\overset{|}{C}}{-}CH_2{-}\overset{\displaystyle HC{=}O}{\overset{|}{C}}{=}CH{-}CH_2{-}CH_2{-}CH_2{-}\overset{\displaystyle O}{\overset{\|}{C}}{-}H}$$

$$\downarrow \quad + \text{ Protein-NH}_2 \quad \downarrow$$

$$\mathrm{H{-}\overset{\displaystyle O}{\overset{\|}{C}}{-}CH_2{-}CH_2{-}CH_2{-}\underset{\displaystyle Protein\text{-}NH}{\underset{|}{C}H}{-}\overset{\displaystyle HC{=}O}{\overset{|}{C}H}{-}CH_2{-}\overset{\displaystyle HC{=}O}{\overset{|}{C}H}{-}\underset{\displaystyle Protein\text{-}NH}{\underset{|}{C}H}{-}CH_2{-}CH_2{-}CH_2{-}\overset{\displaystyle O}{\overset{\|}{C}}{-}H}$$

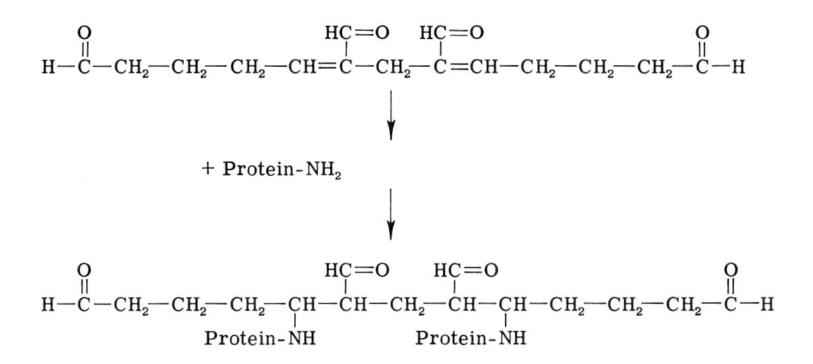

Glutaraldehyde has been used extensively to cross-link and stabilize protein crystals.[68-70] The following example illustrates this use. Crystals of carboxypeptidase A, obtained at pH 7.5 at low ionic strength, are fragile, insoluble in deionized water, and soluble in 1 *M* NaCl. When a suspension of these crystals was kept in a 6% aqueous glutaraldehyde solution, the crystals became yellow and, after 3 hours, were very resistant to mechanical breakage.[68] Less extensive treatment, such as exposure for 30 minutes to 0.1% glutaraldehyde also produced crystals with significantly increased mechanical resistance and lowered solubility in 1 *M* NaCl. The X-ray diffraction patterns of treated and untreated crystals were very similar but not identical. When propionaldehyde was used instead of glutaraldehyde in parallel experiments, there was no increase in mechanical strength of the crystal, nor did the solubility in 1 *M* NaCl show any change. In the reaction with soluble proteins, extensive intermolecular cross-linking takes place, and glutaraldehyde

[68] F. A. Quiocho and F. M. Richards, *Proc. Nat. Acad. Sci. U.S.* **52**, 833 (1964).

[69] W. H. Bishop and F. M. Richards, *J. Mol. Biol.* **33**, 415 (1968).

[70] G. N. Reeke, J. A. Hartsuck, M. L. Ludwig, F. A. Quiocho, T. A. Steitz, and W. N. Lipscomb, *Proc. Nat. Acad. Sci. U.S.* **58**, 2220 (1967).

appears to be the preferred reagent whenever high molecular weight polymers are the desired products.[71,72]

Miscellaneous

A number of other bifunctional or polyfunctional reagents which do not readily fit in the above classes of compounds have also been used and will be briefly listed here. Among these are two monofunctional protein reagents, which rather unexpectedly have been found to cause cross-linking in proteins. One of these is tetranitromethane (see [44] in this volume), an efficient reagent for nitration of tyrosine. Several cases of polymer formation after treatment of proteins with tetranitromethane have been reported,[73,74] but the chemical basis for the cross-link formation is not understood. In the case of the other reagent, diethyl pyrocarbonate, a reasonable mechanism of the cross-link formation has been proposed. This reagent appears to act simply as a condensing agent, facilitating amide bond formation between lysine ε-amino groups and the carboxyl groups of aspartic or glutamic acid in the protein.[75] This reaction is analogous to the condensation reactions in which carbodiimides are used to induce cross-linking.[39] In this connection other approaches which have been used to introduce cross-links into proteins without the use of formal bifunctional reagents should also be mentioned. By reacting proteins with thiolactones[76] or with mercaptosuccinic anhydride,[77] it has been possible to introduce new sulfhydryl groups into the protein. Subsequent air oxidation to disulfides gives rise to a large number of cross-links.

Polydiazotized polymers of aromatic amino acids[78] and potassium nitrosyldisulfonate[79] are other examples of bifunctional reagents which have been used successfully to cross-link proteins.

For the sake of completeness, reagents of the type

Substrate—R-X
(substrate analog)

will be mentioned here although in a strict sense they do not belong

[71] A. F. S. A. Habeeb and R. Hiramoto, *Arch. Biochem. Biophys.* **126**, 16 (1968).
[72] A. Scheiter and A. Bar-Eli, *Arch. Biochem. Biophys.* **136**, 325 (1970).
[73] R. J. Doyle, J. Bello, and O. A. Roholt, *Biochim. Biophys. Acta* **160**, 274 (1968).
[74] R. W. Boesel and F. H. Carpenter, *Biochem. Biophys. Res. Commun.* **38**, 678 (1970).
[75] B. Wolf, J. A. Lesnaw, and M. E. Reichmann, *Eur. J. Biochem.* **13**, 519 (1970).
[76] R. Benesch and R. E. Benesch, *Proc. Nat. Acad. Sci. U.S.* **44**, 848 (1958).
[77] I. M. Klotz and V. H. Stryker, *Biochem. Biophys. Res. Commun.* **1**, 119 (1959).
[78] F. A. Anderer and H. D. Schlumberger, *Immunochemistry* **6**, 1 (1969).
[79] R. Consden and J. A. Kirrane, *Nature (London)* **218**, 957 (1968).

to the group of bifunctional protein reagents as defined at the outset. Only one functional group is present to form a covalent bond with the protein, and the second portion of the reagent, although interacting strongly with the active site, is only a directing group, favoring the formation of a covalent bond near the active site. Notable examples of the application of this type of reagent are the reaction of L-1-tosylamido-2-phenylethyl chloromethyl ketone with chymotrypsin,[80] the reaction of 4-(iodoacetamido)salicylic acid with lactice acid dehydrogenase (in this case the salicylate is a competitive inhibitor analog),[81] and the "affinity labeling" of anti-benzenearsonic acid antibodies with *p*-(arsonic acid)benzenediazonium fluoroborate.[82] The search for such active site-specific reagents is a very active area of research today (see [58] and [59] in this volume). The natural extension of these single reactive group "bifunctional" reagents to compounds which in addition to the affinity group contain two reactive groups capable of covalent bond formation ("trifunctional reagents") has already been mentioned in connection with the bifunctional alkyl halides.[12] Another variation on the same theme is bifunctional reagents containing two affinity groups and no reactive group. An example of this type is a series of compounds, bis-*N*-biotinylpolymethylenediamines which were used to "measure" the distances between the 4 biotin binding sites in avidin.[83]

[80] G. Schoellman and E. Shaw, *Biochemistry* **2**, 252 (1963); cf. this volume [58].

[81] B. R. Baker, W. W. Lee, E. Tong, and L. O. Ross, *J. Amer. Chem. Soc.* **83**, 3713 (1961).

[82] L. Wofsy, H. Metzger, and S. J. Singer, *Biochemistry* **1**, 1031 (1962).

[83] N. M. Green, *Biochem. J.* **104**, 64P (1967).

Section IX

Specific Modification Reactions

[58] Site-Specific Reagents for Chymotrypsin, Trypsin, and Other Serine Proteases[1]

By Elliott Shaw

Chymotrypsin and Related Serine Proteases

Variations in the structure of TPCK (I), the reagent initially devised as a substrate-like alkylating agent for chymotrypsin,[2] have led to inhibitors that either inactivate very rapidly or will otherwise increase the usefulness of this class of agents in structural work with serine proteases, for example, by inactivation of enzymes insensitive to TPCK.

The essentiality of an α-substituent for rapid and exclusive alkylation of histidine in chymotrypsin has been shown.[3] Without it, alkylation of methionine-192 takes place,[4] leading to chymotrypsin derivatives with residual hydrolytic activity.

The nature of the N^{α} substituent has a considerable influence on the speed of inactivation of chymotrypsin; moreover, a bromomethyl ketone is more effective than a chloromethyl ketone.[3] Thus, the second-order rate constant for alkylation by ZPBK[5] (II, X = Br) is about 100 times that for TPCK at pH 7 (see the table).

The substituent on the α-amino group may interact with chymotrypsin at the binding site for the N-terminal portion of a normal polypeptide substrate and, therefore, variations in this portion of the reagent are of added interest. For this purpose, the benzyloxycarbonyl derivatives ZPCK[6] and ZPBK[5] (II) are of particular value as synthetic intermediates since they can be deblocked, without loss of the halomethyl ketone grouping, to the amino compound (III), which is stable when not exposed to alkaline conditions. Acylation of the side chain in (III) is possible, leading to reagent modifications such as the tripeptide chloromethyl ketone (IV) that would not be otherwise accessible.[7,8] These derivatives have a 2-fold potential value. In the case of γ-chymotrypsin,

[1] Research carried out at Brookhaven National Laboratory under the auspices of the U.S. Atomic Energy Commission.

[2] G. Schoellmann and E. Shaw, *Biochem. Biophys. Res. Commun.* **7**, 36 (1962).

[3] E. Shaw and J. Ruscica, *Arch. Biochem. Biophys.* **145**, 484 (1971).

[4] K. J. Stevenson and L. B. Smillie, *Can. J. Biochem.* **48**, 364 (1970).

[5] E. Shaw and J. Ruscica, *J. Biol. Chem.* **243**, 6312 (1968).

[6] E. Shaw, Vol. XI, p. 684.

[7] K. Morihara and T. Oka, *Arch. Biochem. Biophys.* **138**, 526 (1970).

[8] J. C. Powers and P. E. Wilcox, *J. Amer. Chem. Soc.* **92**, 1782 (1970).

RATE OF INACTIVATION OF CHYMOTRYPSIN BY SOME ACTIVE SITE-DIRECTED AGENTS

Agent	pH	°C	k (M^{-1} sec^{-1})	Solvent
Methyl *p*-nitrobenzene-sulfonate (V)	6.84 7.92	25	0.25[a] 0.45	0.1 *M* phosphate, 9.9% acetonitrile
TPCK (I)	7.0	25	7.7[b]	0.05 *M* phosphate, 10% methanol, 10% dimethyl sulfoxide
ZPCK (II, X = Cl)	7.0	25	69.0[b]	0.05 *M* phosphate, 10% methanol
ZAGPCK (IV)	7.0	40	100[c]	0.05 *M* Tris·HCl, 10% dioxane
ZPBK (II, X = Br)	7.0	25	790[b]	0.05 *M* phosphate, 10% methanol, 10% dimethyl sulfoxide
DFP	7.0	25	45[d]	
Phenylmethanesulfonyl fluoride	7.0	25	248[d]	

[a] Y. Nakagawa and M. L. Bender, *Biochemistry* **9,** 259 (1970).
[b] E. Shaw and J. Ruscica, *Arch. Biochem. Biophys.* **145,** 484 (1971).
[c] K. Morihara and T. Oka, *Arch. Biochem. Biophys.* **138,** 526 (1970).
[d] A. M. Gold, Vol. 11, p. 707.

N-acetyl-L-alanyl-L-phenylalanyl chloromethyl ketone produced inactivation in the crystalline enzyme[8,9] and may provide information of subsites. (The crystal form used in studies of α-chymotrypsin leaves little space for α substituents in bound substrates.[10]) A second potential lies in the possibility of obtaining specific alkylation of the active centers of other serine proteinases which are more resistant than chymotrypsin to chloromethyl ketones derived from a single amino acid, a difficulty encountered in the cases of subtilisin and elastase, as discussed below.

A different type of site-specific alkylating agent for chymotrypsin is provided by methyl *p*-nitrobenzenesulfonate (V) which inactivates chymotrypsin by methylation of His-57 at N-3 of the imidazole ring.[11] In contrast to reagents such as TPCK, (V) leaves only a small substituent, the methyl group, attached at the active center. The rates of inactivation of chymotrypsin by a selected group of reagents are listed in the table.

[9] D. M. Segal, D. R. Davies, G. H. Cohen, J. C. Powers, and P. E. Wilcox, *Fed. Proc. Fed. Amer. Soc. Exp. Biol.* **29,** 336 (1970).
[10] T. A. Steitz, R. Henderson, and D. M. Blow, *J. Mol. Biol.* **46,** 337 (1969).
[11] Y. Nakagawa and M. L. Bender, *Biochemistry* **9,** 259 (1970).

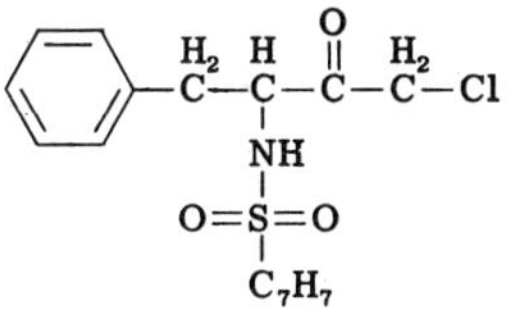

(I)

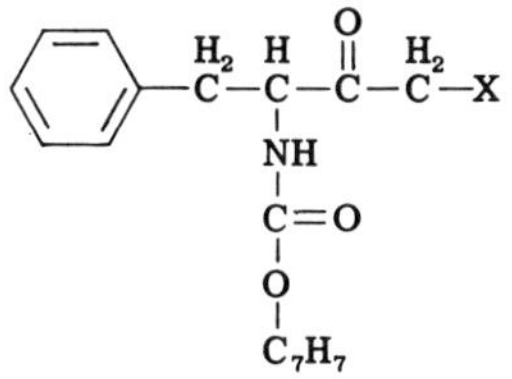

(II)

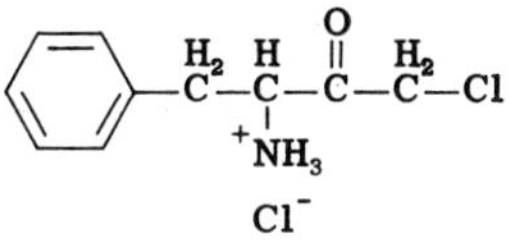

(III)

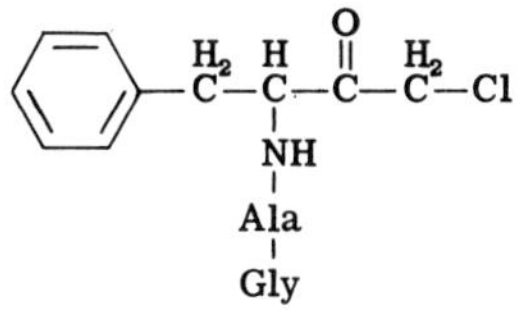

(IV)

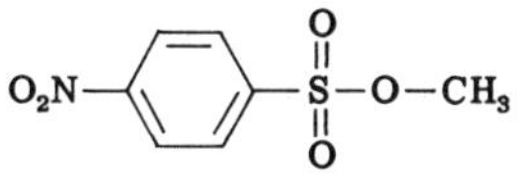

(V)

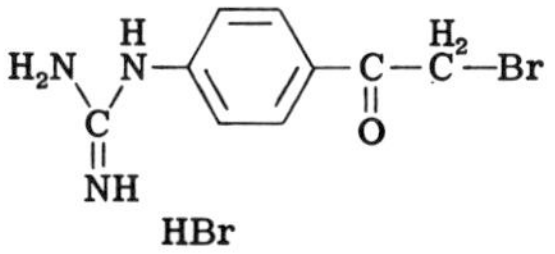

(VI)

Subtilisin and Other Neutral Serine Proteases

The bacterial protease, subtilisin, is inert to TPCK and TLCK although it has esterase activity on the related simple ester substrates. However, a specific alkylation by the more reactive phenylalanine-derived reagent ZPBK (II, X = Br) was demonstrated,[5] permitting identification of the active center residue His-64.[12]

The demonstration by Morihara and Oka[7] that dipeptide and tripeptide chloromethyl ketone derivatives are rapid alkylating agents of subtilisin reflects the importance of occupying subsites. For the most effective inhibitor, ZAGPCK (IV), a second-order rate constant of 25 $M^{-1}sec^{-1}$ was observed at 40°, pH 7.

Elastase and the α-lytic protease of *Sorangium* sp. are serine proteases catalytically similar to chymotrypsin and trypsin but inert to chloromethyl ketones derived from *N*-tosylglycine and *N*-tosyl-L-valine

[12] F. S. Markland, E. Shaw, and E. L. Smith, *Proc. Nat. Acad. Sci. U.S.* **61,** 1440 (1968).

as well as to the bromomethyl ketone of the latter.[13] No conclusion about the role of histidine could be drawn from these negative experiments. The observations made with subtilisin with respect to the importance of having a di- or tripeptide derivative may apply to elastase and other enzymes as well.

Trypsin and Related Serine Proteases

An improved synthesis of TLCK is particularly useful for preparing radioactive forms of the inhibitor.[14] Attempts to obtain chloromethyl ketones of ornithine and arginine derivatives met with difficulty due to cyclization reactions, but a low yield of N^{α}-*p*-nitrobenzyloxycarbonyl arginyl chloromethyl ketone was obtained.[14] This is a highly effective trypsin inactivator, but a useful synthesis is lacking. On the other hand, active site-directed inhibitors could be based on the competitive inhibitors benzamidine and phenylguanidine, that is, *p*-amidinophenacyl bromide and *p*-guanidinophenacyl bromide.[15] The latter, GPB (VI), apparently forms a complex with trypsin in which the bromine is not near His-46 but closer to Ser-183, which thus becomes etherified during the inactivation.[15]

The use of active site-directed reagents for proteolytic enzymes has been reviewed recently.[16]

TLCK has generally been found to inactivate trypsin from a variety of species including dogfish,[17] shrimp,[18] pig,[19] beef,[20] and man,[21] among others.[16] On the other hand, certain macromolecular inhibitors, such as the soybean trypsin inhibitor and ovomucoid, are without effect on human trypsin[21] and thus appear to be less reliable than TLCK as a trypsin inhibitor.

Enzyme Titrants

A variety of reagents which are useful for titrating chymotrypsin, trypsin, and other serine proteases have been developed, some of which

[13] H. Kaplan, V. B. Symonds, H. Dugas, and D. R. Whitaker, *Can. J. Biochem.* **48**, 649 (1970).

[14] E. Shaw and G. Glover, *Arch. Biochem. Biophys.* **139**, 298 (1970).

[15] D. Schroeder and E. Shaw, *Arch. Biochem. Biophys.* **142**, 340 (1971).

[16] E. Shaw, *Physiol. Rev.* **50**, 244 (1970).

[17] H. Neurath, R. A. Bradshaw, and R. Arnon, *in* "Structure-Function Relationships of Proteolytic Enzymes" (P. Desnuelle, H. Neurath, and M. Ottesen, eds.), p. 113. Munksgaard, Copenhagen, 1970.

[18] B. J. Gates and J. Travis, *Biochemistry* **8**, 4483 (1969).

[19] R. A. Smith and I. E. Liener, *J. Biol. Chem.* **242**, 4033 (1967).

[20] E. Shaw, M. Mares-Guia, and W. Cohen, *Biochemistry* **4**, 2219 (1965).

[21] J. Travis and R. C. Roberts, *Biochemistry* **8**, 2884 (1969).

are demonstrably active site directed. Appropriate methods are summarized elsewhere in this series.[22,23]

Preparation and Use of Selected Reagents

Synthesis of PCK, the Chloromethyl Ketone Derived from L-*Phenylalanine.*[3] The benzyloxycarbonyl derivative L-ZPCK[6] (1.0 g) and phenol (0.5 g) were heated in trifluoroacetic acid (5 ml) for 30 minutes. The solvent was removed under reduced pressure, and the residue was taken up in ethanolic HCl containing an excess of hydrogen chloride to ensure conversion of the product to the hydrochloride salt. After removal of the solvent, the residue was triturated with anhydrous ether, and the crystalline product was collected by filtration. Recrystallization from absolute alcohol and ether yielded 0.49 g, 70%, with mp 169–170°.

The bromomethyl ketone (PBK) has been similarly prepared.

*A Peptide Chloromethyl Ketone: Benzyloxycarbonyl-*L-*Alanylglycyl-*L-*Phenylalanine Chloromethyl Ketone* (*ZAGPCK, IV*).[7] Peptide chloromethyl ketones containing a C-terminal phenylalanyl chloromethyl ketone residue can be prepared from PCK (III) by the mixed anhydride method.[7,8] The intermediate PCK is obtainable from ZPCK (II, X = Cl) as described above, by catalytic hydrogenation,[7] or by treatment with hydrogen bromide in acetic acid.[8]

A mixed anhydride was prepared from Z-Ala-Gly (5.7 mmoles) and ethyl chloroformate (5.7 mmoles) in methylene chloride (15 ml) below 5°. The resultant solution was added to a solution of PCK (5.75 mmoles) in methylene chloride to which an equivalent of triethylamine had been added in the cold. The reaction mixture was washed with aqueous hydrochloric acid and sodium bicarbonate; the residue from the organic layer was crystallized from ethyl acetate and petroleum ether to provide a product, mp 160–162°.

Inactivation of Chymotrypsin, Subtilisin, and Other Serine Proteinases with Chloromethyl Ketone Reagents

In general this class of reagent is not soluble in water. Consequently a concentrated stock solution is prepared in an organic solvent; methanol, dioxane, dimethyl sulfoxide, or acetonitrile have commonly been used. The reagent solution is then added to the buffered enzyme solution to achieve the desired final concentration. Chymotrypsin and subtilisin tolerate relatively higher concentrations of certain organic solvents (i.e., 20%). Suitable controls are desirable over the period of observation. The reagents are not stable above pH 7.5–8.0.

[22] F. J. Kezdy and E. T. Kaiser, Vol. XIX, p. 3.

[23] T. Chase and E. Shaw, Vol. XIX, p. 20.

Preparation of Methyl *p*-Nitrobenzenesulfonate[11]

p-Nitrobenzenesulfonyl chloride (0.1 mole) in ether (75 ml) was stirred in a flask provided with a reflux condenser and cooling bath to maintain a temperature of 20°. Sodium methoxide (0.1 mole) in methanol (100 ml) was added dropwise with stirring. After 2.5 hours, the suspension was poured into water (300–500 ml) and shaken with chloroform. The chloroform extract was washed with sodium bicarbonate and dried over sodium sulfate; the filtrate was taken to dryness. The residue was crystallized from ligroin; the product melted at 91.5–92.5°.

Inactivation of Chymotrypsin with Methyl *p*-Nitrobenzenesulfonate[11]

α-Chymotrypsin ($3.65 \times 10^{-4}\ M$) in 0.1 *M* sodium phosphate buffer, pH 7.93, was incubated with methyl *p*-nitrobenzenesulfonate ($2.86 \times 10^{-3}\ M$) at 25° with a final acetonitrile concentration of 9.9%. Active center titration with *N-trans*-cinnamoylimidazole indicated 90% inactivation in about 30–35 minutes.

Improved Synthesis of L-TLCK[14]

N^{ϵ}-Benzyloxycarbonyl-L-TLCK[6] (2.5 g) was heated in trifluoroacetic acid (10 ml) at 90° for 20 minutes. The trifluoroacetic acid was removed under reduced pressure in a rotary evaporator. The residue was dissolved in excess ethanolic hydrogen chloride and taken to a thick syrup once more under reduced pressure. To remove benzyl trifluoroacetate, the oily product was stirred with several portions of anhydrous ether (50 ml), which were discarded. At this point, the TLCK could be slowly crystallized from a concentrated solution in absolute ethanol in 70% yield.

p-Guanidinophenacyl Bromide Hydrobromide[15]

p-Guanidinoacetophenone hydrochloride[24] (2.6 g, 10 mmole) was suspended in 20 ml of glacial acetic acid containing 13% HBr, and a solution of bromine in acetic acid (0.75 ml in 2.5 ml) was added to the vigorously stirred suspension. Within 10 minutes, the bromine color was discharged and solution was complete. After 2 hours, water was added and the mixture was taken to dryness; recrystallization of the residue from acetonitrile-ether gave 3.2 g (9.6 mmoles) of *p*-guanidinophenacyl bromide hydrobromide, mp 181–183°. The reagent is water soluble.

[24] H. King and I. M. Tonkin, *J. Chem. Soc.* p. 1063 (1946).

[59] Site-Specific Reagents for Triose Phosphate Isomerase and Their Potential Applicability to Aldolase and Glycerol Phosphate Dehydrogenase[1]

By Fred C. Hartman

An important advance in the selective chemical modification of ligand-binding sites of biological macromolecules was the introduction of affinity labeling.[2-4] This technique involves designing a chemical reagent that resembles the ligand and thus possesses an affinity for the ligand-binding site. Affinity of the reagent for a specific site increases the likelihood of covalent modification of a group within that site as compared to analogous groups located elsewhere in the macromolecule. Although affinity labeling has been most extensively exploited in the characterization of enzymatic active sites, it has also been used successfully to label antigenic sites of antibodies,[3] and in the future will possibly offer a more rational approach to drug design.[4,5]

Site-Specific Reagents for Triose Phosphate Isomerase

General Considerations

The active site of triose phosphate isomerase can be labeled with haloacetol phosphates[6-11] and glycidol phosphate.[12,13] The former are structurally similar to the substrate dihydroxyacetone phosphate; the latter may be more closely related to the enolate of dihydroxyacetone phosphate, a proposed intermediate in the enzyme-catalyzed intercon-

[1] Research, conducted in author's laboratory, sponsored by the U.S. Atomic Energy Commission under contract with Union Carbide Corporation.

[2] B. R. Baker, "Design of Active-Site-Directed Irreversible Enzyme Inhibitors." Wiley, New York, 1967.

[3] S. J. Singer, *Advan. Protein Chem.* **22,** 1 (1967).

[4] E. Shaw, *Physiol. Rev.* **50,**(2), 244 (1970).

[5] B. R. Baker, *Annu. Rev. Pharmacol.* **10,** 35 (1970).

[6] F. C. Hartman, *Biochem. Biophys. Res. Commun.* **33,** 888 (1968).

[7] F. C. Hartman, *J. Amer. Chem. Soc.* **92,** 2170 (1970).

[8] F. C. Hartman, *Biochem. Biophys. Res. Commun.* **39,** 384 (1970).

[9] F. C. Hartman, *Biochemistry* **10,** 146 (1971).

[10] A. F. W. Coulson, J. R. Knowles, and R. E. Offord, *Chem. Commun.* **1,** 7 (1970).

[11] A. F. W. Coulson, J. R. Knowles, J. D. Priddle, and R. E. Offord, *Nature (London)* **227,** 180 (1970).

[12] I. A. Rose and E. L. O'Connell, *J. Biol. Chem.* **244,** 6548 (1969).

[13] S. G. Waley, J. C. Miller, I. A. Rose, and E. L. O'Connell, *Nature (London)* **227,** 181 (1970).

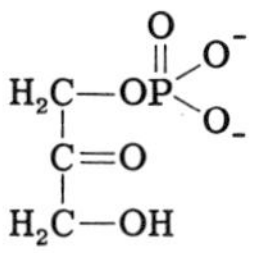

$H_2C{-}OP(=O)(O^-)_2$, $C{=}O$, $H_2C{-}OH$

Dihydroxyacetone phosphate

$H_2C{-}OP(=O)(O^-)_2$, $C{=}O$, $H_2C{-}X$ X = Cl, Br or I

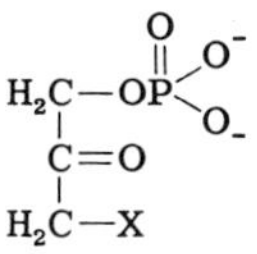

3-Haloacetol phosphates

$H_2C{-}OP(=O)(O^-)_2$, $C{-}O^-$, $\|$ $HC{-}OH$

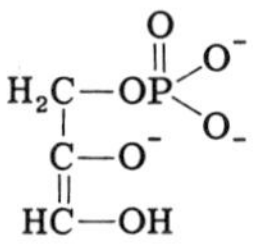

Enolate of dihydroxyacetone phosphate

$H_2C{-}OP(=O)(O^-)_2$, HC, H_2C (epoxide, $-O-$)

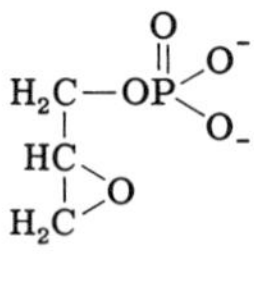

Glycidol phosphate

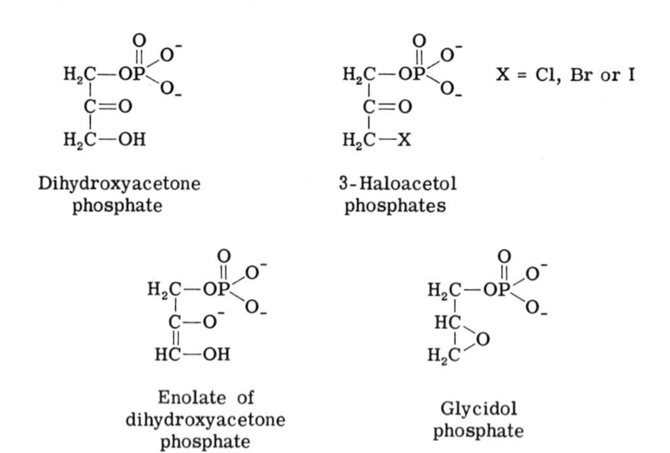

version of dihydroxyacetone phosphate and glyceraldehyde 3-phosphate. Both types of reagents inactivate triose phosphate isomerase by esterification of a glutamyl γ-carboxylate at the active site. All the usual

$$\text{enzyme}{-}(CH_2)_2{-}\overset{O}{\overset{\|}{C}}O^- + XCH_2\overset{O}{\overset{\|}{C}}CH_2OPO_3H_2 \rightarrow$$

$$\text{enzyme}{-}(CH_2)_2{-}\overset{O}{\overset{\|}{C}}OCH_2\overset{O}{\overset{\|}{C}}CH_2OPO_3H_2 + X^-$$

criteria indicative of active-site modification are fulfilled[9]: (1) With large molar excesses of reagents, the kinetics of inactivation are pseudo first-order. (2) One mole of reagent per mole of catalytic subunit is incorporated with absolute specificity for a single glutamyl residue. (3) Substrate and competitive inhibitors protect against inactivation. (4) Triose phosphate isomerases from a variety of species are inactivated. (5) Amino acid sequences adjacent to the labeled residue are identical in the rabbit muscle and chicken muscle enzymes.[9,11,13]

Indirect evidence suggests a catalytically functional role for the reactive glutamyl residue. The rate of inactivation of triose phosphate isomerase by haloacetol phosphates is extremely high (see the table), but the reagents do not react with free glutamic acid.[14] (These inactivation rates are in the same range as those for the inactivation of cholinesterases by certain dialkyl phosphorofluoridates.[15]) The only demonstrable reactivity of haloacetol phosphates with model compounds is toward mercaptans.[14] If the γ-carboxylate, which seems to be involved in catalysis, is the essential group with pK 6.5–7.5 detected in kinetic

[14] F. C. Hartman, *Biochemistry* **9**, 1776 (1970).
[15] See Vol. 11 [81].

BIMOLECULAR RATE CONSTANTS FOR THE INACTIVATION OF TRIOSE PHOSPHATE ISOMERASES BY HALOACETOL PHOSPHATES AND GLYCIDOL PHOSPHATE

	$k_{2nd} \times 10^{-2}$ (M^{-1} sec^{-1})			
Species	Glycidol phosphate	Chloroacetol phosphate[b]	Bromoacetol phosphate[b]	Iodoacetol phosphate[b]
Escherichia coli		11		
Spinach		16		
Mouse liver		19		
Rabbit muscle	0.043[a]	23	26	2.3
Human whole blood		27		
Bakers' yeast		53		

[a] Calculated from data of I. A. Rose and E. L. O'Connell, *J. Biol. Chem.* **244,** 6548 (1969). Conditions: pH 7.4 and 15°.
[b] Data from F. Hartman, *Biochemistry* **10,** 146 (1971). Conditions: pH 6.5 and 2°.

studies,[16,17] the danger of attempting to identify a residue from an observed pK is again very apparent.

As triose phosphate isomerases from various species become available, site-specific reagents will provide a convenient method for determining the subunit molecular weight or the number of subunits in those cases in which the molecular weight has been calculated by an alternative method. The extreme rapidity of the reaction between chloro- or bromoacetol phosphate allows measurement of the active-site concentration by direct titration (Fig. 1); the triose phosphate isomerase concentration in crude extracts has been measured with chloroacetol phosphate, which demonstrates the minimal reaction of the reagent with other proteins.

The ability of chloroacetol phosphate rapidly and selectively to inactivate triose phosphate isomerase in mixtures of proteins permits the removal of triose phosphate isomerase activity in commercial preparations of certain enzymes, such as fructose diphosphate aldolase and glycerol phosphate dehydrogenase. Also, the selective inhibition of a given enzyme is sometimes advantageous in metabolic studies. For example, during investigations on the biosynthesis of alkyl glyceryl ethers by microsomal preparations, the presence of triose phosphate isomerase precludes determining whether dihydroxyacetone phosphate or glyceraldehyde 3-phosphate is the obligatory precursor. After inactivation of the enzyme with chloroacetol phosphate, however, the identi-

[16] I. A. Rose, *Brookhaven Symp. Biol.* **15,** 293 (1962).
[17] R. Wolfenden, *Biochemistry* **9,** 3404 (1970).

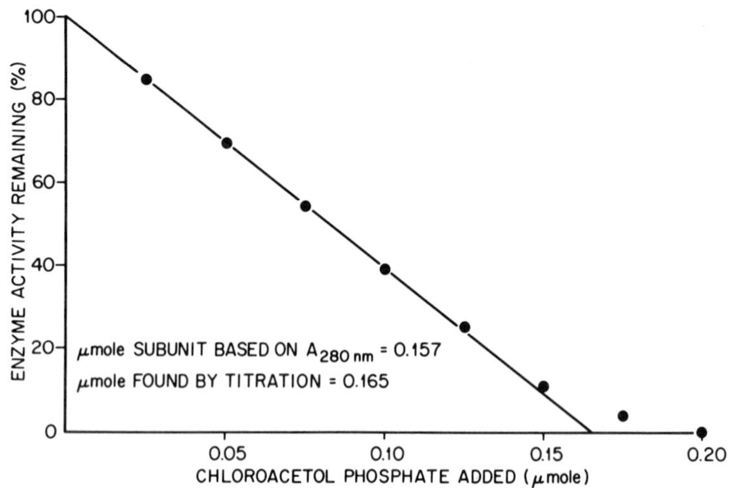

FIG. 1. Titration of triose phosphate isomerase with chloroacetol phosphate. Eight 5-μliter samples of 5 m*M* reagent were added in succession to a solution (1 ml) containing 4.2 mg of rabbit muscle enzyme in 0.1 *M* imidazolium chloride (pH 6.5). Fifteen minutes after each addition, 1-μl aliquots were diluted with imidazolium buffer containing 1 m*M* β-mercaptoethanol and assayed for enzymatic activity.

fication of dihydroxyacetone phosphate as the precursor of alkyl glyceryl ethers becomes unequivocal.[18–21] Recently, a potential use of haloacetol phosphates in cancer chemotherapy has been suggested.[22]

Extent of Incorporation

The most accurate and convenient method of measuring the number of moles of reagent incorporated per mole of protein is with radioactive reagents. [^{32}P]Haloacetol phosphates,[10,23] [^{14}C]bromoacetol phosphate,[10] [^{32}P]glycidol phosphate,[12] and [^{3}H]glycidol phosphate[12] have been prepared. Reasons for choosing ^{32}P-labeled reagents are that the isotope is not introduced until the last step in the synthesis, and autoradiograms of peptide maps, which provide information valuable in the characterization of the modified protein, can be prepared with short

[18] R. L. Wykle and F. Snyder, *Biochem. Biophys. Res. Commun.* **37**, 658 (1969).
[19] F. Snyder, B. Malone, and M. L. Blank, *J. Biol. Chem.* **245**, 1790 (1970).
[20] G. A. Rao, M. F. Sorrels, and R. Reiser, *Lipids* **5**, 762 (1970).
[21] L. E. Puleo, G. A. Rao, and R. Reiser, *Lipids* **5**, 770 (1970).
[22] T. P. Fondy, G. S. Ghangas, and M. J. Reza, *Biochemistry* **9**, 3272 (1970).
[23] F. C. Hartman, *Biochemistry* **9**, 1783 (1970).

exposure times due to the high energy of ^{32}P decay. The disadvantage of this isotope is its short half-life (14 days), which necessitates repeated synthesis of the reagent if an investigation is to continue during long time periods. [^{14}C]Bromoacetol phosphate and [^{3}H]glycidol phosphate are particularly attractive because of the long half-lives of ^{14}C and ^{3}H. Of the reagents considered, [^{14}C]bromoacetol phosphate is the most costly to prepare, and [^{32}P]glycidol phosphate is the most easily prepared.

There are two alternative methods of determining extent of incorporation of haloacetol phosphates into triose phosphate isomerase. The inactivated enzyme may be treated with [^{3}H]sodium borohydride, which results in a stoichiometric incorporation of tritium during the reduction of the carbonyl group of the reagent moiety to a hydroxyl group.[9] The reduction is quantitative, and insignificant cleavage, if any, of the protein-reagent bond occurs. [^{3}H]Sodium borohydride with high specific activities is commercially available at relatively low cost. Accurate determination of the specific activity of sodium borohydride requires using it to reduce a well-characterized organic aldehyde or ketone, whose corresponding alcohol can be quantitated.[24]

The incorporated haloacetol phosphate may also be quantitated by base-labile phosphate determinations (incubation of the labeled protein in 1 N sodium hydroxide for 10 minutes at room temperature followed by inorganic phosphate assays). This method is less accurate and requires larger quantities of protein than radioisotopic methods.

Regardless of the reagent and method used to quantitate incorporation, one must be aware of the lability of the protein-reagent linkage and, in the case of haloacetol phosphates, the lability of the phosphate group. After inactivation of the protein, excess reagent should be removed under mild conditions. Gel filtration at pH 7.9 resulted in the loss of about 35% of the radioactivity incorporated into triose phosphate isomerase inactivated with [^{32}P]bromoacetol phosphate,[10] whereas dialysis at 4° and pH 6.0 caused no loss of phosphate from the incorporated acetol phosphate moiety.[9] Freeze-drying of the inactivated enzyme also removes the phosphate group but does not result in cleavage of the glutamyl ester bond. Sodium borohydride reduction of the acetol ketone group in the modified enzyme stabilizes both the phosphate group and the glutamyl ester. Without reduction of the carbonyl, migration of the acetol moiety from the glutamyl carboxylate to an adjacent tyrosyl hydroxyl group reportedly can occur.[11] Therefore, before at-

[24] H. E. Conrad, J. R. Bamburg, J. D. Epley, and T. J. Kindt, *Biochemistry* **5**, 2808 (1966).

tempting any rigorous characterization of triose phosphate isomerase inactivated with haloacetol phosphates, the carbonyl should be reduced.

Identification of Modified Residue

Because chloroacetol phosphate inactivates triose phosphate isomerases from all species that have been tested, the reagent will be valuable in comparative studies of the active sites. The observation of inactivation does not necessarily mean that in all cases inactivation results from esterifying a glutamyl γ-carboxylate. Esterification is implied if the reagent-protein bond is labile toward base and hydroxylamine, but lability studies cannot distinguish glutamate and aspartate esters. Amino acid analyses are of no value because esters do not survive conditions used for hydrolyzing proteins. Theoretically, one can analyze the intact, inactivated enzyme for esters by reduction to an alcohol or conversion to a hydroxamate, which in turn can give rise to an amine, but these methods are complicated by the potential reactivity of glutamine and asparagine.[25] An ester in an intact protein may also fail to react, at least quantitatively, due to inaccessibility to the reagent. Attempts to use these methods on chloroacetol phosphate inactivated triose phosphate isomerase were unsuccessful.

The only unequivocal method of verifying the presence of an ester is apparently to degrade (by proteolytic digestion) the modified enzyme, isolate the labeled peptide, and subsequently assay the peptide for ester. One should bear in mind the likelihood of protease-catalyzed hydrolysis of esters during digestion of the labeled enzyme. In the case of triose phosphate isomerase inactivated by chloroacetol phosphate, a 2-hour digestion at 40° with 1% by weight of trypsin or chymotrypsin liberated only about 10% of the reagent moiety, but these proteases used in succession resulted in total hydrolysis of the ester. The labeled peptide is amenable to purification by ion-exchange chromatography because of its highly acidic phosphate group.

Potential Site-Specific Reagents for Fructose Diphosphate Aldolases and Glycerol Phosphate Dehydrogenase

A single reagent capable of covalently modifying binding sites for the same ligand on enzymes that catalyze different reactions would be useful in determining whether these sites have structural similarities. Any derivative of dihydroxyacetone phosphate containing a functional group reactive toward amino acid side chains is potentially a site-specific reagent for all enzymes that bind dihydroxyacetone phosphate. In addi-

[25] See Vol. 11 [74].

tion to triose phosphate isomerase, enzymes that catalyze reactions involving dihydroxyacetone phosphate are fructose diphosphate aldolases, glycerol phosphate dehydrogenase, and glycerol kinase. Glycidol phosphate does not inactivate the latter three enzymes.[12] Under physiological conditions chloro- and bromoacetol phosphate do not inactivate rabbit muscle aldolase or glycerol phosphate dehydrogenase, but the reagents are substrates for the dehydrogenase.[26] Iodoacetol phosphate inactivates rabbit muscle aldolase[23] and glycerol phosphate dehydrogenase[26] by oxidation of sulfhydryl groups, but the reaction is not confined to the dihydroxyacetone phosphate binding site. Preliminary experiments suggest that chloroacetol phosphate does react selectively with the active site of yeast fructose diphosphate aldolase.[27]

Dihydroxyacetone phosphate is itself a site-specific reagent for class I fructose diphosphate aldolases; the methodology for active-site labeling has appeared in an earlier volume of this series.[28]

Synthesis of Reagents

1-O-Benzoyl-3-chloro-2-propanone[14]

During 2 hours, benzoyl chloride (140.6 g, 1 mole) is added dropwise to a vigorously stirred solution of 3-chloropropanediol (110.5 g, 1 mole) in 600 ml of anhydrous pyridine at —10 to —20°. After the addition is completed, the reaction mixture is kept at 4° for 4 hours, and the excess pyridine is then removed by concentration at 40° with a rotary evaporator. The product is dissolved in chloroform (500 ml), and the resulting solution is washed at 4° in succession with two 1-liter portions of 1 *N* sulfuric acid, saturated aqueous sodium bicarbonate, and water. The chloroform layer is dried over anhydrous sodium sulfate, filtered, and concentrated to yield a syrupy residue (205 g), about 75% of which is 1-*O*-benzoyl-3-chloropropanediol. This benzoate is oxidized to the corresponding ketone with dimethyl sulfoxide containing dicyclohexylcarbodiimide (DCC).[29] To a precooled (4°), ether solution (400 ml) containing the benzoate (50 g), dimethyl sulfoxide (20 ml), pyridine (3 ml), and DCC (90 g), 3 ml of trifluoroacetic acid is added with stirring. Thirty minutes after initiation of the oxidation, 20 g of oxalic acid in 30 ml of methanol is added to the reaction mixture to decompose excess DCC. After an additional 30 minutes, dicyclohexylurea

[26] S. L. Hackette and F. C. Hartman, unpublished data (1969).

[27] F. C. Hartman, I. L. Norton, Y. Lin, and R. D. Kobes, *Fed. Proc. Fed. Amer. Soc. Exp. Biol.* **30**, 1157 (1971).

[28] See Vol. 11 [77].

[29] K. E. Pfitzner and J. G. Moffatt, *J. Amer. Chem. Soc.* **87**, 5661 (1965).

is removed by filtration. The filtrate is extracted three times with 200-ml portions of cold, saturated sodium bicarbonate, dried over sodium sulfate, and concentrated to yield crystalline 1-*O*-benzoyl-3-chloro-2-propanone. The crystalline mass is dissolved in 300 ml of boiling cyclohexane, and insoluble material is removed by gravity filtration. Cooling at room temperature for 6 hours gives 29 g of crystals, which are then dissolved in 600 ml of ether; traces of insoluble dicyclohexylurea are removed by suction filtration through Celite. The filtrate is concentrated to dryness, and the product is again crystallized from cyclohexane (500 ml) to give 23.5 g (44% from chloropropanediol) of analytically pure 1-*O*-benzoyl-3-chloro-2-propanone, mp 93.5–95.5°.

3-Chloroacetol [^{32}P]Phosphate Dimethyl Ketal Biscyclohexylammonium Salt[14]

1-*O*-Benzoyl-3-chloro-2-propanone (20 g) is converted to its dimethyl ketal by incubation for 24 hours at room temperature in a mixture composed of methanol (200 ml), trimethyl orthoformate (200 ml), and concentrated sulfuric acid (2 ml). The reaction mixture is neutralized with solid sodium bicarbonate, filtered, and concentrated to dryness. The remaining syrup is dissolved in chloroform (200 ml); the solution is then extracted with 250 ml of cold, saturated sodium bicarbonate, dried, and concentrated to yield 24 g of the ketal as a thin syrup.

The ketal (24 g) in 200 ml of methanol is treated with 27 ml of 4 *N* sodium hydroxide for 6 hours at room temperature to remove the benzoyl blocking group. Methanol is removed by concentration, and the remaining aqueous mixture is extracted three times with 100-ml portions of ether. The dried extracts are concentrated to give 12.5 g of the alcohol as a slightly viscous, light yellow liquid.

3-Chloro-1-hydroxy-2-propanone dimethyl ketal (1 g, 6.47 mmoles) in 20 ml of tetrahydrofuran is added during 1 hour to an efficiently stirred solution at —2° of 40 ml of tetrahydrofuran containing [^{32}P]phosphorus oxychloride (1.53 g, 10 mmoles; specific radioactivity of 0.15 mCi/mmole; New England Nuclear Corporation) and pyridine (1.6 g, 20 mmoles). Precautions are taken to exclude atmospheric moisture. Thirty minutes after the addition is completed, the reaction mixture is poured over 50 g of ice with vigorous mixing. Pyridine is then removed from the solution with 30 meq of Bio-Rad AG50W (H^+) resin. Resin is removed by filtration; the filtrate is neutralized (pH 8.5) with cyclohexylamine and concentrated to dryness. Remaining traces of water are removed by two concentrations from 50 ml of ethanol. The residue is slurried with isopropanol (50 ml), which dissolves cyclohexylammonium chloride. The insoluble cyclohexylammonium salts of inorganic phosphate and the desired compound are collected on a Büchner funnel and air-dried. The phosphate salts are then refluxed with 50 ml of ethanol containing 0.1 ml of cyclohexylamine; insoluble cyclohexylammonium phosphate is removed by gravity filtration. Upon cooling of the filtrate in an ice-bath, the product crystallizes; recrystallization is effected by dissolving in 10 ml of water and adding acetone to turbidity, followed by cooling at 4°. The chloroacetol [^{32}P]phosphate dimethyl ketal biscyclohexylammonium salt (1.1 g, 33% from 1-*O*-benzoyl-3-chloro-2-propanone) thus obtained is pure, as judged by paper chromatography and elemental analysis; its specific radioactivity is 290,000 cpm/μmole.

If the ^{32}P label is not required, the phosphorylation can be achieved in 80% yield with diphenylchlorophosphate. The subsequent removal of the phenyl groups by platinum-catalyzed hydrogenation does not cleave the alkyl chloride.

Chloroacetol [^{32}P]Phosphate[14]

A 0.05 *M* solution of chloroacetol [^{32}P]phosphate dimethyl ketal biscyclohexylammonium salt (111 mg in 5 ml of water) is stirred for 5 minutes with 1 meq of Bio-Rad AG50W (H^+) resin, filtered, and incubated at 40°. Chloroacetol phosphate formation is monitored by base-labile phosphate (inorganic phosphate after base treatment minus inorganic phosphate before base treatment) appearance. Base-labile phosphate and chloroacetol phosphate concentrations are equivalent, as the dimethyl ketal is stable to base; and dihydroxyacetone phosphate, whose phosphate group is labile, is not formed during hydrolysis of the ketal (no liberation of chloride ion). Periodically 0.02-ml aliquots are diluted to 0.5 ml; 0.4 ml of the diluted sample is assayed for inorganic phosphate,[30] and 0.1 ml is treated at room temperature for 10 minutes with 1 ml of 1 *N* sodium hydroxide, neutralized with 1 *N* hydrochloric acid, and then assayed for inorganic phosphate. After 40 hours, the conversion of the ketal to the ketone is about 80% complete; some inorganic phosphate (7% of the total phosphate) is formed. The reaction mixture is neutralized to pH 4.5 with solid sodium bicarbonate and frozen. No decomposition of chloroacetol phosphate is detected after 3 months.

Iodo- and Bromoacetol Phosphate[14]

These reagents can be prepared by a series of reactions analogous to those described for the synthesis of chloroacetol phosphate. 1-*O*-Benzoyl-3-iodo-2-propanone is obtained from the corresponding chloro com-

[30] B. B. Marsh, *Biochim. Biophys. Acta* **32**, 357 (1959).

pound by treatment with sodium iodide in acetone. Bromopropanediol[31] is obtained from epibromohydrin.

[^{14}C]Bromoacetol Phosphate[10]

This reagent is prepared by the following series of reactions:

$$\mathrm{Br^{14}CH_2COOH} \xrightarrow{\mathrm{ClC(=O)C(=O)Cl}} \mathrm{Br^{14}CH_2C(=O)Cl} \xrightarrow{\mathrm{CH_2N_2}} \mathrm{Br^{14}CH_2C(=O)CHN_2}$$

$$\mathrm{Br^{14}CH_2C(=O)CHN_2} \xrightarrow{\mathrm{H_3PO_4} + \mathrm{BF_3}} \mathrm{Br^{14}CH_2C(=O)CH_2OPO_3H_2}$$

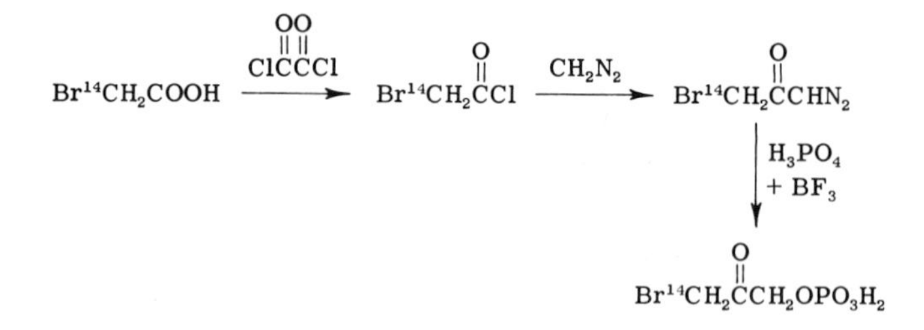

Glycidol Phosphate[12]

This reagent is prepared by phosphorylation of glycidol with phosphorus oxychloride.

Labeling of the Active Site of Rabbit Muscle Triose Phosphate Isomerase with Chloroacetol Phosphate[9]

Rabbit muscle triose phosphate isomerase is isolated by a published procedure[32] or obtained commercially (Sigma Chemical Co. and Boehringer Mannheim Corp.). Protein concentration is determined from the absorbancy at 280 nm using an $\epsilon_{1\,\mathrm{cm}}^{1\%}$ of 13.1.[32] Enzymatic activity is determined spectrophotometrically from NADH oxidation by coupling the isomerase and glycerol phosphate dehydrogenase reactions.[33] The assay mixture (3 ml) contains 0.15 m*M* NADH, 1 m*M* DL-glyceraldehyde 3-phosphate, 28 μg of α-glycerol phosphate dehydrogenase (all three materials from Sigma Chemical Co.), 0.3 m*M* EDTA, and 20 m*M* triethanolamine hydrochloride (pH 7.9). The reaction is initiated by the addition of triose phosphate isomerase (5–15 ng). One unit of activity is the conversion of 1 μmole of D-glyceraldehyde 3-phosphate to dihydroxyacetone phosphate per minute and represents a decrease in $A_{340\,\mathrm{nm}}$ of 2.07.

Triose phosphate isomerase (150 mg, 5.66 μmoles of catalytic subunit) in 10 ml of 0.1 *M* sodium bicarbonate (pH 8.0) containing 1 m*M* EDTA is treated with 0.5 ml of 0.04 *M* chloroacetol [^{32}P]phosphate. After 5

[31] S. Winstein and L. Goodman, *J. Amer. Chem. Soc.* **76,** 4368 (1954).

[32] I. L. Norton, P. Pfuderer, C. D. Stringer, and F. C. Hartman, *Biochemistry* **9,** 4952 (1970).

[33] See Vol. I [57].

minutes, less than 0.01% of the initial enzymatic activity remains, and the excess reagent is decomposed by adding β-mercaptoethanol to a final concentration of 0.01 *M*. The extent of incorporation is measured after removal of excess reagent from a 0.5 ml aliquot by dialysis at 4° against 0.01 *M* sodium phosphate (pH 6.0). The remainder of the reaction mixture is reduced with sodium borohydride and carboxymethylated before further characterization of the inactivated enzyme. After cooling the solution to 4° and adding 1 drop of *n*-octyl alcohol to prevent excessive foaming, the carbonyl group of the incorporated reagent is reduced by a 30-minute treatment with sodium borohydride (0.01 *M*). To the reaction mixture is then added 10 ml of 8 *M* guanidine hydrochloride containing 0.1 *M* sodium phosphate (pH 8.0), followed immediately by 0.7 ml of 1 *M* sodium iodoacetate (pH 8.0) to carboxymethylate protein sulfhydryl groups. Twenty minutes later, arabinose (100 mg) is added to the solution to oxidize the excess borohydride; the reaction mixture is then made 0.1 *M* in β-mercaptoethanol and dialyzed exhaustively against water, followed by 0.1 *M* ammonium bicarbonate (pH 8.0). The dialyzed solution (34 ml) contains 135 mg of protein.

Author Index

Numbers in parentheses are reference numbers and indicate that an author's work is referred to, although his name is not cited in the text.

A

B

C

D

E

H

N

Q

R

S

T

Subject Index

A

B

C

D

E

H

M

N

O

P

U

V

W

Z